Fundamental Concepts of

Inorganic Chemistry

Volume 4

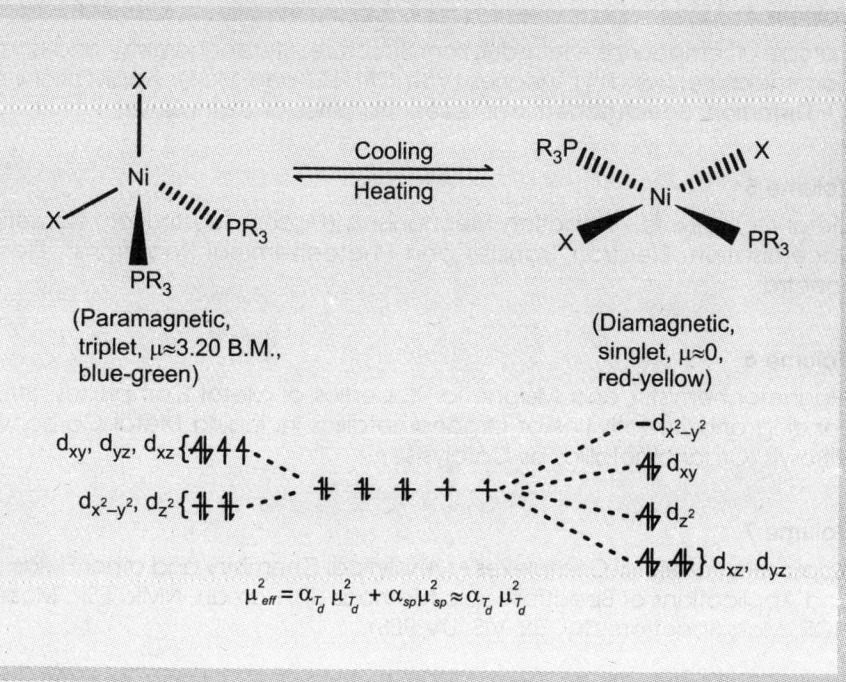

Fundamental Concepts of **Inorganic Chemistry**

Contents of Volumes 1–7

Fundamental Concepts of

Inorganic Chemistry

Volume 4

Asim K. Das
MSc (Gold Medalist, CU), PhD (CU), DSc (Visva Bharati)

Professor of Chemistry
Visva Bharati University, Santiniketan 731235
West Bengal (India)

Mahua Das
MSc (CU), PhD (Visva Bharati)

Former Research Associate, Department of Chemistry
Visva Bharati University, Santiniketan 731235
West Bengal (India)

CBS

CBS Publishers & Distributors Pvt Ltd

New Delhi • Bengaluru • Chennai • Kochi • Kolkata • Mumbai
Bhopal • Bhubaneswar • Hyderabad • Jharkhand • Nagpur • Patna • Pune
Uttarakhand • Dhaka (Bangladesh)

Fundamental Concepts of
Inorganic Chemistry
Volume 4

ISBN: 978-81-239-2351-2

Copyright © Authors and Publisher

First Edition: 2014

Reprint: 2015, 2016, 2018, 2019

Published by Satish Kumar Jain and produced by Varun Jain for

CBS Publishers & Distributors Pvt Ltd

4819/XI Prahlad Street, 24 Ansari Road, Daryaganj, New Delhi 110 002, India.

Ph: 23289259, 23266861, 23266867   Website: www.cbspd.com

Fax: 011-23243014   e-mail: delhi@cbspd.com; cbspubs@airtelmail.in.

Corporate Office: 204 FIE, Industrial Area, Patparganj, Delhi 110 092

Ph: 4934 4934   Fax: 4934 4935   e-mail: publishing@cbspd.com; publicity@cbspd.com

Branches

- **Bengaluru:** Seema House 2975, 17th Cross, K.R. Road,
  Banasankari 2nd Stage, Bengaluru 560 070, Karnataka
  Ph: +91-80-26771678/79   Fax: +91-80-26771680   e-mail: bangalore@cbspd.com
- **Chennai:** 7, Subbaraya Street, Shenoy Nagar, Chennai 600 030, Tamil Nadu
  Ph: +91-44-26680620, 26681266   Fax: +91-44-42032115   e-mail: chennai@cbspd.com
- **Kochi:** 42/1325, 1326, Power House Road, Opposite KSEB Power House,
  Ernakulam 682 018, Kochi, Kerala
  Ph: +91-484-4059061-65   Fax: +91-484-4059065   e-mail: kochi@cbspd.com
- **Kolkata:** 6/B, Ground Floor, Rameswar Shaw Road, Kolkata-700 014, West Bengal
  Ph: +91-33-22891126, 22891127, 22891128   e-mail: kolkata@cbspd.com
- **Mumbai:** 83-C, Dr E Moses Road, Worli, Mumbai-400018, Maharashtra
  Ph: +91-22-24902340/41   Fax: +91-22-24902342   e-mail: mumbai@cbspd.com

Representatives

• **Bhopal** 0-8319310552	• **Bhubaneswar** 0-9911037372	• **Hyderabad** 0-9885175004	• **Jharkhand** 0-9811541605
• **Nagpur** 0-9421945513	• **Patna** 0-9334159340	• **Pune** 0-9623451994	• **Uttarakhand** 0-9716462459
• **Dhaka (Bangladesh)** 01912-003485			

Printed at India Binding House, Greater Noida, UP, India

A tribute to the

Father of Indian Chemistry
Acharya Prafulla Chandra Ray (1861–1944)
on the occasion of his
151st birth anniversary

श्रद्धावान् लभते ज्ञानम्
One who has shraddha acquires knowledge

Preface

We do endeavour. He gives the strength and patience.

The present volumes 4–7 are in continuation of the existing title *Fundamental Concepts of Inorganic Chemistry*. The classnotes and valuable suggestions received from the esteemed readers have been shaped in these volumes.

These volumes cover the structure and bonding (VBT, CFT, and MOT), stability and reactivity, spectral and magnetic properties of metal complexes in depth. Kinetics and reaction mechanisms of ligand substitution, electron transfer and photochemical reactions have been included. Magnetochemistry and organometallic chemistry have been covered. Applications of different spectroscopic techniques (Raman, IR, NMR, ESR, Mossbauer, UV-VIS, UV-PES, etc.) have been discussed to widen the utility of the series. In developing the present extension, we have taken all the measures to retain the basic features of the existing title.

In preparing the manuscript, we have freely consulted the books and reviews of the earlier authors and have borrowed their ideas whenever it has been required. These sources are listed and acknowledged at the end of the text. We are grateful and indebted to these authors. In reality, we have picked up flowers from these gardens to prepare the garland to worship the goddess of learning.

We are extremely thankful and grateful to Mr SK Jain, Managing Director, CBS Publishers & Distributors, for his continued support. We are thankful to Mr YN Arjuna, Senior Director, Publishing, Editorial and Publicity, and the DTP staff for taking the trouble in processing the manuscript.

In spite of our best efforts, some mistakes and misconceptions might have crept in for which we beg to be excused. Constructive criticism and suggestions are always welcome to better the presentation.

Asim K. Das
Mahua Das

Contents

Crown Ether Complexes: Structure; Spherands: The Preorganized Macrocycles; Cryptands –
Multicyclic Macrocycles: Football Ligands; Sepulchrate; Application of the Macrocyclic Ligands
Related to the Crown Ethers; Crown Ether Related Macrocycles and Supramolecular System
and Molecular Recognition; Ionophores Resembling the Crown Ethers and Naturally Occurring
Antibiotics: Alkali Metal Complexes of the Naturally Occurring Ionophores

Contents of Other Volumes

Volume 5

Volume 6

Volume 7

CHAPTER 1

Introduction and Language of the Chemistry of Complex Compounds

1.1 DISCOVERY AND HISTORICAL BACKGROUND

Coordination Chemistry now covers a substantial portion of Inorganic Chemistry and many other branches of chemistry like Analytical Chemistry, Organic Chemistry, Environmental Chemistry, Industrial Chemistry, etc. Extensive researches are going on in the field of coordination chemistry and several hundreds of research papers are being published every month.

The concept of coordination chemistry is of fairly recent origin. In fact, this branch of chemistry was not known to the *Arabian Chemists* of the Middle Ages and the *Alchemists*. Coordination compounds came into use long before their characterization but the actual date when these came into use is not well traced in history. The use of **red madder** (a dye) by Alexander the Great in a battle field probably laid a foundation stone to its origin. *Madder* is the root of the plant, *Rubia tinctorum*, found in the various parts of Europe and Asia Minor. This **red dye** is nothing but the coordination compounds of Ca^{2+} and Al^{3+} (present in clay) with the vegetable product hydroxyanthraquinone present in the madder. These naturally occurring coordination compounds are the representative examples of *alizarin dye*.

Alexander the Great, probably was the first to initiate the *chemical warfare* in the history by using the above mentioned red dye of coordination compounds in a battle with the Persian force. In the battle, the Persian force was much larger than the Macedonian force under Alexander the Great. On the second day of the battle, it was found that the garments of the Macedonian soldiers were blood stained and this sight led the Persian force to conclude that most of the opposition force had been wounded in the fighting of first day and the few who received medical care had come back to the field. Actually, the garments were intentionally red stained by the dye. Thus the Persian force was misguided. This wrong interpretation of the Alexander's trick made the Persian force defeated as they did not take the adequate caution and measure on the second day of battle.

In the ancient days, the **Indian painters** practised the use of metal salts with the vegetable extracts for painting fabrics and walls. The colour was due to complexation of the metal ions with the constituents of vegetable extracts.

Libavius noticed (1597) the formation of **blue colour** on the addition of a copper salt into an aqueous ammonia solution. It is now known that it is due to $[Cu(NH_3)_4]^{2+}$. **Aureolin** *i.e.* $K_3[Co(NO_2)_6].6H_2O$ was used as a **yellow pigment** in the ancient times. The most authenticated coordination compounds in practical use involved the **discovery of prussian blue** (which is now known as $KFe[Fe(CN)_6]$) by the artist's colour maker Diesbach in Berlin in 1704. The accidental discovery of the prussian blue appeared as a by-product when animal waste and soda were heated in an iron pot. Prussian blue was used as a *blue pigment*. Here it is interesting to mention that the metal centre iron was involved in the **accidental discoveries of different types of coordination compounds** which in turn paved the way for the development of new branches of coordination chemistry. These are:

- **prussian blue** (used as a blue pigment by the artists in Germany in the beginning of 18th century);
- **iron carbonyls** (in 1891)
- **iron phthalocyanins** (in 1926)
- **iron cyclopentadienyls** (in 1951)

Other early recorded reports of coordination compounds are: **yellow prussiate of potash**, (*i.e.* potassium ferrocyanide) prepared by Macquer in 1753 by treating prussian blue with alkali; **prussic acid** from yellow prussiate of potash by Schell (1783); **potassium chloroplatinate** (1760) involved in the refining of platinum.

Transition metals are the most potential candidates for complex formation

---

However, the accidental isolation of the orange coloured **cobalt-ammine compound, $CoCl_3.6NH_3$ by Tassaert in France in 1798** from an ammonical solution of cobalt chloride ($CoCl_2$) exposed to air was the **landmark** in the age of understanding of coordination compounds. *Isolation of this new compound from the combination of two already saturated compounds appeared as a vexing problem to the chemists at that time.* In fact, it had to wait for about 100 years to rationalise the observation of Tassaert.

Within about 50 years after the discovery of cobalt-ammine compound $CoCl_3.6NH_3$ by Tassaert, **different coordination compounds of cobalt-ammines** of different compositions were synthesized by different laboratories. The coordination compounds were not only confined to the metal centre cobalt but extended for many other metal centres. These are:

- **Red prussiate of potash** (*i.e.* potassium ferricyanide, $K_3[Fe(CN)_6]$) (in 1822);
- **Zeise's salt** having the composition $PtCl_2.KCl.C_2H_4$ (1827); (**first organometallic compound** of transition metals).
- **Magnus' green salt** having the composition $PtCl_2.2NH_3$ (1828); (now formulated as $[Pt(NH_3)_4][PtCl_4]$).
- **Nitroprusside** (1849)
- **Different cobalt-ammine compounds** by E. Fremy (1851-52), and *colour code names* of these compounds introduced by Fremy (*cf.* Table 1.1.1).

Table 1.1.1 Colour code names of different cobalt(III)-ammines (*cf.* Sec. 1.15.1).

Cobaltammines	Colour	Colour code name
A. $CoCl_3.6NH_3$	Orange-yellow	Luteocobaltic chloride
B. $CoCl_3.5NH_3.H_2O$	Red	Roseocobaltic chloride
C. $CoCl_3.5NH_3$	Purple	Purpureocobaltic chloride
D.* $CoCl_3.4NH_3$	Green	Praseocobaltic chloride
	Violet	Violeocobaltic chloride

* *Praseo* (*i.e.* green) compound is actually the *trans*-isomer, *i.e.* *trans*-$[CoCl_2(NH_3)_4]Cl$ while the *violeo* (*i.e.* violet) compound is the *cis*-isomer, *i.e.* *cis*-$[CoCl_2(NH_3)_4]Cl$. The *cis*-isomer was not isolated upto 1907.

Though the proper understanding of the behaviour and properties of the coordination compounds started late in the nineteenth century with the arrival of **Werner's Theory**, many propositions appeared earlier to cope with the problem. Many of which are of mere historical significance and most of them were to be rejected. But, definitely some propositions prepared the background for Werner's success. For example, the proposal made by *Claus* in 1850s was strongly criticized and rejected at that time but accepted about after 40 years in Werner's theory with some modifications. This theory ultimately recognized Werner by the **Nobel Prize in 1913** and **it was the first example of Noble Prize in Inorganic Chemistry**. Claus proposed that in combination with the cobalt chloride, ammonia lost its basic character and it became passive. He also proposed that the factors which governed the number of ammonia molecules to be associated would also be applicable for the salt hydrate formation. However, these statements of Claus could not be supported by the facts and other examples and this was the limitation with Claus. Similarly, the **Lewis concept of dative bond formation** is found to sound the Graham's theory which was proposed much earlier to explain the behaviour of coordination compounds.

1.2 PROPERTIES OF COBALTAMMINES

Before to discuss any theory to explain the behaviour of coordination compounds, it is required to know the properties of cobaltammines which were known as the representatives examples of coordination compounds at that time.

$CoCl_3.6NH_3$ **(luteocobaltic chloride)** was isolated in the following way by Tassaert.

$$CoCl_2 \text{ in aqueous } NH_3 \xrightarrow[\text{(colour changed)}]{\text{exposed to air}} \underset{\text{(crystal)}}{CoCl_3 \cdot 6NH_3 (\downarrow)}$$

In the above reaction, Co(II) is oxidised to Co(III) by the oxygen present in air.

In $CoCl_3.6NH_3$, the *NH_3 molecules are so tightly bound* that treatment with conc. H_2SO_4 removes all the chlorides as HCl but **no NH_3 is lost**.

$$CoCl_3 \cdot 6NH_3 \xrightarrow{\text{Conc. } H_2SO_4} Co_2 (SO_4)_3 \cdot 12NH_3$$

$CoCl_3.6NH_3$ does not lose any NH_3 molecule even when it was boiled in aqueous HCl. It again indicates that the *NH_3 molecules are tightly bound in the compound*.

$$CoCl_3 \cdot 6NH_3 \xrightarrow[\text{aqueous HCl}]{\text{boiling in}} \textbf{No loss of } NH_3$$

Properties of other cobaltammines are also similar to those of $CoCl_3.6NH_3$. It is illustrated below.

$$\underset{\textbf{(Roseocobaltic chloride)}}{CoCl_3 \cdot 5NH_3, H_2O} \xrightarrow{\Delta(100°C)} \underset{\textbf{(Purpureocobaltic chloride)}}{CoCl_3 \cdot 5NH_3} \xrightarrow{H_2SO_4} CoClSO_4 \cdot 5NH_3$$

Precipitation of AgCl from the aqueous solution of cobaltammines indicates the **presence of free Cl⁻ in solution**. The observations are:

(A) $CoCl_3 \cdot 6NH_3 \xrightarrow{AgNO_3} 3AgCl (\downarrow)$ *(i.e.* 3 Cl⁻ remain free per molecule)

(B) $CoCl_3 \cdot 5NH_3 \cdot H_2O \xrightarrow{AgNO_3} 3AgCl (\downarrow)$ *(i.e.* 3 Cl⁻ remain free per molecule)

(C) $CoCl_3 \cdot 5NH_3 \xrightarrow{AgNO_3} 2AgCl (\downarrow)$ *(i.e.* 2 Cl⁻ remain free per molecule)

(D) $CoCl_3 \cdot 4NH_3 \xrightarrow{AgNO_3} 1AgCl (\downarrow)$ *(i.e.* 1 Cl⁻ remains free per molecule)

(E)* $CoCl_3 \cdot 3NH_3 \xrightarrow{AgNO_3}$ No AgCl *(i.e.* No Cl⁻ remains free in solution)

(* At the early date, $CoCl_3.3NH_3$ was not isolated.)

Precipitation of AgCl from the **freshly prepared solutions** of cobaltammines can indicate the number of free Cl^- ions present in solution.

(**Note:** To estimate the number of free Cl^- ions present in solution, always **freshly prepared solutions** should be used. On standing or heating, other Cl^- ions may be also slowly precipitated. This depends actually on the **ease of solvolysis of the compounds**. If the compounds are *kinetically inert* then this experiment is only meaningful as in the cases of Co(III) and Pt(IV) compounds. This aspect will be discussed at the appropriate place.)

Scheme 1.2.1 illustrates the properties of different cobalt(III)-ammine compounds.

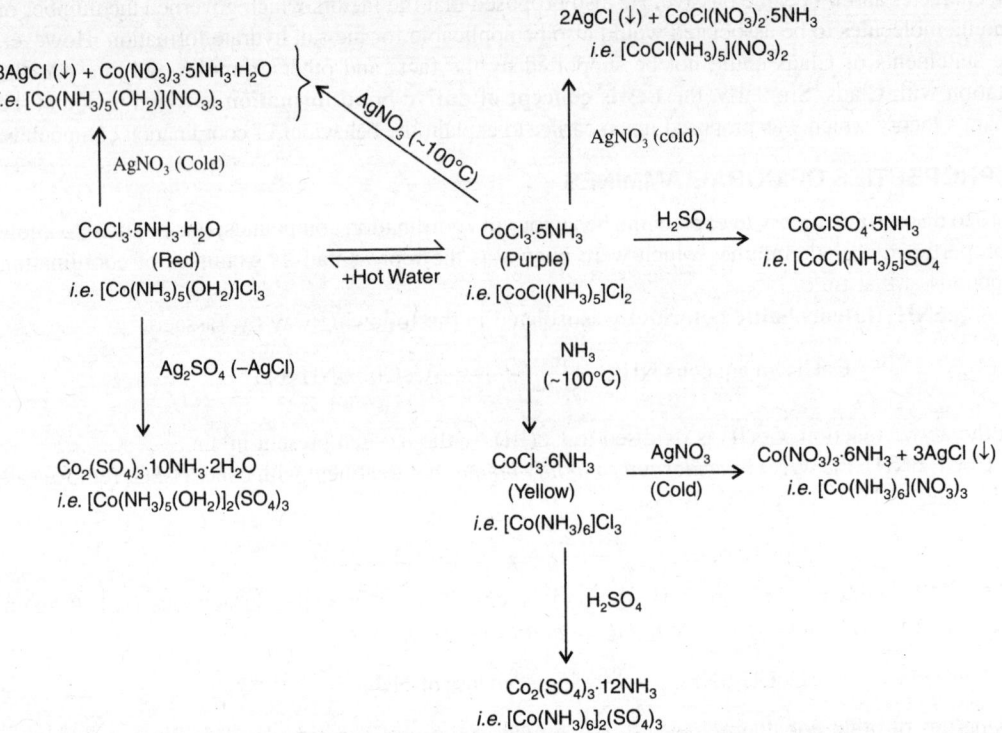

Scheme 1.2.1 Interconversion of cobalt(III)-ammine complexes illustrating their properties.

Molar conductivity of a dilute solution of a compound indicates the number of ions available per molecule in solution. It happens so, as the electrical conductance of a solution depends on the number of ions (*i.e. charge carriers*) present in solution when the other factors remain the same or comparable. This is illustrated for the *dilute solutions* (*ca.* 10^{-3} M) in water for different types of electrolytes.

Type of electrolyte	No. of ions in formula	Λ_M (Ohm^{-1} cm^2 mol^{-1})
1–1 (for NaCl) (*i.e.* **uni-univalent**)	1 + 1 = 2	~ 120
2–1 (for $CaCl_2$, $BaCl_2$) (*i.e.* **bi-univalent**)	1 + 2 = 3	~ 260
3–1 (for $LaCl_3$, $CeCl_3$) (*i.e.* **ter-univalent**)	1 + 3 = 4	~ 400

The Λ_M (in Ohm^{-1} cm^2 mol^{-1}) values for the dilute (0.001 M) solutions of CoCl$_3$.6NH$_3$, CoCl$_3$.5NH$_3$.H$_2$O, CoCl$_3$.5NH$_3$ and CoCl$_3$.4NH$_3$ (*praseo-*) are approximately 432, 413 and 261 and 100 respectively. It indicates the number of ions per molecule as 4, 4, 3 and 2 respectively. The mode of ionization can also be supported by the *cryoscopic measurements (e.g.* depression of freezing point*)*.

Table 1.2.1 Physical properties of cobaltammines.

Compound	Λ_M (Ohm^{-1}cm^2 mol^{-1})	No. of ions per molecule	No. of ionic Cl$^-$ per molecule for AgCl ppt.
CoCl$_3$.6NH$_3$ (Luteocobaltic Chloride)	~ 432	1 + 3 = 4	3
CoCl$_3$.5NH$_3$* (Purpureocobaltic Chloride)	~ 261	1 + 2 = 3	2
CoCl$_3$.5NH$_3$.H$_2$O (Roseocobaltic Chloride)	~ 413	1 + 3 = 4	3
CoCl$_3$.4NH$_3$ (Praseocobaltic Chloride)	~ 100	1 + 1 = 2	1

* It undergoes slowly aquation to increase the Λ_M value.

A Vexing Problem on the Concept of Valence at the Early Date

A stable metal salt like MCl$_x$ chemically combines with the stable molecules like yNH$_3$ to produce a new compound of composition MCl$_x$(NH$_3$)$_y$. **How is it possible?** This was the vexing question. The answer came out from the systematic studies on such compounds in the period 1875–1915 by the **Danish chemist S.M. Jörgensen** (1837–1914) and the **Swiss chemist Alfred Werner** (1866–1919).

1.3 BLOMSTRAND-JORGENSEN'S CHAIN THEORY OF METAL-AMMINE COMPOUNDS

Kekule's proposition (1858) of *tetracovalence* of carbon and *catenation property of carbon* for forming the chain structure of higher hydrocarbons laid a rewarding foundation stone for the advancement of organic chemistry. But extrapolation of this idea in coordination chemistry gave nothing but a negative displacement of inorganic chemistry by nearly half a century.

By following the analogy of catenation in carbon compounds, Blomstrand had put forward the *chain theory* that was later accepted and extended by S.M. Jorgensen to explain the properties of cobalt-ammines and platinum-ammines. Comparing the organic chain like R—CH$_2$—CH$_2$—CH$_2$—, they proposed the structures of ammine complexes (Fig. 1.3.1) assuming the valence of nitrogen as 5 and the valence of cobalt as 3.

To explain the reactivity and other properties (discussed in Sec. 1.2) of different cobalt-ammines, it was suggested that the **instantly reactive chlorides** (to be precipitated as AgCl by the addition of AgNO$_3$ to the freshly prepared aqueous solutions of the ammine compounds) were situated at the end of (NH$_3$)$_x$ chain but the **unreactive chlorides** were directly attached to the cobalt(III) centre (*cf.* Fig. 1.3.1).

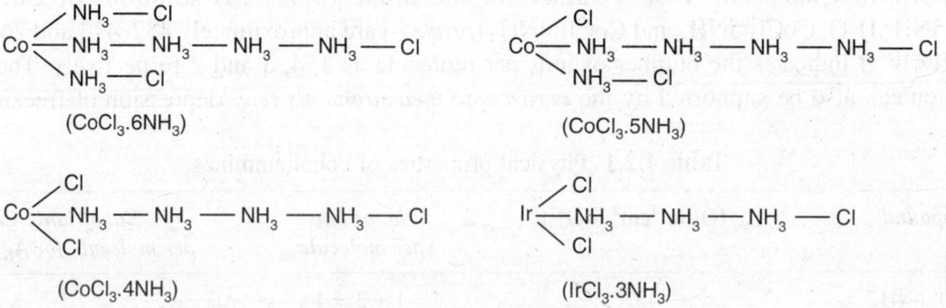

Fig. 1.3.1 Chain structures of different ammine complexes.

● According to **Blomstrand-Jorgensen's chain structure**, in $CoCl_3.6NH_3$ all the three chlorides are away from the metal centre (*i.e. these are less tightly bound*) and these are readily precipitated as AgCl by the addition of $AgNO_3$. This is in agreement with the experimental fact (*cf.* Table 1.2.1).

● In the chain structure of $CoCl_3.5NH_3$, it is evident that out of the three chlorides, two chlorides are reactive with respect to $AgNO_3$ while the third chloride which is directly bound to the metal centre (*i.e. this chloride is relatively more tightly bound*) is unreactive with respect to AgCl precipitation. This is also in agreement with the experimental observation (*cf.* Table 1.2.1).

● In the same way, the chain structure of $CoCl_3.4NH_3$ indicates that one chloride is reactive while the other two chlorides are unreactive with respect to the AgCl precipitation reaction by the treatment with $AgNO_3$ solution. This prediction was also supported by the experimental fact.

● At that time, Jorgensen's school could not prepare the next member $CoCl_3.3NH_3$ of the cobalt-ammine series to verify their prediction from their chain theory. However, they were able to prepare the analogous iridium compound, $IrCl_3.3NH_3$ (*cf.* Ir is a heavier congener of Co in the periodic table). The chain structure of $IrCl_3.3NH_3$ predicts that out of the three chlorides, only one chloride which is held at the end of $(NH_3)_x$-chain should give the precipitate of AgCl on the addition of $AgNO_3$. **But, this compound did not give any precipitate of AgCl and thus the chain theory appeared to be incorrect.** Later on, this deficiency of chain theory was proved in many other metal complexes.

Fate of Jorgensen !

We should remember that Jorgensen's contribution to shape the present form of coordination chemistry can never be narrowed down. As an experimental chemist, he was second to none. **He did a lot of painstaking experiment to characterize the coordination compounds known at his time.** In fact, Werner proposed his revolutionary theory largely based on Jorgensen's data just keeping the limitations of the chain theory in mind. Perhaps, if Jorgensen were not prejudiced by the catenation theory of carbon utilized in explaining the higher hydrocarbons, he would have earned the same results and fame won by Werner.

The *unfortunate influence of organic chemistry* misdirected the school of Jorgensen. But, inorganic chemistry is immensely grateful to **Arrhenius** for his famous **ionisation theory** (1887) that saved the inorganic chemistry from its wrong course induced by the contemporary development in organic chemistry. The ionisation theory was rightly applied to the coordination compounds known at that time to understand their structural features. In fact, by using the principle of ionisation theory of Arrhenius, the coordination chemists started to make a fresh journey in the right direction. This journey earned the Noble Prize in chemistry from the contribution of Werner to understand the coordination compounds.

1.4 WERNER'S THEORY OF COORDINATION COMPOUNDS

Alfred Werner (1866–1919)

He was born in Switzerland. He studied chemistry in Karlsruhe and then in Zurich. His research interest was in understanding of valence. In his doctoral thesis (1890), he explained the structural properties of some nitrogenous compounds by extending the Van't Hoff's theory of tetrahedral carbon atom to nitrogen atom.

To understand the bonding and structural features of coordination compounds, Werner used the principle of *ionisation theory* proposed by Arrhenius in 1887 and the principle of *structural chemistry including stereochemistry*. Based on these findings, he proposed his famous theory known as **Werner's coordination theory** in 1893 at the age of twenty seven. At the same year, he became a full Professor of Chemistry at the university of Zurich. Then for the next 20 years, he proposed and characterized different coordination compounds to perfect and substantiate his famous theory. He first prepared the *optically active purely inorganic compounds* that further helped him to substantiate his theory which ultimately earned the Noble Prize in Chemistry 1913. This was the **first Noble Prize in the branch of Inorganic Chemistry.** Werner passed away in 1919 only at the age of fifty three.

Noble Prizes in Inorganic Chemistry

Alfred Werner (1913) (coordination theory); G. Wilkinson and E. O. Fischer (1973) (ferrocene – organometallic chemistry); William Lipscomb (1976) (borane chemistry); H. C. Brown (1979) (borane chemistry); Roald Hoffman (1981) (organometallic chemistry); Henry Taube (1983) (inorganic reaction mechanism).

1.4.1 Basic Postulates of Werner's Coordination Theory

In the famous *Werner's coordination theory* (1893), it was proposed that the valence of an atom (specially the metal centre) does not remain necessarily confined to a fixed small number as expected in classical idea but *it may get extended over the entire surface of the atom and these valencies are of different strength.*

The basic postulates of the theory can be formulated as follows:

(i) *Metal centre shows two types of valence.* Depending on the properties and strength of the valencies showed by the metal centre (acting as the central atom), the valencies can be classified in two groups: **primary valence or principal valence and secondary valence or auxiliary valence or residual valence**. After satisfaction of the primary valence (*i.e.* ionic charge) of the metal ion, the metal ion still possesses some residual binding capacity that gives the secondary valence.

(ii) The **primary valence** is utilized to form the ionic bonds while the **secondary valence** is utilized to form the coordinate covalent bonds in space.

(iii) The **primary valence** is utilized *to bind the anions ionically* and these anions are ionisable in solution while the **secondary valence** is utilized *to bind the neutral or anionic groups through the coordinate covalent bonds in space*. The groups bound by the secondary valence are not ionisable.

(iv) **Satisfaction of the primary valence** leads to the ordinary molecules like $CoCl_3$, NH_3, etc. while satisfaction of both the primary and secondary valence leads to the coordination compounds like $CoCl_3 \cdot 6NH_3$; $CoCl_3 \cdot 5NH_3$, etc.

 (v) For a particular metal centre, **the secondary valence is generally fixed and it gives the coordination number of the metal centre.**

 (vi) **The secondary valencies are directed in space around the metal centre to constitute the coordination sphere** (more correctly, 1st coordination sphere). The *directional property* of the secondary valence (*i.e.* coordinate covalent bonds) gives *a definite geometrical shape of the coordination sphere.*

(vii) To determine the structure (*i.e.* geometry) of the complex species (*i.e.* coordination sphere), Werner considered the **number of isolated isomers** for a certain geometry. Different geometries for a certain coordination number can produce different numbers of geometrical isomers. From the actual number of geometrical isomers, the geometry was predicted. In some cases, possibility of optical isomerism was considered by Werner to determine a particular geometry. All these aspects have been discussed in detail in Sec. 1.4.5.

The above postulates are explained and illustrated in the next sections.

(**Note:** The **primary valence is nondirectional** and it forms the ionic bonds. Thus, it has nothing to do with the geometrical shape of the complex.

It may be noted that the concept of coordinate bond was popularized by Sidgwick in 1923. Thus, at the time of Werner, the concept of coordinate bond formation by the secondary valence was not well understood.)

1.4.2 Characteristic Features of the Primary and Secondary Valencies

● **Primary Valence:** These are having the following features:

  (i) satisfied ionically by the **anions only**;

 (ii) supposed to satisfy the chemical equivalence of the metal centre;

(iii) anionic moieties are held by the **nondirectional ionic bonds and these are ionizable;**

 (iv) given by the oxidation state of the metal centre; it is **fixed** for a particular metal ion.

● **Secondary valence:** These are having the following features:

  (i) satisfied by the anions and/or neutral molecules; by the coordinate covalent bonds;

 (ii) the **anions** bound by the coordinate covalent bonds simultaneously satisfy **both the primary** and **secondary valence**;

(iii) the moieties held by the secondary valencies are nonionizable;

 (iv) the **coordinate covalent bonds are of directional property** and spatial orientation of these bonds provides a fixed geometry around the metal centre;

  (v) the number of such coordinate covalent bonds gives the **coordination number** of the metal centre; for example: 4-coordinated complexes are either planar or tetrahedral, 6-coordinated complexes are generally octahedral; it **may vary** even for a particular metal ion;

 (vi) the number of such coordinate covalent bonds gives the measure of secondary valence.

(vii) the moieties bound by the coordinate covalent bonds to satisfy the secondary valence constitute the **coordination sphere** which is shown within [ ] and the moieties bound by primary valencies are kept outside the [ ].

(**Note:** The idea of coordinate covalent bond was visualized by G. N. Lewis (1916) and N. V. Sidgwick (1923) after the Werner's theory.)

Illustration:

● The coordination sphere produced by the secondary valencies is generally referred to as **complex entity** that may be neutral, anionic or cationic.

● The moieties held by the secondary valencies within the coordination sphere are described as the **ligands**.

● The anions bound by the secondary valency within the coordination sphere satisfy both the primary and secondary valence simultaneously. **Such anions are generally considered as nonionizable while the ions held ionically outside the coordination sphere by the primary valence are ionizable.** The anions held by the secondary valence within the coordination sphere may ionize (*i.e.* may come out from the coordination sphere) in solution if the metal centre is **sufficiently liable** (a kinetic term). However, for the *inert centre* like Co(III), Pt(IV), etc. the anions held by the secondary valence do not ionize rapidly (*i.e.* the process occurs only slowly). This aspect will be discussed later on at the appropriate place.

● The number of primary valence and secondary valence are given for different complexes in Table 1.4.2.1 for illustration. These are also illustrated in Table 1.4.3.1.

● For a particular metal ion, the **primary valence is fixed** (given by the ionic charge or oxidation state of the metal ion) but the **secondary valence may vary from complex to complex**. In fact, depending on the nature of ligands, a particular metal centre may adopt different geometries and coordination numbers (*i.e.* different secondary valencies). For example, Ni^{2+} may adopt tetrahedral, square planar and octahedral complexes depending on the nature of the ligands. This aspect will be discussed at the appropriate place.

Table 1.4.2.1 Illustration of primary and secondary valencies.

Complex Species	Primary valence	Oxidation state of the metal	Secondary valence (= coordination number)
$[Co(NH_3)_6]^{3+}$	3	+3	6
$[CoCl_2(NH_3)_4]^+$	3	+3	6
$[Fe(CN)_6]^{3-}$	3	+3	6
$[Fe(CN)_6]^{4-}$	2	+2	6
$[FeCl_4]^-$	3	+3	4
$[PtCl_2(NH_3)_4]^{2+}$	4	+4	6
$[PtCl_2(NH_3)_2]^0$	2	+2	4
$[Ni(NH_3)_6]^{2+}$	2	+2	6
$[Ni(CO)_4]^0$	0	0	4
$[Ni(CN)_4]^{2-}$	2	+2	4

1.4.3 Formulation of the Cobalt(III)-Ammine Complexes by Werner's Theory

Physical and chemical properties of the complexes have been discussed in Sec. 1.2. The number of chlorides per molecule precipitated by the addition of $AgNO_3$ to the freshly prepared aqueous solutions gives the idea of the number of *free chlorides* (*i.e.* the number of *ionizable chlorides*). This is also further substantiated from the measurement of molar conductance of the compounds in dilute aqueous solutions (*cf.* Table 1.2.1). For the formulation of the coordination compounds, Werner also considered the geometrical structures of the coordination spheres. This led to the development of **stereoisomerism** in coordination compounds. This aspect immensely helped him to establish his theory (*cf.* Sec. 1.4.5).

(i) **Luteocobaltic chloride** ($CoCl_3.6NH_3$) produces 4 ions (predicted from Λ_M) and 3 free Cl^- ions per molecule (predicted from the precipitation of AgCl) in an aqueous solution.

According to Werner's theory, in the compound, three Cl^- ions are *ionically bound by the primary valence*. The NH_3 molecules which are not lost even by the treatment of concentrated H_2SO_4 (*cf.* Sec. 1.2) are bound to the metal centre by the *secondary valence*. Thus, $6NH_3$ molecules are bound by the coordinate covalent bonds (which are basically covalent bonds) with the cobalt(III) centre to produce the coordination sphere. This coordination sphere is shown within [ ] and the moieties bound by the primary valence are kept outside the [ ] in the formulation of the compound. Thus, the compound $CoCl_3.6NH_3$ is to be represented by $[Co(NH_3)_6]Cl_3$ that ionizes as follows:

$$\left[Co\left(NH_3\right)_6\right]Cl_3 \rightleftharpoons \left[Co\left(NH_3\right)_6\right]^{3+} + 3Cl^- \text{ (i.e. ter-univalent electrolyte)}$$

$CoCl_3 \cdot 6NH_3$:

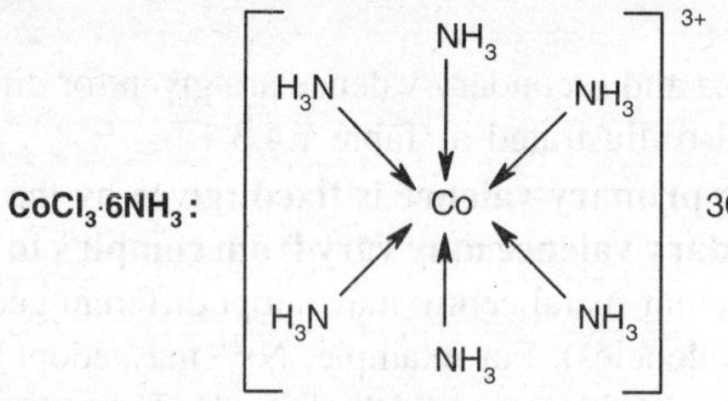

It was proved by Werner that the coordination sphere represents an **octahedral geometry**. (**Note:** Presently the bonds within the coordination sphere are not denoted by ←. These are simply denoted by — as practised for the covalent bonds. In fact, the coordinate bond or dative bond is only a special type of covalent bond and here the bonding electron pair is donated by the donor, *i.e.* ligand, and the electron pair is shared between the combining centres as usual. *Thus, both the coordinate and ordinary covalent bonds are the 2c-2e bonds.*).

(ii) The properties of **roseocobaltic chloride** ($CoCl_3.5NH_3.H_2O$) (*cf.* Table 1.2.1) indicate that it is a *ter-univalent electrolyte producing three free Cl^- ions per molecule*. Thus, according to Werner's theory, these $3Cl^-$ ions are ionically bound by the primary valence while the 6 secondary valencies are satisfied by $5NH_3$ and $1H_2O$ molecules. Thus the compound can be formulated as follows:

$$\left[Co\left(NH_3\right)_5\left(OH_2\right)\right]Cl_3 \rightleftharpoons \left[Co\left(NH_3\right)_5\left(OH_2\right)\right]^{3+} + 3Cl^-$$

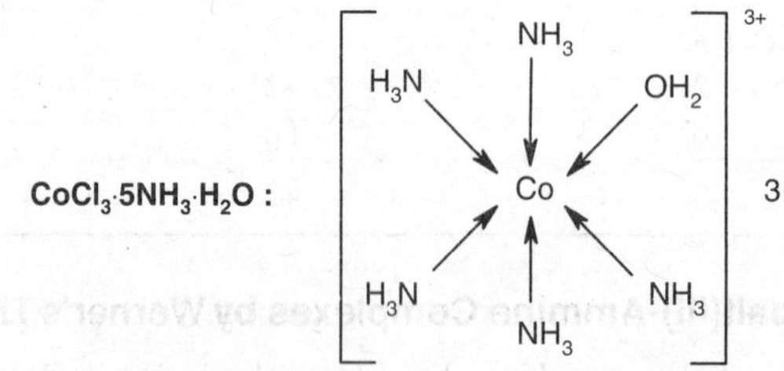

$CoCl_3 \cdot 5NH_3 \cdot H_2O$:

(iii) The properties of **purpureocobaltic chloride** ($CoCl_3.5NH_3$) (*cf. Table 1.2.1*) indicate that it is a *bi-univalent electrolyte producing two free Cl^- ions per molecule* in solution. Thus according to Werner's theory, out of the three Cl^- ions, two Cl^- ions are ionically bound by the primary valence and the other Cl^- ion is present in the coordination sphere and it is bound to the metal centre by a coordinate covalent bond (*i.e.* by a secondary valence). *The Cl^- ion present in the coordination sphere satisfies both one primary valence and one secondary valence simultaneously.* This Cl^- ion does not readily ionize (*cf.* Co(III)-centre in the present compound is highly inert). The other 5 secondary valencies are satisfied by $5NH_3$ molecules. Thus the compound can be formulated as follows:

$$\left[CoCl(NH_3)_5\right]Cl_2 \rightleftharpoons \left[CoCl(NH_3)_5\right]^{2+} + 2Cl^-$$

$CoCl_3 \cdot 5NH_3$:

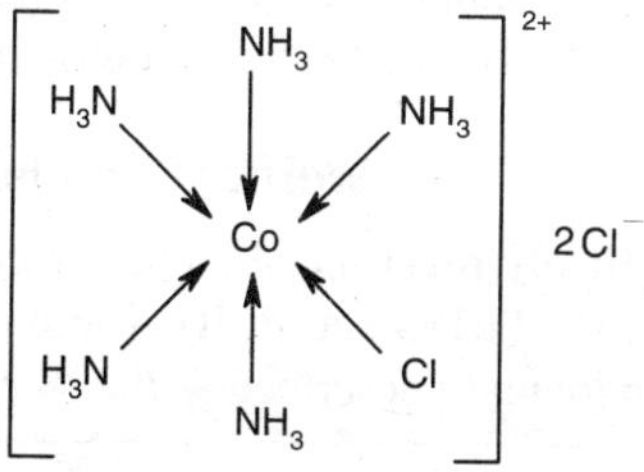

$2Cl^-$

(iv) The properties of **praseocobaltic chloride** ($CoCl_3.4NH_3$) (*cf.* Table 1.2.1) indicate that the compound is a *uni-univalent electrolyte producing one free* Cl^- *ion per molecule in solution*. This can be explained by Werner's theory by considering the presence of $2Cl^-$ ions within the coordination sphere to satisfy two primary and two secondary valencies simultaneously. The third Cl^- is ionically bound by the primary valence. The $4NH_3$ molecules are present in the coordination sphere to satisfy the residual 4 secondary valencies. Thus the compound can be represented as follows:

$$\left[CoCl_2(NH_3)_4\right]Cl_2 \rightleftharpoons \left[CoCl_2(NH_3)_4\right]^+ + Cl^-$$

$CoCl_3 \cdot 4NH_3$: (*trans*-form)

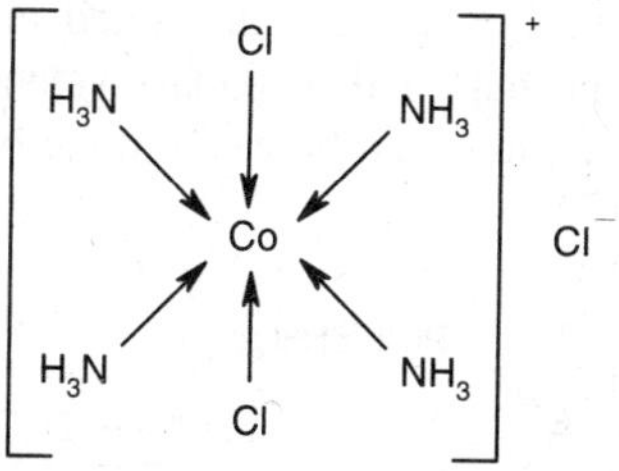

Cl^-

Table 1.4.3.1 Formulation of cobalt(III) - ammine compounds.

Compound***	Formulation	Primary Valence	Secondary Valence
Luteocobaltic Chloride ($CoCl_3 \cdot 6NH_3$)	$[Co(NH_3)_6]Cl_3$	3 by 3 Cl^- ions	6 by 6 NH_3 molecules
Roseocobaltic Chloride ($CoCl_3 \cdot 5NH_3.H_2O$)	$[Co(NH_3)_5(OH_2)]Cl_3$	3 by 3 Cl^- ions	6 by 5 NH_3 and 1 H_2O molecules
Purpureocobaltic Chloride ($CoCl_3 \cdot 5NH_3$)	$[CoCl(NH_3)_5]Cl_2$	3 by 3 Cl^- ions*	6 by 5 NH_3 molecules and 1 Cl^- ion*
Violeo- and praseocobaltic Chloride ($CoCl_3 \cdot 4NH_3$)	$[CoCl_2(NH_3)_4]Cl$	3 by 3 Cl^- ions**	6 by 4 NH_3 molecules and 2 Cl^- ions**

* One Cl^- present within the coordination sphere satisfies both one primary valence and one secondary valence simultaneously.
** 2 Cl^- ions present within the coordination sphere satisfy both two primary valencies and two secondary valencies simultaneously. It can produce the *cis*- (violeo-) and *trans*- (praseo) isomers.
*** The compound, $CoCl_3 \cdot 3NH_3$ can be formulated as $[CoCl_3(NH_3)_3]^0$ which is a nonelectrolyte. It can produce the *fac-mer* isomerism.

(**Note:** Later on, it was proved that the green coloured compound $CoCl_3.4NH_3$ is actually the *trans*-isomer. The corresponding *cis*-isomer is violet coloured and its coloured code name is *violeocobaltic chloride*. Here it is worth mentioning that the *cis*- and *trans*-isomers of $[CoCl_2(en)_2]^+$ also show the similar colour variations, *i.e.* *cis*-$[CoCl_2(en)_2]^+$ (**violet**) and *trans*-$[CoCl_2(en)_2]^+$ (**green**).

Werner Complexes

Werner studied the **kinetically inert** metal-ammonia complexes, *i.e.* metal-ammines. The metal centres were mainly Co(III), Pt(IV) and Pt(II). These classical metal-ammines and their related complexes are very often lovingly described as the *Werner Complexes*.

1.4.4 Formulation of a Series of Platinum(IV)-Ammine Complexes by Werner's Theory

The series of kinetically inert Pt(IV)-ammine complexes are:

$PtCl_4.6NH_3$; $PtCl_4.5NH_3$; $PtCl_4.4NH_3$; $PtCl_4.3NH_3$; $PtCl_4.2NH_3$; $PtCl_4.NH_3.KCl$; $PtCl_4.2KCl$.

The variation of molar conductivity (Λ_M) of the above series of the compounds is shown in Fig. 1.4.4.1. It shows that $PtCl_4.2NH_3$ is a nonelectrolyte having the minimum conductance (close to zero). Thus it is reasonable to consider the structure of the compound as: $[PtCl_4(NH_3)_2]^0$ where no moiety exists outside the coordination sphere.

Thus the measurement of molar conductance of the compounds of Pt(IV) indicates the nature of the compounds as electrolytes. These are illustrated in Table 1.4.4.1. From the mode of ionisation, the moieties to satisfy the primary and secondary valencies can be recognised. From this knowledge, the formulation of the compounds can be easily done by using the Werner's theory (*cf.* Table 1.4.4.1).

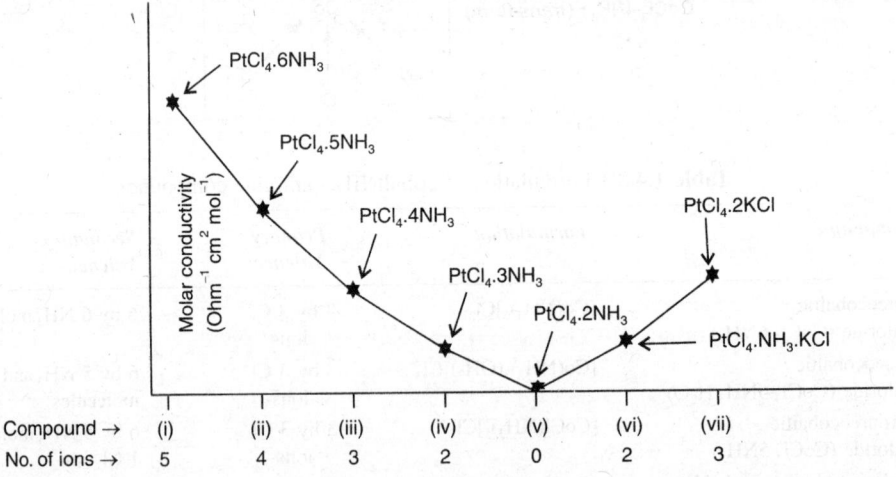

Fig. 1.4.4.1 Molar conductivity of a series of platinum(IV)-ammine compounds (For numbering of the compounds, *see* Table 1.4.4.1).

From the conductance data, formulation of the compounds has been done. Formulation of these compounds indicates the number of Cl^- ions that remain outside the coordination sphere. Such Cl^- ions are bound ionically by the primary valencies only and these are to be precipitated by the addition of $AgNO_3$ to the freshly prepared solutions of the compounds. This prediction is also supported by the experimental fact for the said Pt(IV)-compounds.

Table 1.4.4.1 Molar conductance (Λ_M) of platinum(IV)-ammine compounds and formulation of the compounds according to Werner's theory.

Compound	Λ_M (Ohm^{-1} cm^2 mol^{-1})	Type of electrolyte	No.of ions* per molecule	Formulation**
(i) PtCl$_4$.6NH$_3$	~ 520	4 – 1	4 + 1 = 5	[Pt(NH$_3$)$_6$]Cl$_4$
(ii) PtCl$_4$.5NH$_3$	~ 400	3 – 1	3 + 1 = 4	[PtCl(NH$_3$)$_5$]Cl$_3$
(iii) PtCl$_4$.4NH$_3$	~ 230	2 – 1	2 + 1 = 3	[PtCl$_2$(NH$_3$)$_4$]Cl$_2$
(iv) PtCl$_4$.3NH$_3$	~ 100	1 – 1	1 + 1 = 2	[PtCl$_3$(NH$_3$)$_3$]Cl
(v) PtCl$_4$.2NH$_3$	~ 5***	Nonelectrolyte	0	[PtCl$_4$(NH$_3$)$_2$]
(vi) PtCl$_4$.NH$_3$.KCl	~ 110	1 – 1	1 + 1 = 2	K[PtCl$_5$(NH$_3$)]
(vii) PtCl$_4$.2KCl	~ 255	2 – 1	2 + 1 = 3	K$_2$[PtCl$_6$]

* This conclusion from the Λ_M values can also be substantiated by the *cryoscopic measurement data*.

** In all cases, primary valence = 4, oxidation state of platinum = +4 and secondary valence = 6. Cl$^-$ ions present within the coordination sphere satisfy both the primary and secondary valencies simultaneously and these are not ionizable and not precipitated by the addition of AgNO$_3$.

*** Theoretically it should be zero. The very small value may be due to an impurity or an insignificant reaction with the solvent producing a very small concentration of the ions.

Complication in the interpretation of conductance data and results of silver halide (AgX) precipitation due to solvation at the kinetically labile centres

It has been assumed that the moieties remaining outside the coordination sphere are ionically bound by the primary valence only and these remain as the free ions in solution. On the other hand, the moieties present within the coordination sphere are bound by the secondary valence and these do not come out from the coordination sphere as the free species. **This assumption is true only for the kinetically inert metal centres.** For the labile centres, the moieties present in the coordination sphere come out due to solvation to complicate the interpretation. For example, the *dark green* **bromopraseo salt**, [CoBr$_2$(NH$_3$)$_4$]Br on dissolution gives a dark green solution which gradually changes in colour to red. This colour change is accompanied with the gradual increase in molar conductance. This has been explained by considering the **aquation,** *i.e.* substitution of bromides by water molecules in the coordination sphere as follows:

$$[CoBr_2(NH_3)_4]Br \xrightarrow{+H_2O} [CoBr(NH_3)_4(OH_2)]Br_2$$

(1–1 electrolyte) (2–1 electrolyte)
(Dark green)

$$\downarrow +H_2O$$

$$[Co(NH_3)_4(OH_2)_2]Br_3$$

(3–1 electrolyte)
(Red)

To avoid the complication from the solvation which is likely to occur in labile centres, Werner used the inert complexes formed by Co(III) (octahedral complexes), Pt(IV) (octahedral complexes), Cr(III) (octahedral complexes) and Pt(II) (square planar complexes). Though at that time, the reason behind the **kinetic lability** and **inertness** was not clearly understood.

1.4.5 Stereochemical Aspects of Coordination Compounds and Werner's Theory

In Werner's theory, it was proposed that the coordination sphere is constituted by the formation of **directional coordinate covalent bonds** *(i.e. secondary valence)*. This predicts the development of stereoisomerism in coordination compounds.

Werner dealt with many coordination compounds of coordination number 6 and 4 to verify the prediction of stereoisomerism in coordination compounds based on his theory.

(A) Coordination number (C.N.) 6: The arrangement of six groups around the metal centre within the coordination sphere can lead to three possible geometries—*planar hexagon, trigonal prism and octahedran.* For a particular geometry, if the groups (*i.e.* ligands) are arranged in different ways then

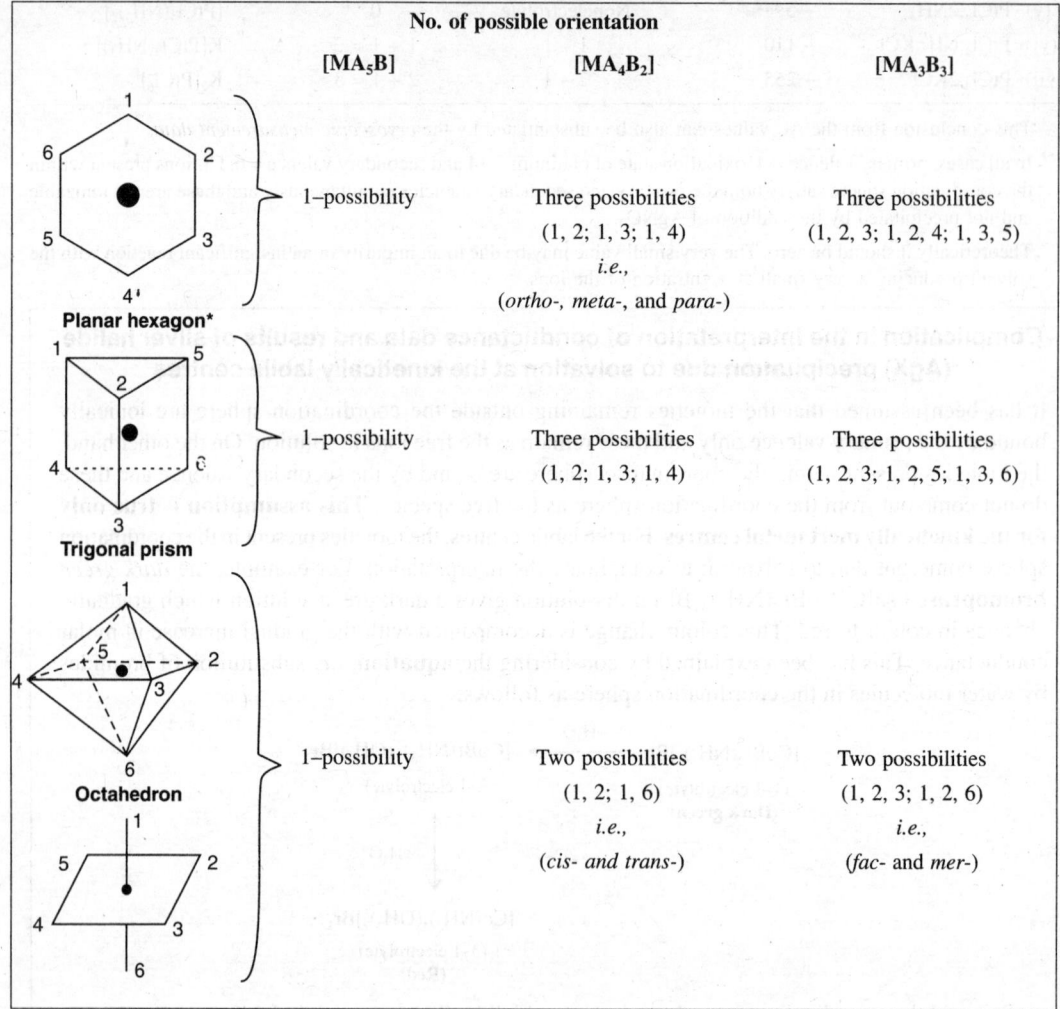

	No. of possible orientation		
	[MA₅B]	[MA₄B₂]	[MA₃B₃]
Planar hexagon*	1–possibility	Three possibilities (1, 2; 1, 3; 1, 4) *i.e.,* (ortho-, meta-, and para-)	Three possibilities (1, 2, 3; 1, 2, 4; 1, 3, 5)
Trigonal prism	1–possibility	Three possibilities (1, 2; 1, 3; 1, 4)	Three possibilities (1, 2, 3; 1, 2, 5; 1, 3, 6)
Octahedron	1–possibility	Two possibilities (1, 2; 1, 6) *i.e.,* (cis- and trans-)	Two possibilities (1, 2, 3; 1, 2, 6) *i.e.,* (fac- and mer-)

* It may also lead to a **pyramidal geometry** that will produce the same number of isomers as in the case of hexagonal planar geometry.

Fig. 1.4.5.1 Number of possible geometrical isomers for coordination number 6 in three cases.
(● denotes the position of metal centre; the numbers 1, 2, 3... denote the positions of the ligands)

Table 1.4.5.1 Number of isomers predicted and actually found for different geometries having coordination number 6.

Complex	Prediction			Observation
	Planar hexagon	Trigonal prism	Octahedron	
$[MA_6]$	1	1	1	1
$[MA_5B]$	1	1	1	1
$[MA_4B_2]$	3	3	2	2
$[MA_3B_3]$	3	3	2	2

(A and B are the monodentate ligands)

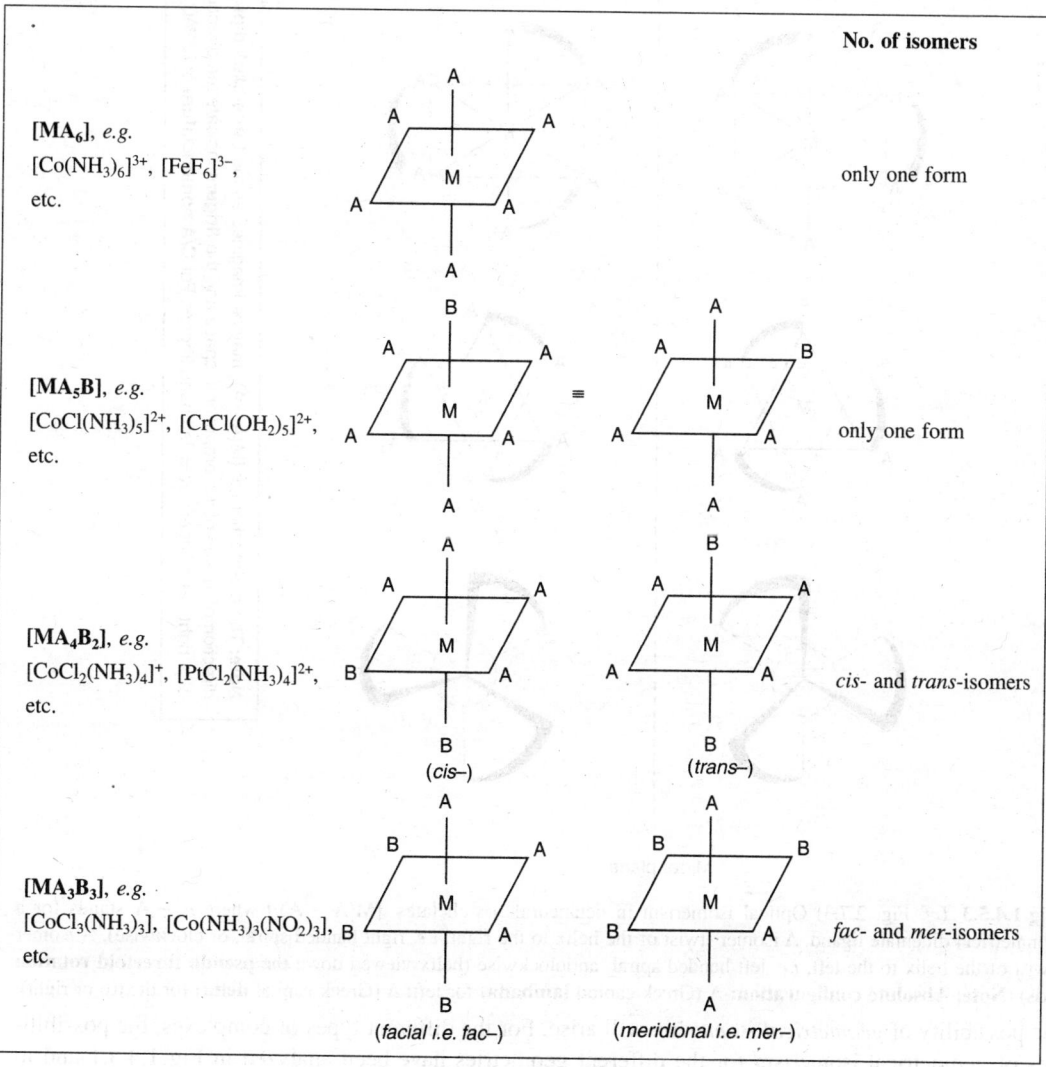

No. of isomers

$[MA_6]$, e.g.
$[Co(NH_3)_6]^{3+}$, $[FeF_6]^{3-}$,
etc.

only one form

$[MA_5B]$, e.g.
$[CoCl(NH_3)_5]^{2+}$, $[CrCl(OH_2)_5]^{2+}$,
etc.

only one form

$[MA_4B_2]$, e.g.
$[CoCl_2(NH_3)_4]^+$, $[PtCl_2(NH_3)_4]^{2+}$,
etc.

cis- and trans-isomers

(cis-)   (trans-)

$[MA_3B_3]$, e.g.
$[CoCl_3(NH_3)_3]$, $[Co(NH_3)_3(NO_2)_3]$,
etc.

fac- and mer-isomers

(facial i.e. fac-)   (meridional i.e. mer-)

Fig. 1.4.5.2 Isomers in the octahedral complexes having the six monodentate ligands (A and B).

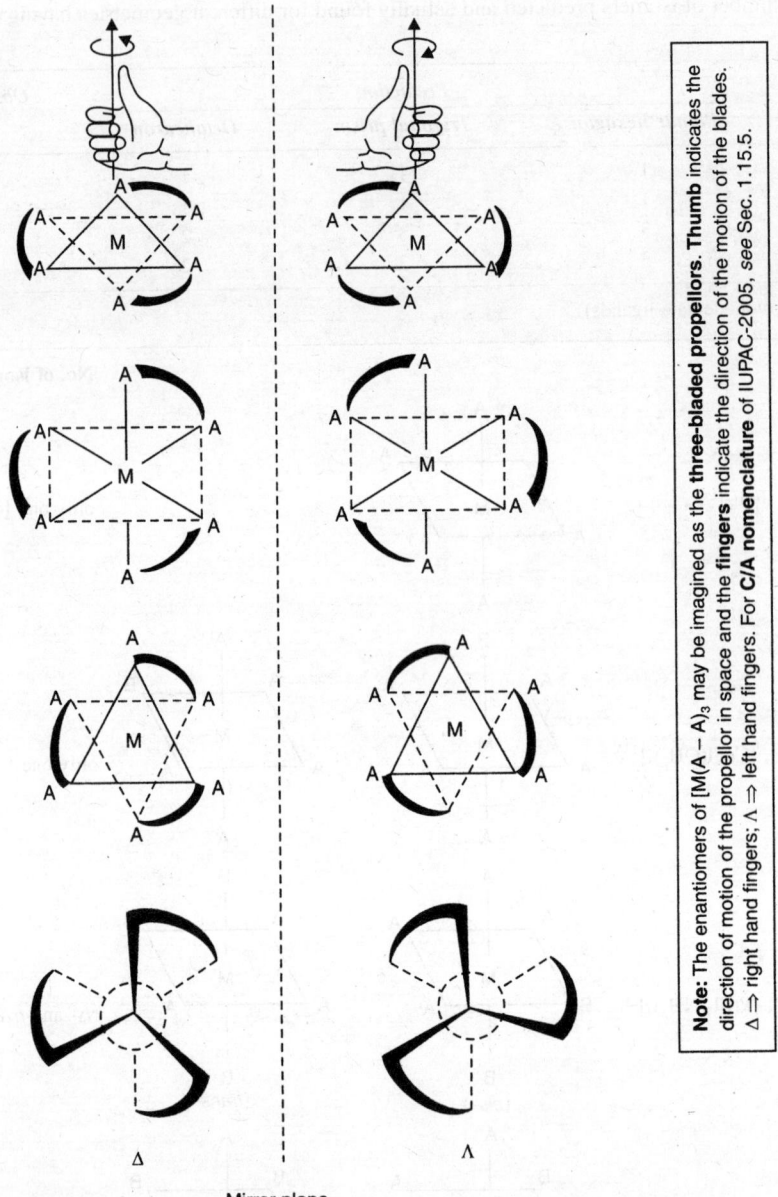

Note: The enantiomers of [M(A – A)₃] may be imagined as the **three-bladed propellors. Thumb** indicates the direction of motion of the propellor in space and the **fingers** indicate the direction of the motion of the blades. For **C/A nomenclature** of IUPAC-2005, see Sec. 1.15.5.
Λ ⇒ right hand fingers; Λ ⇒ left hand fingers.

Δ   Λ

Mirror plane

Fig 1.4.5.3 (*cf.* Fig. 2.7.3) Optical isomerism in octahedral *tris*-chelates, [M(A – A)₃] where A – A stands for a symmetrical didentate ligand. Δ isomer (twist of the helix to the right, *i.e.* right handed spiral, or clockwise), Λ isomer (twist of the helix to the left, *i.e.* left handed spiral, anticlockwise (helix viewed down the **pseudo threefold rotation axis**) (**Note: Absolute configuration: Λ** (Greek capital **lambada**) for left; Δ (Greek capital **delta**) for dextro or right).

the possibility of *geometrical isomerism* will arise. For the different types of complexes, the possibilities of geometrical isomerism for the different geometries have been analyzed in Fig 1.4.5.1 and in Table 1.4.5.1.

(**Note:** *fac-* and *mer-* denote the facial and meridional respectively; *cis* (Latin) meaning *on the same side; trans* (Latin) meaning *across; cf,* E (*entgegen* meaning opposite), Z (*zusammen* meaning together) nomenclature in organic chemistry.)

● **[MA$_6$] and [MA$_5$B] type complexes:** For [MA$_6$] (*i.e.* all identical ligands), no isomerism is expected for the three possible geometries. It also happens so for [MA$_5$B] that has 5 identical ligands and the sixth ligand is different.

● **[MA$_4$B$_2$] type complex:** For the complex [MA$_4$B$_2$], the octahedral structure predicts two isomers while the other two possible geometries predict three isomers in each case. In fact, for the complex [CoCl$_2$(NH$_3$)$_4$]Cl, two isomers (*violeo*-salt, *i.e. cis*-isomer where two chlorides are at the adjacent positions like 1, 2 or 1, 3 or 1, 4 or 1, 5 or 2, 3 or .... and the *praseo*-salt, *i.e. trans*-isomer where two chlorides are at the *trans*-axial positions or diagonally opposite positions like 1, 6 or 2, 4 or 3, 5).

● **[MA$_3$B$_3$] type complex:** For the complex [MA$_3$B$_3$], there are two possible isomers in an octahedral arrangement while for the other two possible geometries, there are three isomers in each case. In the octahedral complex, three identical ligands can be positioned at the corners of a trigonal face (*i.e.* 1, 2, 3 or 1, 4, 5 or 1, 3, 4 or 2, 3, 6, or ...) or along the meridian (*i.e.* 1, 2, 6 or 1, 3, 6 or 1, 4, 6 or 2, 3, 4 or 2, 4, 5, or ...). These isomers are described as the *facial* or *fac-* and meridional or *mer*-isomers respectively. In fact, for the complex [CoCl$_3$(NH$_3$)$_3$], the two isomers are established. The isomers in the octahedral complexes are shown in Fig. 1.4.5.2.

Based on the above results, Werner concluded that for coordination number 6, geometry of the coordination sphere is octahedral. Later on, this prediction was supported from the X-ray structure determination.

● **[M(A – A)$_3$] type complex:** At the time of Werner, complexes of the type [M(A – A)$_3$] where A– A stands for a *symmetrical didentate ligand* like ethylenediamine, H$_2$N—CH$_2$—CH$_2$—NH$_2$ were known. For the octahedral geometry, it should have two *optical isomers, i.e.* Δ and Λ isomers for the right and left-handed spirals (Fig. 1.4.5.3). Their mirror images are *nonsuperimposable.*

In fact, for the complex [Co(en)$_3$]Cl$_3$ (*en* stands for the didentate ligand ethylenediamine), the optical isomerism was established. It was questioned that the observed optical isomerism might be due to the carbon centres present in the ligand. Werner prepared the following purely inorganic complex compound

$$\left[Co\left(\underset{\underset{H}{O}}{\overset{\overset{H}{O}}{<}}Co(NH_3)_4 \right)_3 \right]^{6+}$$

(well known as **hexol complex**) of the type [M(A – A)$_3$] without any carbon atom. **This purely inorganic compound was resolved into the optical isomers.** Thus it was answered to the questioned optical activity of [Co(en)$_3$]Cl$_3$.

> **Note: Other two optically active complexes devoid of any organic moiety are:**
> (i) *cis*-[Rh(HNSO$_2$NH)$_2$(OH$_2$)$_2$]$^-$ (Mann, 1933) of *cis*-[M(A – A)$_2$B$_2$] type
> (ii) [Cr(HPO$_3$)$_3$]$^{3-}$ (Podlaha and Ebert, 1960) of [M(A – A)$_3$] type.

(B) Coordination number (C.N.) 4: For the four coordinated complexes, there are two possible geometries - *tetrahedron* and *square planar.*

● **[M(A)(B)(C)(D)] type complex:** [M(A)(B)(C)(D)] (having all the 4 ligands A, B, C, D different) should give *two optical isomers* if it is a tetrahedral complex. If the complex is a square planar one, then it should produce *three geometrical isomers but no optical isomer* (Fig. 1.4.5.4).

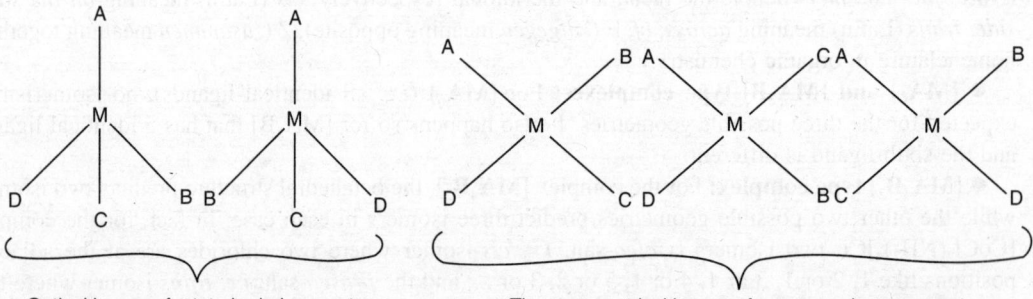

Optical isomers for tetrahedral geometry   Three geometrical isomers for square planar geometry

Fig. 1.4.5.4 Possible isomers of [M(A)(B)(C)(D)] for the tetrahedral and square planar arrangements.

In fact, both the tetrahedral and square planar geometries are known for different metal centres. For platinum(II), the complexes are square planar and in fact, three isomers are known for the complexes like $[Pt(NH_2OH)(NH_3)(NO_2)(py)]^+$, $[PtBrCl(NO_2)(NH_3)]^-$, $[PtBrCl(NH_3)(py)]$.

● **$[MA_4]$ and $[MA_3B]$ type complexes:** For the complexes of the type **$[MA_4]$** and **$[MA_3B]$**, no stereoisomerism is possible for both the tetrahedral and square planar configurations of the complexes.

● **$[MA_2B_2]$ type complexes:** For the complexes of the type $[MA_2B_2]$, there is no stereoisomerism for the tetrahedral structure, but there are *two geometrical isomers* (*cis-* and *trans-*) for the square planar structure. Werner showed that for the complexes like $[Pt^{II}Cl_2(NH_3)_2]$, and $[Pd^{II}Cl_2(NH_3)_2]$, there are two isomers. This proved that the said Pt(II) and Pd(II) complexes are square planar rather than tetrahedral.

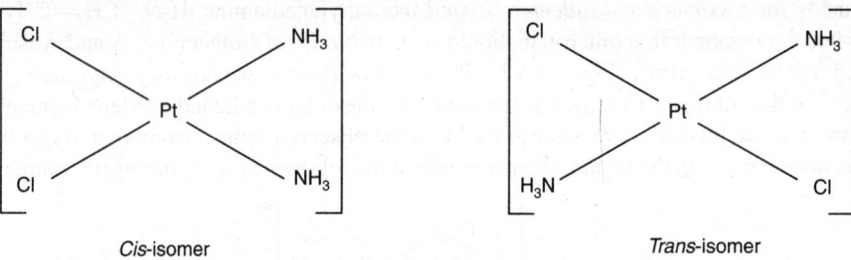

Cis-isomer   Trans-isomer

Fig. 1.4.5.5 *Cis-, trans-* isomerism in the square planar complexes of the type $[MA_2B_2]$.

● **$[MA_2BC]$ type complex:** The complex $[MA_2BC]$ can produce the *cis-* and *trans-* forms for the square planar arrangement but it shows no isomerism for the tetrahedral arrangement.

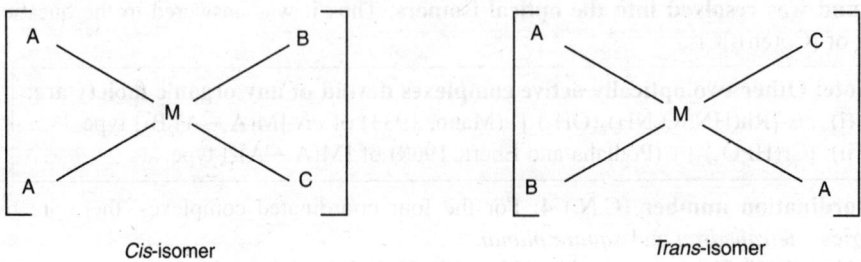

Cis-isomer   Trans-isomer

Fig. 1.4.5.6 *Cis-* and *trans-* isomers of the square planar complex $[MA_2BC]$.

The above results for the mondentate ligands are summarized in Table 1.4.5.2.

Table 1.4.5.2 Number of stereoisomers predicted for coordination number 4 (Square pyramidal geometry not considered).

Complex	Square planar geometry	Tetrahedral* geometry
[MA$_4$]	No isomerism	No isomerism
[MA$_3$B]	No isomerism	No isomerism
[MA$_2$B$_2$]	cis- and trans-isomers	No isomerism
[MA$_2$BC]	cis- and trans-isomers	No isomerism
[MABCD]	three geometrical isomers but no optical isomerism	two optical isomers but no geometrical isomerism

* It may be noted that for the tetrahedral geometry, **no geometrical isomerism is possible** because the relative position of each ligand is the same.

Werner studied mainly the platinum(II) and pladium(II) complexes of C.N. 4 and in such cases, the experimental observations agreed with the predictions made for the square planar geometry. **Thus, he concluded that the Pt(II) and Pd(II) complexes are square planar**. Later on, these were supported by X-ray studies.

Different stereoisomers for the square planar complexes may arise when the ligands are didentate or polydentate. This aspect has been discussed separately in Chapter 2.

(**Note: Werner's elucidation of structure on the basis of the number of isomers isolated:** This method of structure determination applied by Werner was not absolutely correct. One may question that Werner could not prepare all the isomers. He concluded just on the basis of the number of isomers which he could isolate. However, later on, it was confirmed by X-ray studies that the 6 coordinated complexes he studied were really octahedral. Thus, his prediction for the compounds was flawless. He ruled out the possibility of **trigonal pismatic geometry for C.N. 6** based on the number of isolated isomers. But in 1965 (*i.e.* about after 70 years of Werner's proposal), this geometry was proved to exist in reality in the complexes like **[Re(S$_2$C$_2$Ph$_2$)$_3$]**. Now several trigonal prismatic complexes are known. These are discussed in Chapter 2.)

Methods of studying the complexes

Werner had a limited scope of physical methods to study the complexes. He had to rely on *the molar conductivity data* to realize the mode of ionization of a complex species. In some cases, additional support could be obtained from the *precipitation of silver halide.* The mode of ionization could also by supported by the *cryoscopic measurements.* For structure determination, he had to rely on the *number of isolated isomers* and *optical activity.*

Now several powerful methods are available for this purpose. These are: measurement of magnetic properties, dipole moment measurement (for nonionic complexes), spectral properties (different types). The most powerful method is the determination of crystal structure by X-ray.

1.4.6 Definitions of the Terms: Complex, Ligand and Coordination Number (C.N.)

(**A**) **Complex and Ligand:** A metal centre (which may be neutral or cationic) can combine with a definite number of neutral molecules (*e.g.* NH$_3$, H$_2$O, CO, PR$_3$, etc.) or anions (*e.g.* halide, pseudohalide, NO$_2^-$, etc.) or cations (*e.g.* NO$^+$, etc., in rare cases) **in some certain directions** to produce *a new species of sufficient stability so that the new species can be characterized and identified.* Such a species generated is called the *complex*. The groups that combine with the metal centre to produce the complex are called the *ligands*. The ligands generally ligate the metal centre through the coordinate covalent bond formation. Thus in terms of Lewis acid-base concept, *the ligands are Lewis bases, the metal centres are Lewis acids and the complexes are* **Lewis acid-base adducts.** The complex formation equilibrium can be represented as follows:

$$M + xL \rightleftharpoons \left[ML_x\right]$$

M stands for the metal centre; L stands for the ligand; charge is not shown for the sake of simplicity. Here the complex is $[ML_x]$.

The ligands are arranged **in certain directions** around the metal centre to give the preferred geometry. These ligands bound through the coordinate covalent bonds produce the *coordination sphere* denoted by [ ].

(B) Coordination number (C.N.) or ligancy of the metal centre: It can be defined in many ways:

● The **number of secondary valencies** gives the coordination number of the metal centre in the given complex.

● The **number of coordinate bonds** which are the **sigma bonds** formed by the ligands within the coordination sphere to determine the geometry of the entity gives the coordination number of the metal centre in the given complex.

It may be noted that some ligands (*e.g.* CO, CN⁻, etc.) can bind the metal centre by both the σ- and π- bonds within the coordination sphere, *but the coordination number is determined by the number of these σ- bonds.* The π- bonds are not considered to determine the C.N.

● For the monodentate ligands, the **number of ligands** gives the coordination number.

● For the didentate and polydentate ligands, the **sum of denticity** shown by the ligands in the given compound gives the coordination number.

C.N. in Complexes and in Crystals

The term C.N. has been used to describe the crystals where it gives the number of nearest neighbours around the given atom or ion in a crystal. This C.N. used in connection with the crystals is different from the C.N. used in the coordination compounds *where the discrete molecular species exist.* In solid crystals, the question of existence of such discrete molecular species does not arise.

● **Heptacity (η^x) vs. denticity:** The metal-ligand bond may not be always the electron pair bond. In some cases, the π- electron cloud may participate in the formation of metal-ligand σ-bond (more correctly μ-bond). It happens so for the ligands like ethylene, benzene, cyclopentadienyl anion, etc. However, in all cases, *the denticity of a ligand is determined by the number of metal-ligand σ-bond that the ligand can form.* When an ethylenic double bond binds the metal centre as in $[Pt(C_2H_4)Cl_3]^-$, η^2–C_2H_4 contributes unity to the observed C.N. (*e.g.* in the given complex, C.N. = 4). When the cyclopentadienyl anion $C_5H_5^-$ (Cp⁻) binds to the metal centre, η^5–Cp⁻ also contributes unity to the C.N., *e.g.* in $[Mo(CO)_3(\eta^5$–Cp)Cl] C.N. = 5; in $[Fe(CO)(COCH_3)(\eta^5$–$C_5H_5)PPh_3]$, C.N. = 4; in $[ReBr_2(CO)_2(\eta^5$– Cp)] C.N. = 5. In $[Cr(\eta^6$–$C_6H_6)_2]$, C.N. is 2 as each η^6–C_6H_6 forms a single σ-bond with the metal centre. In these complexes, the 2e-donor ligands are Cl⁻, Br⁻, CO, and η^2–C_2H_4 while η^6–C_6H_6 or η^5–Cp⁻ is a 6e-donor. But each of them contributes unity to determine the total C.N.

Thus it is evident that heptacity (denoted by η^x, x = 2, 5, 6, etc.) is not necessarily the denticity of such ligands. Here it may be pointed out that some authors have argued that heptacity is the denticity in some cases to explain the stuctural features as in the distorted tetrahedral structure of $[Fe(CO)(COCH_3)(\eta^5$–$C_5H_5)PPh_3]$ (*cf.* Fig 2.8.1).

- In the complexes like **metal-olefin and metal-alkyne**, the denticity of the ligand is to be judged by the number of σ-*bonds produced* by the ligand with the metal centre. This is illustrated below.

$$\left[PtCl_4\right]^{2-} + C_2H_4 \longrightarrow \left[Pt\left(\eta^2 - C_2H_4\right)Cl_3\right]^- + Cl^-$$

Here, η^2–C_2H_4 replaces one Cl^- and forms one metal-ligand σ-bond. Thus, it is a **monodentate ligand**. But in the following complex, the olefin η^2–$C_2(CN)_4$ forming two sigma bonds acts as a **didentate chelating ligand**.

$$\begin{bmatrix} Ph_3P \diagdown \quad \diagup C(CN)_2 \\ \quad\quad Pt \\ Ph_3P \diagup \quad \diagdown C(CN)_2 \end{bmatrix}$$

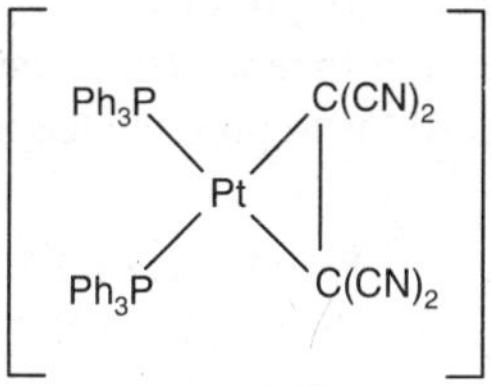

- **Alkyne** can also act as a **monodentate** as well as a **didentate ligand**.

$$\left[Mn(CO)_3\left(C_5H_5\right)\right] + RC\equiv CR \xrightarrow[\text{(UV)}]{hv} \left[Mn(CO)_2\left(C_2R_2\right)\left(C_2H_5\right)\right] + CO$$
$$\text{(\textbf{monodentate } R_2C_2)}$$

$$\left[V(CO)_4\left(C_5H_5\right)\right] + RC\equiv CR \xrightarrow[\text{(UV)}]{hv} \left[V(CO)_2\left(C_2R_2\right)\left(C_5H_5\right)\right] + 2CO$$
$$\text{(\textbf{didentate } R_2C_2)}$$

Replacement of one CO group by R_2C_2 indicates the monodentate character of R_2C_2 while replacement of two CO groups supports the didentate character of R_2C_2. To show the didentate character of R_2C_2, the following two sets of MOs (assuming x-axis as the molecular axis) of R_2C_2 are involved for bonding with the metal centre.

$$\left(\pi_{p_y} + \pi^*_{p_y}\right) \text{ and } \left(\pi_{p_z} + \pi^*_{p_z}\right)$$

The bonding electron pair form a π–BMO is donated to the metal centre and electron cloud is received back into the corresponding π*–ABMO. To act as a monodentate ligand, R_2C_2 utilizes only one set of MO.

- η^2–O_2^{2-}, η^2–NO_3^-, η^2–CO_3^{2-}, etc. acting as the didentate chelating ligands maintain the *heptacity equals to denticity*.

1.4.7 Defects of Werner's Theory

(i) **Nature of bonding:** It cannot explain the nature of bonding, *i.e.* metal-ligand interaction within the coordination sphere. Consequently, it fails to explain the *magnetic properties, spectral properties, thermodynamic stabilities and kinetic stabilities (i.e. lability and inertness) of the complexes*.

(ii) **Structure determination from the number of isolated isomers (*cf.* Sec 1.4.5):** This method applied by Werner is not absolutely correct. One may not be able to isolate an isomer and based on this, it cannot be concluded that the isomer does not exist. This aspect has been discussed.

(iii) **Preference to a certain coordination number and stereochemical configuration:** At the time of Werner, most of the complexes known were of either 4 C.N. or 6 C.N. But Werner's theory could not explain the preference among these complexes for the 4- and 6- coordination numbers.

For C.N. 4, some complexes are tetrahedral while some complexes are square planar. This aspect cannot be understood from the Werner's theory.

Even a particular metal centre may adopt **different coordination numbers and different geometries** in different conditions. It is illustrated by the following examples.

$[Ni(CO)_4]$ (C.N. = 4, tetrahedral); $[Ni(CN)_4]^{2-}$ (C.N. = 4, square planar); $[Ni(NH_6)_6]^{2+}$ (C.N. = 6, octahedral).

This aspect cannot be understood from Werner's theory.

1.5 DOUBLE SALTS AND COMPLEX SALTS: PERFECT AND IMPERFECT COMPLEXES

A large number of *molecular compounds* can be produced through the combination of simpler *molecular inorganic salts* which are capable of their independent existence. The constituent simple inorganic salts combine in the stoichiometric proportions to produce the molecular compounds. The examples of such compounds are:

(a)	**Potash alum:**	$K_2SO_4, Al_2(SO_4)_3.24H_2O$	
(b)	**Mohr's salt:**	$(NH_4)_2SO_4.FeSO_4.6H_2O$	
(c)	**Carnalite:**	$KCl.MgCl_2.6H_2O$	Double salts
(d)	**Chrome alum:**	$K_2SO_4.Cr_2(SO_4)_3.24H_2O$	
(e)	**Tutton Salt:**	$M_2SO_4.CuSO_4.6H_2O$ (M = Na, K).	
(f)	**Potassium ferrocyanide:**	$4KCN.Fe(CN)_2$ or $K_4[Fe(CN)_6]$	
(g)	**Potassium ferricyanide:**	$3KCN.Fe(CN)_3$ or $K_3[Fe(CN)_6]$	
(h)	**Potassium chloroplatinate:**	$2KCl.PtCl_4$ or $K_2[PtCl_6]$	Complex salts
(i)	**Prussian blue:**	$KCN.Fe(CN)_2.Fe(CN)_3$ or $KFe[Fe(CN)_6]$	

Depending on the behaviour of such salts produced by the combination of simple inorganic salts in the stoichiometric proportions, these can be classified into two groups — **double salts and complex salts.** In the above examples, the compounds (a) to (e) are the double salts while the last four compounds (f) to (i) are the complex salts.

(A) **Double Salts or Lattice Compounds:** A double salt is an aggregate of two simple inorganic salts in a certain proportion and the aggregate simply splits into the constituent ions in solution. *It indicates that the aggregate maintains the individuality of the constituent salts.*

Thus a double salt shows its distinct entity in the solid crystalline state but it loses its identity in solution.

Potash alum represented by $K_2SO_4.Al_2(SO_4)_3.24H_2O$ is an example of double salt and it ionizes in solution as follows:

$$K_2SO_4 \cdot Al_2(SO_4)_3 \cdot 24H_2O \xrightarrow{\text{water}} 2K^+ + 4SO_4^{2-} + 2Al^{3+} + 24H_2O$$

(All the ions remain as the solvated ones).

In the solution, existence of these ions can be proved chemically.

Mohr's salt, $(NH_4)_2SO_4.FeSO_4.6H_2O$ is a double salt and it ionizes as follows:

$$(NH_4)_2SO_4 \cdot FeSO_4 \cdot 6H_2O \xrightarrow{\text{water}} 2NH_4^+ + Fe^{2+} + 2SO_4^{2-} + 6H_2O$$

Other examples of double salts also ionize in the same way.

● *Double salts are obtained by crystallization from the solutions containing the constituent simpler salts.* Packing of the constituent ions in the crystal lattice **maintains the stoichiometric composition** of the mixed salt. This is why, the double salts are described as the **lattice compounds**. This fixed stoichiometric composition is a characteristic property of a true compound. **Thus, a double salt is not a simple mechanical mixture of its constituent salts rather it is a true compound.**

- Double salts are different from the **mixed crystals** formed by the *isomorphous crystals* as the mixed crystals do not maintain the fixed composition, a characteristic property of the compound. The composition of a mixed crystal can vary depending on the relative concentrations of the constituent salts in the solution from which the mixed crystal is isolated through crystallization. For example, $KMnO_4$ and $KClO_4$ are isomorphous and they can produce a continuous series of solid solutions (*i.e.* mixed crystals) with one another. (*See* Vols. 2 and 3 of Fundamental Concepts of Inorganic Chemistry.)

(B) **Complex Salts:** In a complex salt, the individual identities of the constituent salts are lost. Thus, a complex salt on being ionized cannot produce all the constituent ions in solution. Instead of the simple constituent ions, a complex salt produces the new complex ions which are totally different from the simple constituent ions.

The behaviour of a complex salt is illustrated by taking the example of potassium ferrocyanide, $K_4[Fe(CN)_6]$ which is obtained by the action of potassium cyanide (KCN) on ferrous sulfate ($FeSO_4$). $K_4[Fe(CN)_6]$ may be considered apparently as $4KCN.Fe(CN)_2$.

$$2KCN + FeSO_4 \longrightarrow Fe(CN)_2 + K_2SO_4, \; 4KCN + Fe(CN)_2 \rightarrow K_4\left[Fe(CN)_6\right]$$

$K_4[Fe(CN)_6]$ ionizes in aqueous solution a follows:

$$K_4\left[Fe(CN)_6\right] \qquad \nearrow 4K^+ + \left[Fe(CN)_6\right]^{4-}$$
$$\left(i.e.,\; 4KCN \cdot Fe(CN)_2\right) \qquad \searrow 4K^+ + 6CN^- + Fe^{2+}$$

The solution of potassium ferrocyanide fails to respond to the **common chemical tests** for CN^- and Fe^{2+} ions. It gives the test for the complex ion $[Fe(CN)_6]^{4-}$ whose property is totally different from those of the constituent ions Fe^{2+} and CN^- ions. *In fact, the complex ion ferrocyanide is stable both in solid and solution phase.*

The complex ions **capable of maintaining their independent existence** in both the solid and solution phase are written within [ ], *e.g.* $[Fe(CN)_6]^{4-}$
Other examples of complex ions:

$$KCN \cdot AgCN, \; i.e. \; K\left[Ag(CN)_2\right] \longrightarrow K^+ + \left[Ag(CN)_2\right]^-$$

$$2KCN \cdot M(CN)_2, \; i.e. \; K_2\left[M(CN)_4\right] \longrightarrow 2K^+ + \left[M(CN)_4\right]^{2-}$$
$$\left(M = Pt, Ni\right)$$

$$3KNO_2 \cdot Co(NO_2)_3, \; i.e. \; K_3\left[Co(NO_2)_6\right] \longrightarrow 3K^+ + \left[Co(NO_2)_6\right]^{3-}$$

$$CoCl_3 \cdot 6NH_3, \; i.e. \; \left[Co(NH_3)_6\right]Cl_3 \longrightarrow \left[Co(NH_3)_6\right]^{3+} + 3Cl^-$$

$$3KCN \cdot CuCN, \; i.e. \; K_3\left[Cu(CN)_4\right] \longrightarrow 3K^+ + \left[Cu(CN)_4\right]^{3-}$$

(C) **Perfect and Imperfect Complexes:** In illustrating the behaviour of a complex ion, it has been said that *the constituent simple ions fail to respond to their common chemical tests.* It is true for the complex ions like $[Fe(CN)_6]^{4-}$, $[Ni(CN)_4]^{2-}$, $[Cu(CN)_4]^{3-}$, $[Co(NH_3)_6]^{3+}$, etc.

On the other hand, some complex ions like $[Cd(CN)_4]^{2-}$ as in $K_2[Cd(CN)_4]$, $[Ni(NH_3)_6]^{2+}$ as in $[Ni(NH_3)_6]Cl_2$, respond to some of the chemical tests for the metal ions and the involved ligands. Thus $K_2[Cd(CN)_4]$ ionizes sufficiently to allow the detection of Cd^{2+} as CdS by passing H_2S. CN^- can also be chemically detected.

$$K_2\left[Cd(CN)_4\right] \rightleftharpoons 2K^+ + \left[Cd(CN)_4\right]^{2-}$$

$$\left[Cd(CN)_4\right]^{2-} \rightleftharpoons Cd^{2+} + 4CN^-$$

$$Cd^{2+} + S^{2-} \longrightarrow CdS \,(\text{yellow precipitate})$$

(**Note:** It may be noted that from the solution of $K_3[Cu(CN)_4]$, Cu^+ cannot be precipitated as Cu_2S.)

Similarly, the complex ion $[Ni(NH_3)_6]^{2+}$ splits to some extent into its constituents to allow the detection of NH_3 in solution. Similarly, the silver ammine complex ion, $[Ag(NH_3)_2]^+$ splits to some extent in aqueous solution into its constituents and Ag^+ can be detected as the precipitate of silver iodide (pale yellow) by the addition of KI solution.

$$\left[Ag(NH_3)_2\right]^+ \rightleftharpoons Ag^+ + 2NH_3 \,; \, Ag^+ + I^- \longrightarrow AgI \,(\text{pale yellow precipitate})$$

Based on the above mentioned behaviour (*i.e.* **tendency to ionize or split into the constituents**) of the complex ions, they were classified as the *perfect* and *imperfect complexes* by the classical chemists. *However, this distinction is arbitrary as the constituents combine in a reversible process with a definite equilibrium constant to produce the complex ion.* In general, the complex formation equilibrium is represented as follows:

$$M + xL \rightleftharpoons \left[ML_x\right], \, K_{stab} = \frac{\left[ML_x\right]}{[M][L]^x}$$

(M stands for the metal centre; L stands for the ligand; charge is not indicated for the sake of simplicity; K_{stab}, **a thermodynamic parameter,** denotes the the the *overall stability constant* of the complex.)

Thus, the dissociation of the complex can be shown as follows:

$$\left[ML_x\right] \rightleftharpoons M + xL, \, K_{instab} = \frac{1}{K_{stab}} = \frac{[M][L]^x}{\left[ML_x\right]}$$

K_{instab} denotes the *instability constant* of the complex and it gives the measure of stability of the complex with respect to its dissociation into the metal and ligands.

If K_{stab} is very high (i.e. very low value of K_{instab}) then the extent of dissociation of the complex into its constituents is very small and such complexes are referred to as the perfect complexes. On the other hand, if K_{stab} is low (i.e. high value of K_{instab}), then the complex dissociates to a significant extent and such complexes are described as the imperfect complexes.

Thus the perfect and imperfect complexes differ only in the degree of stability constant and **there is no sharp line of demarcation between them and it is a matter of relativity only**.

$[Cd(CN)_4]^{2-}$ and $[Cu(CN)_4]^{3-}$ differ in their stability constants. K_{stab} of $[Cd(CN)_4]^{2-}$ is much less than that of $[Cu(CN)_4]^{3-}$. Thus $[Cd(CN)_4]^{2-}$ produces a much higher concentration of the free metal ion than that for $[Cu(CN)_4]^{3-}$. It allows the seperation of cadmium as CdS by H_2S from the mixture of $K_2[Cd(CN)_4]$ and $K_3[Cu(CN)_4]$. Thus in this reaction (*i.e.* precipitation by H_2S in neutral solution), $K_2[Cd(CN)_4]$ acts as an imperfect complex while $K_3[Cu(CN)_4]$ acts as a perfect complex. *However, the classical labeling of perfect and imperfect complexes depends on the* **sensitivity of the chemical test** *used to detect the constituent ions.*

Thus with respect to a particular chemical test of relatively low sensitivity, a complex may appear as a perfect complex but it may appear as an imperfect complex when subjected to a chemical test of higher sensitivity (*cf.* Sec. 1.10).

The stability constants of the complexes $K_2[MCl_4]$ (M = Zn, Cu, Pt) differ significantly.

$$\left[MCL_4\right]^{2-} \rightleftharpoons M^{2+} + 4Cl^-, \ K_{instab} = \frac{\left[MCl_4^{2-}\right]}{\left[M^{2+}\right]\left[Cl^-\right]^4}$$

K_{stab} runs as: $Pt^{2+} \rangle\rangle Cu^{2+} \rangle\rangle Zn^{2+}$ *i.e.* K_{instab} runs as: $Pt^{2+} \langle\langle Cu^{2+} \langle\langle Zn^{2+}$. In fact, $[ZnCl_4]^{2-}$ breaks down into its constituents almost completely while $[PtCl_4]^{2-}$ remains almost unchanged and $[CuCl_4]^{2-}$ shows an intermediate behaviour. Thus the **perfectness in the complexes** increases in the sequence:

$$\left[ZnCl_4\right]^{2-} \langle\langle \left[CuCl_4\right]^{2-} \langle\langle \left[PtCl_4\right]^{2-}$$

(**Note:** In terms of the stability constant, the double salts may be considered to have the **vanishingly small stability constants**. Consequently the double salts ionize completely into its constituents.)

The instability constants which indicate the extent of dissociation of a complex ion into its constituents are given in Table 1.5.1 *Higher value of K_{instab} indicates the higher extent of dissociation (i.e. more imperfectness).*

Thermodynamic and kinetic stability of a complex

Here it should be mentioned that **ease of dissociation of a complex depends on both the thermodynamic stability and kinetic stability** (*i.e.* lability and inertness). The factors governing the kinetic stability of a complex will be discussed later (*cf.* Chapters 4 and 5).

Table 1.5.1 Instability constants of some representative complex ions.

Dissociation equilibrium	K_{instab} (25° C)
$\left[Ag(NH_3)_2\right]^+ \rightleftharpoons Ag^+ + 2NH_3$	6.8×10^{-8}
$\left[Ag(CN_2)\right]^- \rightleftharpoons Ag^+ + 2CN^-$	1.0×10^{-21}
$\left[Cu(NH_3)_4\right]^{2+} \rightleftharpoons Cu^{2+} + 4NH_3$	4.6×10^{-14}
$\left[Cu(CN)_4\right]^{3-} \rightleftharpoons Cu^+ + 4CN^-$	5.0×10^{-28}
$\left[Zn(NH_3)_4\right]^{2+} \rightleftharpoons Zn^{2+} + 4NH_3$	2.6×10^{-10}
$\left[Zn(CN)_4\right]^{2-} \rightleftharpoons Zn^{2+} + 4CN^-$	2.0×10^{-17}
$\left[Cd(NH_3)_4\right]^{2+} \rightleftharpoons Cd^{2+} + 4NH_3$	2.5×10^{-7}
$\left[Cd(CN)_4\right]^{2-} \rightleftharpoons Cd^{2+} + 4CN^-$	1.4×10^{-17}
$\left[CdI_4\right]^{2-} \rightleftharpoons Cd^{2+} + 4I^-$	5.0×10^{-7}
$\left[HgI_4\right]^{2-} \rightleftharpoons Hg^{2+} + 4I^-$	5.0×10^{-31}
$\left[Hg(SCN)_4\right]^{2-} \rightleftharpoons Hg^{2+} + 4SCN^-$	1.0×10^{-22}
$\left[Co(NH_3)_6\right]^{3+} \rightleftharpoons Co^{3+} + 6NH_3$	2.1×10^{-34}

Note: In the present day's literature, the terminology describing the perfect and imperfect complexes is not of any significance. The stability constants (*cf.* Chapter 4) describe the **thermodynamic stability** of the complexes.

1.6 DEFINITION OF INNER SPHERE COMPLEX, OUTER SPHERE COMPLEX AND SECOND COORDINATION SPHERE COMPLEX (*i.e.* SUPERCOMPLEX)

(A) **Inner and Outer Sphere Complexes:** The ligands bound to the metal centre by the coordinate covalent bonds determine the geometrical shape of a complex. These ligands lie within the coordination sphere (more correctly **1st coordination sphere**).

Thus the **inner sphere complexes** are produced by the ligands coordinating to the metal centre. Such inner sphere complexes may be ionic or neutral depending on the conditions. If the inner sphere complex is ionic (*i.e.* charge bearing) then it can hold the oppositely charged ions outside the coordination sphere. This association between the inner sphere complex and the oppositely charged ions produces an **outer sphere complex** where the ligands within the coordination sphere remain unchanged.

$$M^{x+} + nL \xrightleftharpoons{K} \underset{(\text{Inner sphere complex})}{\left[ML_n\right]^{x+}} \quad ;$$

(K or β_n denotes the stability constant or formation constant of the complex).

$$\left[ML_n\right]^{x+} + Y^{m-} \xrightleftharpoons{K_{o.s.}} \underset{\textbf{Ion Pair (Outer sphere complex)}}{\left[ML_n\right]^{x+} \cdot Y^{m-}} \quad ;$$

($K_{o.s.}$ denotes the **outer-sphere association constant**)

Thus $[Co(NH_3)_6]^{3+}$ can hold an halide (X^-) ion to produce an **ion-pair association** outside the coordination sphere.

$$\left[Co(NH_3)_6\right]^{3+} + X^- \rightleftharpoons \underset{(\text{Outer sphere complex})}{\left[Co(NH_3)_6\right]^{3+} \cdot X^-} \quad \textbf{(Ion pair)};$$

$$K_{o.s.} = \frac{\left[\left(\left[Co(NH_3)_6\right]^{3+} \cdot X^-\right)\right]}{\left[Co(NH_3)_6^{3+}\right]\left[X^-\right]}$$

The outer sphere association constant ($K_{o.s.}$) for such an ion-pair formation can be theoretically calculated by *Fuoss's equation* (*cf.* Debye-Huckel theory on electrolytes).

Besides the **ion-pair interaction**, other types of interaction like **ion-dipole interaction**, (*i.e.* ionic inner sphere complex and neutral but dipolar molecules), **dipole-dipole interaction** (*i.e.* neutral complex and neutral dipolar molecules) can also produce the outer sphere complex. The examples are:

$[Co(NH_3)_6]^{3+}.H_2O$ **(ion-dipole interaction)**;
$[CoCl_3(NH_3)_3].H_2O$ **(dipole-dipole interaction.)**

Among these different types of interaction, the electrostatic interaction causing the ion-pair association is the most important one. The attractive force depends on the product of the charges of the ions (*cf.* Coulombic interaction). *It increases with the increase of the product of the charges on the ions and decreases with the increase of size of the ions.* However, the charge effect is more important than the size effect because sizes of the different complex ions are, in general, more or less comparable. The outer sphere association constants ($K_{o.s.}$) for some representative **ion-pair associations** are given in Table 1.6.1.

(**Note:** ● Formation constant or stability constant (Chapter 4) of an inner sphere complex due to the direct metal-ligand interaction is much higher than that of the outer sphere association constant that involves the weaker force.)

● The concept of outer sphere complex is exceedingly important in understanding the *kinetic aspects of metal complexes* (Chapter 5). From the kinetic data, $K_{O.S.}$ may be calculated in some cases.)

Table 1.6.1 Outer sphere association constants ($K_{O.S.}$) for some outer sphere complexes.

Ion-pair	$K_{O.S.}$ (M^{-1})	Ion-pair	$K_{O.S.}$ (M^{-1})
$[Co(NH_3)_6]^{3+}.Cl^-$	74	$[Ni(OH_2)_6]^{2+}.SCN^-$	10.0
$[Co(NH_3)_6]^{3+}.Br^-$	46	*$[Ni(OH_2)_6]^{2+}.NH_3$	1.5
$[Co(NH_3)_6]^{3+}.I^-$	17	*$[Ni(OH_2)_6]^{2+}.py$	1.3
$[Co(en)_3]^{3+}.Cl^-$	52	$[CoCl(NH_3)_5]^{2+}. CH_3CO_2^-$	5
$[Co(en)_3]^{3+}.SO_4^{2-}$	2.8×10^3	$[CoCl(NH_3)_5]^{2+}. N_3^-$	13
$[CoCl(NH_3)_5]^{2+}.Cl^-$	10.0		
$[CoCl(NH_3)_5]^{2+}.SO_4^{2-}$	1.6×10^3		

* Ion-dipole interaction.

Binary and Mixed Ligand Complexes

In the coordination sphere, if only one type of ligand excluding the solvent molecules is present, then the complex is described as a **binary complex or homoleptic complex**, *e.g.* $[Co(NH_3)_6]^{3+}$, $[Co(NH_3)_5(OH_2)]^{3+}$, etc. If there are different types of ligands (excluding the solvent) within the coordination sphere, then the complex is described as a **mixed ligand complex or heteroleptic complex.** *The ligands may differ either by their nature, optical configuration, ionization state or their binding sites.* If there are two types of ligands, the complex is referred to as the **ternary complex**. Similarly, a **quarternary complex** possesses three different types of ligands within the coordination sphere. Examples of the ternary complexes are:

$$[CoCl(NH_3)_5]^{2+}, [Cu(bpy)(en)]^{2+}, [Cu(bpy)(ox)]^0, [Cu(his)(Hhis)]^+, \text{etc.}$$

(B) **Second Coordination Sphere Complex:** It has been already mentioned that outside the *coordination sphere* (*i.e.* inner-sphere), ions or dipolar molecules may get held to produce the outer-sphere complex. In the inner-sphere (more correctly, *first coordination sphere*), the ligands are oriented in some specified directions to provide a definite geometrical structure but in the outer-sphere, the ions or dipoles are held electrostatically relatively loosely and their arrangement generally *does not maintain any definite structural feature.* **Thus in the inner sphere complex, there is a strict structural order while in the outer sphere complex, their is no such strict structural order.**

In some cases, certain macrocyclic ligands are held in the outer sphere (*more correctly, second coordination sphere*) *to maintain a definite structural order* confirmed by the X-ray crystallographic data. Such outer sphere complexes are described as the **second coordination sphere complexes** or *supercomplexes*. An example of supercomplex is:

$$[Rh(cod)(NH_3)_2]. \text{ dibenzo-24-crown-8}$$

cod stands for cyclooctadiene. The macrocyclic crown ether (present in the outer sphere, *i.e.* 2nd coordination sphere) interacts with the NH_3 ligands (present in the 1st coordination sphere) through hydrogen bonding by using the ether–O atoms of the macrocyclic ligand. **Such H–bonding interaction between the ligands (present in 1st coordination sphere) and the species (ions or dipoles) present in the outer sphere leads to second coordination sphere complex (*i.e.* supramolecular association).**

1.7 CLASSIFICATION OF LIGANDS

The ligands remaining within the coordination sphere ligate the metal centre to determine the stereochemistry of the complex. *The ligands are the Lewis bases, the metal centres are the Lewis acids, and the complexes are the Lewis acid-base adducts.* The ligands can be classified on different bases. These are:

- organic and inorganic nature of the ligands;
- hard-soft character of the ligands;
- nature of the donor electrons on the ligands;
- number of binding sites (*i.e.* denticity) of the ligands;
- mode of binding of the ligands to the metal centre;
- charge bearing properties of the ligands;

 (i) Depending on the inorganic and organic nature of the ligands, they can be classified as the **organic** and **inorganic ligands.** The examples are given below:

   Organic ligands: oxalate (*ox^{2-}*) ($C_2O_4^{2-}$); ethylenediamine (*en*) (H_2N—CH_2—CH_2—NH_2); 2,2'-dipyridyl or 2,2'-bipyridyl (*dpy* or *bpy*); pyridine (*py*); etc.

   Inorganic ligands: H_2O, ^-OH, NH_3, halide, pseudohalide, carbonate, $[Co(NH_3)_4(OH)_2]^+$ (*see hexol* complex in Sec. 1.4.5), etc.

(ii) Depending on the **hard-soft character** of the ligands, they can be classified as the *hard ligands and soft ligands.* The hard ligands prefer to coordinate with the hard metal centres while the soft ligands prefer to coordinate with the soft metal centres. The examples of some typical hard and soft ligands are given below:

   Hard ligands: NH_3, H_2O, ^-OH, F^-, etc.

   Soft ligands: CO, CN^-, C_2H_4, SCN^-, R_3P, etc.

   The ligands having the intermediate character are described as the **border line ligands**. For details, *see* the HSAB theory (Vol. 3 of Fundamental Concepts of Inorganic Chemistry).

Replacement of H$^+$ by M^{n+} during ligation

In terms of deprotonation, there are **two types of functional groups** acting as the binding sites.

● **Deprotonation** is generally needed from the following functional groups (say, —X—H) for showing the ligating property. Thus in such cases, 'H' of the functional group is replaced by M. Examples: —CO_2H (carboxylic), —SO_3H (sulfonic), —OH (enolic and phenolic), =NOH (oxime), etc.

$$—X—H + M^{n+} \longrightarrow —X—M^{(n-1)+} + H^+$$

Sometimes, —OH (alcoholic) and =N—OH (oxime) functional groups may coordinate the metal centre **without deprotonation.**

● In the following functional groups, for the coordinating property, deprotonation does not occur.

amines (primary, secondary and tertiary), R_3P, $\rangle C = O$, —O—(ether), —S—(thioether), pyridine, etc.

● The binding sites are the Brönsted bases and they may get protonated in an acidic condition and then complexation is disfavoured as it will need the deprotonation from the protonated basic site.

(iii) The ligands can be classified as **cationic, anionic and neutral** based on their charge bearing properties. Examples of anionic and neutral ligands are abundant as they can act as the good Lewis bases. On the other hand, the examples of cationic ligands are rare as these *electron* *deficient species* are the poor Lewis bases. In addition to this, the *electrostatic repulsion* with the positively charged metal centre disfavours the approach of the cationic ligands.

Anionic: Halide, ^-OH, CO_3^{2-}, $C_2O_4^{2-}$, etc.

Neutral: NH_3, dpy or bpy, en, py, H_2O, etc.

Cationic: NO^+ as in $[Co(diars)_2(NO)]^{2+}$, $[Cr(CN)_5NO]^{3-}$, etc.; $[Co(NH_3)_4(OH)_2]^+$ (used by Werner to prepare a pure inorganic complex, *hexol*, optically active); $(CH_3)_3^+N(CH_2)_2NH_2$ denoted by L as in $[ML_6](ClO_4)_8$ (M = Ni, Co); $NH_2-NH_3^+$ (hydrazinium).

(iv) Ligands can be classified based on the **nature of donor electrons** of the ligands. Some ligands possess the lone pair of electrons that can be donated to the metal centre for the σ-bond formation. Some ligands possess no lone pair of electrons but π-electrons for donation to the metal centre for the σ-bonding. Such σ-bonds are now described as the **μ-bonds** (*see* Vols. 2 and 3 of Fundamental Concepts of Inorganic Chemistry).

● The ligands (*e.g.* alkenes, alkynes, benzene, etc.) having the π-electron cloud for σ-donation to the metal centre to produce the ligand → metal σ-bond (more correctly **μ-bond**, as such bonds lack in the symmetry element $C_∞$) are always having some suitable vacant antibonding π-molecular arbitals (*i.e.* π*—MO). These vacant π*—MOs can receive back the electron cloud from the metal centre through the π-bonding (*see* Fig. 10.7.8.1, Vol. 2 of Fundamental Concepts of Inorganic Chemistry). Such ligands are called the **π-acid ligands**.

● CO and CN^- also use the π*—MOs for showing the metal → ligand π-bonding. PR_3 and S-donor ligands can use the vacant *d*-orbital (or σ*—MO in the case of PR_3) on the donor atom for the metal → ligand π-bonding.

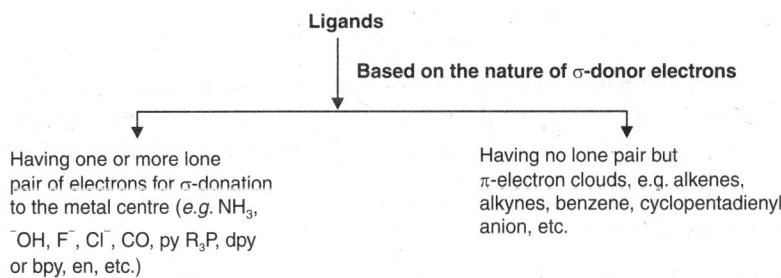

1.7.1 Classification of Ligands Based on the Nature of Metal-Ligand Bonding

Ligands can be classified on the basis of the nature of metal-ligand bonding interaction.

● The **π-acid ligands can stabilize the low oxidation state of the metal centre**. The accumulated electrons on the metal centre from the ligand → metal σ-donation, produces an unfavourable situation for the low oxidation state of the metal centre as the electropositive metal centre does not like the accumulation of much electron cloud. The π-acidic character (*i.e.* removal of electron cloud from the metal centre through the metal → ligand π-back bonding) of the π-acid ligand can balance the unfavourable situation developed from the σ-donation of the ligands.

● For the π-acid ligands, both the ligand → metal σ-donation and metal → ligand π-back bonding go on simultaneously. **These two bondings (*i.e.* σ- and π-bonding) generally act in a synergistic fashion**, *i.e.* if the σ-donation is favoured then the π-acceptance is also favoured.

● The π-acid ligands are generally the **strong field ligands**.

● Ligands having more than one lone pair may utilize the lone pair of lower energy (degeneracy of the lone pairs may be removed through splitting) for σ-donation. While the other lone pairs of relatively higher energy may participate in π-donation. **Such π-donor ligands (*e.g.* ⁻OH) are the weak field ligands**.

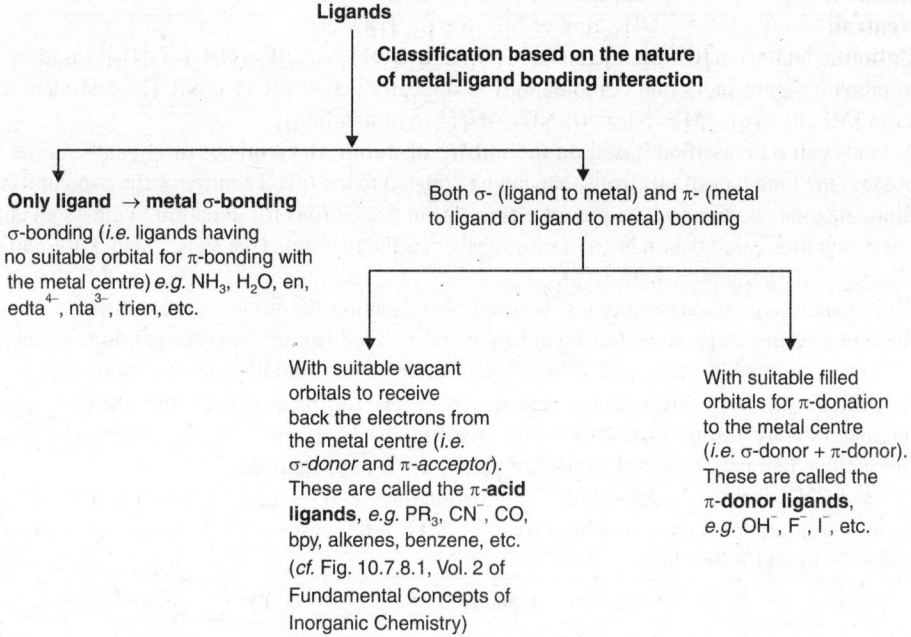

- **The pure σ-donor ligands and the ligands with the π-donor properties can stabilize the higher oxidation state of the metal centre.** *Such ligands are generally the weak field ligands.* The above aspects of metal-ligand bonding are discussed in detail at the appropriate places.

1.7.2 Classification of the Ligands Based on the Number of Donor Sites and Coordinating Behaviour

This classification is on the basis of the number of donor sites (present in a particular ligand) which can participate (simultaneously) in making the metal-ligand linkage. *The number of such donor sites determines the* **denticity** *of the ligands.* The ligands with one donor site are called the **monodentate ligands**; the ligands with *two such donor sites which simultaneously participate in making two different*

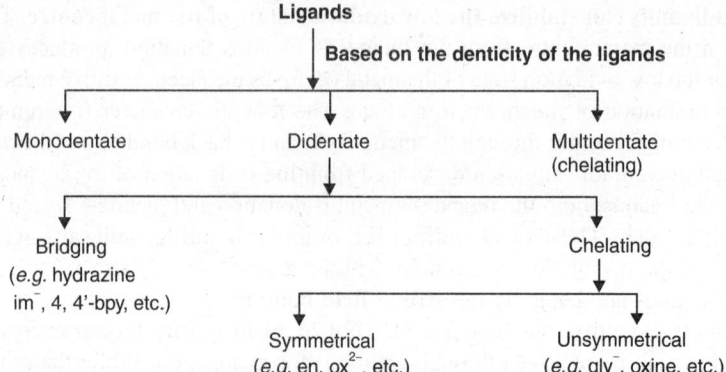

metal-ligand linkages are called the **didentate ligands** (*i.e.* **bidentate ligands**); the ligands having such donor sites more than two are described as the **polydentate or multidentate** ligands. The multidentate ligands are *tridentate, tetradentate, pentadentate, hexadentate,* etc. for the donor sites three, four, five and six respectively.

The multidentate ligands including the didentate ligands are described as the *chelating ligands* (pronounced as *key-lating,* derived from the Greek word *khele* meaning **crab's claw**) as they grasp the metal centre between two or more donor sites.

(**Note:** To express the denticity of a ligand, both **Latin** (*i.e. uni-, bi-, quadri-,* etc.) and **Greek** (*i.e. mono-, di-, tetra-,* etc.) prefixes were previously used. Thus, a ligand with four donor sites was described either as *quadrientate* (with Latin stem) or *tetradentate* (with Greek stem). Presently, only Greek prefixes, *i.e. mono-, di-, tri-, tetra-, penta-, hexa-, hepta-,* etc. are used to express the denticity of the ligands).

- (**σ-donor**) **ligands:** H^-, NH_3, etc. The single lone pair (without any π-electron and vacant orbital) on the donor site is used for σ-donation. These are the **weak field ligands**.

- (**σ-donor + weak π donor**) **ligands:** H_2O, NH_2^-, R_2O, R_2S, R_3PO (O-donor), R_3AsO (O-donor), ROH, etc.

There are two lone pairs on the donor site. Out of these two lone pairs, one is involved in σ-donation $(L \xrightarrow{\sigma} M)$ to the metal centre and the other may act as a weak π-donor $(L \xrightarrow{\pi} M)$. These are the **weak field ligands**.

- (**σ-donor + good π–donor**) **ligands:** O^{2-}, OH^-, NH^{2-}, F^-, Cl^-, Br^-, I^-, S^{2-}, etc.

There are three or four lone pairs on the donor site. During the interaction, these are split into two sets: one lone pair becomes of lower energy while the other lone pairs become of higher energy. The lone pair of lower energy is involved in σ-bonding $(L \xrightarrow{\sigma} M)$ while lone pairs of higher energy participate in π-bonding $(L \xrightarrow{\pi} M)$. **These are the very weak field ligands.**

- (**σ donor + π acceptor**) **ligands** (*cf.* Fig. 10.7.8.1, Vol. 2 of Fundamental Concepts of Inorganic Chemistry): CO, CN^-, C_2H_4, NO^+, PR_3, R_3AS, PF_3, py, dpy or bpy, phen, alkenes, alkynes, benzene, etc. The lone pair or π-bonding electron cloud is involved in the $L \rightarrow M$ σ-bonding while the vacant orbital (d-orbital or σ*-MO or π*-MO) participates in the $M \rightarrow L$ π-bonding. Such ligands are the **π-acid ligands**. These are the **strong field ligands**.

A. Monodentate Ligands: A monodentate ligand can occupy only *one coordination site in the coordination sphere of a complex.* They generally possess one donor site. Sometimes, the ligands having more than one donor sites or binding site can also act as the monodentate ligands. For example, hydrazine $(H_2\ddot{N} — \ddot{N}H_2)$ bears two donor sites but these two donor sites cannot be used to occupy two coordination sites of the coordination sphere of a metal centre. If *it acts as a didentate chelating ligand, a three membered chelate ring is to be produced.*

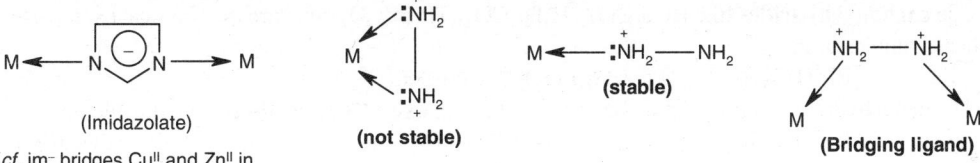

(Imidazolate)

(*cf.* im⁻ bridges Cu^{II} and Zn^{II} in **superoxide dismutase enzyme**)

(**not stable**)

(**stable**)

(**Bridging ligand**)

Such a **strained three membered chelate ring** is not stable. Moreover, in the chelate ring produced by hydrazine, the adjacent N-sites bear the positive charge in the same coordination sphere. This **electrostatic repulsion** creates an unfavourable situation. This is why, it acts **as a monodentate ligand**. However, it can act **as a bridging ligand**.

Some representative examples of the monodentate ligands are given in Table 1.7.2.1

Table 1.7.2.1 Some representative examples of the monodentate ligands.

Formula (donor site bold)	Name of the ligand when present in complex	Formula (donor site bold)	Name of the ligand when present in complex
Neutral molecules		**Anions****	
H$_2$**O**	aqua	**F**$^-$	fluorido
NH$_3$	ammine	**Cl**$^-$	chlorido
CH$_3$**N**H$_2$	methylamine	**Br**$^-$	bromido
CO	carbonyl	**I**$^-$	iodido
NO	nitrosyl	**O**H$^-$	hydroxido
CS	thiocarbonyl	**C**N$^-$	cyanido-C
C$_5$H$_5$**N**	pyridine (py)	C**N**$^-$	cyanido-N (*i.e.* isocyanido)
PPh$_3$	triphenyl phosphane	**O**CN$^-$	cyanato-O
AsR$_3$	trialkyl arsane	OC**N**$^-$	cyanato-N (*i.e.* isocyanato)
(C$_2$H$_5$)$_2$**O**	diethyl ether	**S**CN$^-$	thiocyanato-S
NF$_3$	nitrogen trifluoride	SC**N**$^-$	thiocyanato-N (*i.e.* isothiocyanato)
C$_2$H$_4$	ethene (ethylene)	**N**O$_2$$^-$	nitrito-N (nitro)
CH$_3$**N**C	methylisocyanide*	**O**NO$^-$	nitrito-O
Cations		**H**$^-$	hydrido
NO$^+$	nitrosylium	**N**$_3$$^-$	azido
NH$_2$**N**$^+$H$_3$	hydrazinium	**O**$^{2-}$	oxido
		CH$_3$CO**O**$^-$	acetato
		N^{3-}	nitrido
		NH$_2$$^-$	amido
		ONO$_2$$^-$	nitrato
		NOS$^-$	nitrosylsulfido

** As per the **IUPAC Recommendations 2005**, all anionic ligands ending with *'ide'* will end with *'ido'*, *e.g. chlorido, cyanido, hydrido* (no exception for boranes), *oxido*, etc. instead of *chloro, cyano, hydro, oxo*, etc. respectively. **However, at this stage, many books do not follow this recommendation strictly.**

Note: Oxyanions like NO$_3$$^-$, RCO$_2$$^-$, CO$_3$$^{2-}$, SO$_4$$^{2-}$, etc. can act as both the monodentate and didentate ligands depending on the conditions (*cf.* Fig. 1.7.2.1). *R—N$^+$≡C$^-$ (alkyl isocyanide).

Hydride as the ligands

The **carbonyl hydrides** like HCo(CO)$_4$, H$_2$Fe(CO)$_4$, HMn(CO)$_5$, etc. are the classical examples. Other examples are:

[Rh(H)(NH$_3$)$_5$]$^{2+}$, [PtCl(H)(PEt$_3$)$_2$], [Co(CN)$_5$(H)]$^{3-}$, etc.

The metal hydride complexes may be formed (through **oxidative addition**) from the dihydrogen complexes if the bound dihydrogen (η^2–H$_2$) is sufficiently active. These aspects will be discussed later on.

Dihydrogen *vs.* Dihydrido Complexes

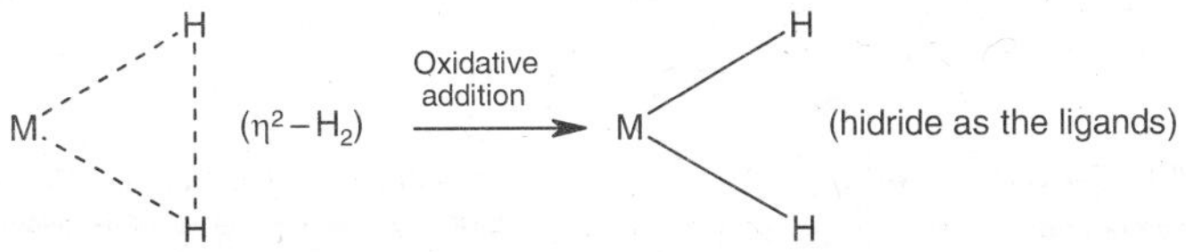

(Dihyrogen complex)		(Dihydrido complex)

$$e.g. \left[Ir^{I}(CO)ClL_2 \right] \left(\textbf{Vaska's complex} \right) + H_2 \longrightarrow \left[Ir^{III}(CO)Cl(H)_2 L_2 \right], L = PPh_3$$

B. Didentate Ligands: The ligands having two different binding sites suitably oriented to occupy simultaneously two coordination sites of *a metal centre are called the didentate ligands.*

Thus these are the didentate chelating ligands. In such cases, chelating rings are formed. The 5 and 6 membered rings are strain-free while the 4 and 7 membered rings experience the strain. Thus the 5 and 6 membered rings are more stable. **Between the 5 and 6 membered rings, the 5 membered ring is more stable.** For example, glycine (gly) and α-alanine (α-ala) form the five membered rings while β-alanine (β-ala) forms a six membered ring. In such cases, the chelate ring formed by α-ala or gly is stabler than the ring formed by β-ala.

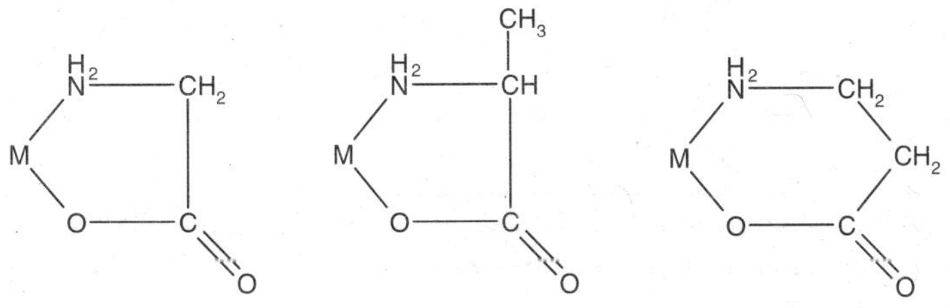

(gly)	(α-ala)	(β-ala)

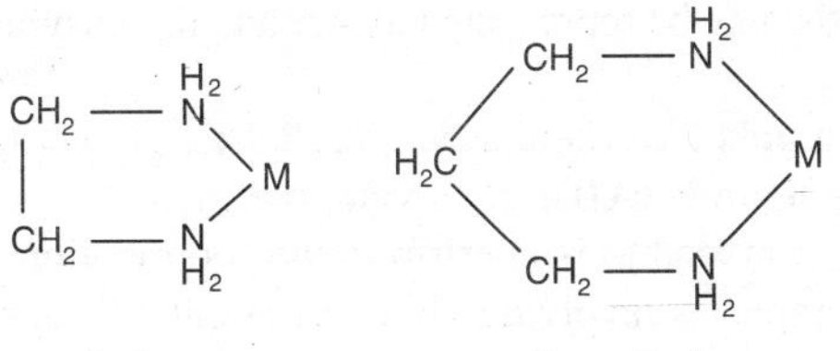

(en)	(tn or tmd)

Because of the same ground, ethylenediamine (en) forms a stabler complex than trimethylenediamine (tn)

"Didentate Bridging Ligands (*e.g.* hydrazine, pyrazine, peroxide, 4,4′-bipyridine) and **Didentate Chelating Ligands** (*e.g.* ethelyene diamine; peroxide; 2,2′-bipyridine; etc.)". This aspect has been critically discussed later.

Three membered chelate rings

In some rare cases, didentate ligands forming the **three membered chelate rings** are known.

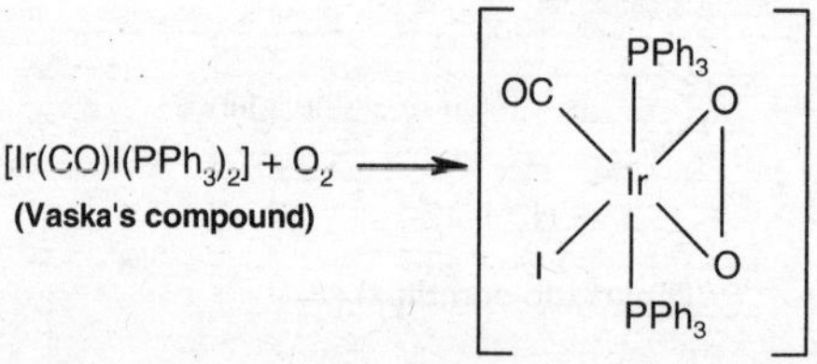

$$[Ir(CO)I(PPh_3)_2] + O_2 \longrightarrow$$
(Vaska's compound)

(*see* Fig. 14.7.5.3, Vol. 3 of Fundamental Concepts of Inorganic Chemistry for details of bonding interaction).

η^2- O_2^{2-} (peroxo group) here acts as a didentate chelating ligand. In **peroxidodisulfatotitanic acid, $H_2[Ti(\eta^2$- $O_2)(SO_4)_2]$**, the peroxide (η^2-O_2^{2-}) makes a such three membered ring. This compound formation is utilized for the detection of Ti(IV) (*see* Chapter 11).

In **[Pt{η^2- $C_2(CN)_4$}$_2$(PPh$_3$)$_2$]**, a **three membered chelate ring** is noticed. Here η^2- $C_2(CN)_4$ acts as a *didentate chelating ligand*.

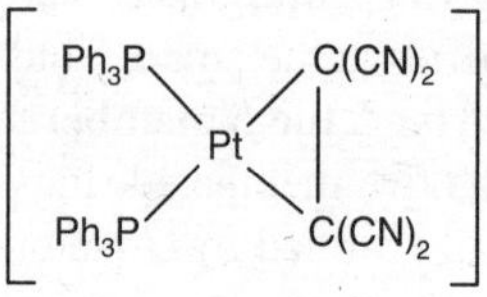

(in the complex, C—C bond length is nearly the same as in ethane).

Another type of **three membered chelate ring** may be as follows:

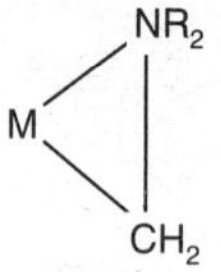

- Both the neutral and anionic didentate ligands are known. Both the binding sites may be neutral (as in en, pn, glyme, diphos, etc.) or anionic (as in ox^{2-}) or one neutral donor site and one anionic site (as in gly, oxine, etc.)
- Depending on the nature of donor sites, the didentate chelating ligands may be *symmetrical* (*i.e.* the binding sites are identical) and *unsymmetrical* (*i.e.* the binding sites are different). The symmetrical didentate ligands may be represented as AA and the unsymmetrical didentate ligands are represented as AB.

Symmetrical didentate ligands (AA): en, acac, pn, diphos, glyme, bpy, phen, diars, etc.

Unsymmetrical didentate ligands (AB): gly, oxine, pic, etc.

The unsymmetrical ligands can lead to isomerism in the coordination complexes. This is illustrated for the square planar Pt(II) complexes involving gly (a representative unsymmetrical ligand) and en (a representative symmetrical ligand).

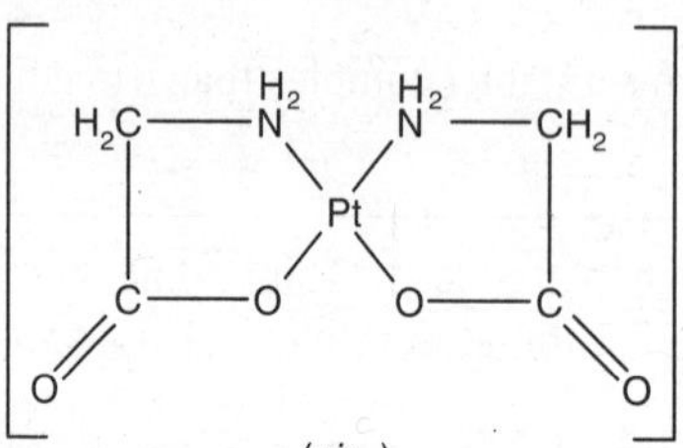

(*cis–*)

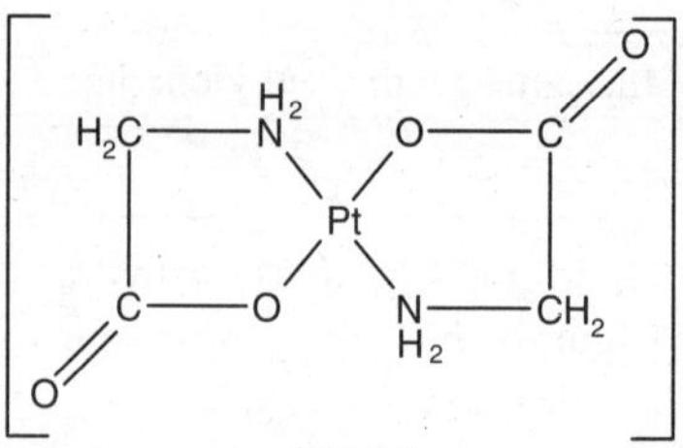

(*trans–*)

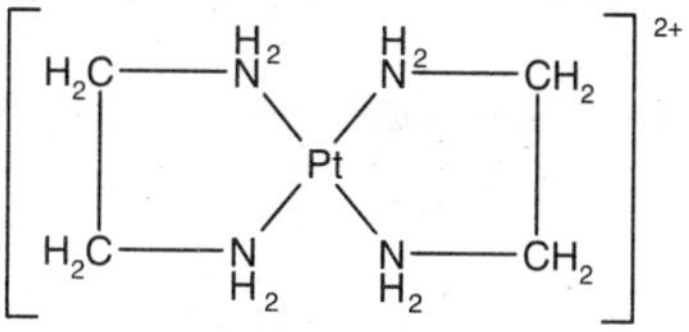

(no isomerism)

- Many ligands need **deprotonation for chelation**. For example, glycine (H_2N—CH_2—CO_2H) acts as a didentate ligand in the form of glycinate (H_2N—CH_2—CO_2^-). It happens so for many other ligands, like oxine, acetylacetone, etc. The deprotonation process may be favoured due to the stability of the complex formed. This aspect is illustrated for *acetylacetone* which exists in the keto and enol forms.

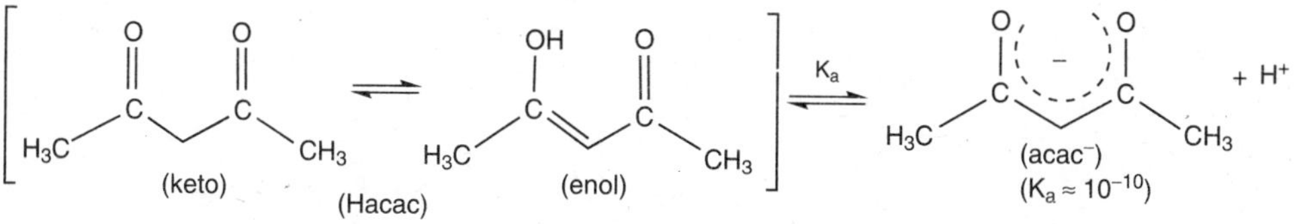

(keto)
(Hacac)
(enol)
(acac⁻)
($K_a \approx 10^{-10}$)

$$Fe^{3+} + 3acac^- \xrightarrow{K_{stab}} \left[Fe(acac)_3\right], \; K_{stab} \approx 10^{26}$$

It is evident that acylacetonate is a very weak acid ($pK_a \sim 10$) but formation of a very stable complex ($K_{stab} \approx 10^{26}$) drives the overall process.

$$Fe^{3+} + 3Hacac \rightleftharpoons \left[Fe(acac)_3\right] + 3H^+$$

In this typical complex, in the chelate ring (6-membered), **aromaticity** is attained by using the metal d-orbital

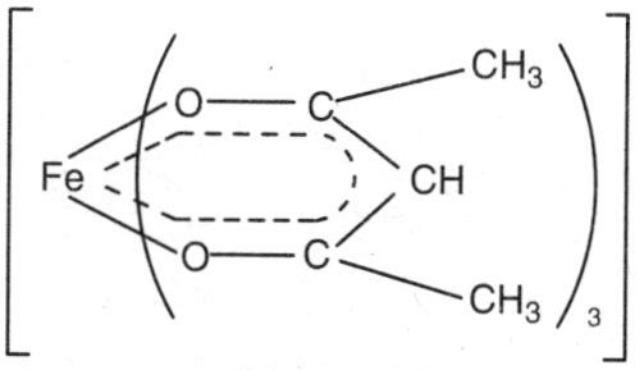

[Fe(acac)₃]

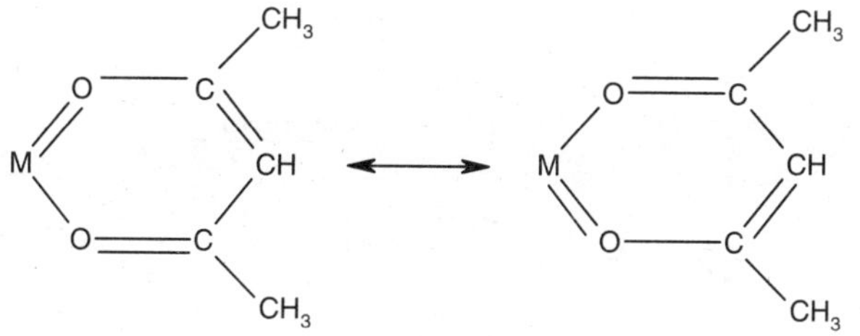

This **resonance** in each chelate ring gives an additional stabilization.

Examples of some representative didentate ligands are shown in Table 1.7.2.2

Table 1.7.2.2 Some representative didentate ligands (*cf.* Sec. 1.15.1).

Name of the ligand when present in complex	Abbreviation (cf. Sec. 1.15)	Formula (donor sites denoted by *)	Donor Sites
Carbonato	—		(O,O)
Nitrato	—		(O,O)
Sulfato	—		(O,O)
Carboxylato	—		(O,O)
Peroxido	—		(O,O)
Ethylenediamine (en) or, 1,2-ethanediamine or, ethane-1,2-diamine	(en)		(N,N)
Propylenediamine or, 1,2-propanediamine or, propane-1,2-diamine	(pn)		(N,N)
Butylenediamine (may be *dl* and *meso*) or, butane-2,3-diamine	(bn)		(N,N)

Isobutylenediamine	(*i*-bn)	CH₂—*NH₂ / H₃C—C—*NH₂ / CH₃	(N,N)
N,N,N′,N′-tetramethyl-ethane-1,2-diamine	(tmen)	H₂C—*NMe₂ / H₂C—*NMe₂	(N,N)
1,2-cyclohexanediamine or, 1,2-diaminocyclohexane	(chxn)	*NH₂ / *NH₂	(N,N)
Trimethylenediamine or, 1,3-diaminopropane or, propane-1,3-diamine	(tn or tmd)	CH—*NH₂ / H₂C / CH—*NH₂	(N,N)
Ethylenediphosphane or, ethane-1,2-diyl*bis*-(phosphane)	(diphos)	CH₂—*PH₂ / CH₂—*PH₂	(P,P)
Ethane-1,2-diyl*bis*-(dimethylphosphane)	dmpe (R = Me)		
Ethane-1,2-diyl*bis*-(diethylphosphane)	depe (R = Et)	CH₂—*PR₂ / CH₂—*PR₂	(P,P)
Ethane-1,2-diyl*bis*-(diphenylphosphane)	dppe (R = Ph)		
Dimethylglycol	(glyme)	CH₂—*O—Me / CH₂—*O—Me	(O,O)
Oxalato	(ox) (H₂ox)#	O=C—*O⁻ / O=C—O⁻*	(O,O)

$$CH_2\!-\!\overset{*}{N}H_2$$
$$H_3C\!-\!C\!-\!\overset{*}{N}H_2$$
$$CH_3$$

N,N,N′,N′-tetramethylethane-1,2-diamine (tmen):
$$H_2C\!-\!\overset{*}{N}Me_2$$
$$H_2C\!-\!\overset{*}{N}Me_2$$

Trimethylenediamine (tn or tmd):
$$CH\!-\!\overset{*}{N}H_2$$
$$H_2C$$
$$CH\!-\!\overset{*}{N}H_2$$

Ethylenediphosphane (diphos):
$$CH_2\!-\!\overset{*}{P}H_2$$
$$CH_2\!-\!\overset{*}{P}H_2$$

Ethane-1,2-diylbis(phosphanes):
$$CH_2\!-\!\overset{*}{P}R_2$$
$$CH_2\!-\!\overset{*}{P}R_2$$

Dimethylglycol (glyme):
$$CH_2\!-\!\overset{*}{O}\diagup Me$$
$$CH_2\!-\!\overset{*}{O}\diagdown Me$$

Oxalato (ox):
$$O\!=\!C\!-\!\overset{*}{O}^-$$
$$O\!=\!C\!-\!O^-_*$$

Name	Abbreviation	Structure	Donor atoms
Biguanide	(Hbig)		(N,N)
Glycinato	(gly) (Hgly)#		(N,O)
8-Hydroxyquinolinato or, oxinato or, 8-quinolinolato	(oxine) (Hoxine)#		(N,O)
Dimethylglyoximato	(Hdmg) (H₂dmg)#		(N,N)
Acetelacetonato or, 2,4-pentanedionate	(acac) (Hacac)#		(O,O)
N,N-diethyldithiocarbamato	(dtc) (Hdtc)#		(S,S)
2,2'-dipyridyl or 2,2'-bipyridyl (in old literature, it is denoted by bipy or dipy)	(dpy, bpy)		(N,N)
1,10-phenanthroline	(phen)		(N,N)

o-Phenylene*bis*-(dimethylarsane) or, bezene-1,2-diyl*bis*-(dimethylarsane)	(diars)		(As, As)
Biuret	—		(O,O)
Xanthato	—		(S,S)

Protonotated form (*i.e.* neutral form) of the ligand; charges of the anionic form not shown (Examples of some other didentate ligands are given in Sec. 1.15).

C. Polydentate Ligands: The ligands having more than two binding *sites* for coordinating simultaneously in a coordination sphere of a metal centre are described as the *polydentate or multidentate ligands*. Depending on the number binding sites (*i.e.* denticity or ligancy), the ligands are *tridentate, tetradentate, pentadentate, hexadentate, etc.* The examples of such ligands are given in Table 1.7.2.3.

D. Schiff's Bases as the Polydentate Ligands (*cf.* Sec. 1.8): Condensation reaction between the carbonyl group $(\rangle C = O)$ and primary amine group $(-NH_2)$ produces the compounds known as *Schiff's bases*. The Schiff's bases properly designed can act as the multidentate ligands. *Some Schiff's bases are stable in free state while some are stable (with respect to hydrolytic cleavage* of the $\rangle C = N$ — linkage) *when complexed.* These are very often prepared by **template reactions** (Sec 1.8).

$$\rangle C = O + H_2N - \xrightarrow[(-H_2O)]{condensation} \rangle C = N-$$

Some examples are:

Glycine + salicylaldehyde $\xrightarrow{-H_2O}$

$$O=C-CH_2$$
$$HO^* \quad ^*N=CH$$
$$H\overset{*}{O}-\bigcirc$$

(tridentate ligand)

(N-salicylideneglycine *i.e.* **H$_2$salgly**)

Glycylglycine + salicylaldehyde $\xrightarrow{-H_2O}$

$$HO-C-CH_2$$
$$H_2C-N^* \quad ^*N=CH$$
$$O=C-\overset{*}{O}H \quad H\overset{*}{O}-\bigcirc$$

(tetradentate ligand)

(N-salicylideneglycylglycine *i.e.* H$_2$salglygly)

en + 2Hacac $\xrightarrow{-2H_2O}$

$$CH_2-CH_2$$
$$H_3C \quad \quad CH_3$$
$$C=N^* \quad ^*N=C$$
$$HC \quad \quad CH$$
$$C-\overset{*}{O}H \quad H\overset{*}{O}-C$$
$$H_3C \quad \quad CH_3$$

(H$_2$acacen)

(tetradentate ligand)

N,N′-ehylene-*bis*(acetylacetoneiminate) or *bis*(acetylacetonato)ethylenediamine or simply H$_2$acacen

To act as a multidentate ligand, OH groups of the Schiff's bases are to be deprotonated. The binding sites are denoted by *.

E. Tripod Ligands (*cf.* Figs. 2.11.7.8-9): These are the tetradentate ligands. The central atom itself acts as a donor site and there are three branches from the central atom and each branch bears a donor site. Such ligands can form the trigonal bipyramidal geometry, tetrahedral geometry and octahedral

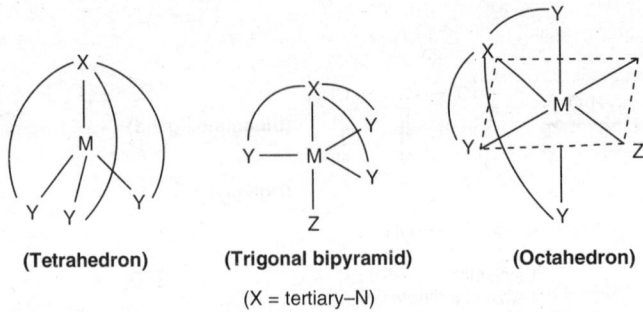

(Tetrahedron)   **(Trigonal bipyramid)**   **(Octahedron)**

(X = tertiary–N)

geometry depending on the flexibility of the branches (*cf.* Sec. 2.11.7, Figs. 2.11.7.8-9, 4.5.2.9). Some representative tripod ligands are:

$$:N(CH_2CH_2\ddot{N}H_2)_3, \quad :N(CH_2CH_2\ddot{P}H_2)_3, \quad :N(CH_2COO^-)_3$$
$$\textbf{(tren)} \qquad\qquad\qquad\qquad \textbf{(nta)}$$

Resonance stabilization in the chelate and electrophilic substitution at the chelate ring

It has already been stated that the **pseudo-aromaticity** in the coordinated acetylacetonate ligand stabilizes the system. **In fact, in this chelate ring, electrophilic substitution at the central C-atom can occur.** This leads to halogenation, acylation, formylation, etc.

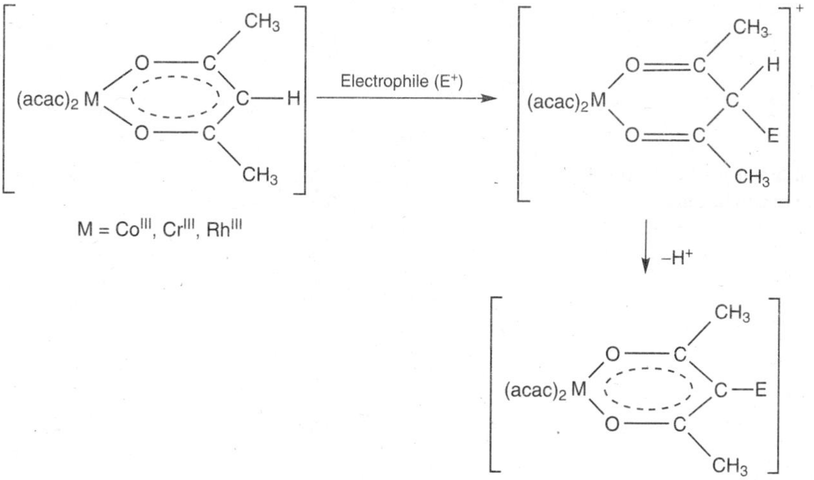

M = Co^{III}, Cr^{III}, Rh^{III}

In the chelated biguanide (bigH) ligand, the **pseudo-aromatic character** is also developed.

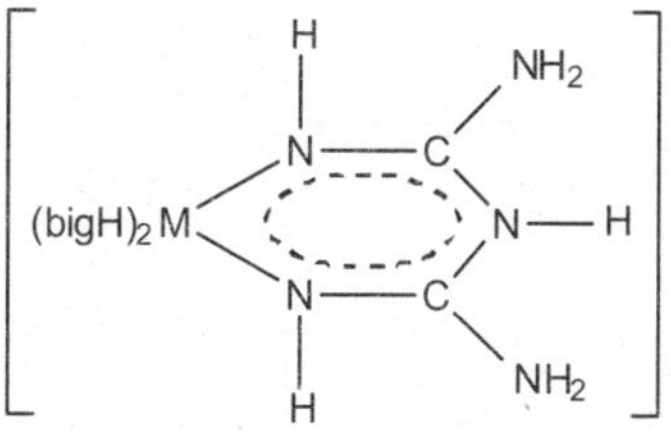

The **extended delocalization of electrons in [Ni(dmgH)$_2$]** stabilizes the system.

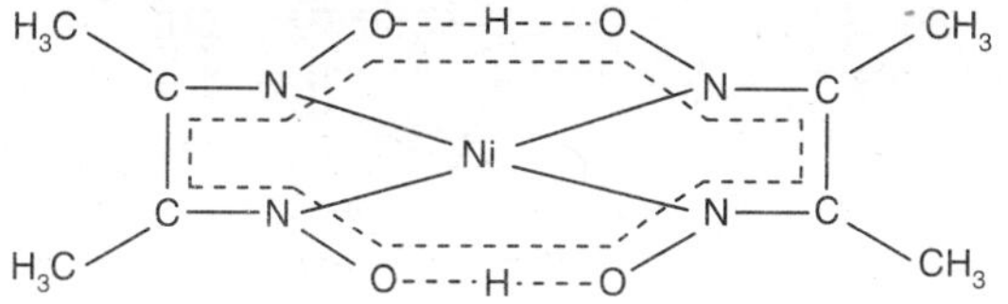

Table 1.7.2.3 Some representative polydentate ligands (*cf.* Sec. 1.15.1).

Name of the ligand when present in complex (cf. Sec. 1.15.1)	Abbreviation	Formula (donor sites denoted by *)	Donor Sites
1,2,3-propanetriamine			(N,N,N) (tridentate)
Diethylenetriamine or, N-(2-aminoethyl)ethane-1,2-diamine	(dien)		(N,N,N) (tridentate)
Iminodiacetato (H₂ida)#	(ida)		(O,N,O) (tridentate)
2,2′,2″-terpyridine or, 2,2′: 6′,2″-terpyridine	(terpy)		(N,N,N) (tridentate)
Glycylglycinato	(glygly) (Hglygly)#		(N,N,O) (tridentate)

Name	Abbreviation	Structure	Donor/Denticity
Cysteinato	(cys) (H₂cys)#		(O,N,S) (tridentate)
Tartrato or, 2,3-dihydroxy-butanedioato	tart	—	(cf. Sec. 1.10)
Histidinato	(his) (Hhis)#		(O,N,N) (tridentate)
Salicylindeneglycinato (Schiff's base)	(salgly) (H₂salgly)#		(O,N,O) (tridentate)
Triethylenetetramine (open linear chain) or, N,N'-bis-(2-aminoethyl)-ethane-1,2-diamine	(trien)	(Open linear chain)	(N,N,N,N) (tetradentate)
tris(2-aminoethyl)amine or, N,N-bis(2-aminoethyl)-ethane-1,2-diamine	(tren)	(tripodal ligand)	(N,N,N,N) (tetradentate)
Nitrilotriacetato or, 2,2',2''-nitrilotriacetato	(nta) (H₃nta)#	(tripodal ligand)	(N,3O) (tetradentate)

Ethylene-*bis*-(salicylaldiminato) or, *bis*(salicylidene)-ethylenediaminato	(salen) (H$_2$salen)#	(Schiff base, en + 2 salicylaldehyde)	(2O,2N) (tetradentate)
o-Phenylene*bis*-(salicylaldiminato)	(saloph) (H$_2$saloph)#		(2O,2N) (tetradentate)
Ethylenediaminediacetato	(edda) (H$_2$edda)#		(2O,2N) (tetradentate)
Ethylenediaminetriacetato			(2N,3O) (pentadentate)
Ethylene*bis*(acetylacetone-iminato) or, *bis*(acetylacetonato)-ethylenediamine	(acacen) (H$_2$acacen)		(2O,2N) (tetradentate)

Tetraethylenepentaamine	(tetraen)		(5N) (pentadentate)

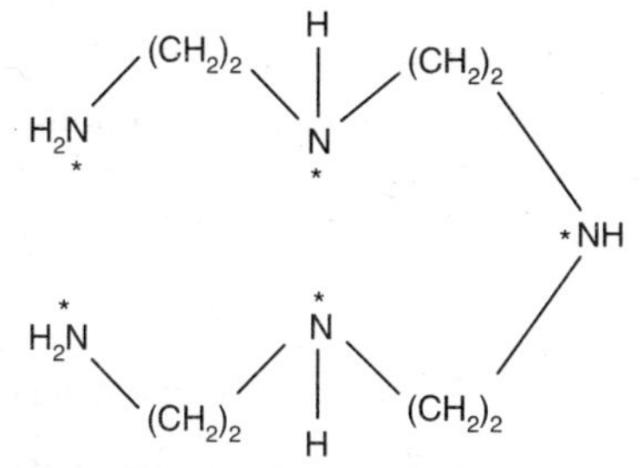

(open linear chain)

Ethylenediaminetetraacetato or, 2,2',2'',2'''-(ehane-1,2-diyldinitrilo)tetraacetato	(edta) (H_4edta)#	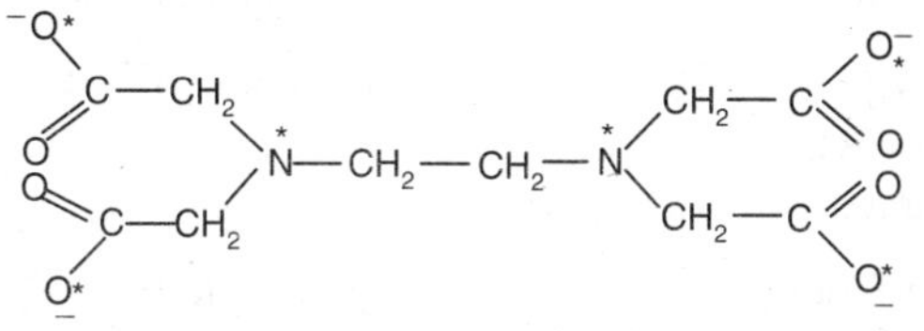	(4O, 2N) (hexadentate)

trans-cyclohexane-1,2-diamine-tetraacetato or, *trans*-2,2',2'',2'''-(cyclohexane-1,2-diyl-dinitrilo)tetraacetato	(cdta) (H_4cdta)#	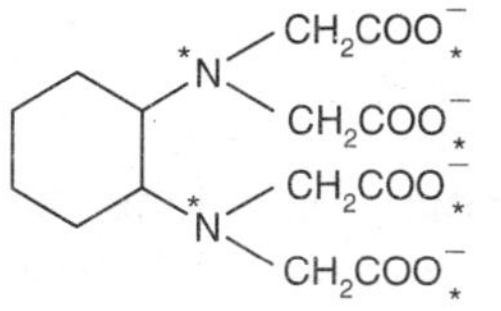	(4O,2N) (hexadentate)

Diethylenetriamine-pentaacetato or, N,N,N',N'',N''-diethylene-triaminepentaacetato	(dtpa) (H_5dtpa)#	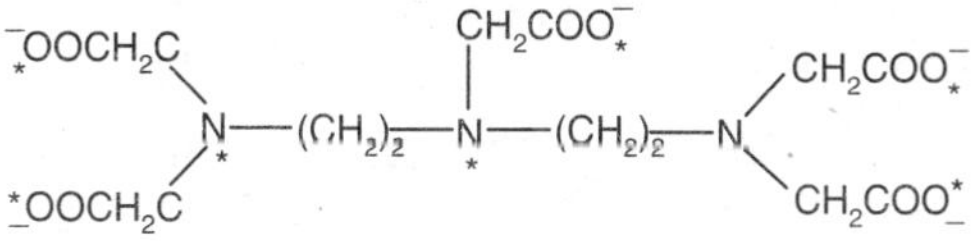	(3N,5O) (octadentate)

Ethylenedibiguanide	H_2endibig	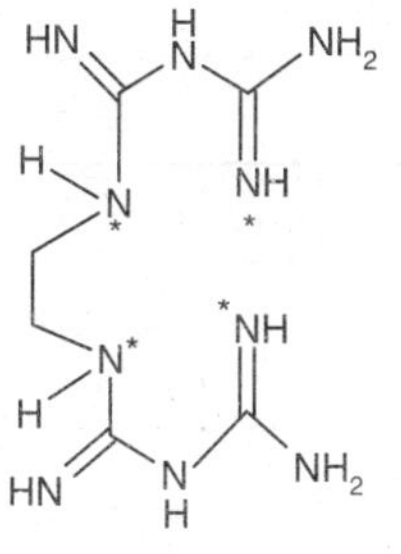	(N,N,N,N) (tetradentate)

Represents the protonated form (*i.e.* neutral form) of the ligand; charges of the anionic form not shown. The **aminopolycarboxylic acids can act as the flexidentate ligands**.

F. Flexidentate Ligands: In some cases for the polydentate ligands (and even for the didentate ligands like oxyanions RCO_2^-, CO_3^{2-}, NO_3^-, SO_4^{2-}, etc.), all the available binding sites of a particular ligand may not participate in making the metal-ligand linkage. In such cases, the denticity (or ligancy) of the ligands depends on the experimental conditions. *This variable denticity character of the ligands is referred to as the flexidentate character of the ligands.*

(i) **Oxyanions:** It has been pointed out that **oxyanions** like carboxylate (RCO_2^-), carbonate (CO_3^{2-}), nitrate (NO_3^-), sulfate (SO_4^{2-}), etc. can act as both the monodentate and didentate ligands depending on the conditions (Fig. 1.7.2.1). This is illustrated below.

$[Co(CO_3)(NH_3)_4]^+$	(didentate $\eta^2\text{-}CO_3^{2-}$, C.N. = 4 + 2 = 6)
$[Co(CO)_3(NH_3)_5]^+$	(monodentate CO_3^{2-}; C.N. = 5 + 1 = 6)
$[U(CO_3)_3O_2]^{4-}$	(didentate $\eta^2\text{-}CO_3^{2-}$; C.N. = 2 + 3 × 2 = 8)
$[Co(NH_3)_5(OSO_3)]^+$	(monodentate SO_4^{2-}; C.N. = 5 + 1 = 6)
$[Co(en)_2(SO_4)]^+$	(didentate $\eta^2\text{-}SO_4^{2-}$; C.N. = 2 × 2 + 2 = 6)
$[U(O_2CR)_3O_2]^-$ (*cf.* $[UO_2(O_2CR)_3]^-$ in old literature)	(didentate $\eta^2\text{-}RCO_2^-$; C.N. = 2 + 3 × 2 = 8)
$[U(NO_3)_2(OH_2)_2O_2]$ (*cf.* $[UO_2(NO_3)_2(OH_2)_2]$ in old literature)	(didentate $\eta^2\text{-}NO_3^-$; C.N. = 2 + 2 × 2 + 2 × 1 = 8)

Ligating behaviour of common axoanions like carbonate, carboxylate, nitrate and sulfate (*see* Sec. 12.1.19 for ir-spectral data)

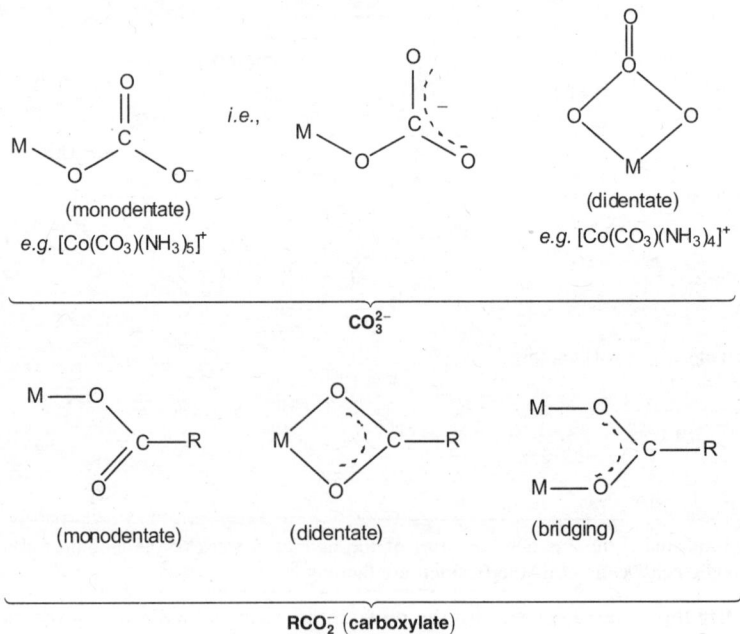

Fig. 1.7.2.1 Flexidentate character of oxyanions. (*contd.*)

(monodentate) (didentate) (bridging)

NO_3^- (nitrate)

(monodentate) (didentate) (bridging)

SO_4^{2-} (Sulfate)

Fig. 1.7.2.1 Flexidentate character of oxyanions.

(ii) **H_4edta:** Flexidentate character of *edta* (fully protonated form H_4edta or edtaH$_4$) is well known. It can show the maximum denticity 6 but in many complexes, it can show denticity less than 6. These are illustrated below:

hexadentate chelation: $[M(edta)]^{2-}$ (M = Ca, Mg); $[M(edta)]^-$ (M = Al, Cr, Co)

pentadentate chelation: $[M(edta)(OH)]^{2-}$ (M = Co, Cr); $[Co(Br)(edta)]^{2-}$

tetradentate chelation: $[M(H_2edta)]$ (M = Pd, Pt)

didentate chelation: $[MCl_2(H_4edta)]$ (M = Pd, Pt)

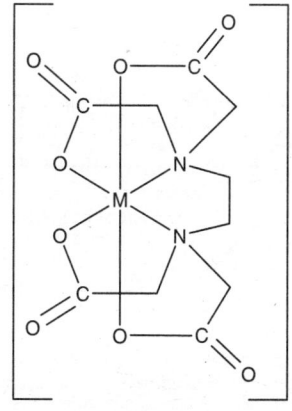

hexadentate edta
(charge is not shown for
the sake of simplicity)

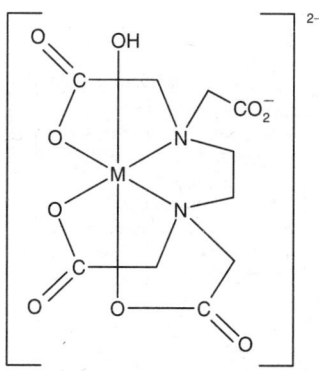

pentadentate edta
(M = Co, Cr)

Fig. 1.7.2.2 (*contd.*)

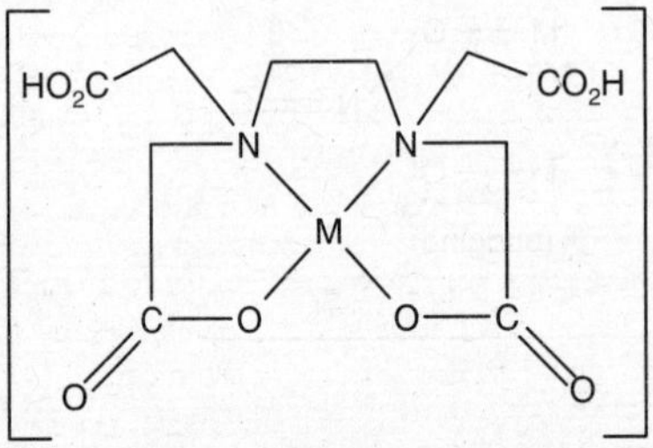

[M(H₂edta)], (M = Pd, Pt)
tetradentate edta

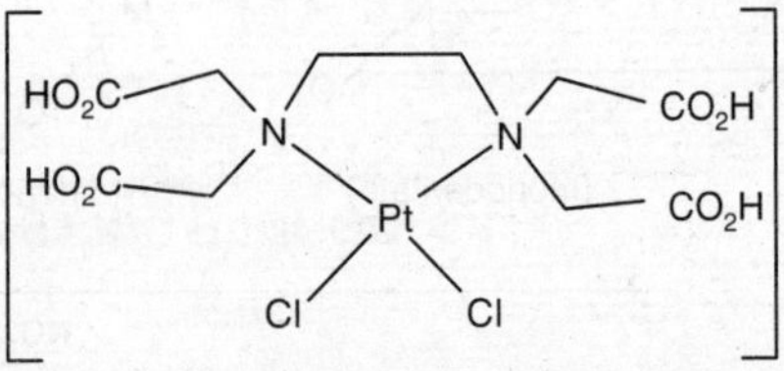

[PtCl₂(H₄edta)]
didentate edta

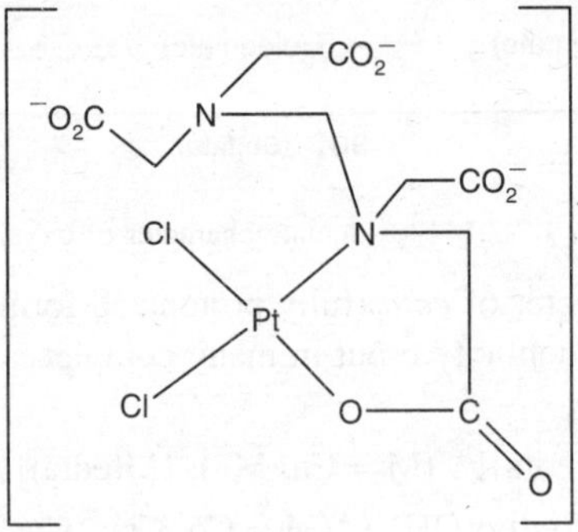

$[PtCl_2(edta)]^{4-}$
didentate edta

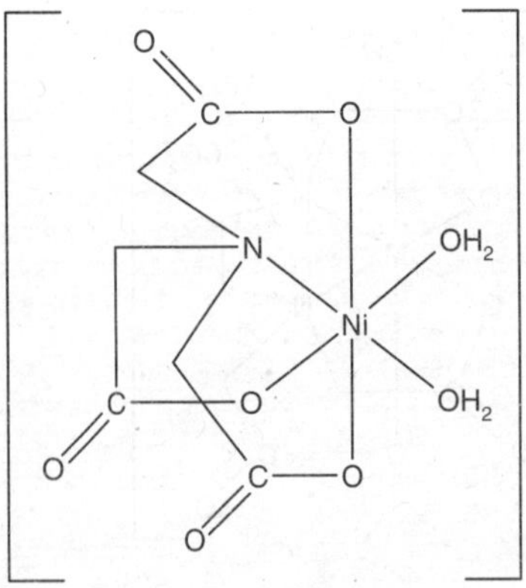

cis-[Ni(nta)(OH₂)₂]⁻
tetradentate nta

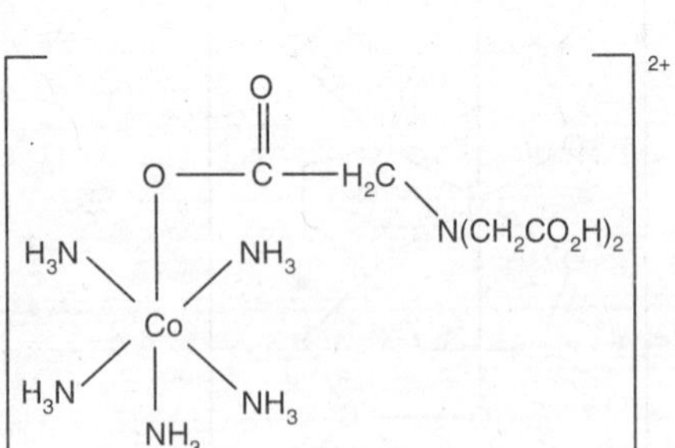

$[Co(H_2nta)(NH_3)_5]^{2+}$
monodentate nta

Fig. 1.7.2.2 Flexidentate and ambidentate character of some aminopolycarboxylates.

Flexidentate character depends on the pH of the solution

The flexidentate character of aminopolycarboxylic acids (*e.g.* H_2ida, H_3nta, H_4edta, etc.), polypeptides (*e.g.* $H_2glygly$), polydentate aminoacids (*e.g.* H_2cys, *i.e.* cysteine, Hhis, *i.e.* histidine) depends largely on the working pH of the solution. At lower pH, very often the ligating sites (*e.g.* $-CO_2^-$, $-S^-$, $-C(O)-N^--$, $-NH_2$, etc.) remain protonated and then they show the lower denticity. This aspect has been illustrated in some examples.

(iii) **Aminopolycarboxylic acids (in general):** Depending on the conditions, these ligands can show the flexidentate as well as the ambidentate character (*cf.* Fig. 1.7.2.2). Like edta (protonated form = H_4edta), other **aminopolycarboxylic acids,** *e.g.* ida (protonated form $idaH_2$ or H_2ida; ida can act maximum as a tridentate ligand), nta (protonated form $ntaH_3$ or H_3nta; nta can act maximum as a tetradentate ligand), etc. can show the flexidentate character (Fig. 1.7.2.2). It is illustrated for nta and *cis*-$[Ni(nta)(OH_2)_2]^-$ and $[Ni(gly)(nta)]^{2-}$, (nta as a tetradentate ligand); $[Co(NH_3)_5(ntaH_2)]^{2+}$ (nta as a monodentate ligand).

(iv) **dien:** It can function as a *tridentate* or as a *didentate* ligand depending on the conditions.

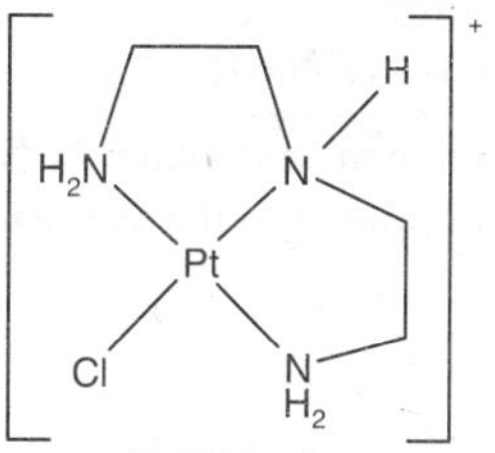

[PtCl(dien)]⁺   [PtCl₂(dien)]
(tridentate chelation)   (*i.e.* **didentate chelation**)

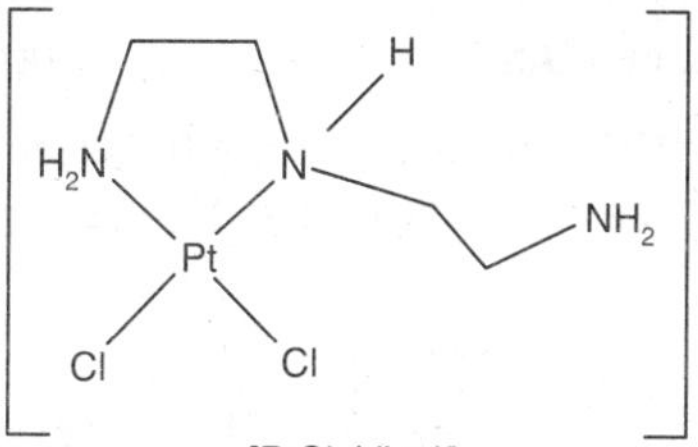

(v) **trien:** Flexidentate character of trien is illustrated below.

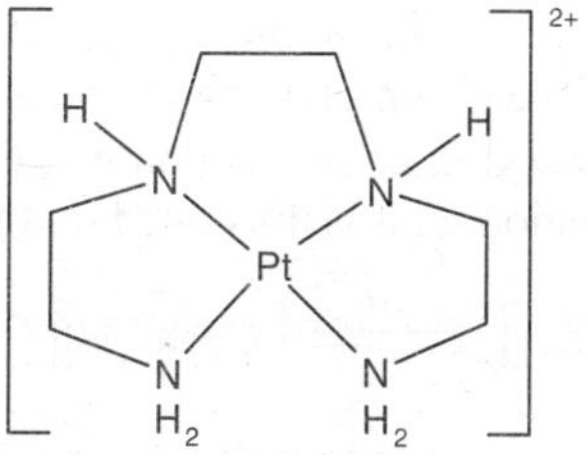

[Pt(trien)]²⁺   [PtCl(trien)]⁺
(tetradentate chelation)   **(tridentate chelation)**

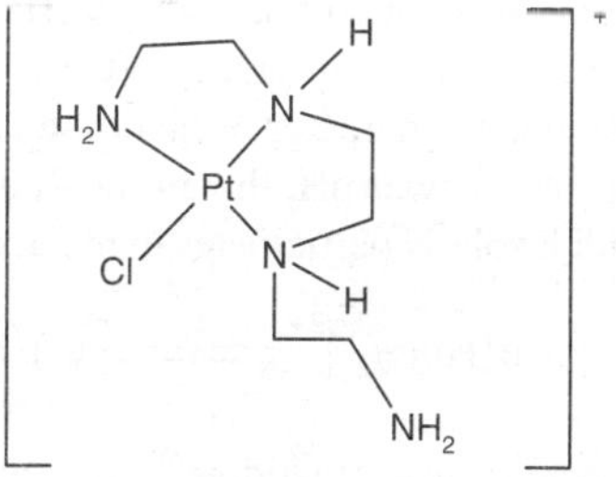

(vi) **H_2cys:** *Cysteine* (H_2cys) can act as a tridentate ligand (S,N,O) and it may act also as a didentate ligand in three different ways (S,N; N,O; S,O; *cf.* **ambidentate character**).

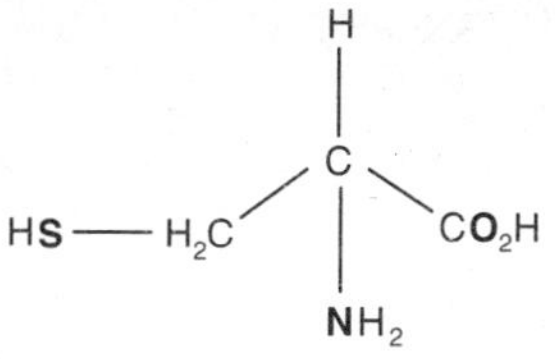

(cysteine, *i.e.* H_2cys, binding sites are bold)

In the dinuclear complex $Na_2[Mo_2(cys)_2O_4].5H_2O$, cysteine acts as a tridentate ligand but on slight acidification, the coordinated —COO^- group is dissociated as —CO_2H (*i.e.* cysteine acts as a didentate ligand).

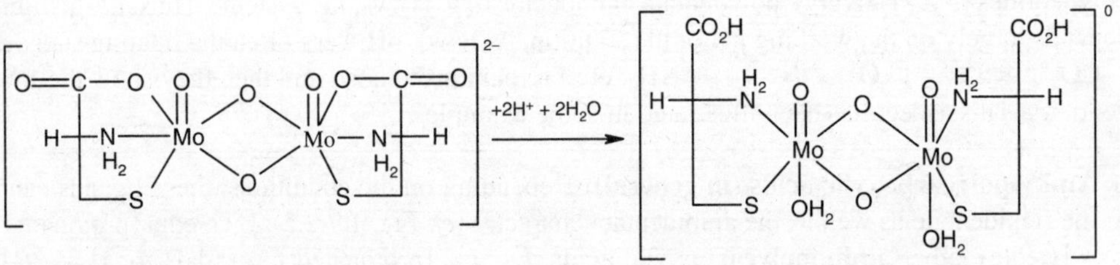

$[Mo_2(cys)_2O_4]^{2-}$
(di-μ-oxido-dioxidodimolybdenum complex)
having **tridentate cysteinate**

(didentate cysteinate)

$[Co(cys)_3]^{3-}$ [*i.e. tris*(cysteinato)cobalt(III)] in which cysteine acts as a **didentate ligand** can show the **linkage isomerism** (*cf.* ambidentate character).

$[Co(cys)_3]^{3-}$, $[Co(cys)_3]^{3-}$
[red, cys(O,S)]   **[green, cys(N,S)]**

(vii) **Hhis:** Histidine (Hhis) can also act as a flexidentate ligand: tridentate (N,N,O), didentate (N,N or N,O) (*cf.* ambidentate character). The N-sites are coming from the imidazole moiety and —NH_2 group.

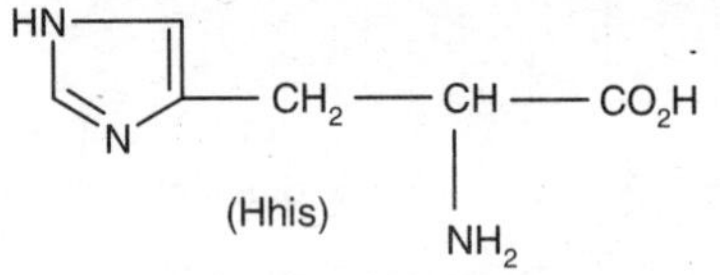

(Hhis)

[Co(L-his)₂], *i.e. bis*(L-histidinato)cobalt(II) where the amino acid functions as a **tridentate ligand**. In *bis*(L-histidine)copper(II) nitrate, *i.e.* [Cu(L-His)₂](NO₃)₂, the amino acid functions as a **didentate ligand**. The flexidentate character of histidine largely depends on pH. The imidazole proton (*i.e.* imH⁺) has $pK_a \approx 6.0$. Thus at lower pH, the amino-N and carboxyl-O act as the donor sites while at relatively higher pH, the imidazole-N participates in metal coordination. This is illustrated for the Cu(II)-complexes

$$\left[Cu(Hhis)_2\right]^{2+} \overset{-H^+}{\rightleftharpoons} \left[Cu(his)(Hhis)\right]^+ \overset{-H^+}{\rightleftharpoons} \left[Cu(his)_2\right]$$
(a ternary complex)

The ligating behaviour of histidine in the copper(II) complex is shown in Fig. 1.7.2.3.

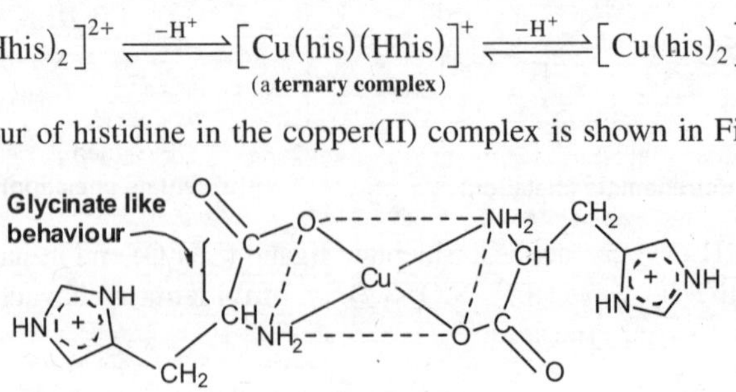

$[Cu(Hhis)_2]^{2+}$

(A)

Fig. 1.7.2.3 (*contd.*)

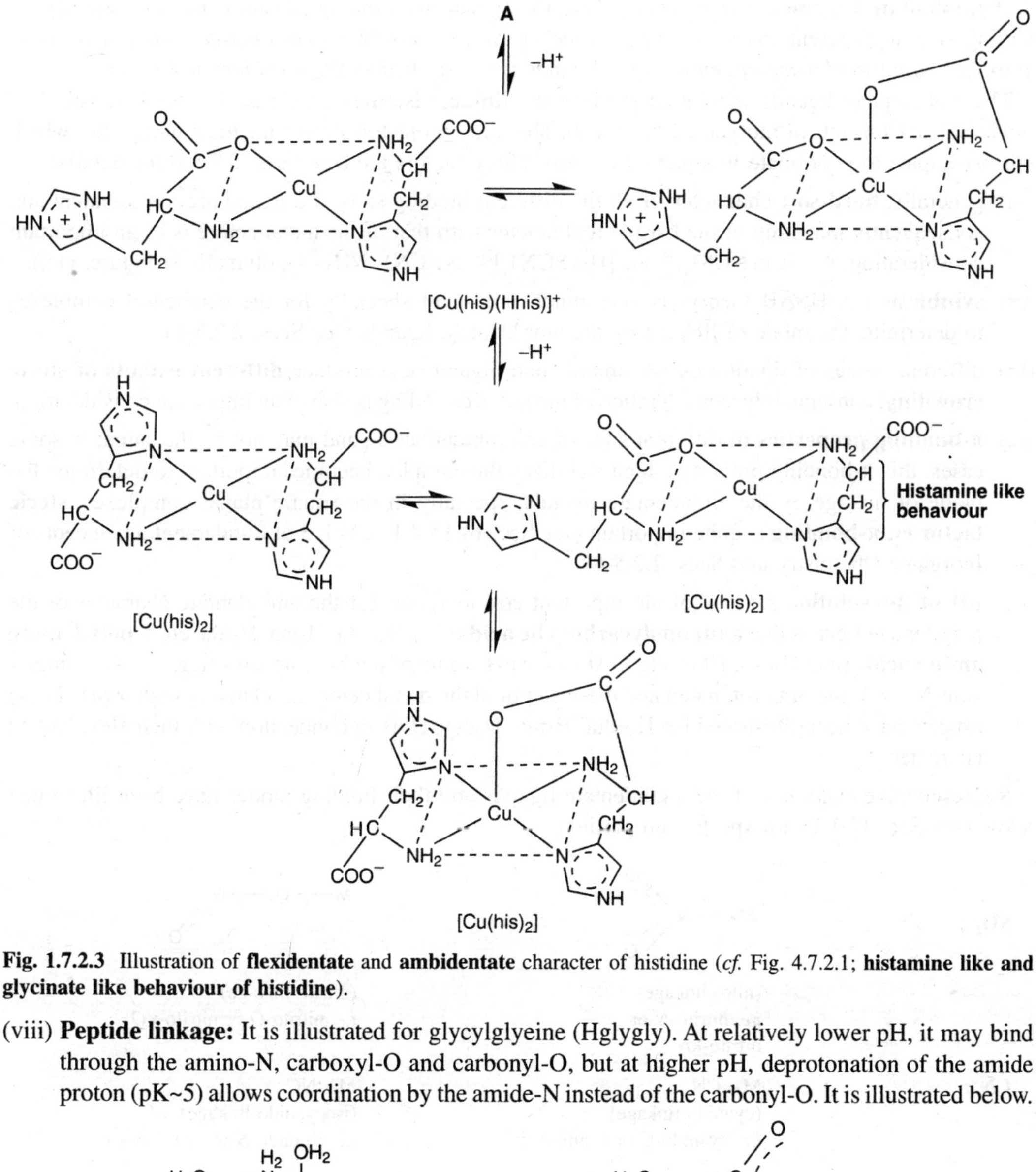

Fig. 1.7.2.3 Illustration of **flexidentate** and **ambidentate** character of histidine (*cf.* Fig. 4.7.2.1; **histamine like and glycinate like behaviour of histidine**).

(viii) **Peptide linkage:** It is illustrated for glycylglyeine (Hglygly). At relatively lower pH, it may bind through the amino-N, carboxyl-O and carbonyl-O, but at higher pH, deprotonation of the amide proton (pK~5) allows coordination by the amide-N instead of the carbonyl-O. It is illustrated below.

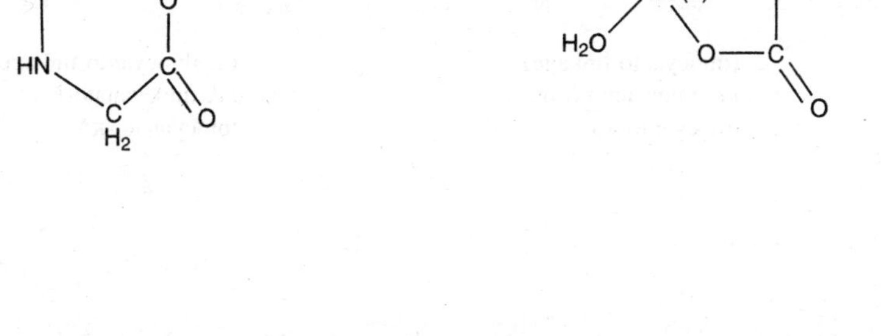

G. Ambidentate Ligands: The ligands having more than one binding site may use different sets of binding sites in different cases, *but all the binding sites generally cannot operate simultaneously to coordinate a particular metal centre.* Such ligands are described as the *ambidentate ligands.*

The ambidentate ligands very often produce the **linkage isomers** (*cf.* Figs. 14.15.4.1-3, Vol. 3 of Fundamental Concepts of Inorganic Chemistry). The different factors determine the binding sites which will participate to coordinate in a particular case. These factors are (*see* Secs. 2.2.5-6 for details):

(i)   generally, **hard-soft character** of all the different binding sites of a ligand are not identical and consequently matching of the hard-soft character with that of the metal centre is of an important consideration. (*cf.* $[Co(NCS)_4]^{2-}$ *vs.* $[Hg(SCN)_4]^{2-}$; $Ni(CN)_2.NH_3$ — polymeric structure, etc.).

(ii) **symbiosis** (*cf.* HSAB theory) is very much important specially for the octahedral complexes to determine the mode of linkage by the ambidentate ligands (*see* Secs. 2.2.5-6).

(iii) different modes of ligation by the ambidentate ligands can produce **different extents of steric crowding**; consequently, consideration of the steric crowding is also of an important consideration.

(iv) **π-bonding properties** of different sites of an ambidentate ligand may not be the same; in some cases, this π-bonding property which stabilizes the complex becomes important to determine the mode of linkage by the ambidentate ligands; specially in the square planar complexes, **steric factor *vs.* π-bonding** is quite important (*see* Sec. 14.15.4.1-3, Vol. 3 of Fundamental Concepts of Inorganic Chemistry and Secs. 2.2.5-6).

(v)   **pH of the solution** may be of an important consideration for the ambidentate character of the polydentate ligands like **aminopolycarboxylic acids** (*e.g.* H_4edta, H_3nta, H_2ida, etc.), **polydentate amino acids** (*e.g.* H_2cys, Hhis, etc.). At lower pH, some of the binding sites (*e.g.* $—CO_2^-$, imidazole-N, etc.), remain protonated and these can bind the metal centre at relatively higher pH. These aspects have been illustrated for H_4edta, H_3nta, H_2cys, Hhis in connection with their flexidentate character.

Representative examples of the ambidentate ligands and their binding modes have been illustrated below (*see* Sec. 12.1.18 for spectral properties).

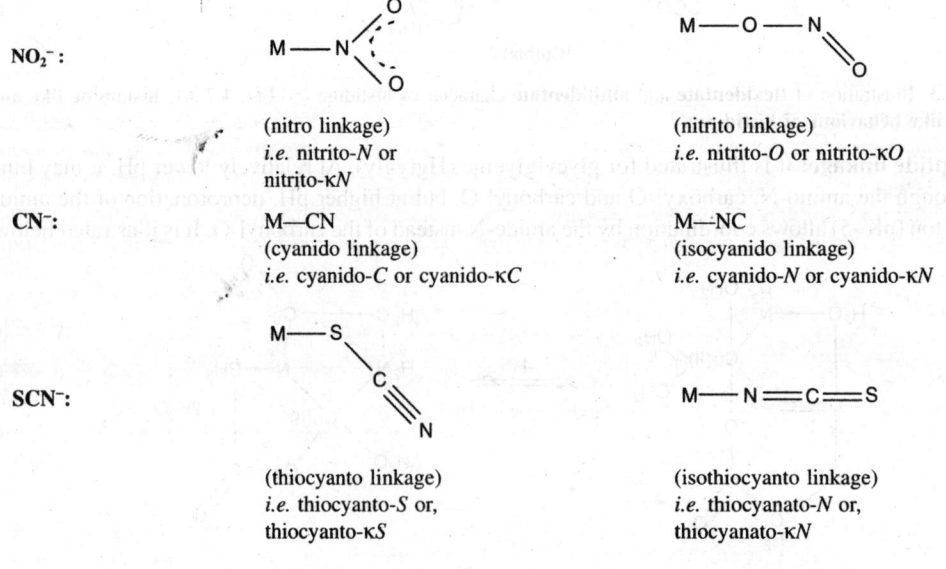

NO_2^- :

(nitro linkage)
i.e. nitrito-*N* or
nitrito-κ*N*

(nitrito linkage)
i.e. nitrito-*O* or nitrito-κ*O*

CN^-:   M—CN
(cyanido linkage)
i.e. cyanido-*C* or cyanido-κ*C*

M—NC
(isocyanido linkage)
i.e. cyanido-*N* or cyanido-κ*N*

SCN^-:

(thiocyanto linkage)
i.e. thiocyanto-*S* or,
thiocyanto-κ*S*

(isothiocyanto linkage)
i.e. thiocyanato-*N* or,
thiocyanato-κ*N*

SeCN⁻:	M—SeCN (selenocyanato-*Se*)	M—NCSe (selenocyanato-*N*)
OCN⁻:	M—OCN (cyanato linkage) *i.e.* cyanato-*O* or cyanato-κ*O*	M—NCO (isocyanato linkage) *i.e.* cyanato-*N* or cyanato-κ*N*
$S_2O_3^{2-}$:	M—SSO₃ (thiosulfato-κ*S*)	M—OSO₂S (thiosulfato-κ*O*)

$C_2S_2O_2^{2-}$

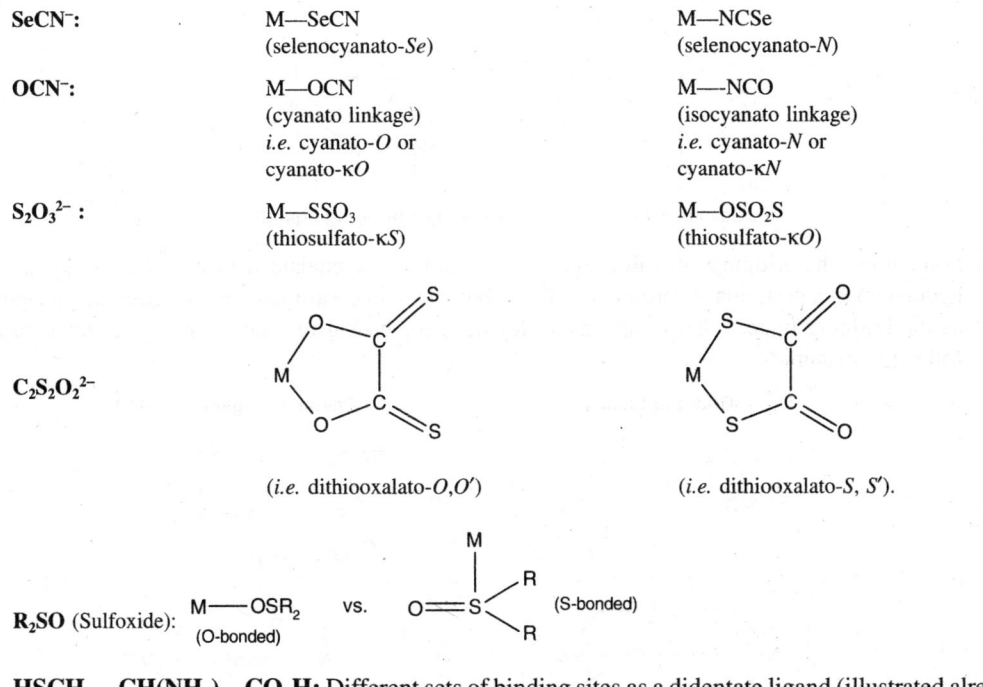

(*i.e.* dithiooxalato-*O,O'*)   (*i.e.* dithiooxalato-*S, S'*).

R₂SO (Sulfoxide):   M——OSR₂   vs.   (O-bonded) ... (S-bonded)
                      (O-bonded)

HSCH₂—CH(NH₂)—CO₂H: Different sets of binding sites as a didentate ligand (illustrated already)
(cysteine)

(S,N), (N,O), (S,O)

Histidine: Different sets of binding as a didentate ligand (illustrated already) (O, Im-N), (O, amino-N, *i.e.* **glycinate like behaviour**), (Im-N, amino-N, *i.e.* **histamine like behaviour**) (*cf.* Fig. 1.7.2.3).

Examples of *linkage isomers* illustrating the ambidentate character of ligands have been discussed in Chapter 2 and in Sec. 12.1.18.

H. Bridging Ligands (*cf.* Sec 1.14): The different ligands having more than one lone pair of electrons can coordinate simultaneously two different metal centres. *Such ligands are described as the bridging ligands.* Sometimes, the bridging ligands may coordinate more than two metal centres (*cf.* μ₂–CO, μ₃–CO in carbonyl clusters; electron deficient M—C bonds).

 (i) Different *monodentate ligands* can act also as the bridging ligands:
     Examples: Cl⁻, F⁻, O²⁻, OH⁻, NH₂⁻, N₃⁻, SCN⁻, CN⁻, (*cf.* μ₂–CO), etc.

 (ii) The *oxyanions* (having the flexidentate character) like CO₃²⁻(carbonate), RCO₂⁻(carboxylate), NO₃⁻(nitrate), SO₄²⁻(sulfate), etc. can act as the bridging ligands. These have been discussed earlier (*cf.* Fig. 1.7.2.1). Carboxylates as the bridging ligands are well documented in the formation of **metal clusters**.

(iii) Sometimes, the didentate and polydentate ligands can act as the bridging ligands. **Oxalate, aminopolycarboxylates**, etc. are the examples in this group. **Peroxide** can act as a didentate chelating ligand in some cases (discussed earlier) but it can act as a bridging ligand in many cases.

(iv) **Hydrazine** can act as a monodentate ligand but it fails to act as a didentate chelating ligand as it will produce a three membered ring which will be in strain. Besides this, in the three membered ring, the positive charges on the adjacent three centres (*i.e.* M, N, N) in the ring will produce an unfavourable situation. However, N₂H₄ can act as a bridging ligand.

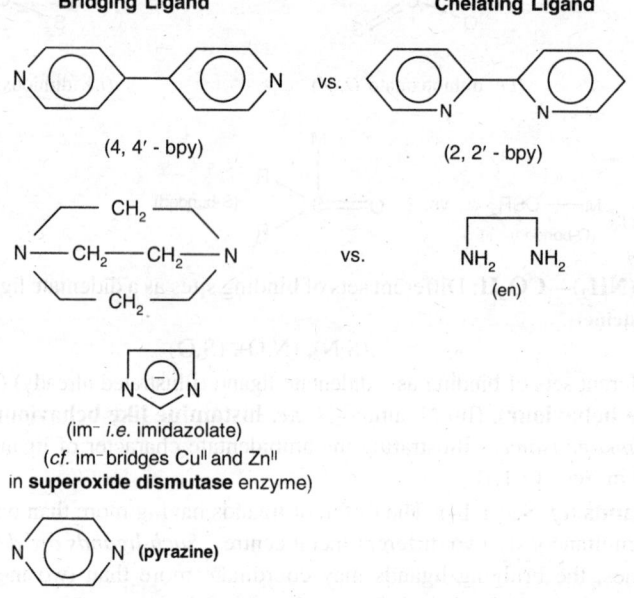

(Unfavoured Situation)   (N₂H₄ as a bridging ligand)

(v) Sometimes, the bridging sites in a ligand do not allow the chelate formation (as in the case of hydrazine) in a particular coordination sphere but they can coordinate two different metal centres as the bridging ligands. This happens so for the geometrical reasons. This is illustrated in the following examples.

Bridging Ligand   **Chelating Ligand**

(4, 4' - bpy)   vs.   (2, 2' - bpy)

vs.   (en)

(im⁻ i.e. imidazolate)
(cf. im⁻ bridges Cuᴵᴵ and Znᴵᴵ
in **superoxide dismutase** enzyme)

(pyrazine)

I. Examples of Complexes having Different types of Bridging Ligands: These are given in connection with the Nomenclature of coordination compounds (Sec. 1.15) and Inner Sphere Electron Transfer Reactions (*see* Chapter 6).

Didentate Bridging Ligand and Didentate Chelating Ligands: The so called monodentate ligands (*e.g.* hydrazine) acting as the bridging ligands are sometimes described as the *didentate bridging ligands* because such ligands show the total denticity two. According to this view, the denticity is determined by the total number of binding sites utilized for binding the metal centres that may be different or single. Thus, hydrazine; 4,4'-bipyridine, etc. can act as the didentate bridging ligands. However, these can never be considered as the *didentate chelating agents* because they fail to form the chelate ring. Thus, 4,4'-bipyridine is a didentate bridging ligand but 2,2'-bipyridine is a didentate chelating ligand which can form a five-membered chelate ring.

Some ligands like peroxide (O_2^{2-}), carboxylates (RCO_2^-), alkyne ($RC \equiv CR$), etc. can act both as the didentate chelating and bridging ligands (*cf.* Secs. 9.11–9.14).

J. Molecular Carbon Dioxide (CO_2), Carbon Disulfide (CS_2), Sulfur Dioxide (SO_2) Oxygen (O_2), Nitrogen (N_2) and Hydrogen (H_2) as the Ligands (*cf.* Secs. 9.11–9.14)

- Nature has utilized **dioxygen as a ligand** in developing her **oxygen uptake metalloproteins** (*e.g.* hemoglobin, myoglobin, hemocyanin, etc.). In many synthetic oxygen carriers, O_2 binds also as a ligand.

- **Vaska's complex** *trans*-[Ir(CO)X(PPh$_3$)$_2$] (X = halide) (*see* Fig. 14.7.5.3, Vol. 3 of Fundamental Concepts of Inorganic Chemistry, for details), [Co(py)(salen)], etc. can **bind O_2 that acts as a ligand**.

- N_2 and CO are isoelectronic. CO is a potential ligand but **N_2 is a very poor ligand** (*cf.* Sec. 9.11.2, Vol. 2 of Fundamental Concepts of Inorganic Chemistry). In fact, Allen and Senoff accidentally prepared (1965) the compound [Ru(NH$_3$)$_5$(N$_2$)]$^{2+}$ where N_2 binds as a ligand. Subsequently, several complexes having N_2 as a ligand have been prepared. Examples are:

  [(H$_3$N)$_5$Ru—N$_2$—Ru(NH$_3$)$_5$]$^{4+}$, [IrCl(N$_2$)(PPh$_3$)$_2$], [Ti(N$_2$)(OR)$_2$], *trans*-[Mo(dppe)$_2$(N$_2$)$_2$]

  (bridging ligand)

  dppe = 1, 2-*bis*(diphenylphosphino)ethane, Ph$_2$P—CH$_2$—CH$_2$—PPh$_2$

- **Dihydrogen complex** (*i.e.* H_2 as a ligand) was first prepared by Kubas in 1985. This historical complex is [W(CO)$_3$(H$_2$)(PR$_3$)$_2$], R = isopropyl group. The lighter congener Mo can also form a similar complex. Now many dihydrogen complexes are known. Some examples are: [M(H)$_2$(H$_2$)(PR$_3$)$_3$] (M = Fe, Ru); [Fe(CO)(H$_2$)(NO)], [Co(CO)$_2$(H$_2$)(NO)].

Ligating behaviour of H_2 and N_2

The ligating behaviour of H_2 (*i.e.* dihydrogen) is quite interesting. It may bind the metal centre and can substitute N_2 (*i.e.* dinitrogen). Thus it indicates that both N_2 and H_2 can act as the weak π-acid ligands evidenced by the lower values of $\nu_{N\equiv N}$ and ν_{H-H} in such complexes (Sec. 12.1.17). If the bound H_2 (*i.e.* $\eta^2 - H_2$) receives sufficient electron in its antibonding MO then it leads to a dihydrido complex through the rupture of the H—H bond. Thus, the **dihydrido complex** is the *oxidative addition of* H_2.

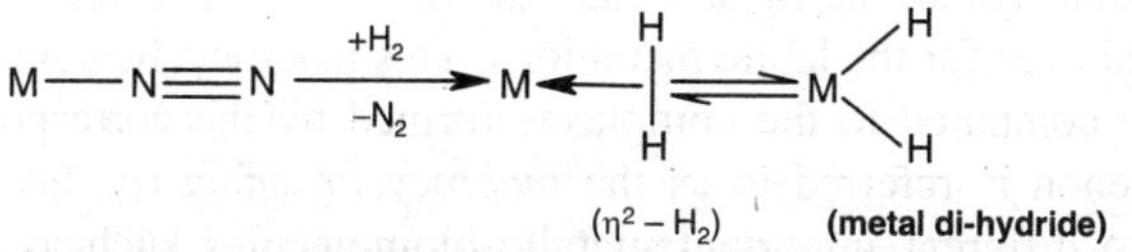

$(\eta^2 - H_2)$ **(metal di-hydride)**

- For **dioxygen (O_2), dinitrogen (N_2) and dihydrogen (H_2)**, there is a similarity in their bonding interaction. From the filled bonding MO, they denote the electron pair and receive back the electron from the metal centre in their vacant antibonding MO. **Thus all of them act as the π-acid ligands**. It is evidenced by the fact that in the complex, the X—X bond length is longer than that in free X_2 (X = H,N,O) and ν_{X-X} decreases in the complex compared to that in free X_2.

The π-acidic property of the ligand (i.e. O_2, N_2 and H_2) makes them chemically more active when complexed. This activation through complexation has been utilized by nature in developing the *oxygenase enzyme, nitrogenase enzyme* (involved in biological nitrogen fixation), and *hydrogenase* enzymes. Thus these complexes are of tremendous importance.

The complexes of H_2, N_2 and O_2 are discussed at a greater length in Secs. 9.11–9.14.

- Ligating behaviour of **CO_2, CS_2 and SO_2** have been discussed in Chapter 9.

K. Compartmental Ligands

Some macrocyclic ligands and open-ended polydentate ligands are big enough with good many donor atoms to accommodate two metal centres side by side. Such ligands are called the compartmental ligands.

(Triketonate)

1.8 MACROCYCLIC LIGANDS (*cf. Sec.* 4.6)

Macrocyclic ligands are the polydentate ligands which *encircle* or *encapsulate the metal centre*. In the cavity of the macrocyclic ligand, the metal centre is sheltered. It provides the *kinetic stability* to the coordination compound even for the labile metal ions. This macrocyclic complexes are also *thermodynamically more stable* compared to the complexes formed by the corresponding linear polydentate ligands. This phenomenon is referred to as the *macrocyclic effect* (*cf.* Sec. 4.5.4). The macrocylic ligands are relevant to different important metallo-biomolecules such as *hemoglobin, myoglobin, cytochromes, ionophores, naturally occurring macrocyclic antibiotics* (*e.g.* valinomycin), *chlorophyll, vitamin* B_{12}, etc. The different types of macrocyclic ligands are discussed below in short.

Characteristics of the macrocycles

- Biologically very important.
- Both thermodynamic and kinetic stability (*macrocyclic effect*).
- Stabilization of higher oxidation states, *e.g.* Cu^{III}, Ni^{III}, etc.

(A) Porphyrins: It is a *tetrapyrrole tetraaza macrocyclic ligand* having four coplanar *pyrrole nitrogen atoms* for binding the metal centre. *There is a conjugated π-system*. These heterocycles are **aromatic** ($4n + 2$ π-electrons). Depending on the ligating sites, these are the N_4-ligands.

The unsubstituted ligand *porphine or porphin,* is a tetrapyrrole connected at the α-carbons by *methylidyne or methine linkages* (=CH—). There are *two dissociable protons* in a porphin ring and it coordinates (*i.e.* encapsulates) as a dinegative anion. The **porphyrins** are the derivatives of the ***porphin.*** The derivative *protoporphyrin IX* (H_2ppIX, 'H_2' indicates the two dissociable NH protons) is bearing specific substitutions on 8 pyrrole positions on the porphin ring. Protoporphyrin IX is used in the *heme group*. The Fe(II)-protophyrin IX unit is called **heme b** which is used in many important biomolecules like *hemoglobin* (Hb), *myoglobin* (Mb), *cytochromes, catalase, peoxidase, cytochrome P-450, NO-synthase, etc*. The porphyrin related macrocyclic ligands are used in chlorophyll and vitamin B_{12}. In chlorophyll, the **chlorin ring** (modified porphyrin ring) coordinates Mg(II). In vitamin B_{12}, the **corrin ring** (modified porphyrin ring) coordinates the cobalt metal centre. In the octahedral geometry, the four pyrrole N-sites provide the *basal plane* and the two *trans*-axial positions are occupied by other ligands.

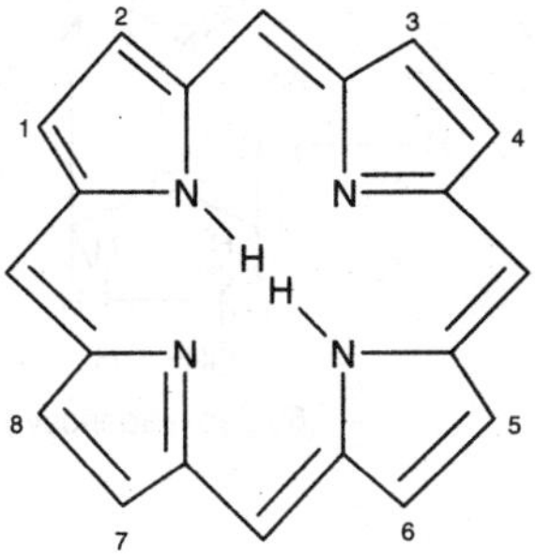

(Porphin)

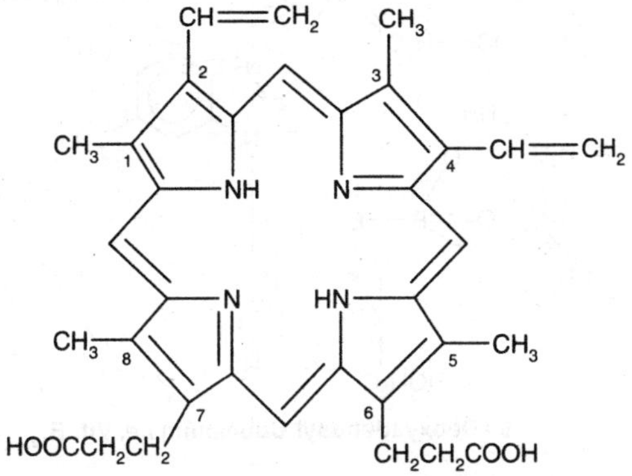

(Protoporphyrin-IX *i.e.* H_2ppIX)

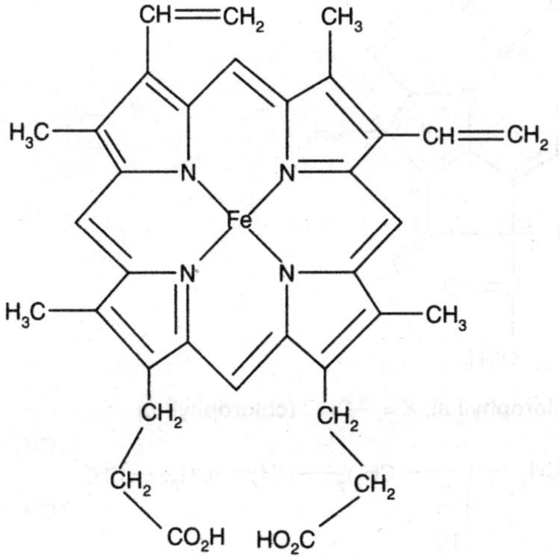

[Fe(II)-ppIX] complex *i.e.* heme-b

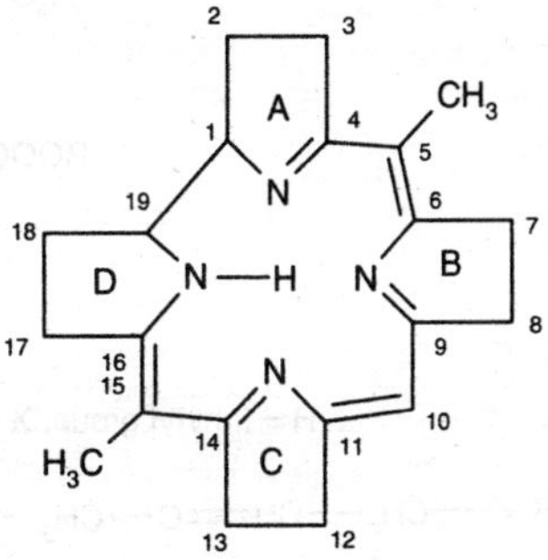

Corrin ring

(Utilised in vit. B_{12})

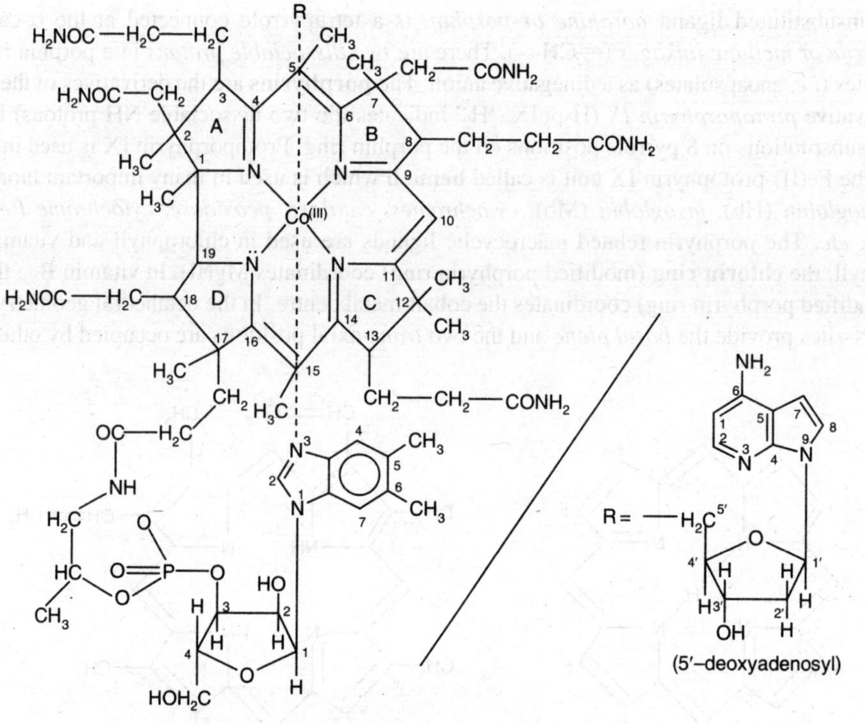

5'- Deoxyadenosyl Cobalamin *i.e.* **Vit. B₁₂**

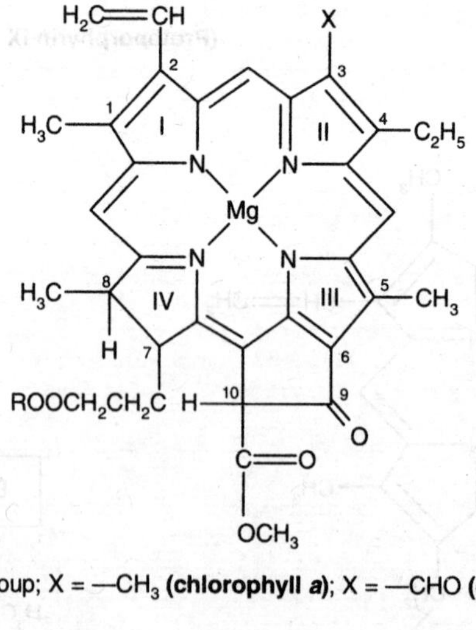

R = (phytyl group; X = —CH₃ (**chlorophyll a**); X = —CHO (**chlorophyll b**)

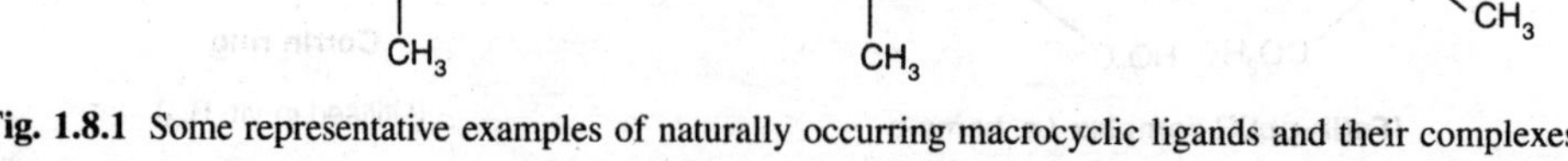

Fig. 1.8.1 Some representative examples of naturally occurring macrocyclic ligands and their complexes.

The **synthetic ligands** mimicking the porphin ring are **octaethylporphyrin** (H_2oep) and **meso-tetraphenylporphyrin** (H_2tpp). 'H_2' denotes the two dissociable protons. The sulfonated derivatives (*i.e.* having the —SO_3H groups as substitution) of pophyrins have been synthesized.

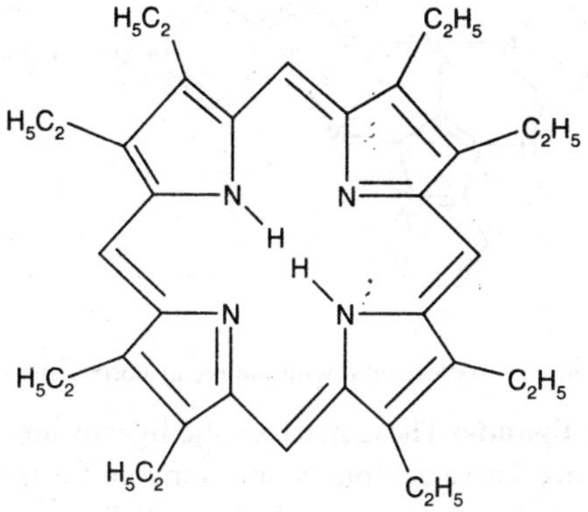

(H₂oep)

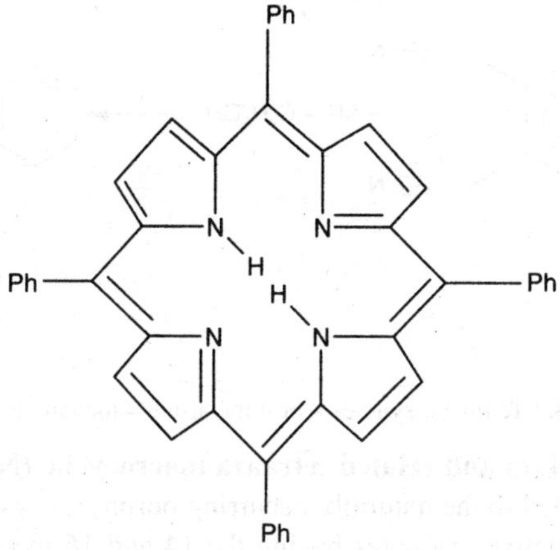

(H₂ttp)

Fig. 1.8.2 Some representative synthetic macrocyclic ligands mimicking the naturally occurring porphyrin.

If the radius of the porphyrin cavity is smaller than the radius of the metal ion, then the metal ion sits out of the plane. *It happens so in the heme-b unit.*

(B) Phthalocyanines (N_4-ligands): These are the synthetic macrocyclic ligands resembling the porphyrins. These are synthesized in the reaction of *phthalonitrile* with the metal halides where the metal centre acts as a *template* (*cf.* template synthesis). The metal-phthalocyanine complexes isoelectronic with the metal-porphyrin complexes are generally described as the **phthalocyanines**. These are thermally stable and used as dyes for their intense colour.

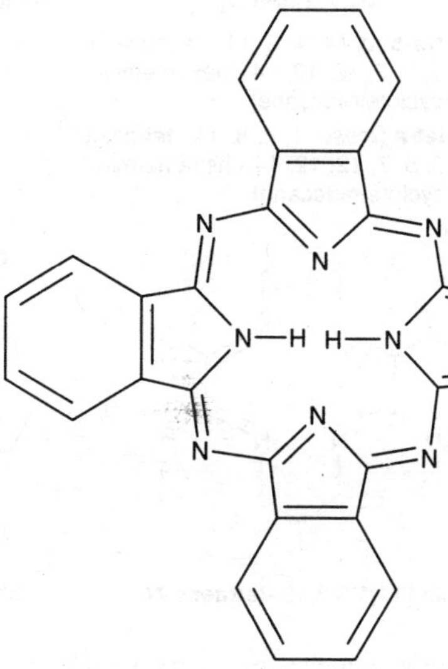

(Phthalocyanine, H₂pc)

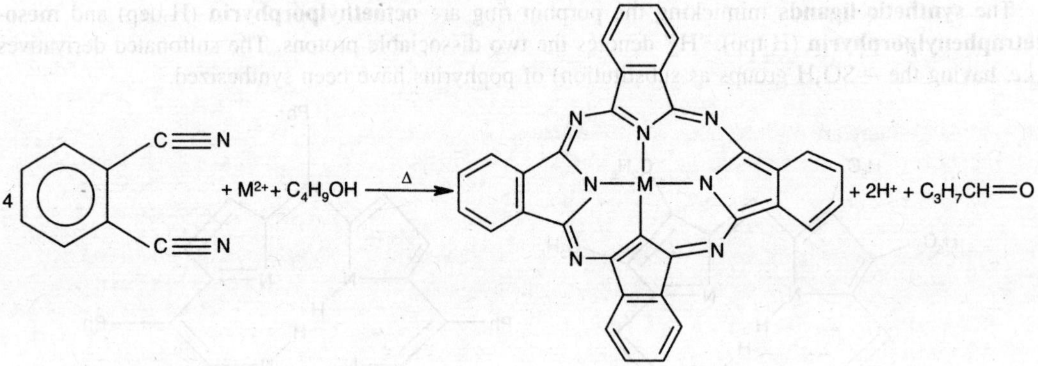

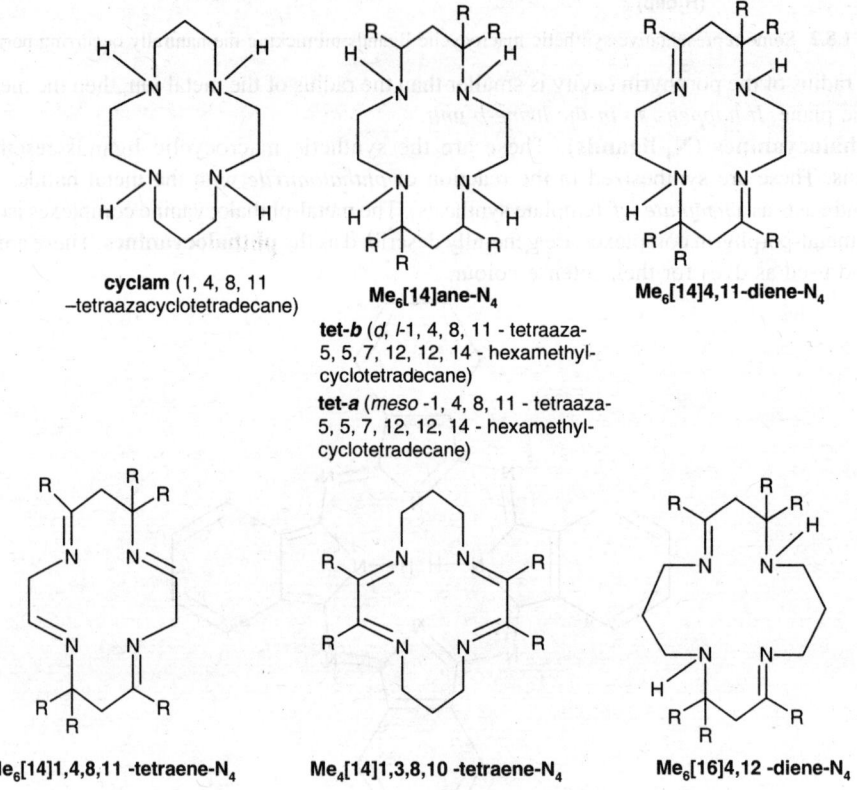

Fig. 1.8.3 Template synthesis of a metal-phthalocyanine complex isoelectronic with the metal-porphyrin complex.

(C) Cyclam and related tetraaza macrocyclic (N_4) ligands: These macrocyclic ligands are structurally related to the naturally occurring porphyrin systems. There are four N-sites for coordination. Some representative examples having the 14 and 16 membered rings are given in Fig. 1.8.4

cyclam (1, 4, 8, 11
–tetraazacyclotetradecane)

Me₆[14]ane-N₄

tet-b (d, l-1, 4, 8, 11 - tetraaza-
5, 5, 7, 12, 12, 14 - hexamethyl-
cyclotetradecane)

tet-a (meso -1, 4, 8, 11 - tetraaza-
5, 5, 7, 12, 12, 14 - hexamethyl-
cyclotetradecane)

Me₆[14]4,11-diene-N₄

Me₆[14]1,4,8,11 -tetraene-N₄   **Me₄[14]1,3,8,10 -tetraene-N₄**   **Me₆[16]4,12 -diene-N₄**

Fig. 1.8.4 Structures of some representative N_4–macrocycles. (R = CH_3).

Naming of the tetraaza macrocyclic ligand

The abbreviated forms for naming such ligands indicate the *nature and number of substitutent groups, ring size, and location of the* C$=$N *bond*. The representation, $Me_6[14]4$, 11-diene-N_4 indicates: there are 6 Me-groups as substituents; 14 membered ring size; 4,11 denote the positions of 'C$=$N' bond. Its full name is: 5,7,7,12,14,14-hexamethyl-1,4,8,11-tetraazacyclotetradeca-4,11-diene.

The N_4-macrocyclic ligands given in Fig. 1.8.4 are the examples where *the conjugated π-systems are absent*. These may be completely saturated or partially saturated. This is the basic difference from the porphyrin systems having the conjugated π-system. The saturated macrocycles are *flexible*. *The resonance stabilization from the conjugated π-system as in phthalocyanines, porphyrins favours the substitution reaction*. Nature has selected the macrocycles enjoying resonance stabilization from the conjugated π-systems. The removal of the active metal from such macrocycles as in the heme group, chlorophyll, etc. is kinetically favoured as the leaving ligand is itself resonance stabilized.

(D) Macrocycles with mixed donor sites: The different macrocycles with N_2O_2; N_2S_2, P_2S_2, etc. donor sites have been prepared.

(E) Crown ethers, cryptands and sepulchrate ligands: These macrocyclic ligands have been discussed separately (*cf.* Sec. 4.6).

Clathro-chelate and Wrapping Effect of the Macrocycles

When the macrocycle is sufficiently **large and flexible**, it can wrap around the metal centre positioned in the hole, *i.e.* three-dimensional cage. Thus the metal centre is encapsulated, *i.e.* wrapped in a three dimensional cage. This phenomenon is called *wrapping effect*. **The flexibility is earned when the chain of the macrocycle is largely saturated.** The resulting complex is called *clathro-chelate* which differs from the simple **clathrate compounds** in which practically no chemical interaction occurs. But in the clathro-chelate, besides the mechanical or physical trapping, metal-ligand (*i.e.* donor sites) bonding interaction exists. *Thus in the clathro-chelate, the metal ion is both physically and chemically arrested within the hole.*

1.9 TEMPLATE SYNTHESIS OF POLYDENTATE AND MACROCYCLIC LIGANDS

Very often, synthesis of different polydentate ligands including the macrocyclic ligands requires the presence of certain metal ions. Formation of such polydentate ligands, specially the macrocyclic ligands passes through the multistep reactions in which the metal ions on being coordinated to the partially formed ligand direct the course of the reaction. *The coordinated metal ion can bring the reactive groups in proximity to facilitate the reaction*. Sometimes, the metal coordination can stabilize the polydentate ligand. The influence of metal ions is described as *template effect*. There are two types of template effect.

(i) **Kinetic template effect:** The course of the reaction is kinetically favoured in presence of the metal ion. In fact, metal coordination to the partially formed ligand or starting reactants brings the reacting groups in a correct position to favour the ligand synthesis reaction kinetically. The metal ion can hold the partially formed ligand.

(ii) **Thermodynamic template effect:** It arises when chelation of the product ligand with the metal centre stabilizes the ligand. In such cases, the reaction product may be a mixture of different compounds and they remain in an equilibrium. *A particular form* (generally, the minor product in absence of template effect) *is thermodynamically less stable and it appears as a minor product in the equilibrium mixture.* Metal ion can stabilize the minor product through coordination. *Thus, the metal ion shifts the position of the equilibrium already existing in the reaction system.*

However, it may not be always possible to determine which of the two effects is more important in a particular synthesis. Both the effects may go on simultaneously.

(A) Examples illustrating the thermodynamic and kinetic template effect

(i) **Synthesis of a tetradentate Schiff base ligand:** It has been already mentioned that very often it is very difficult to identify the type of template effect playing the major role; and sometimes both the effects may go on simultaneously. The following examples illustrate the cases where *the thermodynamic template effect is the main driving force of the reaction.*

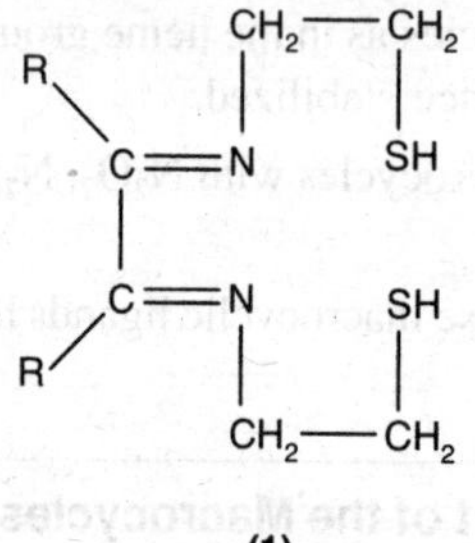

(1)

Schiff Base

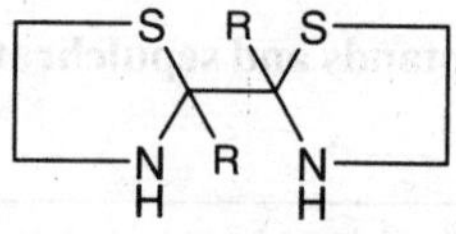

(2)

Thiazoline

To prepare the tetradentate ligand, Schiff base (**1**), it is quite reasonable to start with the α-diketone and 2-aminoethanethiol ($NH_2CH_2CH_2SH$) in the condensation reaction. But the attempt fails and it gives a mixture of the product *thiazoline* (**2**) and the tetradentate ligand (**1**) whose concentration is small. In fact, both the amine ($-NH_2$) and mercapto ($-SH$) groups act as the **competitive nucleophiles.** The nucleophilic attack by the $-SH$ group on the carbonyl carbon

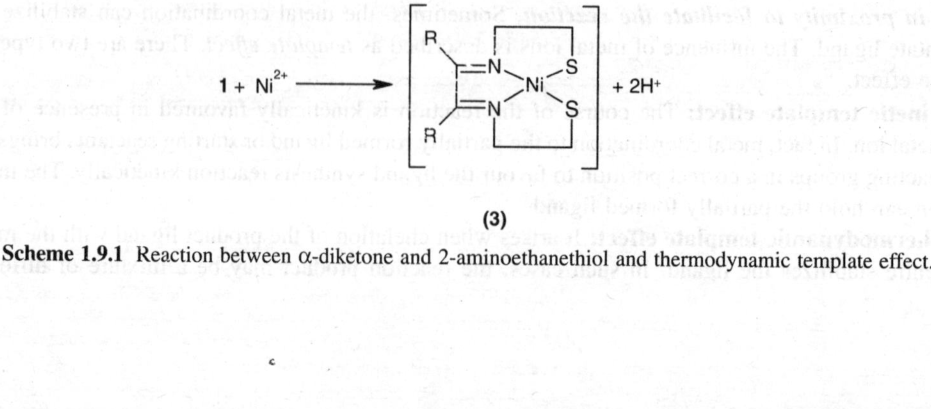

Scheme 1.9.1 Reaction between α-diketone and 2-aminoethanethiol and thermodynamic template effect.

gives the product thiazoline (**2**) while the nucleophilic attack by the —NH_2 group gives the desired tetradentate ligand (**1**). However, if the reaction is carried out in presence of Ni^{2+}, the desired tetradentate ligand is stabilized as a Ni(II)-complex and it becomes the main product. The other possible product **2** is not suitable for complexation with Ni(II). Thus, in presence of Ni^{2+}, the product **2** is not stabilized.

Thus, the equilibrium between **1** and **2** is driven towards **1** due to the favoured complexation between **1** and Ni^{2+}. It is a case of **thermodynamic template effect**.

(ii) **Synthesis of a macrocyclic ligand from a Schiff base ligand:** In the reaction, between the Schiff's base complex (**3**), *i.e.* 2,3-butanedione-*bis*(mercaptoethylamine)-nickel(II) and dibromo-*o*-xylene, the metal ion holds the reactive groups in a suitable position to allow the macrocyclic complex (**4**) formation. This is an example of **kinetic template effect**.

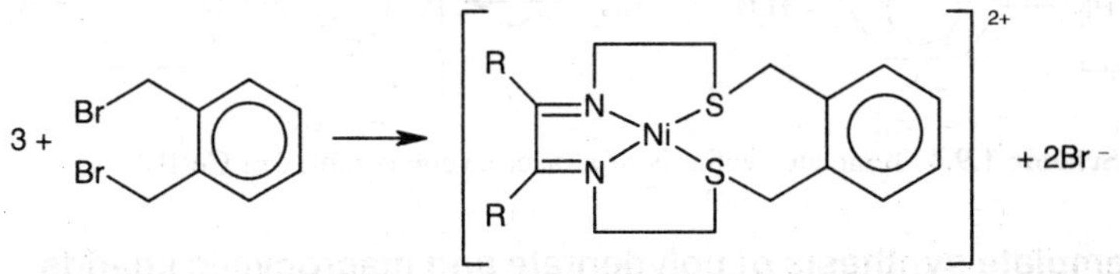

(**4**)

Scheme 1.9.2 Synthesis of a macrocyclic complex of Ni(II) through the kinetic template effect.

In the above reaction, the coordinated thiol groups in the complex **3** act as the nucleophiles to undergo alkylation giving rise to the synthesis of a macrocyclic ligand that remains chelated with the Ni(II) centre.

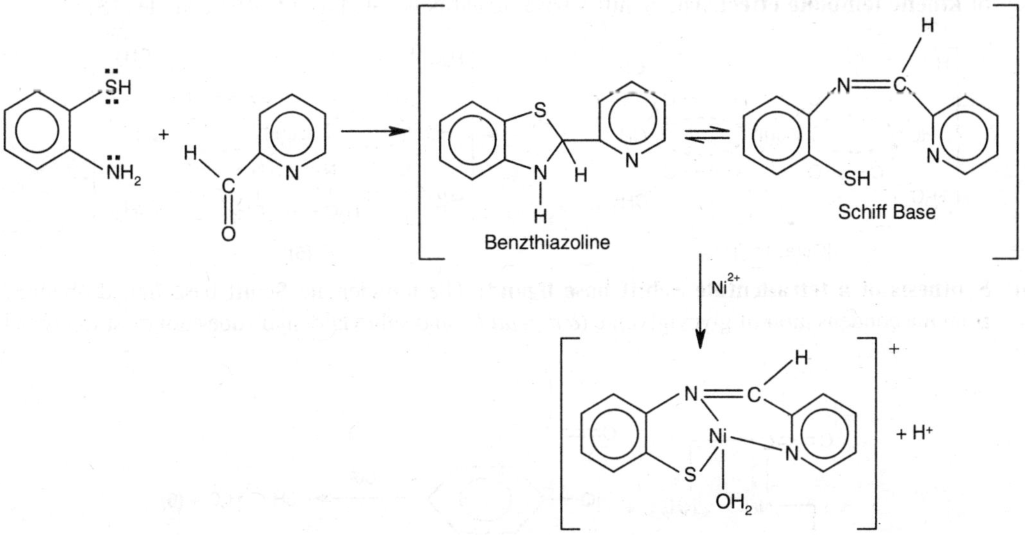

Scheme 1.9.3 Synthesis of a Schiff base complex through the template effect.

(iii) **Synthesis of a tridentate Schiff base ligand: Thermodynamic template effect** can be illustrated for the synthesis of the tridentate Schiff base ligand from the condensation of *o*-aminothiophenol with pyridine-1-carboxaldehyde. Again the competitive nucleophilic action of the —NH_2 and —SH group can produce the *Schiff's base* and *benzthiazoline* respectively and they remain in an equilibrium. The Schiff base as a tridentate ligand can be stabilized in presence of the metal ions like Ni^{2+} to favour its formation.

(iv) **Synthesis of a macrocyclic complex:** The following example of synthesis of a macrocyclic complex of Cu(II) illustrates *both the thermodynamic and kinetic template effect.* For condensation reaction, the reactive carbonyl groups are placed at the appropriate positions. In absence of the metal ion, the above reaction leads to a undesirable polymeric product.

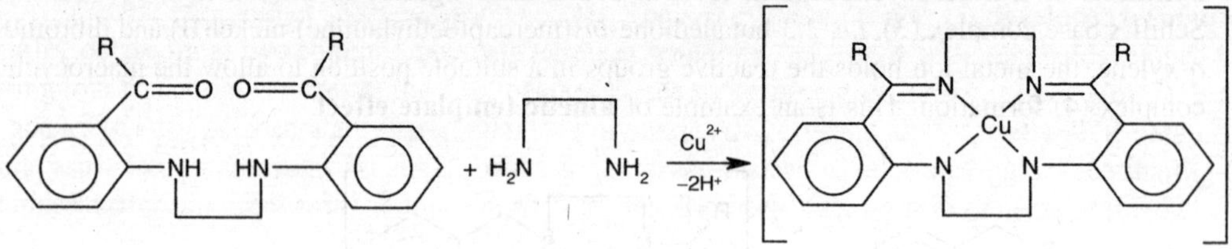

Scheme 1.9.4 Template synthesis of a macrocyclic complex of Cu(II).

(B) Examples of template synthesis of polydentate and macrocyclic ligands

(i) **Synthesis of a tetradentate Schiff base ligand:** Formation of the Schiff base, tetradentate ligand from the condensation of the acetylacetone and ethylenediamine is catalysed by Cu(II). In the Cu(II)-acetylacetonate complex (bearing a pseudo-aromatic character), the nucleophilic attack by the —NH_2 groups of en on the carbonyl carbon is facilitated due to the metal coordination *that makes the carbonyl carbon centres more electron deficient.* Thus it may be considered as a case of **kinetic template effect.** The Schiff's base ligand is isolated as a Cu(II) complex (**5**).

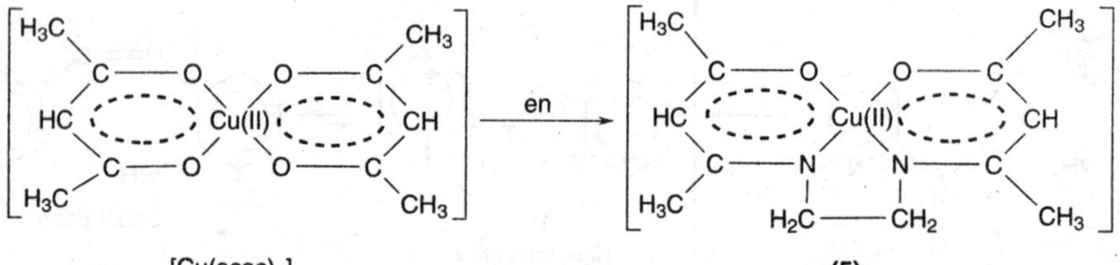

[Cu(acac)₂]   en   (**5**)

(ii) **Synthesis of a tetradentate Schiff base ligand:** The tetradentate Schiff base ligand obtained from the condensation of glycylglycine (*a dipeptide*) and salicylaldehyde does not exist freely but

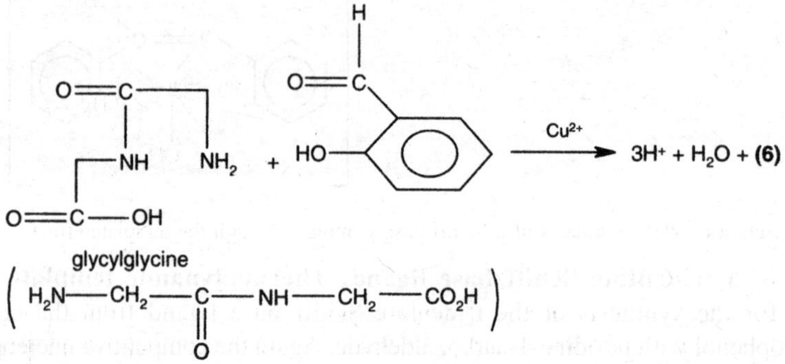

$$Cu^{2+} \rightarrow 3H^+ + H_2O + (6)$$

glycylglycine

$$\left(H_2N\text{---}CH_2\text{---}\overset{\overset{\displaystyle O}{\|}}{C}\text{---}NH\text{---}CH_2\text{---}CO_2H \right)$$

when it is synthesized in presence of Cu^{2+}, the ligand gets stabilized in a complex (**6**). The free ligand experiences a hydrolytic cleavage. *Thus it is an example of* **thermodynamic template effect.**

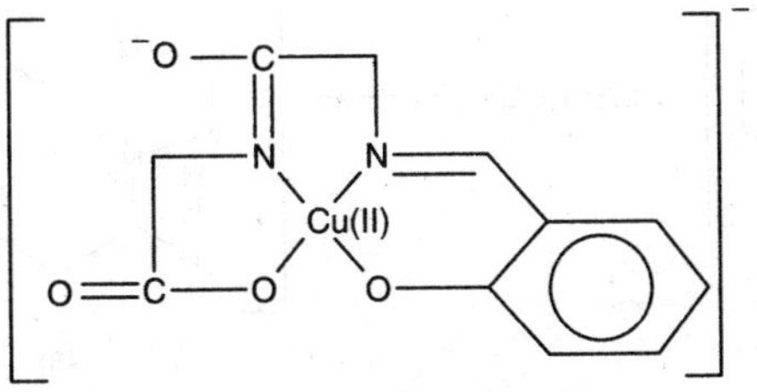

(6)

N-salicylideneglycylglycinatocuprate(II)

In the same way, **aqua-N-salicylideneglycinatocopper(II) complex** (**7**) involving the tridentate Schiff base ligand is obtained through the condensation of salicylaldehyde and glycine.

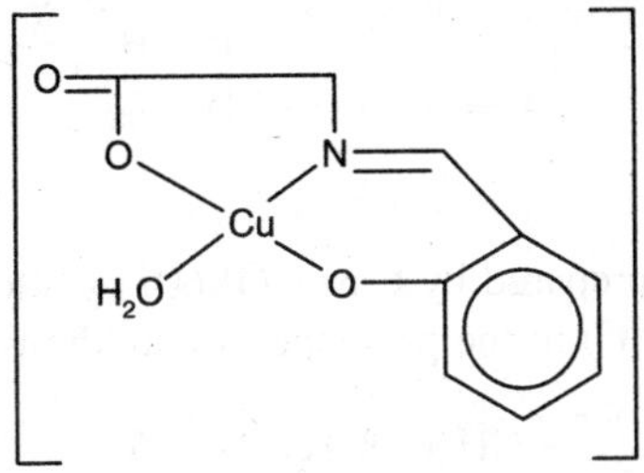

(7)

Labile metal centres used in template synthesis – why?

In template synthesis, generally the labile metal centres like Cu^{II}, Ni^{III}, Fe^{II}, etc. are used. It is due to the following reasons:
(i) Some of the ligands may be required to be completely dislodged and the dislodged ligand may catalyse the reaction as in *Curtis Synthesis* of the macrocyclic-N_4 ligand in the reaction of $[Ni(en)_3]^{2+}$ with acetone; (ii) the functional groups (*e.g.* $—NH_2$, $\rangle C = O$, etc.) already coordinated may be required to be temporarily dissociated for their participation in the reactions as in the Schiff's base formation; (iii) after formation of the desired ligand, removal of the coordinated metal centre is required to isolate the free ligand, if it can freely survive.
These conditions are easily satisfied for the labile metal centres. The inert centres like Co(III), Cr(III), cannot satisfy these requirements.

(iii) **Synthesis of tetraaza-macrocyclic ligands:** The *template synthesis* of the macrocyclic *tetraaza* ligand through the condensation of the coordinated didentate or tetradentate amines with the carbonyl compounds (*e.g.* acetone) is illustrated below.

$$\left[Ni(en)_3 \right]^{2+} + \longrightarrow \left[Ni(en)_2 \right]^{2+} + \text{en}$$

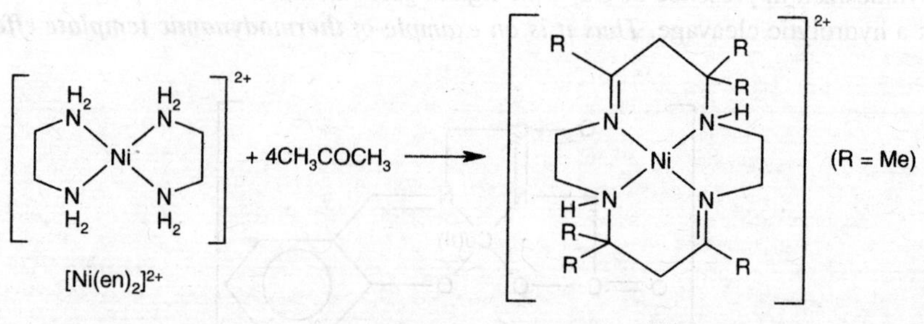

$$[\text{Ni(en)}_2]^{2+} + 4\text{CH}_3\text{COCH}_3 \longrightarrow \quad (R = Me)$$

(8)

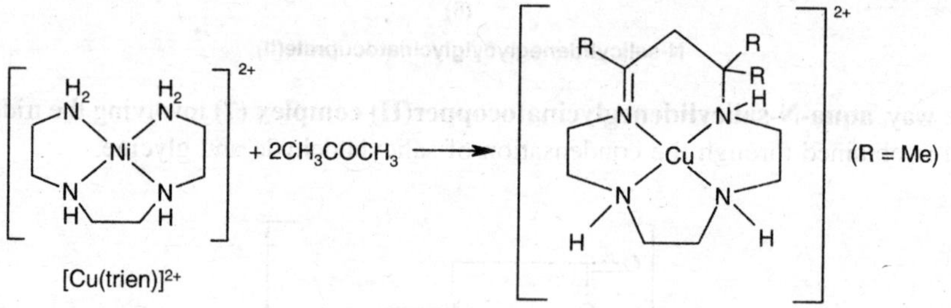

$$[\text{Cu(trien)}]^{2+} + 2\text{CH}_3\text{COCH}_3 \longrightarrow \quad (R = Me)$$

(9)

The macrocyclic complex was prepared by Curtis (1960) by allowing $[\text{Ni(en)}_3]^{2+}$ to react with *dry acetone* in a sealed tube (at ~ 100°C) in the presence of anhydrous CaSO_4

$$\left[\text{Ni(en)}_3\right]^{2+} + \text{CH}_3\text{COCH}_3 \longrightarrow \mathbf{8}$$

The reaction is catalysed by base and retarded by water. *In **Curtis reaction**, probably one dissociated ethylenediamine (en) molecule from* $[\text{Ni(en)}_3]^{2+}$ *catalyses the reaction as a base.* The probable reaction path is outlined in Scheme 1.9.5.

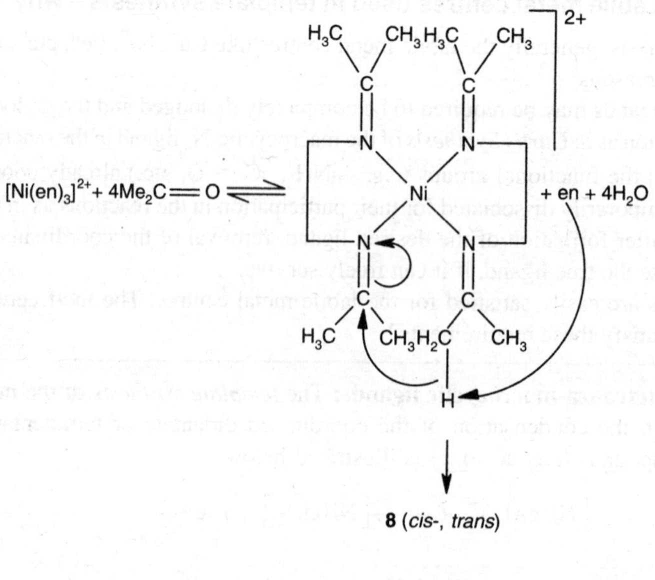

$$[\text{Ni(en)}_3]^{2+} + 4\text{Me}_2\text{C}{=}\text{O} \rightleftharpoons \qquad + \text{en} + 4\text{H}_2\text{O}$$

8 (*cis-, trans*)

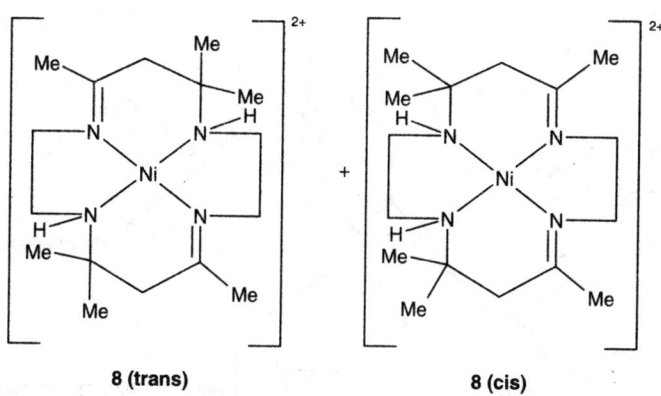

8 (trans) **8 (cis)**

Scheme 1.9.5 Template synthesis (**Curtis synthesis**) of a tetraaza macrocyclic complex of Ni(II) in the reaction of $[Ni(en)_3]^{2+}$ with dry acetone.

The first step leading to condensation of $[Ni(en)_2]^{2+}$ with acetone is a reversible one and presence of water shifts the equilibrium towards the left. To favour the reaction, anhydrous $CaSO_4$ is used for the removal of water. This reaction occurs if the diamine complexes of the liable metal centres like Cu(II) and Ni(II) are used. *If the inert complexes like $[Co(en)_3]^{2+}$ are used then the reaction does not occur.* It suggests the dislodgement of one *en* molecule during the reaction and CH_3COCH_3 probably reacts with the free —NH_2 group *which is temporarily dissociated from the labile metal centre.* **This condition is satisfied only for the labile metal complexes.**

In the next step, proton abstraction from a CH_3 group by the base (*i.e.* released *en* acting as a catalyst) produces a carbanion that makes a nucleophilic attack on the carbon centre of the C=N linkage as shown in Scheme 1.9.5. This nucleophilic attack is favoured by the coordinated metal ion in two ways:

the 'C=N' linkage experiencing the nucleophilic attack is kept in proximity with the nucleophile (i.e. carbanion); coordination by the N-site of the C=N linkage to the metal centre makes the C-centre more electron deficient (i.e. more sensitive towards the nucleophilic attack).

This illustrates the *kinetic template effect.*

In the same way, formation of the macrocyclic Cu(II)-complex (**9**) in the reaction of $[Cu(trien)]^{2+}$ with acetone can be understood.

The complex **8** can exist in *cis-* and *trans-* forms. The metal complex can be reduced to give the saturated amine complex from which the tetradentate macrocyclic ligand can be isolated by the treatment of CN^- which removes Ni^{2+}. This free ligand is **cyclam.**

The complex **8** may be oxidized to give the macrocycle with several double bonds.

(iv) **Synthesis of a macrocyclic complex:** Condensation of **1,2-dicarbonyl** compounds with **1,2-diamines** gives **glyoxalene** derivatives (6-membered ring). But the same condensation reaction with 1,3-diamines gives the polymeric product instead of the 7-membered ring because of its inherent instability. However, in presence of the **labile metal ions** like Ni(II), Cu(II), Pd(II), etc. strongly preferring the square planar coordination, macrocyclic complexes are produced. These are illustrated in Scheme 1.9.6.

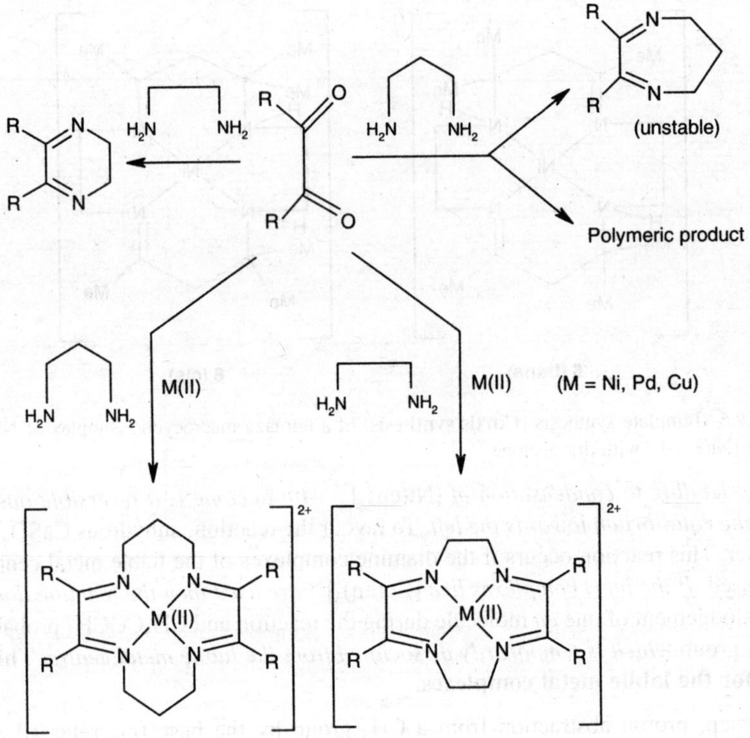

Scheme 1.9.6 Template synthesis of macrocyclic complexes from the condensation of 1,2-dicarbonyl compounds with 1,2-diamines and 1,3-diamines.

(v) **Synthesis of a macrocyclic complex:** The condensation reaction of 2,6-diacetylpyridine with the polyamines in presence of the **labile metal ions** like Ni^{2+}, can produce the macrocyclic ligand. The ring size of the macrocyclic ligand depends on the nature of the polyamine used. This is illustrated below.

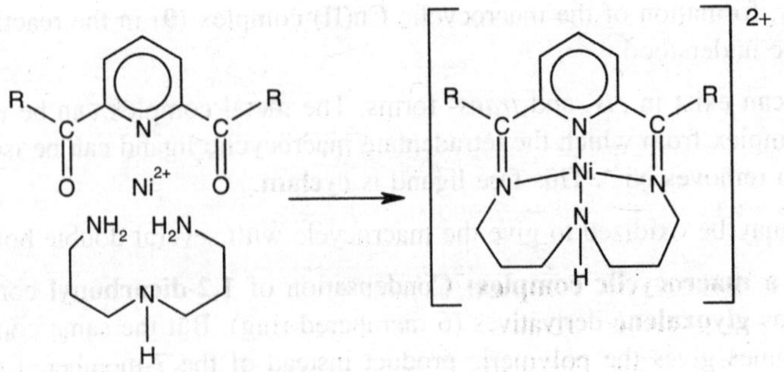

(vi) **Synthesis of a macrocyclic complex:** *Bis*(dimethylglyoximato)nickel(II) can react with boron halides (BX_3) to produce a macrocyclic complex of Ni(II).

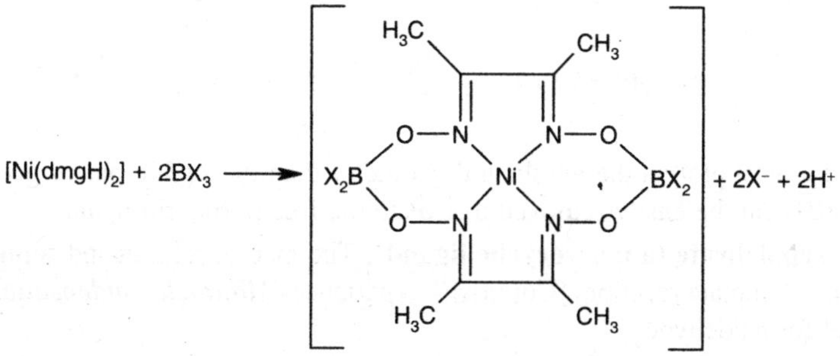

$$[\text{Ni(dmgH)}_2] + 2BX_3 \longrightarrow \quad + 2X^- + 2H^+$$

This cyclisation is an example of *kinetic template effect* as in the case of formation of the complex (**4**).

Template synthesis in nature

Nature probably uses the principle of template synthesis for its macrocyclic ligands like porphyrins. In laboratory, by using the template reaction, the macrocyclic bioligands have been synthesised. However, **Woodward's elegant synthesis of chlorophyll did not use the template reaction.**

(vii) **Synthesis of the macrocyclic complexes:** Nickel(II) complexes of the planar tetradentate macrocyclic ligand TAAB (= tetrakisanhydroaminobenzaldehyde) can be synthesized through the self condensation of *ortho*-aminobenzaldehyde in presence of Ni(II). A tridentate macrocyclic ligand complex is also partly formed from the self condensation of three molecules of *ortho*-aminobenzaldehyde. These complexes can be separated by selective crystallization of their perchlorate salts. These free macrocyclic ligands cannot survive from the hydrolytic cleavages. Only complexation can stabilize these (*i.e.* **thermodynamic template effect**).

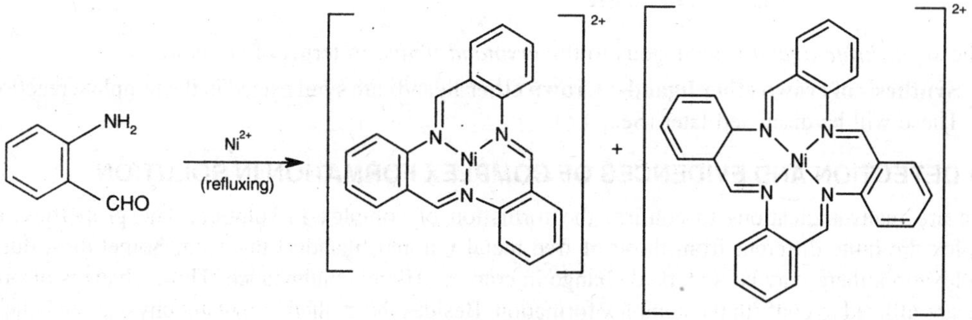

In absence of the complexing metal ion, several other products including the polymeric product are obtained. *It illustrates the operation of thermodynamic template effect in the synthesis of the macrocyclic ligand.*

(viii) **Synthesis of phthalocyanines:** Phthalocyanines (*cf.* Fig. 1.8.3) are synthesized in the template reactions.

$$4C_6H_4(CN)_2 + Ni^{2+} + C_4H_9OH \longrightarrow C_3H_7CHO + 2H^+ + Ni\,(II) - \text{phthalocyanine complex}$$
(*o*-isomer)

(ix) **Synthesis of porphyrines: Porphyrin** can be synthesized in the following template reaction.

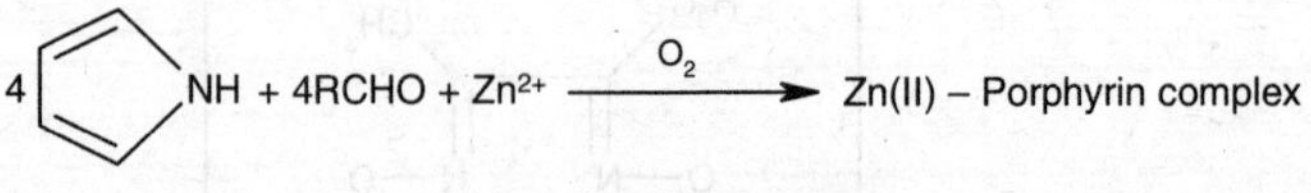

$$4 \bigg\langle \bigg\rangle NH + 4RCHO + Zn^{2+} \xrightarrow{\ O_2\ } Zn(II) - Porphyrin\ complex$$

Zn(II) (a d^{10} system) prefers the tetrahedral geometry than the square planar geometry and consequently, Zn(II) can be easily removed to isolate the free porphyrin ligand.

(x) **Synthesis of sepulchrate (a macrocyclic ligand):** The macrocyclic ligand **sepulchrate** (sep) is synthesized in a template reaction. $[Co(en)_3]^{3+}$ experiences **Mannich condensation** in presence of ammonia and formaldehyde.

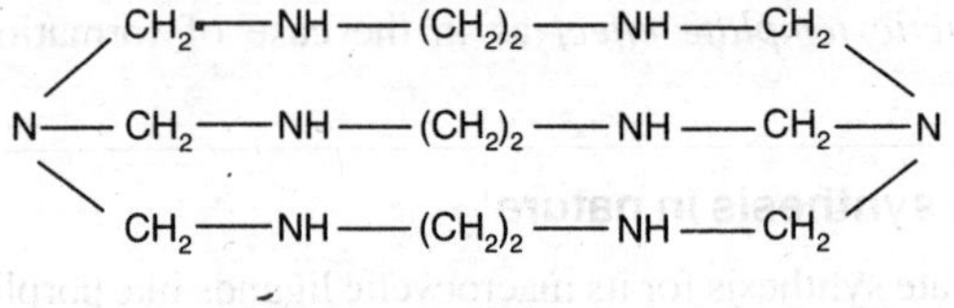

(sep)

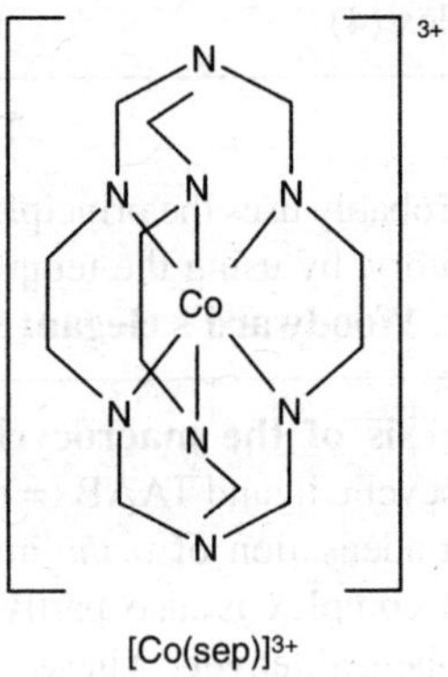

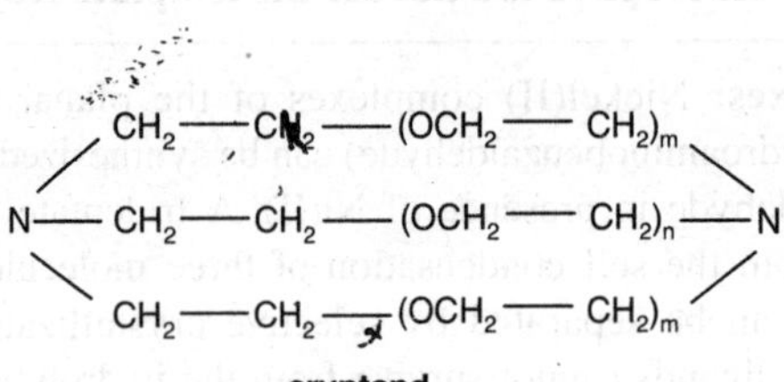

cryptand

(n = m = 2,   C-222;)
(n = 1, m = 2,   C-221)

$[Co(sep)]^{3+}$

The *sepulchrate ligand* is analogous to the *cryptand ligand* in terms of structure.

(xi) **Synthesis of crown ether ligands: Crown ether** ligands are synthesized in the template reactions. These will be discussed later (Sec. 4.6).

1.10 DETECTION AND EVIDENCES OF COMPLEX FORMATION IN SOLUTION

There are many indications to confirm the formation of complex in solution. The properties of a complex are quite different from those of free metal ion and ligand(s) involved. Sometimes, during complexation, there may be a marked change in colour, pH and conductance. These changes of properties are utilized to confirm the complex formation. Besides these, there are many physical and chemical methods to detect the complex formation.

(1) Detection of complex formation by chemical tests: The chemical properties of the ligand and metal ion present within a complex species are quite different from their properties in free condition. These are illustrated in the following examples.

(i) Cu^{2+} is precipitated as hydroxides (*i.e.* $Cu(OH)_2$) in alkaline condition but in presence of organic amines or tartaric acid (a representative α-hydroxy acid) in **Fehling's solution**, it is not precipitated as $Cu(OH)_2$ even in alkaline condition. It indicates that Cu(II) remains as a complexed species. In fact, the **Cu(II)-complexes** of the said ligands (*i.e.* amines, α-hydroxy acids, etc.) are so stable that the concentration of free Cu^{2+} ion obtained from the dissociation of the complexes is insufficient to exceed the solubility product of $Cu(OH)_2$. Similarly, in **Tollen's** reagent ($AgNO_3$ + excess NH_4OH), precipitation of $AgOH$ is prevented due to the $[Ag(NH_3)_2]^+$ complex formation.

(ii) Fe(II) is not precipitated as $Fe(OH)_2$ in alkaline condition in presence of edta. In fact, $[Fe(edta)]^{2-}$ complex is so stable that $[Fe^{2+}]$ generated from the dissociation of the complex is insufficient to exceed the solubility product of $Fe(OH)_2$.

(iii) Fe(III) is not precipitated as $Fe(OH)_3$ even in alkaline condition in presence of tartaric acid due to the **Fe(III)-tartarato** complex formation.

Ligating behaviour of hydroxo carboxylic acids like tartaric acid, HO_2C—$CH(OH)$—$CH(OH)$—CO_2H (*see* Sec. 1.15.3J) and citric acid, HO_2C—CH_2—$C(OH)(CO_2H)$—CH_2—CO_2H

For the hydroxo carboxylic acids, both the hydroxo and carboxylate groups participate in complexation. For the α-hydroxo carboxylic acid like tartaric acid, following type of chelate may be formed (*see* Sec. 1.15.3J).

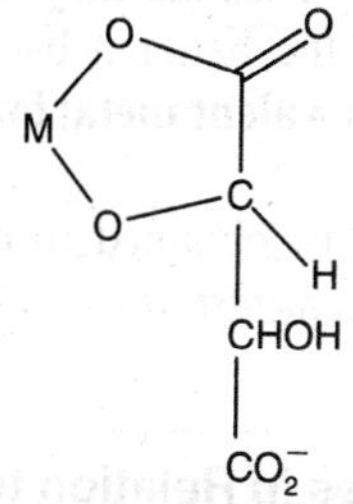

[M(tart)], tart stands for *tartarato* or 2,3-dihydroxy-butanedioato

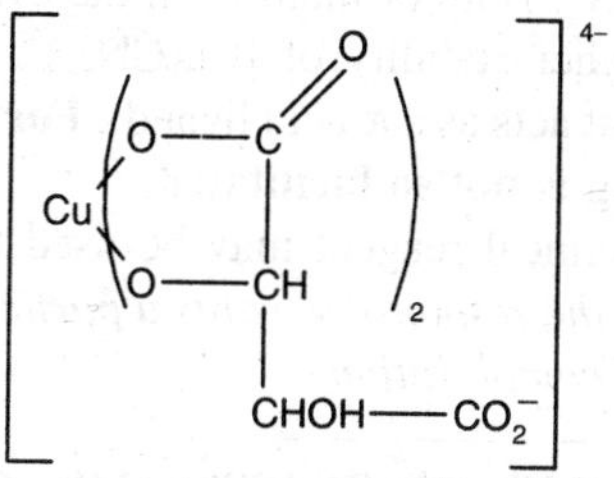

Cu^{II}-tartarato complex (present in **Fehling's solution**)

As a bridging ligand, it can form the binuclear complex and polynuclear complexes. In Fehling's solution, it may form monomeric (shown above), dimeric and even polymeric complexes by using the tartarate bridges. In the dimer, each Cu^{II}-centre is in a square planar geometry by using two tartarate bridges. The bridging property of the tartarate ligand is illustrated for the Sb(III) complex **'tartar emetic'** which is used as an *emetic medicine* (causing vomiting). Each Sb-centre is in a *distorted trigonal bipyramidal geometry* with a lone pair at an equatorial position.

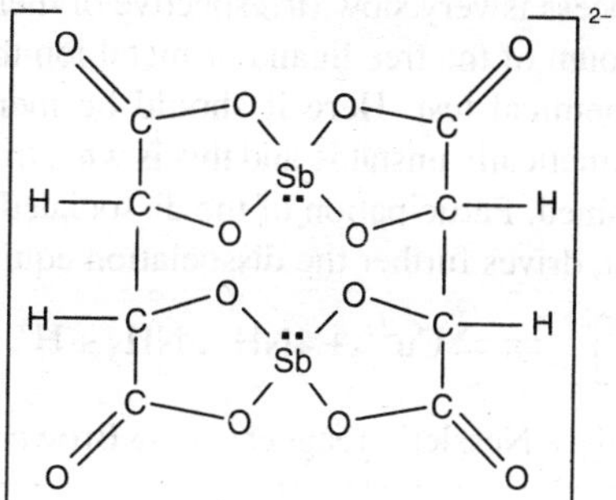

$[Sb_2(C_4H_2O_6)_2]^{2-}$

(iv) If a **mixture of Cu²⁺ and Cd²⁺** ions is treated with the KCN solution and then H_2S is passed through the solution, Cd^{2+} is precipitated as CdS but copper is not precipitated (*see* Secs. 14.17.3, 16.4.4, Vol. 3 of Fundamental Concepts of Inorganic Chemistry).

Apparently it may be assumed that Cd^{2+} does not complex with CN^- under the condition as the property of free Cd^{2+} is yet maintained in terms of precipitation as CdS. The property of free Cu^{2+} or Cu^+ is not manifested in the solution in terms of precipitation by H_2S. Thus, it appears that copper is strongly complexed with CN^- under the condition.

In reality, under the condition, in the solution, both the metal centres remain as complexes $[Cd(CN)_4]^{2-}$ and $[Cu(CN)_4]^{3-}$ (the Cu^I-complex is formed after the reduction of Cu^{II} by CN^-). From the dissociation of $[Cd(CN)_4]^{2-}$, the available $[Cd^{2+}]$ is sufficient to exceed the solubility product of CdS. On the other hand, the available $[Cu^+]$ from the dissociation of $[Cu(CN)_4]^{3-}$ is not sufficient to exceed the solubility product of Cu_2S.

Thus, response to the present chemical test (*i.e.* precipitation as sulfide) depends on two factors: **stability constant of the complex and solubility product of the sulfide**. Thus based on the above observation, it cannot be concluded that Cd^{2+} does not complex with CN^- while Cu^+ forms a complex with CN^-. Both of them form the cyanido complexes but they differ only in stability constants. The higher stability of $[Cu(CN)_4]^{3-}$ is due to the better π-back bonding (metal to π^*–MO of CN^- that acts as a π-acid ligand). **For the higher valent metal ion, *i.e.* Cd²⁺, this type of π-back bonding is not so facilitated.**

A more sensitive chemical reagent may be used to detect Cu^+ obtained from the dissociation of $[Cu(CN)_4]^{3-}$. *Thus from the response towards a particular chemical test, it may not always be possible to conclude the case of complexation.*

Thermodynamic and Kinetic Stability of the Complexes in Relation to Chemical Tests

Thermodynamic stability of a complex species gives the measure of the concentrations of the metal ion and ligand available from the dissociation of the complex under consideration. Whether thus the available concentration of the metal ion or ligand is sufficient or not for the response to a particular chemical test depends on the **sensitivity of the chemical test**. This aspect has been illustrated for Cu^{2+} and Cd^{2+}.

In this regard, **not only the thermodynamic stability but also the kinetic stability of the complex is of an important consideration**. For example, from the complex $[Cu(NH_3)_4]^{2+}$ (Cu^{II}- a kinetically labile centre), all the characteristic properties of NH_3 can be proved by chemically but it cannot happen so for the kinetically inert complexes like $[Co(NH_3)_6]^{3+}$, $[Cr(NH_3)_6]^{3+}$. In the inert complexes, the dissociation process is very slow (irrespective of thermodynamic stability). Hence to accumulate a significant amount of the free ligand or metal ion through dissociation, it takes a long time to invalidate the chemical test. Here it should be mentioned that $[Cu(NH_3)_4]^{2+}$ is thermodynamically stable but kinetically unstable and this is why, from this complex, the chemical tests for Cu^{2+} and NH_3 are obtained. Participation of the dissociated species, *i.e.* Cu^{2+} and NH_3 in some reactions of chemical test, drives further the dissociation equilibrium of the complex.

$$\left[Cu(NH_3)_4\right]^{2+} \rightleftharpoons Cu^{2+} + 4NH_3 \,;\, NH_3 + H^+ \longrightarrow NH_4^+$$

$$\text{or } NH_3 + \text{Nessler's reagent} \longrightarrow \text{Brown ppt.}$$

---

(2) Detection of complex formation by dissolution of insoluble precipitates: It is illustrated in the following cases:

(i) **Solubilisation of AgCl** can be carried out in the presence of excess aqueous NH_3 solution, or concentrated HCl or KCN solution.

$$AgCl(s) + 2NH_3 \rightleftharpoons \left[Ag(NH_3)_2\right]^+ + Cl^-$$

$$AgCl(s) + HCl \rightleftharpoons \left[AgCl_2\right]^- + H^+$$

$$AgCl(s) + 2KCN \rightleftharpoons \left[Ag(CN)_2\right]^- + 2K^+ + Cl^-$$

Actually the dissolution depends on the relative magnitudes of the *instability constant* (K_{instab}) of the complex and *solubility product of the precipitate*. Assuming the complex to be labile, K_{instab} determines the amount of metal ion (*i.e.*, Ag^+ in the present example) obtained through the dissociation of the complex at a particular time.

$$\left[AgL_2\right] \rightleftharpoons Ag^+ + 2L, (L = Cl^-, CN^-, NH_3; \text{ charge is not shown for the sake of simplicity})$$

$$[Ag^+] = K_{instab} [AgL_2]/[L]^2$$

For the precipitation of AgCl, $[Ag^+]$ obtained from the dissociation of the complex $[AgL_2]$ should be sufficient to exceed K_{sp}, *i.e.*

$$\left[Ag^+\right]\left[Cl^-\right] \geq K_{sp} \text{ (AgCl will remain insoluble)}$$

$$\left[Ag^+\right]\left[Cl^-\right] \leq K_{sp} \text{ (AgCl will be solubilised)}$$

Here it is evident that dissolution of the precipitate through complexation is favoured for the higher K_{sp} of the precipitate and lower K_{instab} (i.e. higher K_{stab} of the complex). It can be illustrated by considering the possibility of solubilization of AgCl, AgBr and AgI in presence of dilute NH_3 solution.

$$K_{instab} = \frac{\left[Ag^+\right]\left[NH_3\right]^2}{\left[Ag(NH_3)_2^+\right]} = 6.8 \times 10^{-8}$$

K_{sp} value : $AgCl(1.5 \times 10^{-10}) > AgBr(7.7 \times 10^{-12}) > AgI(1 \times 10^{-16})$

It is evident, AgCl offers the most favourable situation for its solubilization while AgI offers the most difficult situation for its dissolution by NH_3. In fact, AgCl goes into solution in dilute NH_3 solution, AgBr goes into solution only in presence of concentrated NH_3 (that reduces $[Ag^+]$ more) and AgI cannot go into solution under any condition in presence of NH_3.

Halido-complex of Ag(I)

Besides the $[AgCl_2]^-$, the mixed halido-complex like $[Ag(Cl)(I)]^-$ can also enhance the solubility of AgCl. In fact, solubility of AgCl increases with the addition of excess NaI that allows the formation of the complex $[Ag(Cl)(I)]^-$.

(ii) **Ag(OH) is also solubilied in presence of excess NH_3.** In preparing *Tollen's reagent*, dilute NH_3 solution is gradually added to $AgNO_3$ solution. At first, the precipitate of Ag(OH) appears and ultimately it goes into solution due to complexation. The involved equilibria are:

$$NH_3 + H_2O \rightleftharpoons NH_4^+ + OH^-$$

$$Ag^+ + OH^- \rightleftharpoons Ag(OH) (\downarrow)$$

$$Ag(OH)(s) + 2NH_3 \rightleftharpoons \left[Ag(NH_3)_2\right]^+ + OH^-$$

(iii) Bi^{3+} and Hg^{2+} are precipitated **in presence of the precipitating reagent KI**, but in presence of excess KI, they go into solution due to complexation.

$$Bi^{3+} + 3I^- \rightleftharpoons BiI_3 (\downarrow), \ BiI_3 (s) + I^- \rightleftharpoons \left[BiI_4\right]^-$$

$$Hg^{2+} + 2I^- \rightleftharpoons HgI_2 (\downarrow); \ HgI_2 (s) + 2I^- \rightleftharpoons \left[HgI_4\right]^{2-}$$

However, these large anionic complexes can be **precipitated by using the large cations** (*e.g.* pyridinium ion, $C_5H_5NH^+$).

The relatively inert metals like Ag, Au are not attacked by the ordinary acids to liberate H_2 gas. **But in presence of CN^-, they can liberate H_2 gas from the acids** and the metals go into solution. This happens so due to complexation.

$$M(s) + 2H^+ \rightleftharpoons 2M(aq)^+ + H_2(\uparrow), \ M = Ag, \ Au$$

$$M(aq)^+ + 2CN^- \rightleftharpoons \left[M(CN)_2\right]^-$$

The H_2 gas liberation represented above is not favoured in absence of CN^- but due to the stable complex formation by $M(aq)^+$ with CN^-, the H_2 gas liberation process becomes favourable.

In fact, because of the formation of stable complex $[M(CN)_2]^-$ (M = Au, Ag), these metals can be attacked by aerial oxygen and they pass into solution (*see* Sec. 16.4.5, Vol. 3 of Fundamental Concepts of Inorganic Chemistry).

$$4M(s) + 8CN^- + O_2 + 2H_2O \rightleftharpoons 4\left[M(CN)_2\right]^- + 4OH^-$$

(iv) The precipitate of **$BaSO_4$ goes into solution** in presence of Na_2H_2edta at a relatively higher pH.

$$BaSO_4 + H_2edta^{2-} \rightleftharpoons \left[Ba(edta)\right]^{2-} + 2H^+ + SO_4^{2-}; 2H^+ + 2OH^- \longrightarrow 2H_2O$$

$[Ba(edta)]^{2-}$ complex is very much stable and $[Ba^{2+}]$ obtained from the dissociation of the complex is insufficient to exceed the solubility product of $BaSO_4$.

Several metal carbonates go into solution in presence of Na_2H_2edta. It is illustrated for $CdCO_3$.

$$CdCO_3 + H_2edta^{2-} \rightleftharpoons \left[Cd(edta)\right]^{2-} + CO_2 + H_2O$$

(v) Some other illustrative examples are given below.

$$Ag^+ + CN^- \rightleftharpoons Ag(CN)(\downarrow) \xrightarrow{CN^-} \left[Ag(CN)_2\right]^-$$

$$2Ag^+ + S^{2-} \rightleftharpoons Ag_2S(\downarrow) \xrightarrow{CN^-} \left[Ag(CN)_2\right]^- + S^{2-}$$

(**Note:** Ag^I forms a **thiosulfato complex, $[Ag(S_2O_3)_2]^{3-}$ but it is less stable than $[Ag(CN)_2]^-$** and in fact, excess thiosulfate cannot solubilize Ag_2S. Moreover, the thiosulfato complex, $[Ag(S_2O_3)_2]^{3-}$ decomposes to generate the precipitate of Ag_2S (black)

$$2\left[Ag(S_2O_3)_2\right]^{3-} + H_2O \longrightarrow 3S_2O_3^{2-} + Ag_2S(\downarrow) + 2H^+ + SO_4^{2-}$$

(vi) $HgO(s) + 4I^- + H_2O \rightleftharpoons \left[HgI_4\right]^{2-} + 2OH^-$

This reaction is more or less quantitative and *it can be used for the preparation of standard solution of alkali* by treating a known amount of HgO (AR-grade) with an excess amount of KI.

(3) Detection of complex formation by observing the change in colour of the solution: Generally the transition metal ions undergo complexation. During such complexation, the d-orbitals are split into different energy levels. The separation among these energy levels for a particular metal ion *is mainly determined by the nature of the ligand and geometry of the complex.* During complexation in aqueous medium, the aqua ligands are replaced by the complex forming ligands. The d-orbital splitting power of the incoming ligand may be significantly different from that of the aqua ligands. In such cases, the absorption of light energy takes place in a different region compared to that for the aqua ligands. If this new absorption occurs within the *visible range,* then the colour of the solution changes. Sometimes, the **change of symmetry** due to complexation can remarkably enhance the *intensity of colour.* Sometimes the appearance of **new charge transfer band** can produce a new colour to the complex. **Thus due to complexation, position of the absorption band in the electromagnetic radiation and intensity of the light absorption may change significantly to cause a remarkable change of colour.** This technique is mainly applicable for the **labile metal centres** that form the complexes immediately.

Some examples of **labile systems** are given below to illustrate the phenomenon.

$$\left[Cu(OH_2)_6\right]^{2+} + 4NH_3 \rightleftharpoons \left[Cu(NH_3)_4(OH_2)_2\right]^{2+} + 4H_2O$$
(light blue)   (deep blue)

$$\left[Cu(OH_2)_6\right]^{2+} + 4Cl^- \rightleftharpoons \left[CuCl_4\right]^{2-} + 6H_2O$$
(light blue)   (green)

$$\left[Co(OH_2)_6\right]^{2+} + 4Cl^- \rightleftharpoons \left[CoCl_4\right]^{2-} + 6H_2O$$
(light pink)   (instense blue)

$$\left[CoCl_4\right]^{2-} + HgCl_2 + 6H_2O \rightleftharpoons \left[Co(OH_2)_6\right]^{2+} + \left[HgCl_4\right]^{2-} + 2Cl^-$$
(intense blue)   (light pink)

$$\left[Ni(OH_2)_6\right]^{2+} + 6NH_3 \rightleftharpoons \left[Ni(NH_3)_6\right]^{2+} + 6H_2O$$
(bright green)   (intensely blue)

$$\left[Bi(OH_2)_6\right]^{3+} + 3tu \rightleftharpoons \left[Bi(tu)_3\right]^{3+} + 6H_2O,$$
(colourless)   (yellow)

(tu stands for thiourea)

$$\left[Fe(OH_2)_6\right]^{3+} + Cl^- \rightleftharpoons \left[FeCl(OH_2)_5\right]^{2+} + H_2O,$$
(colourless)   (yellow)

$$\left[Fe(OH_2)_6\right]^{3+} + SCN^- \rightleftharpoons \left[Fe(OH_2)_5(SCN)\right]^{2+} + H_2O,$$
(colourless)   (red colour)

$$\left[Fe(OH_2)_5(SCN)\right]^{2+} + 6F^- \rightleftharpoons \left[FeF_6\right]^{3-} + SCN^- + 5H_2O,$$
(red colour)   (colourless)

This remarkable colour change during complexation is used as an important tool to identify the metal ions in quantitative analysis. In fact, **the colour of a particular complex is a characteristic property of the complex** and it depends only on the nature of the metal ion and ligand under consideration at a particular reaction condition.

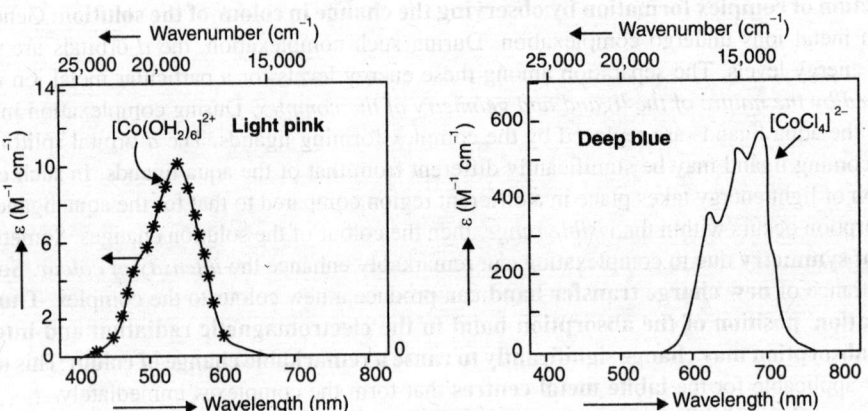

Fig. 1.10.1 Comparison of the electronic absorption spectra of octahedral $[Co(OH_2)_6]^{2+}$ and tetrahedral $[CoCl_4]^{2-}$ complexes. (ε stands for molar extinction coefficient).

(**Note:** ● In the tetrahedral complex, *i.e.* $[CoCl_4]^{2-}$, the Lapporate forbidden *d-d* transition is more allowed than that in the octahedral complex, *i.e.* $[Co(OH_2)_6]^{2+}$. This makes the **higher intensity of the colour of** $[CoCl_4]^{2-}$.

● Substitution of H_2O by Cl^- (which is a **weaker field ligand**) causes the splitting (measured by 10Dq) to decrease. Moreover, in moving from the octahedral to the tetrahedral geometry, the splitting further decreases. Thus, because of both the *ligand field effect* and *geometry effect* in $[CoCl_4]^{2-}$, **the absorption maxima shift towards the higher wavelength.**

● The colour change of Co(II) has been utilized in indicating the presence of moisture or dehydrated condition. In presence of a **dehydrating agent** (or simple heating causing the removal of water), the crystal hydrate of $CoCl_2$ changes colour from light pink to blue. On hydration, the reverse process occurs.

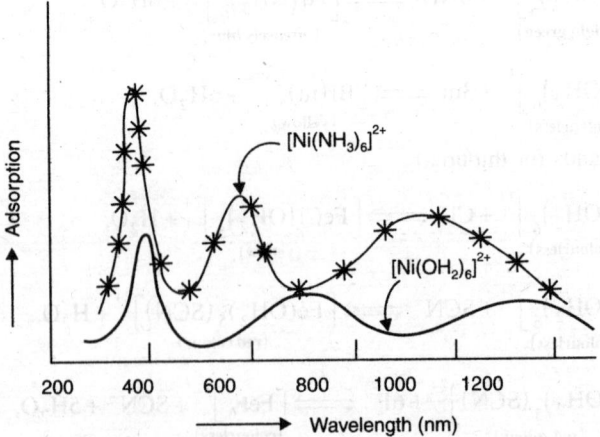

Fig. 1.10.2 Comparison of the electronic absorption spectra of $[Ni(OH_2)_6]^{2+}$ and $[Ni(NH_3)_6]^{2+}$ complexes.

(**Note:** Substitution of H_2O by NH_3 (relatively stronger field ligand) leads to the higher 10Dq value that causes the absorption band (*d-d* transition) to shift towards the shorter wavelength. It changes the colour: bright green to blue.)

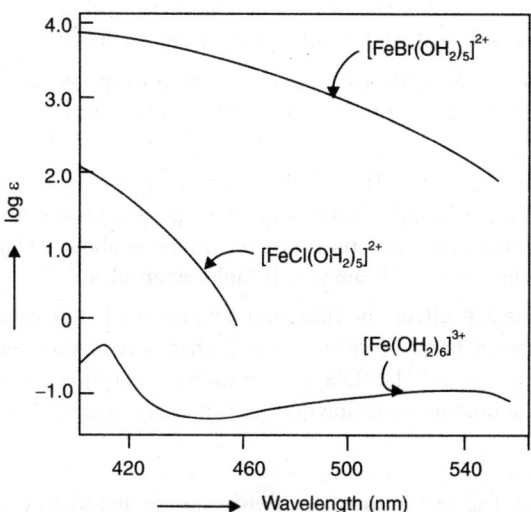

Fig. 1.10.3 Comparison of the electronic absorption spectra of different Fe(III)-species in aqueous solution. $[Fe(OH_2)_6]^{3+}$ (almost colourless, **forbidden d-d transition**) while the halido-complexes are coloured (due to the ligand to metal **charge transfer band**). ε stands for the molar extinction coefficient.

(**Note:** In the above complexes, the charge transfer band arises from the ligand to metal (*i.e.* LMCT). Thus with the increase of polarisability and oxidisability of the ligand, the charge transfer process becomes more favoured. This explains the shifting of the LMCT absorption band towards the longer wavelength in passing from $[Fe(OH_2)_6]^{3+}$ to $[FeCl(OH_2)_5]^{2+}$ to $[FeBr(OH_2)_5]^{2+}$.)

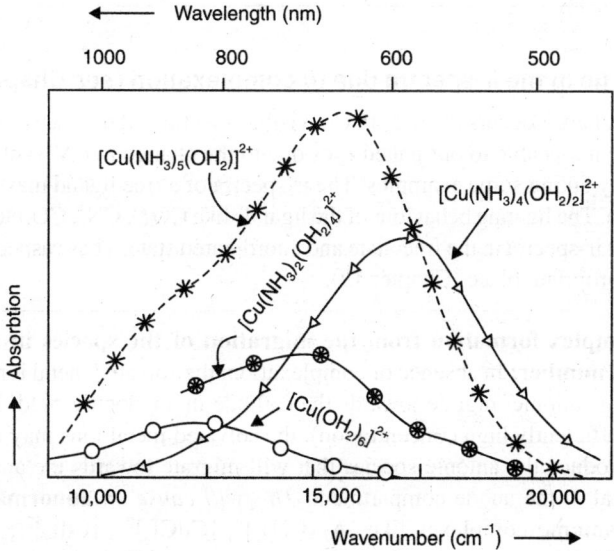

Fig. 1.10.4 Comparison of the electronic absorption spectra of $[Cu(OH_2)_6]^{2+}$ and its different ammine complexes in aqueous solution.

(**Note:** ● Anhydrous $CuSO_4$ is colourless. The ligand field strength of SO_4^{2-} is so weak that the corresponding $d\text{-}d$ transition occurs in the infra-red region of the spectrum.

● In the aqua complex, *i.e.* $[Cu(OH_2)_6]^{2+}$, the maximum absorption occurs at about 800 nm ($\varepsilon \approx$ **10 $M^{-1}\,cm^{-1}$**). Light is absorbed in the range 580–900 nm (*i.e.* yellow-orange-red region) and it **looks light blue.**

● Gradual substitution of H_2O by NH_3 (which is a relatively stronger field ligand) shifts the absorption band from the far red region to the middle of the red region of the spectrum. This shifting continues upto $[Cu(NH_3)_4(OH_2)_2]^{2+}$ that shows the absorption maximum at about 600 nm ($\varepsilon \approx$ **55 $M^{-1}\,cm^{-1}$**). Here the strong absorption occurs above 450 nm (*i.e.* it **looks deep blue**).

● Because of the strong J.T. effect, the fifth ammonia molecule ligates weakly (*i.e.* weak bonding) and it causes the absorption band to shift in the higher wavelength region compared to that of $[Cu(NH_3)_4(OH_2)_2]^{2+}$. In fact, $[Cu(NH_3)_5(OH_2)]^{2+}$ shows its absorption maximum at about 700 nm. In aqueous media, the sixth ammonia molecule does not bind significantly and $[Cu(NH_3)_6]^{2+}$ exists only in liquid ammonia.)

Lability vs. Inertness: The **instantaneous colour change** due to complexation occurs only for the **labile metal ions** and such reactions are mainly used in *analytical chemistry*. However, for the inert **metal ions** like Cr^{3+}, Co^{3+} the colour change also occurs in a similar way but slowly.

$$\left[Cr(OH_2)_6\right]^{3+} + Cl^- \rightleftharpoons \left[CrCl(OH_2)_5\right]^{2+} + H_2O, \text{(slow reaction)}$$
$$\qquad \text{(Blue violet)} \qquad\qquad\qquad \text{(Pale green)}$$

$$\left[CrCl(OH_2)_5\right]^{2+} + Cl^- \rightleftharpoons \textit{Trans-}\left[CrCl_2(OH_2)_4\right]^+ + H_2O, \text{(slow reaction)}$$
$$\qquad \text{(Pale green)} \qquad\qquad\qquad\qquad \text{(Dark green)}$$

Change in the ir-spectra due to complexation (*see* Chapter 12)

The characteristic change occurs in the both UV-visible spectra and ir-spectra due to complexation. The colour change noticeable to our naked eyes due to the change in UV-visible spectra has been illustrated above by taking some examples. The ir-spectra of a free ligand may remarkably change due to coordination. The ligating behaviour of the ligands like CO_3^{2-}, CN^-, CO, etc. can be understood by comparing their ir-spectra in the free-state and coordinated state. These aspects will be discussed in detail at the appropriate place (Chapter 12).

(4) Detection of complex formation from the migration of the species in an electric field and abnormal transport number: In absence of complexation, the solvated metal ions (*e.g.* $[M(OH_2)_x]^{n+}$ in aqueous media) being cationic migrate towards the cathode in an electric field. But in presence of the anionic ligands (at sufficiently high concentration), the solvated metal ions may undergo complexation with the anions to produce the anionic species that will migrate towards the anode. It will lead to an accumulation of metal in the anode compartment. *This will cause an* **abnormal transport number.** Formation of such anionic complexes like $[Ag(CN)_2]^-$, $[CdCl_4]^{2-}$, $[CdI_4]^{2-}$, $[ZnCl_4]^{2-}$, $[HgCl_4]^{2-}$, $[CoCl_4]^{2-}$, etc. can lead to abnormal transport number. As a matter of fact, development of such abnormal transport number is a common phenomenon for the concentrated halide solutions of heavy metals. It is illustrated for CdI_2 solution.

During the electrolysis of a concentrated solution of CdI_2, concentration of Cd(II) around the cathode does not increase and concentration of I^- around the cathode decreases sharply. At about 0.25 mol dm^{-3} CdI_2 solution at 20° C, the transport number of Cd^{2+} becomes close to zero and at higher concentration, it becomes negative (*i.e.* net decrease in the concentration of Cd(II) at the cathode compartment). This can be explained by considering the following mode of ionization of CdI_2.

$$CdI_2 \rightleftharpoons Cd^{2+} + 2I^- \text{ (in dilute solution)}$$

$$2CdI_2 \rightleftharpoons Cd^{2+} + CdI_4^{2-} \text{ (in concentrated solution)}$$

Due to the tendency of such metal ions to undergo complexation, their simple halides combine among themselves to form the **autocomplexes** as in the case of CdI_2. Formation of such autocomplexes is pronounced only in concentrated solution. On dilution, the autocomplexes break down to generate the simple ions.

$$\left[CdI_4\right]^{2-} \rightleftharpoons Cd^{2+} + 4I^- \text{ (on dilution)}$$

For CdI_2, the normal transport numbers are found at a concentration below 0.002 mol dm^{-3}.

(5) Detection of complex formation from the distribution study: When the metal ions form the electrically **neutral inner metallic complexes** with the chelating agents like *acetylacetone* (Hacac), *dimethylglyoxime* (H_2dmg), *8-hydroxyquinoline* (Hoxine), *diethyldithiocarbamic acid* (Hdtc), *diphenylcarbazide, picolinic acid, etc.,* the metal complexes are precipitated from their aqueous solution **due to the increased molecular weight, low hydration energy of the electrically neutral species, and hydrophobic environment around the metal ion.** *Such complexes are very often soluble in organic solvents.*

Thus, when the metal ions in aqueous solutions are treated with the chelating agents (suitable for inner complex formation) taken in a suitable organic solvent (immiscible with water), the neutral metal complexes produced are preferably distributed in the organic phase. For example, when an ammoniacal aqueous solution of Ni(II)-salt is shaken with H_2dmg taken in chloroform, the chloroform layer becomes orange-red. This technique is called *solvent extraction*.

$$\text{Ni(II)-salt} \xrightarrow{\text{H}_2\text{dmg (in CHCl}_3)} \text{Red coloured chloroform layer.}$$
(in ammoniacal
aqueous solution)

Similarly, Pb(II) can be extracted by *diphenylcarbazide* in CCl_4 layer. **Sometimes, the ion pairs experiencing a high electrostatic attractive force can be extracted in an organic phase.** For example, $H^+[FeCl_4]^-$ (ion pair) can be extracted in diethyl ether.

(6) Detection of complex formation from the conductance measurement: It has been already illustrated (Sec. 1.4.3, 1.4.4) that molar conductivity measurement of a solution (at infinite dilution) can tell us the mode of ionization of the compound. In fact, for the **kinetically inert and thermodynamically stable complexes**, the molar conductivity measurement data can be used to characterize a complex. This aspect has been already illustrated (Sec. 1.4.3, 1.4.4) for the Co(III)-ammine and Pt(IV)-ammine complexes.

If during complexation, the number of conducting ions changes, then the electrical conductivity changes. If during complexation, the number of ions decreases then the conductivity decreases as in the following case:

$$M^{n+} + xCl^- \rightleftharpoons \left[MCl_x\right]^{(n-x)+}$$

If during complexation, the number of H^+ or OH^- ions changes, then the electrical conductivity changes drastically. **It happens so because of the abnormally high ionic mobilities (and consequently ion conductance) of the H^+ and OH^- ions.** Thus in the following case, the conductivity increases remarkably.

$$M^{n+} + xLH \rightleftharpoons [ML_x]^{(n-x)+} + xH^+$$

Representative examples of LH are: Hgly, H_4edta, Hoxine, H_2dmg, Hacac, etc.

If the released protons are captured by the bases, then the conductivity will not increase. These aspects are illustrated in the following example.

● **$CuSO_4$ or $CuCl_2$ solution is treated with glycine.**

$$Cu^{2+} + 2\underset{\text{(Hgly)}}{H_2N-CH_2-CO_2H} \rightleftharpoons [Cu(gly)_2]^0 + 2H^+$$

i.e.
$$(Cu^{2+} + 2Cl^-) + 2Hgly \rightleftharpoons [Cu(gly)_2]^0 + 2H^+ + 2Cl^-$$

Cl^- and SO_4^{2-} are the weak Brönsted bases and they do not capture the released H^+ ion. Thus, in such cases, the released protons make the solution more acidic due to complexation. *The conductivity increases remarkably as in the above complexing reaction where one Cu^{2+} ion is replaced by two H^+ ions* (*cf.* the complex is neutral). The **ion conductance of H^+ is abnormally high**.

● **$Cu(OAc)_2$ solution is treated with glycine.**

$$Cu^{2+} + 2\underset{\text{(Hgly)}}{H_2N-CH_2-CO_2H} \rightleftharpoons [Cu(gly)_2]^0 + 2H^+$$

$$2H^+ + 2^-OAc \rightleftharpoons 2AcOH$$

Thus, the overall reaction can be written as follows:

$$(Cu^{2+} + 2^-OAc) + 2Hgly \rightleftharpoons [Cu(gly)_2]^0 + 2AcOH$$

Acetate (^-OAc) is a fairly strong Brönsted base and by capturing the released protons, it produces the weak acid, AcOH (*i.e.* acetic acid) that remains predominantly in the unionized forms. *Thus, in this case, due to complexation, the number of conducting ion decreases and consequently the conductance decreases.*

(7) Detection of complex formation from the pH-metric study: Sometimes, during complexation, pH of the medium changes depending on the nature of the ligand:

$$M^{n+} + xLH \rightleftharpoons [ML_x]^{(n-x)+} + xH^+$$

LH stands for the ligands having the coordinating sites —OH, —COOH, —N $=$ CH—OH (oxime group). The representative examples of such ligands are: amino acids, phenols, enols, aminopolycarboxylic acids, oximes, etc. In the above complex forming reactions, with the progress of the reaction, pH of the solution decreases.

$$M^{n+} + xNH_3 \rightleftharpoons [M(NH_3)_x]^{n+}$$

$$M^{n+} + xen \rightleftharpoons [M(en)_x]^{n+}$$

In the above reactions, the basic ligands like NH_3, en, etc. are incorporated in the coordination sphere. Consequently, acidity of the resultant solution gradually increases (*i.e.* pH decreases) with the progress of the complexation reaction.

(8) Detection of complex formation from the potentiometric study: Due to complexation, the redox potential of a couple may change depending on the relative stabilities of the complexes formed by the oxidized and reduced states of the metal ion under consideration (*see* Sec. 16.4.4, Vol. 3 of Fumdamental Concepts of Inorganic Chemistry for details). This can be understood from the Nernst equation. Let us consider the Co^{3+}/Co^{2+} system.

$$\left[Co(aq)\right]^{3+} + e \rightleftharpoons \left[Co(aq)\right]^{2+}, \; E = E_1^0 + \frac{0.06}{1} \log \frac{\left[Co(aq)\right]^{3+}}{\left[Co(aq)\right]^{2+}}, \; \left(at\ 25°C, E_1^0 = 1.8V\right)$$

The complexation reactions may be represented as follows:

$$\left[Co\,(aq)\right]^{3+} + xL \rightleftharpoons \left[Co^{III}(L)_x\right] + aq; \; K_{Co(III)} = \frac{\left[Co^{III}(L)_x\right]}{\left[Co\,(aq)^{3+}\right]\left[L\right]^x}$$

$$\left[Co\,(aq)\right]^{2+} + xL \rightleftharpoons \left[Co^{II}(L)_x\right] + aq; \; K_{Co(II)} = \frac{\left[Co^{II}(L)_x\right]}{\left[Co\,(aq)^{2+}\right]\left[L\right]^x}$$

Thus in presence of the complexing ligand L, the potential of the couple changes as follows:

$$E_{comp} = \left\{ E_1^0 - 0.06\log \frac{K_{Co(III)}}{K_{Co(II)}} \right\} + 0.06\log \frac{\left[Co^{III}(L)_x\right]}{\left[Co^{II}(L)_x\right]}$$

$$= E_2^0 + 0.06\log \frac{\left[Co^{III}(L)_x\right]}{\left[Co^{II}(L)_x\right]}$$

- $E_2^0 > E_1^0$ (= standard potential for the $\left[Co(aq)\right]^{3+}/\left[Co(aq)\right]^{2+}$ couple) when $K_{Co(II)} > K_{Co(III)}$.

- $E_2^0 < E_1^0$, when $K_{Co(III)} > K_{Co(II)}$.

E_2^0 is the standard potential for the $\left[Co^{III}(L)_x\right]/\left[Co^{II}(L)_x\right]$ couple.

i.e. $\left[Co^{III}(L)_x\right] + e \rightleftharpoons \left[Co^{II}(L)_x\right]$, $E_{comp} = E_2^0 + 0.06\log \frac{\left[Co^{III}(L)_x\right]}{\left[Co^{II}(L)_x\right]}$

The above predictions are illustrated below for the Fe(III)/Fe(II) and Co(III)/Co(II) redox couples (*see* Sec. 16.4.4, Vol. 3 of Fundamental Concepts of Inorganic Chemistry).

Redox Couple	E° (reduction potential in V)	Redox Couple	E° (reduction potential in V)
[Fe(aq)]³⁺/[Fe(aq)]²⁺	0.76	[Co(aq)]³⁺/[Co(aq)]²⁺	1.8
[Fe(CN)₆]³⁻/[Fe(CN)₆]⁴⁻	0.36	[Co^{III}(edta)]⁻/[Co^{II}(edta)]²⁻	0.6
[FeF₆]³⁻/[FeF₆]⁴⁻	0.40	[Co^{III}(phen)₃]³⁺/[Co^{II}(phen)₃]²⁺	0.42
[Fe(phen)₃]³⁺/[Fe(phen)₃]²⁺	1.14	[Co^{III}(en)₃]³⁺/[Co^{II}(en)₂]²⁺	−0.26
		[Co^{III}(CN)₆]³⁻/[Co^{II}(CN)₅]³⁻	−0.83

● It is evident that **CN⁻, F⁻** form the stabler complex with the oxidized form of the Fe(III)/Fe(II) couple while phen forms the stabler complex with the reduced form of the couple. In the Co(III)/Co(II) system, all the ligands shown above from the stabler complex with the oxidized form of the couple.

● Because of the formation of very stable cyanido complex of Cu(I), the formal potential of the Cu(II)/Cu(I) increases sufficiently. It causes the reduction of Cu(II) to $[Cu(CN)_4]^-$ in presence of CN^- and the concomitant oxidation of CN^- to cyanogen gas, *i.e.* $(CN)_2$. **Thus, Cu(II) is unstable in presence of CN⁻.** However, in presence of the N-donor π-acid ligands like 1, 10-phenanthroline, the reduction of Cu(II) is arrested. The five coordinate complexes like $[Cu(CN)(phen)_2]^{2+}$ are stable.

Conclusion: In presence of the complex forming ligands, the redox potential of the couple may change drastically. *This is a strong indication for complex formation.* It may be noted that $Co(aq)^{3+}$ is powerful oxidizing agent while the corresponding cyanido-complex is a very poor oxidizing agent (in fact, the Co^{II}-cyanido complex is a powerful reducing agent.)

(9) **Detection of complex formation from the polarographic study:** It has been shown that due to complexation, the E° value (*i.e.* standard reduction potential) of a redox couple changes. Consequently, the *half-wave potential* $(E_{1/2})$ in the polarographic curve changes due to complexation. *Thus, shifting of half-wave potential is an indication of complexation (see Sec. 16.2.2.1, Vol. 3 of Fundamental Concepts of Inorganic Chemistry).*

1.11 INNER METALLIC COMPLEXES OR INNER COMPLEXES

Some chelating ligands possess both the *salt forming sites* (*i.e.* anionic coordinating sites) *and neutral coordinating sites* and they can form 5 or 6 membered ring at the metal ion centre. Such chelating ligands can simultaneously satisfy both the **primary valence** (*i.e.* charge) and **secondary valence** (*i.e.* coordination number). *The complexes where the coordination number and charge of the metal ions*

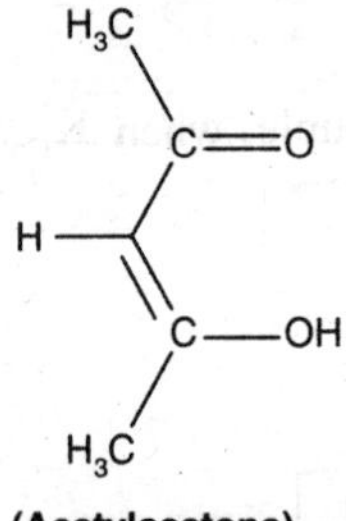

(Acetylacetone)
(enol-form)

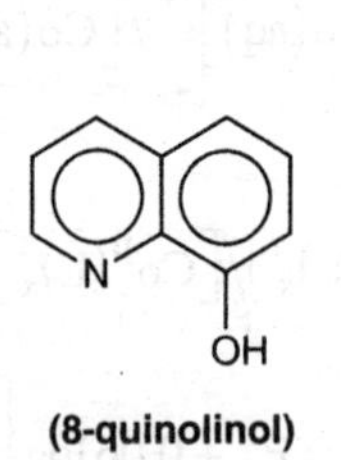

(8-quinolinol)

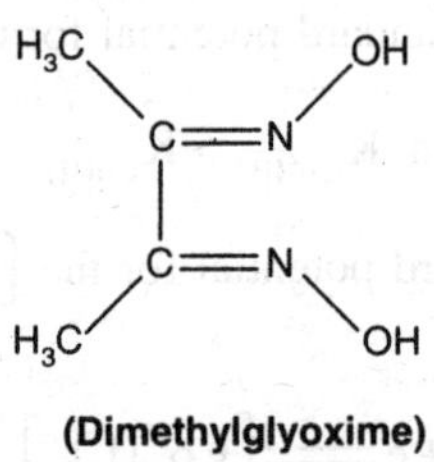

(Dimethylglyoxime)

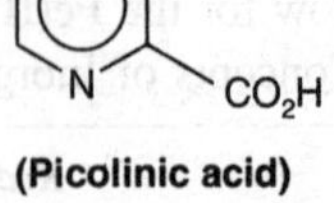

(Picolinic acid)

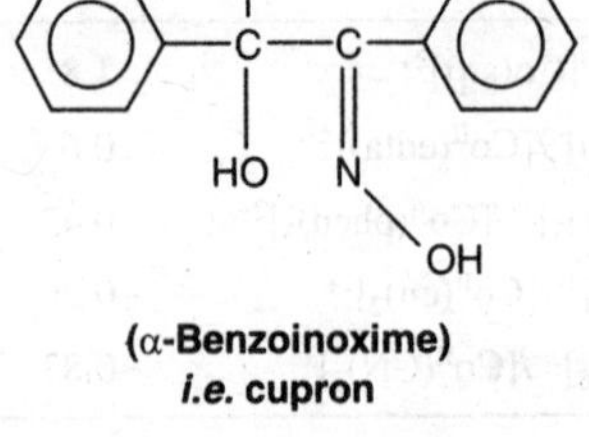

(α-Benzoinoxime)
i.e. cupron

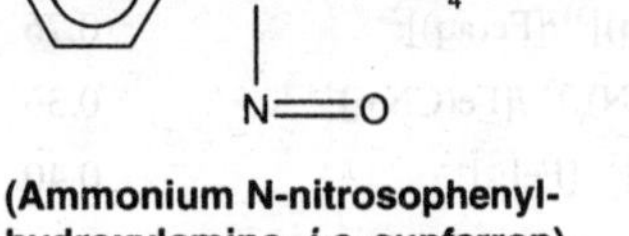

(Ammonium N-nitrosophenyl-
hydroxylamine, *i.e.* cupferron)

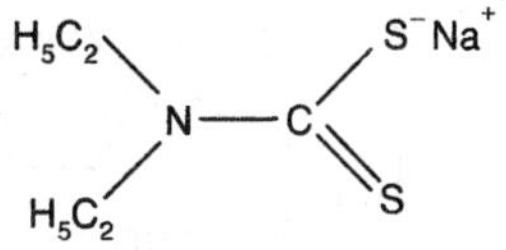

(Sodium N, N-diethyldithio-carbamate)

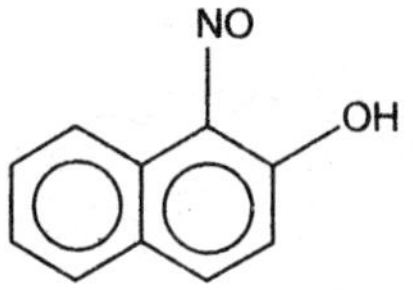

(α-Nitroso-β-naphthol)

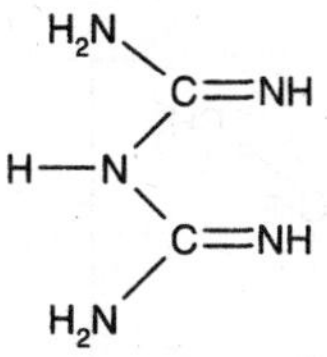

(Biguanide)

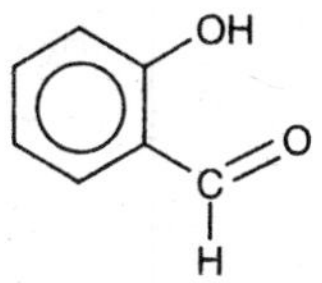

(Salicylaldehyde)

Fig. 1.11.1 Some ligands commonly used in the formation of inner complexes.

are simultaneously satisfied by the chelating ligands are called the inner metallic or simply inner complexes.

The chelating agents commonly used for the inner complex formation are given in Fig. 1.11.1.

Depending on the degree of charge neutralisation, the inner complexes are classified into different groups.

(i) **Inner complexes of first order:** In such complexes, both the coordination number and charge of the metal ion are simultaneously satisfied fully by the coordinating ligands having neutral and anionic

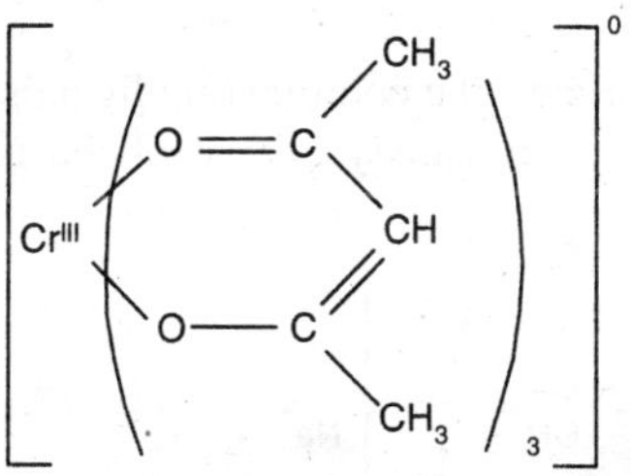

$[Cr(acac)_3]^0$
Tris(acetylacetonato)chromium(III)

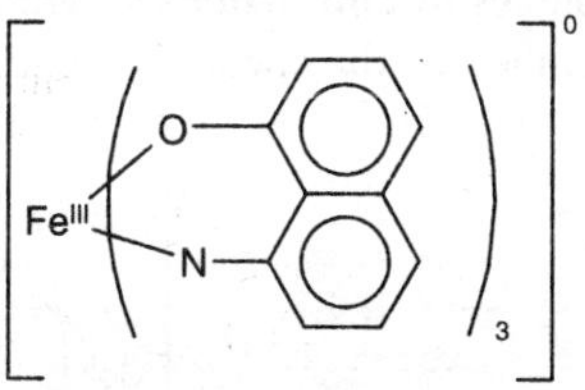

$[Fe(oxine)_3]^0$
Tris(8-quinolinolato)iron(III)

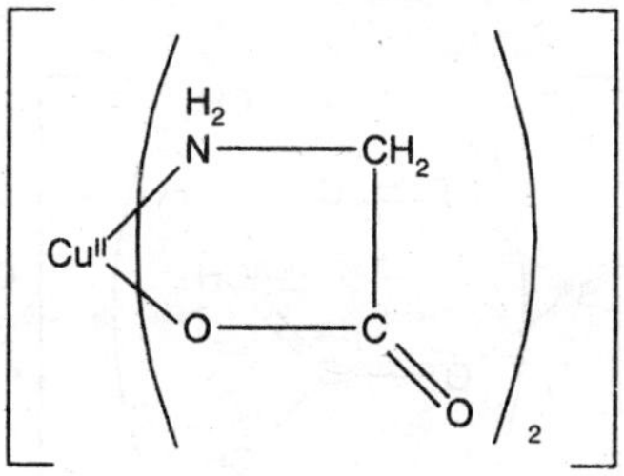

$[Cu(gly)_2]^0$
Bis(glycinato)copper(II)

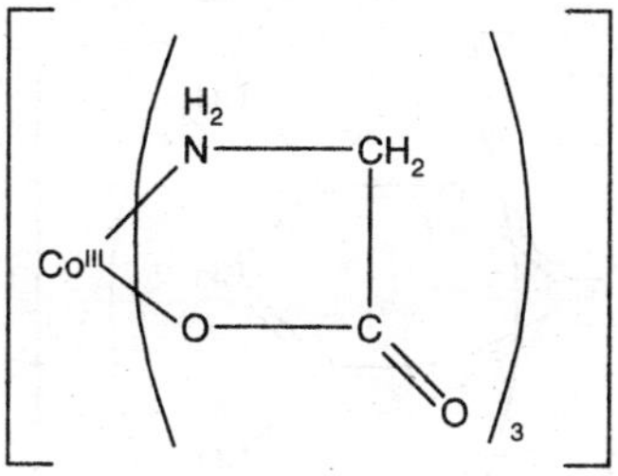

$[Co(gly)_3]^0$
Tris(glycinato)cobalt(III)

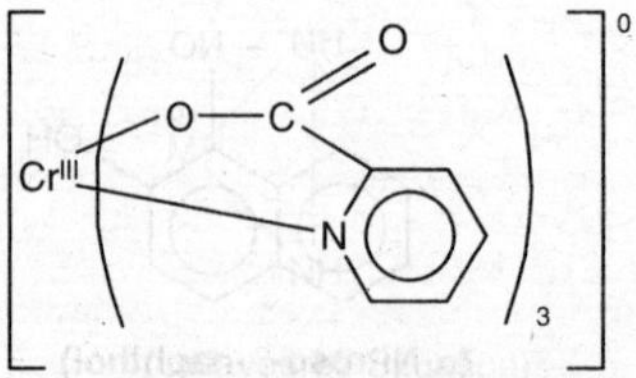

Tris(picolinato)chromium(III) 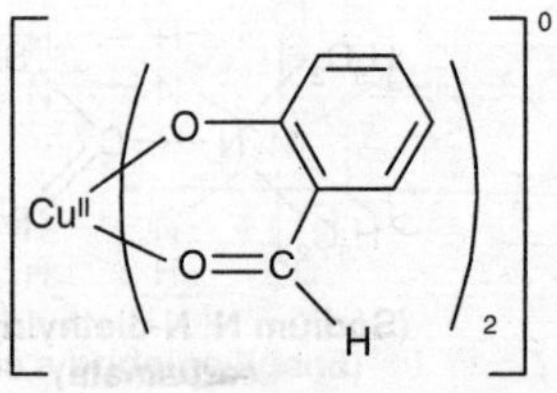   **Bis(salicylaldehydato)copper(II)**

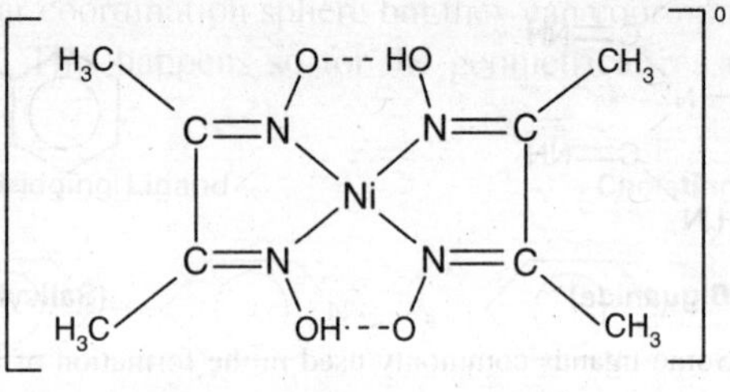

[Ni(dmgH)$_2$]

i.e. **Bis(dimethylglyoximato)nickel(II)**

Fig. 1.11.2 Representative examples of some inner complexes of first-order.

donor sites leading to the formation of neutral complexes. As the complexes are electrically neutral, they are sparingly soluble in water *but highly soluble in organic solvents. These are also characterised by low melting and boiling point.* Examples of such complexes are given in Fig. 1.11.2.

The inner metallic complexes are extremely important in analytical chemistry for the estimation of metal ions.

(ii) **Inner complexes of 2nd order (Werner):** Sometimes, the coordinating ligands satisfy the desired coordination number of the metal ion but cannot exactly satisfy charge of the metal ion (*i.e.* total

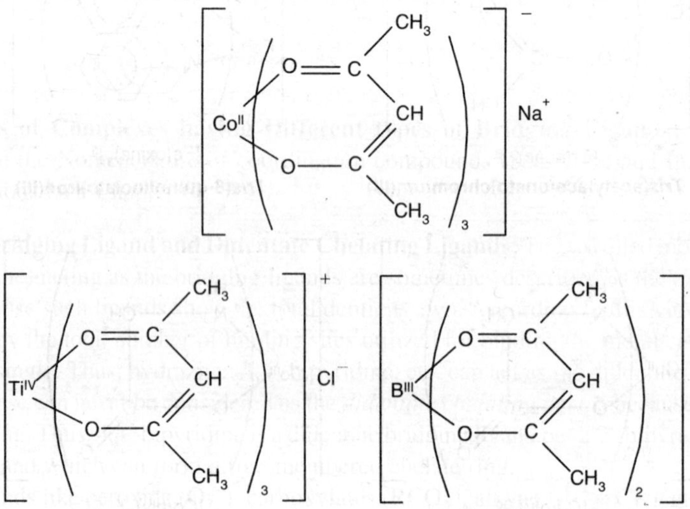

Fig. 1.11.3 Representative examples of some inner complexes of 2nd order.

number of anionic charges is not equal to the charge of the metal ion); consequently such complexes are ionic (*i.e.* **not electronically neutral**). Such charged complexes are described as the **inner complexes of second order (Werner)**. Some examples are: $Na^+[Co^{III}(acac)_3]^-$; $[Ti^{IV}(acac)_3]^+Cl^-$; $[B(acac)_2]^+X^-$ ($X = ClO_4, FeCl_4^-$); $Na^+[Co^{III}(acac)_2(NO_2)_2]^-$; $K^+[Pt^{II}(acac)Cl_2]^-$, etc.

(iii) **Inner complexes of 3rd order (P. Ray)**: In some inner complexes, the coordinated ligands possess some free basic or acidic groups which may undergo protonation or deprotonation depending on the experimental condition to change the charge bearing properties of the complex. *Such complexes were described by P. Ray as the inner complexes of 3rd order.* It is illustrated in the following examples.

(a)

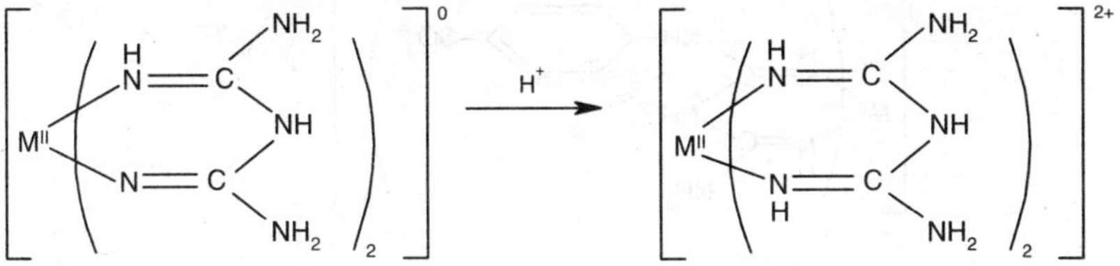

Bis(biguanidato)metal(II)   Bis(biguanide)metal(II) ion

$$i.e. \left[M\left(big\right)_2\right]^0 + 2H^+ \longrightarrow \left[M\left(bigH\right)_2\right]^{2+}$$

bis(biguanidato)metal(II)   bis(biguanide)metal(II) ion

$$\left[Cu\left(big\right)_2\right] \xrightarrow{HCl} \left[Cu\left(bigH\right)_2\right]Cl_2$$

In big^- (*i.e.* deprotonated biguanide), the negative charge is delocalised as in the case of acetylacetonate ($acac^-$).

(big⁻)   (acac⁻)

(b)

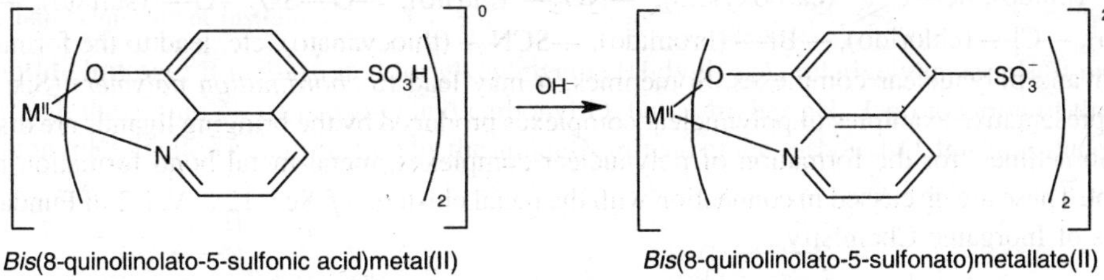

Bis(8-quinolinolato-5-sulfonic acid)metal(II)   Bis(8-quinolinolato-5-sulfonato)metallate(II)

$$i.e. \left[M^{II}\left(oxine\text{-}5\text{-}SO_3H\right)_2\right]^0 \xrightarrow{+OH^-} \left[M^{II}\left(oxine\text{-}5\text{-}SO_3^-\right)_2\right]^{2-}$$

(c) $\left[Ni\left(dmgH\right)_2\right]^0 \xrightarrow[-H^+]{OH^-} \left[Ni\left(dmg\right)_2\right]^{2-}$

Thus $[M(big)_2]^0$, $[M(oxine-5-SO_3H)_2]$, $[M(dmgH)_2]$, etc. are the examples of inner metallic complexes of 3rd order.

(iv) Inner complexes of the fourth order (P. Ray): In some cases, the coordinated ligands possess an equal number of basic and acidic sites and they remain as zwitter ions within the coordination sphere. Such coordinated ligands (and consequently the complexes) undergo protonation in acidic condition to produce the cationic complexes while in basic condition they form the anionic complexes. *Such* **ampholytic complexes** *were described as the inner complexes of fourth order by Ray.* It is illustrated by considering the complexes of phenylbiguanide–*p*–sulfonic acid below.

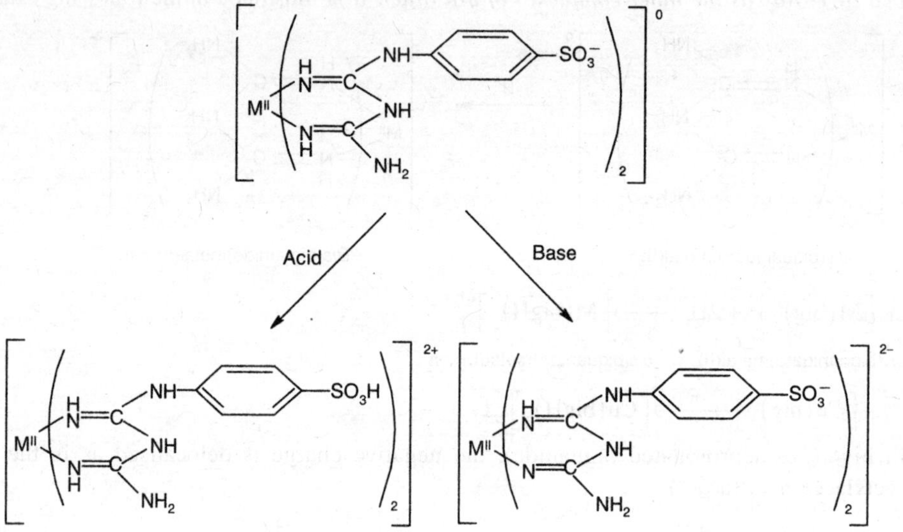

1.12 POLYNUCLEAR OR BRIDGED COMPLEXES

Bridging ligands are denoted by μ-prefix, *e.g.* μ_2–CO (bridging two metal centres), μ_3–CO (bridging three metal centres). Commonly known bridging ligands generally act as the μ_2–bridging ligands. The bridging ligands like: —OH— (hydroxido), —O— (oxido), —O—O— (peroxido), —NH$_2$— (amido), —NH— (imido), R—C$\underset{}{\overset{}{\lessgtr}}$ (carboxylato), —NO$_2$— (nitrito), —O—SO$_2$—O— (sulfato), —CN— (cyanido), —Cl— (chlorido), —Br— (bromido), —SCN— (thiocyanato), etc. lead to the formation of dinuclear and polynuclear complexes. Sometimes, it may lead to *coordination polymers* (Sec. 1.14). Some representative examples of polynuclear complexes produced by the bridging ligands are discussed below. Sometimes, for the formation of polynuclear complexes, metal-metal bond formation may be important. These are discussed in connection with the metal clusters (*cf.* Sec. 12.8, Vol. 2 of Fundamental Concepts of Inorganic Chemistry).

(a) Hydroxido (—OH—) and Oxido (—O—) Bridged Complexes:

(i) Di- and polynuclear hydroxido-aqua complexes: During the base hydrolysis of metal ions, very often the heavy metal ions form the polynuclear hydroxides. The mononuclear units (octahedral or tetrahedral) may join together along their edges or faces or apexes. *Joining through the apexes, edges and faces leads to bridging by one, two and three 'OH' groups respectively.* These are illustrated in Fig. 1.12.1 for the dinuclear complexes.

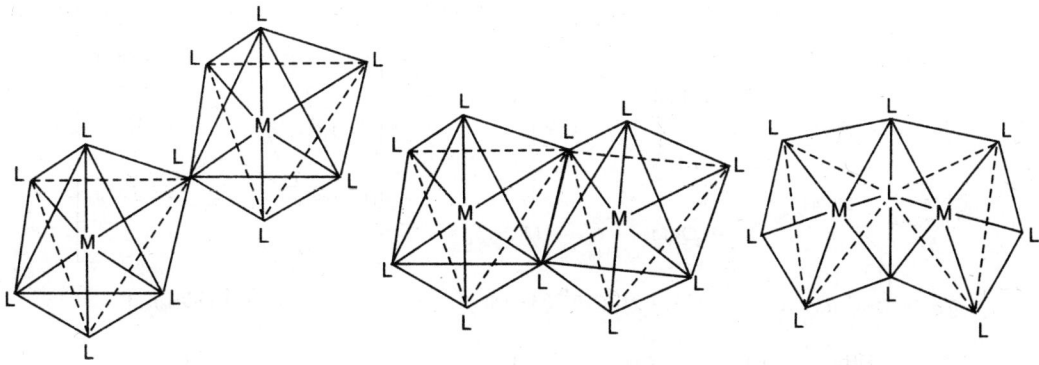

(i) Corner shared octahedra
(one bridging group)

(ii) Edge shared octahedra
(two bridging groups)

(iii) Face shared octahedra
(three bridging groups)

Fig. 1.12.1 Structure of binuclear complexes formed through the sharing of one vertex (i), one edge, *i.e.* two *cis*-vertexes (ii), and one triangular face, *i.e.* three vertexes of a trigonal face (iii) of octahedral units, ML_6 where L acts as a bridging ligand (*e.g.* L = O, OH) (The nonbridging ligands may be the other ligands like H_3N, H_2O).

● For the **bivalent metal ions** like Cu^{2+}, Be^{2+}, Zn^{2+}, Cd^{2+}, Hg^{2+}, Ni^{2+}, Pb^{2+}, etc. the dinuclear complex having one bridging OH group may be represented as follows:

$$\left[(H_2O)_{n-1}M \overset{H}{\underset{}{-O-}} M(OH_2)_{n-1}\right]^{3+} \quad i.e. \quad M_2(OH)^{3+}$$

● For the **trivalent metal ions** like Al^{3+}, Fe^{3+}, In^{3+}, Sc^{3+}, Cr^{3+}, Co^{3+}, etc. the dinuclear complex having two bridging OH groups may be represented as follows:

$$\left[(H_2O)_{n-2}M \underset{\underset{H}{O}}{\overset{\overset{H}{O}}{<}} M(OH_2)_{n-2}\right]^{4+} \quad i.e. \quad M_2(OH)_2^{4+}$$

● For the **tetravalent metal ions** like Ce^{4+}, Th^{4+}, the dinuclear complex may involve three bridging OH groups.

$$\left[(H_2O)_{n-3}M \underset{\underset{H}{O}}{\overset{\overset{H}{O}}{\underset{}{-\overset{H}{O}-}}} M(OH_2)_{n-3}\right]^{5+} \quad i.e. \quad M_2(OH)_3^{5+}$$

The number of 'OH' bridges between the metal centres increases with the increase of positive charge on the metal centre. The higher valent aqua-metal ions are the stronger Brönsted acids, i.e. in the 'M—OH_2' moiety, the 'M—O' bond becomes stronger and the 'O—H' bond becomes weaker. This higher Brönsted acid strength favours the deprotonation process to generate the more metal bound 'OH' groups that may participate in bridging the metal centres.

● The polynuclear hydroxido-aqua complexes may be like these $M_3(OH)_3^{3+}$, $M_4(OH)_4^{4+}$, $M_n(OH)_{2n}^{n+}$, etc.

i.e. $M_3(OH)_3^{3+}$ (M = Ni)   *i.e.* $M_4(OH)_4^{4+}$ (M = Ni)   *i.e.* $[Zr_4(OH)_8(OH_2)_{16}]^{8+}$

i.e. $M_n(OH)_{2n}^{n+}$ (M = Cr^{3+})   *i.e.* $[Be_3(OH)_3(OH_2)_6]^{3+}$

- The polynuclear structure of $M_2O_3.nH_2O$ (variable composition, M = Al, Cr) can be represented as follows:

- On dehydration, the polynuclear complexes with the OH groups are converted into the **oxol** derivative having the oxido-bridges.

(ol-derivative) $\xrightarrow{-nH_2O}$ (**oxol**-derivative)

(ii) **Hydroxido- and oxido-bridged di- and polynuclear Co(III) and Cr(III) complexes:** Some examples are given below.

Hydroxido- and oxido-bridged ammine and ethylenediamine complexes of Cr(III) are well known.

$$\left[(H_3N)_5Cr \overset{H}{\underset{}{-}} O - Cr(NH_3)_5\right]^{5+} \underset{+H^+}{\overset{+OH^-}{\rightleftharpoons}} \left[(H_3N)_5Cr - O - Cr(NH_3)_5\right]^{4+}$$

(**Red,** *more paramagnetic,*   (**Blue,** *less paramagnetic,*

bent hydroxido-bridge)   *almost linear oxido-bridge*)

● The hydroxido- and oxido-bridged dinuclear ammine complexes of Cr(III) are in a deprotonation-protonation equilibrium but their properties are remarkably different. *The oxido-bridged complex with a linear bridge leads to the better spin-pairing (i.e.* superexchange process) *via the filled p-orbitals of the bridging oxygen.* It actually establishes the **3c–4e bonding.**

The above **dinuclear Cr(III)-complexes** can experience aquation (*i.e.* substitution of NH_3 by H_2O) very slowly. These were studied extensively by Werner.

$$\left[(H_3N)_5Cr \overset{H}{\underset{}{-}} O - Cr(NH_3)_5\right]^{5+} \underset{H^+}{\overset{OH^-}{\rightleftharpoons}} \left[(H_3N)_5Cr - O - Cr(NH_3)_5\right]^{4+}$$

(**Rhode ion**)   (**Basic rhode ion**)

(slowly) $\Big\downarrow$ $\begin{array}{l} +H_2O, \\ -NH_3 \end{array}$

$$\left[(H_3N)_5Cr \overset{H}{\underset{}{-}} O - Cr(NH_3)_4(OH_2)\right]^{5+} \underset{H^+}{\overset{OH^-}{\rightleftharpoons}} \left[(H_3N)_5Cr \overset{H}{\underset{}{-}} O - Cr(NH_3)_4(OH)\right]^{4+}$$

(**Erythro ion**)   (**Basic erythro ion**)

● **Dinuclear Co(III)-amine complexes:**

$$2[Co(NH_3)_4(OH)(OH_2)]SO_4 \xrightarrow{100-110°C} \left[(H_3N)_4Co\underset{\underset{H}{O}}{\overset{\overset{H}{O}}{<}}Co(NH_3)_4\right](SO_4)_2 + 2H_2O$$

Similarly, $\left[(en)_2Co\underset{\underset{H}{O}}{\overset{\overset{H}{O}}{<}}Co(en)_2\right]^{4+}$ may be obtained.

● $\left[CoCl(NH_3)_4(OH_2)\right]^{2+}$ on being treated with alkali forms the tetranuclear species (called **hexol**) with a central **CoO₆ octahedron** moiety which is hydroxido-bridged with three Co(NH₃)₄ units. The tetranuclear complex is **optically active having no C-centre.**

$$\left[Co\left(\underset{\underset{H}{O}}{\overset{\overset{H}{O}}{<}}Co(NH_3)_4\right)_3\right]^{6+} \quad \textbf{(Hexol)}$$

A similar tetranuclear complex of Cr(III) is also known. The corresponding linear trinuclear hydroxido-bridged complex is:

$$(L = NH_3, M = Co \text{ or } Cr)$$

(b) Oxido-Bridged Inorganic Polymers (from the p-block Elements)

Silicate, silicone, $(SiO_2)_x$, poly-phosphates, disulfate, etc. are the representative examples.

These oxido-bridged polymeric compounds are well known. Here we shall discuss in short the features of *silicones* which are the organo-silicon polymers having R_2SiO or $RSiO_2$ as the building units (R = CH_3, Ph, etc.). These are called **siloxanes** in general and the commercial product having R = CH_3 is called **silicone**. Controlled hydrolysis of R_2SiCl_2 or $RSiCl_3$ followed by dehydration produces **siloxanes**.

$$nR_2SiCl_2 \xrightarrow[-2nHCl]{2nH_2O} n\left[R_2Si(OH)_2\right] \xrightarrow{-nH_2O} \begin{array}{ccc} R & R & R \\ | & | & | \\ -Si-O-Si-O-Si- \\ | & | & | \\ R & R & R \end{array} \ i.e. \ (O-SiR_2-O)_n$$

$$nRSiCl_3 \xrightarrow[-3nHCl]{3nH_2O} n\left[RSi(OH)_3\right] \xrightarrow{-H_2O} \text{Cross-linked polymer.}$$

Crosslinked polymers

A linear **siloxane polymer** having an exceptionally high thermal stability

Hydrolysis of R_2SiCl_2 can produce the following **cyclic polymer**.

Depending on the degree of polymerization, the siloxanes may be oils, greases, rubbers or resins. *These polymers are water repellants* because of the orientation of the hydrophobic R-groups on the surface. Replacement of some Si-centres by Al, Ti or Sn can enhance the thermal stability.

Siloxane polymers having the metal centres along with the silicon centres

(c) Amido (—NH$_2$—) and Imido (—NH—) Bridged Complexes

● Co(NH$_2$)$_3$ can exist as a polymeric compound through the —NH$_2$— (*i.e.* amido) bridges. It is *amphoteric*. It can react with the both *ammono acids* (e.g. NH$_4$Cl) and *ammono bases* (e.g. KNH$_2$) to produce the cationic complex, [Co(NH$_3$)$_6$]$^{3+}$ and anionic complex, [Co(NH$_2$)$_6$]$^{3-}$ respectively.

(Protonation on the — NH$_2$ groups) | + NH$_4^+$ *i.e.*   [Co(NH$_2$)$_3$]$_x$ + NH$_2^-$ (Replacement of the bridging — NH$_2$ groups)

[Co(NH$_3$)$_6$]$^{3+}$ [Co(NH$_2$)$_6$]$^{3-}$

i.e.
$$\left[Co\left(NH_3\right)_6\right]^{3+} \underset{NH_4^+}{\overset{NH_2^-}{\rightleftharpoons}} Co\left(NH_2\right)_3 \underset{NH_4^+}{\overset{NH_2^-}{\rightleftharpoons}} \left[Co\left(NH_2\right)_6\right]^{3-}$$

● Ammonolysis of [Co(NH$_3$)$_6$]$^{3+}$ can produce a dinuclear complex having two —NH$_2$— bridges.

$$2\left[Co\left(NH_3\right)_6\right]^{3+} + 2NH_3 \rightleftharpoons 2\left[Co\left(NH_2\right)\left(NH_3\right)_5\right]^{2+} + 2NH_4^+$$

● The dinuclear complex, possesses both the hydroxido and amido bridges.

● In the following example, both the 'NH$_2$' and 'NO$_2$' groups acts as the bridging ligands.

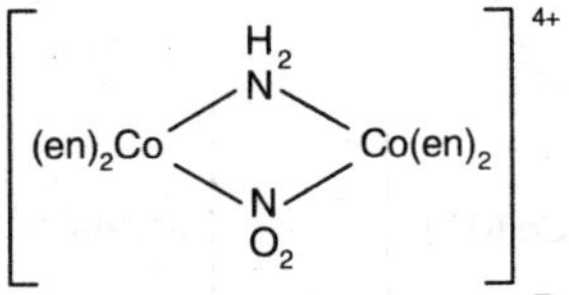

● **Vortmann's sulfate** (a mixture of salts) is a classical example of the amido bridged Co(III) complex. It is prepared by the air oxidation of ammoniacal solution of Co(NO$_3$)$_2$ followed by the neutralization with dilute H$_2$SO$_4$ in cold condition. The major constituents of the mixture are:

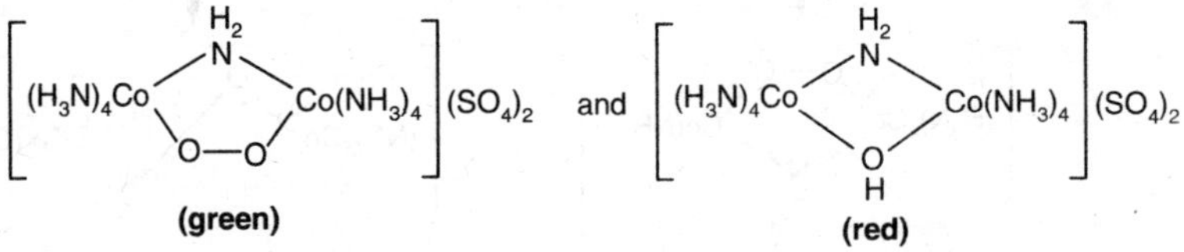

The green salt was formulated by Werner as a *peroxido-bridged complex of Co^{III} and Co^{IV}*. However, *esr data* indicate that the unpaired electron is delocalized over the both Co-centres and the O–O bond length (128 pm) is close to that of O$_2^-$ (*i.e.* superoxide). Thus, it is reformulated as a **superoxido-bridged complex of the Co^{III}- centres**.

Starting from Vortmann's sulfate, different binuclear complexes have been prepared. Some representative examples are:

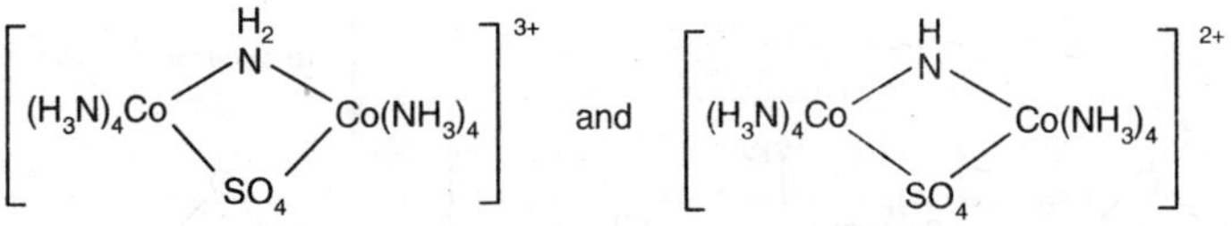

(d) Peroxido (—O—O—) and Superoxido (—O—O—) Bridged Dinuclear Complexes

Co(II)- ammine or -cyanido complexes experience an *oxidative addition* by O$_2$ to give a *transient* Co^{IV} complex that reacts with another Co(II)-complex to produce a *peroxido-bridged dinuclear complex* of Co(III). The peroxido-bridged dinuclear complex of Co(III) can be easily oxidized (in 1e-step) to give a *superoxido-bridged dinuclear complex* of Co(III) centres. It happened so in **Vortmann's green salt**.

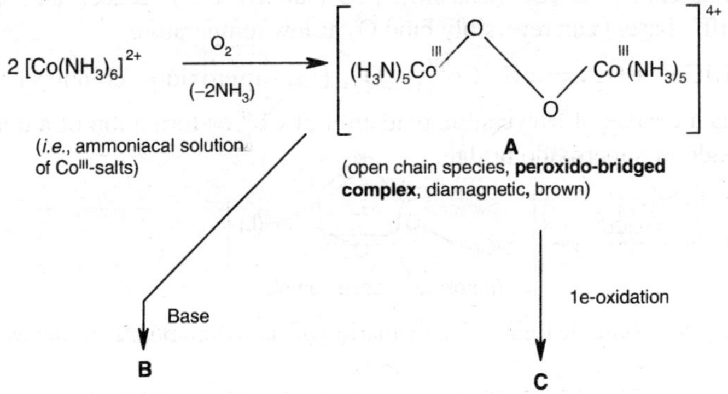

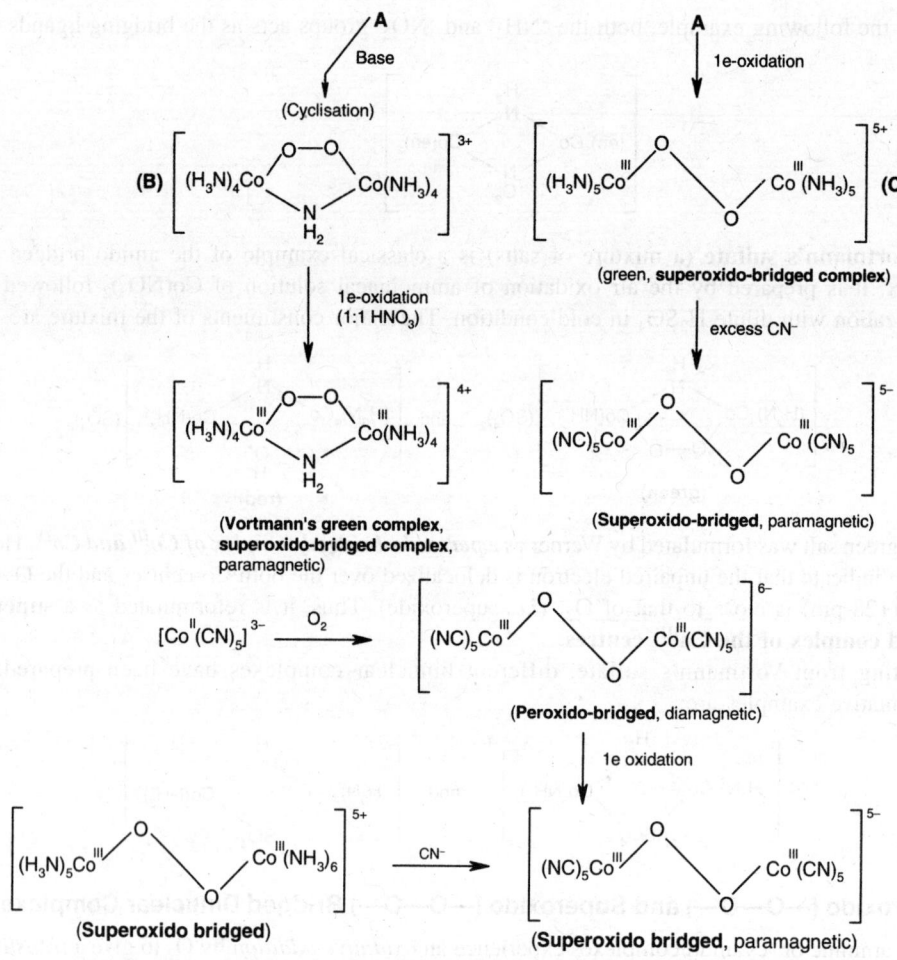

(Note: In all the above mentioned dinuclear complexes, Co(III) remains in low spin state, *i.e.* t_{2g}^6).

Co(II)–Complexes: Oxygen Carriers

Some Co(II)-complexes (*e.g.* [Co^{II}(acacen)], [Co^{II}(salen)], etc.) (acacen and salen denote the tetradentate Schiff's bases) can reversibly bind O_2 at low temperature.

$$(L)Co^{II} + O_2 \rightleftharpoons (L)Co^{III} - O_2^- \quad (i.e. \text{ superoxido} - \text{complex})$$

However, there is a chance of irreversible oxidation of Co^{II} or formation of a dinuclear complex having the peroxido or superoxido bridge.

i.e. $(L)Co^{III} - O_2^- \xrightarrow{+ (L)Co^{II}} \left[(L)Co^{III} \diagdown \overset{O}{\underset{O}{\diagup}} \diagdown Co^{III}(L) \right]$

(peroxido-bridged complex)

However, by selecting a suitable ligand (L), formation of such dinuclear complexes can be sterically prevented.

In **hemocyanin** (an O_2- uptake protein), in the oxy-hemocyanin, there is a peroxido bridge between the Cu(II)- centres.

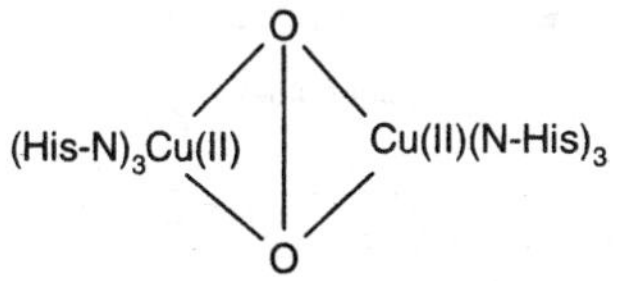

$Cu^{II}(\mu-\eta^2:\eta^2-O_2^{2-})Cu^{II}$ $Cu^{II}(\eta\text{-}1,2\text{-}O_2^{2-})Cu^{II}$

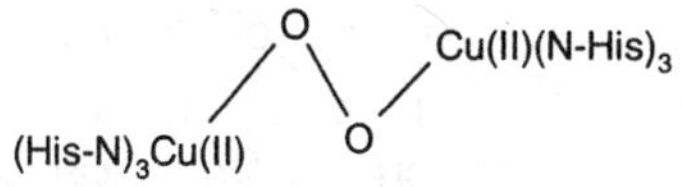

(Oxyhemocyanin)

$Cu^{II}\left(\mu-\eta^2:\eta^2-O_2^{2-}\right)Cu^{II}$ also exists in **tyrosinase** (*see* Bioinorganic Chemistry by A.K. Das).

Dinitrogen-Bridged Dinuclear Complexes

In biological and abiological nitrogen fixation processes, formation of dinitrogen (N_2) bridged complexes may play an important role. The well known example is:

$$\left[(H_3N)_5Ru - N \equiv N - Ru(NH_3)_5\right]^{4+}$$

Here, N_2 acts as a bridging ligand and performs as a π-acid ligand evidenced by lengthening of the nitrogen-nitrogen bond. Thus the ligated N_2 is activated.

(e) Halido-bridged Complexes

These are well known in many cases. Some representative examples are discussed below.

The halido-bridged complexes are very often stable in gaseous and solid state and they get destroyed in aqueous solution due to aquation.

● **$MnCl_2.2H_2O$ *vs.* $MnCl_2.4H_2O$:** The crystal hydrate having relatively lower number of water molecules undergoes dimerisation or polymerization using the anions as bridging ligands. It is illustrated.

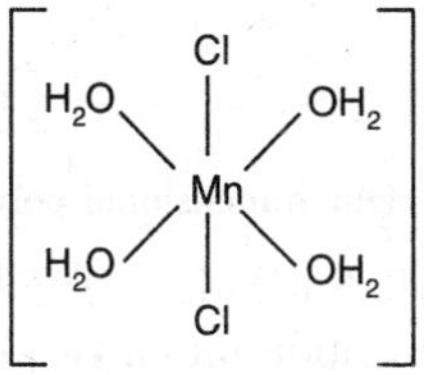

$MnCl_2.4H_2O$ $MnCl_2.2H_2O$

Monomeric form   **Polymeric form**

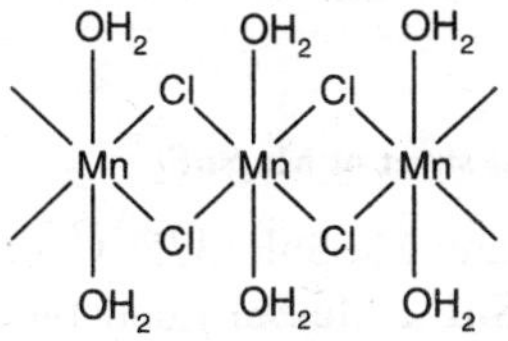

In both the cases, the Mn(II)-centre maintains the octahedral geometry.

(**Note:** $MnSO_4.4H_2O$ produces a dinuclear species using two sulfate groups as the bridging ligands and both the Mn(II)-centres maintain the octahedral geometry. This aspect has been discussed later).

● **$CuCl_2.2H_2O$ (green):** The crystal hydrate contains the polymeric structure.

$$\text{CuCl}_2 \cdot 2\text{H}_2\text{O} \xrightarrow{\text{Water}} [\text{Cu(OH}_2)_6]^{2+} + \text{Cl}^-$$

CuCl$_2$.2H$_2$O
(Green crystal)

(Light blue)

The colour change, green to blue is due to higher crystal field splitting power of H_2O than that of Cl^- as the ligands.

● **SnX$_4$:** SnF$_4$ is solid which sublimes at about 700° C. SnCl$_4$ and SnBr$_4$ are liquids (b.p. 114° C and 207° C respectively); SnI$_4$ is solid (m.p. 144° C). The properties of SnCl$_4$, SnBr$_4$ and SnI$_4$ can be explained by considering the covalent character and molar mass. SnF$_4$ exists in a polymeric form where the octahedral SnF$_6$ units unite by sharing the edges.

● **Me$_3$SnX (X = Cl, Br, I):** The Lewis acid strength sequence: Me$_3$SnF ⟩ Me$_3$SnCl ⟩ Me$_3$SnBr influences their tendencies to undergo polymerization through the halido-bridges in solid state. Me$_3$SnF is extensively aggregated leading to an infinite linear polymer while Me$_3$SnCl and Me$_3$SnBr exist as the molecular monomers.

2-dimensional sheet of Me$_2$SnF$_2$ **One dimensional polymeric chain of Me$_3$SnF**

The **melting points** are: Me$_3$SnF (375° C, decomp.), Me$_3$SnCl (42° C) and Me$_3$SnBr (27° C).

● **Me$_2$SnF$_2$ vs. MeSnF$_3$:** Me$_2$SnF$_2$ undergoes polymerization into a **two dimensional polymeric sheet** while MeSnF$_3$ forms a three dimensional infinite structure.

Note: In the polymeric forms of MeSnX$_2$, Me$_2$SnX$_2$ and Me$_3$SnX, the octahedral geometry around Sn is attained through the sp^3d^2 hybridisation. This **d-orbital participation** (cf. Secs. 9.13.5, 10.7, Vol. 2 of Fundamental Concepts of Inorganic Chemistry) in bonding is favoured most for X = F. This concept also explains their Lewis acid strength (cf. Sec. 14.9.5, Vol. 3 of Fundamental Concepts of Inorganic Chemistry).

● **Crystalline FeCl$_3$:** FeCl$_3$ is polymeric in which octahedral geometry around each Fe(III) centre is maintained by the bridging Cl$^-$ ligands. In water, it breaks down; in organic solvent and during sublimation, it forms the dimer.

(**Note:** Crystalline $FeCl_3$ adopts the **layer structure** while FeF_3 (which is more ionic) adopts the ReO_3-structure in which all the apexes of **'FeF_6' octahedra are shared**. In both cases, halides act as the bridging ligands. $FeCl_3$ lattice melts at $308°$ C while FeF_3 lattice sublimes at $1000°$ C. FeF_3 does not dissolve in water while $FeCl_3$ dissolves easily in water.)

• **Crystalline AlX_3:** AlF_3 is the most ionic compounds among the halides of Al(III). AlF_3 adopts the polymeric structure having the coordination lattice of ReO_3 type. $AlCl_3$ adopts the layer type lattice (*cf.* polymeric FeF_3 – ReO_3 type lattice; polymeric $FeCl_3$ - layer type lattice). Thus in both crystalline $AlCl_3$ and AlF_3, octahedral geometry around Al(III) is maintained by the bridging halides but the degree of polymerisation is relatively less in $AlCl_3$. In AlF_3 crystal, all the corners of AlF_6 octahedra are shared. $AlBr_3$ and AlI_3 form the dimers only in solid state. This structural difference can explain their melting point sequence.

	AlF_3	$AlCl_3$	$AlBr_3$	AlI_3
	(polymeric, ReO_3 type)	(polymeric layer type)	(dimeric)	(dimeric)
m.p.	~$1290°$ C	~$192°$ C	~$98°$ C	~$190°$ C.

Solid AlF_3 is not soluble in water and it is quite unreactive. Solid $AlCl_3$ is polymeric but on melting it forms the dimers Al_2Cl_6 and in the vapour phase, the dimers predominate. On melting of $AlCl_3$, the coordination number drops from 6 to 4. **It leads to an abrupt increase in volume** (*i.e. abrupt decrease in density*).

> ### High melting points of some fluoro-compounds (*cf.* Sec. 10.6.3, Vol. 2 of Fundamental Concepts of Inorganic Chemistry)
>
> It may be noted that the fluoro-compounds very often form the polymeric aggregation that may lead to *one-dimensional chains, two-dimensional network, layer type lattice or three-dimensional network depending on the condition. Consequently, they very often record the very high melting points.* This aspect has been illustrated for SnF_4, Me_3SnF_2, Me_2SnF_3, AlF_3, BeF_2, FeF_3, etc.

● **Polymeric fluoridoaluminates** (*e.g.* $M_2^IAlF_5$, M^IAlF_4, AlF_3): In these polymeric fluoridoaluminates, the 'AlF_6' octahedra join **at some common apexes**. The AlF_5^{2-} **stoichiometry** is developed by uniting the AlF_6 octahedra at two common opposite vertexes while the AlF_4^- **stoichiometry** is developed by sharing four equatorial vertexes of the AlF_6 octahedra (see **structure of SnF_4**, Sec. 12.4.7, Vol. 7). If all the apexes of AlF_6 octahedra are shared then the polymeric AlF_3 crystal (**ReO_3-type**) will be obtained. In all cases, **corner sharing of the AlF_6-units is preferred** instead of edge sharing or face sharing to minimise the repulsion between the adjacent Al^{3+}-centres. **Discrete AlF_6-units exist in cryolite Na_3AlF_6**.

Trans-bridged chain of AlF₅-stoichiometry

AlF_4-stoichiometry of 2D-layer structure; sharing of 4-equatorial apexes

$(AlF_3)_x$ (ReO₃-type 3D-lattice through sharing of all 6 apexes)

● **Solid BeX_2 and fluoberyllates:** BeF_2 (solid) can adopt different structures just like SiO_2 by using the —F— bridges. In fact, like SiO_2, it can be easily virtified. Polymeric $BeCl_2$ attains the fibrous structure. Because of the difference in structure, $(BeF_2)_x$ shows a higher melting point than $(BeCl_2)_x$ (cf. m.p. 800°C *vs.* 450°C). **Fluoberyllates**, *i.e.* **fluoridoberyllates** ($[BeF_3]^-$) can polymerize while BeF_4^{2-} exists as the molecular ion.

Chain lattice of BeF₃⁻ stoichiometry as in KBeF₃

BeF_4^{2-} ion

● **Polymeric $(SbF_5)_n$:** The highly viscous liquid SbF_5 is consisting of the linear polymers in which each Sb-centre is octahedrally surrounded by the fluorides. In it, there are three types of F in the ratio 2:2:1.

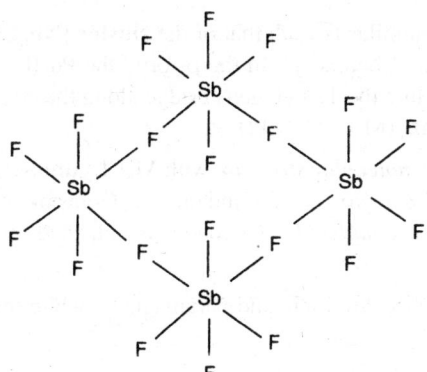

(cis-bridged $[SbF_6]$ **octahedral)**

(Two equivalent axial F;
two equivalent equatorial F;
two equivalent bridging F, *i.e.*,
one bridging F per Sb-atom)

(Linear polymer of liquid SbF$_5$; liquid VF$_5$ also adopts the similar structure)

The solid SbF$_5$ consists of **cyclic tetramers**.

AsF$_5$ (gas): (m.p. $-80°C$; b.p. $-53°C$)
SbF$_5$ (liquid): (m.p. $8°C$; b.p. $142°C$)
Monomeric AsF$_5$ is structurally similar to PF$_5$ but SbF$_5$ is associated in all phases.

(Cyclic tetrameric units of solid SbF$_5$; solid NbF$_5$ and TaF$_5$ also adopt the similar structure)

● **Polymeric** $(MX_3)_n$: The examples are: $(MoI_3)_n$, $(MoBr_3)_n$, etc.

It may be noted that $[Cr(NH_2)_3]_n$, $[Co(NH_2)_2(OH)]_n$ also possess the similar structure.

● **Polymeric PdCl$_2$ and PtCl$_2$:** They can exist in α- and β forms. The α-form of PdCl$_2$ is comparable with that of $(BeCl_2)_x$. (*cf.* BeCl$_4$ *units are tetrahedral;* PdCl$_4$ *or* PtCl$_4$ *units are square planar*). The structure of α-PtCl$_2$ is uncertain.

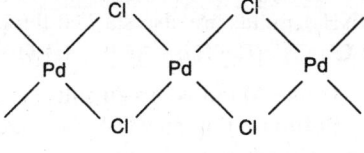

α-PdCl$_2$ **(Polymeric chain)**

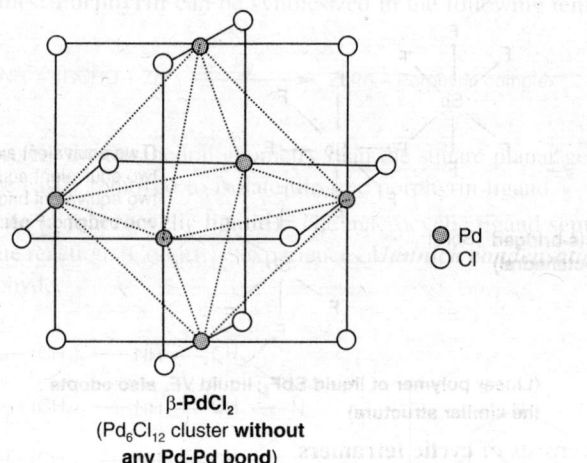

β-PdCl₂
(Pd₆Cl₁₂ cluster **without**
any Pd-Pd bond)

The β-form has got a structural similarity with that of the **cluster $[Nb_6Cl_{12}]^{2+}$** (*See* Ch. 12, Vol. 3 of Fundamental Concepts of Inorganic Chemistry). In the β-form, the Pd(II) or Pt(II) centres occupy the vertexes of a regular octahedron while the 12 halogens bridge along the edges of the octahedron. Thus, the β-form consists of **M_6Cl_{12} units** (M = Pd or Pt).

β-PdCl₂ or β-PtCl₂ produces the *molecular structure* with **M_6Cl_{12} units** stabilized by halogen bridges not by M–M bonds (*see* Sec. 12.8.3, Vol. 3 of Fundamental Concepts of Inorganic Chemistry for bonding and structure in details). The *molecular behaviour* is indicated by the solubility of Pt₆Cl₁₂ in benzene.

It may be noted that PdF₂ and NiX₂ are ionic and paramagnetic while the polymeric PdCl₂ or PtCl₂ is diamagnetic.

(**Note:** Generally, the Pt-centres form the cluster compounds more easily than the analogous Pd-compounds).

● **Halido-bridged Pt(II)-ammine complexes:** These are illustrated below.

$$
\begin{bmatrix}
H_3N & Cl & NH_3 \\
\quad \diagdown \quad \diagup \quad \diagdown \quad \diagup \quad & & \\
\quad Pt \qquad Pt \quad & & \\
\quad \diagup \quad \diagdown \quad \diagup \quad \diagdown \quad & & \\
H_3N & Cl & NH_3
\end{bmatrix}^{2+}
$$

[PtBr₃(NH₃)₂] remains in a polymeric state in which *trans*-[Pt^{IV}Br₄(NH₃)₂] and [Pt^{II}Br₂(NH₃)₂] units are stacked through the bromido-bridges. In **Wolfram's red salt,** *i.e.* PtCl₃(EtNH₂)₄, 2H₂O, *trans*-[Pt^{IV}Cl₂(EtNH₂)₄] and [Pt^{II}(EtNH₂)₄] units are also stacked through the chlorido bridges. It thus represents [Pt^{II}(EtNH₂)₄]$^{2+}$[*trans*-(μ-Cl)₂-Pt^{IV}(EtNH₂)₄]$^{2+}$(Cl⁻)₄·4H₂O.

(**Note:** In the polymeric compounds like **Magnus' green salt,** *i.e.* [Pt(NH₃)₄]$^{2+}$[PtCl₄]$^{2-}$, **KCP,** *i.e.* K₂[Pt(CN)₄]Br₀.₃, etc. the **direct Pt—Pt interaction** prevails. These are discussed in Sec. 12.8.7, Vol. 3 of Fundamental Concepts of Inorganic Chemistry).

[PtBr$_3$(en)] also remains in a polymeric chain like the corresponding ammine complex.

● **R$_2$AuX (X = Cl⁻, Br⁻):** These stable organometallic compounds (having the Au–C bonds) are dimeric.

● **AuX (X = Cl, I):** In solid state, their polymeric forms produce the zig-zig chain structures. (X = Cl, I)

(For X = I, Au—Î —Au ≈ 73°)

● **NiCl$_2$.2H$_2$O; NiBr$_2$.2H$_2$O; CdCl$_2$.2NH$_3$:** They produce the chain structure through the halido-bridges (—X—)

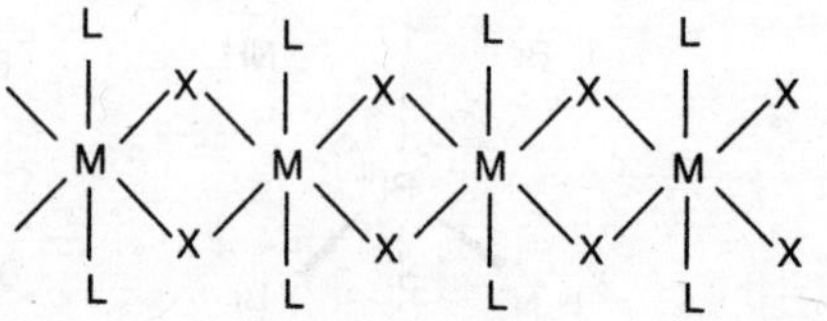

(X = halogen; M = Ni, Cd; L = NH$_3$, OH$_2$)

Thus the octahedral geometry of the metal centre is maintained in the chain structure. [CoCl$_2$(py)$_2$]$_x$ also possesses the similar chain structure.

(f) Cyanido (—C≡N) Bridged Complexes

● [Ni(CN)$_4$]$^{2-}$ on being reduced by Na–Hg forms [Ni^I(CN)$_3$]$^{2-}$ that remains actually as a dimer.

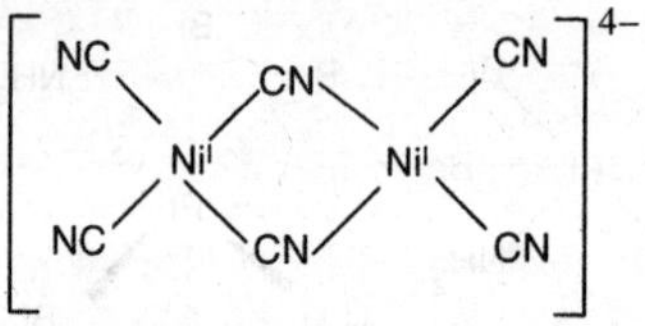

● **AuCN, AgCN:** They produce a linear polymer through the —CN— bridge.

$$— C \equiv N — M — C \equiv N — M — C \equiv N —$$

● **K[Cu(CN)$_2$]:** It is a polymeric compound where the anion produces a polymeric chain by using the bridging properties of CN. In the polymeric chain, Cu(I) is trigonally coordinated (N,C,C) as follows:

$$— CN — Cu(CN) — CN — Cu(CN) — CN — Cu(CN) — CN —$$

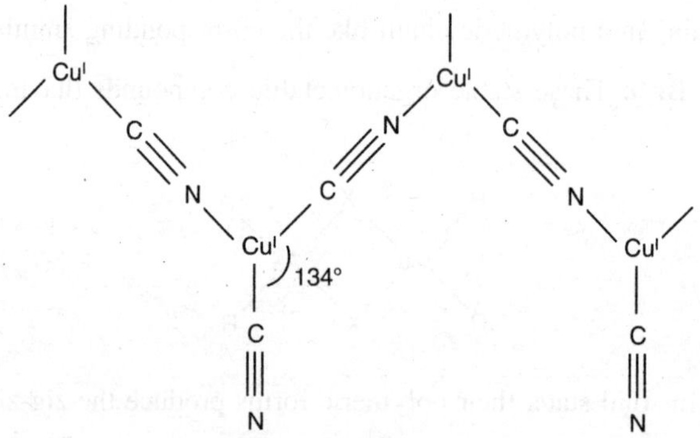

Polymeric chain of Cu(CN)$_2$

● **Pd(CN)$_2$ and Ni(CN)$_2$:** They produce two dimensional planar polymeric networks. The polymeric structure of [Ni(CN)$_2$(NH$_3$)]$_n$ (see Fig. 13.2.2.2, Vol. 3 of Fundamental Concepts of Inorganic Chemistry) is closely related with that of [Ni(CN)$_2$]$_n$. In solid [Ni(CN)$_2$]$_n$, each Ni(II) is 4 coordinated (2C, 2N) but in [Ni(CN)$_2$(NH$_3$)]$_n$ half of the Ni(II) centres are **4 coordinated** (4C, **square planar,** diamagnetic) and the remaining Ni(II) centres are **6 coordinated** (6N, **octahedral,** paramagnetic). In fact, when Ni(CN)$_2$ is crystallised in presence of NH$_3$ and C$_6$H$_6$ (as the guest species), it adopts this special structure **(template effect)** to accommodate the guest. A similar template effect operates to give the special crystal structures of ice to accommodate the guest species (see Sec. 13.2.2, Vol. 3 of Fundamental Concepts of Inorganic Chemistry).

M(2N, 2C)
(Polymeric forms of M(CN)$_2$, M = Ni, Pd)

Ni(4C), Ni(6N)
Polymeric form of Ni(CN)$_2$.NH$_3$ trapping the benzene molecule within the cage. *i.e.,* **Clathrate compound (Hoffman type)** of Ni(CN)$_2$.NH$_3$ and C$_6$H$_6$. (See Sec.13.2.2, Vol. 3 of Fundamental Concepts of Inorganic Chemistry)

(830 pm; Guest)

(**Note:** Different coordinating environments around the Ni(II) centres in the polymeric structure of Ni(CN)$_2$.NH$_3$ can be rationalized. In this polymeric structure there are two types of Ni(II)-centres — one centre is coordinated by 4 C-sites of the CN$^-$ ligands while the other type is coordinated by the 4 N-sites of the bridging CN$^-$ ligands and two NH$_3$ ligands. *The C-sites are softer and less electronegative than the N-sites.* The structures Ni^{II}—C$_4$ and Ni^{II}—N$_6$ are in confirmity with the **symbiotic theory of HSAB principle.**

Template effect in presence of NH$_3$ gives a special structure of [Ni(CN)$_2$]$_x$ where there are *two types of 4 coordinated Ni(II) centres* (4C vs 4N) and this structure accommodates NH$_3$ as the ligands and benzene (or similar molecule) as the guests. According to the *symbiotic theory* of HSAB principle, NH$_3$ (*i.e.* N-donor ligands) coordinates the metal centre which is already coordinated to the N-sites. In other words, the metal centre which receives less electron from the existing ligands in [Ni(CN)$_2$]$_x$ **(template crystal)** will take up the additional ligands (*i.e.* NH$_3$). The metal centre receives more electron from the Ni—C bond than from the Ni—N bond as the C-centre is less electronegative than the N-centre. This is why, the Ni(—N)$_4$ centre of [Ni(CN)$_2$]$_x$ will preferably bind the additional NH$_3$ ligands to produce the polymeric structure of Ni(CN)$_2$.NH$_3$.

● **AuR$_2$CN:** It produces a tetramer where Au(III) is possessing the square planar geometry.

(**Note:** AuR$_2$X (X = Cl, Br) remains in a dimeric state by two bridging X)

● **AgCN:** In solid state, it forms a polymeric linear chain structure.

$$-Ag - C \equiv N - Ag - C \equiv N-$$

● **Iron-cyanido complex:** (i) The iron-cyanido complexes are of much importance in this regard. Prussian blue, Turnbull's blue and other related compounds have the similar structures.

$$K^+ + Fe^{2+} + \left[Fe(CN)_6\right]^{3-} \longrightarrow \textbf{Turnbull's blue}$$

$$K^+ + Fe^{3+} + \left[Fe(CN)_6\right]^{4-} \longrightarrow \textbf{Prussian blue}$$

Both the blues have the similar bridged network structure having the $-Fe^{II} - C \equiv N - Fe^{III}$ segments. Fe^{II}— and Fe^{III}— centres are positioned at the corners of the cubes. The cubes are stacked to produce the cubic lattice where the requisite amount of K^+ ions occupy the centres of alternates cubes to give the composition $KFeFe(CN)_6$. Water molecules may also occupy the cubic holes. Thus, prussian blue and other related materials adopt the **three dimensional polymeric structure.**

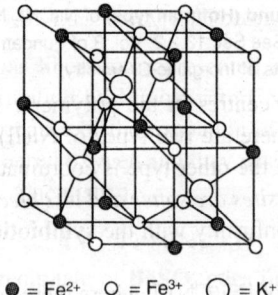

● = Fe^{2+} ○ = Fe^{3+} ◯ = K^+

Structure of polymeric prussian blue $KFe^{II}Fe^{III}(CN)_6$

In Prussian blue, the starting material possesses the Fe^{II}— CN linkage while in Turnbull's blue the starting material bears the Fe^{III}— CN linkage. Thus it is reasonable to assume that in Prussian blue it should have the linkage $-Fe^{II} - C \equiv N - Fe^{III}$ — while Turnbull blue should contain the linkage $-Fe^{III} - C \equiv N - Fe^{II} -$. **But the linkage $-Fe^{II} - C \equiv N - Fe^{III}$ — is stabler than the $-Fe^{III} - C \equiv N - Fe^{II}$ — linkage**. In 'C≡N', the C-end is a strong field ligand which can generate a high *cfse (crystal field stabilization energy)* for the $Fe^{II}(d^6)$ – **low spin complex** (t_{2g}^6). The N-end is a relatively weak field ligand that forms a **high-spin complex** with the Fe^{III} centre ($t_{2g}^3 e_g^2$) which is resistant to some extent to spin pairing due to the high exhange energy. Besides these, the **softer C-end** matches better with the Fe(II)-centre while the relatively **harder N-end** matches better with the relatively harder Fe(III)-centre. Thus, the C-end preferably links with the Fe^{II}-centre. **The linkage $-Fe^{III} - C \equiv N - Fe^{II}$ — initially present in Turnbull blue isomerizes slowly to the $-Fe^{II} - C \equiv N - Fe^{III}$ — linkage which is thermodynamically stabler.**

(ii) $M^ICu^{II}Fe^{III}(CN)_6$, $M^IRu^{II}Fe^{III}(CN)_6$ **(ruthenium purple)** are also structurally similar to Prussian blue and in the Prussian blue structure, Fe^{II} (d^6) is replaced by the bivalent metal ions like Cu^{II} (d^9), Ru^{II} (d^6). M^I- ions are required to balance the charge and they occupy the centres of the cubes.

In $Fe^{III}Fe^{III}(CN)_6$ (**Berlin green**), and $K_2Fe^{II}Fe^{II}(CN)_6$ (**Everitt's salt**), the bridges $-Fe^{III} - C \equiv N - Fe^{III} -$ and $- Fe^{II} - C \equiv N - Fe^{II} -$ exist respectively and the Fe-centres lie at the corners of the cubic array. K^+ ions (required to balance the charge) occupy the centres of the cubes. Fe^{II} or Fe^{III} coordinating with the C-end adopts the low spin state while Fe^{II} or Fe^{III} coordinating with the N-end adopts the high spin state.

(iii) $K^+ + \left[Cr(CN)_6\right]^{3-} + Fe^{2+} \longrightarrow KFe^{II}Cr^{III}(CN)_6$ (**Red compound**)

$$\downarrow \text{standing}$$

$$KCr^{III}Fe^{II}(CN)_6 \text{ (Green compound)}$$

In fact, the red and green compounds are having the same composition $KFe^{II}Cr^{III}(CN)_6$. The red compound contains the linkage $-Fe^{II} - N \equiv C - Cr^{III} -$ while the green compound (*which is thermodynamically stabler*) possesses the linkage $-Fe^{II} - C \equiv N - Cr^{III} -$. The red and green compounds are the **linkage isomers.** The higher thermodynamic stability of the green isomer arises due to the higher *cfse* (crystal field stabilization energy) of the isomer having the low-spin Fe^{II} (t_{2g}^6) which coordinates with the stronger field donor site 'C' of the ambidentate linkage. On the other hand, Fe^{II} remains as a high-spin centre ($t_{2g}^4 e_g^2$) in the red compound where Fe^{II} links with the weaker donor sites 'N' of 'CN'. Cr^{III}- a d^3 system remains as t_{2g}^3 in both the isomers.

(iv) **Colour of the prussian blue and related compounds** (*cf.* Sec. 1.13): These are considered as the compounds of *mixed valence state and the charge transfer occurs from metal centre to metal centre* mediated by the bridging cyanido ligand. The compounds like $Fe^{III}[Fe^{III}(CN)_6]$ (**Berlin green**), $K_2Fe^{II}[Fe^{II}(CN)_6]$ (**Everitt's salt**) contains the metal centres in one oxidation state (*i.e.* these are not mixed valence state complex) and consequently the **metal to metal charge transfer (MMCT) band** cannot arise in such cases. *In fact, these compounds are not intensely coloured.*

(g) 'SCN'—Bridged Complexes

$-S - C \equiv N-$ bridged polynuclear complexes are well known. The polymeric structure of AgSCN produces a zig-zag chain.

(h) Sulfato (—O—SO₂—O—) Bridged Complexes

SO_4^{2-} as a *bridging ligand* is known in many cases.

(i) Chromic sulfate solution depending on the pH and temperature can produce different species having SO_4^{2-} as a didentate bridging ligand, and as a monodentate ligand.

$$\left[Cr(OH_2)_6\right]^{3+} \rightleftharpoons \left[Cr(OH)(OH_2)_5\right]^{2+} + H^+$$
(**Violet**)

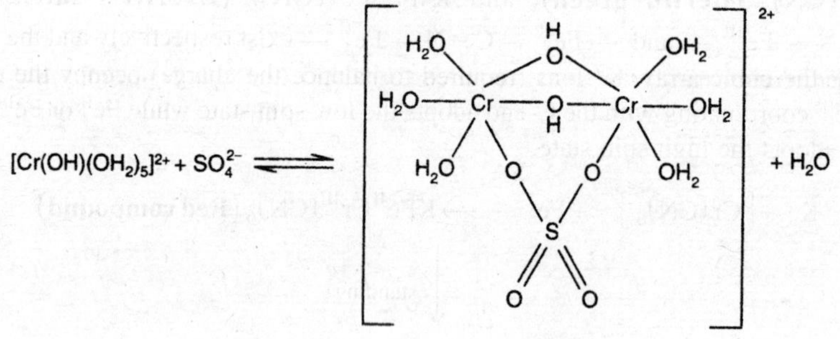

$[Cr(OH)(OH_2)_5]^{2+} + SO_4^{2-} \rightleftharpoons$

(Green)

$[Cr(OH)(OH_2)_5]^{2+} + 3SO_4^{2-} \rightleftharpoons$

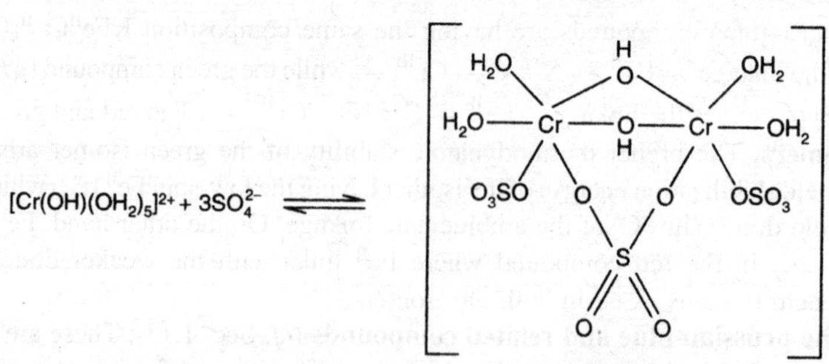

(Green)

Note: ● $Cr_2(SO_4)_3.18H_2O$ (**A**) (**Violet**) and $Cr_2(SO_4)_3). 6H_2O$ (**Green**) (**B**).

B (dissolved in water, green solution) $\xrightarrow{\text{BaCl}_2}$ No precipitate of $BaSO_4$

It indicates that in **B**, SO_4^{2-} ions are present in the coordination sphere as the ligands. It may be noted that $Cr(III)$ (t_{2g}^3) is **kinetically inert**.

B (Solution) $\xrightarrow[\text{standing}]{\text{On prolonged}}$ Violet solution $\xrightarrow{\text{BaCl}_2}$ All the sulfates
Green　　　　　　　　　　　　　　　　　　　　　　　　　　　　　are precipitated
or,

A ⟶↑

The violet colour is due to $[Cr(OH_2)_6]^{3+}$. Thus **A** is reasonably formulated as $[Cr(OH_2)_6]_2(SO_4)_3.6H_2O$.

● In the spectrochemical series, H_2O is given a higher position than that of SO_4^{2-}. Thus, aquation of the sulfato complex (*i.e.* **B**) changes the colour from green to violet.

● **Violet chrome alum**, $K_2SO_4.Cr_2(SO_4)_3.24H_2O$ is made by **A**, K_2SO_4 and water of crystallization.

$Cr_2O_7^{2-} + Fe^{II}$ (or HSO_3^-) + H_2SO_4 ⟶ colour of the final solution ?

$Cr(VI)$ (d^0-system) in $Cr_2O_7^{2-}$ or CrO_4^{2-} is finally reduced to $Cr(III)$ (d^3-system). If $[Cr(OH_2)_6]^{3+}(t_{2g}^3)$ is produced, the colour should be violet. During the reduction of $Cr(VI)$ to $Cr(III)$, the step-wise reduction produces the *labile intermediates* like $Cr(IV)$, $Cr(V)$ that may take up SO_4^{2-} as the ligands. These sulfato intermediates are finally reduced to the sulfato-$Cr(III)$ species that looks green. It may be mentioned that $Cr(III)$ (t_{2g}^3) being **kinetically inert** cannot take up the sulfate ligand after its production. The ligand enters at the labile intermediates produced during the reduction of $Cr(VI)$ to $Cr(III)$.

(ii) In the crystal structure of $MnSO_4.4H_2O$, $CuSO_4.5H_2O$, (*see* Fig. 13.1.6.3, Vol. 3 of Fundamental Concepts of Inorganic Chemistry), etc. SO_4^{2-} exists as a bridging ligand.

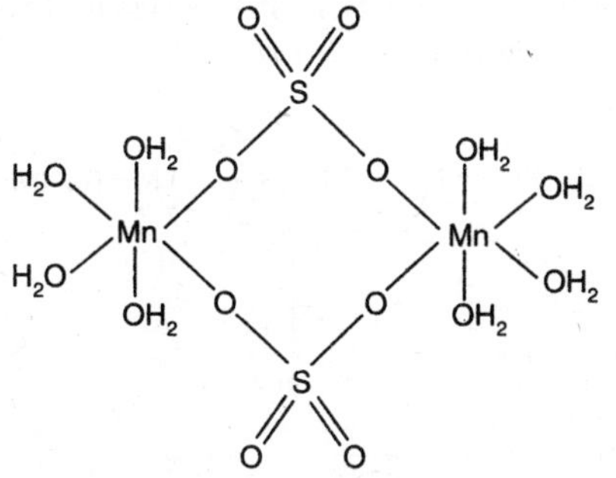

MnSO_4.4H_2O

(i) Halido and Carboxylato Bridged Metal Clusters

These are discussed in Sec. 12.8, Vol. 3 of Fundamental Concepts of Inorganic Chemistry. Carboxylato bridged important biomolecules are: hemerythrin, leucine aminopeptidase, methane monoxygenase, etc. (*see* Bioinorganic Chemistry by A.K. Das).

(j) Bridging Carbonyls and Polynuclear Carbonyls

These are discussed in Sec. 3.3.4 and 9.4; and Sec. 12.8.5, Vol. 3 of Fundamental Concepts of Inorganic Chemistry.

(k) Acetylacetonato Bridged Polynuclear Complexes

The polymeric structures of $[Ni(acac)_2]$ and $[Co(acac)_2]$ have been discussed in Secs. 2.11.6 and 8.26.2.

1.13 MIXED VALENCE COMPLEXES: CONJUGATED BRIDGES AND INTERVALENCE CHARGE TRANSFER

In the polynuclear complexes, oxidation states of the metal centres may be different. The common examples are: Pb_3O_4, Fe_3O_4, Co_3O_4, $PtBr_3(NH_3)_2$ (a combination of $[Pt^{II}Br_2(NH_3)_2]$ and $[Pt^{IV}Br_4(NH_3)_2]$ through the —Br— bridges), prussian blue (a cyanido bridged complex). Here we shall discuss the mixed valence state complexes bridged by the ligands that can allow the electron transfer by their vacant π-orbitals. Such ligands are:

$C \equiv N^-$, $N = N = N^-$, 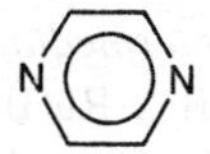 (pyrazine *i.e.* pyz)

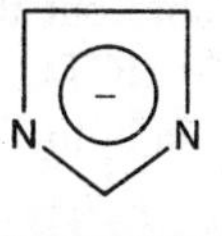

(imidazolate)

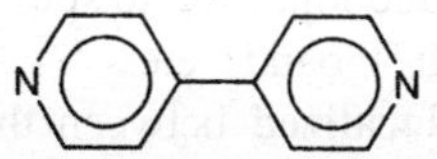

(4, 4'-bipyridine)

● $[(H_3N)_5Ru—(\mu\text{-pyz})—Ru(NH_3)_5]^{5+}$ (**called Creutz-Taube ion**): The complex shows a very low energy absorption band (1050 nm) in the near infra-red (IR) region. This is also an intense band. This is believed to be due to the electron transfer (photon promoted) from Ru(II) to Ru(III) via the bridging ligand. This is called the **metal-metal or intervalence charge transfer band (MMCT or ICT band)**.

This photon-promoted electron transfer leads to an exchange of positions between the Ru(II) and Ru(III) centres. Here the Ru-centres are ordinarily **indistinguishable** and both the Ru—N distances are the same. If Ru(II) and Ru(III) were fixed at their respective positions, the different Ru^{II}—N and Ru^{III}—N distances would have made the ion *unsymmetrical*.

$$\left[(H_3N)_5Ru^{II}-N\bigcirc\!\!-\!\!\bigcirc N-Ru^{III}(NH_3)_5\right]^{5+}$$

$$\downarrow h\nu$$

$$\left[(H_3N)_5Ru^{III}-N\bigcirc\!\!-\!\!\bigcirc N-Ru^{II}(NH_3)_5\right]^{5+}$$

● In **Creutz-Taube ion**, *i.e.* $[(H_3N)_5Ru(\mu\text{-pyz})Ru(NH_3)_5]^{5+}$ there is an extensive delocalization of the electron over the both metal centres through the bridging ligand, but by modifying the ligand environment the electron can be **localized** (*i.e.* different oxidation states can be trapped - described as the **trapped valence**). Such examples are:

$$\left[(bpy)_2ClRu-N\bigcirc N-Ru(bpy)_2Cl\right]^{3+}$$

$$\left[(H_3N)_5Ru^{III}-N\bigcirc N-Ru^{II}(bpy)_2Cl\right]^{4+}$$

● **Prussian blue:** The intense colour is due to electron transfer from Fe(II) to Fe(III) via the CN-bridge (*i.e.* **MMCT bond**).

A. Robin-Day Classification of Mixed Valence Complexes (*cf. Adv. Inorg. Chem. Radiochem.,* **10**, 247-422, 1968): These are of three types. These may be illustrated by taking the example [Ru—X—Ru]$^{5+}$, (X is neutral).

Class-I: In [Ru—X—Ru]$^{5+}$, Ru(III) and Ru(II) centres are *fully localized* and it happens so if they are bridged by a saturated ligand or they are not connected.
Example: Pb_3O_4 ($\equiv 2PbO + PbO_2$) is a mixture of Pb(II) and Pb(IV).

Class-II: It describes an *intermediate character* between the fully localized and delocalized states of electrons. In [Ru—X—Ru]$^{5+}$, Ru(II) and Ru(III) centres are identifiable but there is an electron interaction between the metal centres *like the intervalence charge transfer* (ICT).
Examples: Prussian blue, Creutz-Taube ion, inverse spinel Fe_3O_4 (Fe^{III} in both T_d and O_h holes; Fe^{II} in O_h holes; black colour indicates the ICT band), etc.

Class III: The electron is **fully delocalized** between the metal centres. In [Ru—X—Ru]$^{5+}$, the oxidation state of each Ru-centre is +2.5 and it is not possible to assign the separate integral oxidation number for the Ru-centres. In such cases, *no intervalence charge transfer absorption is possible*.
Example: $[Tc_2Cl_8]^{3-}$ (average oxidation state of Tc = +2.5; direct metal-metal bonding; no bridging ligand); $[Ta_6Cl_{12}]^{3+}$ cluster (with average oxidation state +2.5 per Ta; bridging ligand is there) (*see* Sec. 12.8, Vol. 3 of Fundamental Concepts of Inorganic Chemistry for details regarding the bonding).

Table 1.13.1 Comparison of the characteristics of mixed-valence compounds (Robin-Day classification).

Property	Class-I	Class-II	Class-III
(i) Ligand Environment	Different environments for the metal centres	More or less similar environments for the metal centres	Identical environment for the metal centres.
(ii) Delocalization of the extra electron	Fully localized on one type of metal centre	Not fully localized (intermediate behaviour)	Fully delocalized over the metal centres.
(iii) d-d spectra	Distinct spectra for the metal centres	Modified spectra	New electronic spectra
(iv) Intervalence charge transfer (ICT) band	At very high energy	Invisible or in the near IR-range	Not observed
(v) Magnetic properties	Characteristics of different oxidation states	Same as in Class-I; but at low temperature it may be modified	New magnetic properties

---

Inner-Sphere Mechanism of Electron Transfer Process (*see* Chapter 6)

In this process of electron transfer, a binuclear complex is formed by using a bridging ligand and then electron transfer occurs from the reducing metal centre to the oxidizing metal centre. Thus, in this process, the bridging ligand connecting the metal centres plays a crucial role.

1.14 COORDINATION POLYMERS

(a) Addition polymers: Coordination polymers are developed through the coordination by the metal ions. Many simple inorganic compounds like AlF_3, FeF_3, $FeCl_3$, BeF_2, $BeCl_2$, $Ni(CN)_2$, $Ni(CN)_2.NH_3$, $Pd(CN)_2$, $PdCl_2$, $PtCl_2$, $AgCN$, $AgSCN$, SnF_4, prussian blue and related compounds remain in polymeric forms in solid state These have been discussed in detail in Sec 1.12. Their polymeric structures are developed through the metal coordination by the suitable bridging ligands. These have been discussed in Sec 1.12. These may be considered as **addition polymers**.

Metal rubeanates (i.e. **metal dithiooxamides***)* are the examples of addition polymer.

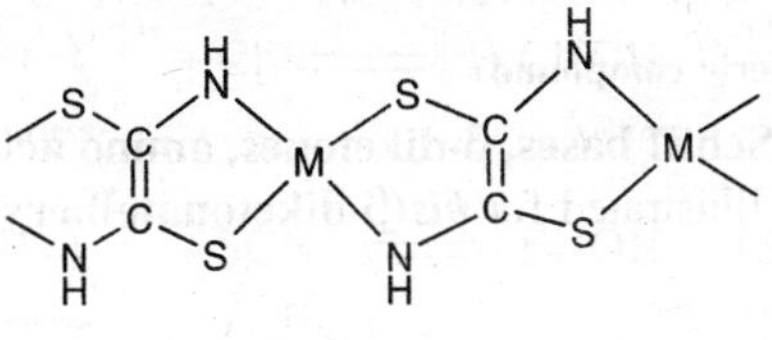

(M = Cu^{II}, Ni^{II}, etc.)

Polymeric form of *bis*(dithiooxamido)metal(II)

2, 5- dihydroxy-*p*-benzoquinone can act as a chelating ligand to give the linear polymeric products with the metal ions like Cu^{2+}, Ni^{2+},

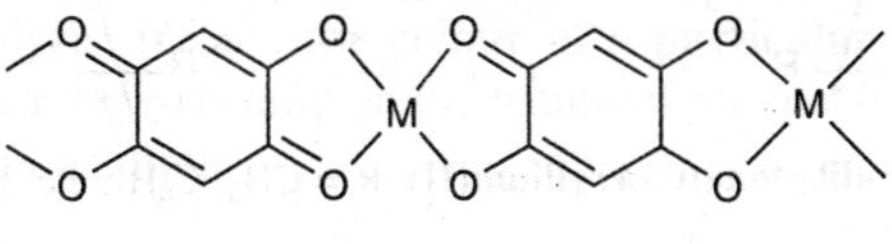

$[Ni(NCS)_2(NH_3)_3]_n$ is a polymeric compound.

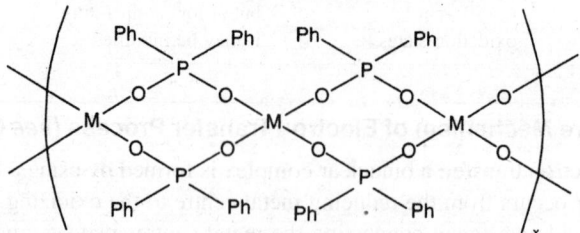

$$[Ni(NCS)_2(NH_3)_3]_n$$

(b) Chain polymers of diphenylphosphinate complexes of Be, Cr, Zn, etc.

Chain polymers of diphenylphosphinate, M = Be, Zn

(c) **Condensation polymerization** of metal chelates leads to condensation between the coordinated ligands present in different coordination spheres. In this polymerization process, coordination spheres of the metal centres remain unchanged. Some representative examples of condensation polymerization involving the metal chelates are given below.

$$HO - Pt^{IV}Cl_2(NH_3)_2 - OH + HO - Pt^{IV}Cl_2(NH_3)_2 - OH$$

$$\downarrow -H_2O$$

$$-O - Pt^{IV}Cl_2(NH_3)_2 - O - Pt^{IV}Cl_2(NH_3)_2 - O -$$

(Polymeric compound)

Bis-chelates involving the ligands like **Schiff bases, β-diketones, amino acids**, etc. may experience the condensation polymerization. This is illustrated for *bis*(β-diketonate)beryllium(II).

(Chain polymer of *bis*(β-diketonate)beryllium(II), R = CH$_3$, C$_6$H$_5$, etc. R' = (—CH$_2$—)$_n$, C$_6$H$_4$)

(d) Sheet polymers of metal-phthalocyanine units: In the usual procedure of phthalocyanine synthesis, if phthalic anhydride is replaced by pyromellitic anhydride, the said sheet polymers may be synthesized.

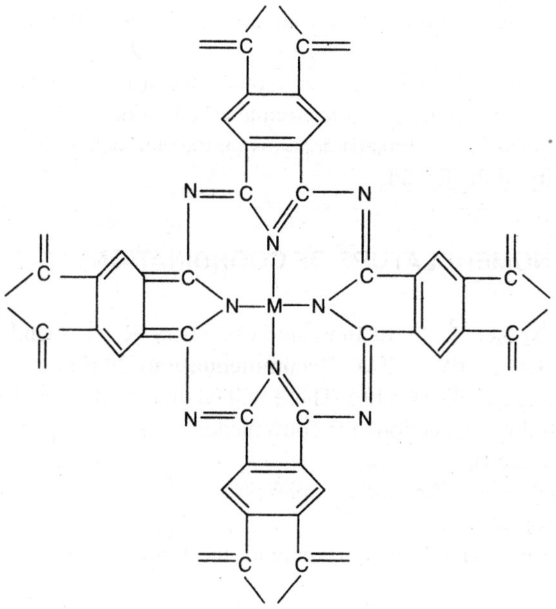

Sheet polymer having the metal – phthalocyanine units 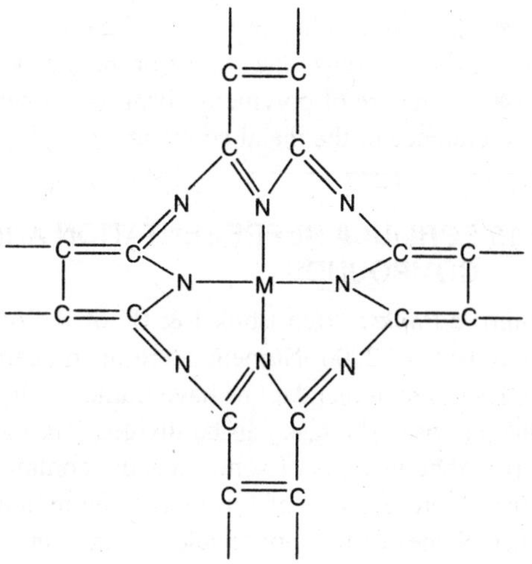   **Sheet polymer coating the metal surface when treated with $C_2(CN)_4$ (M = Fe, Cu, Ni)**

Mn(II)-pthalocyanine undergoes polymerization through the oxido bridges in the presence of oxygen in pyridine solution. Sheet polymers similar to metal-phthalocyanine are considered to be formed on the surface of iron, copper or nickel when $C_2(CN)_4$ vapour is allowed to react with the metal surface.

(d) Polymeric Schiff's bases can be prepared through the condensation of pyridine-2, 6-dialdehyde and diamines (*e.g.* ethylenediamine, hexamethylenediamine, etc.).

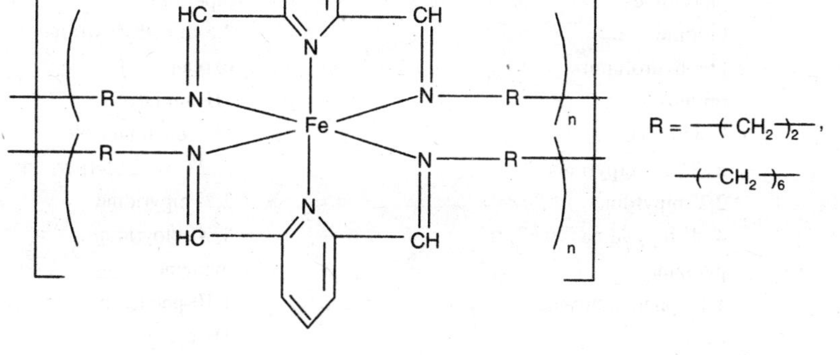

$R = -(CH_2)_2-$,

$-(CH_2)_6-$

(Metal derivatives of polymeric Schiff's bases—thermally stable and ferromagnetic)

General features of the coordination polymers

● If the metal-ligand bond is significantly ionic in character, then because of the electrostatic attraction within the coordination polymer, the flexibility and plasticity of the polymer is significantly reduced. Thus, for a better plasticity, there should be a little ionic character in the metal-ligand bond.

● Thermal stability of the coordination polymers is increased, if the anionic sites of the ligands can just neutralize the charge of the metal centre (*i.e.* inner metallic complexes of 1st order).

● The nature of polymeric forms (*i.e.* linear, sheet, cross-linked) depends on the stereochemical preference of the metal centre and functionality of the ligand.

1.15 FORMULA REPRESENTATION AND NOMENCLATURE OF COORDINATION COMPOUNDS

(*Source Books:* "Red Book I & II" of IUPAC, Nomenclature of Inorganic Chemistry, Recommendations 1990 & 2000; Nomenclature of Inorganic Chemistry, **IUPAC Recommendations 2005**).

These recommendations have framed some rules for this purpose. These will be discussed below for this purpose. The task can be divided into the following sections for convenience.

 (i) Abbreviations of some ligands (commonly used);
 (ii) Representation of formula of the mononuclear coordination complexes;
 (iii) Names of the mononuclear coordination complexes;
 (iv) Formulae and names of the dinuclear and polynuclear complexes with the bridging ligands or metal-metal bondings.

1.15.1 Abbreviations of Some Organic Ligands Commonly Used (*cf.* Table 1.7.2.2, 3)

Abbreviation	Ligand name	Systematic name
	(a) **Hydrocarbons**	
cod	cyclooctadiene	1,5-cyclooctadiene or cycloocta-1,5-diene
cot	cyclooctatetraene	1,3,5,7-cyclooctatetraene or, cycloocta-1,3,5,7-tetraene
Cp	cyclopentadienyl	cyclopentadienyl
	(b) **Heterocycles**	
py	pyridine	pyridine
pip	piperidine	piperidine
lut	lutidine	2,6-dimethylpyridine
thf	tetrahydrofuran	oxolane
Him	imidazole	1H-imidazole
Hbzim	benzimidazole	1H-benzimidazole
terpy	2, 2', 2''-terpyridine	2, 2' : 6', 2''-terpyridine
bpy	2,2'-bipyrdine	2,2'-bipyridine
4, 4'-bpy	4, 4'-bipyridine	4, 4'-bipyridine
pyz	pyrazine	pyrazine
phen	1,10-phenonthroline	1,10-phenonthroline
Hpz	pyrazole	1H-pyrazole
picoline	α-picoline	2-methylpyridine

nia	nicotinamide	3-pyridinecarboxamide or pyridine-3-carboxamide
isn	isonicotinamide	4-pyridinecarboxamide
Hpic	picolinic acid	2-pyridinecarboxylic acid
Hnic	nicotinic acid	3-pyridinecarboxylic acid

(c) Amino acids and amino alchohols

Hgly	glycine	aminoacetic acid
α-Hala	α-alanine	2-aminopropionic acid
β-Hala	β-alanine	–
Hea	ethanolamine	2-aminoethanol
H_2dea	diethanolamine	2,2′-iminodiethanol
H_3tea	triethanolamine	2,2′,2″-nitrilotriethanol

(d) Diketones

Hacac	acetylacetone	2,4-pentanedione
Hba	benzolacetone	1-phenyl-1,3-butanedione
Hfta	trifluoroacetylacetone	1,1,1-trifluoro-2,4-pentanedione
Hhfa	hexafluoroacetylacetone	1,1,1,5,5,5-hexafluoro-2,4-pentanedione

(e) Aminopolycarboxylic acids

H_2ida	iminodiacetic acid	–
H_3nta	nitrilotriacetic acid	2, 2′, 2″-nitrilotriacetic acid
H_4edta	ethylenediaminetetraacetic acid	(1,2-ethanediyldinitrilo)tetraacetic acid or (ethane-1,2-diyldinitrilo)-tetraacetic acid
H_4cdta	*trans*-1,2-cyclohexanediaminetetraacetic acid	*trans*-(1,2-cyclohexanediyldinitrilo)-tetraacetic acid
H_5dtpa	N,N,N′,N″,N″-diethylenetriaminepentaacetic acid	–

(f) Schiff's bases

H_2salen	*bis*(salicylidene)ethylenediamine or N,N′-ethylene*bis*(salicylideneimine)	–
H_2acacen	*bis*(acetylacetone)ethylenediamine or N,N′-ethylene*bis*(acetylacetoneimine)	–
H_2saldien	*bis*(salicylidene)diethylenetriamine	–
H_2salgly	salicylideneglycine	–

(g) Other ligands

en	ethylenediamine	1,2-ethanediamine or ethane-1,2-diamine
dien	diethylenetriamine	N-(2-aminoethyl)-1,2-ethanediamine
pn	propylenediamine	1,2-propanediamine
tn	trimethylenediamine	1,3-propanediamine
tmen	N,N′,N′,N′-tetramethylethylenediamine	N,N,N′,N′-tetramethyl-1,2-ethanediamine
trien	triethylenetetramine	N,N′-*bis*(2-aminoethyl)-1,2-ethanediamine
tren	*tris*(2-aminoethyl)amine	N,N-*bis*(2-aminoethyl)-1,2-ethanediamine

chxn	1,2-diaminocyclohexane	–
Hthsc	thiosemicarbazide	–
dabco	triethylenediamine	1,4-diazabicyclo[2.2.2]octane
depe	1,2-bis(diethylphosphino)ethane	1,2-ethanediylbis(diethylphosphane)
dppe	1,2-bis(diphenylphosphino)ethane	1,2-ethanediylbis(diphenylphosphane)
diars	o-phenylenebis(dimethylarsine)	1,2-phenylenebis(dimethylarsone) or benzene-1,2-diylbis(dimethylarsane)
dmso	dimethyl sulfoxide	–
tu	thiourea	–
ur	urea	–
dmf	dimethylformamide	N,N-dimethylformamide
Hbig	biguanide	–
HEt$_2$dtc	diethyldithiocarbamic acid	–
tcne	tetracyanoethylene	–
H$_2$dmg	dimethylglyoxime	2,3-butanedionedioxime
H$_2$mnt	maleonitriledithiol	2,3-dimercapto-2-butenedinitrile

Names of Coordination Compounds based on the names of discoverers

cis-[PtCl$_4$(NH$_3$)$_2$]:	Cleve's salt
[PtCl(NH$_3$)$_3$]Cl:	Cleve's triammine
$trans$- K[Co(NH$_3$)$_2$(NO$_2$)$_4$]:	Erdmann's salt
K$_3$[Co(NO$_2$)$_6$]:	Fischer's salt
[Co(NH$_3$)$_3$(NO$_2$)$_3$]:	Gibb's salt
[Pt(NH$_3$)$_4$][PtCl$_4$]:	Magnus' green salt (First Pt^{II}-ammine salt discovered)
[PtCl(NH$_3$)$_3$]$_2$ [PtCl$_4$]:	Magnus' pink salt
(NH$_4$)[Cr(NCS)$_4$(NH$_3$)$_2$]: ($trans$-isomer)	Reinecke's salt
K[Pt(η^2-C$_2$H$_4$)Cl$_3$]:	Zeise's salt
Pt(EtNH$_2$)$_4$Cl$_3$ · 2H$_2$O:	Wolfram's red salt
[(H$_3$N)$_5$ Ru(μ-pyz)Ru(NH$_3$)$_5$]$^{5+}$:	Creutz-Taube ion
[PtCl$_2$(NH$_3$)$_2$]:	{ Peyrone's chloride (cis-isomer) Reiset's chloride ($trans$-isomer)

See text for Vortmann's sulfate, Vaska' compound, Lifschitz salt.

Colour code names of cobalt ammine salts (*cf.* Table 1.4.3.1) (Fremy's proposal)

$trans$-[Co(NH$_3$)$_4$(NO$_2$)$_2$]$^+$ (yellow):	croceo-salt
cis-[Co(NH$_3$)$_4$(NO$_2$)$_2$]$^+$ (brown):	flavo-salt
[Co(NH$_3$)$_6$]$^{3+}$ (yellow):	luteo-salt
$trans$-[CoCl$_2$(NH$_3$)$_4$]$^+$ (green):	praseo-salt
[CoCl(NH$_3$)$_5$]$^{2+}$ (purplish red):	purpureo-salt
[Co(NH$_3$)$_5$(OH$_2$)]$^{3+}$ (rose-red):	roseo-salt
cis-[CoCl$_2$(NH$_3$)$_4$]$^+$ (violet):	violeo-salt

1.15.2 Formula Representation of Coordination Compounds (IUPAC 2005)

Here it should be mentioned that the IUPAC-2005 recommendations differ from the earlier recommendations in many aspects. Presently, most of the books available in the market have followed the earlier recommendations. However, here we shall follow the IUPAC 2005 recommendations.

(A) Formula representation of the mononuclear complexes

(a) The **order of symbols** within the coordination formula is:

- *"Central metal atom symbol followed by the ligand symbols in **an alphabetical order** (irrespective of their charge bearing properties)"*. (IUPAC 2005)
- "No space left in writing the components of a complex species".

(*cf.* **Earlier Recommendation:** "Central metal atom symbol followed by the anionic ligands then by the neutral ligands; within each category of ligands, the ligands are listed in an alphabetical order".)

Alphabetical order of the ligand symbols: (i) The ligands are listed in an alphabetical order according to the first atomic symbols of their **line formulae** or **abbreviations.** (ii) If required, formula of different ligands are ordered *alphanumerically*, *i.e.* according to the order of the atomic symbols and right subscripts to these. (iii) A *single letter symbol* precedes a *two letter symbol* with the same initial letter, *e.g.* CO precedes Cl. (iv) Sequence of the ligands does not depend on the charge of the ligands.

The above rules are illustrated in the following examples.

- Order of the inorganic like Cl^-, H_2O, NH_3, NO_3^-, SO_4^{2-}, OH^- is determined by the alphabetical order of the letters C, H, N, N, S, O respectively.
- For CH_3CN, MeCN and NCMe (line formulae of the of the same substance), the order is determined by the alphabetical order of the letters C, M and N respectively.
- For H_2O and OH_2, they are ordered under the letters H and O respectively.
- The following *N*-containing ligands are ordered as:

  N^{3-} (nitride), NH_2^-, NH_3, NO_2^-, NO_3^-, $N_2O_2^{2-}$ (hyponitrite), N_3^- (azide)
- For O^{2-}, OH^-, OH_2 and NH_3 the order is: NH_3, O^{2-}, OH^-, OH_2

Note: For formula representation, OH_2 (not H_2O) is used to convey the mode of ligation. If H_2O and NH_3 are considered, then H_2O precedes NH_3; but if OH_2 and NH_3 are considered, then NH_3 precedes OH_2. Thus we can write: $[Cr(H_2O)_4(NH_3)_2]^{3+}$ as per the earlier recommendation, but presently it is recommended as: $[Cr(NH_3)_2(OH_2)_4]^{3+}$.

- SCN^- and NCS^- are ordered under S and N respectively.

  Note: It cannot be written as CNS^- or SNC^-.

(b) To convey more structural information by formulae, the ligand formula should be written in such a way that **the donor atom symbol is written closest to the central atom symbol.** This recommendation is practised when it is possible as in the case of coordinated water, *i.e.* $[Ni(OH_2)_6]^{2+}$ but not $[Ni(H_2O)_6]^{2+}$

(c) Formula of a coordination entity (whether charged or not) is **enclosed in square brackets.** Formulae of the polyatomic ligands and ligand abbreviations are **enclosed in parentheses.**

(d) For the ionic species, cation is followed by the anion and no space is left between the representations of the cation and anion.

(e) For the charged complex species (shown without the counterion), the **charge** is indicated as a **right hand superscript** outside the enclosing square bracket [ ]. The **oxidation number** of the metal may be indicated by a roman numeral placed as a right hand superscript on the elemental symbol.

(f) Isomers (*e.g. cis-trans; fac-mer, etc.*) and stereochemical descriptions (Sec. 1.15.5) may be indicated.

Illustrative Examples (IUPAC 2005)

$[CoCl_3(NH_3)_3]$

$[Fe(OH)(OH_2)_5]^{2+}$
$[Co(NH_3)_5(N_3)]SO_4$
(*cf.* $[Co(N_3)(NH_3)_5]SO_4$
as per the earlier IUPAC
recommendations)

$[Pt(\eta^2-C_2H_4)Cl_3]^-$
(*cf.* $[PtCl_3(\eta^2-C_2H_4)]^-$
as per the earlier IUPAC
recommendations)
$[Pt(\eta^2-C_2H_4)Cl_2(NH_3)]$
$[Co(NH_3)_6][Cr(CN)_6]$

$[Co^{III}(NH_3)_5(ONO)](ClO_4)_2$
fac–$[Co(NH_3)_3(NO_2)_3]$
$[Hg(CHCl_2)Ph]$
trans–$[PtCl_2(NH_3)_2]$

$[Co(NH_3)_6]Cl(SO_4)$
$Fe_4[Fe(CN)_6]_3$
$[Co^{III}(CN)_5H]^{3-}$
$K_2[Os^{VI}Cl_5N]$
$[PtCl(NH_2CH_3)(NH_3)_2]Cl$

$[CoCl(NH_3)_4(NO_2)]Cl$
$[Ru(NH_3)_5(N_2)]^{2+}$
$[Ir(CO)Cl(PPh_3)_2]$
$[RhI_2(Me)(PPh_3)_2]$
$[Zr(\eta^5-C_5H_5)_2Me_2]$

$[Fe(CCPh_2)(CO)_4]$
$[U(acac)_2O_2]$ (*cf.* not $[UO_2(acac)_2]$)
$[Cr^{III}Cl_2(NH_3)_2(OH_2)_2]^+$
$[PtBrCl(NH_3)(NO_2)]^-$
(*cf.* $[PtBrCl(NO_2)(NH_3)]^-$
as per the earlier IUPAC
recommendations)
$[V(acac)_2O(py)]$ (*cf.* not $[VO(acac)_2(py)]$)
$[PtCl_2\{P(OEt)_3\}]$
$[Pt^{II}Cl_2(NH_3)(py)]$
$[Cr(en)F_2(NH_3)_2]^+$
By using the formula of en, it
is written as:
$[CoF_2(NH_2CH_2CH_2NH_2)(NH_3)_2]^+$
$[CoBr_2(en)(NH_3)_2]^+$
$K[Os^{VIII}(N)O_3]$
$[Fe^0(CO)_5]$
$[Mn^{-I}(CO)_5]^-$
cis–$[PtCl_4(NH_3)_2]$
$[Ru(CO)ClH(PMe_2Ph)_3]$
$[CuCl_2\{O = C(NH_2)_2\}_2]$
$[Fe(CNCH_3)_6]SO_4$
$[Cr^{III}(NCS)_4(NH_3)_2]^-$
$[NiBr_2(Me_2PCH_2CH_2PMe_2)]$
$[FeH(H_2)(Ph_2PCH_2CH_2PPh_2)_2]^+$
$[Co(CO)_4H]$ (not $[HCo(CO)_4]$)
$[Mn(CO)_5H]$
$[Fe(\eta^5-C_5H_5)_2]$
$[W(\eta^5-C_5H_5)_2H_2]$
$[Ti(\eta^5-C_5H_5)_2Cl_2]$
$[Ti(CO)_2(\eta^5-C_5H_5)_2]$
$[Ti(\eta^4-C_8H_8)(\eta^8-C_8H_8)]$

Significance of the η^2–, η^4–, η^5–symbols will be discussed later.

(B) Formula representation of the di- and polynuclear compexes (*cf.* Sec. 1.15.4)

(a) **Order of the metal centres:** For the polynuclear complexes (*i.e.* more than one metal centre), order of the metal atoms is decided by their **relative electronegativities** as follows (IUPAC 2005):

"*the more electronegative centre comes later, i.e. the least electronegative (i.e. the most electroposi-tive) element is listed first.*"

(**Note:** According to the earlier IUPAC recommendation, the central metal atoms are listed in an alphabetical order.)

Working rule based on the electronegativity for ordering of the central atoms: In the periodic table (having 18 groups), if we make a journey from the element F (the most electronegative one) as shown by an arrow in Fig. 1.15.2.1, the element which appears later is less electronegative (**by convention**). Thus it is concluded as follows:

Table 15.2.1.1 Sequence of elements as the central atoms. Arrow starts from F and the element reached later is listed first.

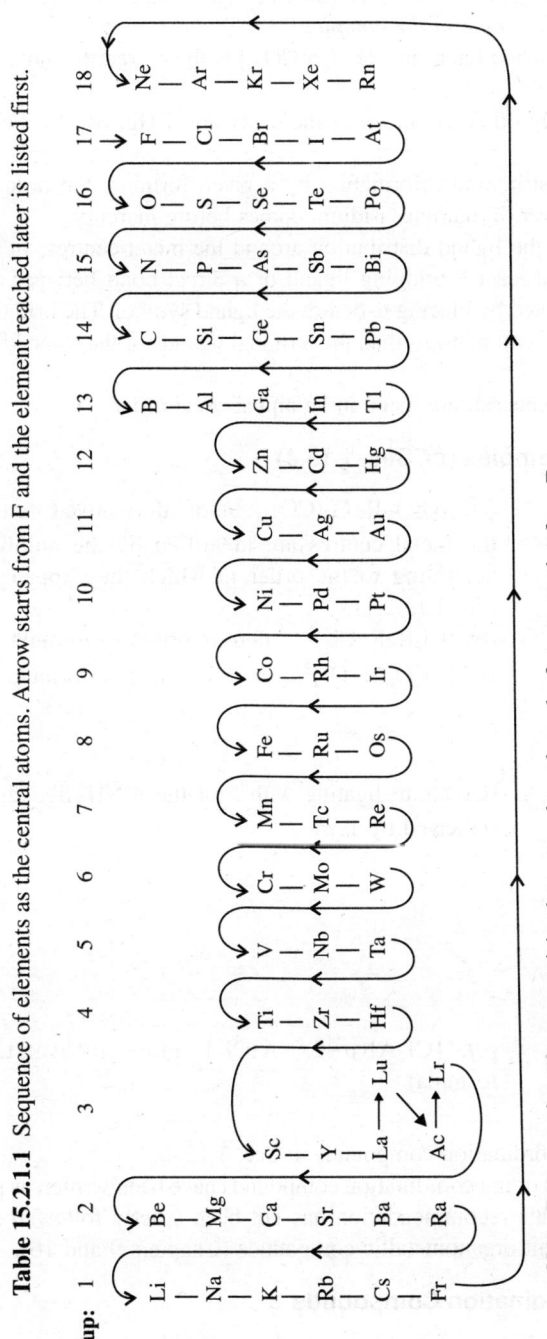

Note: By convention, electronegativity decreases along the path of arrow starting from F.

"the element that appears later in the path of the arrow (illustrated in Fig. 1.15.2.1) in traversing from F, is listed first both in the formula and name of the complex."

Illustration: Between Co and Re, Re is reached later, thus $[ReCo(CO)_9]$ is the correct formula but $[CoRe(CO)_9]$ gives the incorrect formula.

For Co and Cr, the order is CrCo. Similarly other examples of the order are: IrHg; MoRu; FeNi; CrFeCu; etc.

Note: ● Sometimes, to convey the more structural information by a given formula, the order is relaxaed, *e.g.* $[ClHgIr(CO)Cl_2(PPh_3)_2]$. However, in naming, iridium comes before mercury.

● Additional rules are required to specify the ligand distribution around the metal centres. In the polynuclear complexes, there must be at least a bridging ligand or a direct bond between the central atoms. **Bridging ligands** are denoted by placing μ-before the ligand symbol. The bridging ligands are placed further away from the central atoms than the terminal ligands of the same kind, *e.g.* $[Cr_2O_6(\mu–O)]^{2-}$.

(b) **Order of the ligands:** The ligands, in general, are listed in an alphabetical order.

Illustrative Examples (*cf.* Sec. 1.15.4)

$[ReCo(CO)_9]$

(*cf.* $[(OC)_5 \overset{1}{Re} \overset{2}{Co}(CO)_4]$: more informative formula; the metal centres are identified by the numbers according to the order in which they appear in Fig. 1.15.2.1)

$[Re_2Cl_8]^{2-}$

(*cf.* $[Cl_4ReReCl_4]^{2-}$: more informative formula.)

$[Cr_2O_7]^{2-}$

(*cf.* $[Cr_2O_6(\mu–O)]^{2-}$: more informative formula.)

$[\{Cr(NH_3)_5\}_2(\mu–OH)]^{5+}$

$[\{Cu(py)\}_2(\mu–O_2CMe)_4]$

$[(H_3N)_5 \overset{1}{Cr} (\mu–OH) \overset{2}{Cr} (NH_2CH_3)(NH_3)_4]^{5+}$

(Cr-centre ligating with 5 of the 9 NH_3 ligands is indicated by 1)

$[Co\{(\mu–OH)_2 Co(NH_3)_4\}_3]^{6+}$

$[(H_3N)_3 \overset{1}{Co} (\mu–NO_2)(\mu–OH)_2 \overset{2}{Co} (NH_3)_2(py)]^{3+}$

$[\{Pt(\eta^2–C_2H_4)Cl(\mu–Cl)\}_2]$

$[\{Co(NH_3)_3\}_2(\mu–NO_2)(\mu–OH)_2]^{3+}$

$[(H_3N)_5Ru(\mu–pyz)Ru(NH_3)_5]^{5+}$

$[Al_2Cl_4(\mu–Cl)_2]$

(*cf.* $[Cl_2Al(\mu–Cl)_2AlCl_2]$: more informative formula)

$[Fe_2(CO)_6(\mu–CO)_3]$

More examples are given in naming the coordination compounds in Sec. 1.15.4.

Note: Though in most of the cases, formulas of the coordination compounds have been written as per the IUPAC recommendation (2005), always this recommendation has not been strictly followed for convenience in writing the formulas of different organometallic compounds (Chapters 9 and 10).

1.15.3 Naming of the Mononuclear Coordination Compounds

(A) **Sequence of the ligands and central metal atom:** Ligand names (determined by the alphabetical order) are followed by the central metal name without giving any space for a particular coordination entity or species.

(**Note:** In formula representation, metal symbol is given at the starting while in naming the metal name appears at the ending).

(**B**) **Central metal atom and ending of the coordination entities:** Central metal atom is named at the end. For the **anionic complexes**, the name of the metal is ended with *ate* while for the cationic and neutral complexes, no distinguishing termination is required (*i.e.* only the name of the metal).

(**C**) **Stock Number:** The **oxidation number** (without the sign) of the metal is indicated in parentheses by the roman numeral and arabic zero after the name of the central metal atom (including the ending 'ate', if applicable). This is called **Stock Number.** No space is left between the Stock Number and name of the metal.

Illustration: For the anionic complexes of trivalent chromium, bivalent nickel, bivalent palladium, trivalent cobalt, trivalent iron, the endings are chromate(III), nickelate(II), palladate(II), cobaltate(III), ferrate(III) respectively.

For the neutral and cationic complexes, the endings are like these platinum(II), nickel(0), cobalt(III), nickel(II).

(**D**) **Ewens-Bassett Number:** It is an alternative to Stock Number. The net charge of the complex is indicated by an arabic number followed by charge sign (called Ewens-Bassett Number) within parenthesis instead of Stock Number. For the neutral complex, nothing to be added. *When the oxidation number of the metal centre is ill-defined, the Ewens-Bassett Number is very much useful* (*e.g.* brown ring complex).

(**E**) **Alphabetical order of the ligands:** In determining the alphabetical order of the ligands, the numerical prefixes (*e.g.* di, tri, bis, tetrakis, etc.) to indicate the number of a particular ligand is ignored. But, if these prefixes are required for naming the ligand then it is considered to determine its alphabetical order. Thus, for 'dichloro' (*i.e.* two Cl^- ligands), the position is determined by the alphabet 'c' while for dimethylamine (*i.e.* Me_2NH) its position is to be determined by the alphabet 'd'.

(**F**) **Number of ligands:** The simple numerical prefixes (*e.g. di, tri, tetra, penta, hexa*, etc.) are used to indicate the number of ligands (generally inorganic ligands) of each kind. The prefixes *di, tri, tetra, penta, hexa* are replaced by *bis, tris, tetrakis, pentakis, hexakis* respectively for the complicated poly-syllabic ligands (generally organic ligands) and in the cases of probable ambiguity. For the prefixes like di, tri, tetra, etc. parentheses are not required but, **for the prefixes bis, tris, tetrakis, etc. parentheses are required.** For example, diammine (meaning $2\ NH_3$), dipyridine (meaning bpy not 2 py), *bis*(pyridine) (meaning 2 py), dimethylamine (meaning Me_2NH not 2 $MeNH_2$), *bis*(methylamine) (meaning 2 $MeNH_2$). It illustrates: use of di-prefix for pyridine or methylamine does not indicate their number 2. Rather, it leads to names of different ligands.

(**G**) **Naming of the ligands:** (i) The naming of **anionic ligands** are ended with 'O'. In general, for the anionic ligands, the usual endings like –ide, –ite, –ate are converted into –ido, –ito, –ato respectively. For example: halide (halido), hydroxide (hydroxido), peroxide (peroxido), oxide (oxido), sulfate (sulfato), nitrite (nitrito), cyanide (cyanido), hydride (hydrido), acetate (acetato), etc.

Note: In the earlier recommendations, halide, hydroxide, oxide, peroxide, hydride were named as halo, hydroxo, oxo, peroxo, hydro respectively. **In IUPAC 2005 recommendations, these exceptions are not granted.**

(ii) Parentheses are used for the inorganic anionic ligands containing numerical prefixes (*e.g.* triphosphato), and the other prefixes like thio–, seleno–, telluro–, etc. of the oxoanions having more than one oxygen atom (*e.g.* thiosulfato), for the substituted organic ligands and for the cases to avoid ambiguity.

(iii) For the common ligands like aqua, ammine, carbonyl, methyl, nitrosyl, etc. parentheses are not required.

(iv) For the neutral and cationic ligands, the ligand names are used without any modification. The exceptions are: *aqua* for H_2O, *ammine* for NH_3, *carbonyl* for CO, *nitrosyl* for NO.

(**Note:** For NH_3, in the spelling, *mm* is to be used while for the organic amines, *m* is used.)

Table 1.15.3.1 Ligand names of some common ligands.

Formula	Systematic Name	Usual Name	Ligand Name
H^-	hydrido	hydride	hydrido
X^-	–	halide	halido (*e.g.* fluorido, chlorido, etc.)
O^{2-}	oxido	oxide	oxido
O_2	dioxygen	dioxygen	dioxygen
O_2^-	[dioxido(1–)]	superoxide	superoxido
O_2^{2-}	[dioxido(2–)]	peroxide	peroxido
H_2O	–	water	aqua
OH^-	hydroxido	hydroxide	hydroxido
N_3^-	–	azide	azido
CN^-	–	cyanide	cyanido
N_2	dinitrogen	dinitrogen	dinitrogen
NH_3	azane	ammonia	ammine
NH_2^-	azanido	amide	amido
CH_3NH_2	methanamine	methylamine	methylamine
NCS^-	–	thiocyanate	thiocyanato-*N*, thiocynato-*S*
CH_3COO^-	ethanoato	acetate	acetato
$C_2O_4^-$	ethanedioato	oxalate	oxalato
N^{3-}	nitrido	nitride	nitrido
NO	–	nitrogen monoxide	nitrosyl
NO_2^-	[dioxidonitrato(1–)]	nitrite	nitrito-*O*, nitrito-*N*
NO_3^-	[trioxidonitrato(1–)]	nitrate	nitrato
SO_4^{2-}	[tetraoxidosulfato(2–)]	sulfate	sulfato
$S_2O_3^{2-}$	[trioxidothiosulfato(2–)]	thiosulfate	thiosulfato
ClO_4^-	[tetraoxidochlorato(1–)]	perchlorate	perchlorato
IO_6^{5-}	[hexaoxidoiodato(5–)]	orthoperiodate	orthoperiodato
$H_2N(CH_2)_2NH_2$ (en)	1,2-ethanediamine or ethane-1,2-diamine	ethylenediamine	ethylenediamine
$CH_3COCHCOCH_3^-$ (acac$^-$)	2,4-pentanedionato	acetylacetonate	acetylacetonato
	8-quinolinolato	8-hydroxyquinolinate	8-quinolinolato
CH_3CONH_2	–	acetamide	acetamide (not acetamido)
$MePH_2$	–	methylphosphine	methylphosphane
PH_3	–	phosphine	phosphane
AsH_3	–	arsine	arsane

* oxinate and oxinato are not recommended.

(H) Naming of the ambidentate or ambident ligands: The ligating or donor site of such ligands is specified with the *itallic letter* after the ligand name.

Illustration: Thiocyanto–*S* means that thiocyanate (NCS^-) coordinates with its *S*-site; thiocyanato-*N* means that thiocyante coordinates with its *N*-site. Similarly, cyanido-*C* and cyanido-*N* are used to specify the donor sites of CN^-. To specify the donor sites, the italic Greek letter κ **(Kappa)** may also be used before the donor site. Thus, nitrito-*N* (*i.e.* nitrito-κ*N*), nitrito-*O* (*i.e.* nitrito-κ*O*), cyanido-*C* (*i.e.* cyanido-κ*C*), cyanido-*N* (*i.e.* cyanido-κ*N*), cysteinato-κ*N*, κ*S*; cysteinato-κ*N*, κ*O*, etc, are used. The Kappa convention is not essential for the monodentate ambidentate ligands as the *donor site symbol* convention is sufficient. **The Kappa convention is quite meaningful for the polyatomic ligands to specify the donor sites.** If there are two or more such ligating sites, a right superscript numerical locant is used on κ to specify the donor site.

Examples: $[Co(NH_3)_5(ONO)]^{2+}$: pentaamminenitrito-κ*O*-cobalt(III) ion.

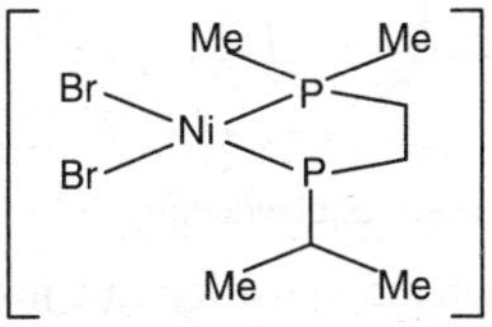

i.e. $[NiBr_2\{(CH_3)_2PCH_2CH_2P(CH_3)_2\}]$

dibromido[1,2-ethanediyl*bis*(dimethylphosphane-κ*P*)]nickel(II) or,
dibromido[ethylenebis(dimethylphosphane-κ*P*)]nickel(II).

(I) Hapto nomenclature symbol to indicate the connectivity between the ligand and metal centre: Unsaturated molecular species, like alkynes, alkenes, cyclopentadienyl (C_5H_5), cyclooctatetraenide anion ($C_8H_8^{2-}$, *i.e.* cot^{2-}), benzene (C_6H_6), 1,3-butadiene (C_4H_6), etc. can bind the metal centres in a complicated fashion. Depending on the condition, the unsaturated organic moieties can bind in different ways by the *C*-sites. The hapto symbol η (*Greek eta*) with a numerical right superscript (to indicate the number of *C*-sites coordinating the metal centre from the ligand) is used.

For the inorganic ligands like NO_3^-, O_2^{2-}, RCO_2^-, etc. the η-symbol may also be used.

Examples:

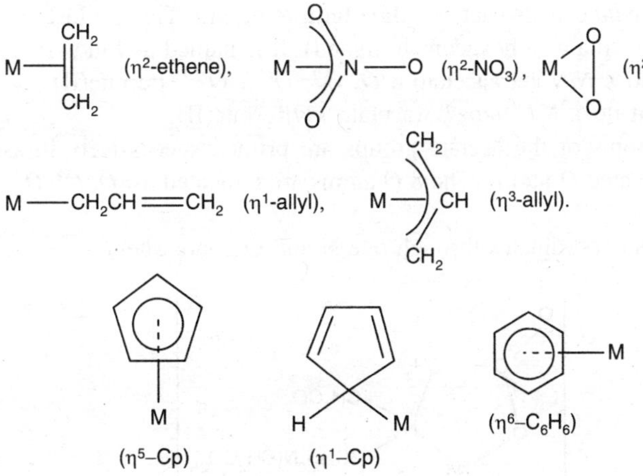

$[Cr(η^3-C_3H_5)_3]$:   *tris*(η^3-allyl)chromium
$[Pt(η^2-C_2H_4)Cl_2(NH_3)]$:   amminedichlorido(η^2-ethene)platinum(II)

$[Fe(\eta^5–C_5H_5)_2]$:   *bis*(η^5-cyclopentadienyl)iron

$[U(\eta^8–C_8H_8)_2]$:   *bis*(η^8-1,3,5,7-cyclooctatetraene)uranium

(**Note:** η^3 is pronounced as '*eta three*' or '*trihapto*'; similarly η^4 as '*eta four*' or '*tetrahapto*', and so on. Sometimes, *h* is also used for the symbol η)

(**J**) **Donor atom symbol in the polydentate ligands:** A polydentate ligand having more than one donor site may act as a flexidentate and also as an ambidentate ligand. In such cases, the donor sites should be indicated by their italicized symbols.

Examples: Dithiooxalate anion can ligate through the *S* or *O* sites and these may be described as dithiooxalate-*S*, *S'* and dithiooxalate-*O*, *O'*. If the ligating sites are different, then these are denoted in an alphabetical order, *e.g.* cysteinato-*N,S*; cysteinato-*N*, *O*; cysteinato-*N*, *O*, *S*.

bis(dithiooxalato-*O*, *O'*)nickel(II)   *bis*(thiooxalato-*S*, *S'*)platinum(II)   (cysteine)

If there are several positions for the same donor site then their positions are indicated by the numerical superscripts. It is illustrated for tartarate or 2,3-dihydroxybutanedioato (*i.e.* tart)

tartarato(3-)-O^1, O^2 tartarato(4-)-O^2, O^3 tartarato(2-)-O^1, O^4

3-, 4- and 2- indicate the number of negative charges on the tartarato ligand

|The polydentate ligand *edta* can also act as a **flexidentate ligand.** This is illustrated in the following examples. **[Fe(edta)]$^{2-}$** (*i.e.* edta as a hexadentate ligand). It is named as follows:

[(ethane-1,2-diyldinitrilo-$\kappa^2 N,N'$)tetraacetato-$\kappa^4 O$, O'', O'''', O'''''']ferrate(II).

or [(ethane-1,2-diyldinitrilo-$\kappa^2 N,N'$)*tetrakis*(acetato-κO]ferrate(II).

Note: The ligated *O*-atoms of the acetate groups are primed successively in ascending order to distinguish from the nonligated *O*-atoms. The 8 *O* atoms are indicated as; *O*, *O'*, *O''*, *O'''*, *O''''*, *O'''''*, *O''''''*, *O'''''''*.

[PdCl$_2$(edta)]$^{4-}$ (*i.e.* edta coordinates through one N and O donor sites)

It is: dichlorido[(ethane-1,2-diyldinitrilo-κN)(acetato-κO)triacetato]palladate(II)

or dichlorido(ethylenediaminetetraacetato-κN, κO)palladate(II)

[PdCl$_2$(edta)]$^{4-}$ (*i.e.* edta coordinates through the both N-sites)

It is: dichlorido[(1,2-ethanediyldinitrilo-κ$^2N,N'$)tetraacetato]palladate(II).

or, dichlorido(ethylenediaminetetraacetato-κ$^2N,N'$)palladate(4−)

It is: [(1,2-ethanediyldinitrilo-κ$^2N,N'$)tetraacetato-κ$^2O,O''$]palladate(2−)

or (ethylenediaminetetraacetato-κ$^2N,N'$-κ$^2O,O''$)palladate(2) or [[ethane 1,2 diyldinitrilo-κ$^2N,N'$)-(N,N'-diacetato-κ$^2O,O''$)(N,N'-diacetato)]palladate(II) **(more informative)**

[Co(edta)(OH$_2$)]$^-$ (*i.e.* edta coordinates through the both N and three O-sites)

It is: aqua[(1,2-ethanediyldinitrilo-κ^2N, N')(tetraacetato-κ^3O, O'',O'''')]cobaltate(III).

or aqua(ethylenediaminetetraacetato-κ$^2N,N'$-κ^3O, O'',O'''')cobaltate(1−)

[N-(2-amino-κN-ethyl)-N′-(2-aminoethyl)-
ethane-1,2-diamine-κ² N,N′]chloridopalladium(II)

[N,N′-bis(2-amino-κN-ethyl)-
ethane-1,2-diamine-κN]chloridopalladium(II)

Illustrative Examples

$Fe_4[Fe(CN)_6]_3$	iron(III) hexacyanidoferrate(II)
	iron(3+) hexacyanidoferrate(4−)
$K_4[Fe(CN)_6]$:	potassium hexacyanido-κC-ferrate(II)
	potassium hexacyanidoferrate(4−)
	tetrapotassium hexacyanido-κC-ferrate(II)
	tetrapotassium hexacyanidoferrate(4−)
$[CoCl_3(NH_3)_3]$:	triamminetrichloridocobalt(III)
$[Co(NH_3)_3(NO_2)_3]$:	triamminetrinitrito-N-cobalt(III)
	triamminetrinitrito-κ³N-cobalt(III)
$[CoCl(NH_3)_4(NO_2)]^+$:	tetraamminechloridonitrito-κN-cobalt(III) ion,
	tetraamminechloridonitrito-κN-cobalt(1+) ion.
$[Co(NH_3)_6]Cl(SO_4)$:	hexaamminecobalt(3+) chloride sulfate
$[Pt(PPh_3)_4]$	tetraKis(triphenylphosphane)platinum(0)
$[FeO_4]^{2-}$	tetraoxidoferrate(VI) ion
$[CoCl(NH_3)_5]^{2+}$	pentaamminechloridocobalt(III) ion
$[Fe(OH_2)_6]^{3+}$	hexaaquairon(III) ion
$[Co(NH_3)_5(N_3)]^{2+}$:	pentaammineazidocobalt(III) ion.
cis-$[Rh(OH_2)_2\{SO_2(NH_2)\}_2]^-$:	cis-diaquabis(sulfamido-κN, κN)rhodate(III) ion.
$[CoCl(NH_3)_3(OH_2)_2]SO_4$:	triamminediaquachloridocobalt(III) sulfate
	triamminediaquachloridocobalt(2+) sulfate
$K_3[Fe(C_2O_4)_3]$:	potassium tris(oxalato)ferrate(III)
$[Co(NH_3)_5(ONO)]Cl_2$:	pentaamminenitrito-κO-cobalt(III) chloride
$[Co(CO_3)(NH_3)_4]NO_3$:	tetraamminecarbonatocobalt(III) nitrate
$[PtCl(NH_2CH_3)_2(NH_3)]^+$:	amminechloridobis(methylamine)platinum(II) ion
$[PtCl_2(NH_3)(py)]$:	amminedichlorido(pyridine)platinum(II)
	amminedichlorido(pyridine)platinum.
$Na[PtBrCl(NH_3)(NO_2)]$:	sodium amminebromidochloridonitrito-κN-platinate(II)
$K[Pt(\eta^2-C_2H_4)Cl_3]$:	potassium trichlorido(η^2-ethene)platinate(II)
	potassium trichlorido(η^2-ethylene)platinate(1−)
$[Pt(\eta^2-C_2H_4)Cl_2(NH_3)]$:	amminedichlorido(η^2-ethene)platinum(II)

(**Note:** The symbol η with a numerical superscript indicates the *heptacity, i.e.* number of the coordinated donor sites in the ligand.)

$[Cr(NCS)_4(NH_3)_2]^-$:   diamminetetrathiocyanato-$\kappa^4 N$-chromate(III) ion.
or, diamminetetrakis(thiocyanato-κN)chromate(III) ion

$Na_2[Fe(CN)_5NO]$:   sodium pentacyanidonitrosylferrate(II)
[if NO^+ is assumed to be present; if neutral NO is assumed then it would be ferrate(III)]
sodium pentacyanido-κC-nitrosyl-κN-ferrate(2−).

$Na_4[Fe(CN)_5NOS]$:   sodium pentacyanido(nitrosylsulfido)ferrate(4−)

$Na_3[Fe(CN)_5CO]$:   sodium carbonylpentacyanidoferrate(3−)

$[Pt(py)_4][PtCl_4]$:   tetrakis(pyridine)platinum(II) tetrachloridoplatinate(II)
terakis(pyridine)platinum(2+) tetrachloridoplatinate(2−)

$[Co(NH_3)_6][Cr(CN)_6]$:   hexaamminecobalt(III) hexacyanidochromate(III).

$[Cr(NH_3)_6]Cl(SO_4)$:   hexaamminechromium(III) chloride sulfate.

cis-$[Co(NH_3)_4(OH_2)_2]^{3+}$:   *cis*-tetraaminediaquacobalt(III) ion.

cis-$[PtCl_2(PPh_3)_2]$:   *cis*-dichloridobis(triphenylphosphane)platinum(II)

$[Co(bpy)(en)_2]Cl_3$:   2,2'- bipyridinebis(ethylenediamine)cobalt(III) chloride.

$[CoCl(en)_2(NO_2)]Cl$:   chloridobis(ethylenediamine)nitrito-κN-cobalt(III) chloride.

$[Zn(OH)_4]^{2-}$:   tetrahydroxidozincate(II) ion.

$[Fe(OH)(OH_2)_5]^{2+}$:   pentaaquahydroxidoiron(2+) ion.

$K_2[SiF_6]$:   potassium hexafluoridosilicate(2−)

$K[CrF_4O]$:   potassium tetrafluoridooxidochromate(V)

$K_2[OsCl_5N]$:   potassium pentachloridonitridoosmate(VI).
potassium pentachloridonitridoosmate(2−)

$[Co(C_2O_4)(en)_2][Co(C_2O_4)_2(OH_2)_2]$:   bis(ethane-1,2-diamine)(oxalato)cobalt(III) diaquabis(oxalato)cobaltate(III)

$[Ni(dmgH)_2]$, *i.e.* $[Ni(C_4H_7O_2N_2)_2]$:   bis(dimethylglyoximato)nickel(II), or bis(2,3-butancdioncdioximato)nickel(II)

cis-$[Ni(nta)(OH_2)_2]^-$:   *cis*-diaquanitrilotriacetatonickelate(II) ion

$[Co(NH_3)_5(SO_4)]^+$:   pentaamminesulfatocobalt(1+) ion.

$[Ru(HSO_3)_2(NH_3)_4]$:   tetraamminebis(hydrogensulfito)ruthenium(II)

$[W(\eta^5-C_5H_5)_2H_2]$:   bis(η^5−cyclopentadienyl)dihydrogentungsten

$[Ti(\eta^5-C_5H_5)_2Cl_2]$:   dichloridobis($\eta^5$−cyclopentadienyl)titanium

$[Ti(CO)_2(\eta^5-C_5H_5)_2]$:   dicarbonylbis($\eta^5$−cyclopentadienyl)titanium

$[Zr(\eta^5-C_5H_5)_2Me_2]$ bis(η^5−cyclopentadienyl)dimethylzirconium

$[NiBr_2(Me_2PCH_2CH_2PMe_2)]$ dibromido[ethane-1,2-diylbis-(dimethylphosphane−κP)]nickel(II)

$[FeCl(S_2CNR_2)_2]$:   chloridobis(*N*,*N*-dialkyldithiocarbamato)iron(III)

$Na[Co(CO)_4]$:   sodium tetracarbonylcobaltate(1−)

$[Fe(CO)_5]$:   pentacarbonyliron(0).

$[Ru(NH_3)_5(N_2)]Cl_2$:   pentaammine(dinitrogen)ruthenium(II) chloride

$[CuCl_2(NH_2CH_3)_2]$:   dichloridobis(methylamine)copper(II)

$[Ti(\eta^2-NO_3)_4]$:   tetra(η^2-nitrato)titanium(IV)

$Na[B(NO_3)_4]$:   sodium tetranitratoborate(III)

$(+)_{589}$-$[Co(en)_3]^{3+}$: $(+)_{589}$-tris(ethylenediamine)cobalt(3+) ion
$(+)_{589}$-tris(ethane-1,2-diamine)cobalt(III) ion

$$\left[\begin{array}{c} O=C-O \quad O-C=O \\ \diagdown \quad Pt \quad \diagup \\ H_2C-N \quad N-CH_2 \\ \quad H_2 \quad H_2 \end{array}\right]$$: *cis*-bis(glycinato-$\kappa N, \kappa O$)platinum(II)

K[AuS(S$_2$)]:   potassium (disulfido)sulfidoaurate(1−)

Cs$_3$[Fe(C$_2$O$_4$)$_3$]:   caesium tris(oxalato)ferrate(3−)

[BF$_4$]$^-$ tetrafluoridoborate(1−) ion

[PF$_6$]$^-$:   hexafluoridophosphate(1−) ion

Ba[BrF$_4$]$_2$:   barium bis(tetrafluoridobromate)

Fe$_4$[Fe(CN)$_6$]$_3$:   iron(III) hexacyanidoferrate(II)

    iron(3+) hexacyanoferrate(4−)

Na$_2$[Fe(CO)$_4$]:   sodium tetracarbonylferrate(2−)

[Mn(CH$_2$CH = CH$_2$)(CO)$_5$]:   (η^1-allyl)pentacarbonylmanganese

[GeF$_4${N(CH$_3$)$_3$}]:   tetrafluorido(trimethylamine)germanium

[WF$_5${N(CH$_3$)$_2$}]:   (dimethylamido)pentafluoridotungsten

[Hg(CHCl$_2$)Ph]:   (dichloromethyl)(phenyl)mecury

[HF$_2$]$^-$:   difluoridohydrogenate(1−)

[H(OH$_2$)$_4$]$^+$:   tetraaquahydrogen(1+) ion

[BH$_2$py$_2$]$^+$:   dihydridobis(pyridine)boron(1+)

[BCl$_2$H$_2$]$^-$:   dichloridodihydridoborate(1−) ion

H[B(C$_6$H$_5$)$_4$]:   hydrogen tetraphenylborate(1−)

Li[AlH$_4$]:   lithium tetrahydridoaluminate(1−)

H$_3$[Fe(CN)$_6$]:   trihydrogen hexacyanidoferrate(3−)

    hexacyanidoferric(III) acid

H$_2$[PtCl$_4$]:   dihydrogen tetrachloridoplatinate(2−)

H$_2$[CrO$_3$(SO$_4$)]:   dihydrogen trioxido(sulfato)chromate(2−)

[CuCl$_2${O = C(NH$_2$)$_2$}$_2$]:   dichloridobis(urea)copper(II)

[Fe(CNCH$_3$)$_6$]$^{2+}$:   hexakis(methyl isocyanide)iron(II) ion.

[Au(C$_2$H$_5$)$_2$(en)]Cl: diethyl(ethylenediamine)gold(III) chloride

[Fe(CH$_3$CO)(CO)$_2$I{PMe$_3$}$_2$]:   acetyldicarbonyliodidobis(trimethylphosphane)iron

[Cr(CO)$_4$(η^4-C$_4$H$_6$)]:   tetracarbonyl(η^4-2-methylene-1,3-propanediyl)chromium, or

    tetracarbonyl(η^4-1,3-butadiene)chromium

[Mo(CO)$_3$(η^7-C$_7$H$_7$)]$^+$:   tricarbonyl(η^7-cycloheptatrienylium)molybdenum(1+) ion

[Ni(PF$_3$)$_4$]:   tetrakis(trifluoridophosphane)nickel(0)

Na$_3$[Ag(S$_2$O$_3$)$_2$]:   sodium bis(thiosulfato−κS)argentate(3−)

[Cr(C$_4$H$_7$O$_2$)$_3$], *i.e.* [Cr(acac)$_3$]:   tris(acetylacetonato)chromium(III)

[W(CO)$_3$(H$_2$)(PPr$_3^i$)$_2$]   tricarbonyl(η^2-dihydrogen)-bis(triisopropylphosphane)tungsten

[PPh$_4$][Li(η^5-C$_5$H$_5$)$_2$]   tetraphenylphosphonium bis(η^5-cyclopentadienyl)lithiate(1−)

[Ru(η^5-C$_5$Me$_5$)$_2$]   bis(pentamethyl-η^5-cyclopentadienyl)ruthenium

[OsEt(NH$_3$)$_5$]Cl   pentaammine(ethyl)osmium(1+) chloride

[Pt{C(O)Me}{Me(PEt$_3$)}$_2$]   acetyl(methyl)-bis(triethylphosphane)platinum

[HgMePh]	methyl(phenyl)mercury
$[Co(edta)]^{2-}$	[(ethane-1,2-diyldinitrilo-$\kappa^2 N,N'$)tetrakis(acetato-κO)]-cobaltate(2−) ion
$[Co(edta)(OH_2)]^-$ (One acetate group remains free)	aqua[(ethane-1,2-diyldinitrilo-$\kappa^2 N,N'$)-tris(acetato-κO)acetato]cobaltate(III) ion
$[ReH_3\{P(C_6H_5)_3\}_5]$:	trihydridopentakis(triphenylphosphane)rhenium(III)
$[U(acac)_2O_2]$ or $[UO_2(acac)_2]$:	bis(acetylacetonato)dioxidouranium(VI)
$[V(acac)_2O(py)]$ or $[VO(acac)_2(py)]$:	bis(acetylacetonato)oxido(pyridine)vanadium(IV)
$K[Os(N)O_3]$:	potassium nitridotrioxidoosmate(1−)
	potassium nitridotrioxidoosmate(VIII)
$(NH_4)_2[VO(ox)_2]$:	ammonium bis(oxalato)oxidovanadate(IV)

tris(8-quinolinolato)iron(III) [or omit (III)]

$[Cr(C_6H_6)_2]$:	bis(benzene)chromium(0)
$[Sb(C_6H_5)Cl_5]^-$:	pentachlorido(phenyl)antimonate(V) ion
$[Pd(bpy)(NCS)_2]$	(2,2′-bipyridine)dithiocyanato-$\kappa^2 N$-palladium(II) or (2,2′-bipyridine)bis(thiocyanato-κN)palladium(II)
$[Pd(bpy)(SCN)_2]$	(2,2′-bipyridine)dithiocyanato-$\kappa^2 S$-palladium(II)
$[Fe(NO)(OH_2)_5]SO_4$:	pentaquanitrosyliron(2+) sulfate

1.15.4 Names and Formulae of the Dinuclear and Polynuclear Complexes with Bridging Ligands or Metal-Metal Bonds (*cf.* 1.15.2B)

(A) Complexes with the Bridging Ligands

- The bridging ligand is indicated by 'μ' placed before the ligand and separated by a hyphen.
- If the bridging ligand appears more than once, then the multiplicative prefixes are used, *e.g.* di-μ-, tri-μ-, etc.
- If the same ligand is present both as the bridging ligand and nonbridging ligand, then the bridging ligand is mentioned first followed by the corresponding nonbridging ligand, *e.g.* μ-chlorido-tetrachlorido..., di-μ-chloridotrichlorido... .
- If a bridging ligand connects two metal centres, then it is simply represented by μ. If the ligand connects more than two metal centres, then it is denoted by μ_n ($n > 2$, denoting the number centres connected by the bridging ligand).

(B) Complexes with the metal-metal bonding

- Metal-metal bonding may be indicated separately in parenthesis at the end followed by the ionic charge (*i.e.* Ewens-Bassett No. within the parenthesis). The involved metals are written by their symbols in italic fonts.

- For the symmetrical dinuclear complexes, the simpler names may be given by using the multiplicative prefixes.

- For the unsymmetrical binuclear complexes, the metal centres are numbered 1 and 2 according to certain priority sequence and the ligating atoms are denoted by Kappa as usual.

- *Priority sequence of the metal centres in the unsymmetrical dinuclear complexes with the metal-metal bond:* If **both the metal centres are the same**, the centre connected to the larger number of ligands is denoted 1 and the other is 2. If both the metal centres bear the same number of ligands, then the centre connecting the larger number of alphabetically preferred ligands is denoted by the number 1.

(C) If the **metal centres are different** then the more metallic centre (*i.e.* less electronegative; *cf.* Fig. 1.15.2.1) is assigned the number 1.

Illustrative Examples

$[Be_4(\mu_4\text{–}O)(\mu\text{–}O_2CMe)_6]$ hexakis(μ-acetato-κO: $\kappa O'$)-μ_4-oxido-*tetrahedro*-tetraberyllium

$[Al_2Cl_4(\mu\text{–}Cl)_2]$ or $[Cl_2Al(\mu\text{–}Cl)_2AlCl_2]$ di-μ-chlorido-tetrachlorido-$1\kappa^2Cl$, $2\kappa^2Cl$-dialuminium di-μ-chlorido-bis(dichloridoaluminum)

$[(NH_3)_5Co\text{—}O_2\text{—}Co(NH_3)_5](NO_3)_4$ μ-peroxido-bis[pentaamminecobalt(III)] nitrate.

$[(NH_3)_5Cr\text{—}OH\text{—}Cr(NH_3)_5]Cl_5$
i.e. $[\{Cr(NH_3)_5\}_2(\mu\text{-}OH)]Cl_5$ μ-hydroxido-bis(pentaamminechromium)(5+) chloride

$[\{PtCl(PPh_3)\}_2(\mu\text{-}Cl)_2]$ di-μ-chlorido-bis[chlorido(triphenylphosphane)]platinum

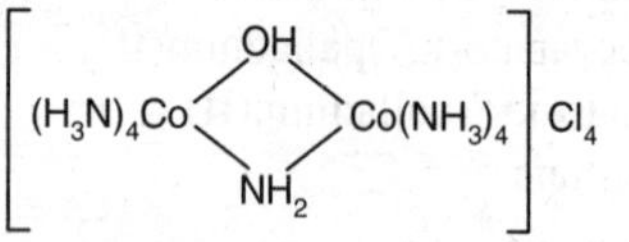

μ-amido-μ-hydroxido-bis[tetraamminecobalt(III)] chloride

i.e. $[\{Co(NH_3)_4\}_2(\mu\text{-}NH_2)(\mu\text{-}OH)]Cl_4$

$[\{Co(NH_3)_3\}_2(\mu\text{-}NO_2)(\mu\text{-}OH)_2]Cl_3$

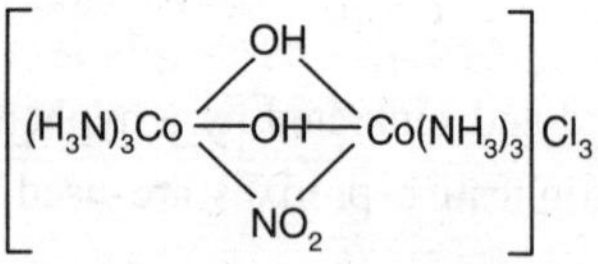

: di-μ-hydroxido-μ-nitrito-κN:κO-bis[triamminecobalt(III)] chloride

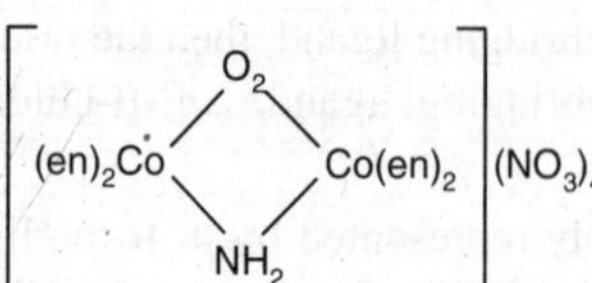

: μ-amido-μ-superoxidobis[bis(ethylenediamine)-cobalt(III)] nitrate

(Assuming the superoxo bridging ligand)

$[\{Ni(\eta^5\text{-}C_5H_5)\}_3(\mu_3\text{-}CO)_2]$ (*see* the structure)   di-μ_3-carbonyl-*cyclo*-tris(η^5-cyclopentadienylnickel)(3 *Ni—Ni*)

1   2
$[(bpy)(H_2O)Cu(\mu\text{-}OH)_2Cu(bpy)(SO_4)]$ aqua-$1\kappa O$-(2,2´-bipyridine-$1\kappa^2N,N'$)-di-μ–hydroxido-(sulfato-$2\kappa O$)dicopper(II)

$[\{Cr(en)_2\}_2(\mu\text{-}NH)(\mu\text{-}OH)]^{3+}$ μ-hydroxido-μ-imido-bis[bis(ethylenediamine)-chromium(III)] ion

$[(OC)_5MnMn(CO)_5]$ { decacarbonyldimanganese(*Mn—Mn*)

i.e. $[Mn_2(CO)_{10}]$ { bis(pentacarbonylmanganese)(*Mn—Mn*)

$[(Ph_3As)Au\overset{2}{—}Mn(CO)_5]$:   pentacarbonyl-1κ⁵*C*-(triphenylarsane-2κ*As*)-goldmanganese(*Au—Mn*)

$[(H_3N)_5\overset{1}{Cr}(\mu\text{-}OH)\overset{2}{Cr}(NH_2CH_3)(NH_3)_4]^{5+}$ { nonoammine-1κ⁵*N*, 2κ⁴*N*-μ-hydroxido-(methylamine-2κ*N*)dichromium(5+) ion

   { nonaammine-μ-hydroxido-(methylamine)-dichromium(5+) ion

$[(H_3N)_3\overset{1}{Co}(\mu\text{-}NO_2)(\mu\text{-}OH)_2\overset{2}{Co}(NH_3)_2(py)]^{3+}$ pentaammine-1κ³*N*, 2κ²*N*-di-μ-hydroxido-μ-nitrito-1κ*N*:2κ*O*-(pyridine-2κ*N*)dicobalt(3+) ion

$[\{Cu(py)\}_2(\mu\text{-}O_2CCH_3)_4]$ tetrakis(μ-acetato-κ*O*:κ*O′*)bis[(pyridine)copper(II)]

$[\overset{1}{Cl}_4\overset{2}{ReReCl}_4]^{2-}$ octachlorido-1κ⁴*Cl*,2κ⁴*Cl*-dirhenate(*Re–Re*)(2−)

$[Re_2Cl_8]^{2-}$ octachloridordihenate(*Re–Re*)(2−)

$[ReCo(CO)_9]$ nonacarbonylrheniumcobalt(*Re–Co*)

$[(OC)_5\overset{1}{Re}\overset{2}{Co}(CO)_4]$ nonacarbonyl-1κ⁵*C*, 2κ⁴*C*-rheniumcobalt(*Re–Co*)

$[\overset{1}{Cl}(H_3N)_4Ni\{\mu\text{-}O_2CMe\}\overset{2}{NiCl}_2(NH_3)_3]$ heptaammine-1κ⁴*N*, 2κ³-trichlorido-1κ*Cl*,2κ²*Cl*-(μ-acetato-1κ*O*:2κ*O′*)dinickel

B_4Cl_4 tetrachlorido-*tetrahedro*-tetraboron(6 *B–B*)

$[Mo_6Cl_8]^{4+}$ (bridging Cl⁻ ions occupying the 8-triangular octahedral faces)   octa-μ₃-chlorido-*octahedro*-hexamolybdenum(4+) ion.

$[Nb_6Cl_{12}]^{2+}$ (bridging Cl⁻ ions occupying the 12 octahedral edges)   dodeca-μ-chlorido-*octahedro*-hexaniobium(2+) ion.

$(NH_4)_3[Re_3Cl_{12}]$ ammonium dodecachlorido-*triangulo*-trirhenate-(3 *Re–Re*)(3−)

$\begin{array}{c}\left(\!\!+\!\!\!—\text{Ag}\!\!—\!\!\text{NC}\!\!—\!\!\!+\!\!\right)_x\end{array}$ *catena*-poly[silver-μ-(cyanido-*N:C*)]

$\begin{pmatrix} & H_3N & \diagup Br \\ & \diagdown Pt — Br — \\ & Br \diagup & \diagdown NH_3 \end{pmatrix}_x$ *catena*-poly[(diamminedibromidoplatinum)-μ-bromido] (Polymeric forms of [PtBr₃(NH₃)₂])

$Cs_3[Re_3Cl_{12}]$ caesium dodecachlorido-*triangulo*-trirhenate(3 *Re–Re*)(3−)

$[Os_3(CO)_{12}]$ *cyclo*-tris(tetracarbonylosmium)(3 *Os–Os*), *cyclo*-dodecacarbonyl-1κ^4C,2κ^4C,

3κ^4C–triosmium(3 *Os–Os*)

i.e.

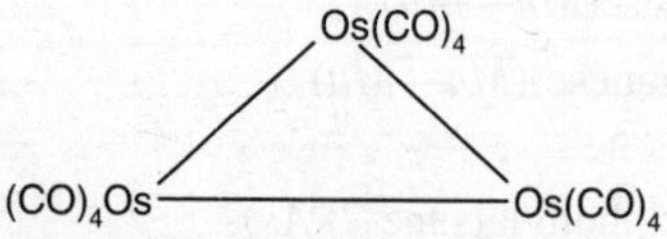

or tris(tetracarbonylosmium)(3 *Os–Os*), or dodecacarbonyl-*triangulo*-triosmium(3 *Os–Os*)

[Fe$_2$(CO)$_9$], *i.e.* [Fe$_2$(CO)$_6$(μ–CO)$_3$]   tri-μ-carbonyl-bis(tricarbonyliron)(*Fe—Fe*)

[(NH$_3$)$_5$Ru(C$_4$H$_4$N$_2$)Ru(NH$_3$)$_5$]$^{5+}$ μ-pyrazine-bis(pentaammineruthenium)(5+) ion
[H$_3$N)$_5$Ru(μ–pyz)Ru(NH$_3$)$_5$]$^{5+}$

(Creutz-Taube ion)

1   2
[Cu(bpy)(H$_2$O)(μ-OH)$_2$Cu(bpy)(SO$_4$)]   aqua-1κO-bis(2,2′-bipyridine)-1$\kappa^2 N'$,N'';2$\kappa^2 N'$,N'-di-μ-hydroxido-{sulfato(2−)-2κO}dicopper(II)

[{Ru(η^5-Cp)}$_3$(CO)$_2$]   di-μ$_3$-carbonyl-*cyclo*-tris(η^5-cyclopentadi-enylruthenium)(3 *Ru–Ru*)

2   1
[Cl(PhNH)$_2$GeGeCl$_3$]   tetrachlorido-1κ^3Cl,2κCl-bis(phenylamido-2κN)-digermanium(*Ge—Ge*)

[Pb$_2$(Et)$_2$], *i.e.* [(Et)$_3$PbPb(Et)$_3$]   hexaethyldilead(*Pb—Pb*)

                         bis(triethyllead)(*Pb—Pb*)

[{Cu(py)$_2$}$_2$(μ-O$_2$CCH$_3$)$_4$]   tetrakis(μ-acetato-κO:$\kappa O'$)bis[(pyridine)copper(II)]

[{Fe(NO)$_2$}$_2${μ-P(C$_6$H$_5$)$_2$}$_2$]   bis(μ-diphenylphosphanido)bis(dinitrosyliron)

[Cr$_3${μ-CH$_3$COO}$_6$(μ$_3$-O)]Br   hexakis(μ-acetato-κO:$\kappa O'$)-μ$_3$-oxido-trichromium(III) bromide.

Na$_2$[(O$_2$)$_2$OCr-O-O-CrO(O$_2$)$_2$]   disodium μ-peroxido-bis(oxidodiperoxidochromate)(2−)

[ClHgIr(CO)Cl$_2$(PPh$_3$)   carbonyl-1κC-trichlorido-1$^2\kappa Cl$, 2κCl-bis(triphenylphosphane-1κP)iridiummercury(*Ir–Hg*)
                         (*cf.* iridium is named before mercury, Fig. 1.15.2.1)

[Cr$_2$O$_7$]$^{2-}$ heptaoxidodichromate(2−)

[Cr$_2$O$_6$(μ–O)]$^{2-}$ μ-oxido-hexaoxidochromate(2−)

[{Ni(η^5–C$_5$H$_5$)}$_3$(μ$_3$–CO)$_2$]   di-μ$_3$-carbonyl-*cyclo*-tris(cyclopentadienylnickel)(3 *Ni–Ni*)

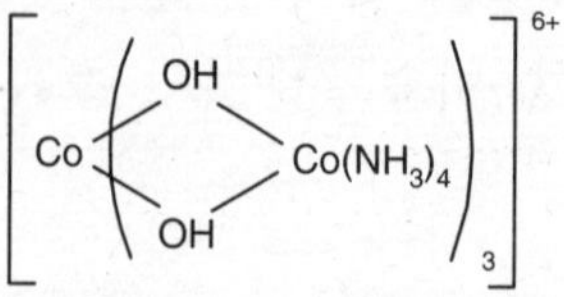

dodecammine-1κ^4N, 2κ^4N, 3κ^4N-hexa-μ-hydroxido-1:4 κ^4O; 2:4 κ^4O; 3:4 κ^4O-tetracobalt(6+) ion

or, tris[tetraammine-μ-dihydroxidocobalt(III)]cobalt(III) ion

[(LiMe)$_4$]   tetra-μ$_3$-methyl-tetralithium

[(OC)$_3$Fe(μ-CO)$_3$Fe(CO)$_3$]   tri-μ-carbonylbis(tricarbonyliron)(*Fe—Fe*)

[Cr$_3$(μ-CH$_3$COO)$_6$(μ$_3$-O)]Cl   hexakis(μ-acetato-κO:$\kappa O'$)-μ$_3$-oxido-trichromium(III) chloride.

[Rh$_3$H$_3${P(OMe)$_3$}$_6$]   trihydridohexakis(trimethylphosphite)trirhodium

1.15.5 Descriptions of Configuration of the Mononuclear Complexes (*i.e.* Coordination Entities)

(A) Polyhedral symbol describing the coordination geometry: It describes the geometrical arrangement of the ligands around the metal centre. It consists of one or more *capital italic letter* (*s*) (derived from the common name of the geometry of the coordination entity) followed by an arabic numeral (with a hyphen) indicating the coordination number. Polyhedral symbols of some common polyhedra are given in Table 1.15.5.1.

Table 1.15.5.1 Polyhedral symbols.

Coordination polyhedron	C.N.	Polyhedral symbol
Linear	2	L–2
Tetrahedron	4	T–4
Square plane	4	SP–4
Square pyramid	5	SPY–5
Trigonal bipyramid	5	TBPY–5
Octahedron	6	OC–6
Trigonal prism	6	TPR–6
Pentagonal bipyramid	7	PBPY–7
Octahedron, face centred	7	OCF–7
Cube	8	CU–8
Square antiprism	8	SAPR–8
Dodecahedron	8	DD–8

(B) Configuration index: It describes the *priorities of the ligands* on the vertices of the coordination polyhedra. It can distinguish the **diastereoisomers**. The digits of configuration index are determined by the **priority numbers** (decided by **CIP procedure** developed by Cahn, Ingold and Prelog for the enantiomeric carbon compounds) of the ligands. *The ligating atom with the highest priority is indicated by the priority number 1,* the next one denoted by the priority number 2, and so on. Ligand priority determination by the CIP rule is illustrated at the end of this section.

By the common terms like *cis-trans, fac-mer,* etc., always the diastereomers cannot be identified. For example, the isomers of [Ma_2b_2] type square planar complexes can be identified by the *cis-trans* terminology, but this terminology appears insufficient to describe the isomers of [*Mabcd*] type square planar complexes. In such cases, configuration index is quite helpful and it has got many advantages. This aspect is illustrated for the square planar, octahedral complexes and for some common geometries.

(a) **Configuration index for the square planar (*SP*–4) systems:** It is represented by **a single digit which is the priority number of the ligating atom *trans* to the ligating atom of the highest priority** (*i.e.* priority number 1). If there are two possibilities, the higher numerical value of priority is taken for the purpose of configuration index. Let us take the [*Mabcd*] type complex having the priority order as follows:

Priority order: a(1) ⟩ b(2) ⟩ c(3) ⟩ d(4) (say)

Here, *the lower numerical value indicates the higher priority.*

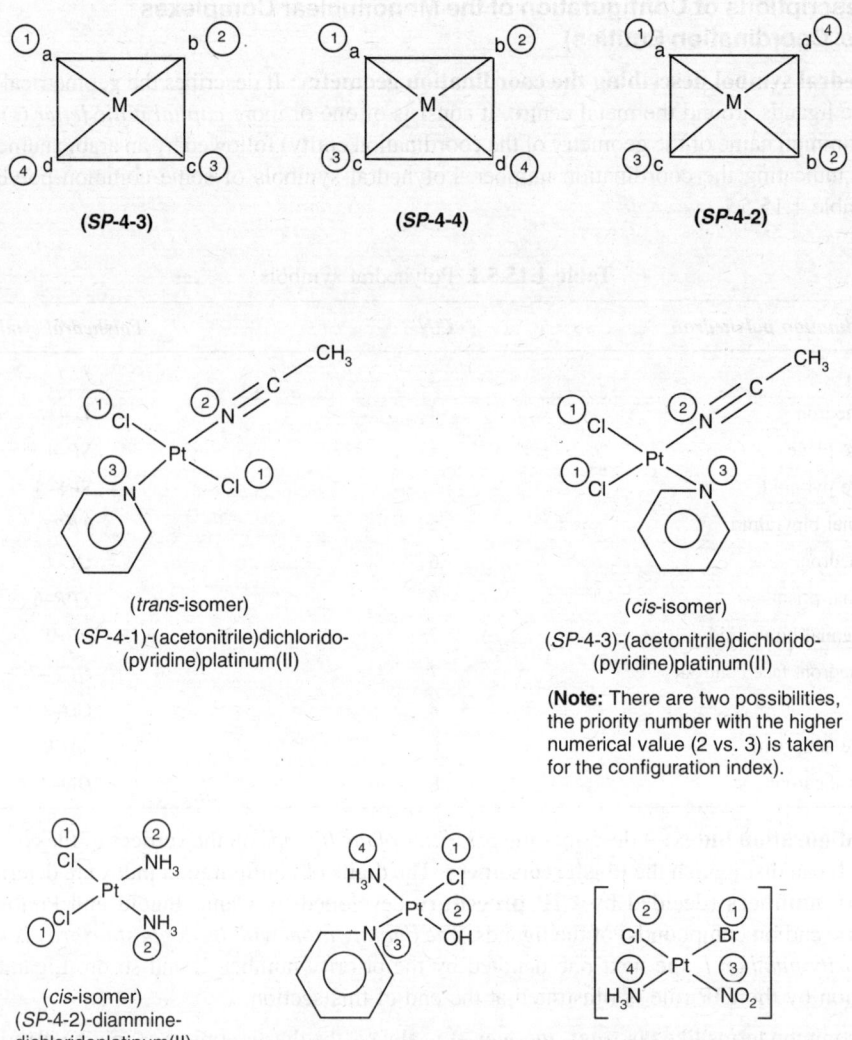

(*trans*-isomer)
(*SP*-4-1)-(acetonitrile)dichlorido-
(pyridine)platinum(II)

(*cis*-isomer)
(*SP*-4-3)-(acetonitrile)dichlorido-
(pyridine)platinum(II)

(**Note:** There are two possibilities,
the priority number with the higher
numerical value (2 vs. 3) is taken
for the configuration index).

(*cis*-isomer)
(*SP*-4-2)-diammine-
dichloridoplatinum(II)

(*SP*-4-3)-(amminechlorido-
hydroxido(pyridine)platinum(II)

(*SP*-4-4)-amminebromidochlorido-
nitrito-κN-platinate(II)

(b) **Configuration index of the octahedral (*OC*-6) systems:** Here the configuration index consists of **two digits**. The first digit gives the priority number of the ligating atom *trans*- to the ligating atom of priority number 1. If there is more than one possibility, then the *trans* ligand with the lowest priority (*i.e.* highest numerical value) is taken as the first digit of the configuration index. These two ligating atoms define the **reference axis** of the octahedron. The second digit of the configuration index is determined by the priority numbers of the ligating atoms lying in the plane (*i.e.* basal plane) perpendicular to the reference axis. This second digit is the priority number of the ligating atom *trans* to the ligating atom with the lowest priority number in the basal plane. This aspect is illustrated in the following examples.

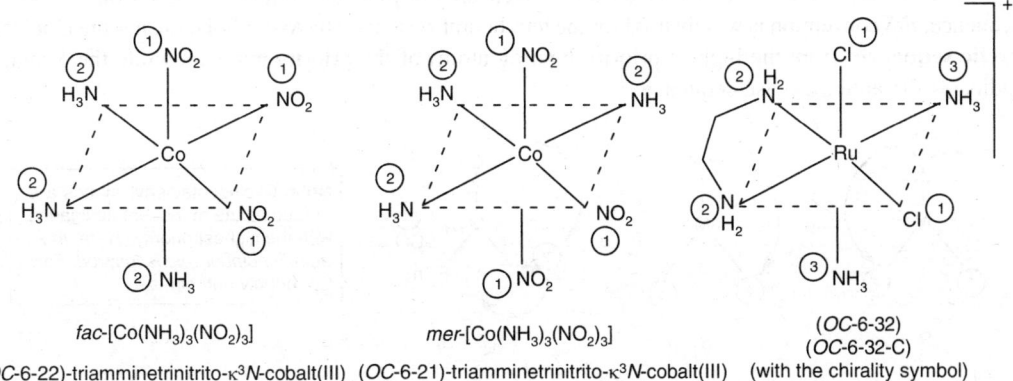

fac-[Co(NH₃)₃(NO₂)₃]

(OC-6-22)-triamminetrinitrito-κ³N-cobalt(III)

mer-[Co(NH₃)₃(NO₂)₃]

(OC-6-21)-triamminetrinitrito-κ³N-cobalt(III)
(cf. ref. axis, 1-1 vs. 1-2; 1-2 is chosen).

(OC-6-32)
(OC-6-32-C)
(with the chirality symbol)

The *fac-mer* isomers of M(AB)₃ can also be expressed by the configuration indexes.

[Co(gly)₃]: For the *fac* and *mer* isomers, the configuration indexes are (*OC*-6-22) and (*OC*-6-21).

(c) **Configuration index of the square pyramidal system (SPY-5):** It consists of **two digits**. The first digit is denoted by the priority number of the ligating atom along the axial direction (*i.e.* C₄ symmetry axis taken as the reference axis). The second digit is given by the priority number of the ligating atom *trans*-to the ligating atom of highest priority (*i.e.* lowest priority number) in the basal plane (*i.e.* plane perpendicular to the reference axis). If there is more than one possibility in the basal plane, then the second digit is given by the highest numerical value of priority.

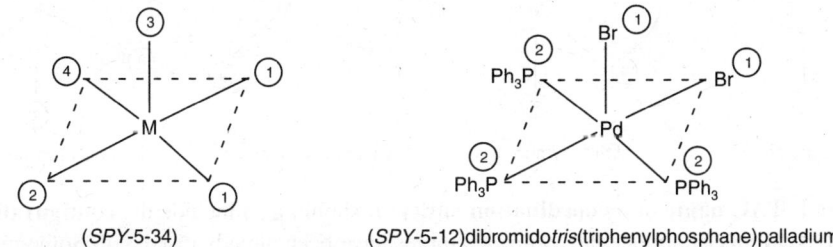

(SPY-5-34)

(SPY-5-12)dibromido*tris*(triphenylphosphane)palladium

(d) **Configuration index of the trigonal planar system (TBPY-5):** It consists of **two digits** given by the priority numbers of the ligating atoms lying in the axial direction (*i.e.* C₃ axis taken as the reference axis). The lower priority number gives the first digit.

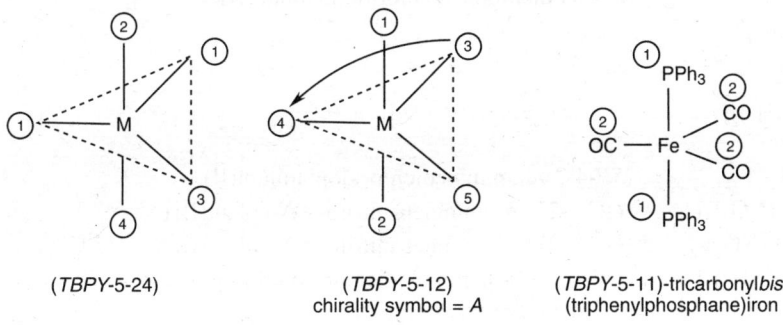

(TBPY-5-24)

(TBPY-5-12)
chirality symbol = A

(TBPY-5-11)-tricarbonyl*bis*-
(triphenylphosphane)iron

(C) Chirality symbols (*C/A* convention) based on the priority sequence: Based on the priority sequence, *R/S* convention is widely used for the tetrahedral systems. The *R*-symbol indicates the **clockwise cyclic sequence** (from the highest priority ligating atom) of the priority numbers while the *S*-symbol indicates the anticlockwise sequence.

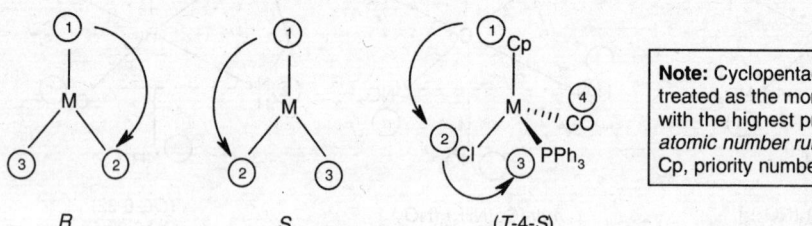

R   S   (*T-4-S*)

> **Note:** Cyclopentadienyl ligands are treated as the monodentate ligands with the highest priority. *Here, the atomic number rule is ignored.* For, Cp, priority number is 1.

The same procedure may be extended for the other geometries with some modifications for the *C/A* convention. In terms of the priority numbers of the ligating atoms, the reference axis is to be defined. **Then the structure is viewed from the axial ligating atom of higher priority.** If the priority numbers of the ligating atoms lying in the plane perpendicular to the reference axis follow a clockwise sequence, then it is described by the *C* symbol and for the anticlock sequence, it is described by the *A*-symbol.

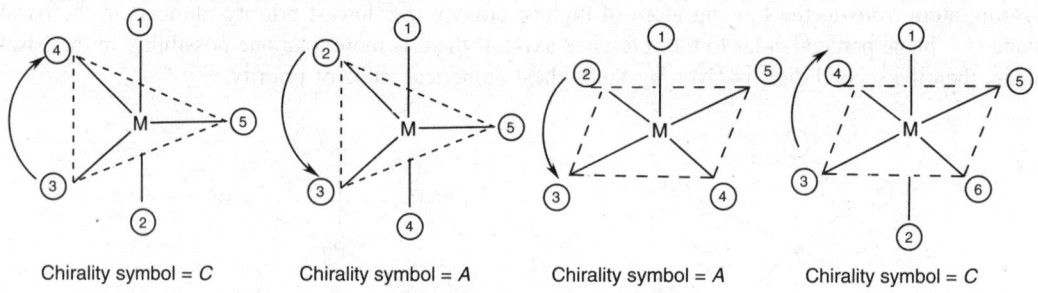

Chirality symbol = *C*   Chirality symbol = *A*   Chirality symbol = *A*   Chirality symbol = *C*

Complete IUPAC name of a coordination entity: It should also include the **configuration index** and **chirality symbols** (if applicable). Some examples have been already discussed. Some more representative examples are given here.

$trans$-$[PtCl_2(NH_3)_2]$: (*SP*-4-1)-diamminedichloridoplatinum(II)

cis-$[PtCl_2(NH_3)_2]$	: (*SP*-4-2)-diamminedichloridoplatinum(II)
fac-$[Co(NH_3)_3(NO_2)_3]$	: (*OC*-6-22)-triamminetrinitrito-$\kappa^3 N$-cobalt(III)
mer-$[Co(NH_3)_3(NO_2)_3]$	: (*OC*-6-21)-triamminetrinitrito-$\kappa^3 N$-cobalt(III)
cis-$[RuCl_2(en)(NH_3)_2]^+$	: (*OC*-6-32C)-diamminedichlorido(ethane-1,2-diamine)ruthenium(III)

(already illustrated)

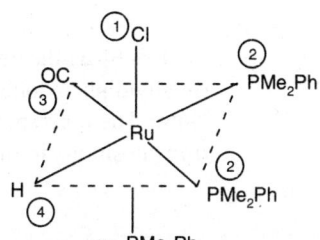

$[Ru(CO)ClH(PMe_2Ph)_3]$:

Its descriptor is: *OC*-6-24-*A*

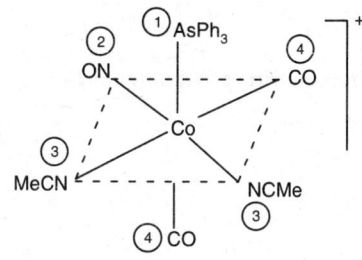

(*OC*-6-24-A)-carbonylchloridohydrido*tris*-
(dimethylphenylphosphane)ruthenium(II)

: (*OC*-6-43)-*bis*(acetonitrile)dicarbonylnitrosyl-
(triphenylarsane)cobalt(1+)

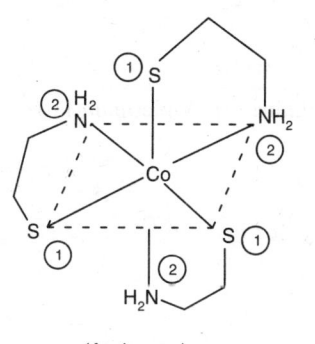

(*fac*-isomer)

: (*OC*-6-22)-*tris*(2-aminoethanethiolato)cobalt(III)
The discriptor for the *mer*-isomer is: *OC*-6-21
For the *fac*- and *mer*- isomers of $[M(gly)_3]$ the discriptors
are *OC*-6-22 and *OC*-6-21 respectively.

$(NH_4)[Cr(NCS)_4(NH_3)_2]$ (**Reneicke salt**)

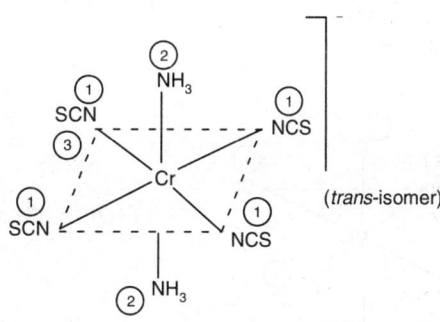

(*trans*-isomer)

Ammonium (*OC*-6-11)-diammine*tetrakis*(thiocyanto-
κ*N*)chromate(III) ion

(**Note:** η^5–Cp is considered as a monodentate ligand with
the **highest priority**).

(*T*-4-*S*)-bromidocarbonyl(η^5–cyclopentadienyl)-
(triphenylphosphane)iron(II)

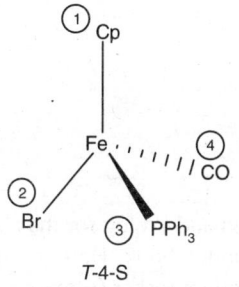

T-4-*S*

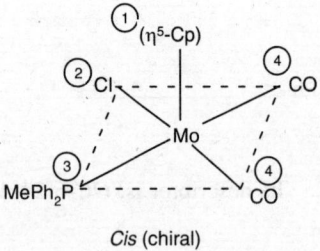

Cis (chiral)

: (*SPY*-5-14-*A*)-dicarbonylchlorido(η^5–cyclopentadienyl)-
(methyldiphenylphosphane)molybdenum(II)
Note: The corresponding *trans*-isomer (*i.e.* CO groups at the
trans positions in the basal plane) is achiral.

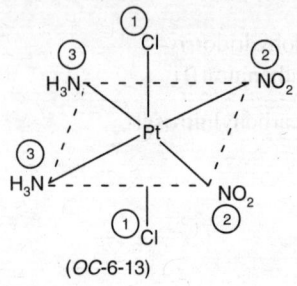

(*OC*-6-13)

: (*OC*-6-13)-diamminedichloridodinitriro-$\kappa^2 N$-platinum(IV)

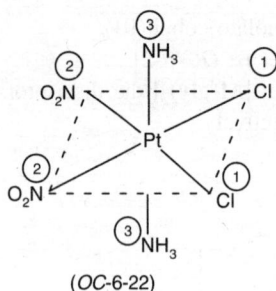

(*OC*-6-22)

: (*OC*-6-22)-diamminedichloridodinitriro-$\kappa^2 N$-platinum(IV)

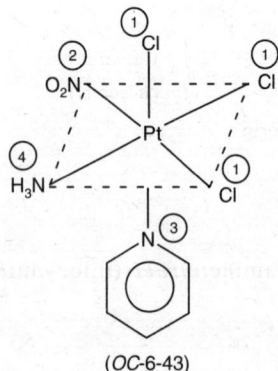

(*OC*-6-43)

: (*OC*-6-43)-amminetrichloridonitriro-κN-pyridineplatinum(IV)

(D) Ligand priority determination (CIP rule developed by Cahn, Ingold and Prelog for the chiral
carbon compounds): This rule is discussed in detail in any standard organic textbook. Here we shall
illustrate the rule. **The priority of a ligating atom is determined by its *atomic number* (atomic mass**

is considered for the isotopes). **Higher atomic number gives higher priority (*i.e.* lower numerical value).** If the atomic numbers of the ligating atoms cannot determine the priority order then moving outwards the ligand structure will be required.

Priority sequence of some ligands is illustrated here.

Br $\rangle$ Cl $\rangle$ PPh$_3$ $\rangle$ NMe$_3$ $\rangle$ NH$_3$ $\rangle$ CO; py $\rangle$ NHEtMe $\rangle$NHMe$_2$ $\rangle$ NH$_2$Me

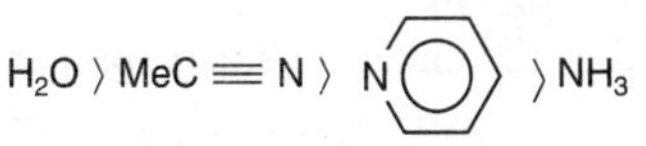

H$_2$O $\rangle$ MeC $\equiv$ N $\rangle$ N$\langle$◯$\rangle$ $\rangle$ NH$_3$

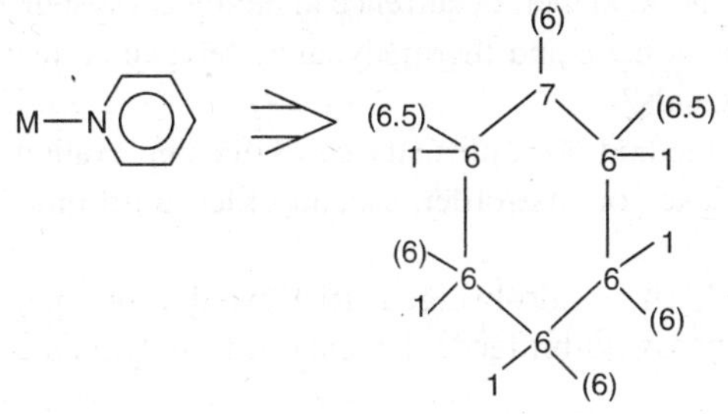

(In terms of atomic number; values within brackets are taken to account for the double bonds; 6.5 value

i.e. $\dfrac{6+7}{2}$ due to the resonating structure)

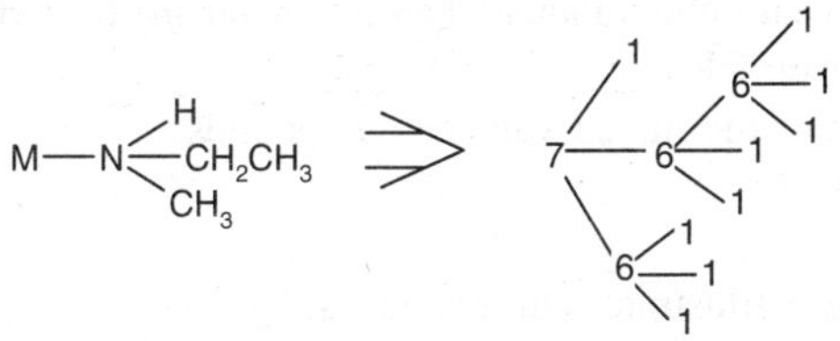

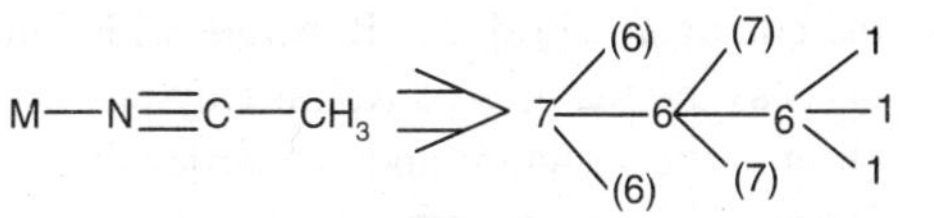

(Proceeding along the branches of the ligand structure allows the priority determination.)

Ref: Nomenclature of Inorganic Chemistry (IUPAC Recommendations 2005), RSC Publishing.

EXERCISE 1

A. General Questions

1. Give the colour code names of the cobalt(III)-ammines and discuss their properties.
2. Discuss the Blomstrand-Jorgensen's Chain Theory of coordination compounds. Mention its limitations.
3. What are the basic postulates of Werner's coordination theory?
4. Discuss the characteristic features of primary and secondary valence of Werner's coordination theory.
5. How did Werner formulate the cobalt(III)-ammine and platinum(IV)-ammine complexes?
6. How did Werner establish the geometry of the coordination compounds of coordination number 4 and 6?
7. What are the possible geometries for the coordination number 4 and 6?
8. Give the examples of optically active coordination compounds which are purely inorganic in nature.
9. Consider the heptacity and denticity of ligands to determine the coordination number with suitable examples.
10. What are the main defects of Werner's coordination theory?
11. Define and illustrate the double salts, complex salts, perfect and imperfect complex.

12. Define and illustrate the terms– inner sphere complex, outer sphere complex, 1st coordination sphere, 2nd coordination sphere, binary complex, mixed ligand complex, heteroleptic and homoleptic complexes.

13. Classify the ligands in terms of the nature of metal-ligand bonding interaction. Give suitable examples.

14. Discuss the properties of representatives examples of ambidentate ligands.

15. Different tautomeric structures of thiosulfuric acid can be supported by the linkage isomers (prepared by P. Ray). Illustrate.

16. Give the representative examples of macrocyclic ligands and their occurrence in biological system.

17. What do you mean by template effect? Illustrate the kinetic and thermodynamic template effect in the synthesis of polydentate and macrocyclic ligands?

18. Discuss the different chemical methods and physical methods to confirm the cases of complexation.

19. Give the characteristic features of the inner complexes of first-order, second-order, third-order and fourth-order.

20. Discuss the structural features of (i) OH-bridged, (ii) oxido-bridged, (iii) cyanido-bridged, (iv) halido-bridged, (v) superoxido-bridged, (vi) peroxido-bridged, dinuclear and polynuclear complexes.

21. What do you mean by mixed valence complexes and intervalance charge transfer medicated by the conjugated bridges? Illustrate with suitable examples.

22. Discuss the features of Robin-Day classification of the mixed valence compounds.

23. Discuss the structure and properties of silicones.

24. Discuss the coordination polymers.

25. What do you mean by the compartmental ligands? Illustrate with examples.

26. Illustrate the polyhedral symbol, configuration index, and chirality symbols (R/S, C/A) as per the IUPAC recommendation.

B. Justify the following statements:

1. Measurement of molar conductivity was utilized by Werner to characterize the cobalt(III)-ammine compounds.

2. In terms of Werner's theory, the anions bound by the primary valence are ionizable but the anions bound by secondary valence are not ionizable.

3. The neutral ligands can only satisfy the secondary valence but the anions may satisfy both the primary and secondary valencies.

4. The secondary valencies are directional to provide the specific geometry of a complex but the primary valencies have nothing to do with the geometry of the coordination compounds.

5. Characterization of the coordination compounds from the measurement of molar conductivity and precipitation of some anions by the precipitating agents (*e.g.* $AgNO_3$, $BaCl_2$, etc.) may be complicated in some cases when the metal centres are labile.

6. $[CoBr_2(NH_3)_4]Br$ is a 1–1 electrolyte but its molar conductivity gradually increases in aqueous solution with a concomitant colour change.

7. From the number of isolated isomers, Werner could establish the geometry of coordination compounds of coordination numbers like 4 and 6.

8. $[M(A–A)_3]$ (where A–A stands for a symmetrical didentate ligand) can produce two optical isomers.

9. Werner's method of elucidation of structure of coordination compounds based on the stereochemical aspects was not free from criticism.

10. The oxy-anions, aminopolycarboxylates, cysteine, histidine, etc. can show the flexidentate character.
11. The concept of coordination number in the coordination compounds is different from that in the ionic solids.
12. In determining the coordination number in coordination compounds, the number of π-bonds formed by the ligands with the metal centre has no role.
13. Heptacity does not necessarily give the denticity of a ligand.
14. Olefins can act both as the monodentate and didentate ligands depending on the conditions.
15. Alkynes can act both as the monodentate and didentate ligands depending on the conditions.
16. Denticity of ligands is of an important consideration to determine the coordination number.
17. Double salts are different from the mixed crystals formed by the isomorphous crystals.
18. The distinction between the perfect and imperfect complexes is arbitrary.
19. Perfect and imperfect complexes can be qualitatively distinguished in terms of their formation constants (*i.e.* stability constants) or dissociation constants (*i.e.* instability constants).
20. In general, there is no strict structural order in the second coordination sphere but in some rare cases, a definite structural order is maintained in the second coordination sphere.
21. The number of cationic ligands is relatively rare.
22. μ-bond is a special type of σ-bond.
23. The π-acid ligands can stabilize the low oxidation states of metals.
24. For π-acid ligands, the σ-and π-bonds in the metal-ligand interactions work in a synergistic fashion.
25. The π-acid ligands are the strong field ligands while the pure σ-donor ligands or the ligands with the π-donor properties are the weak field ligands.
26. Hydrazine can act as a monodentate and as a didentate bridging ligand but not as a didentate chelating ligand.
27. Oxyanions (*e.g.* RCO_2^-, CO_3^{2-}, NO_3^-, NO_2^-, SO_4^{2-}, $S_2O_3^{2-}$) can show the flexidentate character.
28. Though the three membered chelate rings are rare, in some cases these are known.
29. Didentate ligands generally from the five and six membered chelate rings.
30. There are two types of didentate ligands- symmetrical and unsymmetrical didentate ligands.
31. Tripodal tetradentate ligands can participate in different types of geometries of coordination sphere.
32. In the chelating ring produced by acetylacetonate, pseudo-aromaticity is maintained.
33. Schiff's bases generally from the polydentate ligands.
34. Electrophilic substitution reaction in the acetelacetonate chelate ring supports the aromaticity in the ring.
35. In the chelated biguanide ligand, the pseudo-aromatic character is maintained.
36. edta, nta, cys, his can show the flexidentate character and their flexidentate character largely depends on pH.
37. Thiosulfate can act as an ambidentate ligand to produce the linkage isomers.
38. M–SCN linkage is bent while M–NCS linkage is linear.
39. Cysteinate as a didentate ligand can produce the linkage isomers like $[Co(cys)_3]^{3-}$(red) and $[Co(cys)_3]^{3-}$ (green).
40. Histidine (Hhis) can produce different isomeric *bis*-complexes of Cu(II) as shown below:

$$\left[Cu(Hhis)_2\right]^{2+} \xrightarrow{-H^+} \left[Cu(his)(Hhis)\right]^+ \xrightarrow{-H^+} \left[Cu(his)_2\right]$$

41. Ambidentate character of a peptide linkage mainly depends on pH.

42. Some didentate ligands can act as the bridging ligands but cannot act as the chelating ligands.

43. Dioxygen (O_2), dinitrogen (H_2) can act as the π-acid ligands.

44. Dihydrogen (η^2–H_2) complexes may lead to dihydrido complexes depending on the extent of back donation.

45. Nature has used many macrocyclic ligands for its biomolecules.

46. Phthalocyanine mimicking the porphyrin ligand can be readily synthesized in a template reaction.

47. Wrapping effect shown by the **flexible macrocyclic ligands** can stabilize the complex.

48. Clathro-chelates differ from the simple clathrates.

49. Kinetic template effect favours the reaction pathway while thermodynamic template effect stabilizes a particular product.

50. In each of the following reactions, there are two possible products but in presence of some metal ions like Ni^{2+}, Cu^{2+}, one particular product is stabilized.

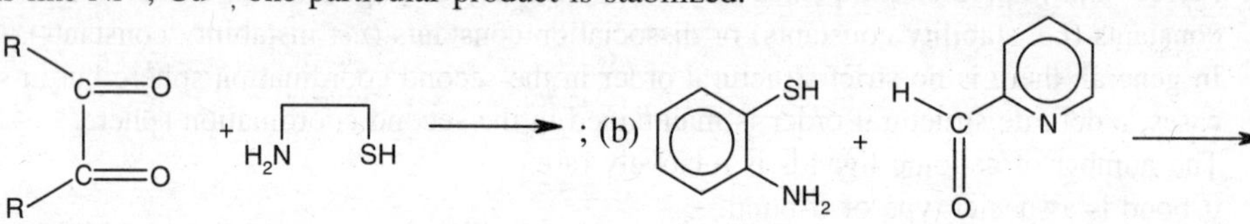

51. There are many template synthesis reactions of the polydentate ligands where both thermodynamic and kinetic effects operate.

52. Synthesis of Schiff's base ligands in the condensation reactions: (i) Hacac + en; (ii) Hglygly + salicylaldehyde; (iii) trien + acetone; (iv) Hgly + salicylaldehyde; (v) 2, 6-diacetyl pyridine + polyamines — the labile metal ions like Ni^{2+}, Cu^{2+} can catalyze the processes.

53. In Curtis synthesis of macrocyclic ligands:

    $[M(en)_3]^{3+}$ + acetone (dry) $\longrightarrow$ Macrocyclic tetraaza ligand

    'M' must be a labile centre and there must be a dehydrating agent like anhydrous $CaSO_4$.

54. In Curtis synthesis of macrocyclic ligands, $[Co(en)_3]^{3+}$ cannot be used instead of $[Ni(en)_3]^{2+}$.

55. In the template synthesis of polydentate and macrocyclic ligands, generally the labile metal centres are used.

56. Both the unsaturated and saturated (*e.g.* cyclam) macrocyclic ligands may be synthesized in Curtis reaction.

57. Sepulchrate (sep) can be prepared by Mannich condensation in a template reaction.

58. Sepulchrate mimicks the cryptand.

59. Fehling's solution is highly alkaline but Cu^{2+} is not thrown out as $Cu(OH)_2$.

60. *Tartar emetic* is a binuclear complex.

61. H_2S can precipitate cadmium but not copper from a mixture of Cu^{2+} and Cd^{2+} in presence of excess CN^-.

62. AgX (X^- = halide) can be solubilised in presence of excess halide or pseudohalide like CN^-.

63. Tollen's reagent is highly alkaline but Ag^+ is not precipitated as $Ag(OH)$.

64. Bi^{3+} or Hg^{2+} is precipitated by KI solution but in presence of excess KI, the precipitate is redissolved; a precipitate may reappear in presence of large cations like pyridinium ion.

65. AgCl is soluble in dilute NH_3 solution, AgBr is soluble in concentrated NH_3 solution, but AgI is insoluble even in concentrated NH_3 solution.

66. H_2S cannot precipitate AgS in presence of excess CN^-.

67. The noble metals like Ag, Au, etc. cannot react with the ordinary acids to liberate the H_2 gas but they can perform the task in presence of CN^-.

68. HNO_3 cannot alone attack the noble metals like Au, Pt but it can attack in presence of HCl or tartaric acid.

69. Thermodynamically, aqua regia cannot be more oxidizing than the concentrated HNO_3, but aqua regia oxidizes the noble metals while the concentrated HNO_3 alone fails to do the task.

70. The insoluble precipitates like $BaSO_4$, $BaCO_3$ can be solubilized in presence of edta.

71. The thiosulfato complex of Ag(I) is unstable.

72. The reaction between HgO(s) and excess KI can be utilized to prepare a standard solution of alkali.

73. Aqueous solution of Co(II) is light pink while in concentrated HCl media, it looks blue; the blue colour becomes again light pink in presence of excess Hg^{2+}, i.e.,

$$Co^{2+} \underset{(\text{Light pink})}{(\text{aq. solution})} \xrightarrow{\text{excess Cl}^-} Blue \xrightarrow{\text{excess Hg}^{2+}} \text{Light pink.}$$

74. The colour changes as follows:

   (i) Anhydrous $CuSO_4$ (Colourless) $\xrightarrow{\text{water}}$ Light blue $\xrightarrow{NH_3}$ Deep blue.

   (ii) Ni^{2+} (aq. solution) (Green) $\xrightarrow{NH_3}$ Blue

   (iii) Fe^{3+} (aq. acidic solution) (Colourless) $\xrightarrow[\text{SCN}^-]{\text{Cl}^-,\text{Br}^- \text{ or}}$ coloured

   (iv) Commercial $CrCl_3 \cdot H_2O$ (Pale green) $\xrightarrow{\text{on standing}}$ Blue-violet

75. The metal ions like Ag^+, Cd^{2+}, Zn^{2+}, Hg^{2+}, etc. can show the abnormal transport number in presence of the complexing halides or pseudohalides.

76. Fe(III) can be extracted in diethyl ether in presence of concentrated HCl.

77. pH of $CuSO_4$ solution falls sharply when titrated with glycine but pH of $Cu(OAC)_2$ solution does not change sharply when titrated with glycine.

78. pH-metric study can support the complexation reaction in some cases.

79. The Co(III)/Co(II) couple is a powerful oxidizing agent in pure aqueous solution but the oxidizing power falls sharply in presence of the ligands like phen, en, CN^-, etc.

80. Oxidizing power of the Fe(III)/Fe(II) couple decreases in presence of CN^- and F^- but it increases in presence of phen.

81. Inner complexes of first-order are of much importance in analytical chemistry.

82. Some inner complexes of first-order may be considered as the inner complexes of 3rd-order.

83. Some inner complexes of first-order may be considered as the inner complexes of 4th-order.

84. In the hydroxido-bridged polymeric compounds, the number of OH-bridges between the metal centres depends on the charge of the metal centres.

85. The properties of the hydroxido- and oxido- bridged dinuclear complexes,

$$[(H_3N)_5Cr \overset{H}{-} O - Cr(NH_3)_5)]^{5+} \text{ and } [(H_3N)_5Cr-O-Cr(NH_3)_5]^{4+} \text{ are different.}$$

86. $Co(NH_2)_3$ actually exists as a polymeric coordination compound and depending on the conditions, $[Co(NH_3)_6]^{3+}$ and $[Co(NH_2)_6]^{3-}$ may be obtained from the polymeric species $\{Co(NH_2)_3\}_x$.

87. In Vortmann's green salt, both the Co-centres are equivalent.

88. (a) Ammoniacal Co(II)-salt $\xrightarrow{O_2}$ A (brown dinuclear complex) $\xrightarrow{i.e.,\ oxidation}$

Vortmann's Green $\xleftarrow{\quad}$ (i) Base | B (green,
Complex $\qquad$ (ii) 1e-oxidation $\qquad$ dinuclear complex)

(b) $[Co(CN)_5]^{3-} \xrightarrow{O_2} A \xrightarrow{i.e.,\ 1e-oxidation} B$

89. Simple Co(II)-compounds are irreversibly oxidized by O_2 but the sterically hindered complexes like $[Co^{II}(py)(acacen)]$ or $[Co^{II}(py)(salen)]$ can bind O_2 reversibly (*Hints:* Steric Control).

90. In crystal, $MnCl_2.4H_2O$ exists as a monomeric one while $MnCl_2.2H_2O$ produces a polymeric form.

91. $CuCl_2.2H_2O$ crystal hydrate looks green but on dissolution in water it gives a light blue colour.

92. SnF_4 is solid while $SnCl_4$ and $SnBr_4$ are liquid at ordinary temperature.

93. The melting points run as: $Me_3SnF \rangle Me_3SnCl \rangle Me_3SnBr$.

94. Me_3SnF forms a one dimensional polymeric chain, Me_2SnF_2 forms a two dimensional polymeric sheet while $MeSnF_3$ forms a three dimensional polymeric network.

95. $FeCl_3$ (solid) adopts a layer structure while FeF_3 (solid) adopts a ReO_3-structure; they also widely differ in melting points and solubilities.

96. Fluoro compounds (*e.g.* AlF_3, FeF_3, SnF_4, BeF_2, etc.) show very high melting points.

97. On melting of $AlCl_3$ crystal, the volume increases abruptly.

98. The modes of polymerization of SbF_5 in solid and liquid state are different. The physical properties of AsF_5 and SbF_5 differ significantly.

99. Solid $PdCl_2$ and $PtCl_2$ can exist in $\alpha-$ and $\beta-$ forms.

100. β-$PdCl_2$ and β-$PtCl_2$ show the molecular behaviour (*e.g.* soluble in benzene).

101. $[PtBr_3(NH_3)_2]$ and $[PtBr_3(en)]$ remain in polymeric forms.

102. $[Cu(CN)_2]^-$ exists in a polymeric chain.

103. $[Ni(CN)_4]^{2-}$ is a monomeric unit while $[Ni(CN)_3]^{2-}$ remains as a dimer.

104. $M(CN)_2$ (M = Pd, Ni) can produce a two-dimensional planar polymeric network in solid state but Prussian blue and related materials produce a three dimensional polymeric structure.

105. AgCN and AuCN can produce the polymeric chains in solid state.

106. $AuR_2(CN)$ produces a tetrameric unit in a solid state while AuR_2X (X = Cl, Br) produces a dimeric unit.

107. $NiX_2.2H_2O$ (X = Cl, Br) and $CdCl_2.2NH_3$ produces the chain structures in solid state.

108. On standing **Turnbull's blue** isomerizes to **Prussian blue**.

109. The polymeric structure of $Ni(CN)_2.NH_3$ can be rationalized by considering the **modified (template effect)** skeleton structure of polymeric $Ni(CN)_2$.

110. In the modified (template effect) polymeric structure of $Ni(CN)_2$, the additional NH_3 ligands coordinate to the Ni(II)-centre which is already coordinated to the N-ends of CN^-.

111. Both the Prussian and Turnbull's blue are highly coloured.

112. Prussian blue is thermodynamically stabler than the isomeric compound Turnbull's blue.

113. Berlin green, *i.e.* $Fe^{III}Fe^{III}(CN)_6$ and Everitt's salt $K_2Fe^{II}Fe^{II}(CN)_6$ are not intensely coloured.

114. $FeSO_4 + K_3[Cr(CN)_6] \xrightarrow{\text{Cold}}$ Red $\xrightarrow{\text{Heating}}$ Green
                                        Compound   Compound

115. AgSCN (solid) forms a polymeric zig-zag chain.

116. $MnSO_4.4H_2O$ forms a dinuclear species in crystal.

117. $Cr(aq)^{3+}$ complex is violet, but the Cr(III) species produced from the reduction of Cr(VI) by Fe(II) in aqueous H_2SO_4 media looks green.

118. $Cr(aq)^{3+}$ complex is violet but the partially substituted sulfato or chlorido complex of Cr(III) generally looks green (*i.e.* aquasulfato–or aquachlorido–complex of Cr(III) is green but on aquation it changes to violet).

119. $Cr_2(SO_4)_3.18H_2O$ is violet while $Cr_2(SO_4)_3.6H_2O$ is green; and from the green salt solution, addition of $BaCl_2$ does not produce any precipitate while from the violet solution, $BaCl_2$ gives the precipitate of $BaSO_4$.

120. Chrome alum, $K_2SO_4.Cr_2(SO_4)_3.24H_2O$ is violet coloured.

121. $[Ni(NCS)_2(NH_3)_2]$ forms a polymeric compound.

122. Metal rubeanates are the examples of addition polymer.

123. $[PtBr_3(NH_3)_2]$ is a combination of $[Pt^{II}Br_2(NH_3)_2]$ and $[Pt^{IV}Br_4(NH_3)_2]$.

124. Creutz-Taube ion, $[(H_3N)_5Ru—N\bigcirc N—Ru(NH_3)_5]^{5+}$ shows a very low energy absorption band.

125. In, $[(H_3N)_5Ru—N\bigcirc N—Ru(NH_3)_5]^{5+}$, there is an extensive delocalization of the electron over the both metal centres (*i.e.* metal centres are equivalent) but in $[(H_3N)_5Ru^{III}—N\bigcirc N—Ru^{II}Cl(bpy)_2]^{5+}$, the electron is localized (*i.e.* the metal centres are non-equivalent).

126. Robin-Day classification of mixed valence complexes depends on the extent of delocalization of charge.

127. Intervalence charge transfer band at low energy is a characteristic feature of class-II (in terms of Robin-Day classification) mixed valence complexes.

CHAPTER 2

Stereochemistry and Structural Aspects of Complex Compounds

2.1 ISOMERISM IN COORDINATION COMPOUNDS AND DIFFERENT TYPES OF ISOMERISM

The compounds having the same chemical composition with different properties due to the structural difference are called **isomers** (*isos* meaning *the same* and *meros* meaning *the parts*).

The structural difference causing the isomerism can arise in two ways — **difference in atom or bond connectivity** (*i.e.* in chemical linkage) and **difference in orientation of the atoms or groups in space.**

There are **three types of isomerism.**

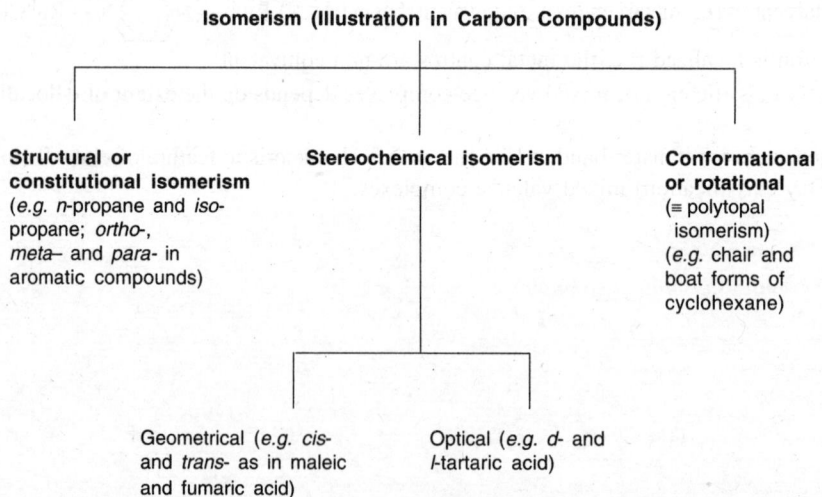

Compared to the isomerism in organic compounds, the **diversity of isomerism in coordination compounds** is of much wider range because of various possibilities of atom connectivity (*i.e.* chemical linkage) and stereochemical configurations in coordination compounds.

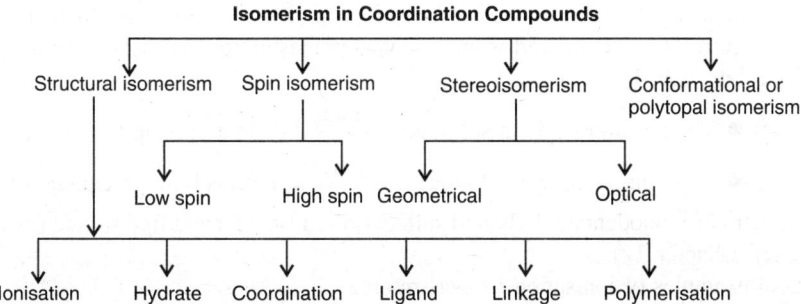

Isomerism in Coordination Compounds

These different types of isomerism in coordination compounds are discussed here.

2.2 STRUCTURAL ISOMERISM

2.2.1 Ionisation Isomerism

It occurs due to the exchange between the ligating ions (present in the coordination sphere) and the ions residing outside the coordination sphere. Thus it leads to **an interchange of the ions between the inner– and outer–coordination sphere.**

● The ions present outside the coordination sphere are held ionically and these only ionise while the ligands present within the coordination sphere are held by Werner's secondary valence and these are not ionised. *Consequently, in the case of ionisation isomerism, the available ions in solution are different.*

● The construction of inner sphere determines the **colour of a coordination compound.** *Thus the ionisation isomerism (i.e. different composition of inner sphere) leads to different colours.*

● Because of the different modes of ionisation in ionisation isomerism, the **molar conductivities** are different for such isomers.

These are illustrated below.

$$\left[CoBr(NH_3)_5\right]SO_4 \rightleftharpoons \left[CoBr(NH_3)_5\right]^{2+} + SO_4^{2-}, \ (2-2 \text{ electrolyte})$$
$$\text{(Violet)}$$

$$\left[Co(NH_3)_5(SO_4)\right]Br \rightleftharpoons \left[Co(NH_3)_5(SO_4)\right]^{+} + Br^{-}, \ (1-1 \text{ electrolyte})$$
$$\text{(Red)}$$

In the **violet isomer,** bromide is a ligand but sulfate is not a ligand while the reverses are true in the **red isomer.**

Preparation of the above violet and red isomers (cf. Scheme 1.2.1)

$$CoBr_2 \xrightarrow[NH_4Br+NH_4OH]{O_2 \ (air)} \left[Co(NH_3)_5(OH_2)\right]Br_3 \xrightarrow{Heating} \left[CoBr(NH_3)_5\right]Br_2$$

$$\left[CoBr(NH_3)_5\right]Br_2 \xrightarrow[-2AgBr]{Ag_2SO_4} \left[CoBr(NH_3)_5\right]SO_4$$

$$\left[Co(NH_3)_5(SO_4)\right]Br \xleftarrow[-BaSO_4]{BaBr_2} \left[Co(NH_3)_5(SO_4)\right]HSO_4 \xleftarrow[-Br_2]{Conc. \ H_2SO_4} \left[CoBr(NH_3)_5\right]SO_4$$
$$\text{(Red)} \qquad\qquad \left(\text{conc. } H_2SO_4 \text{ oxidises } Br^- \text{ to } Br_2\right) \qquad\qquad \text{(Violet)}$$

The molar conductance values of the violet isomer, petaamminebromidocobalt(III) sulfate (2–2, i.e. bi-bivalent electrolyte) and the red isomer pentaamminesulfatocobalt(III) bromide (1–1, i.e. uni-univalent electrolyte) are different.

- Violet compound (in solution) $\xrightarrow{\text{BaCl}_2}$ White precipitate of $BaSO_4$

- Red compound (in solution) $\xrightarrow{\text{AgNO}_3}$ Pale yellow precipitate of AgBr.

- The ir-spectra of monodentatedly bound sulfate (in red isomer) and free sulfate (in violet isomer) are different (cf. Chapter 12).

Other sets of examples of ionisation isomerism are:

(i) $[CoCl_2(NH_3)_4]NO_2$ and $[CoCl(NH_3)_4(NO_2)]Cl$; trans-$[CoCl_2(en)_2]NO_2$ (green) and trans-$[CoCl(en)_2(NO_2)]Cl$ (red)

(ii) $[CoCl(en)_2(NO_2)]Br$, $[CoBr(en)_2(NO_2)]Cl$ and $[CoBrCl(en)_2]NO_2$

(iii) $[CoCl_2(NH_3)_4]NO_2$ and $[CoCl(NH_3)_4(NO_2)]Cl$

(iv) $[PtCl_2(NH_3)_4]Br_2$, $[PtBrCl(NH_3)_4]BrCl$ and $[PtBr_2(NH_3)_4]Cl_2$

Difference in the properties for the ionisation isomers

The ionisation isomers differ in the properties like colour, molar conductance, chemical tests for the respective ions, ir-spectra, etc.

Note: The phenomenon of ionisation isomerism exists based on the assumption that the ions held (through the primary valence) outside the coordination sphere are only ionised in solution, and the ions present within the coordination sphere satisfying both the primary and secondary valence of the central metal ion simultaneously do not ionise. **This assumption is strictly true only for the kinetically inert complexes of the metal ions like Co(III) (low spin), Cr(III), Pt(IV), etc.** For the **labile metal complexes,** the ions from the inner sphere come out rapidly through solvation and *in such cases, the ionisation isomerism cannot be characterised based on the presence of ions in solution.* However, other studies (e.g. ir-spectra) in **solid state** can distinguish the ionisation isomers.

2.2.2 Hydration Isomerism

It occurs due to the **different modes of distribution of water molecules between the inner sphere and outer sphere**.

- The water molecules present as ligands within the coordination sphere are much more tightly bound than the water molecules present outside the coordination sphere. In fact, the water molecules residing outside the coordination sphere are loosely bound and *easily lost* by heating (ca. 100°C) or over the dehydrating agent like concentrated H_2SO_4. The ligand water molecules need a much higher temperature (ca. 150–170°C) to be lost.

- Because of the difference in the composition of the inner sphere in such isomerism, it leads to different colours of different isomers.

- These are basically **a special type of ionisation isomerism** when it occurs due to the exchange of H_2O molecules and ions between the inner sphere and outer sphere. In such cases, the different hydration isomers produce different ions in solution. It leads to difference in molar conductance values, chemical tests for the ions in solutions and spectral properties, etc.

These are illustrated for **chromic chloride hydrate** producing the isomers.

(i) $[Cr(OH_2)_6]Cl_3$;   (ii) $[CrCl(OH_2)_5]Cl_2 \cdot H_2O$;   (iii) $[CrCl_2(OH_2)_4]Cl \cdot 2H_2O$
      Violet   **Blue-green**   **Dark-green.**

                                               (It may have the *cis*– and *trans*- isomers)

(i) **Violet isomer:** No loss of water over conc. H_2SO_4; all the chlorides are precipitated by the addition of $AgNO_3$ solution to the freshly prepared solution; molar conductance of the freshly prepared solution is comparable to that of 3-1, *i.e.* ter-univalent electrolyte;

(ii) **Blue-green isomer:** Loss of one molecule of water over conc. H_2SO_4; two chlorides are readily precipitated by $AgNO_3$; molar conductance of the freshly prepared solution is comparable to that of 2–1, *i.e.* bi-univalent elctrolyte.

(iii) **Dark-green isomer:** Loss of two molecules of water over conc. H_2SO_4 and one-third of the chlorides is precipitated by $AgNO_3$; molar conductance of freshly prepared solution is comparable to that of 1–1, *i.e.* uni-univalent electrolyte.

What is commercial $CrCl_3 \cdot 6H_2O$? What is the colour of $Cr_2(SO_4)_3$ solution?

Commercial $CrCl_3 \cdot 6H_2O$ is constituted by the above mentioned three isomers. The predominant form is *trans*–$[CrCl_2(OH_2)_4]Cl \cdot 2H_2O$. These isomers can be separated by using the cation exchange resins followed by elution with dil. $HClO_4$ of different concentrations. Green solution of commercial $CrCl_3 \cdot 6H_2O$ on standing for a prolonged period is gradually converted into violet solution through aquation generating $[Cr(OH_2)_6]Cl_3$. See Sec. 1.12 (O) for the **colours of chromic sulfate solution.**

Other examples of hydration isomerism are:
  (i) $[CoCl_2(NH_3)_4]Cl \cdot H_2O$ and $[CoCl(NH_3)_4(OH_2)]Cl_2$
  (ii) $[Cr(en)_2(OH_2)_2]Br_3$ and $[CrBr(en)_2(OH_2)]Br_2 \cdot H_2O$

2.2.3 Coordination Isomerism

● When the salts are produced by the cationic complex and anionic complex, **the metal centers between the complex entities may be interchanged** to give the coordination isomerism.

    Thus it causes the exchange of metal centers between the coordination spheres leading to the coordination isomerism.

Coordination Position Isomerism

In the binuclear and polynuclear complexes, interchange of the ligands between/among the metal centers can lead to *a special type of coordination isomerism* that may be described as **coordination position isomerism**. It is illustrated in some typical binuclear complexes.

● Sometimes the **interchange or redistribution of ligands** between the coordination spheres can lead to coordination isomerism. The examples are:

(i) $[Pt(NH_3)_4][PdCl_4]$ and $[Pd(NH_3)_4][PtCl_4]$
(ii) $[Co(NH_3)_6][Cr(CN)_6$ and $[Cr(NH_3)_6][Co(CN)_6]$ Exchange of two different metal
(iii) $[Co(NH_3)_6][Cr(ox)_3$ and $[Cr(NH_3)_6][Co(ox)_3]$ centers between the coordination spheres.
(iv) $[Cu(NH_3)_4][PtCl_4]$ and $[Pt(NH_3)_4][CuCl_4]$

(v) $[Pt^{II}(NH_3)_4][Pt^{IV}Cl_6]$ and $[Pt^{IV}Cl_2(NH_3)_4][Pt^{II}Cl_4]$ Redistribution of
(vi) $[Cr(NH_3)_6][Cr(SCN)_6]$ and   ligands between the
     $[Cr(NH_3)_4(SCN)_2][Cr(NH_3)_2(SCN)_4]$ coordination spheres.

2.2.4 Ligand Isomerism

It arises due to the **isomeric nature of the ligands**. The isomeric ligands are illustrated below:

(i) **Monodentate amines:** $N(CH_3)_3$, $CH_3CH_2CH_2NH_2$, (*n*-pram) CH_3CHCH_3 (*iso*-pram)
               (trimethylamine)   (*n*-propylamine) NH_2
                                                             (isopropylamine)

(ii) **Didentate amines:**
           H_3C——CH——NH_2 and CH_2——NH_2 / CH_2 \ CH_2——NH_2
                CH_2——NH_2 *d*-pn and *l*-pn (by considering the
                 (pn)   (tn)   optical isomers in the ligand)

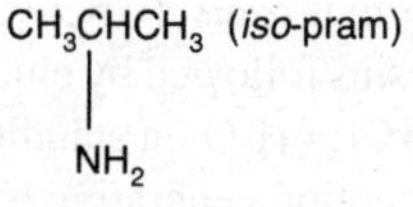

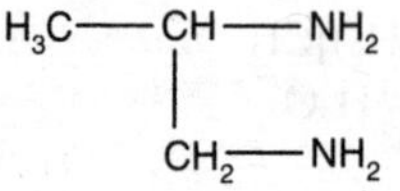

(iii) **Didentate amino acids:**   H——C——NH_2 and
                             CH_3 (above) H_2C \ NH_2
                             CO_2^- H_2C \ CO_2^-
                          (α-ala)   (β-ala)

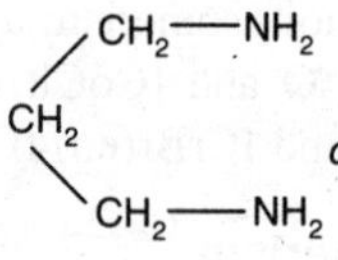

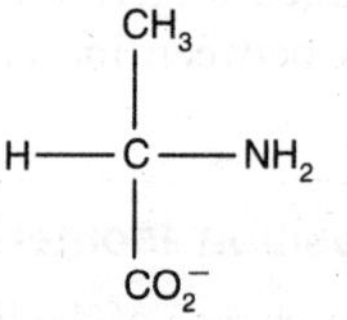

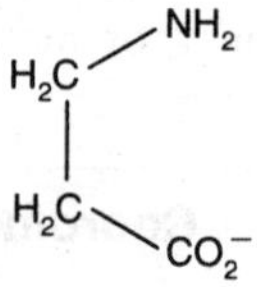

(iv) **Didentate amines:** H_3C——CH——NH_2 (bn) and H_3C——C——NH_2 (*i*-bn)
                       H_3C——CH——NH_2 CH_2——NH_2 (top) CH_3 (bottom)

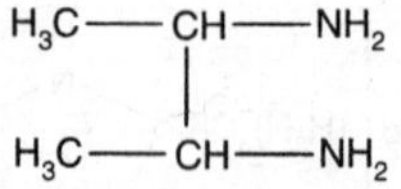

(v) **Tetradentate amines:**   HN——$(CH_2)_2$——NH   and   N——$(CH_2)_2$——NH_2
                      $(CH_2)_2$——NH_2 H_2N——$(CH_2)_2$ $(CH_2)_2$——NH_2
                                                  $(CH_2)_2$——NH_2
                             (trien)   (tren)

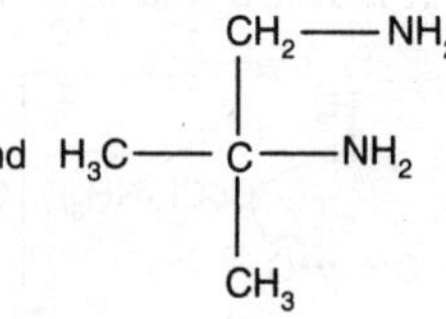

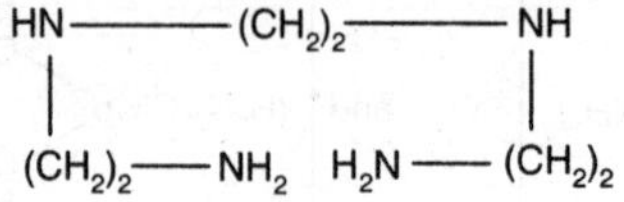

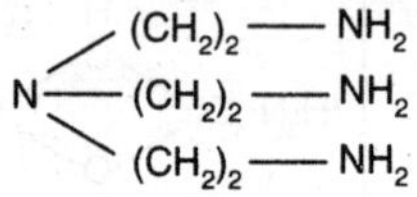

(vi) **Isomeric pyridine carboxylic acids as the didentate ligands:**

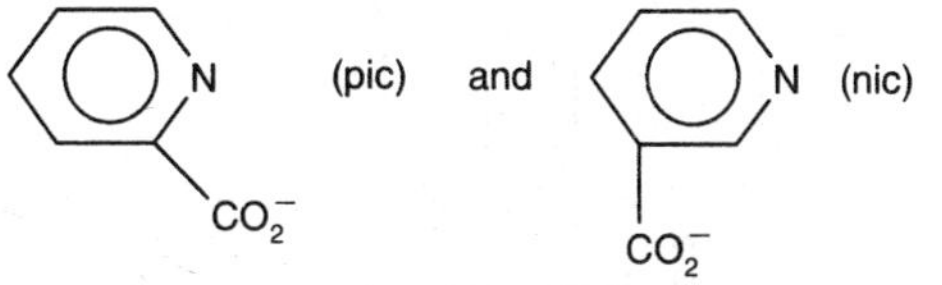

(picolinate *i.e.*
pyridine-2-carboxylate)

(nicotinate *i.e.*
pyridine-3-carboxylate)

(vii) **Isomeric olefinic ligands:**

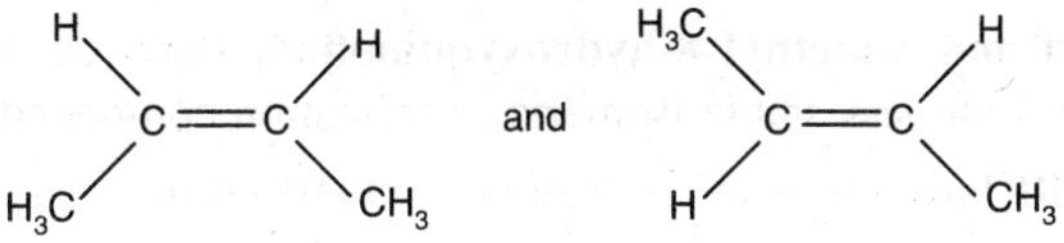

(*cis*-butene)

(*trans*-butene)

(viii) **Isomeric oxine-derivatives:**

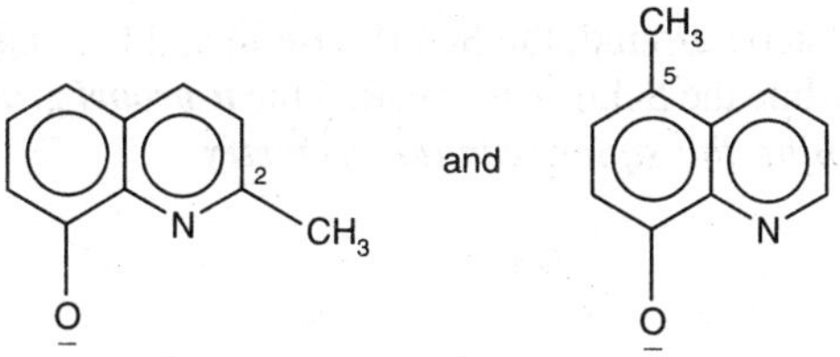

and

(ix) **Isomeric Schiff-bases arising from the isomeric amines:**

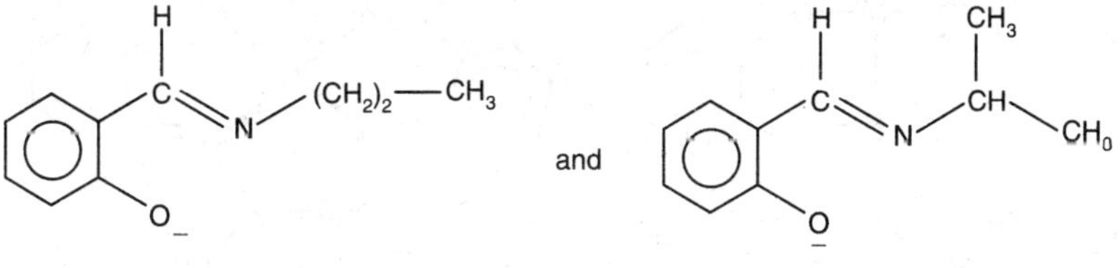

and

(Schiff-base anions obtained from the condensation of salicylaldehyde and *n*-propylamine or *iso*-propylamine)

(x) **Optical isomerism in ligands:** Optical isomers of the ligand, *e.g.* *d*-pn and *l*-pn; *d*-tartarate, *l*-tartarate and *meso*-tartarate; etc.

Examples of coordination compounds showing ligand isomerism.

$[Co(\alpha\text{-ala})_3]$ and $[Co(\beta\text{-ala})_3]$; $[Ag(pic)_2]^-$ and $[Ag(nic)_2]^-$

$[CoCl_2(pn)_2]$ and $[CoCl_2(tn)_2]$; $[Co(bn)_2Cl_2]^+$ and $[Co(i\text{-bn})_2Cl_2]^+$; etc.

● **tren** and **trien** (*cf.* Fig. 4.5.2.9): Both are the tetradentate ligands but tren **(a tripod ligand)** favours sterically the tetrahedral geometry while trien favours sterically the square planar complexes. *These preferences are based on the flexibility of the ligands.* This is why, the metal centers like Zn^{2+} (d^{10} system, *i.e.* no *cfse*) which prefere the tetrahedral geometry prefer the tren ligand; while the metal centers like Cu^{2+} which prefer the square planar geometry (high *cfse*) prefer the trien ligand. Thus:

$[Zn(tren)]^{2+}$ » $[Zn(trien)]^{2+}$; $[Cu(trien)]^{2+}$ » $[Cu(tren)]^{2+}$ **(stability order)**.

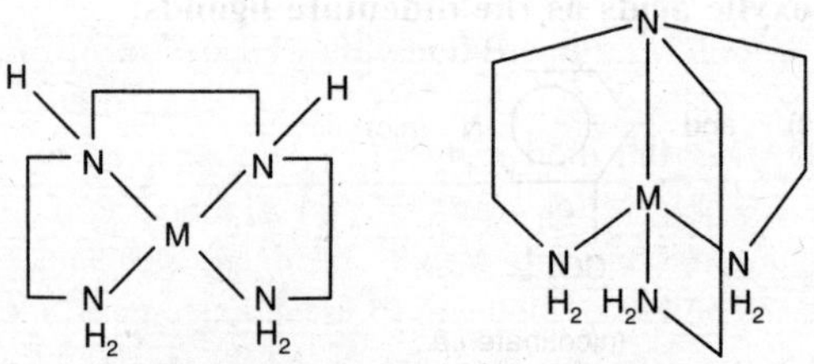

[M(trien)]²⁺   **[M(tren)]²⁺**

● **2-methyl and 5-methyl 8-hydroxyquinoline:** Between these isomeric ligands, the 2-methyl derivative forms the less stable isomeric coordination compounds because of the steric ground.

● The **Schiff base** given in (ix) can be synthesised in a *template reaction, i.e.*

$$bis\,(\text{salicylaldehydato})\,\text{nickel}\,(\text{II}) \;+\; \underset{(n\text{-Pr}-\text{NH}_2/iso\text{-Pr}-\text{NH}_2)}{\text{propylamine}} \longrightarrow \begin{array}{l}\text{Immine Schiff-base}\\ \text{complex of Ni}\,(\text{II}).\end{array}$$

Because of the steric ground, the Schiff base ligand bearing the *isopropyl group prefers the tetra-hedral geometry* while the Schiff base bearing the *n-propyl group does not experience any such steric crowding in attaining the square planar geometry.*

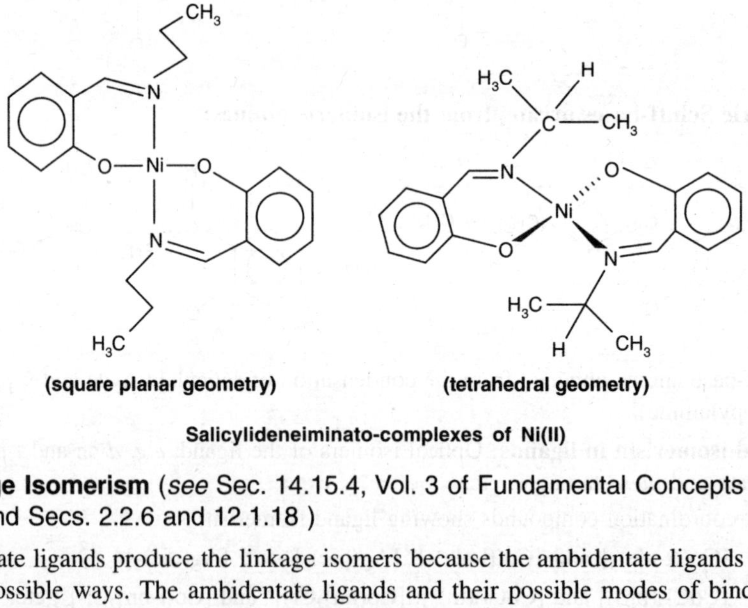

(square planar geometry)   **(tetrahedral geometry)**

Salicylideneiminato-complexes of Ni(II)

2.2.5 Linkage Isomerism (*see* Sec. 14.15.4, Vol. 3 of Fundamental Concepts of Inorganic Chemistry and Secs. 2.2.6 and 12.1.18)

The ambidentate ligands produce the linkage isomers because the ambidentate ligands can coordinate in different possible ways. The ambidentate ligands and their possible modes of binding have been discussed in 1st Chapter. Some representative **ambidentate ligands** are: NO_2^-, SCN^-, CN^-, $SeCN^-$, $S_2O_3^{2-}$, SO_3^{2-}, $(CH_3)_2SO$, etc.

The phenomenon of linkage isomerism is illustrated in the following examples.

(i) **Nitro- (*i.e.* nitrito-κ*N*) and nitrito (*i.e.* nitrito-κ*O*) isomers:**

Linkage isomers from the ambidentate ligand NO_2^- (nitrito-κ*N* and nitrito-κ*O*) are illustrated below.

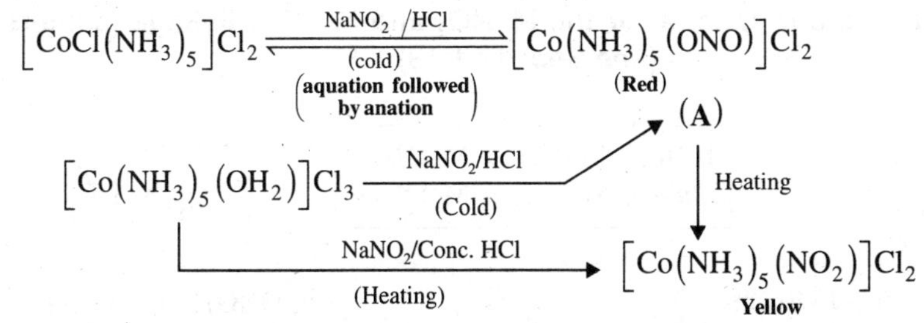

$$\left[\text{CoCl}(NH_3)_5\right]Cl_2 \underset{(\text{cold})}{\overset{NaNO_2 / HCl}{\rightleftharpoons}} \left[\text{Co}(NH_3)_5(ONO)\right]Cl_2$$

$$\begin{pmatrix}\text{aquation followed}\\\text{by anation}\end{pmatrix} \qquad \text{(Red)}$$

$$\qquad\qquad\qquad\qquad\qquad\qquad\qquad\qquad \textbf{(A)}$$

$$\left[\text{Co}(NH_3)_5(OH_2)\right]Cl_3 \xrightarrow[\text{(Cold)}]{NaNO_2/HCl}$$

$$\xrightarrow[\text{(Heating)}]{NaNO_2/\text{Conc. HCl}} \left[\text{Co}(NH_3)_5(NO_2)\right]Cl_2$$

Heating

Yellow

The interconversion in solid state may be also photochemically carried out.

$$\left[\text{Co}(NH_3)_5(NO)_2\right]^{2+} \overset{h\nu}{\rightleftharpoons} \left[\text{Co}(NH_3)_5(ONO)\right]^{2+}$$

The red isomer is a **kinetically controlled product** while the yellow isomer is **thermodynamically more stable** because of the **symbiotic effect.** When $[\text{Co}(NH_3)_5(^*OH_2)]^{3+}$ is treated with NO_2^-, the labelled oxygen (O^*) is found to be completely incorporated in the nitrito (O), i.e. nitrito-κO, isomer. In fact, for the formation of nitrito (O) isomer, from the aqua complex, the '**Co — O' bond is not ruptured and this is why, the process is fairly fast.**

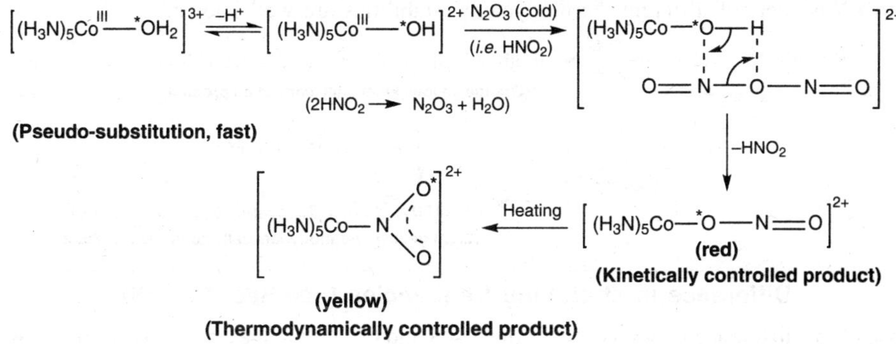

$$\left[(H_3N)_5\overset{III}{Co}{-\!\!-}^*OH_2\right]^{3+} \overset{-H^+}{\rightleftharpoons} \left[(H_3N)_5\overset{III}{Co}{-\!\!-}^*OH\right]^{2+} \xrightarrow[(i.e.\ HNO_2)]{N_2O_3\ (cold)} \left[(H_3N)_5Co{-}^*O{-}H \cdots \right]^{2+}$$

$$(2HNO_2 \rightarrow N_2O_3 + H_2O)$$

(Pseudo-substitution, fast)

$$\begin{bmatrix}(H_3N)_5Co{-}N\overset{O^*}{\underset{O}{\diagdown}}\end{bmatrix}^{2+} \xleftarrow{\text{Heating}} \begin{bmatrix}(H_3N)_5Co{-}^*O{-}N{=}O\end{bmatrix}^{2+}$$

$$\text{(red)}$$
$$\text{(Kinetically controlled product)}$$

$$-HNO_2$$

(yellow)
(Thermodynamically controlled product)

During isomerisation of the nitrito (O) isomer to the thermodynamically more stable nitrito (N) isomer, **the 'Co — O' bond is to be ruptured and this is why the process is slow.** From the isotope labelling experiment, it has been established that the isomerisation process goes on through an **intramolecular rearrangement** as shown below.

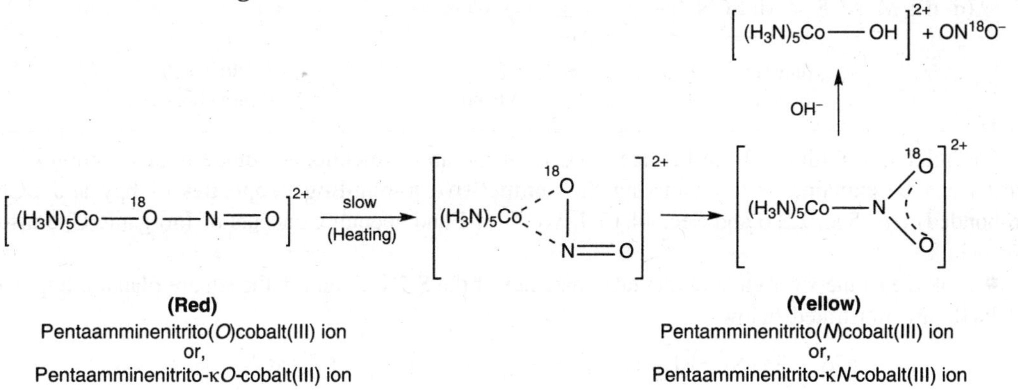

$$\left[(H_3N)_5Co{-\!\!-}OH\right]^{2+} + ON^{18}O^-$$

$$OH^-$$

$$\begin{bmatrix}(H_3N)_5Co{-\!\!-}\overset{18}{O}{-}N{=}O\end{bmatrix}^{2+} \xrightarrow[\text{(Heating)}]{\text{slow}} \begin{bmatrix}(H_3N)_5Co\overset{\overset{18}{O}}{\underset{N{=}O}{\diagdown}}\end{bmatrix}^{2+} \longrightarrow \begin{bmatrix}(H_3N)_5Co{-}N\overset{\overset{18}{O}}{\underset{O}{\diagdown}}\end{bmatrix}^{2+}$$

(Red)
Pentaamminenitrito(*O*)cobalt(III) ion
or,
Pentaamminenitrito-κ*O*-cobalt(III) ion

(Yellow)
Pentamminenitrito(*N*)cobalt(III) ion
or,
Pentaamminenitrito-κ*N*-cobalt(III) ion

Difference in the ir-spectra for the M–NO₂ and M–ONO linkage isomers
(*see* Sec. 12.1.18)

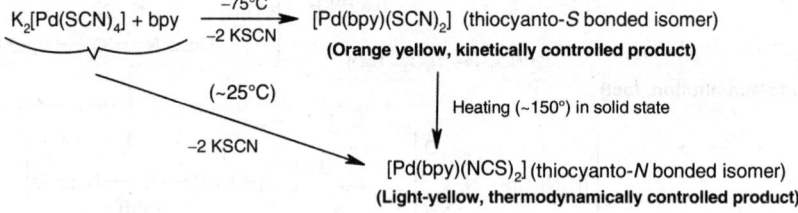

| 10Dq: | M–NO$_2$ > M–ONO |
| λ(d–d): | M–NO$_2$ < M–ONO |

$v_s(NO_2) \approx 1310 - 1340$ cm^{-1}

$v_a(NO_2) \approx 1370 - 1450$ cm^{-1}

$\Delta v \approx 60 - 110$ cm^{-1}

$v_s(ONO) \approx 1040 - 1100$ cm^{-1}

$v_a(ONO) \approx 1420 - 1480$ cm^{-1}

$\Delta v \approx 400$ cm^{-1}

cf. For **free NO$_2^-$ ion:** v_s (symmetric stretch) = 1250 cm^{-1}, v_a (antisymmetric stretch) = 1335 cm^{-1}.
Higher band separation (Δv) for the nitrito-isomer is due to the inequality in the two N–O bond lengths. ρ_W (NO$_2$) (wagging mode) $\approx 620 - 640$ cm^{-1} for the nitro-isomer and it is absent for the nitrito-isomer.

(ii) **Linkage isomers from the ambidentate ligand SCN⁻:**

● The isomeric compounds [Pd(bpy)(SCN)₂] (*i.e.* thiocyanato-κS-isomer) and [Pd(bpy)(NCS)₂] (*i.e.* thiocyanto-κN-isomer) of different thermodynamic stabilities are well known.

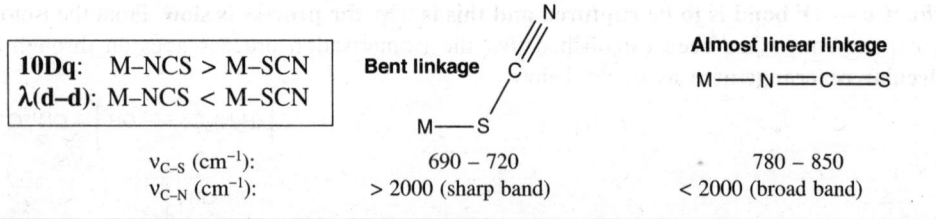

Difference in stretching frequencies (*see* Sec. 12.1.18)

Because of the difference in bond order in the C – S and C – N linkages, for two different possible modes of binding of SCN⁻, the *ir* – stretching frequencies of the linkages differ.

| 10Dq: | M–NCS > M–SCN |
| λ(d–d): | M–NCS < M–SCN |

Bent linkage

Almost linear linkage

M——N══C══S

M——S

	Bent linkage	Almost linear linkage
v_{C-S} (cm^{-1}):	690 – 720	780 – 850
v_{C-N} (cm^{-1}):	> 2000 (sharp band)	< 2000 (broad band)

The relative stabilities of the linkage isomers of the above mentioned square planar complexes of Pd(II) can be explained by considering the **competitive π–bonding properties** of bpy and SCN⁻ (*S*-bonded) (*see* Sec. 2.2.6 and Sec. 14.15.4, Vol. 3 of Fundamental Concepts of Inorganic Chemistry for details).

● Linkage isomers for the ambidentate character of the SCN⁻ ligand in the square planar complexes of Pt(II) are illustrated below:

$$Cis - \left[Pt(NCS)_2(NH_3)_2 \right] \rightleftharpoons Cis - \left[Pt(NH_3)_2(SCN)_2 \right]$$

$$Cis-\left[Pt\left(PR_3\right)_2\left(SCN\right)_2\right] \rightleftharpoons Cis-\left[Pt\left(NCS\right)_2\left(PR_3\right)_2\right]$$

The relative stabilities of the above mentioned linkage isomers can be explained by considering the π–bonding effect of the ligands and steric effect. (*see* Sec. 2.2.6 and Sec. 14.15.4, Vol. 3 of Fundamental Concepts of Inorganic Chemistry, for details).

Mode of Linkage by SCN⁻ and HSAB Principle (*cf.* Sec. 2.2.6; Sec. 14.15.4, Vol. 3 of Fundamental Concepts of Inorganic Chemistry)

In terms of the HSAB principle, the relatively harder centers like Cr(III), Fe(III), Co(III), etc. should preferably bind through the N–end of SCN⁻ while the softer centers like Hg(II), Pd(II), Pt(II), etc. should prefer the S–site of SCN⁻ ligand. Here, it should be mentioned that this prediction may not always work because of many other factors like steric factor, π–bonding factor, electronic factor, cfse, etc. These aspects have been discussed separately later (*cf.* Sec. 2.2.6).

● Many other linkage isomers involving the SCN⁻ ligand are well known, *e.g.*

$$\left[Co\left(NH_3\right)_5\left(SCN\right)\right]^{2+} \rightleftharpoons \left[Co\left(NCS\right)\left(NH_3\right)_5\right]^{2+}$$

$$\left[Co\left(CN\right)_5\left(NCS\right)\right]^{3-} \rightleftharpoons \left[Co\left(CN\right)_5\left(SCN\right)\right]^{3-}$$

$$\left[Rh\left(NH_3\right)_5\left(SCN\right)\right]^{2+} \rightleftharpoons \left[Rh\left(NCS\right)\left(NH_3\right)_5\right]^{2+}$$

$$\left[Zn\left(SCN\right)_4\right]^{2-} \rightleftharpoons \left[Zn\left(NCS\right)_4\right]^{2-}$$

$$\left[Co\left(SCN\right)_4\right]^{2-} \rightleftharpoons \left[Co\left(NCS\right)_4\right]^{2-}$$

$$\left[Hg\left(NCS\right)_4\right]^{2-} \rightleftharpoons \left[Hg\left(SCN\right)_4\right]^{2-}$$

$$\left[Pd\left(NCS\right)_4\right]^{2-} \rightleftharpoons \left[Pd\left(SCN\right)_4\right]^{2-}$$

$$\left[Cd\left(NCS\right)_4\right]^{2-} \rightleftharpoons \left[Cd\left(SCN\right)_4\right]^{2-}$$

Symbiosis and HSAB principle

The relative stability of above mentioned linkage isomers can be explained by considering the *matching of hard-soft character and symbiotic effect.*

(iii) **Linkage isomers from the ambidentate ligand CN⁻:**

CN⁻ can bind through the C–end or N–end to produce the linkage isomers.

● The **prussian blue** and related compounds are the classical examples. In prussian blue, the stable form possesses the following type of linkage.

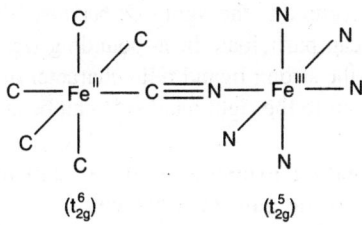

Preference for the C–end (**strong field ligand site**) of CN^- by the Fe(II)–center (d^6) can be explained by considering the gain of *cfse*. This aspect has been already discussed in Chapter 1 (*see* 1.12).

● A similar linkage isomerism in the polymeric compound having the metal centers Fe(II) and Cr(III) connected by the bridging ligand CN^- has been characterised (*cf.* Sec. 1.12).

$$FeSO_4 + K_3\left[Cr(CN)_6\right] \xrightarrow[(-K_2SO_4)]{\text{Cold Condition}} KFe^{II}Cr^{III}(CN)_6$$
$$\text{(Brick red solid)}$$

$$\Big\downarrow \begin{array}{l}\text{Heating}\\ (\sim100°C)\end{array}$$

$$KCr^{III}Fe^{II}(CN)_6$$
$$\text{(Dark green)}$$

Both the brick red and dark green solids have the polymeric structure similar to the skeleton structure of prussian blue. In the brick red solid (**kinetically controlled product**), the strong field donor ends (*i.e.* C–end) of the CN^- ligands octahedrally coordinate the Cr^{III}– center (d^3 system) while in the dark green solid (**thermodynamically controlled product**), the Fe^{II}–center (d^6–system) is octahedrally coordinated by the C–ends of the CN^- ligands.

In both the forms, Cr^{III} remains as t_{2g}^3 but Fe^{II} attains the *low-spin state* ($t_{2g}^6 e_g^0$) in the dark green product due to the ligation by the **strong field donor sites (*i.e.* C–ends) of the CN–ligands** while Fe^{II} remains in *high-spin state* ($t_{2g}^4 e_g^2$) in the brick red compound due to ligation by the weak field donor sites (*i.e.* N-ends). **Thus the gain of cfse makes the green product thermodynamically more stable**. This aspect has been discussed in Chapter 1 (Sec. 1.12). Many other examples of the linkage isomers produced by the CN–bridges have been discussed in Sec. 1.12.

● The relative stabilities of the following isomers of Co(III) can be explained by considering the concept of gain of *cfse* and *symbiosis*.

$$\left[Co(CN)_5 (NC)\right]^{3-} \rightleftharpoons \left[Co(CN)_6\right]^{3-}$$

The above equilibrium is also favoured to the right side *because of the π–bonding interaction*. When CN^- binds through the C–end, it can participate in π–bounding (*metal to ligand*) but the N–site is a non–π–bonding site. Besides this, the strong ligand field character of the C–end leads to a higher *cfse*. Thus the above favoured equilibrium to the right hand side can be considered as the joint effect of the following factors:

(i) **symbiotic effect; (ii) π–bonding property of the C–end of CN⁻; (iii) strong ligand field character of the C–end leading to the gain of more cfse.**

● Now let us consider the equilibrium between the given linkage isomers.

$$cis\text{-}\left[Co(CN)_2(trien)\right]^+ \rightleftharpoons cis\text{-}\left[Co(NC)_2(trien)\right]^+$$

Symbiotic effect favours the equilibrium towards the right side while the π-bonding effect favours to the left side. In terms of cfse, the equilibrium is also favoured to the left side.

(iv) Linkage isomers from the thiosulfate ($S_2O_3^{2-}$) ligand:

● The purple coloured pentaamminethiosulfatocobalt(III) salts prepared by **P. Ray** is actually a mixture of the following two isomeric compounds (*see* Sec. 6.17.1, Vol. 1 of Fundamental Concepts of Inorganic Chemistry).

$[(H_3N)_5Co\text{—}O\text{—}S_2O_2]^+$ and $[(H_3N)_5Co\text{—}S\text{—}SO_3]^+$

monodentate $S_2O_3^{2-}(\kappa O)$ monodentate $S_2O_3^{2-}$ (κS)

     i.e. η^1–O)   *i.e.* η^1–S)

    (*ca.* 90%)   (*ca.* 10%)

● Other linkage isomeric coordination compounds from $S_2O_3^{2-}$ were also prepared by **P. Ray**.

$[(NC)_5Co\text{—}O\text{—}S_2O_2]^{4-}$ and $[(NC)_5Co\text{—}S\text{—}SO_3]^{4-}$

     (Yellow) (η^1–O)   **(Red) (η^1–S)**

● In fact, thiosulfuric acid ($H_2S_2O_3$) remains in an equilibrium mixture of the two isomeric forms (*i.e.* tautomers)

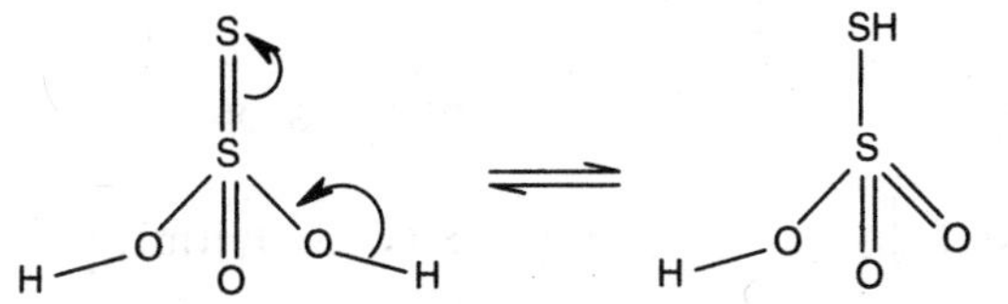

(prototropic tautomerism)

The structure of thiosulfate should be represented as follows.

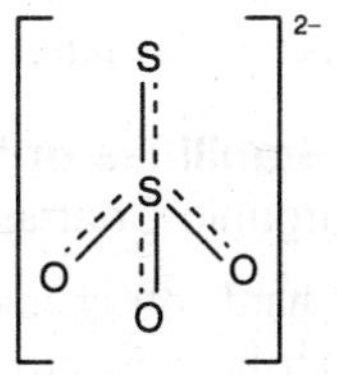

● Existence of such isomeric coordination compounds stands as an indirect proof for the existence of two tauotomeric forms of thiosulfuric acids.

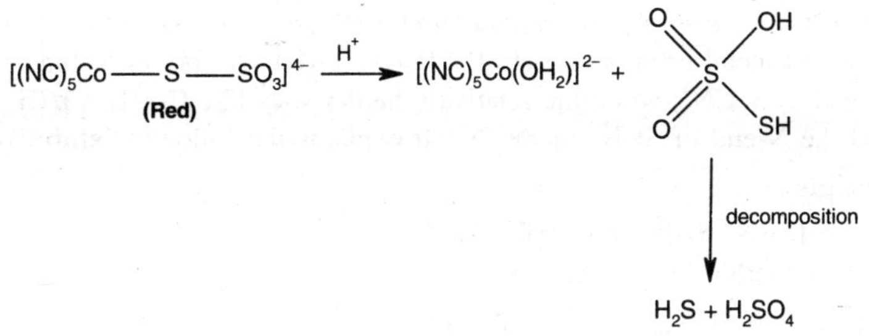

$[(NC)_5Co\text{—}S\text{—}SO_3]^{4-} \xrightarrow{\ H^+\ } [(NC)_5Co(OH_2)]^{2-} +$

     (Red)

decomposition

$H_2S + H_2SO_4$

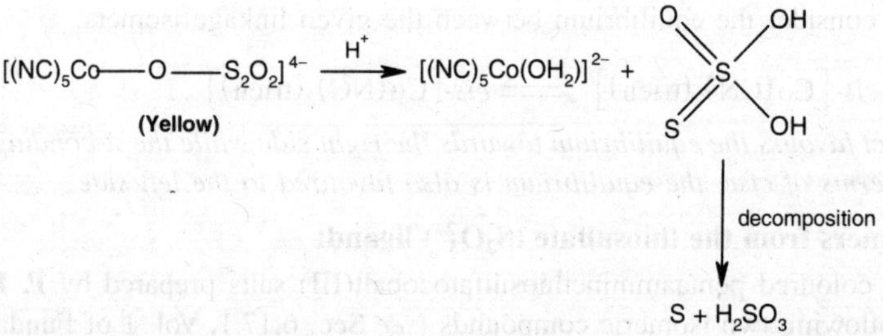

$$[(NC)_5Co{-\!-}O{-\!-}S_2O_2]^{4-} \xrightarrow{\ H^+\ } [(NC)_5Co(OH_2)]^{2-} +$$

(Yellow)

(structure of thiosulfuric acid, with S bearing O, OH, S, OH groups)

decomposition ↓

$$S + H_2SO_3$$

Acid catalysed decomposition of the linkage isomeric compounds as outlined above gives the different decomposition products from the coordinated thiosulfato moiety. **It indirectly supports the two different structures of thiosulfuric acid**.

● It may be mentioned that thiosulfate can act also as a didentate ligand in two possible ways, *i.e.* η^2–O, O and η^2–S, O giving rise to two linkage isomers.

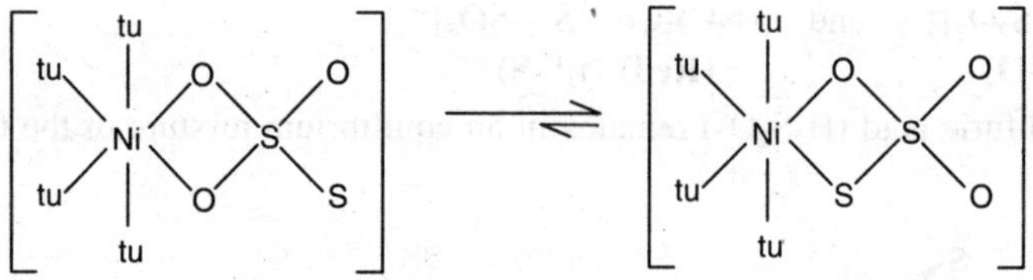

$$S_2O_3^{2-}\ (\eta^2 - O,\ O) \qquad\qquad S_2O_3^{2-}\ (\eta^2 - S,\ O)$$

$$\left[Ni\!\left(\eta^2\text{-}O,O - S_2O_3\right)(tu)_4 \right] \qquad \left[Ni\!\left(\eta^2\text{-}S,O - S_2O_3\right)(tu)_4 \right]$$

(v) Linkage isomers from the flexidentate ligands:

The flexidentate ligands like *edta, cysteinate, histidinate, dithiooxalate*, etc. can produce the linkage isomers. These have been discussed in Chapter 1 (*cf. See* 1.7.2). In such cases, the nature of linkage isomer depends on pH and hard-soft character matching.

2.2.6 Factors Affecting the Relative Stabilities of the Linkage Isomers (*cf.* Sec. 14.15.4, Vol. 3 of Fundamental Concepts of Inorganic Chemistry)

The important factors are: (i) matching of hard-soft character as expected from the HSAB principle; (ii) symbiotic effect; (iii) steric factors; (iv) π-bonding properties of the ligands; (v) *cfse* consideration.

A. Hard-soft character matching and symbiotic effect

If other factors are less important then the HSAB principle works well, *i.e.* soft metal centers prefer the soft donor sites and hard metal centers prefer the hard donor sites. *In terms of symbiosis, like ligands prefer the like ligands, i.e. hard and soft ligands prefer to group together separately.*

This is why, the soft metal centers like Hg(II), Pd(II), Pt(II), etc. prefer to bind with the *S*-end of SCN⁻ and the Se-end of SeCN⁻, while the relatively harder sites like Co(II), Cr(III), Fe(III), Co(III), etc. prefer to bind the N-end of SCN⁻ or SeCN⁻. It explains the following **stability orders**.

HSAB principle

$[M(SCN)_4]^{2-} > [M(NCS)_4]^{2-}$ (M = Hg, Pd, Pt),

$[M(SeCN)_4]^{2-} > [M(NCSe)_4]^{2-}$ (M = Hg, Pd, Pt).

$[Hg(SCN)_4]^{2-} > [Cd(SCN)_4]^{2-} > [Zn(SCN)_4]^{2-}$

$[Co(NCS)_4]^{2-} > [Co(SCN)_4]^{2-}$

$[Cr(OH_2)_5(SCN)]^{2+} < [Cr(NCS)(OH_2)_5]^{2+}$

$cis–[M(NH_3)_2(SCN)_2] > cis–[M(NCS)_2(NH_3)_2]$, (M = Pt, Pd).

Symbiotic principle

$[Co(NH_3)_5(NO_2)]^{2+} > [Co(NH_3)_5(ONO)]^{2+}$

$[Co(CN)_5(NCS)]^{3-} < [Co(CN)_5(SCN)]^{3-}$

$[M(NCS)(NH_3)_5]^{2+} > [M(NH_3)_5(SCN)]^{2+}$,(M = Co, Rh)

$[Co(CN)_5(NC)]^{3-} < [Co(CN)_6]^{3-}$

$Rh(SCN)_6]^{3-} > [Rh(NCS)(SCN)_5]^{3-}$

$[Fe(CO)_2(Cp)(SeCN)] > [Fe(CO)_2(Cp)(NCSe)]$

Fe^{II}-center prefers the harder N-donor site of $SeCN^-$ as in $[Fe(NCSe)_4]^{2-}$ but when **softened by Cp⁻ and CO**, it prefers to bind through the softer end of $SeCN^-$ as in **[Fe(CO)₂(Cp)(SeCN)]**. Cp⁻ and CO being enriched with the π-electron cloud can act as the soft ligands.

It is evident from the above examples that in terms of symbiotic principle, irrespective of the hardness of the metal center, the harder ligands group together and similarly the softer ligands group together. Thus $[Rh(SCN)_6]^{3-}$, $[Fe(NCSe)_4]^{2-}$, etc. are stable.

π-Bonding and Steric Control in the Octahedral Linkage Isomers

Sometimes, binding by the π-bonding ends of the ambidentate ligands may require more space (**angular volume**) as in SCN^-, $SeCN^-$, etc. (*cf.* Sec. 12.1.18; Fig. 14.15.4.2, Vol. 3 of Fundamental Concepts of Inorganic Chemistry). In such cases, the steric crowding by the other ligands may have some control to dictate the mode of binding by the ambindentate ligands. This explains the following stability orders:

$[Co(NCS)(NH_3)_5]^{2+} > [Co(NH_3)_5(SCN)]^{2+}$ (for the bulky NH_3 ligands, steric factor more important)

$[Co(CN)_5(SCN)]^{3-} > [Co(CN)_5(NCS)]^{3-}$ (for the less bulky CN^- ligands, π-bonding more important)

It may be noted that the above given stability orders are also expected from the symbiotic effect. These aspects have been discussed in detail latter.

● In the polymeric structure of **Ni(CN)₂ · NH₃** (*cf.* Sec. 1.12), 'CN' bridges the Ni(II)–centers producing the coordinating sites $Ni(C)_4$ and $Ni(N)_4$ in the skeleton of the **rearranged (template effect) polymeric structure of Ni(CN)₂.** The NH_3 ligands coordinate the $Ni(N)_4$ center giving rise to the $Ni(N)_6$ octahedral symmetry while the other center $Ni(C)_4$ maintains the square planar symmetry. In terms of the linkage isomerism, we have the stability order;

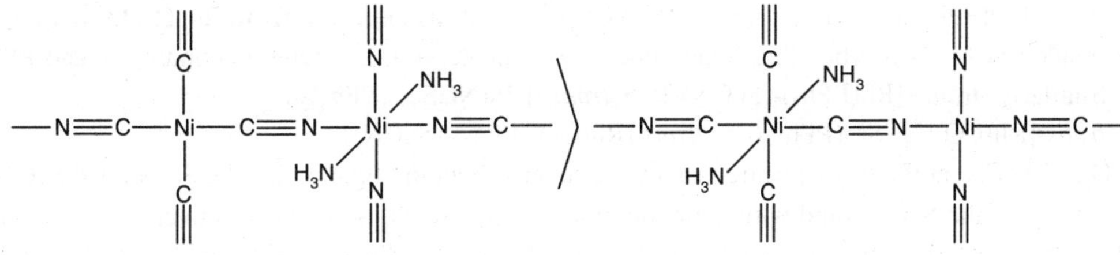

The above stability order is quite expected from the *principle of symbiotic theory.*

Here there is another point of consideration. Ni(II) center gets less electron from the 'Ni—N' bond than from the Ni—C bond as the C–end is less electronegative than the N–end of the CN^- ligand. *Thus the electronegativity of the metal center having already the coordination sphere Ni(C)$_4$ is less than that of the Ni(N)$_4$ center.* This is why, the Ni(N)$_4$ center preferably binds with the additional NH$_3$ ligands.

● The flexidentate ligand **dithiooxalate** ($C_2S_2O_2^{2-}$) can produce the linkage isomers.

(*i.e.*, dithioxalato - *O, O'*)   (*i.e.*, dithioxalato - *S, S'*)
when M is hard   when M is soft

● The linkage isomers formed by **thiosulfate can** also be rationalised in terms of HSAB principle.

M is relatively softer.   M is relatively harder.

B. Effect of π-bonding and competitive π-bonding interaction in the square planar complexes: In the octahedral complex, the π-bonding effect is relatively less pronounced than that in the square planar complex. **This is why, the simple prediction of the HSAB principle or symbiotic principle may not always work in the square planar complexes.** This is illustrated (in terms of the stability order) in some representative examples of the square planar complexes.

● *cis*–[Pt(NH$_3$)$_2$(SCN)$_2$] > *cis*–[Pt(NCS)$_2$(NH$_3$)$_2$] (**contradicting the symbiotic principle**)
(S–bonded SCN$^-$ is a π–bonding ligand while N–bonded SCN$^-$ and NH$_3$ are the non-π-bonding ligands)

● *cis*–[Pt(NCS)$_2$(PR$_3$)$_2$] > *cis*–[Pt(PR$_3$)$_2$(SCN)$_2$]   (**contradicting the HSAB principle**)
(PR$_3$ is a better π-bonding ligand than the S–bonded SCN$^-$ ligand).

● [Pd(SCN)$_4$]$^{2-}$ > [Pd(NCS)$_4$]$^{2-}$ (**as expected from the HSAB principle**)
(S–bonded SCN$^-$ is a π–bonding ligand, but N-bonded SCN$^-$ is a non-π-bonding ligand).

● [Pd(bpy)(NCS)$_2$] > [Pd(bpy)(SCN)$_2$] (**contradicting the HSAB principle**)
(bpy is a better π–bonding ligand than the S–bonded SCN$^-$ ligand)

● *cis*–[Pd(NCS)$_2$(XPh$_3$)$_2$] > *cis*–[Pd(SCN)$_2$(XPh$_3$)$_2$](**contradicting the HSAB principle**)
(X = As, P) (XPh$_3$ is a better π-bonding ligand than the S-bonded SCN-ligand)

● *trans*–[Pd(PPh$_3$)$_2$(SCN)$_2$] > *trans*–[Pd(NCS)$_2$(PPh$_3$)$_2$] (**as expected from the HSAB principle**)
(S-bonded SCN$^-$ as a π-bonding ligand does not compete with the better π-bonding ligand PPh$_3$)

● Similarly, *trans*–[Rh(PPh$_3$)$_2$(SeCN)$_2$]$^-$ > *trans*–[Rh(NCSe)$_2$(PPh$_3$)$_2$]$^-$

● *trans*-[Rh(CO)(NCSe)(PPh$_3$)$_2$] > *trans*-[Rh(CO)(PPh$_3$)$_2$(SeCN)]

(CO and SeCN$^-$ at the *trans*-positions; CO is a better π-bonding ligand than the Se-bonded SeCN$^-$)

When SCN$^-$ and SeCN$^-$ bind with their soft ends, *i.e.* S-end and Se-end respectively, they can at as the π-acceptor ligands. This π-bonding (metal to ligand) enhances the metal-ligand bonding interaction and thus stability of the system increases. In the square planar complexes of the specially soft metal centers (*e.g.* Pd^{II}, Pt^{II}), if two different π–*bonding ligands* are at the *trans*-positions, then they mutually compete for the same metal *d*-orbital for π-bonding. **The better π-bonding ligand tends to monopolise the metal *d*-orbital and consequently it encourages the *trans*-ambidentate ligand to bind through the non-π-bonding site.**

Symbiotic effect *vs.* Competitive π-bonding effect (*i.e.* Antisymbiotic effect)

The above competitive π-bonding interaction works against the symbiotic theory (*i.e.* accumulation of the soft ligands). This is why, the competitive π-bonding interaction has been described as the **antisymbiotic effect**. *In general, in the square planar complexes, the antisymbiotic effect is more important than the symbiotic effect while in the octahedral complexes, the symbiotic effect is more important than the* **antisymbiotic effect** *(i.e. competitive π-bonding interaction). In the tetrahedral complexes, the antisymbiotic effect is less pronounced than that in the square planar complexes but it is more pronounced than that in the octahedral complexes.* Thus the importance of **antisymbiotic effect** runs in the sequence:

square planar >> tetrahedral > octahedral

The **competitive π-bonding** effect leading to an antisymbiotic effect is illustrated in Fig. 2.2.6.1.

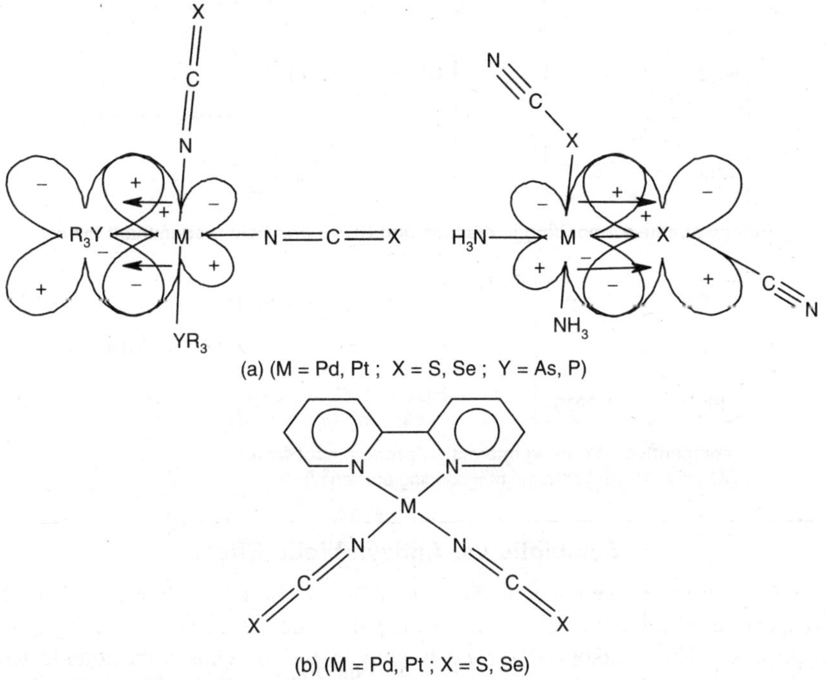

(a) (M = Pd, Pt ; X = S, Se ; Y = As, P)

(b) (M = Pd, Pt ; X = S, Se)

Fig. 2.2.6.1 Effect of the competitive π-bonding to determine the ligating behaviour of SCN$^-$ and SeCN$^-$ in the square planar complexes of Pd(II) and Pt(II). (a) Metal → Ligand π–bonding in *cis*–[M(YR$_3$)$_2$(NCX)$_2$] and *cis*–[M(NH$_3$)$_2$(XCN)$_2$]. (b) Structure of [M(bpy)(NCX)$_2$].

Considering the competitive π-bonding effect between the *trans*-ligands in the square planar complexes, the relative stabilities of the given linkage isomers can be explained. The **efficiency of π-bonding property** runs in the following order.

$$PR_3 > SCN^- (S); \quad bpy > SCN^- (S); \quad A_SR_3 > SCN^-(S); \quad CO > SeCN^- (Se)$$

Thus in terms of the **competitive π-bonding effect** *vs.* **noncompetitive π-bonding effect**, the relative stabilities of the following linkage isomers can be explained.

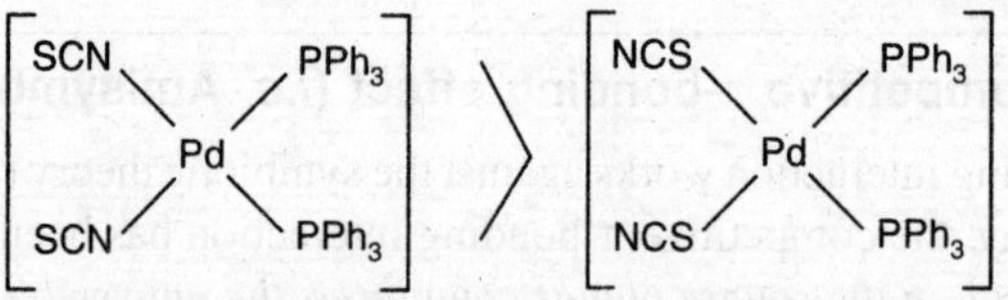

$$\left[\begin{array}{cc} SCN & PPh_3 \\ & Pd \\ SCN & PPh_3 \end{array}\right] > \left[\begin{array}{cc} NCS & PPh_3 \\ & Pd \\ NCS & PPh_3 \end{array}\right] \quad (cis\text{-configuration})$$

**(competitive π-bonding interaction predominant factor;
PPh$_3$ > SCN$^-$ (S) in terms of π-bonding efficiency)**

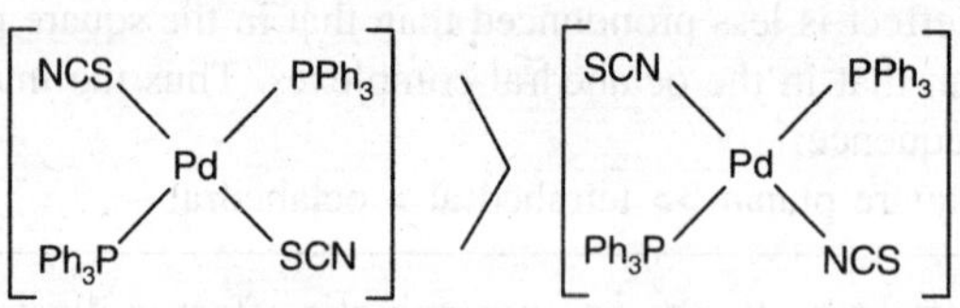

$$\left[\begin{array}{cc} NCS & PPh_3 \\ & Pd \\ Ph_3P & SCN \end{array}\right] > \left[\begin{array}{cc} SCN & PPh_3 \\ & Pd \\ Ph_3P & NCS \end{array}\right] \quad (trans\text{-configuration})$$

(noncompetitive π-bonding interaction and HSAB principle predominant factor)

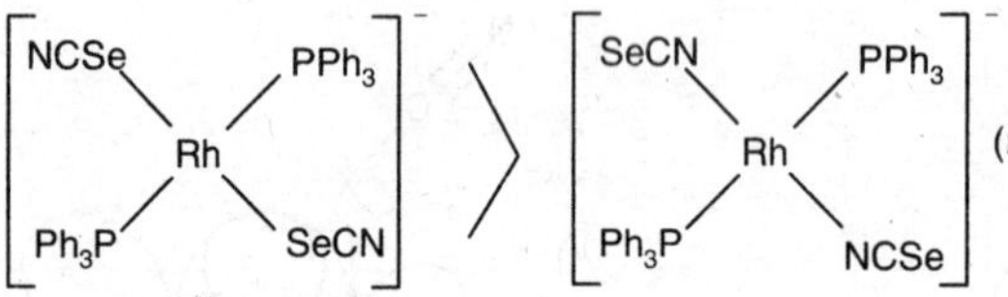

$$\left[\begin{array}{cc} NCSe & PPh_3 \\ & Rh \\ Ph_3P & SeCN \end{array}\right] > \left[\begin{array}{cc} SeCN & PPh_3 \\ & Rh \\ Ph_3P & NCSe \end{array}\right] \quad (trans\text{-configuration})$$

(noncompetitive π-bonding interaction and HSAB principle predominant factor)

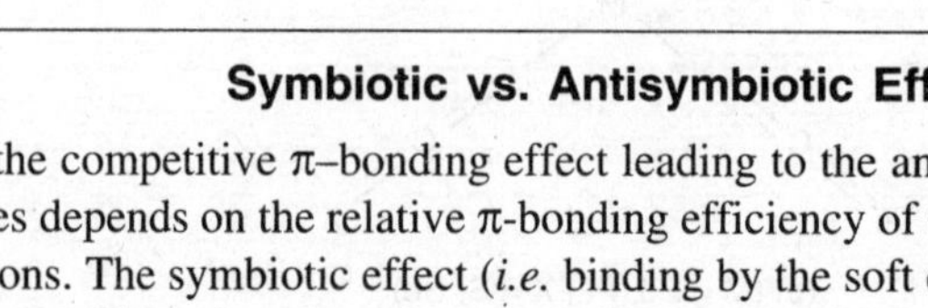

$$\left[\begin{array}{cc} OC & PPh_3 \\ & Rh \\ Ph_3P & NCSe \end{array}\right] > \left[\begin{array}{cc} OC & PPh_3 \\ & Rh \\ Ph_3P & SeCN \end{array}\right] \quad (trans\text{-configuration})$$

**(competitive π-bonding interaction predominant factor;
CO > SeCN$^-$ (Se) in terms of π-bonding efficiency)**

Symbiotic vs. Antisymbiotic Effect

Importance of the competitive π–bonding effect leading to the antisymbiotic effect in the square planar compexes depends on the relative π-bonding efficiency of the competing ligands placed at the *trans*-positions. The symbiotic effect (*i.e.* binding by the soft ends of the ambidentate ligands in the present cases) is opposed by the competitive π-bonding effect. *When these two opposing effects are balanced, both the possible linkage isomers will have the comparable stability.* But, if the π-bonding properties of the competing *trans* ligands widely differ (*i.e.* antisymbiotic effect is highly predominant) then the isomer accumulating the soft ligands may not be isolated.

● The example given in Fig. 2.2.6.2 illustrates the effect of π-bonding (both competitive and noncompetitive).

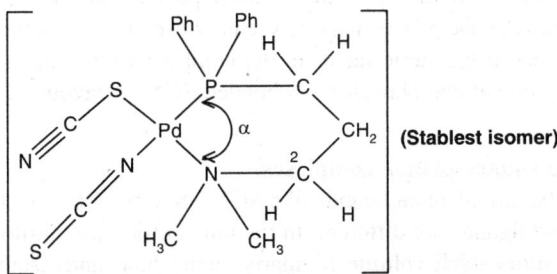

Fig. 2.2.6.2 Illustration of the effect of **competitive (P vs. S.) and noncompetitive (N vs. S.) π-bonding interaction.**

In the above example, the more efficient π-bonding P-donor site encourages the *trans*-ligand to coordinate by its non-π-bonding N-site; while SCN⁻ coordinates through its π-bonding S-site whose *trans*-position is occupied by the non-π-bonding N-site of the didentate ligand. Thus there is a *competitive π-bonding* effect between the *trans*-positions occupied by SCN⁻ and P (of the didentate ligand) and there is a noncompetitive π-bonding interaction between the other *trans-positions* occupied by S (of SCN⁻) and N (of the didentate ligand).

Note: In the example of Fig. 2.2.6.2, the $\widehat{P—Pd—N}$ angle is sufficiently high to cause the steric crowding (cf. Fig. 2.2.6.5) between the substituents on N or P with the adjacent ambidentate ligands. Thus SCN⁻ should bind through the N-site, *i.e.* linear form, in terms of steric crowding. This coordination through the N-end of SCN⁻ at the *cis*–position of the N-donor site bearing two Me-groups is justified by both steric factor and competitive Pd—P π–bonding (at the *trans*-position of N–bonded SCN⁻). *On the other hand, more steric crowding on P (of the didentate ligand) suggests the adjacent SCN⁻group to bind again through the N-end but the noncompetitive π-bonding interaction allows it to bind through the S-end.* The analysis in terms of steric crowding is discussed in the following section.

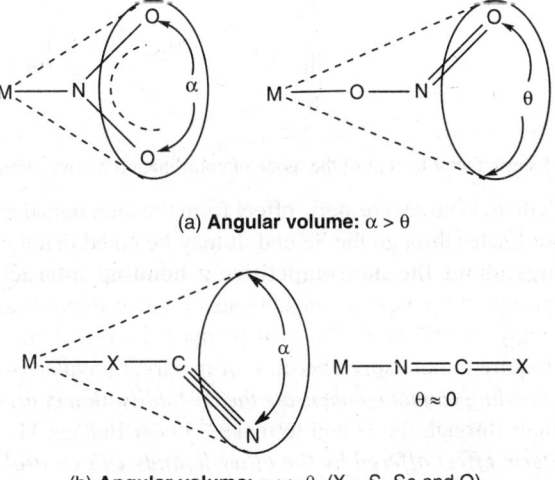

(a) **Angular volume:** $\alpha > \theta$

(b) **Angular volume:** $\alpha \gg \theta$, (X = S, Se and O)

Fig. 2.2.6.3 Steric crowding for different modes of ligation by the ambidentate ligands, *e.g.* (a) M—NO₂⁻ vs. M—ONO⁻; (b) M—SCN vs. M—NCS; M—SeCN vs. M—NCSe or M—OCN vs. M—NCO (*cf.* Sec. 12.1.18).

Here it may be mentioned that the **aryl substituted phosphine is a weak π-accepter ligand** (*cf.* electron donation from the phenyl ring into the vacant *d*-orbital of P reduces the π-acceptor property of PAr_3) and if there is no strong steric factor in the competitive π-bonding interaction between the S-bonded SCN^- and aryl substituted phosphine, S-bonded SCN^- is favoured. This is illustrated in Fig. 2.2.6.5.

C. Steric effects in the square planar complexes:

Steric requirements of the ambidentate ligands like SCN^-, $SeCN^-$, NO_2^- for the two possible ways of linkages for each of these ligands are different. In the nitro-nitrito, *i.e.* nitrito (*N*)–nitrito (*O*) isomers, the nitro isomer requires **more steric volume**. Similarly, in the thiocyanto–isothiocyanato, *i.e.* thiocyanto (*S*)–thiocyanto (*N*) and selenocyanto–isoselenocyanto, *i.e.* selenocyanto (*Se*)–selenocyanto (*N*) isomers, the steric requirements are more for the S-bonded SCN^- and Se–bonded $SeCN^-$. In the same way, the OCN^- (*O*) linkage requires more space than the OCN^- (*N*) linkage. These are illustrated in Fig. 2.2.6.3.

If the steric crowding within the coordination sphere is sufficiently high, then **stability order** of the linkage isomers follows the following sequences.

$$M—ONO > M—NO_2; \quad M—NCS > M—SCN;$$

$$M—NCSe > M—SeCN; \quad M—NCO > M—OCN.$$

Steric factor to control the mode of linkage by the ambidentate ligand is illustrated in the example given in Fig. 2.2.6.4.

Steric factor vs. noncompetitive π-bonding:

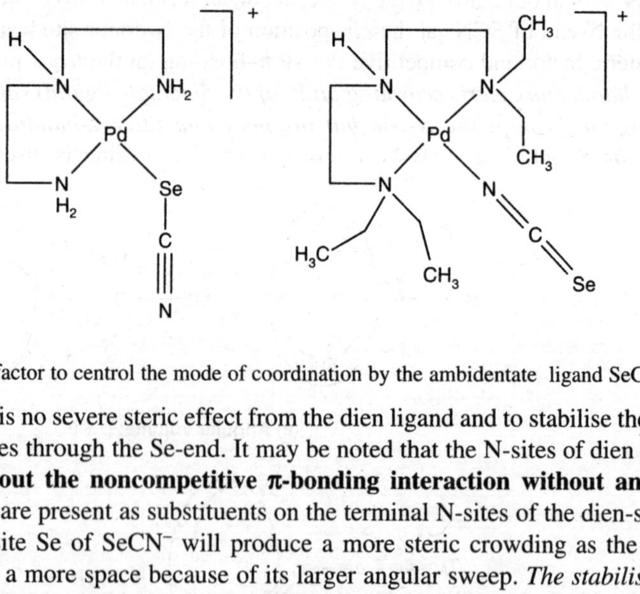

Fig. 2.6.2.6.4 Effect of steric factor to centrol the mode of coordination by the ambidentate ligand $SeCN^-$.

In $[Pd(dien)(SeCN)]^+$, there is no severe steric effect from the dien ligand and to stabilise the system by π-bonding, $SeCN^-$ coordinates through the Se-end. It may be noted that the N-sites of dien are non-π-bonding. **Thus it brings about the noncompetitive π-bonding interaction without any steric hindrance.** When ethyl groups are present as substituents on the terminal N-sites of the dien-skeleton, the linkage by the π-bonding site Se of $SeCN^-$ will produce a more steric crowding as the binding through the Se-end will require a more space because of its larger angular sweep. *The stabilisation to be earned through the π-bonding cannot compensate the destabilisation caused by the steric crowding.* To avoid this, $SeCN^-$ ligates through the N-end to make a linear linkage M—N≡C≡Se that needs a less space. ***Thus, the steric effect offered by the other ligands can control the mode of linkage by the ambidentate ligands specially for the square planar complexes.***

Steric control to determine the relative stability of the linkage isomerism is illustrated in the following example.

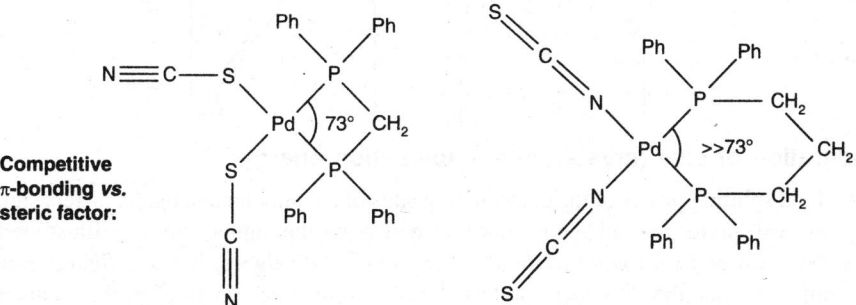

$X = SCN, SeCN, n = 1, 2, 3$

As the value of n increase, the angle α opens (*i.e.* increases) and consequently the angle β decreases to cause a more steric repulsion with the ambidentate ligands like $SeCN^-$ or SCN^-. Thus, the increase of n favours the coordination by the linear form of $SeCN^-$ or SCN^- (*i.e.* N-bonded). In fact, for $n = 1$ (*i.e.* $Ph_2P-CH_2-PPh_2$) the angle α is about $73°$ and SCN^- binds through the S-end (*i.e.* bent mode of binding) to earn the stabilisation through π-bonding as usual. If n is increased to 3 (*i.e.* $Ph_2P-(CH_2)_3-PPh_2$), the angle α ($\sim 90°$) opens sufficiently to increase the steric crowding and then SCN^- binds through the N-end to get relief from the steric crowding.

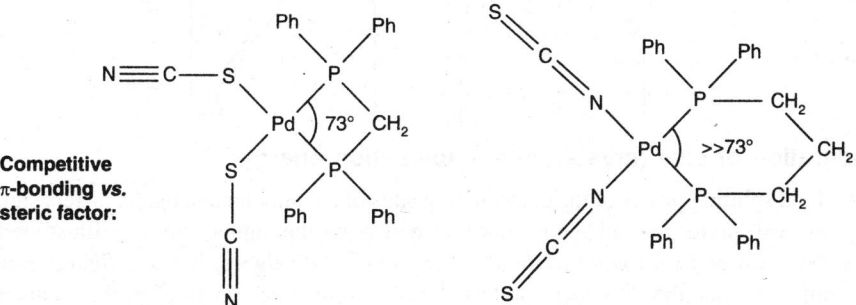

Fig. 2.2.6.5 Effect of steric factors to control the mode of ligation by SCN^-.

Here, it should be noted that the aryl-alkyl substituted phosphines (given in the above examples) are more weakly π-bonding (*cf.* electron pushing mesomeric effect of the phenyl ring into the vacant d-orbital of P) compared to the S-bonded SCN^-. This is why, when α-is small (*i.e.* less steric crowding), SCN^- coordinates through the S-end. On the other hand, when α is high (*i.e.* more steric crowding), SCN^- binds in a linear form (*i.e.* N-bonded) and **the P-center participates in the π-bonding interaction** in this competitive π-bonding interaction. In the case of **noncompetitive π-bonding interaction**, SCN^- may coordinate through the S-end even when partially disfavoured by the steric crowding (*cf.* Fig. 2.2.6.2).

Tune between the steric factor and π-bonding effect in the square planar complexes for the ambidentate ligands

Figs. 2.2.6.2 and 2.2.6.4-5 illustrate the balance among the different factors like **steric factor, competitive π-bonding effect, noncompetitive π-bonding effect**. Figs. 2.2.6.2 and 2.2.6.5 are important to compare. In Fig. 2.2.6.5, steric factor and competitive π-bonding are present while in Fig. 2.2.6.2, steric factor, competitive π-bonding and noncompetitive π-bonding are present. In Fig. 2.2.6.4, steric factor and non-competitive π-bonding are present.

D. Steric control in the octahedral complexes to determine the stabilities of the linkage isomers

The stabilities of the linkage isomers follow the sequence:

$$[Co(NCS)(NH_3)_5]^{2+} > [Co(NH_3)_5(SCN)]^{2+}$$
$$[Co(CN)_5(SCN)]^{3-} > [Co(CN)_5(NCS)]^{3-}$$

This has been explained in terms of the **symbiotic theory**. However, the **steric factor** also predicts the same order. The *pyramidal NH_3 ligand needs more space than the linear ligand CN^-*. Thus presence of 5 NH_3 molecules in the coordination sphere causes a more steric crowding to encourage the binding of SCN^- in the linear form, *i.e.* by the N-end while in the cyanido complex, the steric crowding is relatively less and then to earn the stabilisation, SCN^- binds through the π-bonding end (*i.e.* S-end, bent mode of binding).

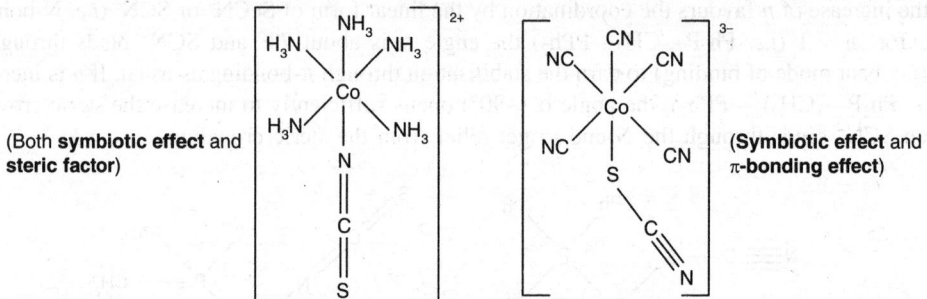

(Both **symbiotic effect** and steric factor)

(**Symbiotic effect** and π-**bonding effect**)

E. Consideration of cfse (crystal field stabilisation energy)

If the crystal field splitting power of the coordinating sites of an ambidentate ligand differs significantly, then the system will prefer the linkage isomer that will bring the higher *cfse*. It is illustrated for CN^- whose *C-end is a strong field ligand while the N-end is a relatively weaker field ligand*. Between the d^5 and d^6 configurations, the d^5-system prefers the high spin state (*i.e.* $t_{2g}^3 e_g^2$ in the octahedral field) because of the higher *exchange energy* while the d^6 system prefers the low spin state because of the high *cfse* (*i.e.* $t_{2g}^6 e_g^0$ in the octahedral field). This is why, in **prussian blue (a cyanido bridged polymeric compound)**, C-end of CN^- coordinates the Fe(II)-center to induce the low spin state (*i.e.* t_{2g}^6) and the N-end of CN^- coordinates the Fe(III)-center to allow the high spin state ($t_{2g}^3 e_g^2$). Here, it may be noted that the harder Fe(III) center coordinates with the harder N-site of the ambidentate ligand CN^-. On the other hand, the softer Fe(II) centre coordinates with the softer C-end of CN^-. In the **prussian blue structure,** Fe(II) may be replaced by Ru(II) (d^6-system) that again links with the C-end to favour the low spin state (t_{2g}^6) because of the same ground. The corresponding compound, $KRu^{II}Fe^{III}(CN)_6$ is called **ruthenium purple** (*cf.* $KFe^{II} Fe^{III}(CN)_6$ prussian blue). In the prussian blue structure, Fe^{III} may be replaced by Cr^{III} (d^3-system) that links with the N-end of CN^- and it possesses the electronic configuration (t_{2g}^3). All these aspects have been discussed in Chapter-1 (*see* 1.12).

F. Effect of pH to determine the coordinating behaviour of the ligands with the donor sites of different basicities

For the ligands like peptide linkage, aminopolycarboxylic acids (*e.g.* H_4edta, H_3nta, etc.), amino acids (*e.g.* H_2cys, Hhis, etc.), stability of the linkage isomers depends on the pH of solution. At a lower pH, the higher basic sites remain protonated and they generally fail to coordinate the metal center. Thus, such sites can act as the ligating sites at a relatively higher pH. The relatively weak basic sites act as the ligating sites at a relatively lower pH. It is illustrated for the **peptide linkage**.

Thus for **peptide**, the 'M–O' linkage is favoured at a relatively lower pH while the 'M–N' linkage is favoured at a relatively higher pH. The similar situation has been illustrated in Sec. 1.7.2 for **histidine** having three coordinating sites imidazole-N, amino-N and carboxylate-O whose basicities are widely different.

These are discussed and illustrated in Chapter 1 (*see* 1.7.2)

2.2.7 Polymerisation Isomerism

Such compounds have the *same empirical molecular formula* but their structure and molecular mass are different. The examples are:

(i) $[PtCl_2(NH_3)_2]$, $[Pt(NH_3)_4][PtCl_4]$, $[Pt(NH_3)_4][PtCl_3(NH_3)]_2$
and $[PtCl(NH_3)_3]_2[PtCl_4]$; (having the empirical formula $[PtCl_2(NH_3)_2]$)

(ii) and (Werner's **hexol** complex)

(iii) $[Co(NH_3)_3(NO_2)_3]$, $[Co(NH_3)_4(NO_2)_2][Co(NH_3)_2(NO_2)_4]$,
$[Co(NH_3)_5(NO_2)][Co(NH_3)(NO_2)_5]$ and $[Co(NH_3)_6][Co(NO_2)_6]$
(having the empirical formula $[Co(NH_3)_3(NO_2)_3]$)

Note: Here the term *polymerisation* is misnomer as there is no repetition of a *simple building block unit* in constructing the said isomers. In the polymers, the simple building block unit is repeated but here no such situation arises.

2.2.8 Some Special Types of Structural Isomerism — Summation Isomerism and Valence Isomerism

Examples of *summation isomerism:*
$[cis-[CoCl(en)_2(H_2NCH_2CH_2Br)](NO_3)_2$
and $[cis-[CoBr(en)_2(H_2NCH_2CH_2Cl)](NO_3)_2$

Examples of *valence isomerism* where the charge of the complex ion is different:

and which is actually

2.3 CONFORMATIONAL ISOMERISM (*cf.* Secs. 7.17.2, 8.26.1-3)
(*cf.* Sec. 2.11 for the mechanism of conformational isomerism)

Conformational isomers are the special types of stereoisomers which can be interconverted without any bond rupture. Thus the conformational isomers have the same ligands and same coordination number but different stereochemistries. The classical example is the isomerism between the *square planar geometry and tetrahedral geometry.* If the tetrahedral geometry is compressed then the square planar geometry is attained. This intramolecular pathway leads to the *flattening movement of the tetrahedron.* This is illustrated in Fig. 2.3.1.

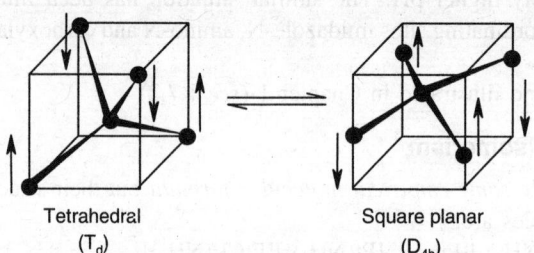

Tetrahedral　　　　　　　　　　　　　　Square planar
(T$_d$)　　　　　　　　　　　　　　　　(D$_{4h}$)

Fig. 2.3.1 Interconversion between a tetrahedral and a square planar geometry; flattening motion of a tetrahedron produces the square planar geometry.

In this conformational change, the bond angle changes, *e.g.* the tetrahedral angle 109° (approx.) becomes 90° in the square planar geometry. Thus, these isomers are also described as **allogons** (*allos* meaning different, *gonia* meaning angle) because these only differ in bond angle.

(a) **[NiX$_2$(PR$_3'$)$_2$] showing the isomerism involving the square planar and tetrahedral geometries:**
The conformation isomerism is well noted in some Ni(II)−complexes involving the square planar and tetrahedral geometry. It is illustrated for **[NiBr$_2$(PEtPh$_2$)$_2$]**

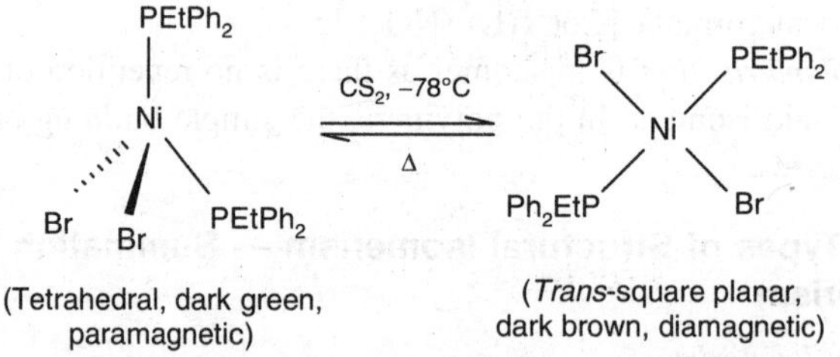

PEtPh$_2$　　　　　　　　　　　　　　　　　　　　　　　Br　　　PEtPh$_2$

Ni　　　　　　　　CS$_2$, −78°C　　　　　　　　　　　Ni

　　　　　　　　⇌

Br　Br　PEtPh$_2$　　　　Δ　　　　　　Ph$_2$EtP　　　　Br

(Tetrahedral, dark green,　　　　　　　　(*Trans*-square planar,
paramagnetic)　　　　　　　　　　　　dark brown, diamagnetic)

Fig. 2.3.2 Polytopal or conformational isomerism in [NiX$_2$(PR$_3'$)$_2$].

In the tetrahedral geometry, the steric crowding is less because of the larger bond angle (compared to that of the square planar geometry, *cf.* 109° *vs.* 90°) but *cfse* is higher in the square planar geometry. *If there is a balance between these opposing factors, then both the isomers can be isolated.*
[NiBr$_2$(PBzPh$_2$)$_2$] (Bz = benzyl group), **[NiX$_2$\{PPh$_2$(CH$_2$Ph)\}$_2$]** can also show the similar isomerism. In their crystals, both the geometries also exist.
The Ni(II) complexes of the formula **[NiX$_2$(PR$_3'$)$_2$]** (X$^-$ = halide, R′ = alkyl or aryl group) in general show the conformational isomerism between the tetrahedral and square planar geometries. If R′ is very bulky then to avoid the **steric crowding**, the tetrahedral geometry is favoured. The **electronic effect** may also stabilise a particular conformation. If the ligand field strength is increased by monitoring the nature of R′-group, **then the higher ligand field strength will favour the square planar geometry.** The better M → P π-back bonding will give the stronger ligand field strength.

That the steric crowding favours the tetrahedral geometry is evidenced from the following observations. Fraction of the tetrahedral isomer increases in the sequence:

$X^- = I^- > Br^- > Cl^-$; $PR'_3 = PAr_3 > PAr_2R > PArR_2 > PR_3$. (Ar = aryl group, R = alkyl group).

Note: The **aryl substituted phosphines** are the relatively weaker π-acceptor ligands because of the **electron pushing mesomeric effect of the phenyl ring** into the vacant *d*-orbital of phosphorus. **Thus PAr₃ being bulkier and weaker as a π-acid ligand compared to PR₃ favours the tetrahedral geometry more.**

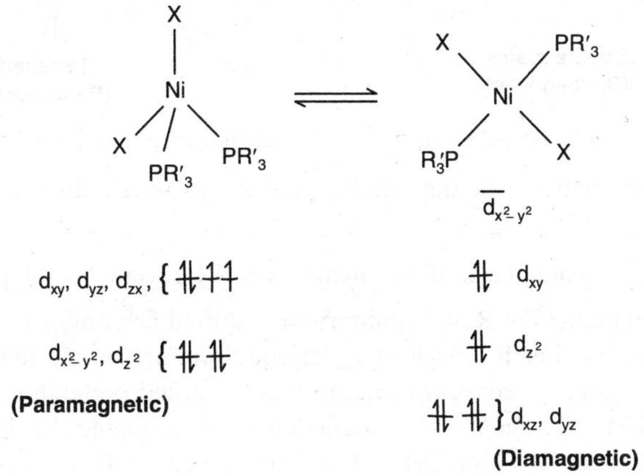

Fig. 2.3.3 Splitting of *d* orbitals in the T_d and D_{4h} geometries

---

Polytopal isomers and fluxional species

The mutual interconversion, tetrahedral ⇌ square planar **does not give the example of fluxional species** though the complexes are *stereochemically nonrigid*. The fluxional species are also *stereochemically nonrigid* but at the same time, **they must be chemically equivalent.** The tetrahedral and square planar structures are **not chemically equivalent** as they give different bond angles, different bond lengths, different colour, different magnetic properties, and different chemical reactivities.

(b) **Four coordinate salicylaldiminato complexes of Ni(II) and Cu(II) showing the isomerism involving the tetrahedral and square planar geometries:** The polytopal isomerism between the square planar and tetrahedral geometries is also noticed in the **salicylaldiminato complexes of Ni(II)**.

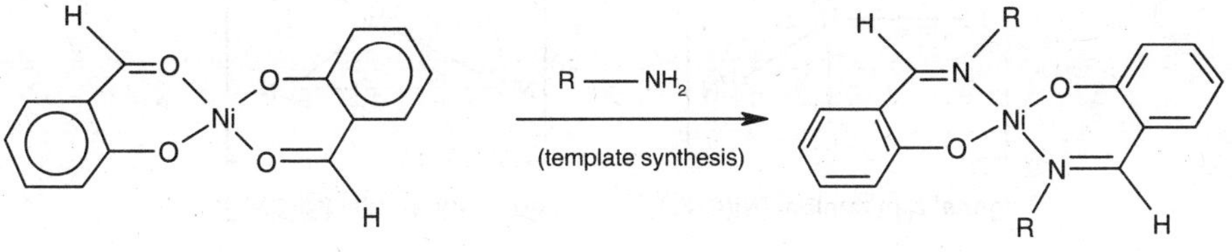

bis(salicylaldehydato)nickel(II)   Salicylaldiminato complex of Ni(II)

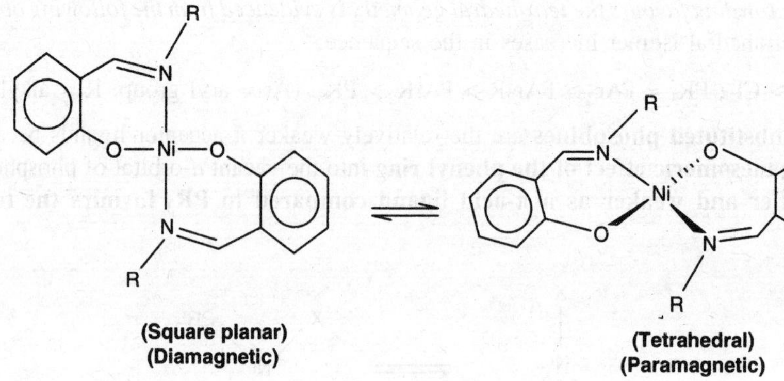

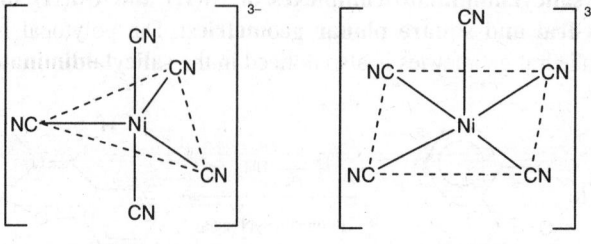

(Square planar)   **(Tetrahedral)**
(Diamagnetic)   **(Paramagnetic)**

Fig. 2.3.4 Polytopal isomerism in salicylaldiminato complexes of Ni(II).

There is an equilibrium between the square planar and tetrahedral forms for $[NiL_2]$ (LH = salicylaldimine).

$$\left[NiL_2\right]\text{(square planar, diamagnetic)} \rightleftharpoons \left[NiL_2\right]\text{(tetrahedral, paramagnetic)}$$

With the increase of bulkiness of R, the equilibrium is shifted favourably to the right hand direction to avoid the steric crowding. For R = *n*-propyl, the equilibrium is in the direction of square planar geometry, while for R = *t*-butyl group, the equilibrium is shifted preferably towards the tetrahedral geometry ($\mu_{obs} \approx 3.2$ B.M., *i.e.* about 95% tetrahedral form assuming 3.3 B.M. for the tetrahedral geometry). If we compare for R = *n*-propyl and isopropyl, then for R = isopropyl, the equilibrium is again favourably shifted to the tetrahedral geometry while for R = *n*-propyl, it is favourably shifted to the square planar geometry. In fact, because of the above equilibrium, **anomalous magnetic moment** depending on the size of R has been noticed. Similar polytopal isomerism in the salicylaldiminato complexes of Cu(II) can also arise.

(c) **Conformational isomerism of 5 and 6 and higher coordinate systems (*cf.* Sec. 2.11):** Besides the conformational isomerism between the tetrahedral and square planar geometries, it can happen in other cases also like trigonal bipyramidal and square pyramidal for 5 coordination number (*cf.* Sec. 2.11.7); octahedral and trigonal antiprism for 6 coordination number (*cf.* Sec. 2.11.8).

● $[Ni(CN)_5]^{3-}$: For 5 coordination number (*see* 2.11.7), there is a well documented example of conformational isomersim for $[Ni(CN)_5]^{3-}$ that exists in the crystal structure of $[Cr(en)_3][Ni(CN)_5]\cdot1.5H_2O$. In the crystal hydrate structure of the given compound, both the square pyramidal and trigonal bipyramidal units of $[Ni(CN)_5]^{3-}$ exist.

Trigonal bipyramidal $[Ni(CN)_5]^{3-}$ **Square pyramidal $[Ni(CN)_5]^{3-}$**

Fig. 2.3.5 Ideal trigonal bipyramidal and square pyramidal structures of $[Ni(CN)_5]^{3-}$. (*cf.* Fig. 2.11.7.4 for the actual structural parameters.)

Probably, stabilities of the trigonal bipyramidal and square pyramidal geometries are comparable. This is why, in the crystal of $[Cr(en)_3][Ni(CN)_5] \cdot 1.5H_2O$, the anion exists in both the forms. Interestingly, if the cation $[Cr(en)_3]^{3+}$ is replaced by $[Cr(tn)_3]^{3+}$ (tn = 1,3-diaminopropane), then the anion $[Ni(CN)_5]^{3-}$ exists in the square pyramidal geometry. This aspect has been discussed in See 2.11.7.

Polytopal Isomerisms (*i.e.* polytopism) and Polytopal Rearrangements

The above mentioned conformational isomers are also described as the polytopal isomers which actually describe the *coordination polyhedra*. Such polytopal isomers can be mutually interconverted through the polytopal rearrangement without the bond rupture. These are illustrated below: square planar $(D_{4h}) \rightleftharpoons$ tetrahedral (T_d); tetrahedral $(T_d) \rightleftharpoons$ distorted tetrahedral $(D_{2d}) \rightleftharpoons$ square planar (D_{4h}); trigonal bipyramidal $(D_{3h}) \rightleftharpoons$ square pyramidal (C_{4v}); dodecahedron $\rightleftharpoons$ cube or square antiprism.

Note: *See* Sec. 2.11 for the mechanism of polytopal isomerism.

● **[CoCl(dppe)₂]⁺:** Another example of isomerism involving the square pyramidal and trigonal bipyramidal shape is a low spin complex of Co(II) (d^7 system) given below (*see* Sec 2.11.7 for details).

$$\left[\text{CoCl(dppe)}_2\right]^+ \rightleftharpoons \left[\text{CoCl(dppe)}_2\right]^+$$

red, square pyramid   green, trigonal bipyramid

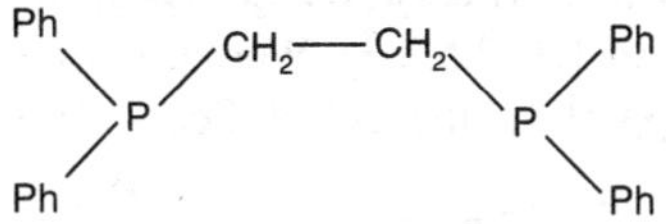

dppe = ethane-1,2-diylbis(diphenylphosphane)

Stereochemical terms

Stereoisomers: They differ only in the spatial arrangement of the ligands or bonded groups.

Enantiomers: They (*i.e.* stereoisomers) are not superimposable on their mirror images. These are optically active.

Diastereoisomers: Such stereoisomers do not have the mirror image relationship, *i.e.* these are not the enantiomers.

Asymmetric: Except C_1, they do not have any other symmetry element.

Dissymmetric: They do not have the S_n-axis (*i.e. rotation-reflection axis*). However, they may or may not possess the C_n axis ($n > 1$). Absence of S_1 is equivalent to the absence of *plane of symmetry* (σ). Similarly, absence of S_2 indicates the absence of *center of symmetry* (*i*).

It indicates that all the asymmetric species are dissymmetric but the reverse is not true.

For the *optical isomers* (*i.e.* enantiomers), they must not have the S_n axis.

Chiral Compound: The optically active compounds (*i.e.* enantiomers) are described as the chiral compounds. The term *chirality* means *handedness*. The enantiomers are of opposite chirality and they are related to each others as *'the right hand is related to the left hand'*.

Optical activity: The chiral compounds can rotate the plane of plane polarised light.

2.4 STEREOISOMERISM IN THE COORDINATION COMPOUNDS

2.4.1 Concept of Stereoisomerism

Due to the difference in the spatial arrangements of the ligands within the coordination sphere, the phenomenon of *stereoisomerism* arises. These are of two types:

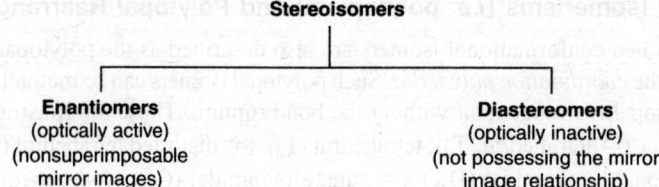

The *geometric isomers* (*i.e.* *cis-trans* isomers) produce a *sub-class of diastereomers.* **A pair of enantiomers belongs to the same geometric isomer.** The enantiomers are *chiral* while the diastereomers are *achiral*. Sometimes, the stereoisomers are classified as **chiral diastereomers** and **achiral diastereomers**. Thus the enantiomers are the *chiral diastereomers*.

Asymmetric carbon center is chiral (*i.e.* 4 different groups). This is also true for the tetrahedral geometry of coordination compounds. **[M(A—B)₂] type tetrahedral complexes (A—B denoting an unsymmetric didentate ligand) are also optically active.** In the octahedral coordination compounds, to produce the chirality, all the groups need not be different. For the monodentate ligands, three types of ligands can produce the chirality as in $[PtCl_2(NH_3)_2(NO_2)_2]$. It is illustrated below for **[MA₂B₂C₂]** having 5 geometric isomers (Figs. 2.4.1.1, 2.5.3.3). Werner considered the stereochemical aspects of coordination number 4 and 6 to support his theory (*cf.* Sec. 1.4.5).

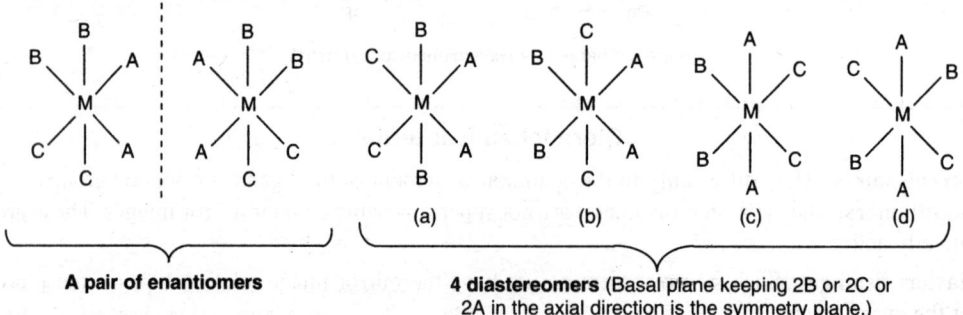

A pair of enantiomers   **4 diastereomers** (Basal plane keeping 2B or 2C or 2A in the axial direction is the symmetry plane.)

Optically active (using only 3 types of the monodentate ligands): *cis*-A₂, *cis*-B₂, *cis*-C₂. **Optically inactive:** (d) *trans*-A₂, *trans*-B₂, *trans*-C₂; (c) *trans*-A₂, *cis*-B₂, *cis*-C₂; (a) *trans*-B₂, *cis*-A₂, *cis*-C₂; (b) *trans*-C₂, *cis*-A₂, *cis*-B₂.

Fig. 2.4.1.1 (*cf.* Fig. 2.5.3.3) 5 geometric isomers for the [MA₂B₂C₂] type octahedral complexes. The geometric isomer having the same ligands at the *cis*-positions is optically active.

Note: *It may be noted that cis-[MA₂B₂C₂] is the only one example of chirality in the octahedral complexes with the minimum number of types of the monodentate ligands. In all other cases, it needs at least one chelate ring.*

2.4.2 Concept of Geometric Isomerism (*cf.* Sec. 1.15.5 for **Configuration Index**)

The geometric isomers differ in the spatial arrangement of the ligands but the difference is not considered in *terms of handedness* (*i.e.* chirality). **This is why, a pair of enantiomers represents the same geometric isomer though their handedness differs.**

For a particular geometric isomer, the given spatial orientation of the ligands determines its configuration. For the interconversion of the geometric isomers, *there is an activation energy barrier.*

The geometric isomers are very often described as the *cis-trans* isomers. The Latin words *cis* and *trans* mean *'on the same side'* and *'across'* respectively. The terms *cis-* and *trans-* were first used by van't Hoff (1874) in explaining the geometric isomers of alkenes, but these are still *in use in square planar and octahedral complexes.* In organic chemistry, E (entegegin meaning *opposite*) and Z (*zusammen* meaning *together*) nomenclature is now in use.

2.5 GEOMETRIC ISOMERISM IN THE COORDINATION COMPOUNDS

2.5.1 Geometric Isomerism in the Complexes of Coordination Number Four
   (*cf.* Sec. 1.15.5 for **Configuration Index**)

Coordination number 4 is given by the tetrahedral and square planar complexes. *In a tetrahedral structure, relative position of each ligand is the same having the bond angle ~109°.* **This is why, in a tetrahedral structure, there is no geometric isomerism.**

Square Planar Complexes: In a square planar structure, relative positions of the ligands are not equivalent. *Consequently, geometric isomerism may arise in the square planar complexes.* The possible cases are discussed in Sec. 1.4.5. The square planar complexes are generally **optically inactive** because the molecular plane is the symmetry plane.

Table 1.4.5.2 gives the possible number of geometric isomers for the different types of square planar complexes involving the monodentate ligands. The results are:

(i) **[MA$_4$]**: no isomer; *e.g.* [Pt(NH$_3$)$_4$]$^{2+}$, [PtCl$_4$]$^{2-}$, [Pt(tu)$_4$]$^{2+}$, [Ni(CN)$_4$]$^{2-}$, etc.

(ii) **[MA$_3$B]**: no isomer; *e.g.* [PtCl(NH$_3$)$_3$]$^+$

(iii) **[MA$_2$B$_2$]**: two isomers; *e.g.* [Pt(NH$_3$)$_2$(tu)$_2$]$^{2+}$, [PtCl$_2$(PR$_3$)$_2$], [PtCl$_2$(NH$_3$)$_2$], (*cis*-and *trans*)

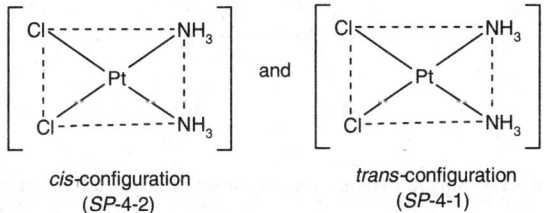

cis-configuration   and   trans-configuration
(SP-4-2)   (SP-4-1)

Fig. 2.5.1.1 *cis-trans*-isomers of [PtCl$_2$(NH$_3$)$_2$]; **configuration indexes**, *i.e.* (SP-4-1), (SP-4-2) are discussed in Sec. 1.15.5.

(iv) **[MA$_2$BC]**: two isomers, *e.g.* [Pt(AsR$_3$)$_2$BrCl], [PtBrCl(PR$_3$)$_2$] (*cis-* and *trans*)

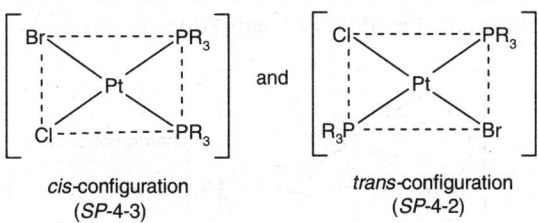

cis-configuration   and   trans-configuration
(SP-4-3)   (SP-4-2)

Fig. 2.5.1.2 *cis-trans*-isomers of [PtBrCl(PR$_3$)$_2$].

(v) **[MABCD]**: There are three *optically inactive geometric* isomers, *e.g.* [PtBrCl(NH$_3$)(NO$_2$)]$^-$, [PtBrCl(NH$_3$)(py)], [Pt(NH$_3$)(NH$_2$OH)(NO$_2$)(py)]$^+$, etc.

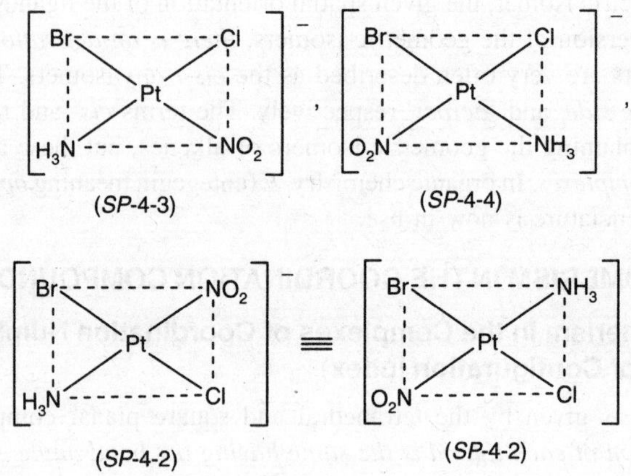

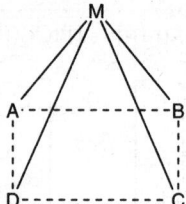

Fig. 2.5.1.3 Three isomers for [Pt(A)(B)(C)(D)] type complexes. **Note:** *cis-trans* terminology is inadequate to describe these isomers but **configuration indexes can describe these** (*cf.* Sec. 1.15.5).

In fact, from the number of isomers, Werner concluded that the complexes like [PtCl$_2$(NH$_3$)$_2$], [Pt(NH$_3$)(NH$_2$OH)(NO$_2$)(py)]$^+$, etc. were square planar. Later, these were supported by X-ray studies. **Note:** For **[MABCD]**, existence of three geometric isomers can also be explained by considering another geometry having the *square pyramidal arrangement of metal-ligand bonds, i.e.* the metal center lies above or below the plane containing the ligands.

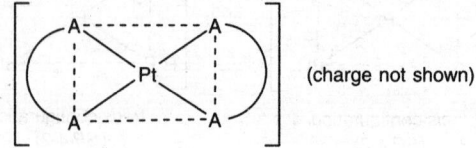

Fig. 2.5.1.4 Square pyramidal arrangement of the metal-ligand bonds.

For the **tetrahedral structure of [MABCD]**, there is no geometrical isomer but the compound is **optically active** (*i.e.* there is a pair of enantiomer).

(vi) **[M(AA)$_2$], [M(AA)B$_2$], [M(AB)C$_2$] and [M(AB)CD]**

For [M(AA)$_2$] type complexes (AA = a symmetrical didentate ligand there is only one possible configuration. It is illustrated for [Pt(en)$_2$]$^{2+}$ and [Pt(ox)$_2$]$^{2-}$, etc.

$$\left[\left(\begin{array}{c} A \text{------} A \\ \diagdown \text{Pt} \diagup \\ A \text{------} A \end{array} \right) \right] \quad \text{(charge not shown)}$$

Fig. 2.5.1.5 Structure of [Pt(AA)$_2$].

For [M(AA)B$_2$] (AA stands for a symmetrical didentate ligand), there is only one configuration. It is illustrated for [PtCl$_2$(en)].

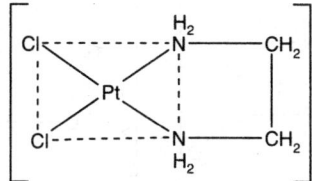

Fig. 2.5.1.6 Structure of [PtCl₂(en)].

Similarly, for [M(AB)C₂], (AB = an unsymmetrical ligand), there is also only one possible geometric configuration as in [PtCl₂(gly)]⁻.

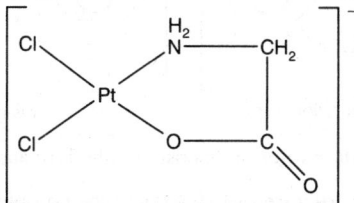

Fig. 2.5.1.7 Structure of [PtCl₂(gly)]⁻.

For [M(AB)CD] (AB stands for an unsymmetrical didentate ligand), there are two possible geometric isomers as in [PtBrCl(gly)]⁻

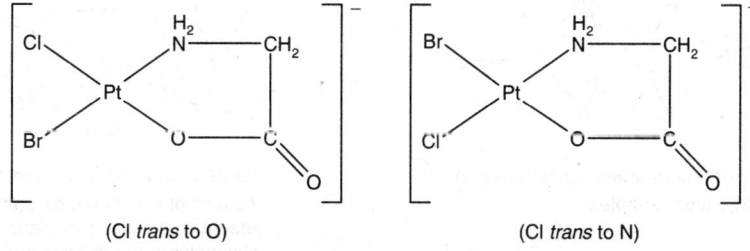

Fig. 2.5.1.8 Geometric isomers for [PtBrCl(gly)]⁻.

(vii) **[M(AB)₂]:** Here AB stands for an unsymmetrical didentate ligand having the ligating sites A and B. It will produce the *cis-trans* isomers.

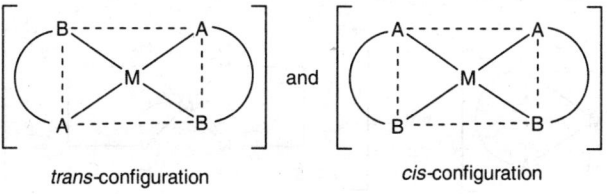

Fig. 2.5.1.9 *cis-trans* isomers of [M(AB)₂].

It is illustrated for **[Pt(gly)₂]** and **[Ag(pic)₂]**

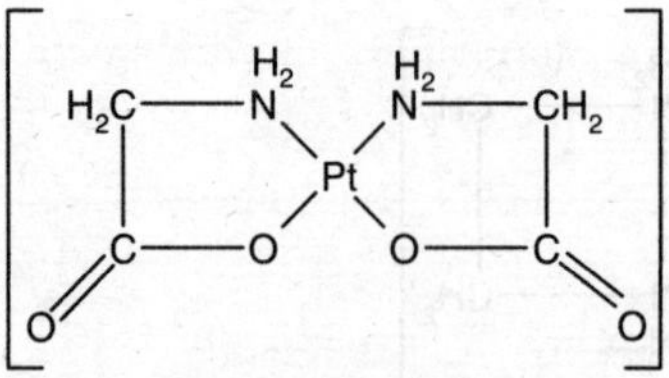

cis-configuration 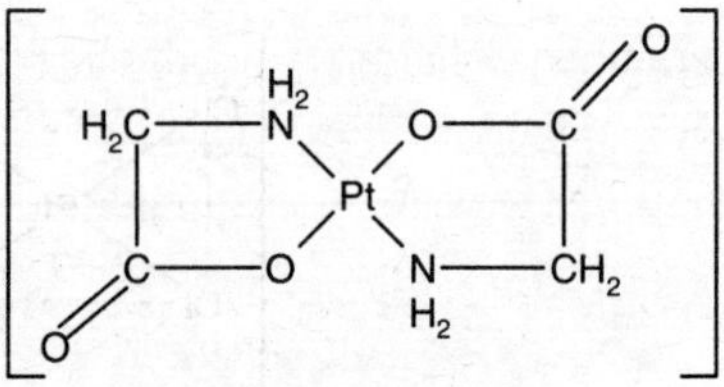   *trans*-configuration

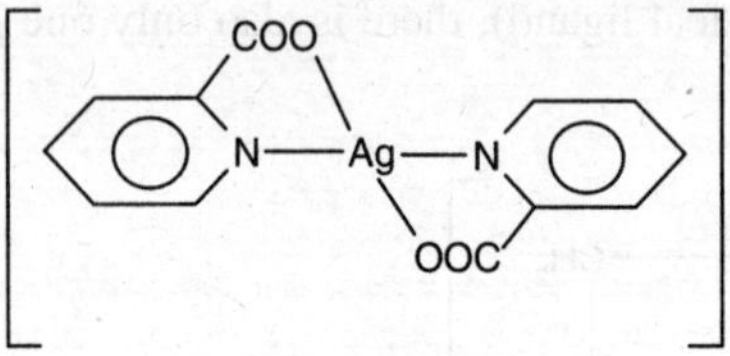

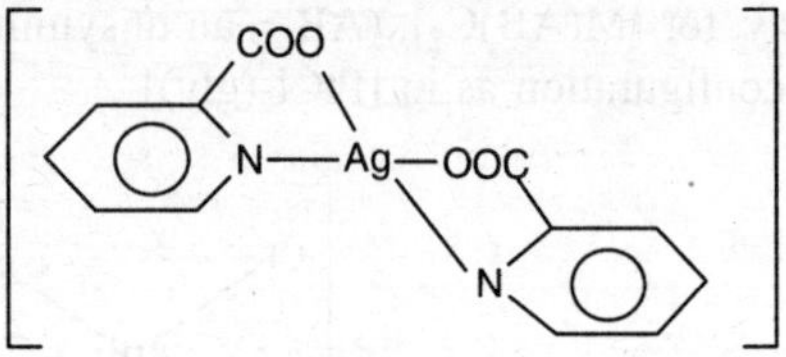

trans-configuration   *cis*-configuration

Fig. 2.5.1.10 *cis-trans* isomerism in [Pt(gly)$_2$] and [Ag(pic)$_2$].

Note: For the square planar structure of [M(AB)$_2$] type complex, there will be two geometric, *i.e.* *cis–* and *trans–* isomers and both of them are *optically inactive*. On the other hand, for the tetrahedral structure, there will be no geometrical isomerism and it will give only one configuration which will be *optically active*. Thus, the tetrahedral structure of [M(AB)$_2$] will produce a *pair of enantiomers.*

A pair of enantiomers for tetrahedral [M(AB)$_2$] type complex

(*cis*)

[M(AB)$_2$] type (both *cis-* and *trans-* **square planar complex, optically inactive**, (existence of plane of symmetry in the molecular plane)

Fig. 2.5.1.11 Optical activity in [M(AB)$_2$] type tetrahedral complex and optical inactivity in [M(AB)$_2$] type square planar complex.

(viii) **[M(AA)$_2$]:** Normally for the symmetrical didentate ligands (denoted by AA), like *en, ox, acac,* etc. there should be no geometric isomerism. This has been already discussed (*cf.* Fig. 2.5.1.5).

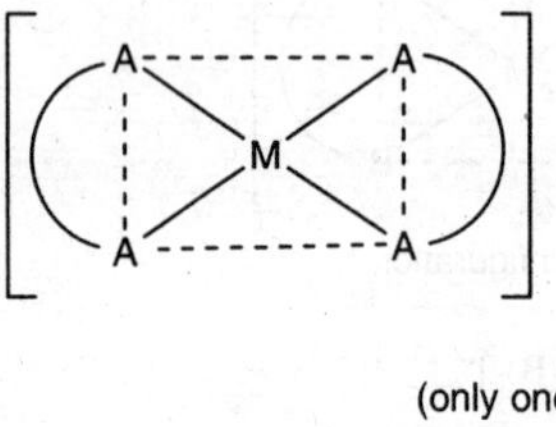

e.g. 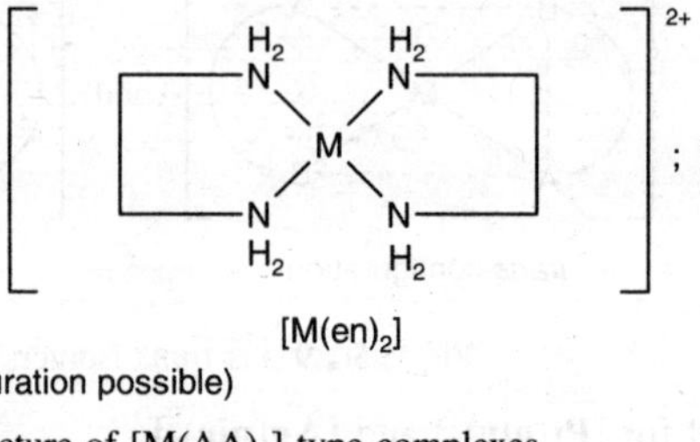   ; M = Pd^{2+}, Pt^{2+}

[M(en)$_2$]

(only one configuration possible)

Fig. 2.5.1.12 Structure of [M(AA)$_2$] type complexes.

Sometimes, the *cis –trans* isomerism may arise from **the difference in spatial arrangement of the substituents in the ligand**. It is illustrated in Fig. 2.5.13.

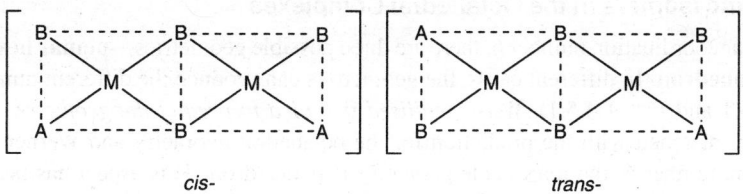

cis-configuration   *trans*-configuration

Fig. 2.5.1.13 Isomerism in [Pt(AA)$_2$] type complexes due to the difference in spatial arrangement of the substituents.

(ix) **[M(AB)$_2$]:** In the planar [M(AB)$_2$] type examples, *e.g.* [Pt(gly)$_2$] the geometric isomerism has developed due to different ligating sites (*e.g.* O, N in gly). In some cases, the unsymmetrical nature of ligand may generate the *cis- trans* isomerism even for the same coordinating sites. It is illustrated for the unsymmetrical glyoximes and β-diketones.

cis-   *trans-*

cis-   *trans-*

Fig. 2.5.1.14 *cis– trans* isomerism depending on the *cis– trans* positions of R$_1$ and R$_2$ in the square planar complexes with unsymmetric glyoximes and β–diketones (*cf.* Fig. 2.5.1.13).

(x) **[M$_2$A$_2$B$_4$]:** It represents a **bridged dinuclear complex** where 'B' acts as a bridging ligand and A, B stand for the monodentate ligands. Such dinuclear complexes can show the *cis– trans* isomers along with an unsymmetrical isomer.

Examples: [Pt$_2$Cl$_4$(PR$_3$)$_2$], [Pt(AsR$_3$)$_2$Cl$_4$]

cis-   *trans-*

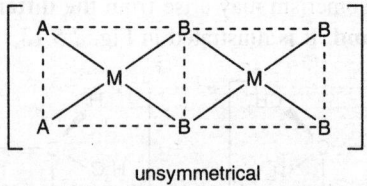

unsymmetrical

Fig. 2.5.1.15 Geometric isomerism in [$M_2A_2B_4$] type complexes (2A ligands coordinate a particular metal centre).

2.5.2 Geometric Isomerism in the Square Pyramidal Complexes of Coordination Number Five

Geometric isomerism is rare in the five coordinate complexes. One typical example is dibromido-dicarbonylcyclopentadienylrhenium(III), *i.e.* [$ReBr_2(CO)_2(\eta^5-C_5H_5)$] in which η^5–Cp occupies the axial position while the other four ligands are placed at the basal plane. At the basal plane, two bromides can be placed either at the *cis*– or *trans*-positions. Both of them are optically inactive.

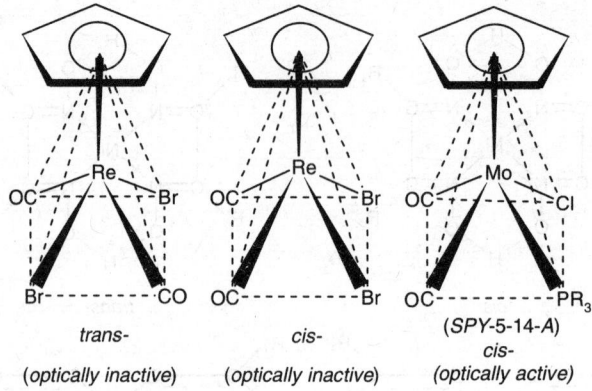

trans-	cis-	(*SPY*-5-14-A)
		cis-
(*optically inactive*)	(*optically inactive*)	(*optically active*)

Fig. 2.5.2.1 Some representative examples of isomerism in square pyramidal complexes.

Similarly, [$Mo(CO)_2(\eta^5-C_5H_5)Cl(PR_3)$] (square pyramidal geometry can have the *cis*– and *trans*–isomers with respect to the positions of CO ligands in the basal plane. It may be noted that in these complexes, η^5– Cp occupies the apical position as usual. In this Mo(II)-complex, the *cis*–isomer is *chiral and optically active* while the *trans*–isomer having a plane of symmetry passing through the η^5– Cp, Cl and PR_3 is optically inactive. The **descriptor of the optically active form** is (*SPY*-5-14-A) (*cf.* Sec. 1.15.5).

Note: It may be argued that η^5– Cp occupies 5 coordination sites. Then such complexes are described as the 9 coordinate complexes.

2.5.3 Geometric Isomers in the Octahedral Complexes

Theoretically, for coordination number 6, there are three possible geometries—**planar hexagon, trigonal prism and octahedron**. In different cases, the geometries can produce the different number of isomers (*cf.* Table 1.4.5.1 and Fig. 1.4.5.1). *Werner utilised this idea to predict the geometry.* The number of isolated isomers matched with the prediction for the octahedral geometry and Werner concluded that for coordination number 6, the reasonable geometry is octahedron. This aspect has been discussed in detail in See. 1.4.5. In fact, even to this date, Werner's prediction appears to be correct in most of the cases. *Only in some rare cases, the trigonal prism geometry has been found for the coordination number 6.*

(i) **[MA$_6$]** and **[MA$_5$B]:** No isomer (A, B for the monodentate ligands).

(ii) **[MA$_4$B$_2$]:** The possible *cis – trans* isomers are shown in Fig. 2.5.3.1.

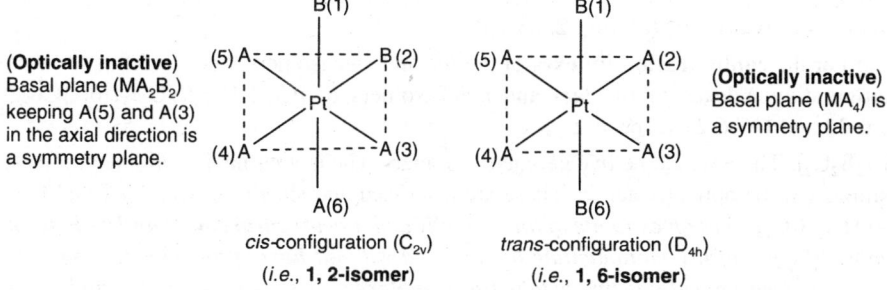

Fig. 2.5.3.1 *cis-trans* isomers for the [MA$_4$B$_2$] type octahedral complexes.

- The *cis–* isomer is generated by placing two B's at the *cis–* positions (*i.e.* adjacent positions), *i.e.* (1, 2), (1, 3), (1, 4), (1, 5), (2, 3), (3, 4), (4, 5), (5, 2), (6, 3), (6, 2), (6, 5) and (6, 4).
- The *trans-*isomer is obtained by placing two B's at the *trans-*positions (*i.e.* along the diagonals), *i.e.* (1, 6), (2, 4), and (3, 5).

The examples are: [CoCl$_2$(NH$_3$)$_4$]$^+$ (*cis–*violet, *trans–*green); [Co(NH$_3$)$_2$(NO$_2$)$_4$]$^-$, (*cis–*yellow-brown, *trans–*yellow)

(iii) **[MA$_3$B$_3$]** (*cf.* **[M(AB)$_3$], [M(AAA)$_2$], [M(AAA)(BB)C]:** If the same (or similar) three ligands occupy **a triangular face** (*i.e.* the same or similar ligands are always at the *cis–*positions), then it leads to a *facial* or *fac–*isomer (equivalent to *cis-*isomer). When the same (or similar) three ligands are distributed in a way so that two of them are at the *trans-*positions, it leads to the *meridional isomer or mer-*isomer (equivalent to *trans–*isomer). In other words, in the *mer-*isomer, the three similar or identical binding sites are placed **along the meridian.**

Examples: [CrCl$_3$(NH$_3$)$_3$], [Co(NH$_3$)$_3$(NO$_2$)$_3$], [RhCl$_3$(py)$_3$], [RuCl$_3$(OH$_2$)$_3$], etc.

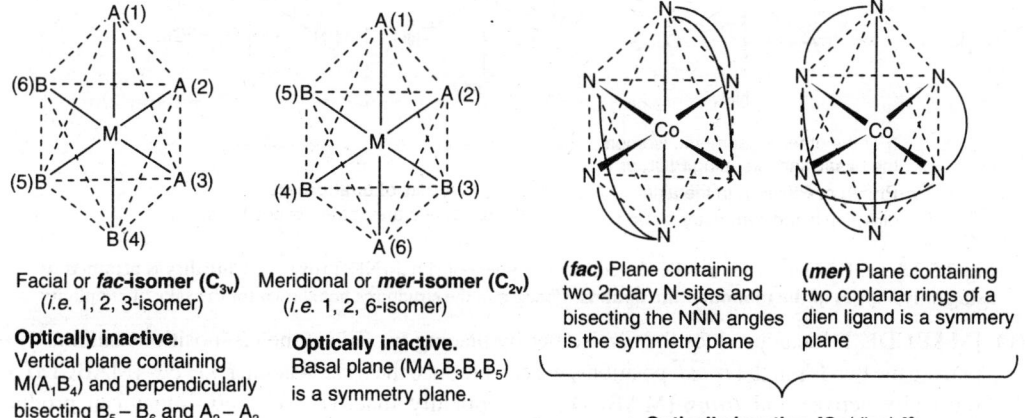

Fig. 2.5.3.2 *fac-mer* isomers in the [MA$_3$B$_3$] type complexes and [M(dien)$_2$], *i.e.* [M(AAA)$_2$] type complexes.

Note: The *fac-mer* isomerism may be generated for the polydentate ligands also, *e.g.* [Co(dien)₂]³⁺ (*fac-*, *mer*), *fac*–[CoCl(dien)(en)]²⁺, *mer*–[CoCl(dpt)(en)]²⁺, *mer*–[CoCl(dpt)(tn)]²⁺, etc. Depending on the *stereochemical positions of the donor sites of the tridentate ligands* like *dien*, *dpt*, the *fac-mer* isomerism arises. tn = H_2N—$(CH_2)_3$—NH_2; dpt = H_2N—$(CH_2)_3$—NH—$(CH_2)_3$—NH_2; dien = H_2N—$(CH_2)_2$—NH—$(CH_2)_2$—NH_2; [Co(dien)₂]³⁺ (*fac- mer*) (Fig. 2.5.3.2)

[M(AB)₃] where AB denotes an unsymmetrical didentate ligand, the *fac-mer* isomerism can develop as in [Co(gly)₃] (*cf.* Fig. 2.5.3.9).

Note: For the **configuration indexes** of the *fac* and *mer* isomers, Sec. 1.15.5 is to be consulted. Configuration indexes of the *fac-* and *mer-* isomers of [Co(NH₃)₃(NO₂)₃] or [Co(gly)₃] are *OC*-6-22 and *OC*-6-21 respectively.

(iv) **[MA₂B₂C₂]:** There would be five geometric isomers. The isomer having all like groups at the *cis*-positions will be optically active. These are discussed and shown in See. 2.4.1 and Fig. 2.4.1.1. *cis*–**[MA₂B₂C₂]** *gives an example of the optically active octahedral compound with the minimum number of types of the monodentate ligands and without having any chelate ring. In fact, for all other known optically active octahedral complexes, there is at least one chelate ring.*

In fact, all the five geometric isomers of [PtCl₂(NH₃)₂(py)₂]²⁺ have been prepared. **Thus the total number of stereoisomers is 6(= 4 + 2).**

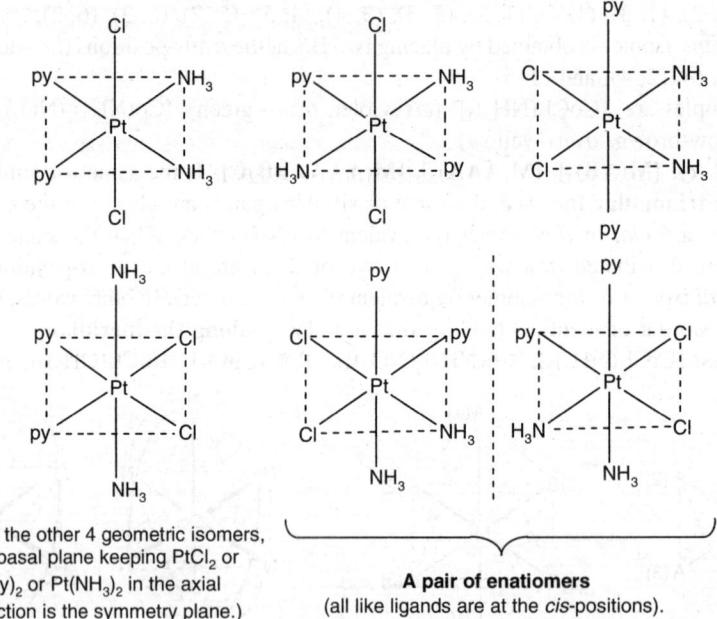

(For the other 4 geometric isomers, the basal plane keeping PtCl₂ or Pt(py)₂ or Pt(NH₃)₂ in the axial direction is the symmetry plane.)

A pair of enatiomers (all like ligands are at the *cis*-positions).

Fig. 2.5.3.3 (*cf.* Fig. 2.4.1.1 for details) 5-geometric isomers of [PtCl₂(NH₃)₂(py)₂]²⁺. **Chirality is attained by using only 3 types of the monodentate ligands**. Charges in the structures not shown for the sake of simplicity.

(v) **[MABCDE₂]:** It can produce the *cis*–isomer by placing two E's at the *cis*–positions; similarly by placing the two E's at the *trans*–positions, we can produce the *trans* – isomer, *i.e. cis*–[MABCD(E)₂] (optically active) and *trans*–[MABCD(E)₂] (optically inactive). The *cis*–iomer can produce 6-pairs of enantiomers having AB, AC, AD, BC, BD and CD at the *trans*-positions respectively. The *trans*-isomer can generate three geometrical isomers having AB, AC and AD at the *trans*-

positions respectively. All these *trans*–isomers are optically inactive. **Thus in total there are 9 geometric isomers.**

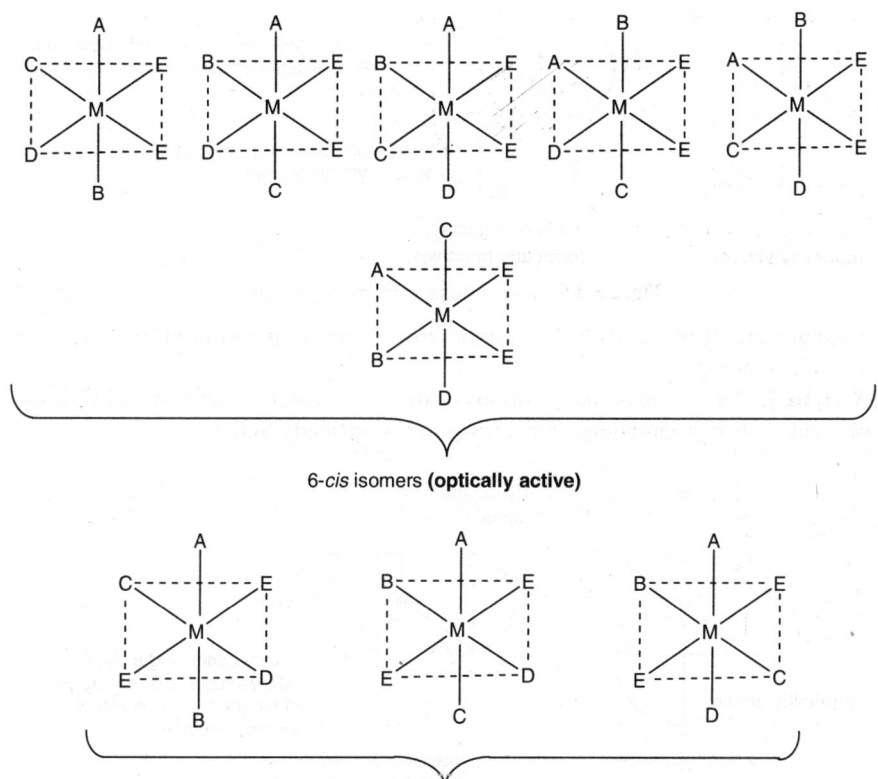

6-*cis* isomers **(optically active)**

(Basal plane assuming E—M—E as the axial direction is the symmetry plane)
3-*trans*-isomers **(optically inactive)**

Fig. 2.5.3.4 9 geometric isomers for [MABCDE$_2$] type octahedral complexes.

cis–[MABCD(E)$_2$]: 6 geometric isomers and all of them are optically active, *i.e.* 6 pairs of enantiomers.

trans–[MABCD(E)$_2$]: 3 geometric isomers and all of them are optically inactive because of the plane of symmetry passing through M(ABCD).

Thus we have: total number of geometric isomers = 6 + 3 = 9; and **total number of stereoisomers = 12 + 3 = 15.**

(vi) **[MABCDEF]:** There are *fifteen geometrical isomers* and each of them is optically active. Thus there are **15 pairs of enantiomers**. The only known example is:

$$[Pt(Br)(Cl)(I)(NH_3)(NO_2)(pn)]$$

(vii) **[M(AA)$_2$B$_2$]:** Here AA stands for a symmetrical didentate ligand (*e.g.* acac, en, ox, etc., and B's are the monodentate ligands). A didentate ligand is considered to occupy only the *cis*–positions and it never can occupy the *trans*–positions. The complex can give the *cis*–*trans* isomerism. The *cis*–compound is optically active.

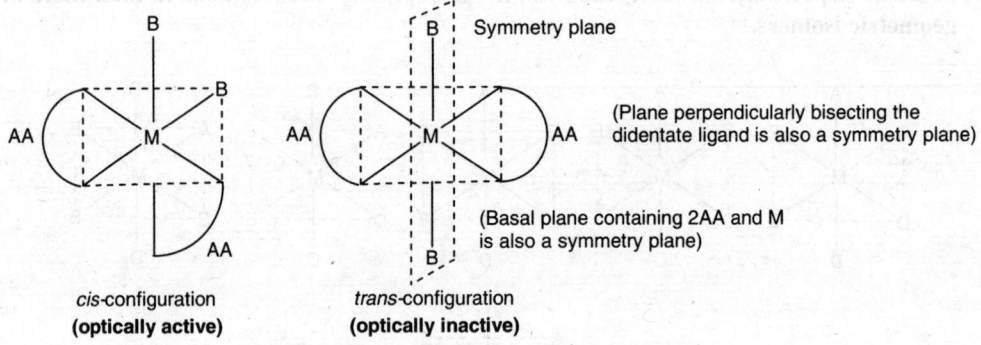

cis-configuration
(optically active)

trans-configuration
(optically inactive)

Fig. 2.5.3.5 Geometric isomers of [M(AA)₂B₂].

The examples are: [CoCl₂(en)₂]⁺ (*cis* – purple, *trans* – green), [Cr(en)₂(NCS)₂]⁺ (*cis* – orange red, *trans* – yellow orange)

(viii) **[M(AA)₂ BC]:** Depending on the positions of the monodentate ligands B and C, it can produce the *cis*- and *trans*-configurations. The *cis*-isomer is **optically active.**

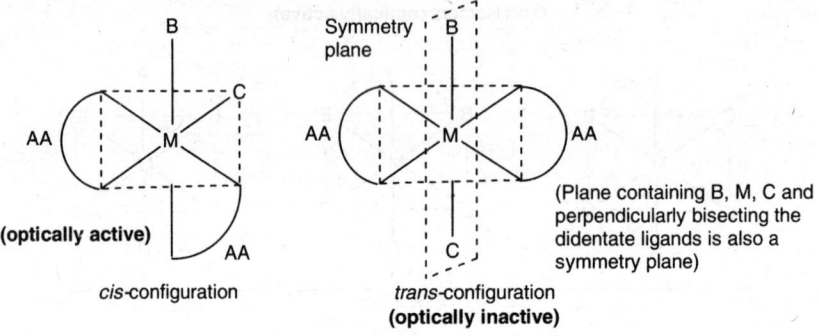

(optically active)

cis-configuration

trans-configuration
(optically inactive)

Fig. 2.5.3.6 Geometric isomers of [M(AA)₂ BC].

Examples: [CoCl(en)₂(NH₃)]²⁺, [CoCl(en)₂(NO₂)]⁺, etc.

(ix) **[M(AA)B₂C₂]:** Depending on the *cis–trans* positions of the monodentate ligands, three configurations may be obtained.

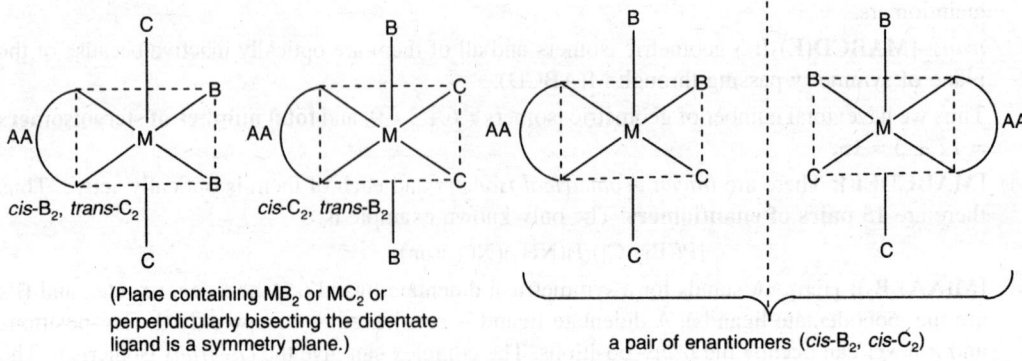

cis-B₂, *trans*-C₂

cis-C₂, *trans*-B₂

(Plane containing MB₂ or MC₂ or perpendicularly bisecting the didentate ligand is a symmetry plane.)

a pair of enantiomers (*cis*-B₂, *cis*-C₂)

Fig. 2.5.3.7 Geometric isomers of [M(AA)B₂C₂].

The examples are: $[CrCl_2(en)(NH_3)_2]^+$, etc.

The configuration in which both the like monodentate groups are mutually at the *cis*–positions, is optically active. Thus we have: **total number of geometrical isomers = 3** and **total number of stereoisomers = 4**

(x) **[MABCD(EE)]:** The didentate ligand (EE) spans two *cis*–positions. Keeping the didentate ligand fixed, distribution of the monodentate ligands produces 6 geometrical isomers and each of them is optically active. Thus it gives 6 pairs of enantiomers.

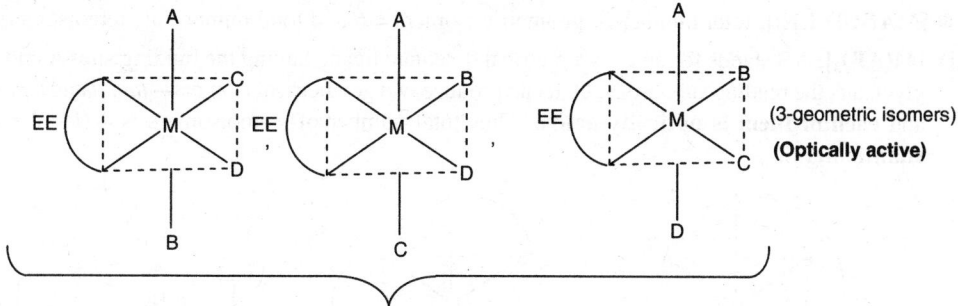

(a) A is placed at an axial position and its *trans* axial position is occupied by B or C or D, *i.e. trans*-axial positions are A, B; A, C; and A, D.

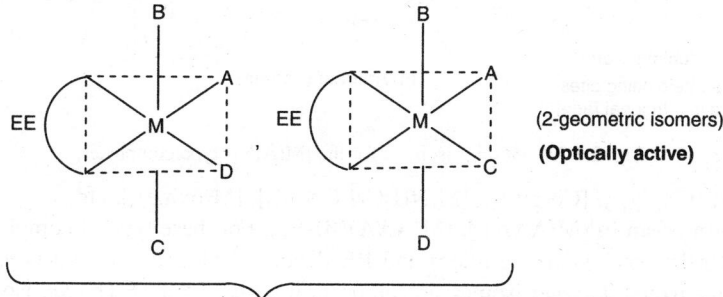

(b) B is placed at an axial position; and the *trans*–axial positions are: B, C; B, D.

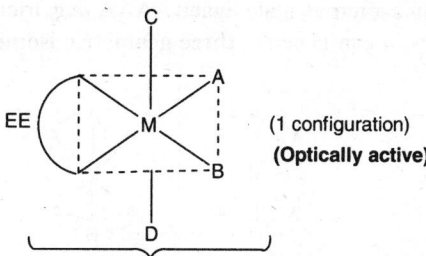

(c) C is placed at an axial position and the *trans*-axial position is C, D.

Fig. 2.5.3.8 Geometric isomers of [MABCD(EE)].

Thus the total number of geometric isomers is $3 + 2 + 1 = 6$ and the total number of stereoisomers is 12 (*i.e.* **6 pairs of enantiomers**).

Note: [MABCDE₂] vs. [MABCD(EE)]: If two monodentate E ligands are replaced by a didentate ligand (EE), then a restriction is introduced as the didentate ligand can only occupy the two *cis*–positions. Thus in [MABCD(EE)], the number of isomer decreases compared to the case of [MABCDE₂] which has been discussed earlier.

● **[MABCDE₂]:** total number of geometric isomers = 3 (for *trans*-positions of 2E) + 6 (for *cis*–positions of 2E) = 9; and total number of stereoisomers = 3 + 6 × 2 = 15, (as the *cis*–isomers are optically active).

● **[MABCD(EE)]:** total number of geometric isomers = 6 and total number of stereoisomers = 12.

(xi) **[M(AB)₃]:** AB stands for an unsymmetrical didentate ligand having the binding sites A and B. For glycinate, the binding sites are N, O. It can produce two geometrical isomers—*facial* and *meridional* and **each of them is optically active.** Thus total number of stereoisomers is 4 (*i.e.* **2 pairs of enantiomers**).

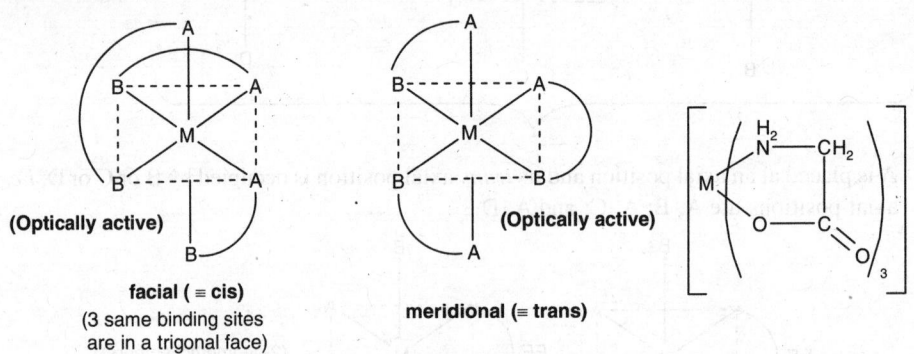

facial (≡ cis)
(3 same binding sites
are in a trigonal face)

meridional (≡ trans)

Fig. 2.5.3.9 *fac-mer* isomers for the [M(AB)₃] type complexes.

Examples: [Cr(gly)₃], [Co(gly)₃], [M(NH₂CH₂CH₂S)₃], [M(oxine)₃], etc.

(xii) *fac-mer* **isomerism in [M(AAA)₂], [M(AAA)(BB)C]:** For these type of complexes where AAA denotes a tridentate ligand (*e.g.* dien) and BB denotes a didentate ligand (*e.g.* en) and C is a monodentate ligand, *fac-mer* isomerism can occur (*cf.* Fig. 2.5.3.2). The *fac*-isomer places three binding sites A, A, A of the tridentate ligand AAA **on a trigonal face** while in the *mer*-isomer, these are placed **along the meridian.**

(xiii) **[M(AAAA)B₂]:** For the linear tetradentate ligand AAAA (*e.g.* trien), depending **on the positions of different chelate rings,** it can generate **three geometric isomers. Example:** [Co(trien)Cl₂]⁺

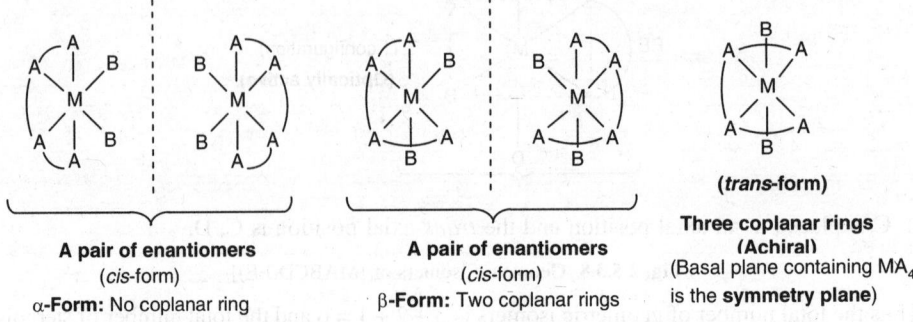

A pair of enantiomers
(*cis*-form)
α-Form: No coplanar ring

A pair of enantiomers
(*cis*-form)
β-Form: Two coplanar rings

(trans-form)

Three coplanar rings
(Achiral)
(Basal plane containing MA₄
is the **symmetry plane**)

Fig. 2.5.3.10 Isomers of the [M(AAAA)B₂] type complexes (AAAA = trien).

- **α-Form:** All the three chelate rings in different planes (*i.e.* no coplanar ring); 2B in *cis*-positions; optically active.
- **β-Form:** Two coplanar rings; 2B in *cis*-positions; optically active.

Trans-Form: Three coplanar rings; 2B in *trans*-positions; optically inactive.

Note: Depending on the conformations of the individual chelate rings, all these 3 geometric isomers have the additional isomers. This aspect is discussed later.

(xiv) **[M(trien)(XY)]** *vs.* **[M(tren)(XY)]:** XY denotes an unsymmetrical didentate ligand (*e.g.* gly). [M(trien)(XY)] can **give only the cis-forms (α- and β)** where XY spans the *trans*-positions of secondary N-atoms in the α-form. With respect to the N-atoms, different isomers are possible (*cf.* Fig. 2.5.3.11). **It cannot have the trans-form** having trien in a plane (*i.e.* three coplanar rings of trien) because then XY will have to occupy the *trans*-positions but it is not possible.

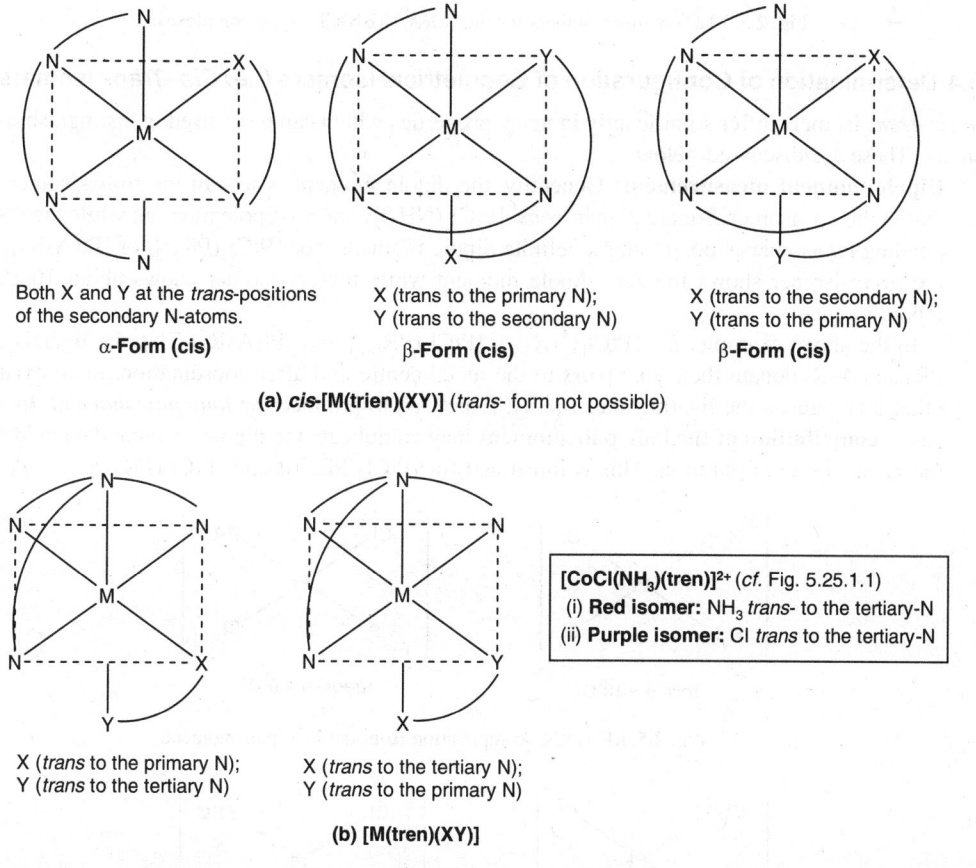

Both X and Y at the *trans*-positions of the secondary N-atoms.

α-Form (cis)

X (trans to the primary N); Y (trans to the secondary N)

β-Form (cis)

X (trans to the secondary N); Y (trans to the primary N)

β-Form (cis)

(a) *cis*-[M(trien)(XY)] (*trans*- form not possible)

[CoCl(NH₃)(tren)]²⁺ (*cf.* Fig. 5.25.1.1)
(i) **Red isomer:** NH₃ *trans*- to the tertiary-N
(ii) **Purple isomer:** Cl *trans* to the tertiary-N

X (*trans* to the primary N); Y (*trans* to the tertiary N)

X (*trans* to the tertiary N); Y (*trans* to the primary N)

(b) [M(tren)(XY)]

Fig. 2.5.3.11 Isomerism in [M(trien)(XY)] and [M(tren)(XY)] (*cf.* Isomerism in [M(N₄)XY], N₄ = trien or tren; X and Y monodentate ligands).

(xv) **[M(AA)(BB)CD]:** The monodentate C and D ligands can be positioned at the *cis*- or *trans*-positions. The two *cis*- forms are chiral while the *trans*-form is achiral. Total number of stereo-isomers = 5 (2 pairs of enantiomers of the *cis*- form and one achiral *trans*-form).

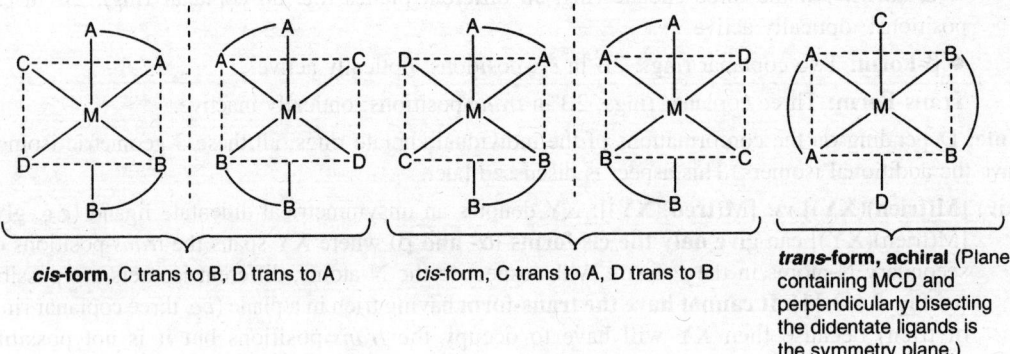

| cis-form, C trans to B, D trans to A | cis-form, C trans to A, D trans to B | **trans-form, achiral** (Plane containing MCD and perpendicularly bisecting the didentate ligands is the symmetry plane.) |

Fig. 2.5.3.12 *cis-trans* isomers for the [M(AA)(BB)CD] type complexes.

2.5.4 Determination of Configuration of Geometrical Isomers (*i.e. Cis–Trans* isomers)

The *cis–trans* isomers differ significantly in many properties which can be utilised in distinguishing the isomers. These are discussed below.

(a) **Dipole moment measurement:** Generally, the dipole moment is less in the *trans*–isomer than that in the *cis*–isomer. Square planar *trans*–$[PtCl_2(NH_3)_2]$ has no dipole moment while the corresponding *cis*-isomer is possessing a definite dipole moment. For $[PtCl_2(PR_3)_2]$ or $[Pt(AsR_3)_2Cl_2]$, the *trans*–isomer shows the zero-dipole moment while the *cis*–isomer shows about 10 Debye dipole moment.

In the above example, *i.e.* $[PtCl_2(NH_3)_2]$, $[PtCl_2(PR_3)_2]$ and $[Pt(AsR_3)_2Cl_2]$, the ligands NH_3, PR_3 and AsR_3 donate their lone pairs to the metal centre and after coordination, there exists no other lone pair on the ligating sites (*i.e.* N, P and As) to produce the *lone pair moment*. In some cases, **contribution of the lone pair moment** may complicate the dipole moment data to identify the geometric configuration. This is illustrated for $[PtCl_2(SEt_2)_2]$ and $[PtCl_2(PR_3)_2]$

$$\left[\begin{array}{c} R_3P \diagdown \qquad \diagup Cl \\ Pt \\ R_3P \diagup \qquad \diagdown Cl \end{array} \right] \qquad \left[\begin{array}{c} Cl \diagdown \qquad \diagup PR_3 \\ Pt \\ R_3P \diagup \qquad \diagdown Cl \end{array} \right]$$

$$(\textit{cis-}\ \mu \approx 10\ D) \qquad\qquad\qquad (\textit{trans-}\ \mu \approx 0\ D)$$

Fig. 2.5.4.1 (a) No complication from the lone pair moment.

$$\left[\begin{array}{c} Et_2\overset{..}{S} \diagdown \qquad \diagup Cl \\ Pt \\ Et_2\overset{..}{S} \diagup \qquad \diagdown Cl \end{array} \right] \qquad \left[\begin{array}{c} Cl \diagdown \qquad \diagup \overset{..}{S}Et_2 \\ Pt \\ Et_2\overset{..}{S} \diagup \qquad \diagdown Cl \end{array} \right]$$

$$(\textit{cis-}\ \mu \approx 9.5\ D) \qquad\qquad\qquad (\textit{trans-}\ \mu \approx 2.5\ D)$$

Fig. 2.5.4.1 (b) Complication from the lone pair moments.

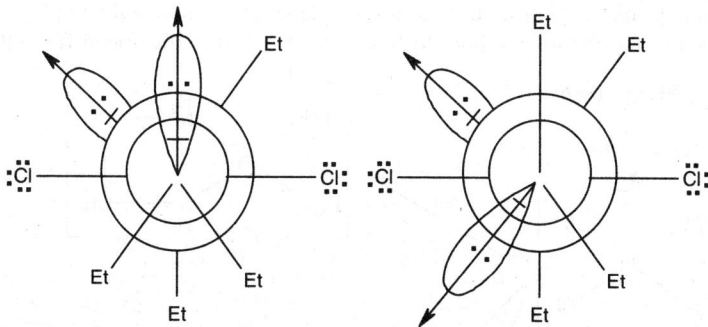

Fig. 2.5.4.2 Different hybrid lone pair moments and bond moments in *trans*–[PtCl$_2$(SEt$_2$)$_2$].

The zero dipole moment for *trans*–[PtCl$_2$(PR$_3$)$_2$] is quite expected but the nonzero dipole moment for *trans*–[PtCl$_2$(SEt$_2$)$_2$] needs an explanation which has been discussed in Sec. 10.5.3 in Vol. 2 of Fundamental concepts of Inorganic Chemistry. Simply it can be said that for the *trans*–configuration, the bond moments and lone pair moments of the *trans*-Cl centres are mutually cancelled but the lone pair moments of the S–centers are not mutually cancelled fully and this gives the origin of **the nonzero value of μ for the *trans*–configuration**. It may be noted that the corresponding *cis*–isomer possesses a higher value of μ.

(b) **Infra-red (ir) spectroscopic study** (*cf.* Sec.12.1): In the *trans*–compounds, the symmetric vibration of the *trans*–ligands cannot produce any dipole moment change. Consequently, this mode of vibration is ir-inactive. On the other hand, for the corresponding *cis*–isomer, both the symmetric and unsymmetric stretchings can produce an appreciable change in dipole moments and consequently both these modes are ir-active. In fact, the vibrational spectrum of *cis*–[PtCl$_2$(NH$_3$)$_2$] or *cis*–[CoCl$_2$(NH$_3$)$_4$]$^+$ is different from that of the corresponding *trans*–isomer.

(c) **Electronic spectra** (*cf.* Chapter 7): For the transition metal complexes, the intensity of a ligand field band (*i.e. d–d* transition) depends on the efficiency with which the Lapporte Selection Rule can be relaxed. In the *cis*–complexes having no center of symmetry, the *d–d* transition is more allowed compared to that in the centro-symmetric *trans*–complexes (for which the *vibronic coupling* can only relax the Lapporte Selection Rule). This is why, in general, for *the cis–complexes, the molar extinction coefficients are higher than those of their corresponding trans–complexes.*

(d) **X-ray studies:** It can most convincingly establish the geometrical configuration of the isomers.

(e) **Chemical method (Grinberg's method):** It is established that a five or six membered chelate forming ligand can only occupy the adjacent *cis*–positions and it can never occupy the *trans* positions in the square planar or octahedral structure. *Thus in a suitable ligand substitution reaction, if such a chelating ligand can replace two unidentate ligands to give a chelated compound then it indicates that the two unidentate ligands were at the cis–positions provided during the substitution reaction, no change in configuration has occurred.* For this purpose, the typical chelating ligands used are:

(oxalic acid, H$_2$ox)   (glycine, Hgly)   (ethylenediamine, en)

This method is applicable for both the square planar and octahedral complexes. This method is also described as Grinberg's method. In Scheme 2.5.4.1, it is illustrated for [PtCl$_2$(NH$_3$)$_2$].

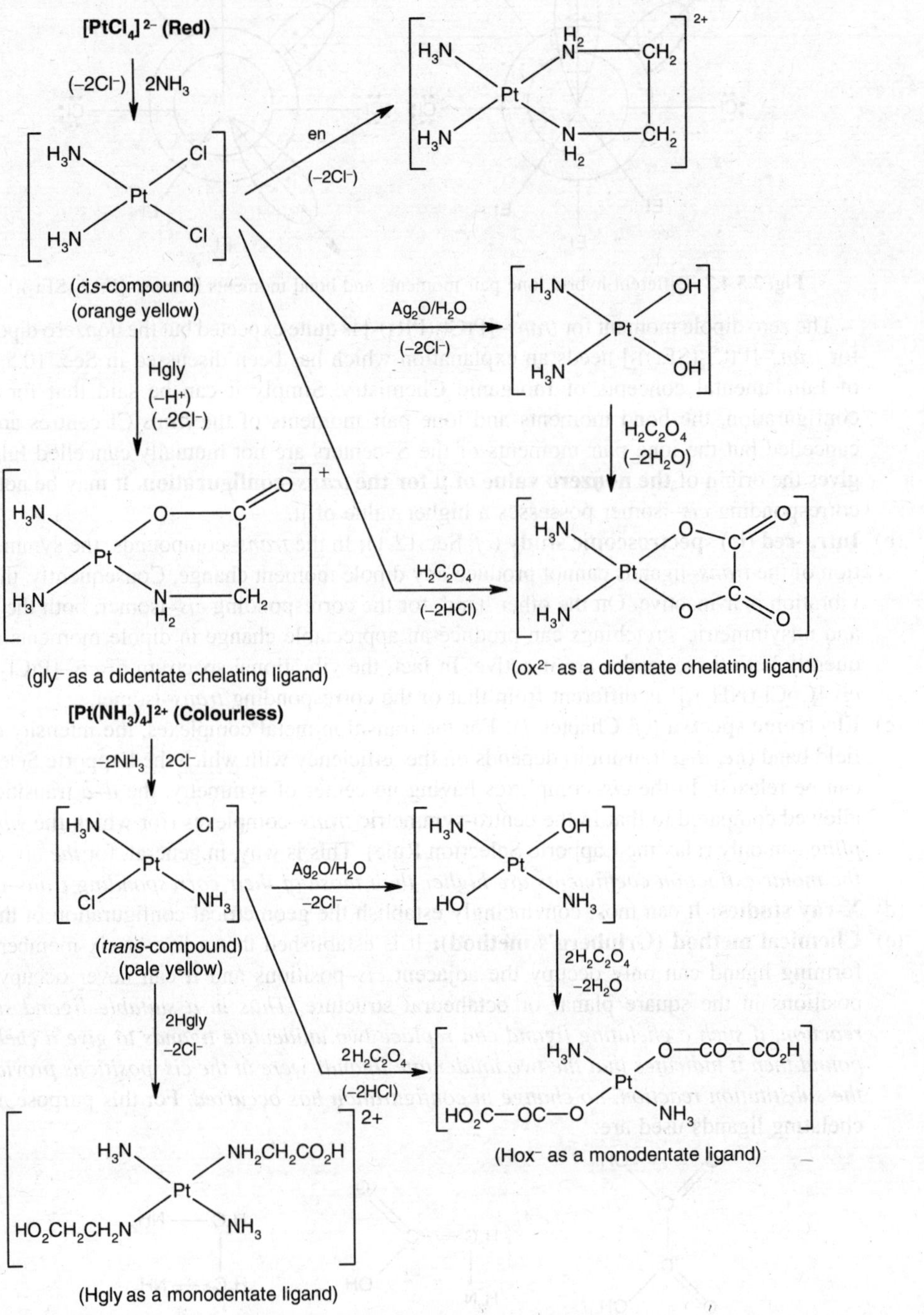

Scheme 2.5.4.1 Identification of the *cis–trans* configurations of [PtCl$_2$(NH$_3$)$_2$] by Grinberg's method.

Grinberg's method can be applied for the octahedral complexes. Scheme 2.5.4.2 illustrates the determination of configurations of $[CrCl_2(en)_2]^+$.

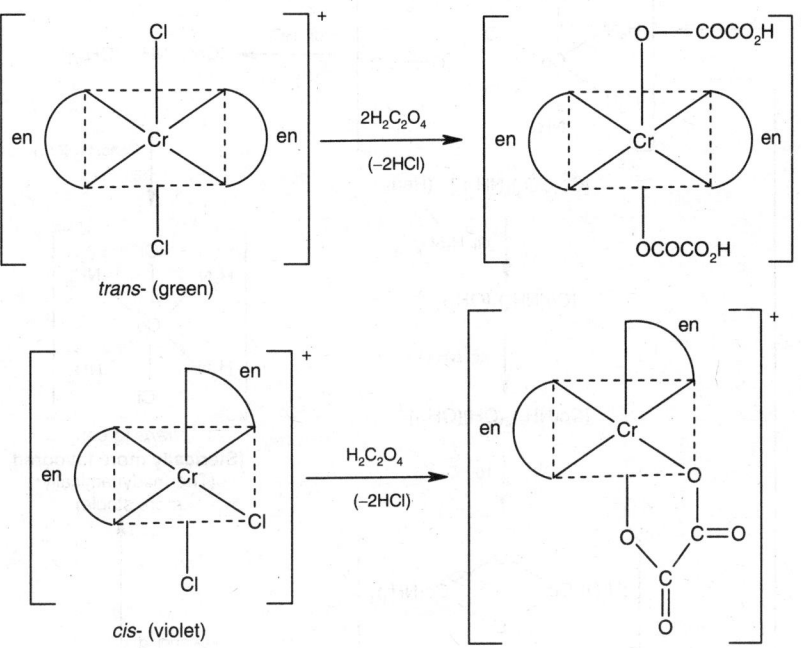

Scheme 2.5.4.2 Identification of the *cis –trans* configurations of $[CrCl_2(en)_2]^+$.

Note: A didentate chelating ligand can substitute two monodentate ligands if present at the *cis*–positions, and this substitution gives the *cis*–product. But replacement of a didentate ligand by two monodentate ligands may not necessarily give the *cis*–product because during this substitution process there may be a change in configuration. It is illustrated in Scheme 2.5.4.3 where substitution of a didentate carbonato ligand (present as a chelating one) by two monodentate ligands can produce both *cis*– and *trans*–compounds depending on the conditions.

At low temperature, treatment of dil. HCl or alcoholic HCl, produces the *cis*-compound through the substitution of the chelated carbonato group by two monodentate ligands. But this on standing isomerises to the *trans*–compound. Direct treatment of HBr or conc. HCl + conc. H_2SO_4 on the carbonato complex gives the *trans*–compound. **It is believed that the bulky halides when positioned at the cis–configuration, the steric hindrance occurs.** *This is why, the trans–configuration is thermodynamically more favoured.* **Br^- is bulkier than Cl^- and this is why, direct treatment with HBr even in mild condition gives the *trans*–isomer** (*cf.* Scheme 2.5.4.3).

(f) **Chemical method (Kurnakov's reaction) for the square planar complexes based on the *trans-effect*:** It is illustrated for $[PtCl_2(NH_3)_2]$ reacting separately with tu (thiourea) and thiosulfate ($S_2O_3^{2-}$) possessing a *trans*–effect much higher than that of Cl^-. The reaction is illustrated in Scheme 2.5.4.4. The observation outlined in Scheme 2.5.4.4 can be explained by considering the *trans*–directing power (*i.e. trans*–effect) of Cl^-, tu and $S_2O_3^{2-}$. Both tu and $S_2O_3^{2-}$ show a higher *trans*–effect than that of Cl^-. When the *cis*–compound is treated with tu, it produces *cis*–$[Pt(NH_3)_2(tu)_2]^{2+}$ first; then it reacts with excess tu to produce $[Pt(tu)_4]^{2+}$ because of the very strong *trans*–labilising effect of the tu groups towards the NH_3 groups present in *cis*–$[Pt(NH_3)_2(tu)_2]$.

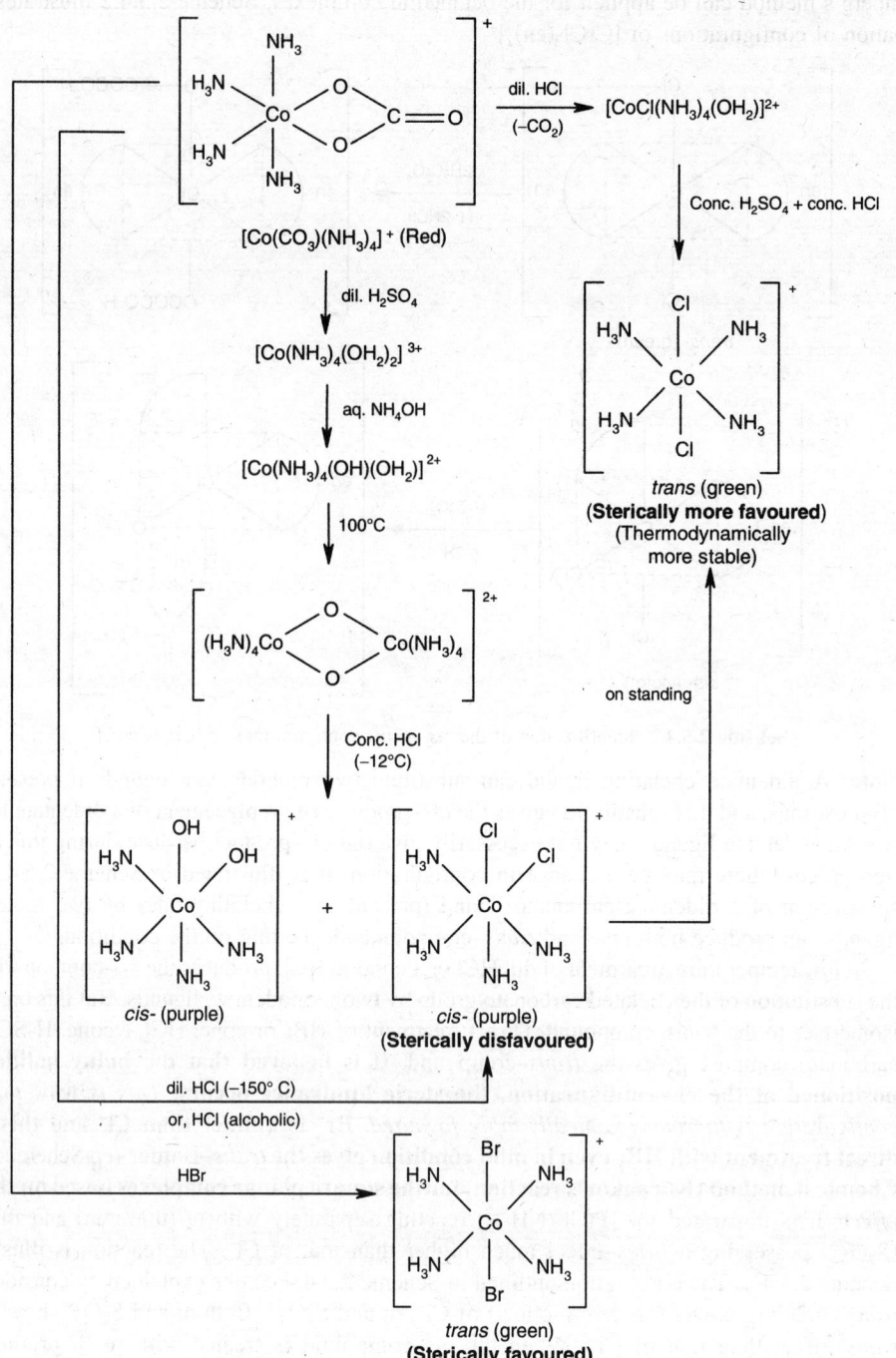

Scheme 2.5.4.3 Change of configuration during the substitution of a didentate ligand by two monodentate ligands.

When, $S_2O_3^{2-}$ is allowed to react with the starting *cis*–compound, it first produces $[Pt(NH_3)_2(S_2O_3)]$ in which the NH_3 groups are labilised by the *trans*–labilising effect of the chelated $S_2O_3^{2-}$. This is why, $[Pt(NH_3)_2(S_2O_3)]$ readily reacts with the excess $S_2O_3^{2-}$ to produce $[Pt(S_2O_3)_2]^{2-}$ through the substitution of NH_3 groups.

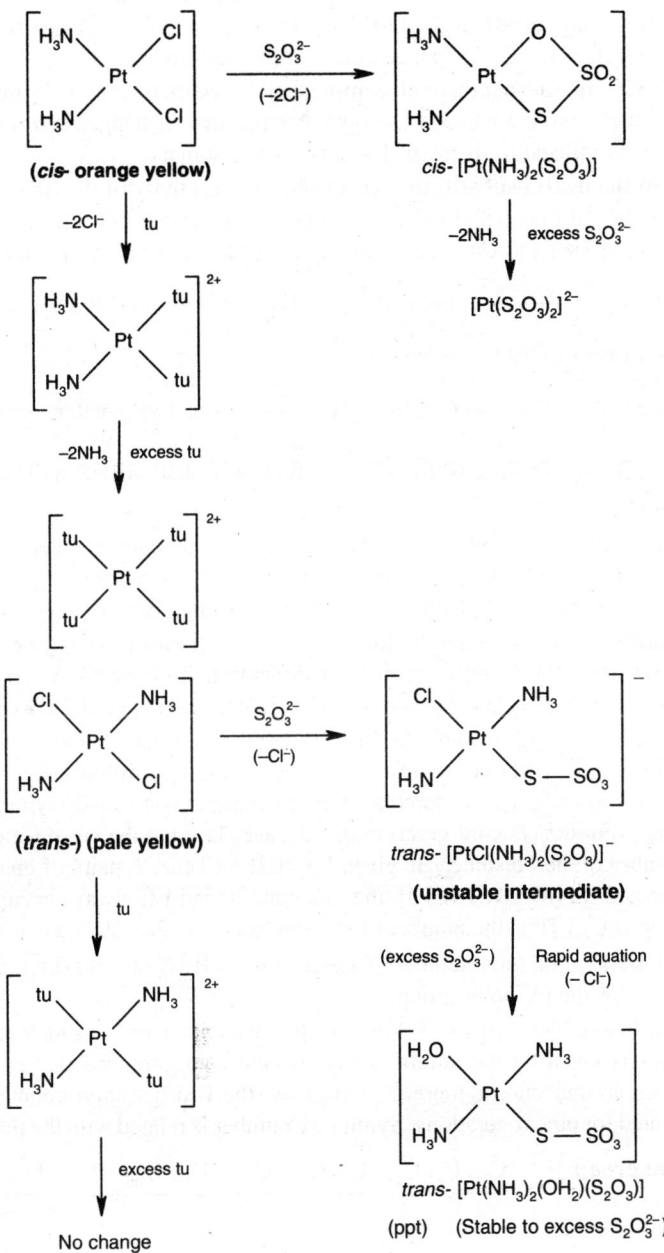

Scheme 2.5.4.4 Identification of the *cis-trans* isomers of $[PtCl_2(NH_3)_2]$ by chemical method (Kurnakov's reaction).

When $trans$–$[PtCl_2(NH_3)_2]$ is allowed to react with tu, it produces first $trans$–$[Pt(NH_3)_2(tu)_2]^{2+}$ which is stable even in presence of excess tu. Because, in $trans$–$[Pt(NH_3)_2(tu)_2]$ the NH_3 ligands are at the cis–positions of tu and thus tu cannot labilise the bound NH_3 groups. When $trans$–$[PtCl_2(NH_3)_2]$ is treated with $S_2O_3^{2-}$, it first produces $trans$–$[PtCl(NH_3)_2(S_2O_3)]^-$ which is unstable with respect to the aquation reaction (*i.e.* replacement of Cl^- by H_2O). **Because, the chlorido group is strongly labilised by the $trans$–$S_2O_3^{2-}$ group.** Thus the product, $trans$–$[Pt(NH_3)_2(OH_2)(S_2O_3)]$ separates out as the final product.

(g) **Optical activity in the octahedral complexes:** Very often, the cis–compounds are optically active while the $trans$–compounds are optically inactive. It happens so for $[CoCl_2(en)_2]^+$. This property may be utilised to distinguish the cis-$trans$-isomers.

(h) **Reactivity in the ligand substitution reaction:** The reactivity of the cis– and $trans$–compounds may differ widely. It may help to distinguish the cis-$trans$ isomers. The reason for this difference will be discussed later (*cf.* Chapter 5). Some illustrative examples are:

$$\left[PtCl_2(NH_3)_2\right] \xrightarrow{\text{aquation}} \left[PtCl(NH_3)_2(OH_2)\right]^+ \xrightarrow{\text{aquation}} \left[Pt(NH_3)_2(OH_2)_2\right]^{2+}$$

Reactivity order : $trans$- $\ggg$ cis-

$$\left[CoCl_2(en)_2\right]^+ \xrightarrow{\text{aquation}} \left[CoCl(en)_2(OH_2)\right]^{2+} ; \textbf{Reactivity order}: \; cis\text{-} > trans\text{-}$$

2.6 ENUMERATION OF TOTAL NUMBER OF STEREO-ISOMERS IN THE OCTAHEDRAL COMPLEXES

Different cases have been illustrated in Sec. 2.5.3. To calculate the total number of isomers, sometimes, becomes very difficult. In such cases, *computer programs* and *application of group theory* are needed. Here we shall discuss some simple methods to calculate the number of stereoisomers.

If there are n different ligands, then the total number of stereoisomers is given **by $n!/\alpha$ where α is the symmetry number of the complex under consideration**. To know the value of α, it is required to know the point group to which the complex belongs. It is illustrated for **[MABCDE$_2$]** and **[MABCD(EE)]**.

In **[MABCDE$_2$]**, if two E ligands are placed at the $trans$-positions, then its point group is D_{4h}. In the $trans$–configurations (*i.e.* positions of two Es fixed), the remaining 4 different ligands can be distributed in different possible ways to give the stereoisomers. Its number is $4!/8 = 3$ (symmetry number is 8 for the D_{4h} point group). For the cis-configuration (*i.e.* 2E are placed at the cis–positions), the point group is C_{2v} and the number of stereoisomers is given by $4!/2 = 12$ (*i.e.* 6 pairs of enantiomers) where the symmetry number is 2. In **[MABCD(EE)]**, the didentate ligand EE always occupies the cis-positions and the point group is C_{2v}. Thus the number of stereoisomers is $12(= 4!/2)$, *i.e.* 6 pairs of enantiomers.

● For **[M(ABCDEF)]**, the total number of stereoisomers is $6!/24 = 30$ (*i.e.* 15 enantiomers); symmetry number is 24 for the O_h point group.

Symmetry number (*cf.* Sec. 10.14.13, Vol. 2 of Fundamental Concepts of Inorganic Chemistry) of a particular species is given by the number of equivalent configurations that could be generated by rotating the species around one or more C_n axis/axes; the configuration counted for one operation should not be counted for other operations. Symmetry number is related with the point group as follows:

Point group:	C_∞, C_1, C_s	C_n, C_{nh}, C_{nv}	D_n, D_{nd}, D_{nh}	$D_{\infty h}$	T_d	O_h
Symmetry number:	1	n	2n	2	12	24

(*See* the present Author's book, Fundamental Concepts of Inorganic Chemistry Vol. 2, for details)

Table 2.6.1 Total number of stereoisomers for different types of octahedral complexes.

Octahedral	Total number	Pairs of enantiomers
[MA$_6$]	1	0
[MA$_5$B]	1	0
[MA$_4$B$_2$]	2	0
[MA$_3$B$_3$]	2	0
[MA$_4$BC]	2	0
[MA$_3$BCD]	5	1
[MA$_2$BCDE]	15	6
[MABCDEF]	30	15
[MA$_2$B$_2$C$_2$]	6	1
[MA$_2$B$_2$CD]	8	2
[MA$_3$B$_2$C]	3	0
[M(AA)(BC)DE]	10	5
[M(AB)$_2$CD]	11	5
[M(AB)$_3$]	4	2

2.7 CHARACTERISTICS OF THE OPTICAL OR MIRROR IMAGE ISOMERS

(a) **Optical activity in terms of interaction with the plane polarised light**

The optical isomers differ only in rotating the *plane of plane polarised light.* If in a pair of enantiomers, one enantiomer rotates the plane of the plane polarised light to the left then the other of the pair will rotate to the right.

The isomer which rotates the plane of the plane polarised light *towards left* (*i.e. anticlockwise*) is called *laevo-rotatory* (*i.e. l*-form) and the one which rotates the plane of the plane polarised light *towards right* (*i.e.* clockwise) is called *dextorotatory* (*i.e. d*-form). Some times, the *d*- and *l*-forms are designated by (+) and (−).

The optical isomers are the nonsuperimposable mirror images to each other. They are called enantiomers (*enantio* = opposite, *morphs* = forms). This phenomenon is called **enantiomorphism.** An equimolar mixture of the enantiomers is optically inactive as the rotation caused by one enantiomer in a particular direction is cancelled by the other enantiomer of the pair. This optical inactivity arises due to the *external compensation* and such a mixture is called the **racemic mixture.**

(b) **Why do the enantiomers interact in different ways with the plane polarised light?**

In light (*i.e. electro-magnetic radiation*), the electric and magnetic fields oscillate perpendicularly with respect to each other and perpendicular to the direction of propagation of light. Light when passed through a **Nicol prism**, in the transmitted light (described as the *plane polarised light*), the electric and magnetic field vectors are confined in perpendicular planes and the **plane of electric field vector is described as the plane of polarisation.** To understand the nature of plane polarised light, we are interested here with the oscillating electric vector in the plane of polarisation. *In fact, the electric vector* (E) *oscillates as a sine-wave with the frequency of light in the plane of polarisation.* Thus, simply, it can be stated that in ordinary light, it vibrates in all possible directions perpendicular to the direction of its propagation, **while in the plane polarised light, the vibration is restricted in one direction.** This restriction is imposed by the atoms of crystals present in Nicol Prism.

The plane polarised light may be considered to be constituted by *two interfering circularly polarised waves* of equal amplitude and rotation. The two components add vectorially to give the *wave of plane polarised light* (*cf.* Fig. 2.7.1).

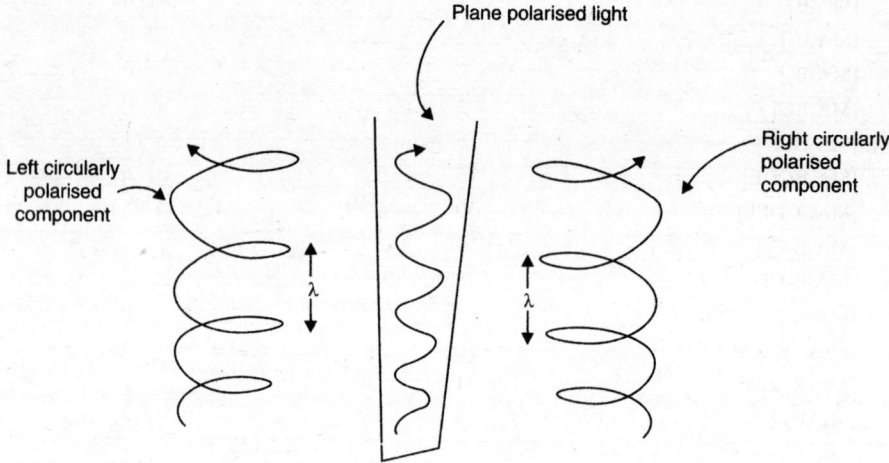

Fig. 2.7.1 Plane polarised light as the vectorial sum of the left and right circularly polarised light which are mutually mirror images to each other.

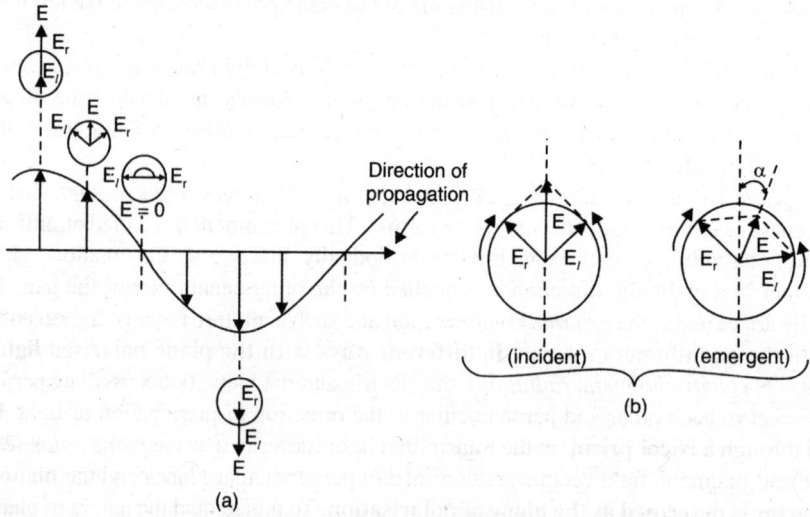

Fig. 2.7.2 (a) Plane polarised light (sine wave) as the resultant of left and right circularly polarised light. The arrows perpendicular to the direction of propagation denotes the **resultant electric vector (E)** generated by the left and right circularly polarised components E_l and E_r respectively at a particular instant. (b) The left circularly polarised light is retarded more in the interaction with the chiral compound (*i.e.* $n_{left} > n_{right}$) causing the rotation of the plane of the plane polarised light by an angle α which is positive or clockwise (*i.e.* dextrorotatory). If $n_{left} < n_{right}$, α becomes negative or anticlockwise (*i.e.* laevorotatory). n denotes the refractive index.

Thus the plane polarised light is considered to be composed of *right and left circularly polarised light*. The electric vector of right circularly polarised component (denoted by E_r) *spirals to the right* along the direction of propagation and the opposite is true for E_1 (the direction is defined by considering the observer looking at the source along the direction of propagation). In fact, the electric vector of a circularly polarised light rotates uniformly about the direction of propagation by 2π angle in each cycle. At any time, E_r and E_1 vectors can add to give the resultant *oscillating electric field* (sine wave) (E) in a plane along the direction of propagation.

It is important to note that the two circularly polarised components are mutually enantiomorphs, i.e. one is the non-superimposable mirror image of the other.

Optical rotation

- Plane of the plane polarised light is rotated due to the difference of velocity of the two circularly polarised components in the interaction with the chiral compound. In other words, the optical rotation is due to the difference in refractive indices, n_{left} and n_{right}.

- If the right circularly polarised component travels faster, then the rotation (α) is positive or clockwise and the optical medium is called **dextrorotatory**. On the other hand, if the left circularly polarised component travels faster, then α is negative or anticlockwise and the optical medium is called **laevorotatory.**

- The angle of rotation (α) in radians per unit optical path length (*i.e.* cm) is given by:

$$\alpha = \frac{\left(n_{left} - n_{right}\right)}{\lambda}$$

- If the rotation is expressed in degrees per decimeter (dm), then it is given by:

$$\alpha = \frac{180 \times 10}{\lambda}\left(n_{left} - n_{right}\right), \quad (\lambda \text{ in dm})$$

(optical path length = 1 dm = 10 cm)

Thus for Na–D light, $\lambda = 589$ nm, it is given by 5.89×10^{-6} dm for use in the above expression.

The two components, *i.e.* left and right circularly polarised light, of the resultant plane polarised light travel *with the same velocity in an optically inactive medium* but when passed through an optically active medium, the velocity of one of the circularly polarised components is retarded compared to the other. **The cause of this relative velocity retardation can be understood in terms of electron density distribution in the chiral molecule**. The interaction between the propagating light wave and electron density of the medium will retard the velocity of light (*i.e.* higher the electron density, lower the velocity, *i.e.* higher the refractive index). *One circularly polarised component experiences a greater electron density as* **the chirality of a molecule leads to such an nonhomogeneous electron density distribution.**

The refractive index (n) depends on the velocity, *i.e.*

$$n = \frac{\text{velocity in vaccum}}{\text{velocity in the given medium}}$$

Thus, for the plane polarsied light passing through a chiral or optically active medium, refractive indices of the two circularly polarised components differ, *i.e.* $n_r \neq n_1$.

In other words, in the interaction between the plane polarised light and optically active medium, one component of the light is slowed down compared to the other. Consequently, after traversing through an optically active compound, the two circularly polarised components of a plane polarised light will be slightly out of phase. This is why, the plane of the plane polarised light is rotated by a certain angle (say α) in the interaction. If the right circularly polarised component is slowed down more, then the plane of polarisation is rotated (in other words, the resultant vector E is rotated) to the anticlockwise when viewed towards the source along the direction of propagation of light.

(c) **Optical rotation (α)**

If a pencil beam of a plane polarised light of wavelength λ is passed through a solution of path length l of an enantiomer, then the plane of the polarised light will be rotated by an angle α, given by:

$$\alpha = [\alpha]_\lambda lc$$

where, $[\alpha]_\lambda$ = specific rotation at the wavelength λ,

           l = path length in decimeter,

           c = concentration in g ml^{-1}

Generally sodium light (D-line, 589 nm) is used.

(d) **Condition of optical activity**

The species to show the optical activity must be nonsuperimposable to its mirror image. Compounds lacking in the *axis of improper rotation* (S_n including $S_1 \equiv \sigma$ and $S_2 = i$) are optically active. These are described as the *chiral compounds*. The term *chirality* means *handedness*. The enantiomers are of *opposite chirality* and they are related to one another as *the right hand is related to the left hand.*

(e) **Resolution of the optical isomers:**

Separation of the dextro- and laevorotatory enantiomers from a *racemic mixture* is called *resolution*. Thus, resolution is the reverse process of *recemization*. If the activation energy in the racemisation process is very low and it can be attained at ordinary working temperature, then the resolution process is of no practical significance. Because, in such cases, the pure enantiomers will undergo racemisation at a fairly high rate.

(i) **Resolution by diastereomeric salt formation:**

The enantiomers being chemically and physically identical cannot be separated by the ordinary methods like fractional crystalisation, fractional distillation, etc. But when the enantiomeric complex ions are allowed to react with the optically active ions of opposite charge to produce the **diastereomeric salts,** these diastereomeric salts differ significantly in solubilities. *Such diastereomeric salts can be separated through fractional crystallisation.* Such diastereomeric salts are the stereoisomers but are not the mirror images to each other, *i.e.* **these salts are not enantiomers.**

If the complex enantionmer is a cation, this is treated with an optically active anion like tartarate, α-bromocamphor-*p*-sulfonate, etc. On the other hand, for the anionic complex

$$\left.\begin{array}{c} d\text{–}M^+ \\ + \\ l\text{–}M^+ \end{array}\right\} + 2d\text{–}A^- \longrightarrow \underbrace{d\text{–}Md\text{–}A + l\text{–}Md\text{–}A}_{\text{A pair of diastereomeric salt.}}$$

Scheme 2.7.1 Formation of diastereomeric salts for the resolution purpose.

enantiomer, it is to be treated with an optically active cation like protonated strychine or brucine. Resolved complex ions like $[Co(en)_3]^{3+}$, $[Co(edta)]^-$, $[Co(ala)_3]$, etc. may also be used as the resolving agents.

The isolated diastereoisomeric salts are treated with suitable non-optically active salts or acids to recover the pure enantiomers, *i.e. d-* and *l-*forms.

The basic steps of resolution process are given in Schemes 2.7.1-2.

Stereoisomers d–M^+ and l–M^+ denote the optically active dextrorotatory and laevorotatory enantiomers of a complex cation; and d–A^- is the optically active salt forming anion used as a resolving agent. The diastereomeric salts obtained can be very often separated through fractional crystallisation because of their different solubilities.

● **Resolution of dl-[Co(en)$_3$]Cl$_3$:** The procedure is illustrated for the resolution of dl–[Co(en)$_3$]Cl$_3$ in Scheme 2.7.2.

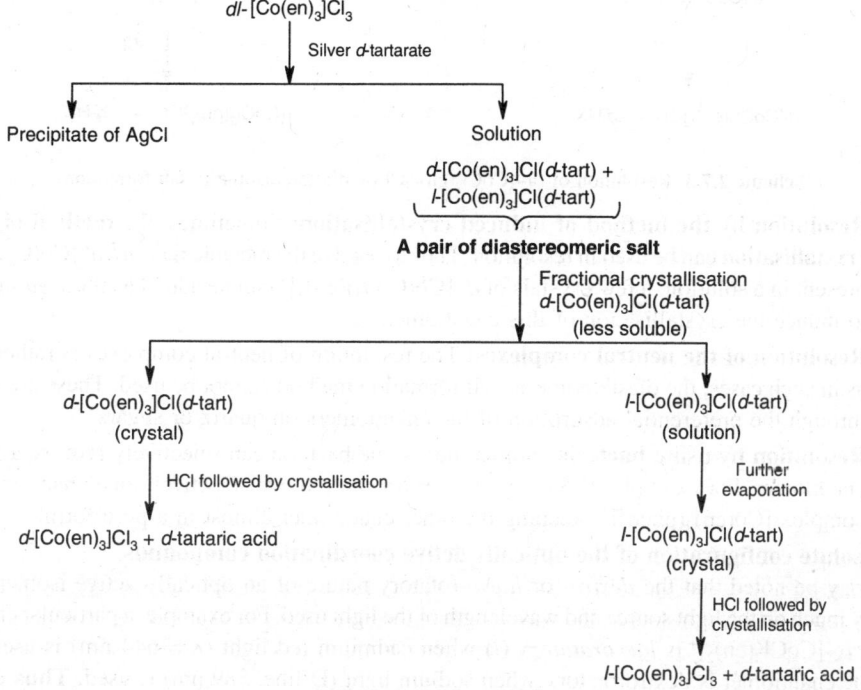

Scheme 2.7.2 Resolution of dl–[Co(en)$_3$]Cl$_3$ by diastereomeric salt formation method.

● **Resolution of dl-cis-[CoCl$_2$(en)$_2$]Cl:** Resolution of dl–cis-[CoCl$_2$(en)$_2$]Cl by ammonium d–α–bromocamphor $-p$–sulfonate, $NH_4(d-C_{10}H_{14}BrO_4S)$ which may be simply represented by $NH_4 d - X$ is schematically shown in Scheme 2.7.3.

● **Resolution of dl-cis[Co(en)$_2$(NO$_2$)$_2$]$^+$:** For the resolution of dl–cis-[Co(en)$_2$(NO$_2$)$_2$]$^+$, the solution is treated with potassium antimony d–tartarate, *i.e.* K^+ salt of d-[Sb$_2(d$–$C_4H_2O_6)_2]^{2-}$, a dinuclear complex (*cf.* Sec. 1.10). The diastereomeric salt, $\{l$–cis-[Co(en)$_2$(NO$_2$)$_2$]$\}_2\{d$–[Sb$_2(C_4H_2O_6)_2]\}$ produced is less soluble and it can be separated through fractional crystallisation and then from the solution $\{d$–cis-[Co(en$_2$)(NO$_2$)$_2$]$\}_2\{d$–[Sb$_2(C_4H_2O_6)_2]\}$ can be separated after further evaporation.

● **Resolution of dl-[Co(ox)$_3$]$^{3-}$:** For resolution of [Co(ox)$_3$]$^{3-}$, protonated brucine or strychine may be used as the resolving agent.

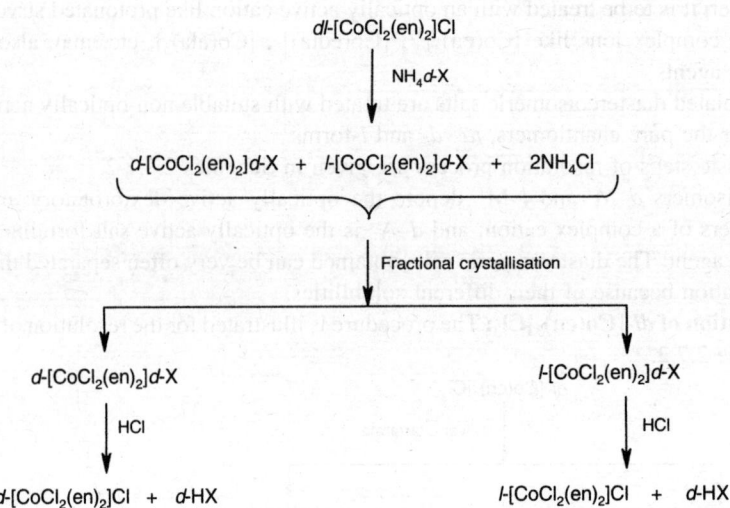

$dl\text{-}[CoCl_2(en)_2]Cl$

$\downarrow$ NH$_4$d-X

$d\text{-}[CoCl_2(en)_2]d\text{-}X$ + $l\text{-}[CoCl_2(en)_2]d\text{-}X$ + 2NH$_4$Cl

Fractional crystallisation

$d\text{-}[CoCl_2(en)_2]d\text{-}X$ $l\text{-}[CoCl_2(en)_2]d\text{-}X$

$\downarrow$ HCl $\downarrow$ HCl

$d\text{-}[CoCl_2(en)_2]Cl$ + $d\text{-}HX$ $l\text{-}[CoCl_2(en)_2]Cl$ + $d\text{-}HX$

Scheme 2.7.3 Resolution of $dl\text{-}[CoCl_2(en)_2]Cl$ by diastereoisomeric salt formation.

(ii) **Resolution by the method of induced crystallisation:** Sometimes, the **method of induced crystallisation** can be used in resolution. Thus to resolve the racemic mixture of $[Co(C_2O_4)(en)_2]^+$ present in a solution, a few crystals of $d\text{-}[Co(C_2O_4)(en)_2]^+$ can be added as the *seeding crystals* to induce the crystallisation of this enantiomer.

(iii) **Resolution of the neutral complexes:** The resolution of neutral complexes is rather difficult as in such cases, the diastereomeric salt formation method cannot be used. These are separated through the preferential adsorption of the enantiomers on quartz or sugars.

(iv) **Resolution by using bacteria:** Sometimes, some bacteria can selectively remove a particular enantiomer. For example, *P. Stutzeri* bacterium can remove selectively one enantiomer of the complex $[Co(en)_2(phen)]^{3+}$ keeping the other enantiomer almost in a pure form.

(f) **Absolute configuration of the optically active coordination compounds**

It may be noted that the *dextro-* or *laevo-*rotatory nature of an optically active isomer depends very much on the light source and wavelength of the light used. For example, a particular enantiomer of $cis\text{-}[CoCl_2(en)_2]^+$ is *laevoratatory* (l) when cadmium red light (λ = 644 nm) is used and the same enantiomer is dextrorotatory when sodium light (D-line, 589 nm) is used. **Thus the d and l designations are totally meaningless if the source of light and its wavelength are not mentioned.** This is why, we need the concept of absolute configuration of the optically active isomer.

To designate the absolute cofigurations of the optically active octahedral coordination compounds, different symbols like D and L; R and S (equivalent to C and A, *cf.* Sec. 1.15.5); P and M or Δ and Λ are used. R and S stand for *rectus* and *simistra* respectively; M and P stand for *minus* and *plus* respectively. Among these, Δ ($\equiv$ L) and Λ ($\equiv$ D) symbols (proposed by T.S. Piper) are very much popular. For the *tris*-(didentate) complexes, *e.g.* $[M(en)_3]^{n+}$, Δ is used for the chelate rings describing a right-handed helix and Λ is used when the chelate rings describe a left handed helix. It is illustrated in Fig. 2.7.3(i) for the optically active octahedral *tris*-chelates and *bis*-chelates. The tetrahedral complexes of the type $[M(A-B)_2]$ where $A-B$ stands for an unsymmetrical didentate ligand, the Δ and Λ configurations for the enantiomers are shown in Fig. 2.7.3(ii). R-S symbols for the tetrahedral complexes have been discussed in Sec. 1.15.5.

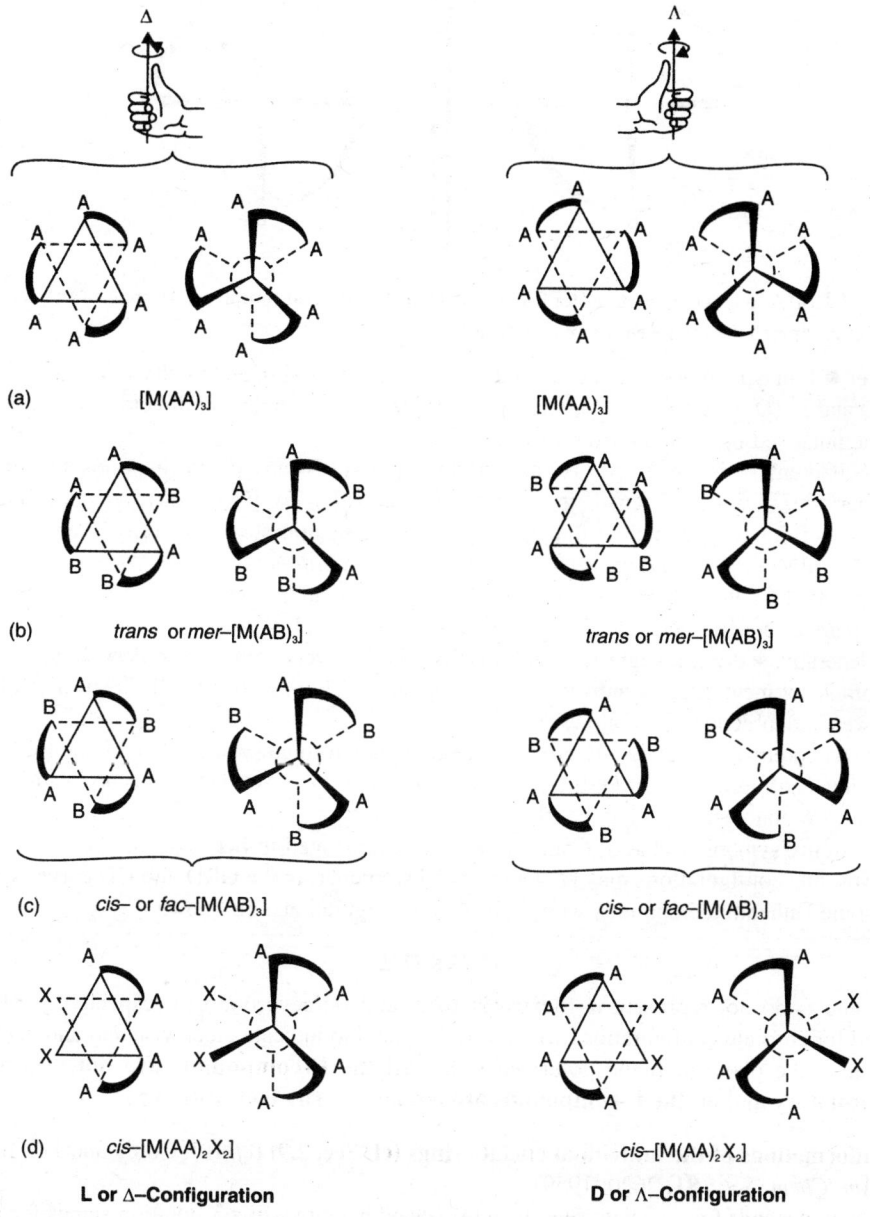

Fig. 2.7.3 (i) Assignment of absolute configurations for (a) $[M(AA)_3]$ type (*e.g.* $[Co(en)_3]^{3+}$, $[Fe(ox)_3]^{3-}$, etc.); (b) *trans* or *mer*-$[M(AB)_3]$ type (*e.g. mer*-$[Co(gly)_3]$); (c) *cis* or *fac*-$[M(AB)_3]$ type and (d) *cis*-$[M(AA)_2X_2]$ type (*e.g. cis*-$[CoCl_2(en)_2]^+$) **octahedral complexes**. (*See* the **Note** in Fig. 1.4.5.3 for determination of Δ and Λ configuration).

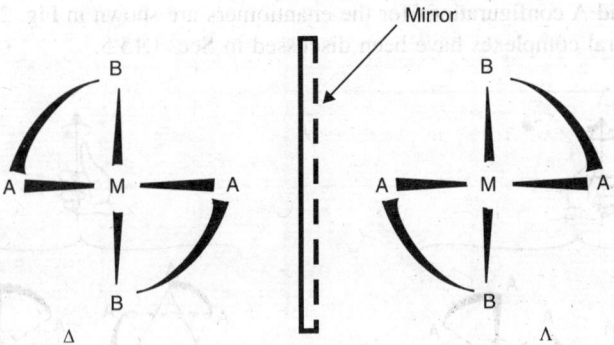

Fig. 2.7.3 (ii) Δ and Λ configurations for [M(A – B)₂] type **tetrahedral complexes**. (Here $\Delta \rightleftharpoons \Lambda$ change can occur through the **intermediate planar form**).

Note: • It may be noted that the absolute configurations designated by the symbols like Δ and Λ or D and L, (D ≡ Λ and L ≡ Δ), etc. do not depend on the angle and direction of rotation of the plane polarised light. Some examples are:

$(+)_{589}$ [Co(en)₃]³⁺ is the Λ or D isomer and thus $(-)_{589}$ [Co(en)₃]³⁺ corresponds to the Δ or L isomer; $(+)_{589}$ *cis*–[CoCl₂(en)₂]⁺ is the Λ isomer and $(-)_{589}$ *cis*–[CoCl₂(en)₂]⁺ is the Δ isomer.

• Here it is interesting to know that the dextrorotatory enantiomers of [Co(en)₃]³⁺ and [Rh(en)₃]³⁺, *i.e.* $(+)_{589}$ [M(en)₃]³⁺ *have the opposite absolute configurations*.

• For the [M(AB)₃] type complex, *e.g.* [Co(gly)₃] there are *fac–* and *mer–*isomers giving rise to two pairs of enantiomers, *i.e. fac–*Δ and *fac–*Λ; *mer–*Δ and *mer–*Λ.

• Generally, sodium D–light (λ = 589 nm) is used and very often in the designation, the wavelength is not mentioned. It indicates λ = 589 (sodium D-line) automatically. Thus optical isomers are very often denoted as follows:

Λ–(+)–[Co(en)₃]³⁺ or D–(+)–[Co(en)₃]³⁺ meaning dextrorotatory at the Na–D line (589 nm) and absolute configuration Λ or D. If may be noted that the 'D' of Na–D line has no relationship with the dextrorotatory or absolute configuration D or Δ.

• Chirality symbols *R* and *S*; *C* and *A* have been illustrated in Sec. 1.15.5.

• Absolute configurations may be determined by comparing the ORD and CD curves with those of some authentic compounds whose absolute configurations are known.

---

d, l vs *D, L*

D– and *L–*do not represent *d*–(dextrorotatory) and *l*–(laevorotatory) respectively. *D–* and *L–*stand for absolute configurations while *d–*and *l–*stand to indicate clockwise and anticlockwise rotation of the plane of plane polarised light. **All the D-compounds are not necessarily dextrorotatory and all the L-compounds are not necessarily laevorotatory.**

(g) **Conformations of the individual chelate rings (cf. Sec. 2.9)** (*cf.* E.J. Corey and J.C. Bailer, Jr., *J. Am. Chem. Soc.,* **81,** 2620, 1959)

δ notation stands for a chelate ring of right-handed helicity while λ notation stands for a chelate ring of left-handed helicity. In terms of δ and λ notation, a particular isomer can be designated. Here we shall illustrate by taking 'en' as the chelate ring forming didentate ligand.

• **M(en) chelate ring conformation:** The **gauche** conformation of M(en) chelate ring may be reasonably compared to the **chair form of cyclohexane.** An etheylenediamine chelate ring (*gauche* *conformation*) is viewed along the C_2-axis of a M(en) chelate ring system, then the C—C bond and the imaginary line joining the two nitrogens (*i.e.* the octahedral edge along which the didentate ligand en spans) are **skewed** and they describe a helix. The *left handed helicity* of the chelate ring conformation is denoted by λ while the *right-handed helicity* stands for the δ–conformation. *In other words, in the chelate ring system, the C—C bond is skewed down to the left of the viewer (viewing through the C_2-axis) for λ–conformation and skewed down to the right for δ–conformation.*

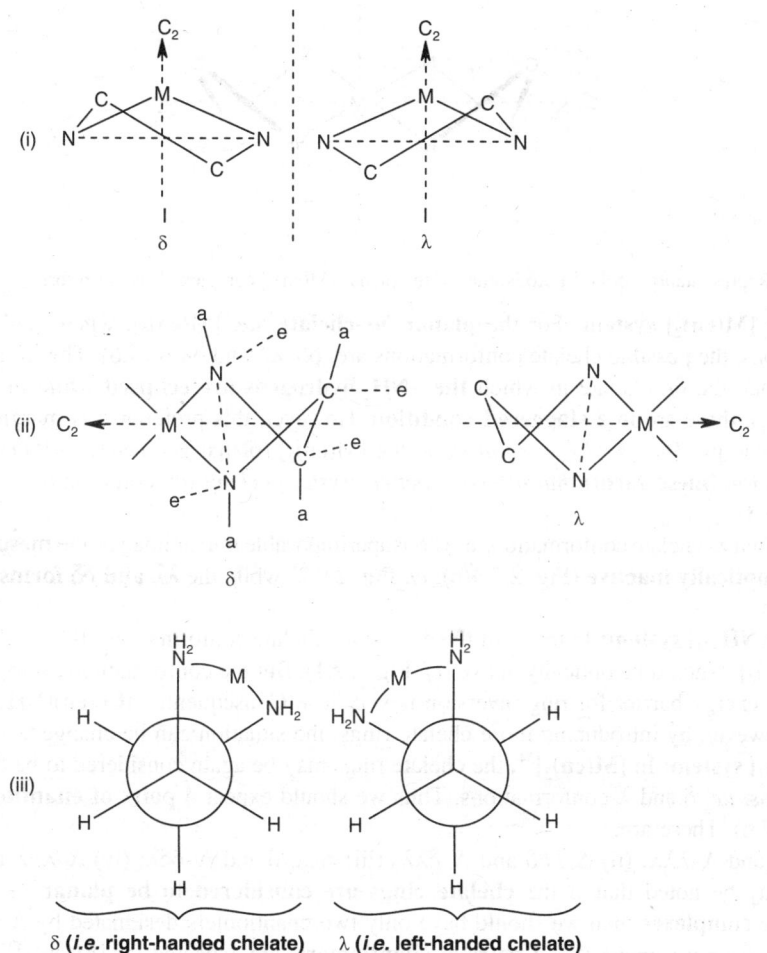

Fig. 2.7.4 (a) Different ways for representation and illustration of the metal-ethylenediamine, *i.e.* M(en) chelate ring conformation. Here, the gauche configuration of ethylenediamine (en) has been considered. N⋯⋯N for the imaginary line lying along the edge of an octahedron. **Fig. 2.7.4(a) illustrates the δ- and λ-chelate ring conformations of the M(en) system.**

● **Interconversion between the δ- and λ- forms of M(en) ring:** The gauche form M(en) is *dissymetric* and *chiral*. The δ- and λ- enantiomeric forms can interconvert through a **symmetric planar conformation** (Fig. 2.7.4b) because the energy barrier is not sufficiently high. *It is obvious that for a planar chelate ring as in the β-diketonate complex, no such enantiomeric ring conformations occurs.*

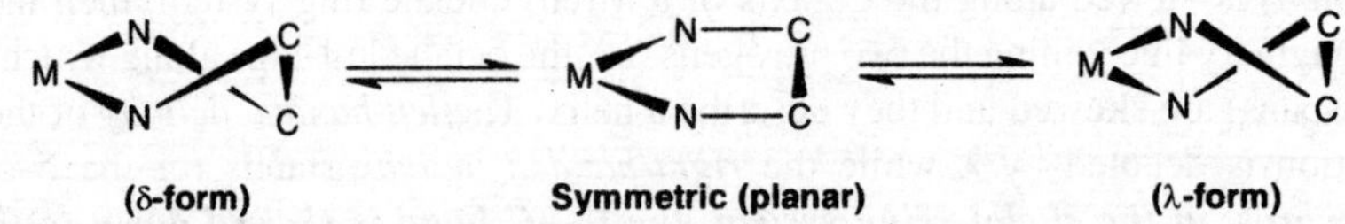

(δ-form)   Symmetric (planar)   (λ-form)

Fig. 2.7.4 (b) Interconversion between the δ-and λ-forms of the M(en) ring system.

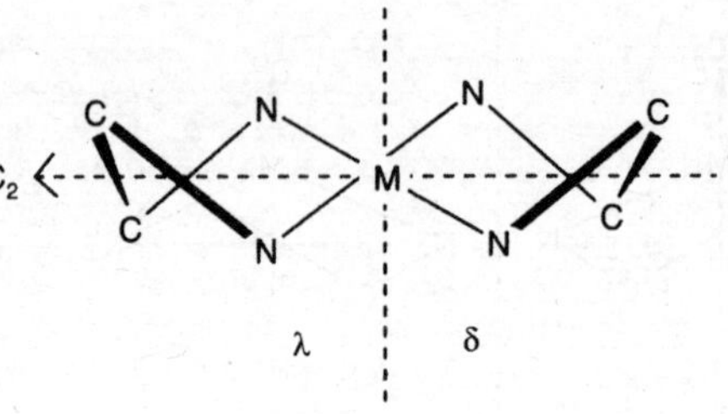

λ   δ

Fig. 2.7.4 (c) Representation of the M–λδ isomer of the planar $[M(en)_2]$ complex. This *meso* form is optically inactive.

● **Planar $[M(en)_2]$ system:** For the **planar *bis*-chelate,** *i.e.* $[M(en)_2]$, where 'en' ligands are in *trans*–positions, the possible chelate conformations are: δδ, λλ and δλ (or λδ). The λλ and δδ chelates are stabler than the δλ chelate in which **the $-NH_2$ hydrogens are eclipsed** while in the λλ and δδ conformations, these are in a **staggered condition**. However, this preference is marginal in terms of stabilisation energy. *This is why, in solid state,* the $[M(en)_2]$ (planar geometry) *unit crystallises in the* λδ *conformation* (**meso form**) *that allows a better crystal packing* (*cf.* contribution of the **statistical entropy effect**).

As the δ– and λ–chelate conformations are nonsuperimposable mirror images, the **meso-diastereomer** λδ– form is **optically inactive** (Fig. 2.7.4(c); *cf.* Fig. 2.9.2) while the **λλ and δδ forms give a pair of enantiomers.**

● **$[M(en)(NH_3)_4]$ system:** In terms of the δ– and λ– chelate conformations, the octahedral complex $[Co(en)(NH_3)_4]^{3+}$ should be optically active (*cf.* Fig. 2.9.1). But the conformational stability is marginal only and the energy barrier for ring inversion is very low. Consequently, $[Co(en)(NH_3)_4]^{3+}$ cannot be resolved. However, by introducing more chelate rings, the situation can be changed.

● **$[M(en)_3]$ system:** In $[M(en)_3]^{3+}$, the chelate rings may be again considered to be of two different conformations, *i.e.* δ and λ conformations. Thus we should expect **4 pairs of enantiomers** (*cf.* Figs. 2.7.5 and 2.7.6). These are:

(i) Δ–δδδ and Λ–λλλ; (ii) Δ–λδδ and Λ–δλλ; (iii) Δ–λλδ and Λ–δδλ; (iv) Δ–λλλ and Λ–δδδ.
(**Note:** It may be noted that if the chelate rings are considered to be planar as in the case of β-diketonate complexes then we should have only two enantiomers designated by Δ and Λ.)

Though 8 distinct isomers (*i.e.* **4 pairs of enantiomers**) are expected for $[M(en)_3]^{3+}$, but in reality, a much smaller number of isomers are obtained. *In fact, the predicted 4 pairs differ in energy because of the difference in C—H and N—H interactions.* Only the stable forms (*i.e.* of low energy) are isolated. *This phenomenon has been described as stereoselectivity.*

Corey and Bailar have discussed the structures of Δ–λλλ and Δ–δδδ forms. In the Δ–λλλ form, the C—C bonds are **parallel** with the pseudo–C_3 axis (passing through the centers of the opposite octahedral faces) and this form is described as the *lel*–**form**. In Δ–δδδ form, the C—C bonds are **oblique** to the pseudo–C_3 axis and this form is described as the *ob*–**form**. It has been estimated that the *lel* form, *i.e.* Δ–λλλ is about 7.5 kJ mol^{-1} (*i.e.* 2.5 kJ per ligand) more stable than the *ob*–form, *i.e.* Δ–δδδ. This higher stability is mainly due to the nonbonded N—H······N—H interactions (*i.e.* H-bonding interaction). It may be noted that for the Λ isomer, $\delta\delta\delta$– form is more stable that the $\lambda\lambda\lambda$–form.

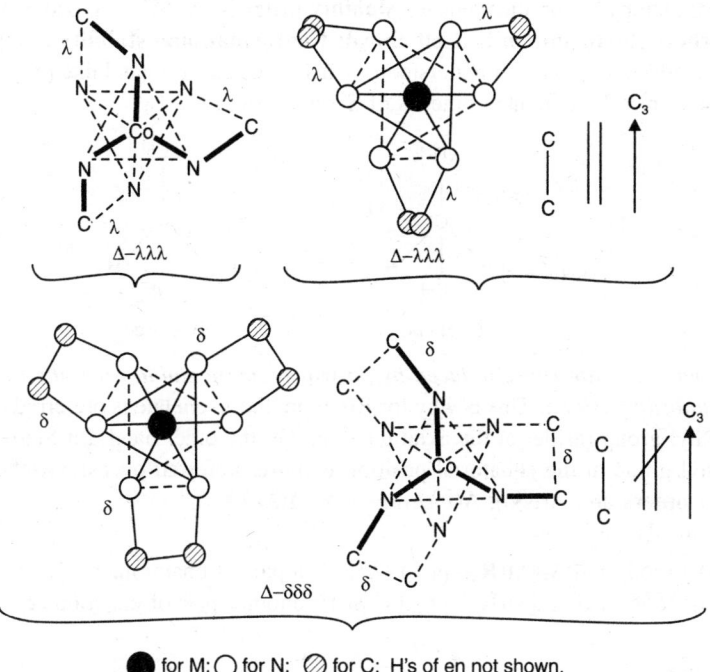

● for M; ○ for N; ⊘ for C; H's of en not shown.

Fig. 2.7.5 The *lel* ($\equiv \lambda\lambda\lambda$) and *ob* ($\equiv \delta\delta\delta$) conformations of Δ–[M(en)$_3$]$^{n+}$ in terms of orientation of the C—C bonds with respect to the C_3–axis.

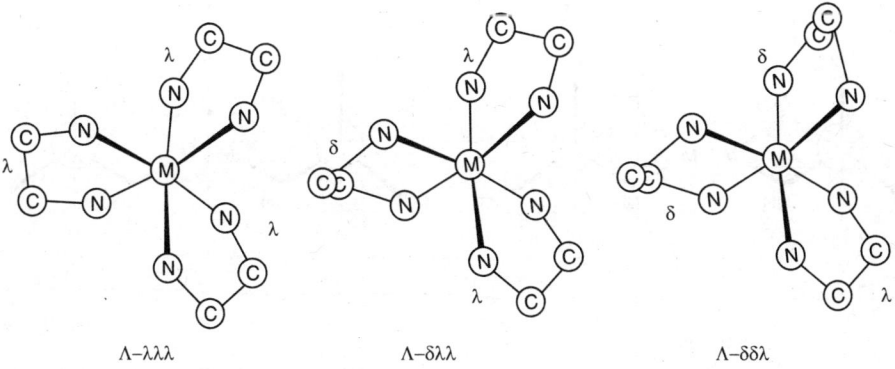

Fig. 2.7.6 Conformations of Λ–[M(en)$_3$]$^{n+}$ (in terms of conformations of the chelate rings). These conformations are observed because of the **statistical entropy effect.** Theoretically expected the **most stable isomer** Λ–$\delta\delta\delta$ is not observed in reality.

The three conformations for Λ–[M(en)$_3$]$^{3+}$ are shown in Fig. 2.7.6. These have been found in [Cr(en)$_3$][Ni(CN)$_5$] · 1.5 H$_2$O and [Cr(en)$_3$][Co(CN)$_6$] · 6H$_2$O.

Here it may be mentioned that in solid state, for the Λ–configuration, the δ–conformation is more stable than the λ–conformation. Because of the **very low energy barrier**, there is a rapid interconversion between the conformations. Due to the **statistical entropy effect**, in solution Λ–$\delta\delta\lambda$ isomer is more abundant than the Λ–$\delta\delta\delta$ isomer. By considering both the enthalpy (difference 2-3 kJ mol^{-1} between the diastereomers) and entropy factor, the predicted **stability order** is: Λ–$\delta\delta\lambda$ > Λ–$\delta\delta\delta$ ≈ Λ–$\delta\lambda\lambda$ >> Λ–$\lambda\lambda\lambda$.

● **[M(pn)$_3$]** where the ligand *pn* is itself chiral: **Conformational stability** of a particular chelate ring leading to stereoslectivity is well understood by using the chiral ligand like propylenediamine, *i.e.* 1, 2-diaminpropane (pn). Its (+) enantiomer is of S-figuration, *i.e.*

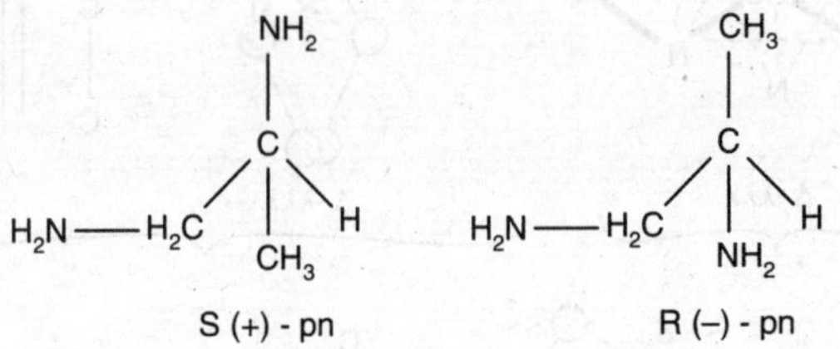

S (+) - pn   R (−) - pn

The five-membered chelate ring produced by pn will be more stable when the bulky substituent is placed at the equatorial position. This is why for R(−)–pn, the λ–chelate is preferred over the δ–chelate in which the methyl group resides at the axial position. On the other hand, for S(+)–pn, the δ-chelate bearing the methyl group at the equatorial position is more preferred over the λ-chelate (Fig. 2.7.7). Thus the **stable isomers** are: $\Lambda(\delta\delta\delta)$, $\Lambda(\lambda\lambda\lambda)$, $\Delta(\delta\delta\delta)$, $\Delta(\lambda\lambda\lambda)$.

For pn, these are shown below:

Λ–[M(S − pn)$_3$] ($\delta\delta\delta$) and Δ–[M(R − pn)$_3$] ($\lambda\lambda\lambda$)}, a pair of enantiomer

Δ–[M(S − pn)$_3$] ($\delta\delta\delta$) and Λ–[M(R − pn)$_3$] ($\lambda\lambda\lambda$)}, another pair of enantiomer

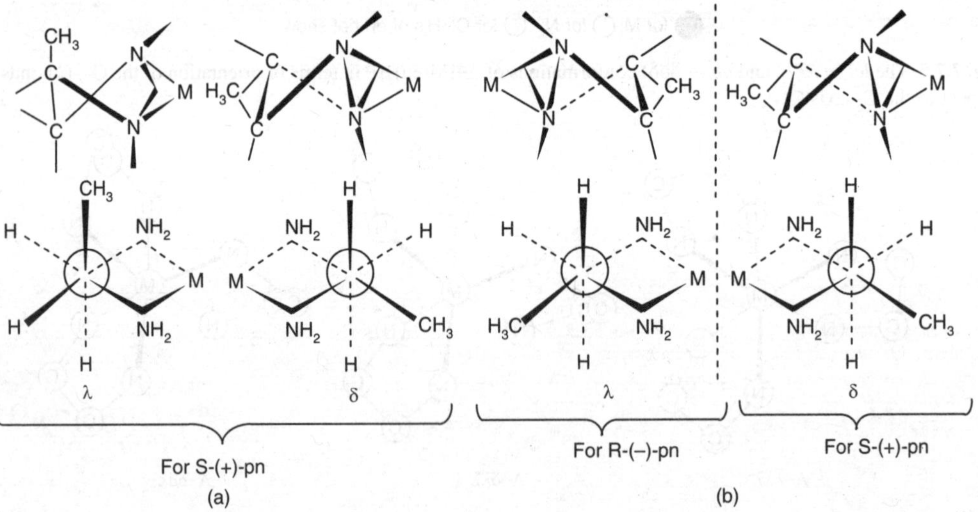

For S-(+)-pn

(a)

For R-(−)-pn   For S-(+)-pn

(b)

Fig. 2.7.7 (a) λ and δ conformations of metal(M) − S-(+) − 1, 2–diaminopropane (pn); Me-group in axial position for λ–conformation and Me-group in equatorial position for δ–conformation. (b) Mirror image conformation of chelate ring of S–pn (d) and R–pn (l) keeping the Me–group in the equatorial position for greater stability.

Note: The imaginary line joining the N-atoms, *i.e.* N······N lies along the edge of the octahedron.

In all these isomers, the Me-groups are at the equatorial positions. It is important to note that R–pn can give two isomers of empirical formula [M(R–pn)$_3$] but **these are not enantiomers. They are two**

diastereoisomers. The same thing happens for $[M(S-pn)_3]$. For $[Co(pn)_3]^{3+}$, **R**-pn and **S**-pn, give two pairs of enantiomers and the specific rotations for the 2-pairs of enantiomers are:

$$\Lambda - \left[Co(S\text{-}pn)_3 \right]^{3+} (\delta\delta\delta) : +24° \quad \Lambda - \left[Co(R\text{-}pn)_3 \right]^{3+} (\lambda\lambda\lambda) : +214°$$
$$\Delta - \left[Co(R\text{-}pn)_3 \right]^{3+} (\lambda\lambda\lambda) : -24° \quad \Delta - \left[Co(S\text{-}pn)_3 \right]^{3+} (\delta\delta\delta) : -214°$$

(h) Optical rotatory dispersion (ORD) and circular dichorism (CD) (cf. R.D. Gillard, *Prog. Inorg. Chem.*, **7**, 215-76, 1966)

● **ORD Curves:** It has been already mentioned that when the plane polarised light interacts with the chiral compounds, velocity of one circularly polarised component is retarded, *i.e.* the two components to constitute the plane polarised light travel with different velocities in this interaction. It gives a difference in the refractive indices of the left (n_{left}) and right (n_{right}) circularly polarised components and consequently, the plane of polarisation of the plane polarised light is rotated by an angle (α–say, which may be positive or negative).

The angle of rotation (α) depends on the wavelength of light. It changes dramatically in the region of absorption band. This effect, *i.e.* change of α with wavelength is called *optical rotatory dispersion* (ORD). Thus the plot of α (angle of rotation) versus wavelength (λ) is called the ORD curve. The *angle of rotation* (α) depends on the difference of refractive indices, *i.e.* $n_{left} - n_{right}$. Thus is why, sometimes, ($n_{left} - n_{right}$) is plotted against λ to obtain the ORD curve. Thus the ORD curve is:

$$\alpha \text{ or } (n_{left} - n_{right}) \text{ versus } \lambda \text{ or } \nu \text{ (frequency)}$$

The abrupt change of α in the vicinity of absorption band is called the Cotton effect (described after the name of the French physicist, A. Cotton who first observed the phenomenon in 1895). In Cotton effect, the sign of rotation changes abruptly at λ_{max} passing through zero.

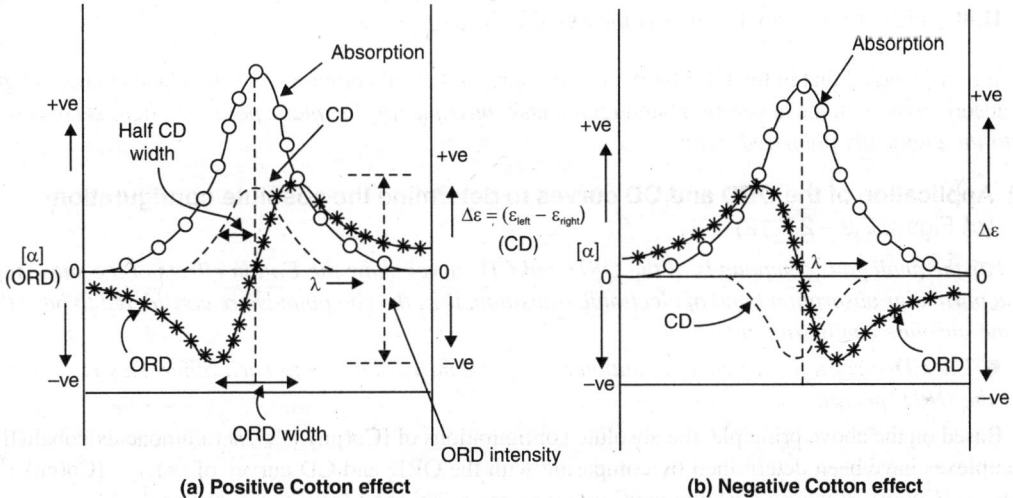

(a) Positive Cotton effect   (b) Negative Cotton effect

Fig. 2.7.8 Illustration of optical rotatory dispersion (ORD) and circular dichroism (CD) for a particular absorption band (*i.e.* electronic transition) of an optical isomer. (a) Representation of a **positive Cotton effect**; (b) Representation of a **negative Cotton effect**.

If the absorption band is approached from the shorter wavelength side, the rotation (α) may decrease to a minimum value and then attain a maximum value passing through zero at λ_{max}. After attaining the maximum value, α again falls with the increase of λ. This is called the **positive Cotton effect.** For the opposite observation, the phenomenon is described as the **negative Cotton effect** (*i.e.* with the increase of wavelength, in the vicinity of absorption band, α increases to a maximum value then decreases to a minimum value passing through zero at λ_{max}). These are illustrated in Fig. 2.7.8. Distance between the maximum and minimum in an ORD curve is called ORD width.

Thus it is evident that the mentioning sign of rotation (that determines the dextrorotatory and laevorotatory character) without the mentioning of wavelength of light is of no meaning.

Enantiomers (of opposite absolute configurations) show the opposite Cotton effects, *i.e.* if one optical isomer shows a positive Cotton effect, then the other isomer shows a negative Cotton effect for the same electronic transition. *The ORD curves of the enantiomers are the mirror images.*

ORD intensity is measured by peak-to-trough height in an ORD curve.

● **CD Curves:** The molar extinction coefficient (ε) of a particular compound depends on its ability to absorb the light via the electronic transition. We have already mentioned that the left and right circularly polarised components of the plane polarised light interact in different ways with a chiral compound. This is why, the molar extinction coefficients of the two components (*i.e.* ε_{left} and ε_{right}) are different in the interaction with a chiral compound. The difference, $\Delta\varepsilon = \varepsilon_{left} - \varepsilon_{right}$ is called the *circular dichorism* which may be positive or negative. This difference arises as in the interaction with a chiral compound, one of the circularly polarised components is absorbed more.

$\Delta\varepsilon$ (= CD) is almost zero at the wavelengths away from the absorption band and the magnitude of $\Delta\varepsilon$ becomes appreciable only at the region of absorption band and it becomes maximum at λ_{max}. Like the ORD curve, the CD curve can show both the *positive Cotton effect* (*i.e.* $\Delta\varepsilon$ is positive) and the *negative Cotton effect* (*i.e.* $\Delta\varepsilon$ is negative). As in the ORD curves, the enantiomers of opposite chirality show the opposite Cotton effects in the CD curves. These are illustrated in Fig. 2.7.8.

Half width or width a CD curve is measured at $\frac{1}{2}|\Delta\varepsilon|_{max}$ value.

It may be noted that in the CD effect, one circularly polarised component of the plane polarised light is absorbed more in the absorption band region and *consequently, the plane polarised light is converted into the eliptically polarised light.*

(i) Application of the ORD and CD curves to determine the absolute configuration
(*cf.* Figs. 2.7.9 – 2.7.12)

● *For the analogous compounds, if the ORD and CD curves show the Cotton effects of the same sign at a particular absorption band of electronic transition, then the compounds are considered to have the same absolute configuration.*

● *The CD-spectra for a pair of enantiomers are the mirror images to each other. This is also true for the ORD spectra.*

Based on the above principle, the absolute configurations of $[Co(pn)_3]^{3+}$, *tris* (aminoacids)cobalt(III) complexes have been determined by comparing with the ORD and CD curves of $(+)_{589} - [Co(en)_3]^{3+}$, $(-)_{589} - [Co(en)_3]^{3+}$ whose absolute configurations are established from the X-ray analysis.

The ORD and CD curves (in the range of 400 – 500 nm, *i.e.* $A_{1g} \rightarrow T_{1g}$ *transition range for the corresponding perfectly octahedral symmetry) of* $\Lambda-[Co(en)_3]^{3+}$, $\Lambda-[Co(S-pn)_3]^{3+}$, $\Lambda-[Co(S-ala)_3]$ are similar (*cf.* Figs. 2.7.9-10). It may be noted that α-alanine can have the *R* and *S*-configurations.

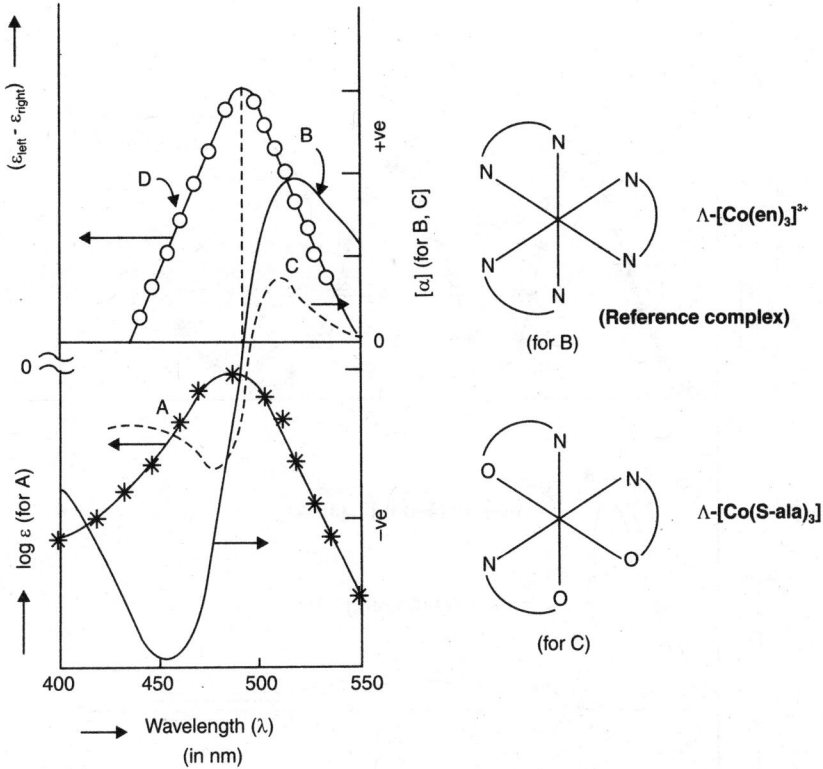

Fig. 2.7.9 (A) Absorption band in the range 400 – 550 nm for the transition $^1A_{1g} \rightarrow {}^1T_{1g}$ (in terms of O_h symmetry) of $[Co(en)_3]^{3+}$. (B) ORD curve for Λ–$[Co(en)_3]^{3+}$; (C) ORD curve for Λ–$[Co(S$–$ala)_3]$ (*mer*–isomer), (D) CD–curve for Λ–$[Co(en)_3]^{3+}$. **(Taking the ORD curve of $[Co(en)_3]^{3+}$ as the reference one, the absolute configuration of $[Co(S$–$ala)_3]$ has been assigned.)**

Note: $[Co(en)_3]^{3+}$ possesses the D_3 symmetry instead of O_h. Thus the electronic transitions are:

$$\left.\begin{array}{l} ^1A_{1g} \rightarrow {}^1T_{1g} \\ ^1A_{1g} \rightarrow {}^1T_{2g} \end{array}\right\} O_h \qquad \left.\begin{array}{l} ^1A_1 \rightarrow {}^1A_2 \text{ and } {}^1E \\ ^1A_1 \rightarrow {}^1A_1 \text{ and } {}^1E \end{array}\right\} D_3$$

(*i.e.* T_{1g} is split into A_2 and E; and T_{2g} is split into A_1 and E on moving from O_h to D_3 symmetry) The splitting of T_{1g} and T_{2g} terms in D_3 symmetry is not pronounced in terms of the electronic absorption spectrum because the corresponding peaks are closed spaced and they merge within the spectral bandwidth. The CD curve shows two Cotton effects (*cf.* Fig. 2.7.10) for the corresponding transition $^1A_{1g} \rightarrow {}^1T_{1g}$ (*i.e.* lower energy band) indicating the splitting of T_{1g} term as expected for the D_3–symmetry of $[Co(en)_3]^{3+}$. Thus in $[Co(en)_3]^{3+}$, for the ligand field transition ($^1A_{1g} \rightarrow {}^1T_{1g}$ in terms of O_h symmetry) at 21,400 cm^{-1}, there are **two CD peaks of opposite Cotton effect at 20,300 and 23,400 cm^{-1}.** The electronic absorption band is broader than the CD peaks. **This is why, in the CD spectrum, the peaks are noticeable but in the ligand field transition band, they merge.** The ORD curve also records a hint for a second Cotton effect but it is not pronounced. For the transition, $^1A_{1g} \rightarrow {}^1T_{2g}$ (*i.e.* higher energy band), there is only one weak Cotton effect. It is due to the fact that the transition $^1A_1 \rightarrow {}^1A_1$ is forbidden and the observed Cotton effect is only due to the transition to $^1A_1 \rightarrow {}^1E$.

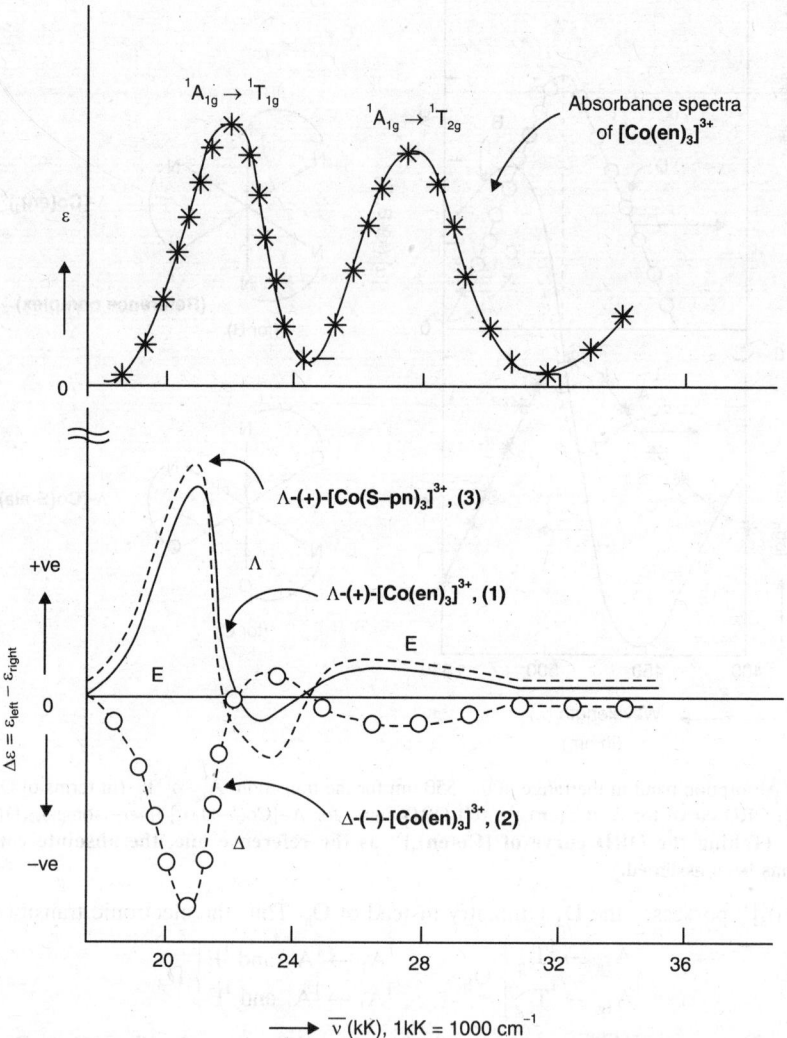

Fig. 2.7.10 Absorption spectrum of $[Co(en)_3]^{3+}$ and comparison of CD spectra of optical isomers $\Lambda-(+)-[Co(en)_3]^{3+}$, $\Delta-(-)-[Co(en)_3]^{3+}$ and $\Lambda-(+)-[Co(S-pn)_3]^{3+}$.

Note: Mirror image relation between the CD spectra of (1) and (2); 1 kK = 1000 cm^{-1}

Note: MCD (magnetic circular dichroism): When a substance is placed in a *magnetic field,* it can rotate the plane of the plane polarised light, *i.e.* ORD and CD effects irrespective of chirality of the substance. These phenomena are called the **Faraday effect.** Only chiral substances can show the Cotton effect but any substance (*i.e.* both chiral and achiral) can show the Faraday effect. In fact, when these are studied in presence of a magnetic field, one component of the magnetic field works in the direction of propagation of the plane polarised light.

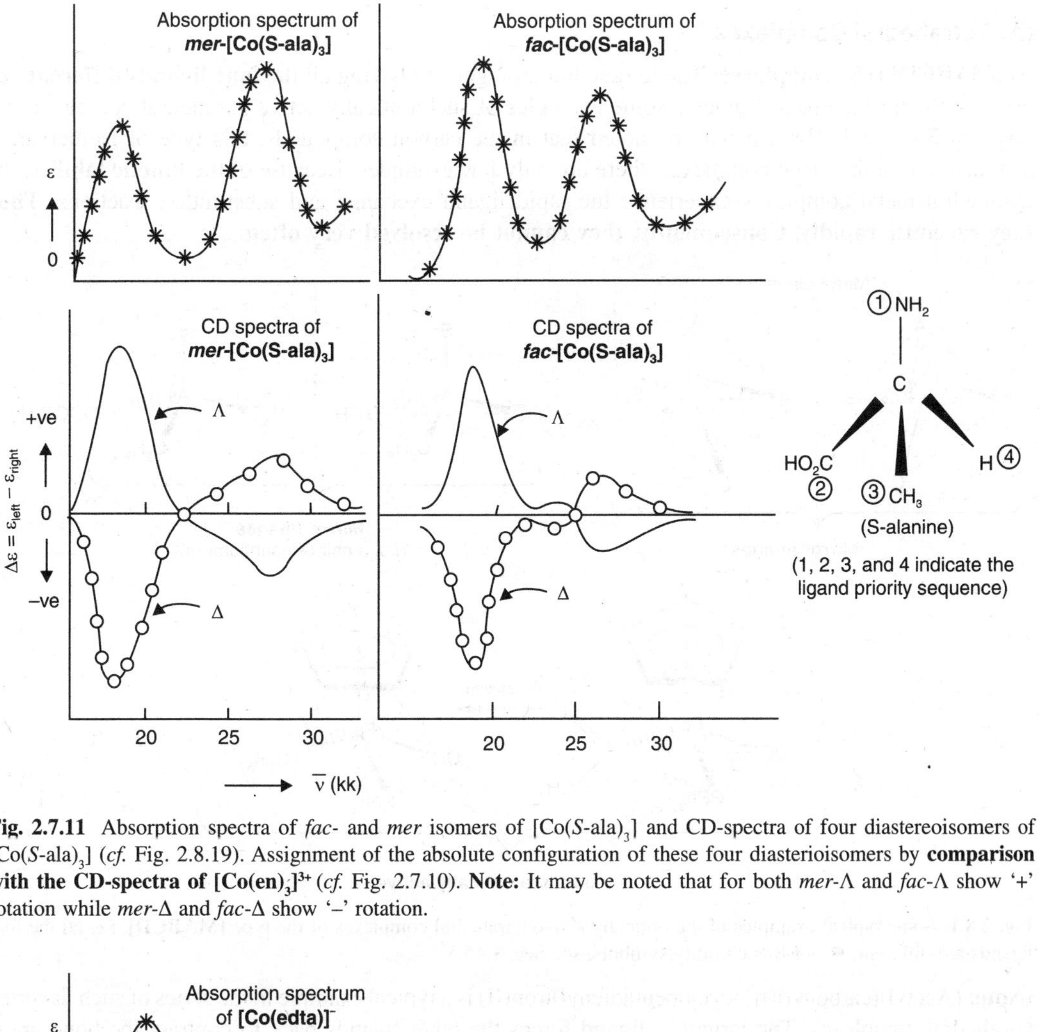

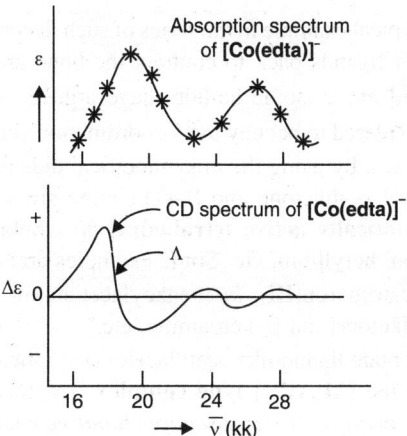

Fig. 2.7.11 Absorption spectra of *fac-* and *mer* isomers of [Co(S-ala)₃] and CD-spectra of four diastereoisomers of [Co(S-ala)₃] (*cf.* Fig. 2.8.19). Assignment of the absolute configuration of these four diasterioisomers by **comparison with the CD-spectra of [Co(en)₃]³⁺** (*cf.* Fig. 2.7.10). **Note:** It may be noted that for both *mer-*Λ and *fac-*Λ show '+' rotation while *mer-*Δ and *fac-*Δ show '−' rotation.

Fig. 2.7.12 Absorption and CD-spectra of [Co(edta)]⁻. The absorption spectrum is similar to that of [Co(en)₃]³⁺. The binding sites in [Co(edta)]⁻ are comparable to those in [Co(ala)₃]. Assignment of absolute configuration of [Co(edta)]⁻ in comparison with the CD-spectra of [Co(en)]³⁺ (Fig. 2.7.10). **Note:** At λ = 546 nm (from a Hg-source), Λ-[Co(edta)]⁻ (as a K⁺-salt) shows '−' rotation.

2.8 OPTICAL ISOMERISM IN THE 4–, 5– AND 6–COORDINATION COMPOUNDS

(A) Tetrahedral Complexes

(i) **[MABCD] type complexes:** The tetrahedral arrangement having all the **four ligands different** can produce the mirror image isomers. Some examples of such optically active tetrahedral complexes are given in Fig. 2.8.1. Here, it may be noted that in the carbon compounds, this type of isomerism is abundant but in the metal complexes, there are only few examples. Because of the **kinetic lability,** the tetrahedral metal complexes experience the rapid ligand exchange and substitution reactions. **Thus they racemise rapidly. Consequently, they cannot be resolved very often.**

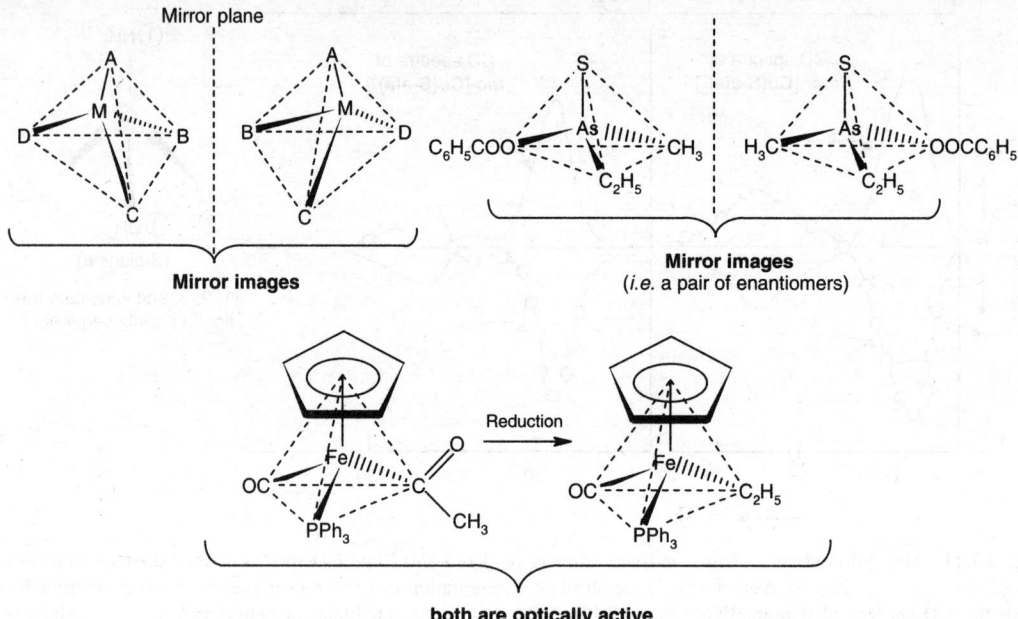

Fig. 2.8.1 Some typical examples of the optically active tetrahedral complexes of the type **[MABCD]**, *i.e.* all the four ligands are different. ● For R/S chirality symbols, see Sec. 1.15.5.

(Note: (Acetyl)(carbonyl)(η^5–cyclopentadienyl)iron(II) is a typical example in the series of such distorted tetrahedral complexes. The large Cp–ligand forces the other ligands back to contract the bond angle from $109\frac{1}{2}°$ to about 90°. To explain this contracted bond angle, some authors have argued such complexes as the **8-coordinate complexes,** *i.e.* η^5–Cp is considered to occupy the 5 coordination sites.)

(b) **[M(AB)₂] type complexes:** In the tetrahedral complexes, by using the unsymmetrical didentate ligands, the chirality may be introduced. The unsymmetrical β–diketone and β–ketoamine are such ligands to produce the optically active compounds. Such **optically active tetrahedral *bis*–chelates** very often involve the non-transition elements such as boron, beryllium, etc. Some examples are:

bis(benzoylpyruvato)beryllium(II), *bis*(salicylaldehydato)boron(III), *bis*(benzoylacetonato)-
beryllium(II), complex of Ni(II) involving β–diketone and β–ketoamine, etc.

It may be noted that by using the simple symmetrical didentate ligand like acetylacetonate, optically active tetrahedral complex cannot be obtained. In general, the **[M(AB)₂] type complex** is optically active when it adopts the tetrahedral geometry. *But it is not optically active in a square planar geometry.*

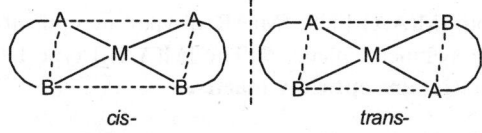

Nonsuperimposable mirror images for the tetrahedral [M(AB)$_2$] type complex.

Mirror images (*i.e.* a pair of enantiomers)
Bis(salicyladehydato)boron(III) ion.

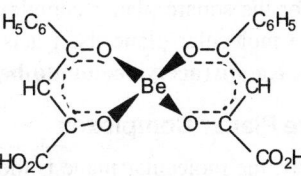

cis-   *trans-*

Square planar [M(AB)$_2$] type complexes,
optically inactive, molecular plane - symmetry plane.

Bis(benzoylpyruvato)beryllium(II)

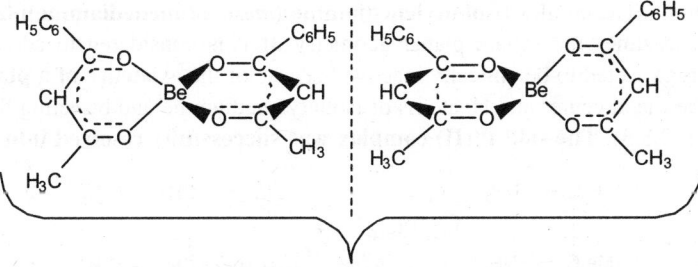

(Tetrahedral complexes)

Mirror images (*i.e.* a pair enantiomers)
Bis(benzoylacetonato)beryllium(II)

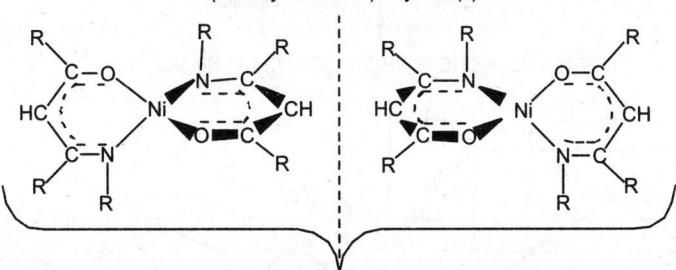

(Tetrahedral complexes)

A pair of enantiomers for β-ketoamine complexes of Ni(II)
(all R's may be the same or different)

Fig. 2.8.2 *(Contd.)*

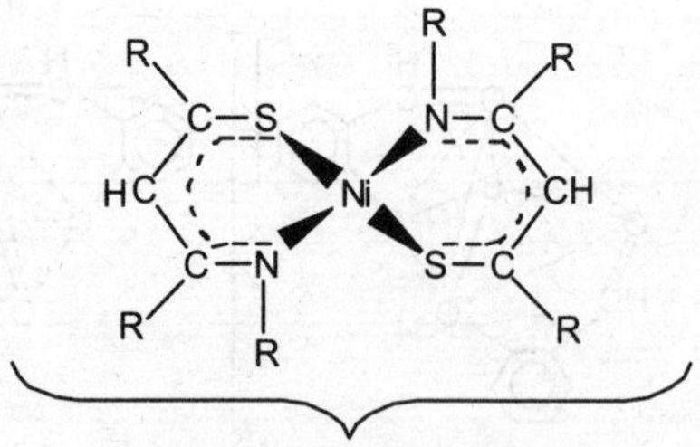

(Tetrahedral complex)

Optically active Ni(II)complex of
β-thioketoamine (all R's may be
the same or different)

Fig. 2.8.2 Optical activity in the $[M(AB)_2]$ type tetrahedral and square planar complexes. Examples of the optically active $[M(AB)_2]$ type tetrahedral complexes where AB stands for an unsymmetrical didentate ligand (Charges not shown).

Note: • For the square planar complexes, both $[M(AA)_2]$ and $[M(AB)_2]$ types are optically inactive because the molecular plane itself acts as the symmetry plane. • **The $[M(AA)_2]$ type tetrahedral complexes**, *e.g. bis*(acetylacetonato)beryllium(II) **are optically inactive.**

(B) Square Planar Complexes

(i) Generally, the molecular plane is the *plane of symmetry* in a square planar complex. In the square planar complexes, the chirality can be introduced by using some special types of didentate ligands. Mills and Quibell synthesised a mixed ligand complex of Pt(II) involving isobutylenediamine and *meso*–stilbenediamine. The comlex **(isobutylenediamine)(*meso*–stilbenediamine)platinum(II)** is optically active if it assumes the square planar geometry. If it is considered to adopt the tetrahedral geometry, then it is expected to be optically inactive because of the **existence of a plane of symmetry** passing through the metal center and N–atoms of isobutylenediamine and bisecting the C—C bond of the other ring (Fig. 2.8.3). **The said Pt(II) complex was successfully resolved into its enantiomers**

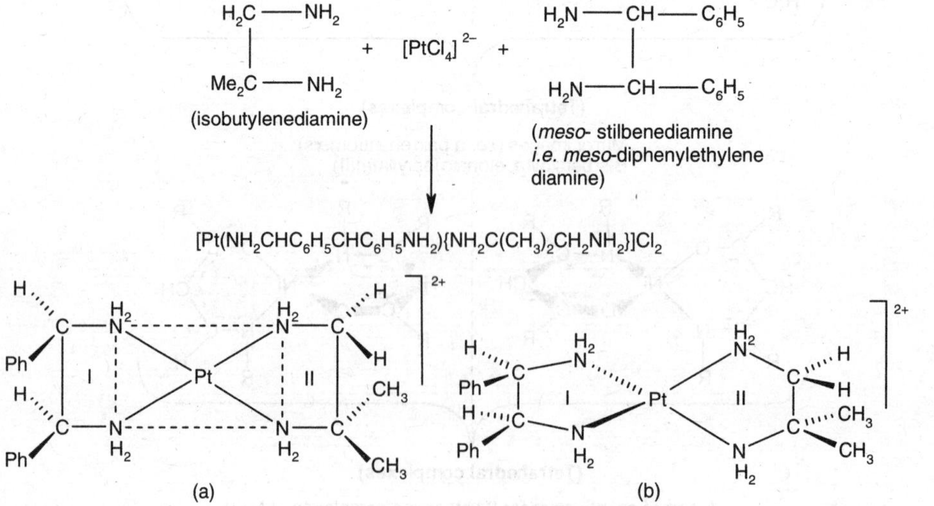

H_2C —— NH_2
|
Me_2C —— NH_2
(isobutylenediamine)

$+$ $[PtCl_4]^{2-}$ $+$

H_2N —— CH —— C_6H_5
|
H_2N —— CH —— C_6H_5
(*meso*- stilbenediamine
i.e. meso-diphenylethylene
diamine)

$[Pt(NH_2CHC_6H_5CHC_6H_5NH_2)\{NH_2C(CH_3)_2CH_2NH_2\}]Cl_2$

(a)
Planar coordination around Pt; **optically active**

(b)
Tetrahedral coordination around Pt; **optically inactive**

Fig. 2.8.3 Possible geometries of (isobutylene diamine)(*meso*–stilbenediamine)platinum(II).

and it conclusively proved the square planarity of the complex. Pd(II) can also form the optically active square planar complex by using these two ligands.

If the complex is argued to be a tetrahedral one in which the ring II lies in the plane of the paper and the ring I lies perpendicular to the plane (*cf.* Fig. 2.8.3). The plane containing the ring II and perpendicular to the ring I and bisecting the C—C bond of ring I is acting as a plane of symmetry. The plane contains the metal center, two N– and two C–atoms of isobutylene diamine and the plane bisects the C—C bond of *meso*–stilbenediamine. This symmetry plane allows the reflection of Ph, Me groups, etc. **Thus, plane of the paper is the plane of symmetry.** In the square planar arrangement, both the rings lie in the same plane, *i.e.* the plane contains all the C–, N–atoms of both rings and the metal center and **this is not a symmetry plane.** It is evident by considering the reflection of Ph-groups. Thus the plane of symmetry is denied. It also does not possess any center of symmetry.

(ii) The complex anion, **[Pd(edta)]$^{2-}$** has been found optically active and it can be resolved into its optical isomers. [Pd(edta)]$^{2-}$ can have two geometrical isomers, *i.e. cis–* and *trans–* isomers but **the cis– isomer cannot be optically active** (Fig. 2.8.4). For the *cis*-isomer, the plane containing Pd and perpendicularly bisecting the H$_2$C—CH$_2$ bond is the mirror plane. The optical activity exists with the *trans*–isomer. In fact, the *cis*–isomer could not be detected.

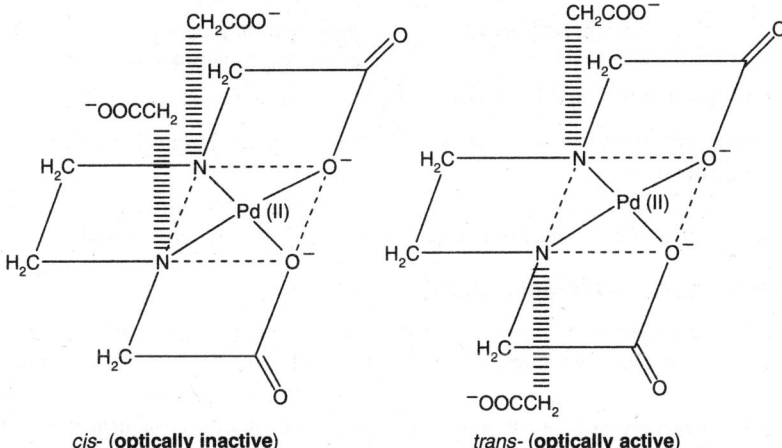

cis- **(optically inactive)**   trans- **(optically active)**

Fig. 2.8.4 Geometrical isomers of [Pd(edta)]$^{2-}$ having the tetradentate edta (2N, 2O) and their optical activities.

(iii) By using the **ligand chirality** (*i.e.* asymmetric ligand), optically active square planar complexes may be synthesised. For example, the compound, (N-methyl-N-ethylglycinato)dinitrito-κ^2N-platinate(II) is optically active due to the existence of the chiral N-center (Fig. 2.8.5).

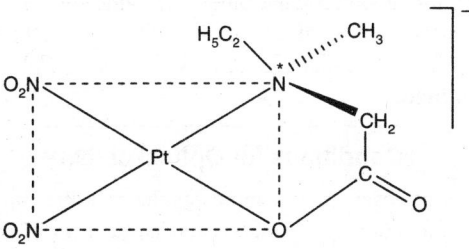

Fig. 2.8.5 Optical activity introduced in the square planar complex (N-methyl-N-ethylglycinato)dinitrito-κ^2N-platinate(II) due to the presence of the asymmetric N–site (denoted by N*).

(C) Optical isomerism in the 5-coordinate complexes (*cf.* Sec. 2.5.2)

In the square pyramidal complexes, optical isomerism has been reported in some rare examples. One such example is a Mo(II) complex, $[Mo(CO)_2(\eta^5–Cp)Cl(PR_3)]$, in which η^5–Cp occupies the axial or apical position and the other 4 ligands are placed at the basal plane. Depending on the position of two carbonyl groups, the complex can have the *cis–* and *trans–* geometric isomers. If there is no chiral center in PR_3, the *cis–* isomer lacking in the S_n axis is optically active while the *trans–*isomer is optically inactive. In the *trans–*compound, the vertical plane passing through the Mo, Cl, P, is the plane of symmetry.

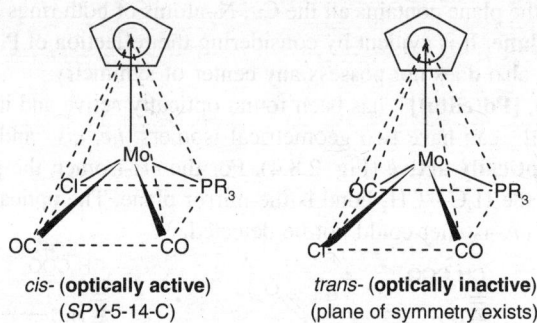

cis- (**optically active**)   *trans-* (**optically inactive**)

(*SPY*-5-14-C)   (plane of symmetry exists)

Fig. 2.8.6 Optical activity in $[Mo(CO)_2(\eta^5–Cp)Cl(PR_3)]$, a distorted square pyramidal geometry.

Another way to introduce the optical activity in the 5–coordinate systems is to use the ligands which are themselves optically active.

Note: Considering η^5–Cp to occupy five coordination sites, such complexes may be argued as the 9 coordinate complexes. For the chirality symbol, Sec. 1.15.5 may be considered.

(D) Octahedral Complexes (*cf.* Sec. 2.5.3)

(a) **[MABCDEF]** type complexes (Fig. 2.8.7) having six different monodentate ligands can produce **15 pairs of enantiomers**. One such complex is: $[Pt(Br)(Cl)(I)(NH_3)(NO_2)(py)]$. But this complex has not been resolved.

(b) **[MA₂B₂CD]** type complex can theoretically produce **6 geometric isomers** in which two are optically active (*cf.* Fig. 2.8.8) having the *cis*-orientations of A_2, B_2 and CD. The optically inactive forms are: *trans*-A_2, *trans*-B_2, *trans*-CD; *trans*-A_2, *cis*-B_2, *cis*-CD; *cis*-A_2, *trans*-B_2, *cis*-CD; *cis*-A_2, *cis*-B_2, *trans*-CD (Fig. 2.8.8).

(c) **[MA₂B₂C₂]** type complex (Fig. 2.8.9) can produce **6 stereoisomers** in which there is a pair of enantiomers. $[PtCl_2(NH_3)_2(NO_2)_2]$ gives **5 geometrical isomers** (Fig. 2.5.3.3) in which one geometrical isomer is optically active. In this geometrical isomer, the same ligands are at the *cis*–positions (Figs. 2.8.9 and 2.4.1.1). *cis*-$[MA_2B_2C_2]$ is the example of optical activity in the octahedral systems having the minimum number of the types of monodentate ligands, *i.e.* **using only three types of monodentate ligands, the chirality is attained.**

Conditions for Optical activity

It has already been stated that absence of S_n indicates the chirality. In practice, for the octahedral complexes, it is very convenient to identify the existence of **center of inversion ($i \equiv S_2$) and plane of symmetry ($\sigma \equiv S_1$)** to determine the optical activity. If σ and i are absent then the complex is optically active.

(d) **[M(AA)B_2CD]** type complexes (Fig. 2.8.10) can produce **4 geometric isomers** in which two are optically active (AA stands for a symmetrical didentate ligand). One hypothetical complex is $[Co(Br)(Cl)(en)(NH_3)_2]^+$.

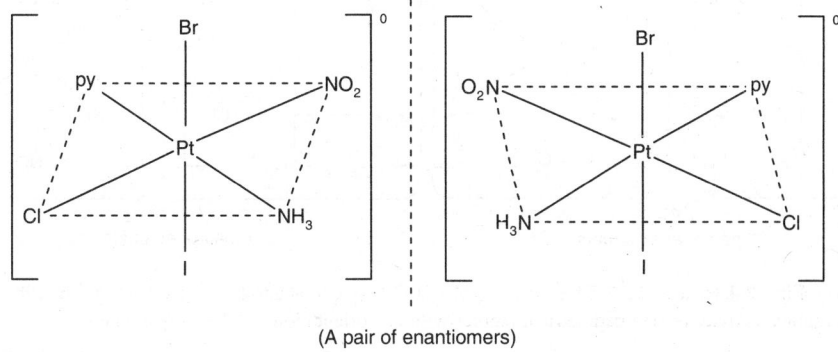

Br
py
NO2
Pt
Cl
NH3
I
0
Br
O2N
py
Pt
H3N
Cl
I
0
(A pair of enantiomers)

Fig. 2.8.7 A pair of enantiomers of the [MABCDEF] type complex, e.g. $[Pt(Br)(Cl)(I)(NH_3)(NO_2)(py)]$. There are total 15 pairs of enantiomers.

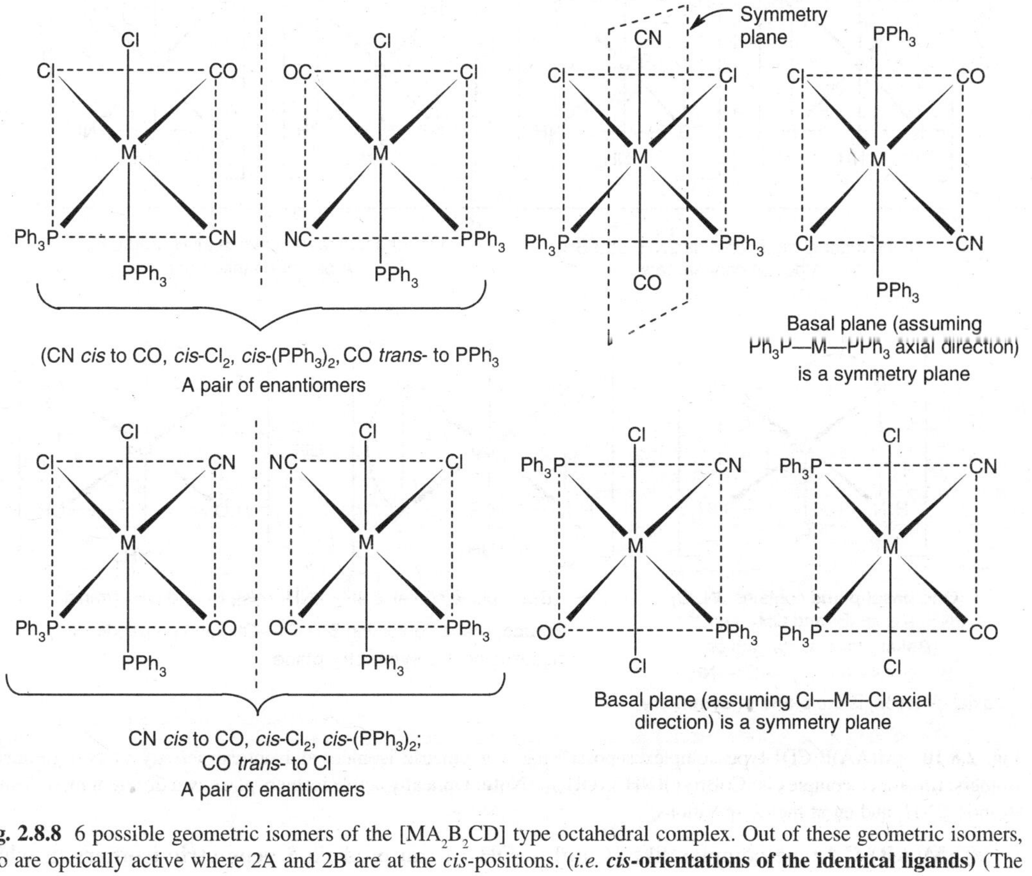

Cl
Cl
CO
M
Ph3P
CN
PPh3
Cl
OC
Cl
M
NC
PPh3
PPh3
(CN cis to CO, cis-Cl2, cis-(PPh3)2, CO trans- to PPh3
A pair of enantiomers
Symmetry plane
CN
Cl
Cl
M
Ph3P
PPh3
CO
PPh3
Cl
CO
M
Cl
CN
PPh3
Basal plane (assuming Ph3P—M—PPh3 axial direction) is a symmetry plane
Cl
Cl
CN
M
Ph3P
CO
PPh3
Cl
NC
Cl
M
OC
PPh3
PPh3
CN cis to CO, cis-Cl2, cis-(PPh3)2;
CO trans- to Cl
A pair of enantiomers
Cl
Ph3P
CN
M
OC
PPh3
Cl
Cl
Ph3P
CN
M
Ph3P
CO
Cl
Basal plane (assuming Cl—M—Cl axial direction) is a symmetry plane

Fig. 2.8.8 6 possible geometric isomers of the $[MA_2B_2CD]$ type octahedral complex. Out of these geometric isomers, two are optically active where 2A and 2B are at the *cis*-positions. (*i.e.* ***cis*-orientations of the identical ligands**) (The illustrated complex is a hypothetical one).

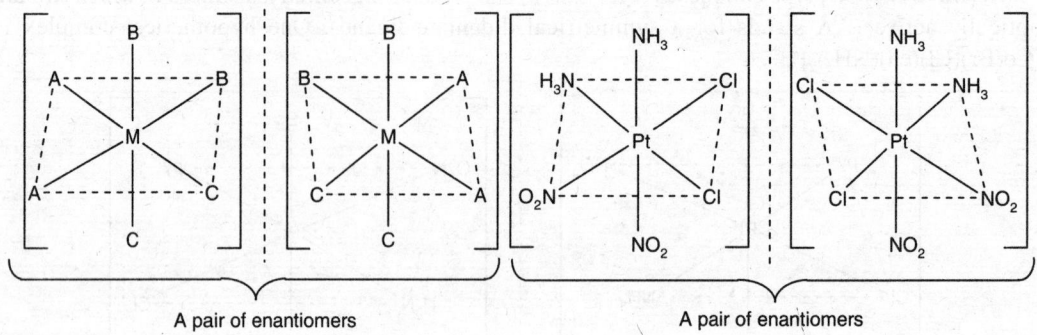

Fig. 2.8.9 (*cf.* **Figs. 2.4.1.1 and 2.5.3.3** for details) [MA$_2$B$_2$C$_2$] type octahedral complex having 5 possible geometrical isomers and optical activity of one geometrical isomer with *cis*–orientations of the same ligands.

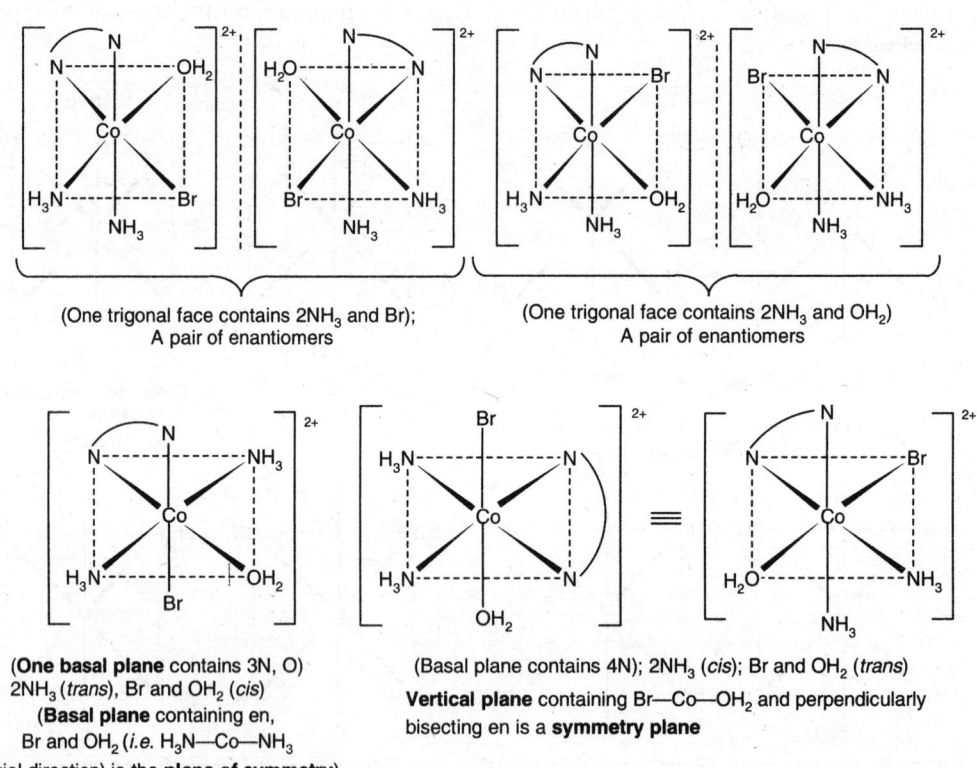

(One trigonal face contains 2NH$_3$ and Br);
A pair of enantiomers

(One trigonal face contains 2NH$_3$ and OH$_2$)
A pair of enantiomers

(**One basal plane** contains 3N, O)
2NH$_3$ (*trans*), Br and OH$_2$ (*cis*)
(**Basal plane** containing en,
Br and OH$_2$ (*i.e.* H$_3$N—Co—NH$_3$
axial direction) is the **plane of symmetry**)

(Basal plane contains 4N); 2NH$_3$ (*cis*); Br and OH$_2$ (*trans*)

Vertical plane containing Br—Co—OH$_2$ and perpendicularly bisecting en is a **symmetry plane**

Fig. 2.8.10 [M(AA)B$_2$CD] type complexes possessing 4 geometric isomers and optical activity of two geometric isomers. Illustrated complex is [CoBr(en)(NH$_3$)$_2$(OH$_2$)]; (**Note:** Optically active isomers place two different monodentate ligands, 2NH$_3$ and en at the *cis*–positions).

(e) **[M(AB)$_2$C$_2$]** type complex like [Cu(gly)$_2$(OH$_2$)$_2$] can produce **5 geometric isomers** in which three *cis*–diaqua isomers are optically active (Fig. 2.8.11).

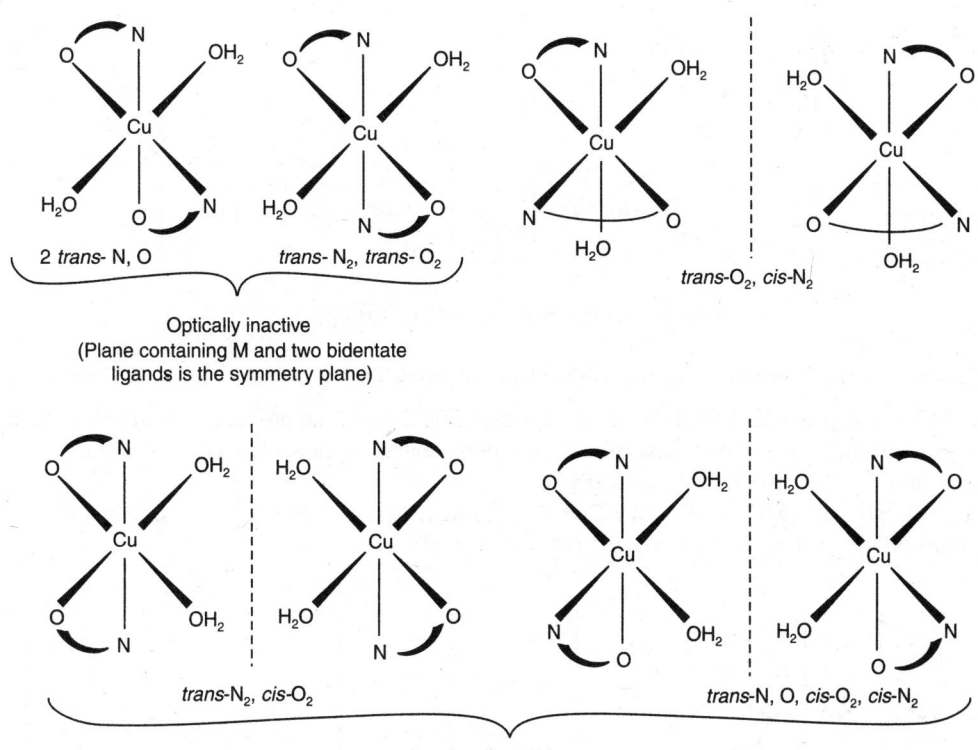

2 trans- N, O

trans- N$_2$, trans- O$_2$

trans-O$_2$, cis-N$_2$

Optically inactive
(Plane containing M and two bidentate
ligands is the symmetry plane)

trans-N$_2$, cis-O$_2$

trans-N, O, cis-O$_2$, cis-N$_2$

3 pairs of enantiomers

Fig. 2.8.11 5 geometric or diastereoisomers for [(M(AB)$_2$ C$_2$)] type complex and optical activity of 3 diastereoisomers out of 5 (Illustrated complex is [Cu(gly)$_2$(OH$_2$)$_2$]).

(f) **[M(AA)B$_2$C$_2$]** type complex like [CoCl$_2$(en)(NH$_3$)$_2$]$^+$, [Co(C$_2$O$_4$)(NH$_3$)$_2$(NO$_2$)$_2$]$^-$, etc. can produce **three geometrical isomers:** *cis*–B$_2$ and *cis*–C$_2$; *cis*–B$_2$ and *trans*–C$_2$; *cis*–C$_2$ and *trans*–B$_2$. Among these, the geometrical isomer having the *cis*–B$_2$ and *cis*–C$_2$ *configuration is optically active* (Fig. 2.8.12).

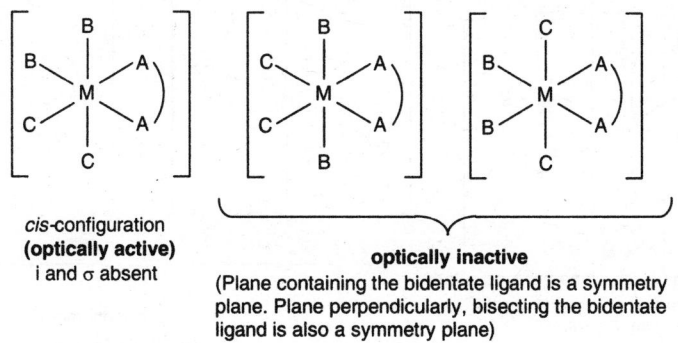

cis-configuration
(optically active)
i and σ absent

optically inactive
(Plane containing the bidentate ligand is a symmetry
plane. Plane perpendicularly, bisecting the bidentate
ligand is also a symmetry plane)

Fig. 2.8.12 *(Contd.)*

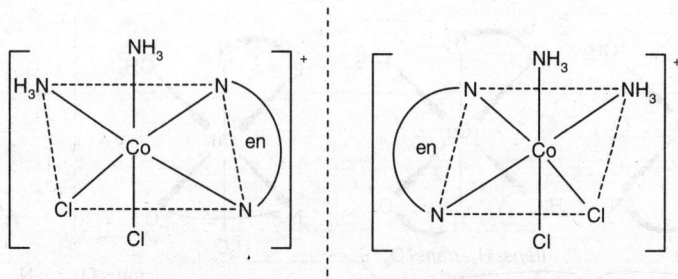

A pair of enantiomers of *cis*-[CoCl$_2$(en)(NH$_3$)$_2$]$^+$

Fig. 2.8.12 Stereoisomerism in the [M(AA)B$_2$C$_2$] type complexes. Illustrated complex is *cis*-[CoCl$_2$en(NH$_3$)$_2$]$^+$.

(g) **[M(AA)$_2$B$_2$] or [M(AA)$_2$BC]** type complexes (Fig. 2.8.13) can produce both the *cis*- and *trans*-isomers depending on the positions of the two monodentate ligands. The *cis*-isomers are optically active while the *trans*-isomers are optically inactive.

The examples are: [MCl$_2$(en)$_2$]$^+$ (M = Co^{III}, Cr^{III}); [MCl$_2$(ox)$_2$]$^{x-}$ (M = Co^{III}, Cr^{III}, Ir^{III}, Ir^{IV}, Rh^{IV}, etc.); [MCl(en)$_2$NH$_3$]$^{2+}$, [M(en)$_2$(NCS)(NH$_3$)]$^{2+}$ (M = Co^{III}, Cr^{III}).

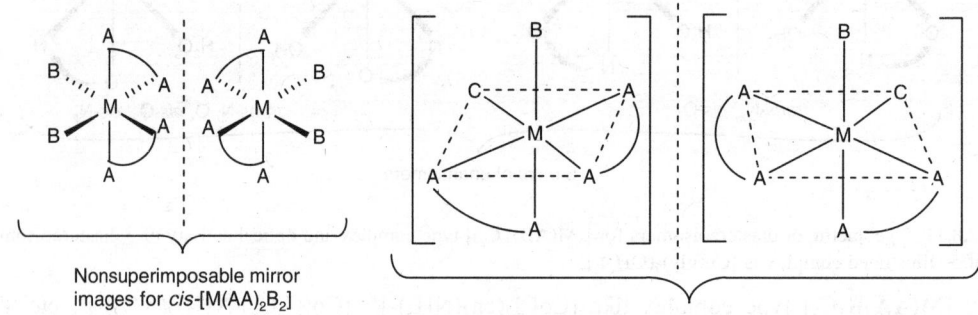

Nonsuperimposable mirror
images for *cis*-[M(AA)$_2$B$_2$]

Nonsuperimposable mirror images for *cis*-[M(AA)$_2$BC]

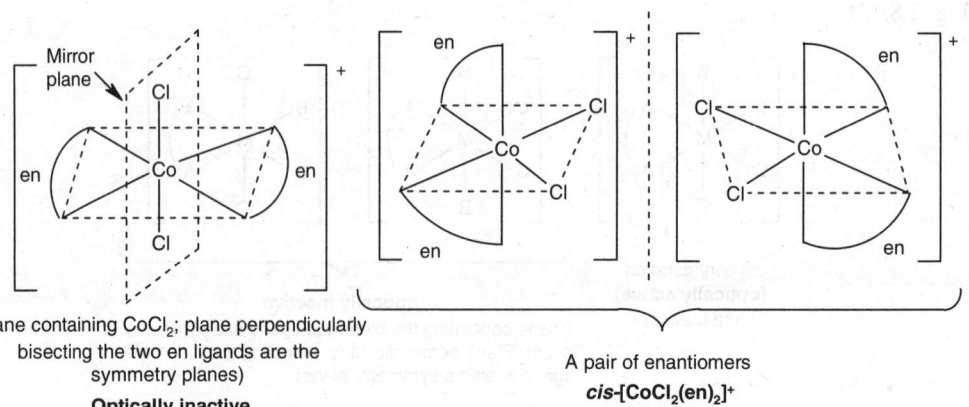

(Plane containing CoCl$_2$; plane perpendicularly
bisecting the two en ligands are the
symmetry planes)
Optically inactive
***trans*-[CoCl$_2$(en)$_2$]$^+$**

A pair of enantiomers
***cis*-[CoCl$_2$(en)$_2$]$^+$**

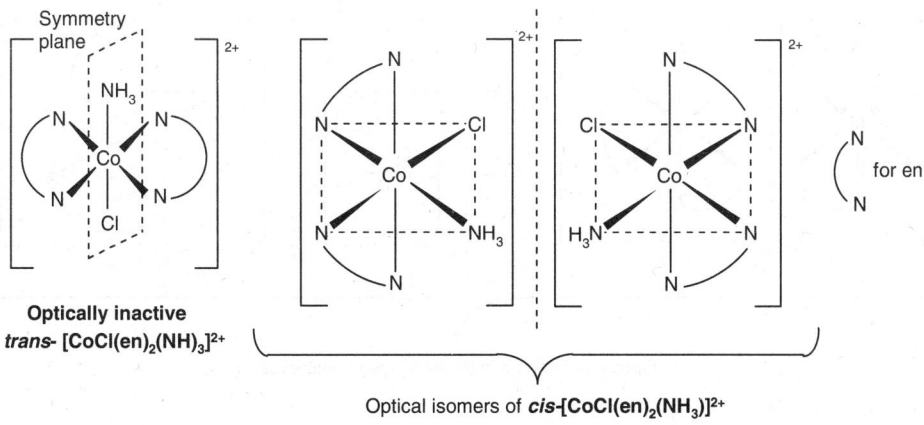

Optically inactive
trans- [CoCl(en)$_2$(NH$_3$)]$^{2+}$

Optical isomers of *cis-*[CoCl(en)$_2$(NH$_3$)]$^{2+}$

Fig. 2.8.13 Illustration of optical activity in the [M(AA)$_2$B$_2$] and [M(AA)$_2$BC] type complexes.

(h) **[M(AA)$_2$(BB)]** type complexes (Fig. 2.8.14) like [Co(CO$_3$)(en)$_2$]$^+$, [Cr(C$_2$O$_4$)(en)$_2$]$^+$, etc. can produce the optical isomers.

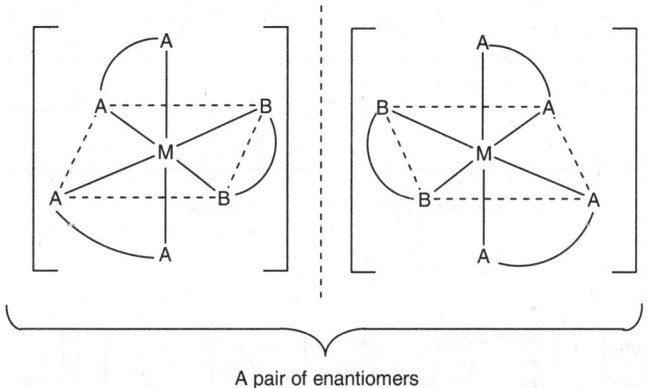

A pair of enantiomers

Fig. 2.8.14 Optical isomers in [M(AA)$_2$(BB)] type complex.

(i) **[M(AA)(BB)CD]** type complexes (*cf.* Fig. 2.5.3.12) can produce both the *cis-* and *trans-* isomers depending on the positions of the monodentate ligands C and D. There are two types of *cis-* isomers (C *trans-* to B and D *trans-* to A; C *trans-* to A and D *trans-* to B) which are optically active while the *trans-* form is optically inactive.

(j) **[M(AAAA)B$_2$]** and **[M(AAAA)BC]** type complexes (*cf.* Fig. 2.5.3.10) where AAAA denotes a linear tetradentate ligands like 'trien' can produce *cis-* (**α and β forms) and *trans-* isomers depending on the planirity of the chelate rings**. The *cis-* forms are optically active while the *trans* form is optically inactive.

(k) **[M(AA)$_3$]** type complexes (Fig. 2.8.15) (D$_3$ symmetry) are optically active. It has got no plane of symmetry and no centre of symmetry.

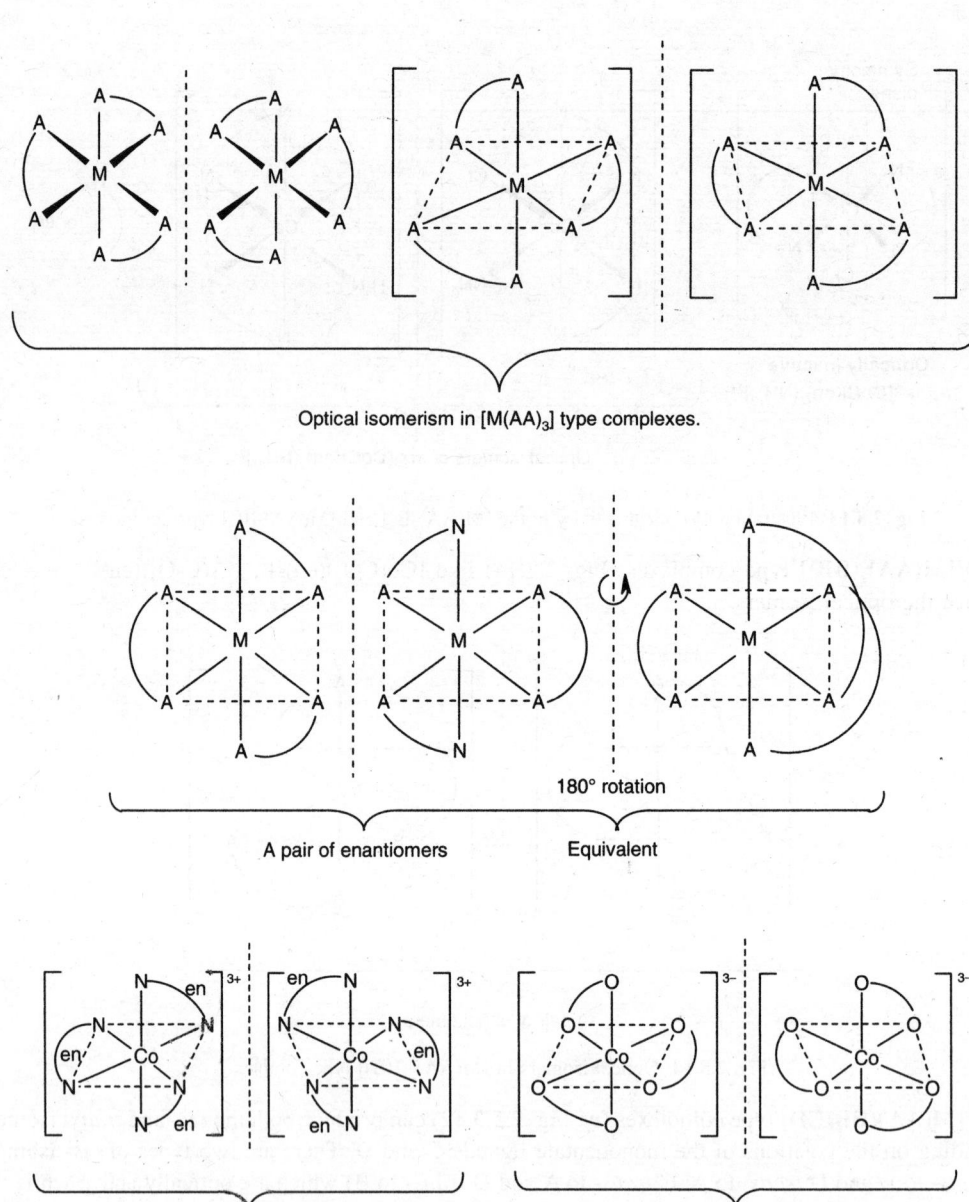

Optical isomerism in [M(AA)₃] type complexes.

A pair of enantiomers   Equivalent

180° rotation

Enantiomers of [Co(en)₃]³⁺   Enantiomers of [Co(C₂O₄)₃]³⁻,

Fig. 2.8.15 Illustration of optical activity in the [M(AA)₃] type complexes.

The **examples** are: $[M(en)_3]^{n+}$ ($M = Co^{III}$, Cr^{III}, Rh^{III}, Ni^{II}, etc.); $[M(bpy)_3]^{2+}$ ($M = Fe^{II}$, Ni^{II}); $[M(C_2O_4)_3]^{3-}$ ($M = Fe^{III}$, Cr^{III}, Co^{III}), $[M(phen)_3]^{2+}$ ($M = Fe^{II}$, Ni^{II}, Ru^{II}), $[Co(bigH)_3]^{3+}$, etc.

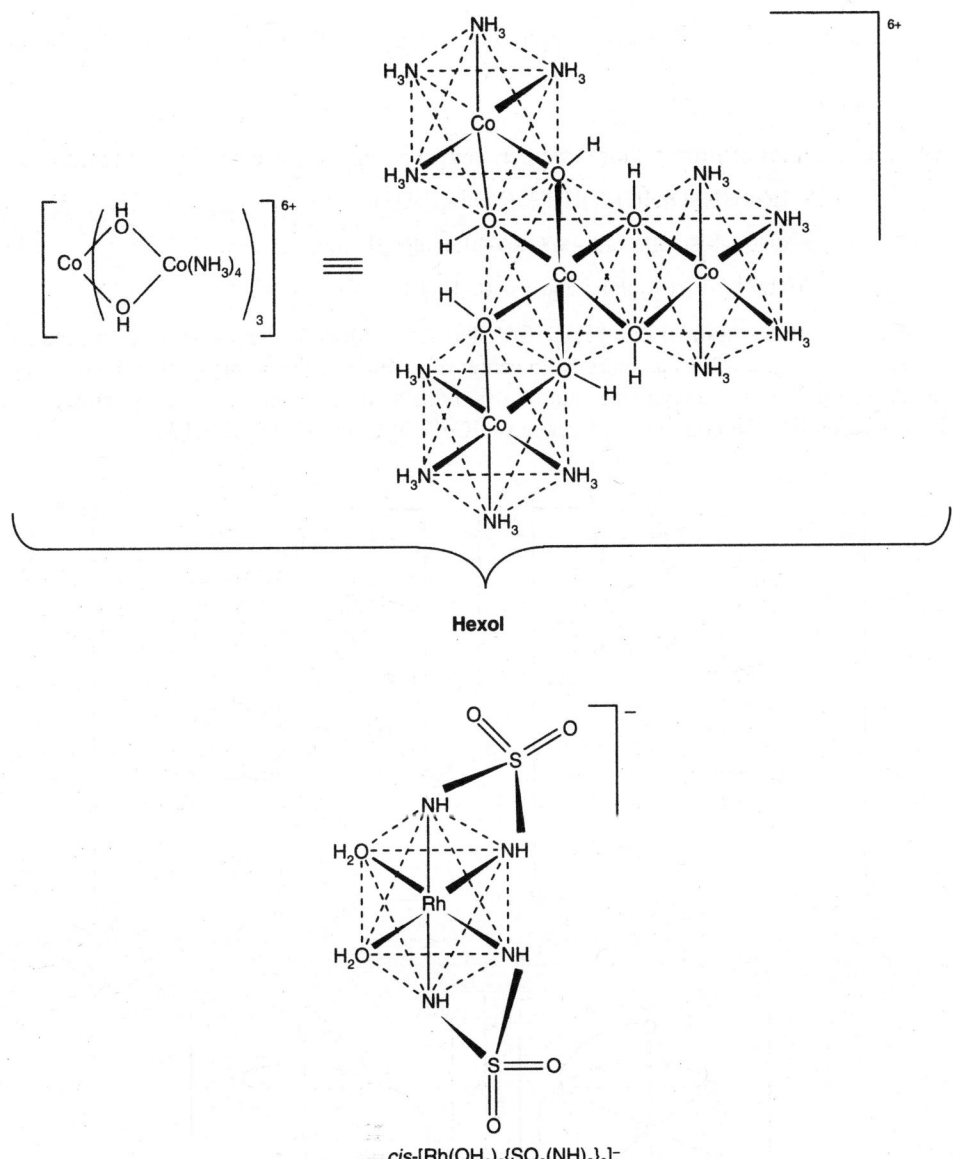

Fig. 2.8.16 Examples of purely inorganic compounds as the optically active compounds.

(l) **Optically active purely inorganic complex:** Werner was able to resolve many complexes of the type [M(AA)₃] and this led him to conclude their octahedral geometry. However, critics of his time argued that the optical activity might be due to the C-center present in the ligands. To answer this question, Werner was able to prepare the optically active compound (purely inorganic compound) having no carbon atom. The compound is:

tris[tetrammine-μ-dihydroxidocobalt(III)]cobalt(III) ion (commonly called *hexol*) in which three

inorganic didentate ligands, $\left[(H_3N)_4Co \begin{smallmatrix} OH \\ \\ OH \end{smallmatrix} \right]^+$ are placed octahedrally around the central Co^{3+} ion. The

complex is shown in Fig. 2.8.16.

Other **two examples of purely inorganic octahedral complexes** having optical activity are:

cis–[Rh(HNSO$_2$NH)$_2$(OH$_2$)$_2$]$^-$ (Mann, 1933)

i.e.   *cis*–diaquabis(sulfamido-κ*N*, κ*N*-)rhohate(III) ion

[Cr(HPO$_4$)$_3$]$^{3-}$ (Podlaha and Ebert, 1960).

(m) **[M(hexadentate ligand)] complex:** [Co(edta)]$^-$, *i.e.* ethylenediaminetetraacetatocobaltate(III) is optically active (Fig. 2.8.17a) as there is no center of inversion and no plane of symmetry. Another hexadentate ligand, 1, 8–*bis*(salicylideneamino)–3, 6–dithiaoctane can form an optically active octahedral complex with Co(III). The complex is denoted by [Co(S$_2$N$_2$O$_2$)]$^+$ (*cf.* Fig. 2.8.17)

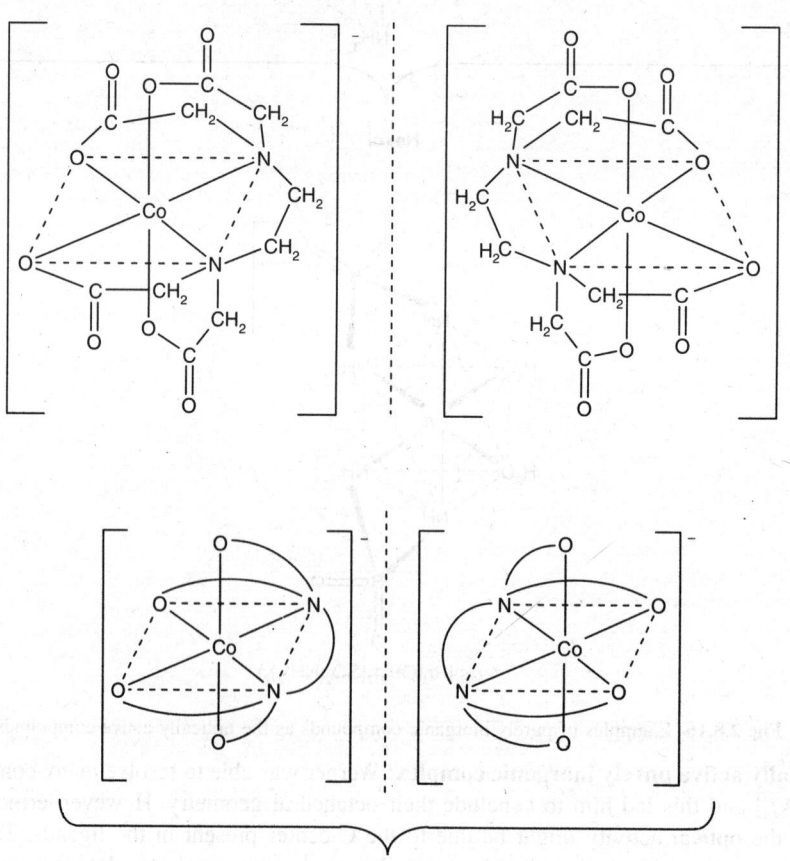

Optical isomers of [Co(edta)]$^-$, *i.e.* [Co(N$_2$O$_4$)]$^-$

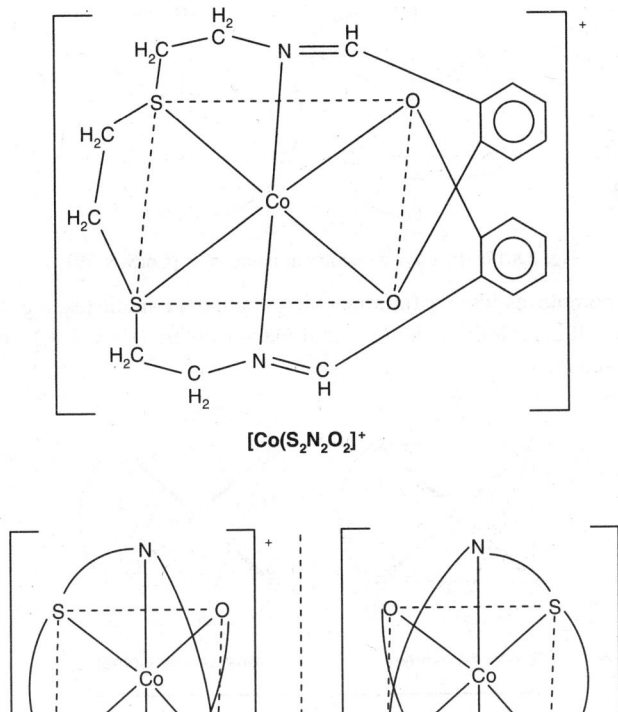

$[Co(S_2N_2O_2)]^+$

Optical isomers of $[Co(S_2N_2O_2)]^+$

Fig. 2.8.17 (a) Optical isomers of Optical isomers of $[Co(edta)]^-$ and $[Co(S_2N_2O_2)]^+$. (see text for the ligand $S_2N_2O_2^{2-}$).

Here it may be pointed out that the said complex $[Co(S_2N_2O_2)]^+$ can give **four geometrical isomers** (*cf.* Fig. 2.8.17b) **and each of these isomers is optically active. The Co(III)-complex with the ligand probably shows the highest molecular rotation among the known coordination compounds.**

$H_2S_2N_2O_2 =$

i.e.   HO   N   S   S   N   OH

(Contd.)

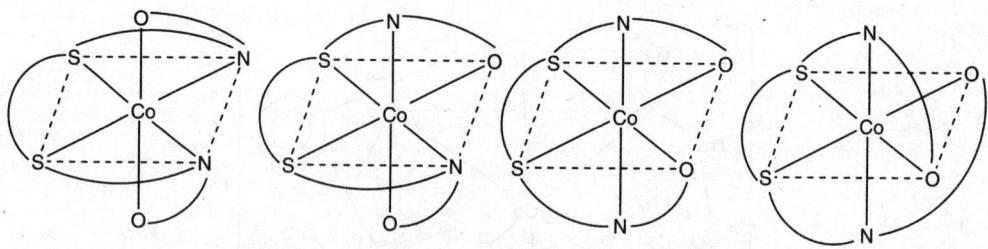

Fig. 2.8.17 (b) Four geometrical isomers of $[Co(S_2N_2O_2)]^+$.

(n) **[M(AB)₃]** type complexes like *tris*(aminoacido)cabalt(III) complexes, *e.g.*$[Co(gly)_3]$, $[Co(ala)_3]$, etc. are optically active. It can produce the *fac–* and *mer–* isomers (Figs. 2.8.18 and 2.5.3.9) and both of them are optically active.

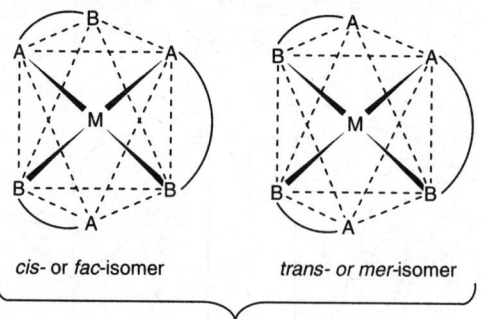

cis- or fac-isomer   trans- or mer-isomer

Fig. 2.8.18 Optically active *fac–* and *mer–* isomers of [M(AB)₃] type complex, *e.g.* [Co(gly)₃].

(o) Optical activity arising from the chirality in chelate ring conformations as in [M(ala)₃], [M(pn)₃], and [M(en)₃]:

● **[M(ala)₃]**: If the amino acid is itself chiral then it may complicate the situation. For **alanine**, there may be **R** and **S** configurations (Fig. 2.8.19). For glycine (which is optically inactive), the *fac–*Δ and *fac–*Λ isomers produce an enantiomeric pair; and similarly, *mer–*Δ and *mer–*Λ produce another pair of enantiomers. **For [Co(S–ala)₃], *fac-*Δ and *fac-*Λ are not the enantiomers but they are the diastereomers.** For, [Co(S–ala)₃], a mirror image cannot be obtained by using S–alanine. It can be obtained by using the R-alanine. Thus for [Co(S–ala)₃], we get four diastereoisomers, *fac–*Δ, *fac–*Λ, *mer–*Δ and *mer–*Λ (*cf.* Fig. 2.7.11) and they can be separated easily as they differ in physical properties.

● For **[Co(pn)₃]³⁺** (*cf.* Fig. 2.7.7) where pn can have the **R–** and **S–** configurations. In fact, because of the **R–** and **S–** configurations of $H_2N - CH_2 - CH - NH_2$ (pn), we can expect **eight optically**
$$| \atop CH_3$$
active isomers. These are:

Δ, Λ–[Co(**R**–pn)₃]³⁺; Δ, Λ–[Co(S–pn)₃]³⁺;

Δ, Λ–[Co(**R**–pn)(S–pn)₂]³⁺; Δ, Λ–[Co(S–pn)(**R**–pn)₂]³⁺.

But thermodynamic stabilities of these all possible isomers widely differ. Consequently, all these isomers are not observed in reality. The development of these isomers from the knowledge of chelate ring conformations (*i.e.* δ, λ) has been discussed earlier (*cf.* Sec. 2.7).

The enantiomeric pairs (*cf.* Fig. 2.7.7) are:

Δ–[Co(**R**–pn)$_3$]$^{3+}$ ($\lambda\lambda\lambda$) and Λ–[Co(**S**–pn)$_3$]$^{3+}$($\delta\delta\delta$)

Δ–[Co(**S**–pn)$_3$]$^{3+}$ ($\delta\delta\delta$) and Λ–[Co(**R**–pn)$_3$]$^{3+}$($\lambda\lambda\lambda$)

The δ and λ denote the chelate ring conformations. These aspects have been discussed in Sec. 2.7 and Fig. 2.7.7.

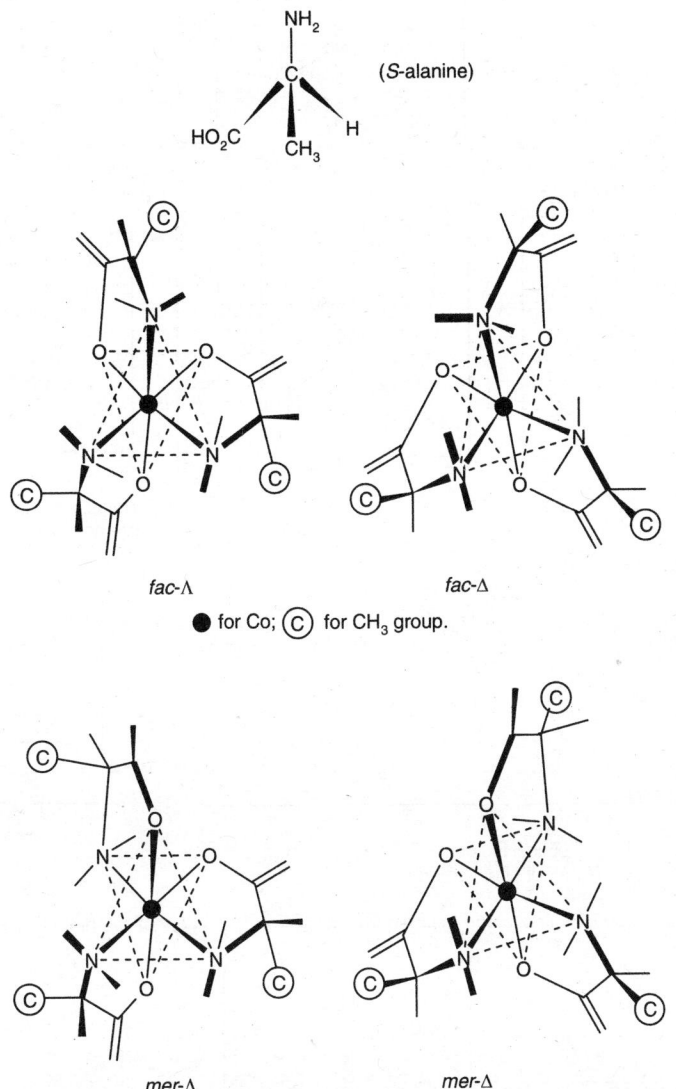

Fig. 2.8.19 Optically active four diastereoisomers of *tris*–(**S**–alaninato)cobalt(III), *i.e.* [Co(S–ala)$_3$].

● For [M(en)$_3$], by considering the δ and λ conformations of the chelate ring, 4 pairs of enantiomers (*cf.* Figs. 2.7.5 and 2.7.6) are possible. (i) Δ-$\delta\delta\delta$ and Λ-$\lambda\lambda\lambda$, the (ii) Δ-$\lambda\delta\delta$ and Λ-$\delta\lambda\lambda$, (iii) Δ-$\lambda\lambda\delta$ and Λ-$\delta\delta\lambda$; (iv) Λ-$\lambda\lambda\lambda$ and Λ-$\delta\delta\delta$. If the chelate ring conformations are not considered then, we can have only two enantiomers designated by Δ and Λ. These aspects have been illustrated in Figs. 2.7.5 and 2.7.6.

(p) **Optical activity in the polynuclear complexes:** Polynuclear complexes possessing more than one chiral centers have been reported (Fig. 2.8.20). The following dinuclear complex can exist in three forms, *i.e.* a pair of enantiomers and one *meso*–form (which is internally compensated).

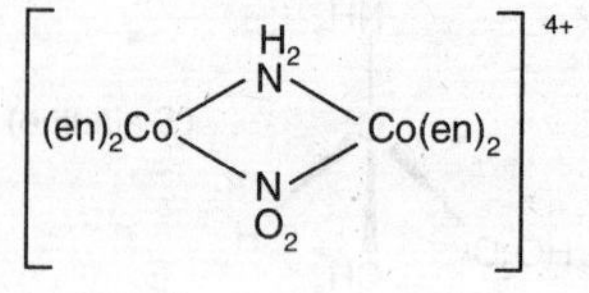

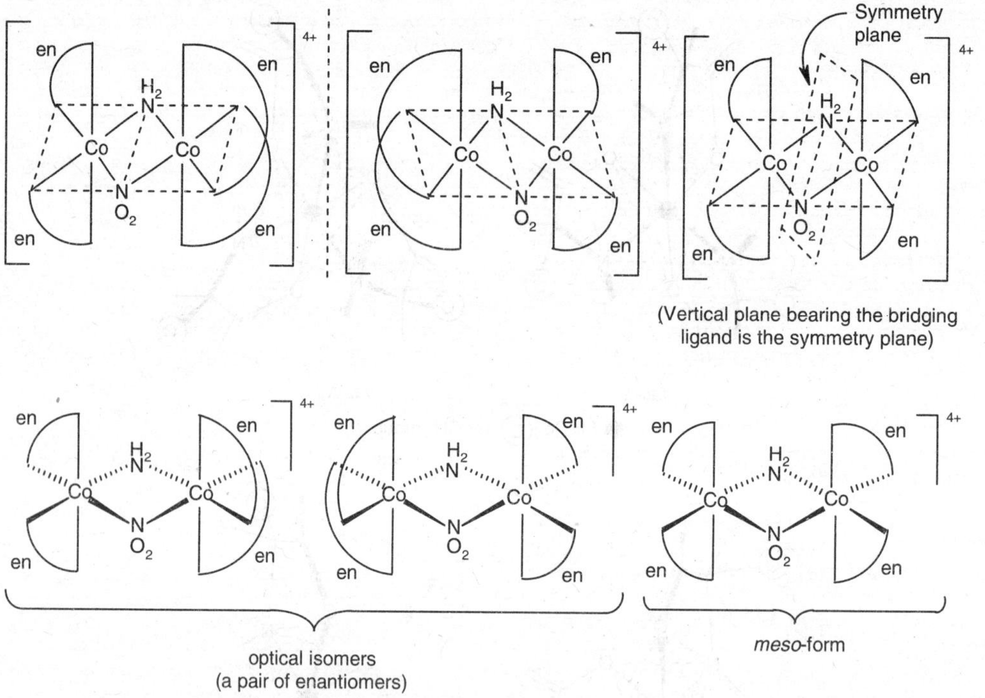

(Vertical plane bearing the bridging ligand is the symmetry plane)

optical isomers
(a pair of enantiomers)

meso-form

Fig. 2.8.20 Optically active and inactive (*i.e. meso*–form) isomers of the binuclear complex,

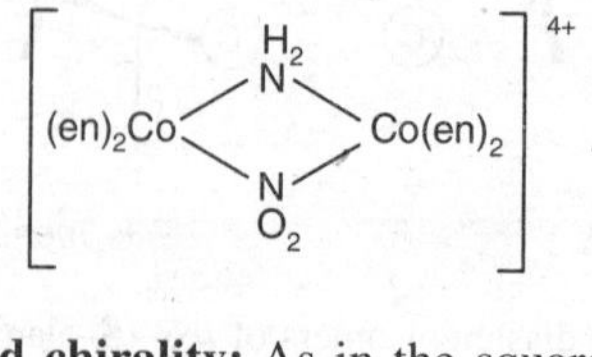

(q) **Optical activity due to ligand chirality:** As in the square planar complexes, the presence of *ligand chirality* can introduce the optical activity in the octahedral complexes. The common example is $[Co(NH_3)_5(LH^{\pm})]^{3+}$ (*cf.* Fig. 2.8.21) where $LH^{\pm}$ stands for the Zwitterionic form of an optically active amino acid like alanine that can exist as R– and S– configurations.

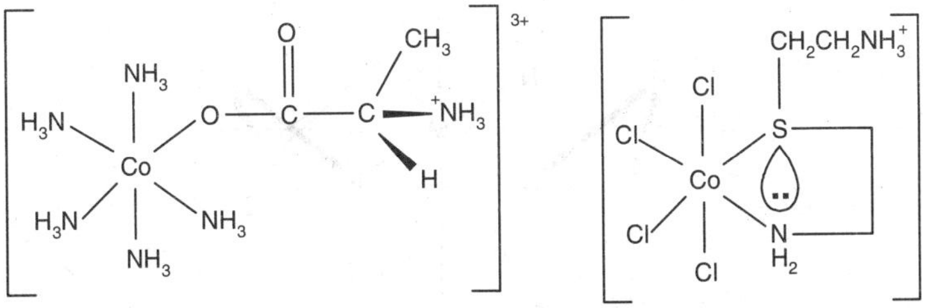

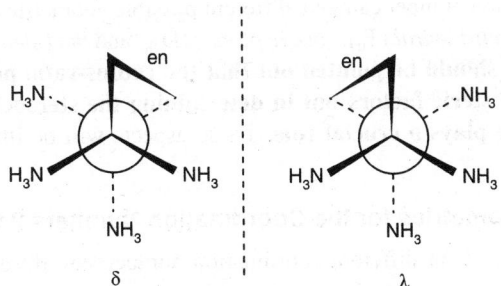

Fig. 2.8.21 Illustration of optical activity due to chirality in a ligand.

In another example (*cf.* Fig. 2.8.21), the **ligated S–site is asymmetric** and the compound is optically active (*cf.* S–site, Fig. 2.8.21). Here the ligated S–site is tetrahedral by considering its one lone pair. Pt(IV) can form the similar compound with the said ligand.

2.9 OPTICAL ACTIVITY FROM THE CHELATE RING CONFORMATION IN SQUARE PLANAR AND OCTAHEDRAL COMPLEXES

It has already been stated that chelate ring conformations can produce different optical isomers (*cf.* Secs. 2.7 and 2.8). The ethylenediamine chelate ring, *i.e.* M(en) in its **gauche** form can be compared with the **chair form of cyclohexane**. In the chelate ring, there is a C_2–axis passing through the metal center and mid-point of the C—C bond (Fig. 2.7.4). The chelate ring can have the **δ– and λ– conformations** which are the nonsuperimposable mirror images. Thus, $[Co(en)(NH_4)_4]^{3+}$ is expected to gave a pair of enantiomers (*cf.* Fig. 2.9.1), but it cannot be resolved because of the very low energy barrier for ring inversion. In fact, **the conformational stability is marginal**. However, by introducing more chelate rings, the situation can be made different. In the square planar complex $[M(en)_2]$ in which the two en-ligends are in the *trans*–positions, there can be a pair of enantionmers (δδ and λλ) and a meso-form (δλ or λδ) (Fig. 2.9.2). These are discussed in Sec. 2.7 and Fig. 2.7.4. The isomers from the chelate ring conformations in the octahedral complexes have been already discussed in Sec. 2.7 (*cf.* Figs. 2.7.4-7 and 2.8.19). Possible isomers in $[M(en)_3]^{3+}$, $[M(ala)_3]$ and $[M(pn)_3]^{3+}$ (considering the R, S, configurations of ala and pn) have been discussed in Secs. 2.7 and 2.8.

Fig. 2.9.1 The δ and λ enantiomers of $[Co(en)(NH_3)_4]^{3+}$.

By considering the chirality in the chelate rings present in **$[M(dien)_2]$** (*cf.* Fig. 2.5.3.2) (which can exist in the *fac-* and *mer* forms without considering the ring conformations), it can generate the optical isomers. Ring conformations in *cis-* (α, β) and *trans-* forms of **$[M(trien)X_2]$** (*cf.* Fig. 2.5.3.10) will also produce the similar results.

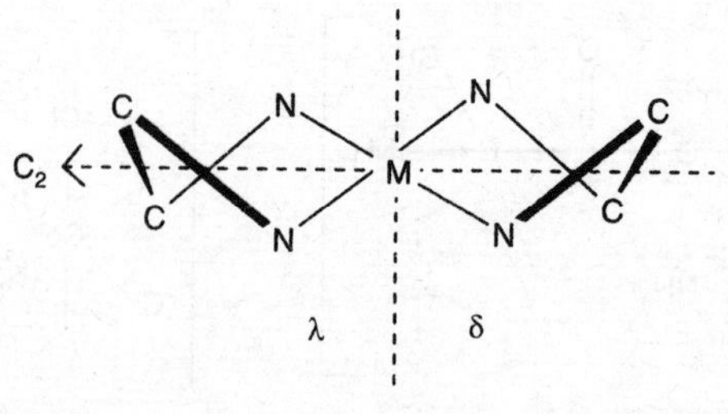

Fig. 2.9.2 M–λδ isomer of the [M(en)$_2$] planar complex. (This *meso* form is optically inactive).

2.10 SPIN STATE ISOMERISM

This aspect has been discussed in detail in Secs. 8.24–25.

2.11 STRUCTURE OF COORDINATION COMPOUNDS AND STEREOCHEMICAL ASPECTS

2.11.1 Different Factors to Control the Stereochemical Preferences

In determining the coordination number (C.N.) and stereochemistry of coordination compounds, many factors depending on the properties of metal center and ligands operate. All these factors cannot be generalised. However, the following general trends are maintained in most of the cases.

(i) To increase the metal-ligand attractive interaction (both covalent and ionic interactions), *the coordination number will tend to be as high as possible*.

(ii) *Crystal field stabilisation energy (cfse)* may direct the preferred stereochemistry.

(iii) The stereochemical orientation of the ligands will occur in a way so that the **ligand-ligand repulsion will be minimum**. Sometimes, the **ligand-ligand bonding interaction** may stabilise a particular geometry (*cf.* trigonal prismatic structure of [M(S$_2$C$_2$R$_2$)$_3$])

(iv) In ionic lattices, the ***radius-ratio*** of the involved cation and anion is very much important. The same principle is also applicable for the coordination compounds, specially for the cases where a particular coordination number can give different possible geometries [*e.g.* C.N. = 4 can give the possible geometries: *tetrahedral* (T$_d$), *square planar* (D$_{4h}$) and *buckeled square planar* or *squashed tetrahedral* (D$_{2d}$)]. **It should be pointed out that the radius-ratio principle mainly works on the consideration of steric factors but in determining the stereochemistry of coordination compounds, the *cfse* plays a crucial role.** These aspects will be illustrated at the appropriate places.

2.11.2 The Possible Geometries for the Coordination Numbers 2 to 7

The possible stereochemistries for different coordination numbers are illustrated below.

C.N.	Possible geomeries
(A) 2 (*i.e.* ML$_2$)	linear (D$_{\infty h}$) and bent (C$_{2v}$).

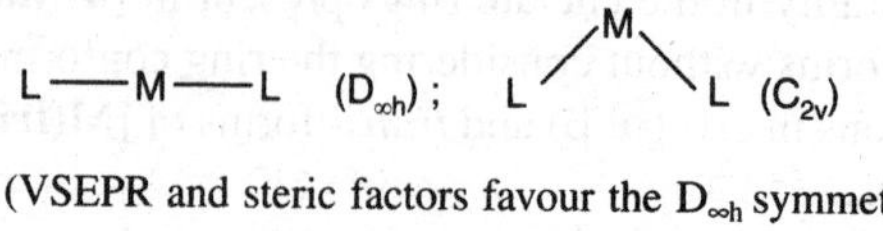

(VSEPR and steric factors favour the D$_{\infty h}$ symmetry).

(B) 3 (*i.e.* ML$_3$)   **planar (symmetrical) (D$_{3h}$), pyramidal (C$_{3v}$) and T-shaped (C$_{2v}$)**

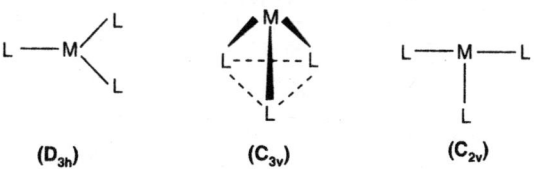

(D$_{3h}$)   **(C$_{3v}$)**   **(C$_{2v}$)**

Fig. 2.11.2.1 Different geometries for C.N. = 3.

C$_{3v}$ is obtained from the D$_{3h}$ by keeping the metal center out of the plane of the ligating sites.

(C) 4 (*i.e.* ML$_4$)   **tetrahedral (T$_d$), square planar (D$_{4h}$), buckeled square planar or squashed**
(very common C.N.)   **tetrahedral (D$_{2d}$)**—an intermediate geometry between the limiting geometries characterised by D$_{4h}$ and T$_d$.

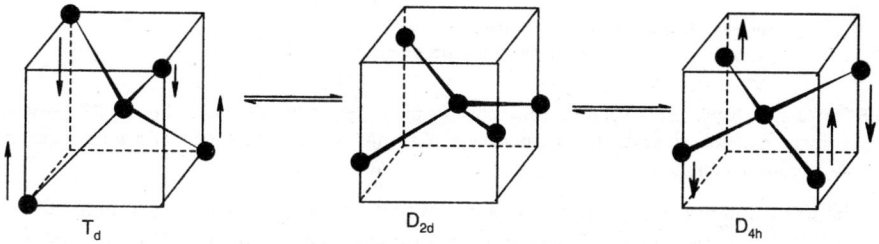

T$_d$ D$_{2d}$ D$_{4h}$

(Metal at the centre of the cube)

Fig. 2.11.2.2 Interconversion among the four coordinate structures (*i.e.* **polytopism**).

Another geometry characterised by C$_{2v}$ symmetry is also possible (specially important for the nontransition central elements, *e.g.* **:**SF$_4$, **:**TeCl$_2$Me$_2$, etc.) when the central element possesses a lone pair. The structure may be described by **disphenoid** or **see-saw** or **teeter-totter** (comparable to children's balancing toy) geometry.

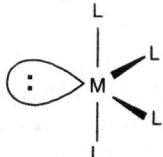

Fig. 2.11.2.3 C$_{2v}$ symmetry of C.N = 4.

(D) 5 (*i.e.* ML$_5$)   **trigonal bipyramidal (D$_{3h}$) and square pyramidal (C$_{4v}$)**
(very common C.N.)

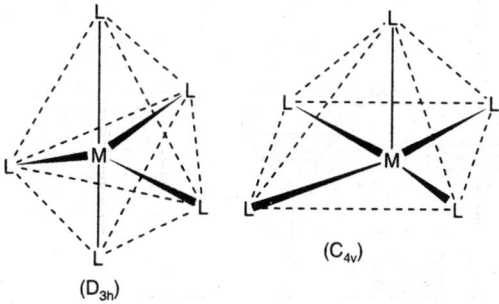

(C$_{4v}$)

(D$_{3h}$)

Fig. 2.11.2.4 Trigonal bipyramidal and square pyramidal geometries for the 5-coordinate complexes.

These two geometries are interconvertible and the geometry having the intermediate structure (C_{2v}) is also known. These easy deformability and interconvertibility (*cf.* Fig. 2.11.7.2, 2.11.7.4) give the **fluxional property** (*i.e.* **stereochemical nonrigidity**).

(E) 6 (*i.e.* ML$_6$)   **Octahedral (O$_h$) (very common); and trigonal prism (D$_{3h}$) (in rare cases).**
(Very Common C.N.)

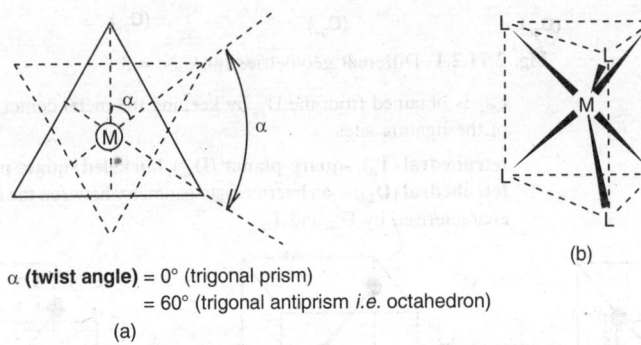

α **(twist angle)** = 0° (trigonal prism)
                 = 60° (trigonal antiprism *i.e.* octahedron)
(a)

Fig. 2.11.2.5 (a) Relation between octahedron and trigonal prism in terms of twist angle (*cf.* Fig. 2.11.8.1); twisting about the C_3-axis, *i.e.* rotation of one triangular face of an octahedron with respect to the parallel triangular face. (b) Trigonal prism for the 6-coordinate complexes.

The octahedral geometry is favoured in general over the trigonal prismatic geometry for 6–coordinate complexes. It is due to the fact: **ligand-ligand repulsion,** *i.e.* **steric crowding is more in trigonal prism than in octahedron.**

Trigonal prismatic geometry arises when the **steric requirement of a polydentate ligand enforces this geometry.** This geometry is also found in the case of $[M(S_2C_2R_2)_3]$ when a set of six sulfur atoms mutually interact to stabilise the system (*cf.* Fig. 2.11.8.4)

The **distortion in the octahedral geometry** can produce different types of structures. These are of mainly three types.

(a) **Tetragonal distortion** (*cf.* Jahn-Teller effect): Contraction (*i.e.* **z–in**) or elongation (*i.e.* **z–out**) along a single C_4–axis giving rise to the D$_{4h}$ symmetry (*cf.* Fig. 3.11.1.1).

(b) **Trigonal distortion:** Contraction or elongation along one of the C_3–axes giving rise to the D$_{3d}$ symmetry.

(c) **Rhombic distortion:** Elongation along one C_4 axis while contraction along another C_4 axis (*i.e.* lengths of the three C_4–axes are unequal) giving rise to D$_{2h}$ symmetry.

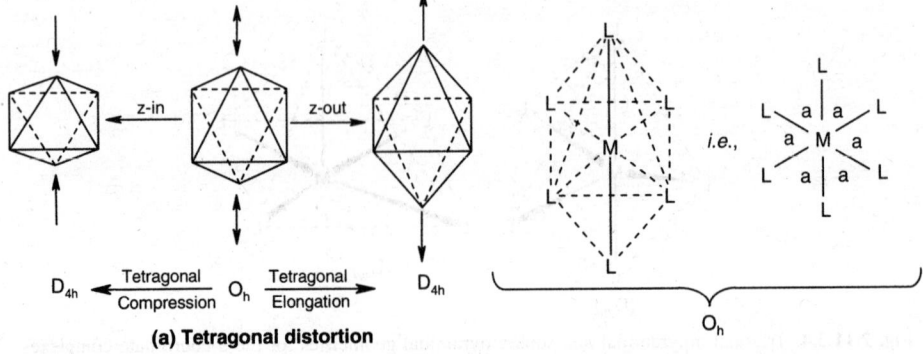

(a) Tetragonal distortion

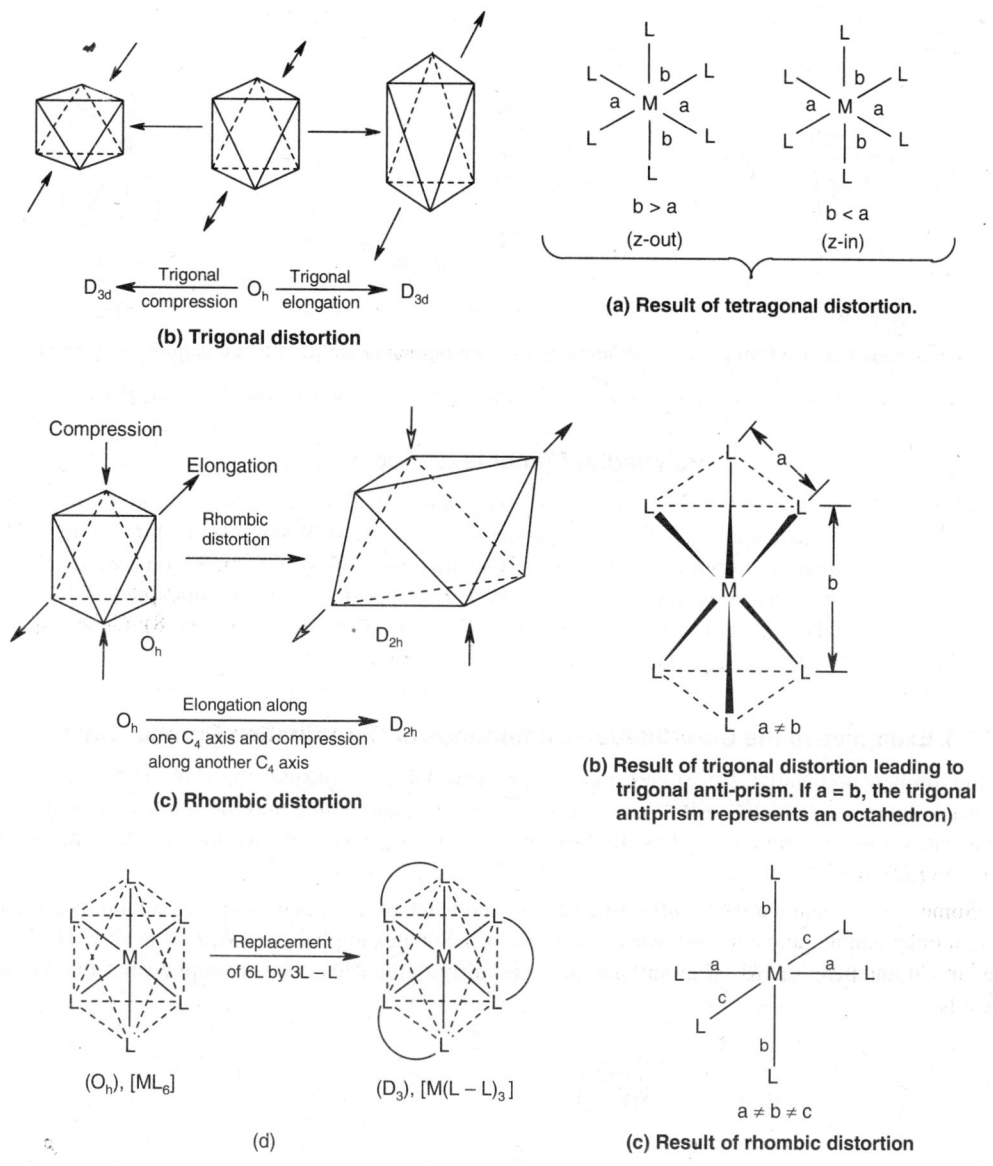

(a) Result of tetragonal distortion.

$b > a$
(z-out)

$b < a$
(z-in)

$D_{3d} \xleftarrow[\text{compression}]{\text{Trigonal}} O_h \xrightarrow[\text{elongation}]{\text{Trigonal}} D_{3d}$

(b) Trigonal distortion

$O_h \xrightarrow[\substack{\text{one } C_4 \text{ axis and compression} \\ \text{along another } C_4 \text{ axis}}]{\text{Elongation along}} D_{2h}$

(c) Rhombic distortion

(b) Result of trigonal distortion leading to trigonal anti-prism. If a = b, the trigonal antiprism represents an octahedron)

$(O_h), [ML_6]$

$\xrightarrow[\text{of 6L by 3L – L}]{\text{Replacement}}$

$(D_3), [M(L – L)_3]$

(d)

$a \neq b \neq c$

(c) Result of rhombic distortion

Fig. 2.11.2.6 Different types of distortions (a) tetragonal distortion, (b) trigonal distortion, (c) rhombic distortion (d) reduction of symmetry by replacement of 6 unidentate ligands (L) by three didentate ligands (L—L) in the octahedral complexes: (d) not a case of distortion; substitution by didentate ligands reduces the symmetry.

Here it may be mentioned that the **trigonal prism structure may experience also a distortion to produce the trigonal antiprism structure.**

(F) 7 (*i.e.* ML₇)   pentagonal bipyramidal (**D_{5h}**), **capped octahedral (C_{3v})**-obtained by placing the 7th ligand at the center of a trigonal face of an octahedron, and **capped trigonal prism (C_{2v})**-obtained by placing the 7th ligand at the center of a rectangular face of a trigonal prism.

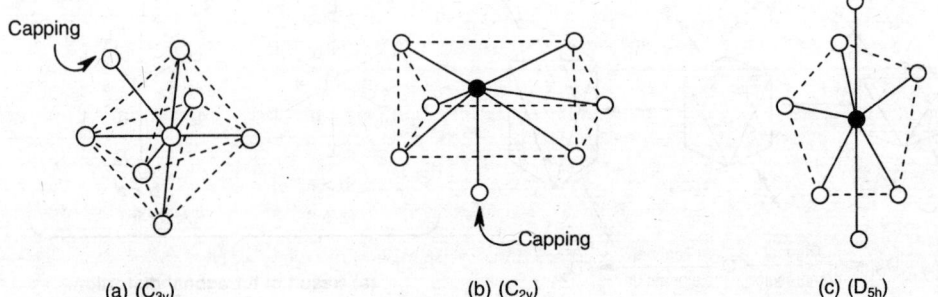

(a) (C₃ᵥ)   (b) (C₂ᵥ)   (c) (D₅ₕ)

(a) **monocapped octahedron (C₃ᵥ)** (b) **tetragonally capped trigonal prism (C₂ᵥ)** (c) **pentagonal bipyramid (D₅ₕ)**

Fig. 2.11.2.7 (*cf.* Fig. 2.11.9.1) Different possible geometries for the 7 coordinate complexes.

Polyhedral Symbols (*cf.* Sec. 1.15.5)

Linear (C.N. 2): *L*-2; Angular (C.N. 2): *A*-2; Trigonal plane (C.N. 3): *TP*-3; Trigonal pyramid (C.N. 3): *TPY*-3; Tetrahedron (C.N. 4): *T*-4; Square plane (C.N. 4): *SP*-4; Trigonal bipyramid (C.N. 5): TBPY-5; Square pyramid (C.N. 5): *SPY*-5; Octahedron (C.N.6): *OC*-6; Trigonal prism (C.N. 6): *TPR*-6; Pentagonal bipyramid (C.N. 7): *PBPY*-7; Octahedron, face monocapped (C.N. 7) = *OCF*-7; Trigonal prism, square face monocapped (C.N. 7): *TPRS*-7; Cube (C.N. 8): *CU*-8; Square antiprism (C.N. 8): *SAPR*-8; Dodecahedron (C.N. 8): *DD*-8.

2.11.3 Examples of the Coordination Compounds of Coordination Number One

In the ion-pair like Na^+Cl^- (in vapour phase), it is reasonable to consider the C.N. of Na^+ as 1. In the transient species like VO^{2+}, BiO^+, etc, the C.N. of the metal center may be argued as 1. But such transient species coordinate readily with the other available ligands to expand the coordination number, *e.g.* $[VO(OH_2)_5]^{2+}$, $[VO(acac)_2]$ (more correctly, $[V(acac)_2O]$ etc.)

Sometimes, by using a **very bulky ligands**, the tendency of the metal center to expand its coordination number can be prevented to restrict the C.N. to 1. Such examples are: $[M(2, 4, 6\text{-}Ph_3C_6H_2)]$ [M^I = Ag^I or Cu^I and there is a M—C bond) and these are fairly stable when they are kept away from the small ligands.

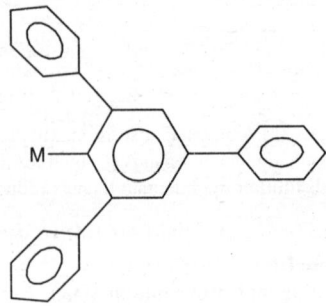

Fig. 2.11.3.1 Structure representation of the organometallic compounds $[M(2, 4, 6\text{–}Ph_3C_6H_2)]$ (M = Cu, Ag) illustrating the C.N. 1.

Coordination Number Zero: What is the C.N. of Na⁺ in zeolite?

In determining the C.N., the distance between the metal center and ligating atom should not exceed their sum of ionic and/or atomic radii. If this interionic distance is larger than their sum of radii, then the surrounding ions are not considered to contribute to determine the coordination number. In zeolites, the fate of alkali metal ions appears so. In fact, the alkali metal ions are present in the cavities of zeolites where only weak electrostatic attraction prevails because of the long interionic distance. Thus, it may be concluded that the **C.N. of alkali metal ions in zeolites is effectively zero.** In fact, such alkali metal ions can be readily exchanged by much larger cations which can fit better in the cavities.

2.11.4 Examples of Coordination Number Two

It is generally found for the Gr. 11(IB) metals of +1 oxidation state characterised by d^{10} configuration, e.g. Cu(I), Ag(I) and Au(I). It also occurs so for a closely related species Hg(II) (a d^{10} system). The examples are:
$[Ag(NH_3)_2]^+$, $[Cu(NH_3)_2]^+$, $[AuCl_2]^-$, $[Hg(CN)_2]$, $[Cl—Ag—NH_2—(CH_2)_2—NH_2—Ag—Cl]$ (en acting as a bridging ligand), $[Ag(CN)_2]^-$, $[Au(PPh_3)_2]^+$, $[AuCl(PEt_3)]$, etc.

The **linear oxo-cations** like $[UO_2]^{2+}$, $[UO_2]^+$, $[PuO_2]^{2+}$, etc. may be considered to have the C.N. 2. However, such oxo-cations tend to ligate with other ligands to increase the C.N., e.g. $[UO_2(NO_3)_2(OH_2)_2]$ (more correctly $[U(NO_3)_2O_2(OH_2)_2]$) (C.N. = 8), $[UO_2(CO_3)_3]^{4-}$ or, $[U(CO_3)_3O_2]^{4-}$ (C.N. = 8), etc. Similarly, Ag^+, Hg^{2+}, etc. also tend to increase the C.N., e.g.

$$\left[Ag(NH_3)_2\right]^+ + 2NH_3 \rightleftharpoons \left[Ag(NH_3)_4\right]^+ \;(\text{C.N.} = 4, T_d)$$

$$\left[Hg(CN)_2\right] + 2CN^- \rightleftharpoons \left[Hg(CN)_4\right]^{2-} \;(\text{C.N.} = 4, T_d)$$

In fact, the MX_2 species very often attains the C.N. 4 in solution through solvation. Pd(0) (a d^{10} system) can also show the C.N. 2 as in $[Pd\{PPh(t–Bu)_2\}_2]$

(a) Some Important Aspects in Connection with C.N. 2

- Hg^{2+}—CN^- system: $\log K_1 = 18$, $\log K_2 = 16.7$, $\log K_3 = 3.83$, $\log K_4 = 3.0$
- Hg^{2+}—I^- system: $\log K_1 = 12.9$, $\log K_2 = 11.0$, $\log K_3 = 3.70$, $\log K_4 = 2.4$

In terms of the step-wise formation constants (i.e., K_1, K_2...), there is a **sudden fall at K_3** for the above mentioned Hg^{2+}–complexes. In fact, at this step, the linear $[HgL_2]$ species converts into the tetrahedral species $[HgL_3(OH_2)]^-$. The *strong preference of Hg^{2+} for the linear geometry* is responsible for this sharp drop. Moreover, *this step is also **entropically disfavoured** because of fixing one solvent molecule within the coordination sphere to provide the tetrahedral geometry.*

$$\left[HgL_2\right] + L^- + H_2O \xrightarrow{K_3} \left[HgL_3(H_2O)\right]^-$$

- Ag^+—NH_3 system: $\log K_1 = 3.15$, $\log K_2 = 3.85$

The higher value of K_2 than that of K_1 goes **against the statistical prediction**. In fact, K_2– step leads to the formation of linear species from a nonlinear species.

$$\left[Ag(NH_3)(H_2O)_n\right]^+ + NH_3 \xrightarrow{K_2} \left[Ag(NH_3)_2\right]^+ + nH_2O$$

$\quad$ (**Nonlinear species**) $\qquad\qquad\qquad$ (**Linear species**)

This step is favoured because of two factors: *strong preference for the linear geometry and **entropic favour** due to the release of the coordinated solvent molecules to provide the required linear geometry.* (*cf.* **Cd^{2+}–Br$^-$ system:** $K_1 \rangle K_2 \rangle K_4 \rangle K_3$; [CdBr$_4$]$^{2-}$ is T$_d$; others are O$_h$)

● [Ag(NH$_3$)$_2$]$^+$ is more stable than [Ag(en)]$^+$. *Apparently,* **it goes against the concept of chelate effect.** In fact, the strong preference for the linear geometry is satisfied in [Ag(NH$_3$)$_2^+$ but not in [Ag(en)]$^+$. The chelating ligand en cannot provide the linear coordination as desired by Ag$^+$.

● In general, the 5-membered chelate rings are more stable than the lower or higher membered rings because of the steric factors. In the case of Ag$^+$, compared to en that forms a five membered chelate ring, the ligands having the similar N-donor sites but capable of forming the larger rings (*i.e.* 6–, 7– or 8– membered rings) can form the more stable complexes. Because, *these larger rings can stretch more towards the linear coordination as desired by Ag$^+$.*

● **Cu(I)– a d^{10} system,** generally prefers the tetrahedral geometry but [CuCl$_2$]$^-$ and [Cu(NH$_3$)$_2$]$^+$ are linear.

● **[Ag(CN)$_2$]$^-$** and **[Au(CN)$_2$]$^-$** are the discrete linear complexes, but solid **K[Cu(CN)$_2$]** possesses a polymeric chain structure of the anion by using the bridging CN$^-$ ligands to provide C.N. 3 (2C,1N) around Cu(I) (*cf.* Fig. 2.11.5.2).

● Most of the metal centers showing the C.N. 2 very often tend to increase the C.N. in presence of suitable ligands. *In this respect, Au(I) is exceptional and it shows a little tendency to increase its C.N. from 2 (see* **"Energetics"** *in bonding model for C.N. 2).*

● In the d^{10}–configuration, there is no *cfse* for both the C$_{2v}$ (bent) and D$_{\infty h}$ (linear) symmetry. Thus the steric factors and VSEPR favour the linear stereo-chemistry.

● Infrared studies on HgCl$_2$, HgBr$_2$ and HgI$_2$ vapour *indicate the very low bending force constants.* This suggests the **very weak directional properties** of the bonding orbitals to provide the C.N. 2.

C.N. 2 for the metal centers other than d^{10} configuration: Ligand imposed C.N. 2

It has been already mentioned that the heavier congeners with d^{10} electronic configurations are the best candidates for showing the C.N. 2. However, by using the **bulky and anionic ligands like [N(SiMe$_3$)$_2$]$^-$, [N(SiMePh$_2$)$_2$]$^-$,** the metal ions like Mn^{2+}(d^5), Fe^{2+}(d^6), Co^{2+}(d^7) and Ni^{2+}(d^8) can be forced to attain the C.N. 2. However, it is not certain whether such complexes are linear (D$_{\infty h}$) or bent (C$_{2v}$).

(b) Bonding Model for C.N. 2

(i) Hybridisation model: The linear coordination can be attained by considering the simple *sp*-hybridisation. The linear coordination is mainly shown by the heavier congeners (Gr. 11 and 12) having the d^{10} configuration. Now the simple question arises regarding the **preference for *sp*–hybridisation over the the *sp^3*–hybridisation.** Apparently, the *sp^3*–hybridisation should bring about more stabilisation through the formation of additional metal-ligand linkages. *In fact, this apparent anomaly cannot be explained by considering the simple sp-hybridisation for the linear coordination.* To explain this, Orgel has proposed the participation of the $(n-1)d_{z^2}$ orbital in attaining the linear coordination. The process is energetically favoured when the energies of the $(n-1)d$, *ns* and *np* orbitals are comparable. *For the $(n-1)d^{10} ns^0 np^0$ electronic* configuration, hybridisation involving the $(n-1)d_{z^2}$, *ns* and *np$_z$* orbitals *followed by the **redistribution of the electron density will remove the electron density from the +z and −z directions** (i.e. regions of the approaching ligands) to stabilise the complex.*

The above mentioned **Orgel concept** can be illustrated by considering the step-wise hydridisation for the sake of simplicity.

Step 1: Hybridisation between $(n-1)d_{z^2}$ (major lobes bearing + signs) and ns–orbitals.

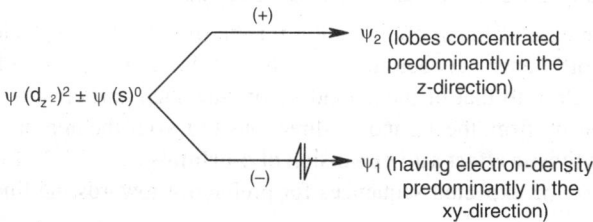

The electrons are **redistributed** as follows:

$$(n-1)d_{z^2}^2\, ns^0 \longrightarrow \psi_1^2 \psi_2^0$$

i.e. shifting of electrons from the d_{z^2} orbital to ψ_1–hybrid orbital (**mainly concentrated in the xy-plane**) reduces the electron density from the $+z$ and $-z$ directions (*i.e.* directions of ligand approach).

Step 2: Hybridisation between ψ_2 and p_z

$$\psi_2 \pm \psi(p_z) \longrightarrow \psi_3 + \psi_4$$

The ψ_3 and ψ_4 are of the same energy and they are projected at 180° to each other as in the case of simple sp–hybridisation. **These ψ_3 and ψ_4 hybrid orbitals are vacant to accept the lone pairs from the approaching ligands.**

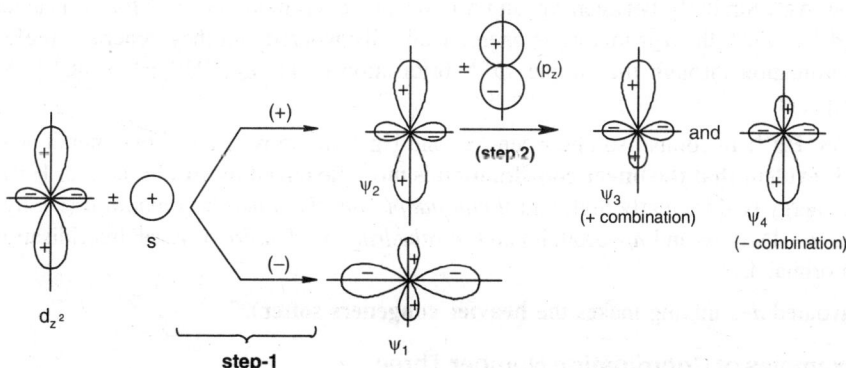

Fig. 2.11.4.1 Hybridisation of $(n-1)d_{z^2}$, ns and np orbitals to produce ψ_1, ψ_3 and ψ_4 hybrid orbitals housing the d_{z^2}– electrons in ψ_1. (Step 1 removes the electron density from the z–direction to ψ_1).

The above Orgel-scheme generates the ψ_3 and ψ_4 orbitals and these are projected in the regions that have been evacuated of electron density at the first step of hybridisation. This favours the metal-ligand linkage formation. If the simple sp-hybridisation is considered (without the participation of $(n-1)d_{z^2}$) for the $(n-1)d^{10}ns^0\, np^0$ electronic configuration, then the electron density in d_{z^2} will repel the approach of the ligands from the $+z$ and $-z$ directions to destabilise the metal-ligand linkages.

(ii) **MOT model:** The above mentioned stepwise hybridisation among the $(n-1)d_{z^2}$, ns and np_z orbitals has been considered just for the sake of simplicity. **In terms of MOT**, combination of the $(n-1)d_{z^2}$, ns and np_z orbitals of the metal center and two ligand orbitals will produce 5 MOs:

two bonding, one nonbonding and two antibonding. In these MOs, 6-electrons (2 from d_{z^2} and 4 from two ligands) are placed in the 2 BMOs and 1 NBMO keeping the 2 ABMOs vacant.

(iii) **Energetics:** The ease of linear coordination for the $(n-1)d^{10} ns^0 np^0$ electronic configuration depends on the energy difference between the $(n-1)d$ orbitals and ns and np orbitals. If the energy of the $(n-1)d$ orbitals is close to that of the ns and np orbitals then participation of the d_{z^2} orbital will remove the electron density from the $+z$ and $-z$ directions to favour the approach of the ligands. Thus from the concept of **relativistic effect and expansion of d-orbitals** (*see* Vol. 1 of Fundamental Concepts of Inorganic Chemistry), the expected sequences for preference towards the linear coordination are:

$$Au^+ > Ag^+ > Cu^+; Hg^{2+} \rangle\rangle Cd^{2+} \rangle Zn^{2+}.$$

● **Hg²⁺ vs. Zn²⁺, Cd²⁺:** Among Hg^{2+}, Cd^{2+} and Zn^{2+}, the heaviest congener Hg^{2+} shows a marked tendency to the form linear coordination while the lighter congeners Cd^{2+} and Zn^{2+} prefer the tetrahedral coordination number. It is due to the fact that in Hg^{2+}, the energy difference between the $(n-1)d$ and ns orbital is minimum. It is evident from the excitation energy (given below in kJ mol^{-1}) for the process $(n-1)d^{10} ns^0 \rightarrow (n-1)d^9 ns^1$:

Ag⁺	Au⁺	Hg²⁺	Cd²⁺	Zn²⁺
462	183	512	966	937

It is evident that in Cd(II) or Zn(II), the excitation energy is much higher than that required for Hg^{2+}. Thus, it is reasonable to expect that for Hg^{2+}, participation of the $(n-1)d_{z^2}$ orbital in hybridisation with the ns and np_z orbitals is more favourable compared to that for its lighter congeners (*i.e.* Zn^{2+}, Cd^{2+}).

● **Au⁺ vs. Ag⁺:** Similarly, between Ag^+ and Au^+, the process is more favoured for the heavier congener Au^+. In Cd^{2+} or Zn^{2+}, the d–s mixing is energetically disfavoured and they generally prefer the tetrahedral coordination through the simple sp^3 hybridisation as in $[Zn(OH_2)_4]^{2+}$, $[ZnCl_4]^{2-}$, $[CdBr_4]^{2-}$, $[Cd(NH_3)_4]^{2+}$, etc.

● **Au⁺ vs. Hg²⁺:** In comparison between Au^+ and Hg^{2+}, in terms of the above mentioned excitation energy, **it is evident that the linear coordination is more favoured in Au⁺.** In fact, for Hg(II), the said excitation energy is sufficiently high and *it can adopt both the linear coordination (i.e.* hybridisation among the $(n-1)d_{z^2}$, ns and np_z orbitals) and *tetrahedral coordination (i.e.* sp^3 hybridisation from the ns and np orbitals).

(**Note:** Favoured d–s mixing makes the **heavier congeners softer**).

2.11.5 Examples of Coordination Number Three

There are only few examples showing the C.N. 3. As in the case of C.N. 2, it is very often shown by the metal centers characterised by the d^{10} electronic configurations. Some common examples are: $[HgI_3]^-$, $[CuCl(tu)_2]$, $[Cu(tu)_3]^+$, $[Cu(SPMe_3)_3]^+$, $[Au(PPh_3)_3]^+$, $[AuCl(PPh_3)_2]$, $[Pt(PPh_3)_3]$ etc.

C.N. 3 for the metal centers other than d¹⁰ configuration

By using the **bulky anionic ligands** like disilylamido $N(SiMe_3)_2^-$, many trivalent metal ions (M^{3+}) of first-transition series (*e.g.* Fe^{3+}, Co^{3+}, Cr^{3+}, etc.) *can be forced to adopt C.N. 3.* The examples are: $[M\{N(SiMe_3)_2\}_3]$, (M = Ti, V, Cr, or Fe). Some divalent cations can produce the complexes with C.N. 3, *e.g.* $[M\{N(SiMe_3)_2\}_2PPh_3]$ (M = Co, Ni). All these trigonal planar complexes are distorted (*cf.* Fig. 2.11.5.1).

The structure of *tris*(trimethylphosphanesulfidio)copper(I) is given in Fig. 2.11.5.1

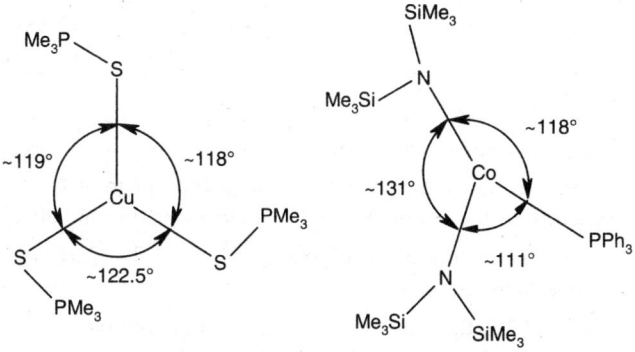

Fig. 2.11.5.1 Strucutres of $[Cu(SPMe_3)_3]^+$ (*cf.* :$\ddot{S}$ ← PMe_3) and $[Co\{N(SiMe_3)_2\}_2\,PPh_3]$.

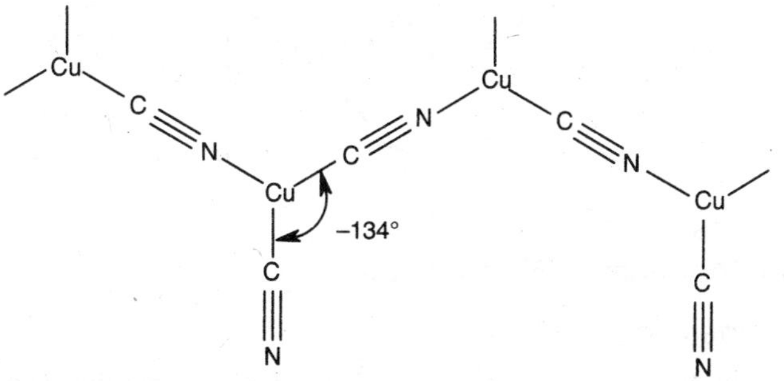

Fig. 2.11.5.2 Polymeric structure of $[Cu(CN)_2]^-$ with the counter ion K^+ or Bu_4N^+.

In **K[Cu(CN)₂]** and **[Bu₄N][Cu(CN)₂]**, the anion forms a polymeric zig-zag chain in which Cu(I) is maintaining C.N. 3 (2C and 1N) (Fig. 2.11.5.2) as follows:

$$—CN—Cu(CN)—CN—Cu(CN)—CN—Cu(CN)—CN—$$

Thus, in the polymeric anion, there are two types of CN⁻ ligands—bridging and nonbridging. Here it may be mentioned that $[Ag(CN)_2]^-$ and $[Au(CN)_2]^-$ maintain the discrete anionic structure with C.N. 2. *Thus it again supports that compared to Cu(I), the heavier congeners, i.e.* **Ag(I) and Au(I)** *are more willing to maintain the linear coordination geometry.* This is also expected in terms of the feasibility of *d–s* mixing (*cf.* relativistic effect, and Sec. 2.11.4) required for their linear coordination.

● **Bonding Scheme in Complexes with C.N. 3:** All such complexes are trigonal planar with a significant extent of distortion (*cf.* Fig. 2.11.5.1). *The bond angles are not equal even for the identical ligands.* It indicates that the simple sp^2–hybridisation (very often found in nontransition elements) of the metal center fails to explain the observation. It is expected that as in the linear geometry, some *d*–orbitals may participate for maintaining the C.N. 3. **It may be sd^2 hybridisation (a trigonal hybridisation).** It has been already mentioned that C.N. 3 is mainly shown by the d^{10} metal centers (*e.g.* Au¹) having low promotional energy for $(n-1)d \rightarrow ns$, i.e. *d–s* mixing is energetically favoured. *This again supports that for giving the C.N. 3, simple sp^2 hybridisation does not operate.* It needs participation of the *d*–orbitals in hybridisation. The possible hybridisation is sd^2.

Here it may be mentioned that for the d^{10} configuration, there is no question of *cfse*. By using the *bulky anionic* disilylamido ligand $N(SiMe_3)_2^-$, the tri-coordinate *binary complexes* of M^{3+} (M = T, V, Cr, Fe) and *mixed ligand complexes* of M^{2+} (M = Co, Ni) produced remain in a *spin-free state*. *cfse* values for all the possible geometries, *i.e.* pyramidal, trigonal planar and T-shaped (*cf.* See. 2.11.2) are more or less comparable. Thus in such cases also, the *cfse* is not a dominant factor and consequently, the steric *factor and VSEPR are the predominant ones to determine the geometry for C.N. 3*. This is why, the observed *trigonal planar geometry* (with some distortions) is not at all surprising.

Note: ● Here it may be noted that many complexes with the apparent C.N. 3 as in $[CuI(PR_3)_2]$, $[Cu(bpy)I]$, etc. exist in reality as the halido-bridged complexes having the effective C.N. 4.

● In $Cs[CuCl_3]$, the anion remains in a polymeric chain $—Cl—Cu(Cl)_2—Cl—Cu(Cl)_2—Cl—$ (infinite single chain) to give the C.N. 4.

● In $K[CuCl_3]$ or $NH_4[CdCl_3]$, the anion exists as the infinite double chains to give the effective C.N. 6 (*i.e.* distorted octahedron).

● In $K_2[Ni(CN)_3]$, the anion exists as a dimeric anion $[Ni_2(CN)_6]^{4-}$ to give the C.N. 4.

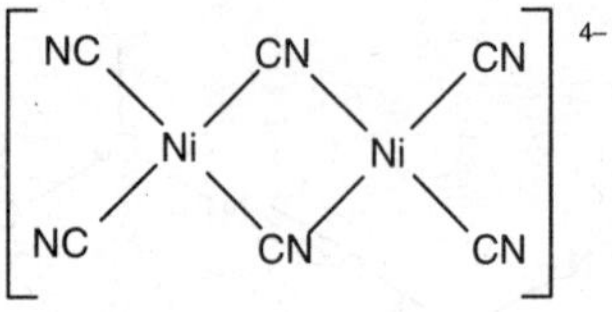

2.11.6 Examples of Coordination Number Four

A. *Possible geometries and polytopism*

The examples of four-coordinate complexes are abundant. The four-coordinate complexes can have the following geometries which are interconvertible (*cf.* Fig. 2.11.2.2).

tetrahedral (T_d) ⇌ squashed tetrahedral or buckled ⇌ square planar (D_{4h})
planar (D_{2d})

If the tetrahedral and square planar geometries are considered to describe the **two limiting situations** then their **intermediate geometry** possesses the D_{2d} symmetry (*cf.* Fig. 2.11.2.2).

The tetrahedral and square planar geometries are the examples of **polytopal isomerism or polytopism.** *In polytopism, the ligands are placed around the metal center in different ways to describe the different coordination polyhedra maintaining the same composition of the complexes (cf.* **difference from fluxionality**).

In the tetrahedral geometry, the steric hindrance is minimum (*i.e.* bond angle ~109°) while it is maximum in the square planar geometry (*i.e.* bond angle 90°). On the other hand, the crystal field stabilisation energy (cfse) for the low spin square planar complexes is much higher compared to that in the tetrahedral complexes which are generally the high-spin complexes. **Thus the strong field ligands favour the square planar geometry while the weak field, and bulky anionic ligands favour the tetrahedral geometry.** The bulky and anionic ligands produce more ligand-ligand repulsion to favour the tetrahedral structure having the higher bond angles. Of course, the d^n-configuration of the metal center plays a crucial role to determine the gain of cfse in moving from the T_d to D_{4h} symmetry. The common examples showing the T_d, D_{2d} and D_{2h} symmetries are given in Table 2.11.6.2.

The three geometries, *i.e.* octahedral (O_h), tetrahedral (T_d) and square planar (D_{4h}) are very much common in coordination compounds and *cfse* in these complexes are compared in Table 2.11.6.1 and the important conclusions are given there.

Table 2.11.6.1 Crystal field stabilisation energy (*cfse*) in the unit of $Dq_{(oct)}$ for the octahedral (O_h), tetrahedral (T_d) and square planar (D_{4h}) complexes (*cf.* crystal field splitting of the *d*-orbitas).

(a) Energy (in the unit of $Dq_{(oct)}$) of different *d*-orbitals in different ligand field symmetries

Geometry	C.N.	$d_{x^2-y^2}$	d_{z^2}	d_{xy}	d_{xz}	d_{yz}
Octahedral	6	+6.0	+6.0	−4.0	−4.0	−4.0
Tetrahedral	4	−2.67	−2.67	+1.78	+1.78	+1.78
Square planar (bonds in the *xy* plane)	4	+12.28	−4.28	+2.28	−5.14	−5.14

(b) *cfse* for different d^n–configurations in weak and strong field (WF and SF) ligands (*cf.* Table 3.14.12.1)

d^n	Ions (examples)	cfse (in $Dq_{(oct)}$) (without considering the Jahn-Teller distortion)					
		Octahedral		Tetrahedral		Square planar	
		WF(h.s)	SF(l.s)	WF(h.s)	SF(l.s)	WF(h.s)	SF(l.s)
d^0	Ca^{2+}, Al^{3+}, Ti^{4+} Cr^{6+}, Mn^{7+}	0	0	0	0	0	0
d^1	Ti^{3+}, V^{4+}	4	4	2.67	2.67	5.14	5.14
d^2	Ti^{2+}, V^{3+}, Mo^{4+}	8	8	5.34	5.34	10.28	10.28
d^3	V^{2+}, Cr^{3+}	12	12	3.56	8.01(1)	14.56	15.42(1)
d^4	Cr^{2+}, Mn^{3+}	6	16(1)	1.78	10.68(2)	12.28	19.7(2)
d^5	Mn^{2+}, Fe^{3+}, Ru^{3+}	0	20(2)	0	8.9(2)	0	24.84(2)
d^6	Fe^{2+}, Co^{3+}, Ru^{2+} Pt^{4+}	4	24(2)	2.67	7.12(1)	5.14	29.12(2)
d^7	Co^{2+}, Ni^{3+}	8	18(1)	5.34	5.34	10.28	26.84(1)
d^8	Ni^{2+}, Cu^{3+}, Pd^{2+}, Pt^{2+}, Au^{3+}	12	12	3.56	3.56	14.56	24.56(1)
d^9	Cu^{2+}, Ag^{2+}	6	6	1.78	1.78	12.28	12.28
d^{10}	Ni^0, Cu^+ Ag^+, Zn^{2+}, Cd^{2+}, Hg^{2+}	0	0	0	0	0	0

● The figures in the parentheses (*cf.* Table 2.11.6.1) indicate the additional number of electrons that must be paired in moving from the weak field (*i.e.* high-spin condition) to strong field (*i.e.* low spin condition) ligand environment. In calculating the net *cfse*, nP must be subtracted from the shown values where n denotes the additional number (with respect to those in free ion) of electron pairs paired and P–denotes the electron pairing energy.

● High-spin square planar and low-spin tetrahedral complexes are rarely found.

Important conclusions from Table 2.11.6.1

(i) **d^0 and d^{10} systems** having no *cfse* in all cases, prefer the **tetrahedral geometry** based on the steric considerations. *e.g.* $[AlCl_4]^-$, $[VOCl_3]$, $[MnO_4]^-$, $[CrO_4]^{2-}$, $[Ni(CO)_4]$, $[MX_4]^{2-}$, $[M(NH_3)_4]^{2+}$, $[M(CN)_4]^{2-}$ (M = Zn, Cd, Hg and X = halogen), $[Cu(CN)_4]^{3-}$ etc.

(ii) **d^5 system for the weak field ligand** (*i.e.* spin-free state) having no *cfse* prefers the **tetrahedral geometry** because of the steric grounds, *e.g.* $[FeCl_4]^-$, $[MnCl_4]^{2-}$, etc.

(iii) **d^7, d^8 and d^9 systems for the strong field ligands** (*i.e.* low spin state) prefer the **square planar geometry** because of the relatively high gain of *cfse* in the square planar geometry, *e.g.* $[Ni(CN)_4]^{2-}$, $[Pt(NH_3)_4]^{2+}$, compared to the other possible geometries.

(iv) **d^{1-6} systems with the strong field ligands** (*i.e.* low spin state) prefer the **octahedral geometries**.

(v) **d^1, d^2, d^6 and d^7 systems with the weak field and bulky ligands** prefer the **tetrahedral geometry** because in such cases, the loss of *cfse* in attaining the tetrahedral structure compared to the other possible geometry like octahedral is not severe (*cf.* Dq value for the weak field ligands is very low). Examples are: $[VCl_4]$ (d^1), $[VCl_4]^-$ (d^2), $[FeCl_4]^{2-}$ (d^6), $[CoCl_4]^{2-}$ (d^7), $[Co(NCS)_4]^{2-}$ (d^7)

(vi) **d^3, d^4, d^8 and d^9 systems with the moderately strong field ligands** prefer the **octahedral geometry**. In such cases, the ligand field strength is not sufficiently high to induce the square planar geometry. However, for the d^9–system, the z–out distortion (Jahn-Teller distortion) can lead to the square planar like geometry.

(vii) **d^7 *vs.* $d^{8,\,9}$: Co^{2+} (d^7) can readily form the tetrahedral complex** while under the identical ligand field environment, Ni^{2+} (d^8) and Cu^{2+} (d^9) do not form the *truly tetrahedral complexes*.

● **$[CuCl_4]^{2-}$ (steric factor, cfse and crystal force):** In moving from the tetrahedral (T_d) to square planar (D_{4h}) geometry through the D_{2d} symmetry (*i.e. flattened tetrahedron*), the steric crowding increases but the gain of *cfse* increases. The relative importance of these two opposing factors depends on the bulkiness of the ligands and ligand field strength. *Exact calculations for $[CuCl_4]^{2-}$ indicate that the D_{2d} symmetry is more stable than the T_d symmetry by about 2 kcal mol^{-1} while the T_d symmetry is more stable than the D_{4h} symmetry by about 18 kcal mol^{-1}.* Thus for $[CuCl_4]^{2-}$, the D_{2d} symmetry is most

Table 2.11.6.2 Representative examples showing the T_d, D_{2d} and D_{4h} symmetries.

Complex	d^n	Structure	Complex	d^n	Structure
$[Ni(CO)_4]$	d^{10}	T_d	$[Cu(NH_3)_4]^{2+}$	d^9	D_{4h}
$[MX_4]^{2-}$	d^{10}	T_d	$[Ni(CN)_4]^{2-}$	d^8	D_{4h}
(M = Zn, Cd, Hg;			$[PtX_4]^{2-}$	d^8	D_{4h}
X = halogen or CN)			(X = halogen, CN)		
$[FeCl_4]^-$	d^5	T_d			
$[MnCl_4]^{2-}$	d^5	T_d	$[Pt(NH_3)_4]^{2+}$	d^8	D_{4h}
$[TiCl_4]$	d^0	T_d	$[CuCl_4]^{2-}$	d^9	D_{4h}
$[GaCl_4]^-$, $[AlCl_4]^-$	d^0	T_d	(with NH$_4^+$ counter ion)		
$[CoCl_4]^{2-}$	d^7	T_d	$[CuCl_4]^{2-}$	d^9	D_{2d}
$[MO_4]^{x-}$	d^0	T_d	(with Cs$^+$ counter ion)		
(M = V^V, Cr^{VI}, Mn^{VII})					
			$[NiCl_4]^{2-}$	d^8	D_{2d}

favoured, but the preference between the D_{2d} and T_d structure is marginal (*ca.* 2 kcal mol^{-1}). This is why, the actual structure of $[CuCl_4]^{2-}$ in crystal depends on the nature of counter ion and **mode of crystal packing** (*cf.* Table 2.11.6.2). In fact, $[CuCl_4]^{2-}$ (D_{2d}) adopts the D_{4h} symmetry under high pressure. In this regard, **Jahn-Teller distortion** in $[CuCl_4]^{2-}$ is also to be considered to describe its structure. It favours the **flattened tetrahedral structure** (*cf.* Table 3.11.3.3, and Sees. 3.11.3-4).

B. Tetrahedral (T_d) and distorted tetrahedral (D_{2d}) complexes (*cf.* Sec. 3.14.12)

Generally the *neutral and anionic* complexes are known to possess the T_d and D_{2d} structures. The T_d and D_{2d} structures are favoured under the following conditions:

(i) **weak field and bulky ligands,** *i.e.* low *cfse* producing and imparting more steric crowding;

(ii) **bulky and anionic ligands** favour the tetrahedral geometry over the octahedral and square planar geometries to minimise the ligand-ligand repulsion.

(iii) the **non-transition metal ions** (*i.e.* d^0) earning no *cfse* (e.g. Al^{3+}, Be^{2+}, B^{3+}, Ga^{3+}, etc.);

Low spin tetrahedral complexes

It is well known that because of the $10Dq_{tet} \ll P$ condition, the tetrahedral complexes are generally the **high-spin complexes**. However, the **low-spin tetrahedral complexes** have been reported in some rare cases, *e.g.*

(i) **[Cr(NO){N(SiMe$_3$)$_2$}$_3$]**
   (distorted tetrahedral of C$_{3v}$ symmetry).

(a complex of Cr^{II} having the d^4 configuration, *i.e.* $e^4 t_2^0$; this leads to the maximum *cfse*.) The ligands are: N(SiMe$_3$)$_2^-$ (disilylamido ligand, bulky and anionic), NO$^+$

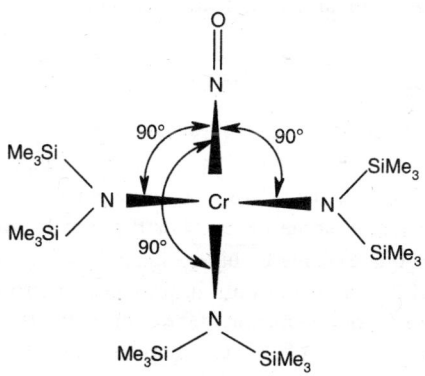

(C$_{3v}$ symmetry *i.e.* distorted tetrahedral)

Fig. 2.11.6.1 Distorted tetrahedral structure of [Cr(NO){N(SiMe$_3$)$_2$}$_3$] with the **low spin configuration.**

(ii) **tetrakis(1–norbornyl)cobalt,** (a complex of Co^{IV} having the d^5 configuration, *i.e.* $e^4 t_2^1$). 1-norbornyl anion is:

$$\overset{-}{:}C \begin{array}{c} CH_2 \!-\! CH_2 \\ \diagup \qquad\qquad \diagdown \\ CH_2 \!-\!\!-\!\!-\! CH \\ \diagdown \qquad\qquad \diagup \\ CH_2 \!-\! CH_2 \end{array} \qquad \textbf{(bulky and anionic ligand)}$$

(iv) d^0 and d^{10} transition metal ions earning no cfse;

(v) d^1, d^2, d^5, d^6, d^7 systems of *first transition series* in presence of the *weak field ligands, i.e.* loss of cfse in attaining the tetrahedral structure instead of the octahedral structure is zero (for d^5) or marginally small.

(vi) Tetrahedral complexes are generally the **high-spin complexes** because of the condition $10 \, Dq_{tet} \ll P$ (pairing energy). For the d^2 and d^7 systems, the *cfse* of the tetrahedral complexes is fairly high.

Thus the tetrahedral structure is preferred when the loss of cfse in attaining the tetrahedral structure (*cf.* OSSE) is either zero or very small and the ligands are bulky and anionic (*i.e.* steric crowding).

● For the **first transition series**, the metal ions of the Groups 8 – 10 (*i.e.* VIII) very often form the tetrahedral complexes in presence of the **weak field ligands**. Among these metal ions, the *tetrahedral complexes of Co(II) are well documented while* **Ni(II) is relatively more reluctant in forming the tetrahedral complexes.**

● For the d^4 (*e.g.* Cr^{2+}) and d^9 (*e.g.* Cu^{2+}) systems, the tetrahedral geometry is disfavoured. They prefer the distorted octahedral geometry (J.T. effect) even in presence of the weak field ligands.

● For the **heavier transition metal ions** of 4d– and 5d–series, even for the weak field ligands, the *cfse* is fairly high (because of the higher splitting power of the heavier congeners). Consequently, the loss of *cfse* in attaining the tetrahedral structure compared to the other possible structures like octahedral or square planar which are in low-spin state is very high. *This is why, in such cases, the tetrahedral structure is disfavoured.*

● **Examples** of complexes having the T_d and D_{2d} structures are given in Table 2.11.6.2.

● **Optical isomerism** in the tetrahedral complexes has been discussed in Sec. 2.8 (*cf.* Figs. 2.8.1, 2.8.2) with suitable examples.

C. Square planar complexes

It is favoured under the following conditions: (i) **high *cfse* generating metal centers**; (ii) **non bulky strong field ligands with the π–bonding properties** (*i.e.*, π-acid ligands).

● **d^8–system:** Square planar geometry is well documented for the d^8–systems like Rh(I), Ir(I), Pd(II), Au(III), Cu(III) and Ni(II) (in some cases). For the metal ions of 4d– and 5d– series of d^8 configuration, formation of the square planar complexes is highly favoured. For the d^8 configuration of first transition series as in Ni(II), it can occur only in presence of strong field ligands.

● **d^9 and low-spin d^7 system:** z-out distortion (Jahn-Teller effect), in the d^9 system can lead to the square planar geometry. Low-spin d^7 (*i.e.* $t_{2g}^6 e_g^1$) system experiencing the z-out distortion can also produce the square planar complexes.

Examples of some square planar complexes (d^8–systems) of some special interest

cis–dammminedichloridoplatinum(II), **(anticancer drug)**	choloridotris(tri–phenylphosphane)-rhodium(I) **(Wilkinson's catalyst)**	trans–carbonylchloridobis-(tri–phenylphosphane)-iridium(I). **(Vaska's compound)**

The d^7, d^8 and d^9 configurations, in general, favour the square planar geometry with the strong field ligands allowing the spin pairing (*cf.* Table 2.11.6.2). This is due to the gain of *cfse* in the square planar geometry compared to other possible geometries.

● **Examples of some common square planar complexes:**

[Ni(CN)$_4$]$^{2-}$, [PtCl$_2$(NH$_3$)$_2$], [Pt(NH$_3$)$_4$]$^{2+}$, [M(bigH)$_2$] (M = Cu, Ni), [Ni(Hdmg)$_2$], [AuCl$_4$]$^-$, etc.

● **Polytopal isomerism** between the square planar and tetrahedral geometry has been discussed separately in the Sections 2.2.4, 2.3. The isomerism in the square planar complexes has been discussed in detail in Secs. 2.5.1 and 2.8 (*cf.* Figs. 2.8.3 – 4)

D. Ligand imposed square planar geometry

● **Planar macrocyclic ligand** (*cf. Sec. 1.8*): Planar fused-ring systems, *e.g.* phthalocyanine (mimiking the porphyrin ligand) produce the planar complexes with different metal centers like Mg(II) (d^0), Be(II) (d^0), Mn(II) (d^5), Fe(II) (d^6), Co(II) (d^7), Ni(II) (d^8), Cu(II) (d^9), Pt(II) (d^8), etc. It is evident that some of these metal ions (*e.g.* Be^{II}, Mg^{II}, Mn^{II}, Co^{II}, etc.) normally favour the tetrahedral geometry rather than the planar geometry. In fact, *the planar fused ring systems enforce the metal center to adopt the square planar geometry.* Mg(II) complex of chlorophyll is a typical example occurring in nature.

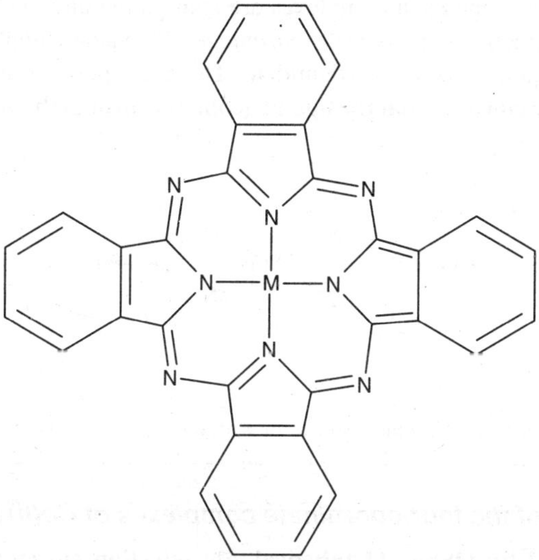

Fig. 2.11.6.2 Square planar complex of M(II) with phthalocyanine, a case of forced configuration.

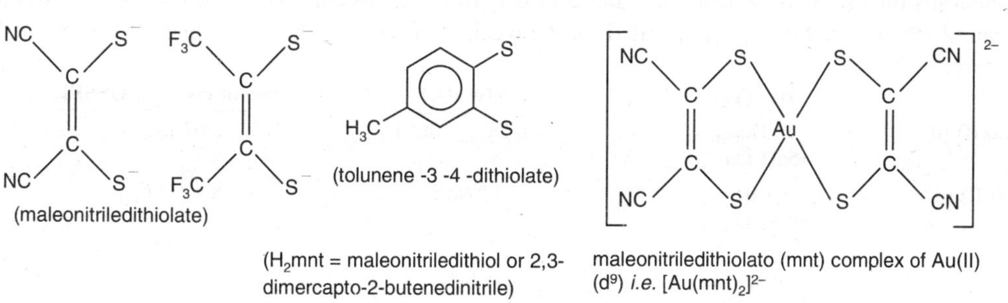

(maleonitriledithiolate)

(tolunene -3 -4 -dithiolate)

maleonitriledithiolato (mnt) complex of Au(II) (d^9) *i.e.* [Au(mnt)$_2$]$^{2-}$

(H$_2$mnt = maleonitriledithiol or 2,3-dimercapto-2-butenedinitrile)

Fig. 2.11.6.3 Some representative 1, 2–dithiolates favouring the square planar complexes.

● **1,2-Dithiolate ligands:** The 1, 2–dithiolates (Fig. 2.11.6.3) favour the square planar geometry for many metal ions like Fe(II) (d^6), Co(II) (d^7), Au(II), (d^9), etc. with respect to other possible geometries like tetrahedron or octahedron. It is suggested (H.B. Gray) that in **such square planar complexes, the p_z orbital of the metal center is fully utilised in π–bonding with the ligand.** *This favoured pi–bonding interaction favours the square planar-geometry.*

● **trien *vs.* tren (*cf.* Sec. 2.2.4):** The tetradentate ligand *tris*(2–aminoethyl)amine, *i.e.* $N(CH_2CH_2NH_2)$ (*tren*) is ideal for giving the tetrahedral geometry but not for providing the square planar structure. The closely related ligand triethylenetetraamine, *i.e.* $H_2N(CH_2)_2NH(CH_2)_2NH(CH_2)_2NH_2$ (*trien*) can form the square planar complex. **The ligand *tren* is more basic than *trien* and thus *tren* is expected to form the more stable complexes than *trien*.** It is found true for Zn(II) (d^{10}) which prefers the tetrahedral geometry than the square planar geometry. On the other hand, for Cu(II) (d^9) which prefers the planar geometry than the tetrahedral geometry, *trien* is found to form the more stable complex in spite of its lower basic strength (*cf.* Sec. 2.2.4).

High-spin square planar complex of Ni(II)

It is well known that for Ni(II) (a d^8–system), only the strong field ligands like CN^- can produce the square planar geometry and such complexes are spin-paired and diamagnetic. But there is a typical example illustrating the high-spin (*i.e.* paramagnetic) planar complex of Ni(II). The ligand is P, P-di-tert-butylphosphinic-N-isopropylamidate. The corresponding Ni(II)-complex is *trans*-planar. **It is reasonable to consider that the ligand geometry favours the square planar geometry.**

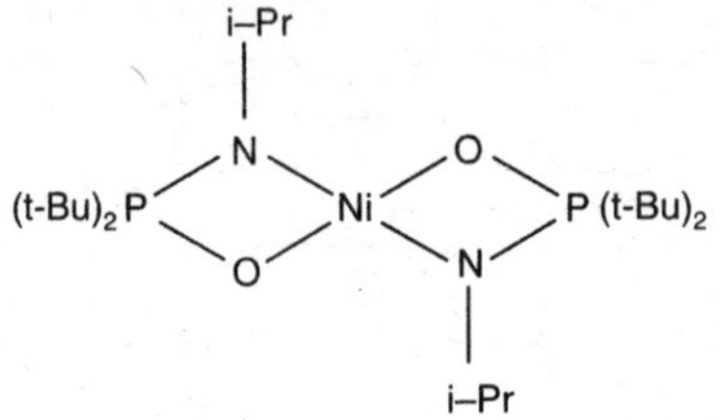

Fig. 2.11.6.4 Structure of a typical square planar complex of Ni(II) with high-spin state.

E. Structural features of the four-coordinate complexes of Co(II) and Ni(II)

(i) **Loss of *cfse* measured by OSSE (Octahedral site selection energy) (Ni^{II} vs. Co^{II}) (cf. Secs. 3.14.12-13):** Co(II) can easily form the tetrahedral structure while Ni(II) is more reluctant to have the tetrahedral structure. It is evident (*cf.* Table 2.11.6.1) from the loss of *cfse* in attaining the T_d structure instead of the O_h structure. It is illustrated for the weak field ligands which are ideal for the tetrahedral structure.

	cfse (T_d)	cfse (O_h)	loss of cfse (*i.e.* OSSE)
Co(II) (d^7):	12 $Dq_{(Tet)}$ ≈5.33 $Dq_{(oct)}$	8 $Dq_{(oct)}$, (h.s.)	2.67 $Dq_{(oct)}$
Ni(II) (d^8):	8 $Dq_{(Tet)}$ ≈3.55 $Dq_{(oct)}$	12 $Dq_{(oct)}$	8.45 $Dq_{(oct)}$

$$\left[\text{Taking } Dq_{(Tet)} \approx \frac{4}{9} Dq_{(oct)} \right].$$

Thus in terms of the loss of the *cfse*, for Co(II), the tetrahedral structure is more favoured. *In other words, because of the steric relaxation, the high-spin tetrahedral complexes are more preferred than the high-spin octahedral complexes for Co(II)*. **For Ni(II), the loss of *cfse* is more and Ni(II) prefers the octahedral geometry rather than the tetrahedral geometry even in presence of the weak field ligands.** In fact, in the case of Ni(II), *only the bulky and anionic weak field ligands* like Cl^- can produce the distorted tetrahedral structure (*i.e.* D_{2d}) under certain conditions.

● **Reluctance of Ni(II) to form the T_d structure:** In fact, Co(II) easily forms the tetrahedral complexes *with the halides and pseudohalides, e.g.* $[CoCl_4]^{2-}$, $[CoBr_4]^{2-}$, etc. But Ni(II) is reluctant to form such tetrahedral complexes. It is illustrated in the following reactions.

$$\underset{\substack{(\text{aq. solution}) \\ \text{light pink}}}{Co(II)} \xrightarrow{\text{Conc. HCl}} \underbrace{\left[Co(H_2O)Cl_3\right]^- \text{ and } \left[CoCl_4\right]^{2-}}_{(\text{deep blue})}$$

$$\underset{(\text{aq. solution})}{Ni(II)} \xrightarrow{\text{Conc. HCl}} \left[Ni(H_2O)_{6-x}Cl_x\right]^{(2-x)+} \quad (x = 1, 2 \text{ even in 12M HCl})$$

$[NiCl_4]^{2-}$ (distorted tetrahedral structure) can be generated in ***nonaqueous and noncoordinating solvents and isolated with large cations like $AsPh_4^+$ or $N(C_2H_5)_4^+$***.

● **β-diketonate complexes of Co(II) and Ni(II)** (*cf.* **Sec. 8.26.2**): These are quite interesting to compare. **[Co(acac)₂]**, *i.e.* bis(acetylacetonato)cobalt(II) can exist both as a monomer and as a tetramer. The monomer is *tetrahedral* while in the tetramer (*cf.* Fig. 2.11.6.5), the Co(II) centers attain the distorted octahedral geometry. If the methyl groups are replaced by the bulky groups, then the corresponding Co(II) complex is stabilised as a ***monomeric one having the tetrahedral structure.***

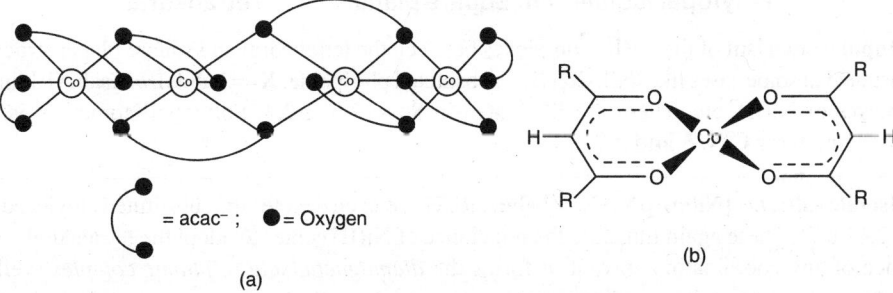

= acac⁻ ; ● = Oxygen

(a)

(b)

Fig. 2.11.6.5 (a) Schematic representation of the tetrameric structure of [Co(acac)₂], *i.e.* bis(acetylacetonato)cobalt(II) indicating the octahedral coordination of Co(II) in the tetramer. (b) **Tetrahedral** monomeric complex of Co(II) when the Me-groups of acac are replaced by the bulky R-groups.

[Ni(acac)₂] is stabilised in a trimeric form in which each Ni(II) center is having the octahedral symmetry (*cf.* Fig. 2.11.6.6). If the steric crowding on the β–diketonate ring is enhanced through substitution then the polymerisation is hindered. **In such cases, it adopts the monomeric square planar geometry rather than the tetrahedral structure.** The square planar geometry is diamagnetic. It has been noticed that when the β–diketonate ring is properly sterically hindered, the corresponding Ni(II)–complex in poor coordinating solvents shows *an anomalous magnetic property due to the following equilibrium.*

$$\text{square planar}\left(\textbf{diamagnetic}\right)\left(\textbf{monomer}\right) \rightleftharpoons \text{octahedral}\left(\textbf{paramagnetic}\right)\left(\textbf{polymer}\right)$$

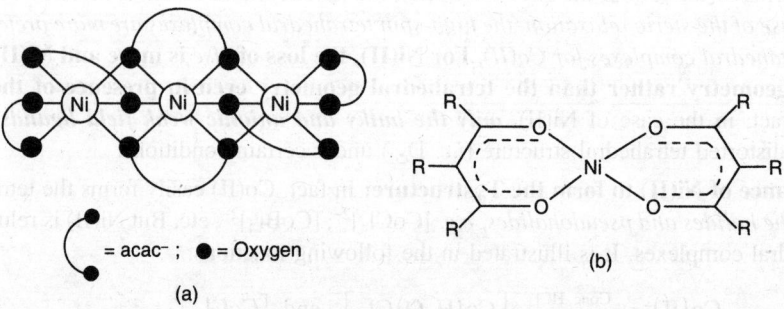

$\mathbf{C}$ = acac⁻ ; ● = Oxygen

(a)   (b)

Fig. 2.11.6.6 (a) Schematic representation of the trimeric structure of [Ni(acac)₂], *i.e.* bis(acetylacetonato)-nickel(II) indicating the octahedral coordination of Ni(II) in the trimer. (b) *Monomeric square planar complex* of Ni(II) with the sterically hindered β-diketonate. (*R* = bulky alkyl groups, all *R*'s may not be the same.)

The trimeric structure of **[Ni(acac)₂]** maintaining the octahedral geometry of Ni(II) is quite stable in the noncoordinating solvents but in the coordinating solvents, it undergoes depolymerisation to produce the monomers of **octahedral structure like [Ni(acac)₂(S)₂]** where S stands for the solvent like H₂O, py, etc.

$$\{[Ni(acac)_2]\}_3 \text{ (polymer)} \xrightarrow{\text{coordinating solvent (S)}} [Ni(acac)_2(S)_2] \text{ (monomer)}$$

Thus the above examples of β–diketonate complexes indicate the willingness of Co(II) to form the tetrahedral structure and unwillingness of Ni(II) to adopt the tetrahedral structure.

Polytopal isomerism: Square planar ⇌ Tetrahedral

Polytopal isomerism of the Ni(II) complexes between the tetrahedral and square planar structures is illustrated in some cases like [NiL₂X₂] (L = substituted phosphine, X⁻ = halide), *bis*(salicylaldimine) complexes of Ni(II), etc. These are illustrated in Secs. 2.3, 2.2.4. Polytopal isomerism is very much common for **C.N. 5 and ≥ 7.**

● **Lifschitz salts,** *i.e.* **[Ni(en)₂]X₂** (X⁻ = halide, RCO₂⁻, etc; en denotes the substituted ethylenediamine) (*cf.* Fig. 2.11.6.7): These again illustrate the reluctance of Ni(II) center to adopt the tetrahedral structure. In absence of any coordinating solvent, it forms the *diamagnetic square planar complex* (yellow-red colours) but in presence of the coordinating solvents, it produces the *paramagnetic octahedral complexes* (blue-green colours) like [Ni(en)₂(S)₂]²⁺ (S = coordinating solvent). *Thus the following equilibrium can explain the anomalous magnetic property and colours of the Lifschitz salts.*

$$\left[Ni(en)_2\right]^{2+} \text{(square planar)} + 2S \rightleftharpoons \left[Ni(en)_2(S)_2\right]^{2+} \text{(octahedral)}$$
(diamagnetic, yellow-red)   **(paramagnetic, blue-green)**

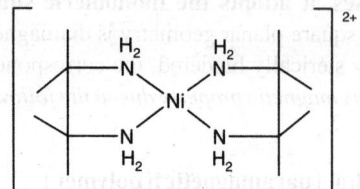

Fig. 2.11.6.7 Square planar structure of the cation in Lifschitz salts, having the substituted ethylenediammine as the ligands for Ni(II) in the cation.

F. Polytopal isomerism leading to square planar (D_{4h}) $\rightleftharpoons$ tetrahedral (T_d):

This aspect has been discussed in detail in Secs. 2.2.4 and 2.3.

G. Four coordinate complexes with C_{2v} symmetry (disphenoid or see-saw geometry)

It is noticed in the complexes of nontransition metals bearing a lone pair as in [:SbCl$_4$]$^-$, [:AsCl$_4$]$^-$, etc. According to VSEPR, in the trigonal bipyramidal geometry, the lone pair occupies one equatorial position and the ligands are placed at the remaining positions of the trigonal bipyramidal geometry (*cf.* Fig. 2.11.2.3). It gives the geometry described by **disphenoid or see-saw**.

2.11.7 Examples of Coordination Number Five (*cf.* Sec. 3.10)

The geometry of 5-coordinate complexes can be explained by considering the trigonal bipyramid (TBP) (D_{3h}) and square pyramid (SP) (C_{4v}) geometries. However, depending on the conditions, they may be distorted or regular.

In terms of number, the 5-coordinate complexes are next to the 4- and 6-coordinate complexes.

A. Characteristics of the trigonal bipyramidal (TBP) and square pyramidal (SP) geometries (*cf.* Sec. 3.10)

● **VSEPR, *cfse* and π-bonding:** In terms of *steric crowding and ligand-ligand repulsions, the trigonal bipyramidal geometry is favoured over the square pyramidal geometry.* This is why, the five coordinate compounds of the *s*– and *p*–block elements (*i.e.* nontransition elements) as the central elements are trigonal bipyramidal as in PF$_5$ (*cf.* VSEPR). Here, it must be mentioned that there is no lone pair on the central atom in PF$_5$. *On the other hand, for the 5-coordinate complexes of transition metals, the π-bonding (if possible) and cfse slightly favour the square pyramidal geometry in some cases over the trigonal bipyramidal geometry.*

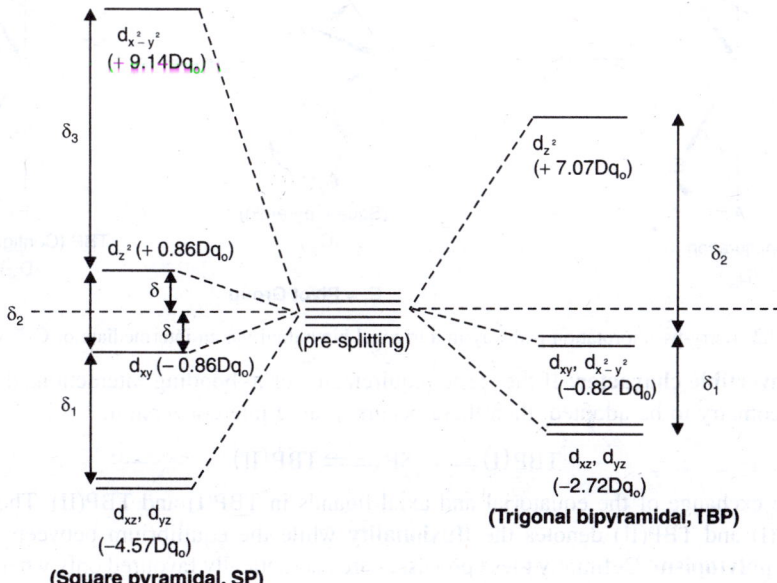

Fig. 2.11.7.1 Splitting of the *d*-orbitals in the square pyramidal (SP) and trigonal bipyramidal (TBP) geometries.

● **VSEPR vs cfse and π-bonding:** The VSEPR/steric factors favour the TBP geometry but the *cfse* and bonding interactions (specially π–bonding) involving the *d*–orbitals marginally favour the SP geometry over the TBP geometry.

● **cfse:** The *crystal field splitting* of *d*–orbitals in the said two geometries are shown in Fig. 2.11.7.1. Energy (in Dq_{oct}) of the *d*-orbitals for the said two geometries are given below.

	$d_{x^2-y^2}$	d_{z^2}	d_{xy}	$d_{xz} = d_{yz}$
Trigonal bipyramidal (D_{3h}):	−0.82	+7.07	−0.82	−2.72
Square pyramidal (C_{4v}):	+9.14	+0.86	−0.86	−4.57

(assuming the equatorial ligands in the *xy*–plane)

In terms of crystal field spiltting, both the geometries give a pair of *d*–orbitals of the lowest energy and one *d*-orbital of very high energy. The d^{5-8} configurations may be either high-spin or low spin. The low-spin complexes of d^{5-8} configuration keep the highest energy *d*–orbital vacant. **The d^{5-8} low spin complexes are favoured for the square pyramidal geometry over the trigonal bipyramidal geometry.** However, the relative gain of *cfse* in the square pyramidal complex is marginally small and the VSEPR/steric factors can compensate this in the trigonal bipyramidal geometry. *This allows the isolation of both isomers.*

It is evident that for d^0, d^5 (high-spin) and d^{10} systems, there is no *cfse* for both the geometries. In such cases, the steric factors/VSEPR favour the TBP geometry.

● **Berry-pseudorotation:** The said two geometries are interconvertible through *Berry-pseudorotation* (without bond breaking).

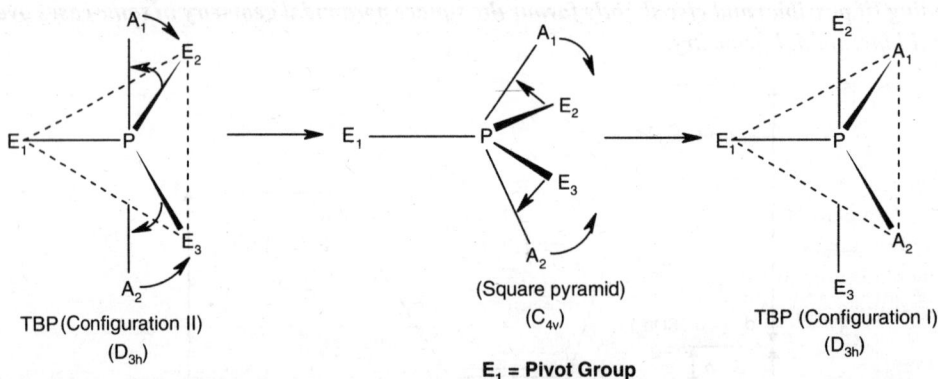

TBP (Configuration II)	(Square pyramid)	TBP (Configuration I)
(D_{3h})	(C_{4v})	(D_{3h})

E_1 = Pivot Group

Fig. 2.11.7.2 Berry-pseudorotation pathway in a trigonal bipyramid via an intermediate of C_{4v} symmetry.

● **Interconvertible character:** If the steric requirements or π–bonding interactions do not enforce a particular geometry to be adopted, then there occurs a rapid interconversion:

$$TBP\,(I) \rightleftharpoons SP \rightleftharpoons TBP\,(II)$$

It leads to the exchange of the equatorial and axial ligands in TBP(I) and TBP(II). The equilibrium between TBP(I) and TBP(II) denotes the **fluxionality** while the equilibrium between TBP and SP represents the **polytopism.** Definitely these processes are energetically favoured only when the energies of the TBP and SP are comparable. *This easy interconvertibility gives the **stereochemical nonrigidity,** i.e. polytopism and fluxional character in the 5-coordinate compounds of phosphorous and silicon.*

• **C.N. 6 vs. C.N. 5:** Thermodynamically, the 6-coordinate complexes are more stable than the 5-coordinate complexes and there is a competition between these preferences. *The 5-coordinate complexes are generally formed by the 1st transition series metals but the heavier congeners (i.e. 2nd and 3rd transition series) form much less number of the 5-coordinate complexes.* This occurs so because of the two reasons:

(i) the **loss of cfse** in attaining the 5–coordinate geometry instead of the 6-coordinate geometry is much higher for the heavier congeners because of their much higher crystal field splitting power;

(ii) the relatively larger size for the heavier congeners can accommodate 6 ligands better, *i.e.* steric crowding is less severe for the heavier congeners.

• **Disproportionation:** Because of the higher thermodynamic stabilities (by considering the electrostatic interactions only) of the 6 coordinate complexes over the 5-coordinate complexes, the 5–coordinate complexes may undergo *disproportionation* to produce the 4-and 6-coordinate complexes, *i.e.*

$$2[MX_5] \rightleftharpoons [MX_4] + [MX_6], \quad \text{(charges not shown for the sake of simplicity)}.$$

The above mentioned disproportionation process is very much favoured in terms of *pure electrostatic interactions only*. The covalent bonding interactions which are quite important in coordinate complexes can balance the situation to allow the isolation of the 5 coordinate complexes.

However, the tendency of above mentioned disproportionation is noticed in many cases. This is illustrated below.

The empirical formula of **[CoCl$_2$(dien)]** (dien = H_2N—$(CH_2)_2$—NH—$(CH_2)_2$—NH_2, a tridentate ligand) apparently suggests a 5-coordinate compound. But it is a salt [Co(dien)$_2$][CoCl$_4$] where the cation [Co(dien)$_2$]$^{2+}$ is octahedral and the anion [CoCl$_4$]$^{2-}$ is tetrahedral, *i.e.*

$$2[CoCl_2(dien)] \longrightarrow [Co(dien)_2][CoCl_4]$$

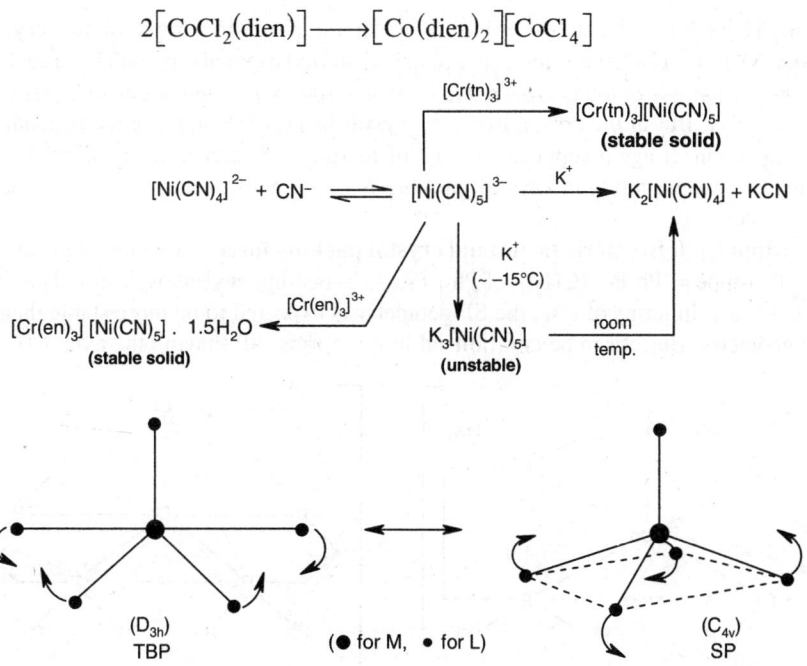

Fig. 2.11.7.3 Interconversion (*i.e.* polytopism) between the trigonal bipyramidal (TBP) and square pyramidal (SP) geometries.

● **Isomeric forms of comparable stabilities:** Isomerisation between the TBP and SP forms have been noticed in some cases. It has been already mentioned that very often, the energies of the two forms differ marginally and consequently they may be isolated in both possible forms. In fact, they are interconvertible with a low activation energy barrier (*cf.* Fig. 2.11.7.2).

(i) **[Ni(CN)₅]³⁻** (*cfse*, **steric factor- and crystal force**): This is illustrated for $[Ni(CN)_5]^{3-}$ for which both the geometries earn the comparable stabilities (*cf.* Sec. 3.10). For the isolation of this large anion in solid crystals, we need the large cations like $[Cr(en)_3]^{3+}$, $[Cr(tn)_3]^{3+}$, etc. K^+ as a counter ion fails to trap this large anion (*cf.* Fig. 2.11.7.3).

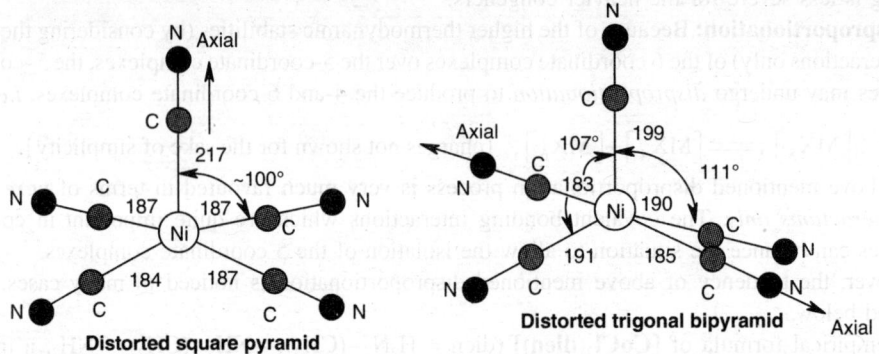

Distorted square pyramid   **Distorted trigonal bipyramid**

Fig. 2.11.7.4 Structural features of two isomeric forms of $[Ni(CN)_5]^{3-}$ found in $[Cr(en)_3][Ni(CN)_5] \cdot 1.5\ H_2O$. (Bond length in pm).

In **[Cr(tn)₃][Ni(CN)₅]**, the anion exists as a square pyramidal one. In the crystal hydrate, **[Cr(en)₃][Ni(CN)₅] · 1.5 H₂O**, the anion exists both in distorted trigonal bipymidal and square pyramidal geometries. *In solution, the anion is expected to exist as a square pyramidal one by considering the cfse and π–bonding effect.* But in the crystal hydrate, **crystal forces** may stabilise the trigonal bipyramidal geometry of the anion. It again supports the fact of marginal difference in stabilities of the said two forms and this is why, the **mode of crystal packing** is important to determine the geometry of the 5-coordinate complex.

(ii) **[CoCl(dppe)₂]⁺** (*cfse*, **steric factor and crystal packing force**): Another interesting complex is: $[CoCl(dppe)_2]^+$, (dppe = Ph_2P—$(CH_2)_2$—PPh_2, *i.e.* 1, 2–bis(diphenyl)phosphane). This is a *low-spin* complex of Co^{2+} (d^7). In terms of *cfse*, the SP geometry is expected to be more stable than the trigonal bipyramidal geometry. But, it can be **crystallised** in two forms: SP (having the axial Cl) - *red crystals*

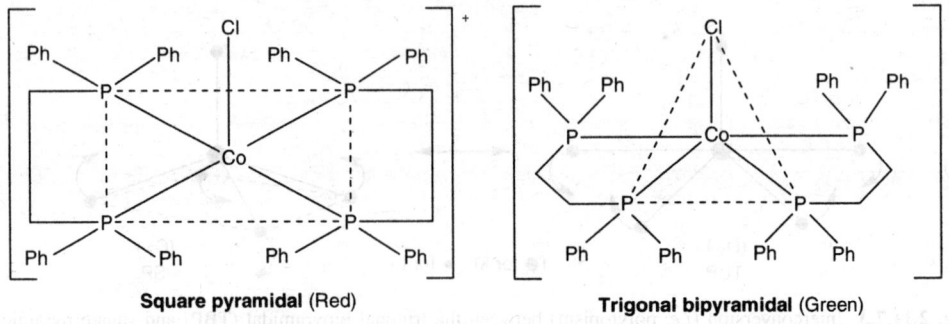

Square pyramidal (Red)   **Trigonal bipyramidal (Green)**

Fig. 2.11.7.5 Structural representation of the TBP and SP forms of $[CoCl(dppe)_2]^+$.

and TBP (two P atoms are in the axial directions, and other two P and Cl⁻ centres in the equatorial positions) - **green crystals** (Fig. 2.11.7.5). Existence of both forms again supports the fact that the energies of the two forms are comparable.

(iii) **[Ni(CN)₂{PPh(OEt)₂}₃]:** It is a typical structure describing the intermediate situation of TBP and SP. *It may be considered as a highly distorted TBP geometry* (CN⁻ ligands are in the axial directions) or *it may be considered as a highly distorted SP geometry.* Better, it is described to possess an intermediate structure.

Examples of complexes contradicting the empirical formula in relation to C.N. 5

● [MCl₅] (M = Nb, Ta, Mo, etc.) maintains the 5-coordinate structure in vapour phase but in solid state they exist as the dimers providing the octahedral structure via the Cl-bridges.

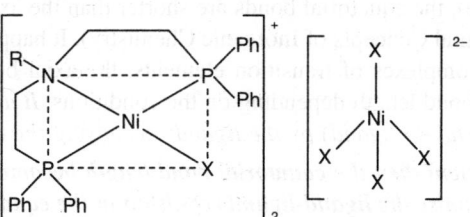

● SbF₅, AlF₅²⁻, etc. produce the octahedral geometry through polymerisation by F-bridges (*cf.* Sec. 1.12e).

● PCl₅ in solid state exists as a salt [PCl₄]⁺[PCl₆]⁻

● **Cs₃[CoCl₅] and (NH₄)₃[ZnCl₅]:** The anions are not the 5–coordinate complexes. They actually exist as the tetrahedral anions [MCl₄]²⁻ and free Cl⁻ ions are trapped in the lattice. It may be noted that **[CdCl₅]³⁻** can maintain the C.N. 5.

● **[CoCl₂(dien)]** actually exists as **[Co(dien)₂][CoCl₄]** where the cation is octahedral and anion is tetrahedral.

● **[Ni(PNP)X₂]** where PNP denotes the tridentate ligand R—N(CH₂CH₂PPh₂)₂, maintains the 5 coordinate structure at low temperature but on heating it produces a new salt [Ni(PNP)X]₂[NiX₄] where the cation is a square-planar moiety and the anion is a tetrahedral moiety.

● **[Ni(Et₄dien)X₂]** is actually a four-coordinate complex, [Ni(Et₄dien)X]X. Interestingly, the corresponding Co(II) complex maintains a 5-coordinate structure.

B. Examples of TBP complexes

[MCl₅]³⁻ (M = Cu, Cd), *e.g.* [Cr(NH₃)₆][CuCl₅]; [Ni(CN)₂{PPh(OEt)₂}₃]

[Ni(CN)₅]³⁻ in [Cr(en)₃][Ni(CN)₅]·1.5H₂O (both the SP and TBP structure of the anion)

(in [Cr(tn)₃][Ni(CN)₅], the anion is SP).

$\left[M(NMe_3)_2Cl_3\right]$ (M = Ti, V, Cr) having the Cl-atoms in the equatorial positions.

$\left[Fe(CO)_5\right], \left[Mn(CO)_5\right]^-, \left[Fe(CO)_4(PR_3)\right], \left[Pt(SnCl_3)_5\right]^{3-}, \left[Co(NCMe)_5\right]^{2+}$

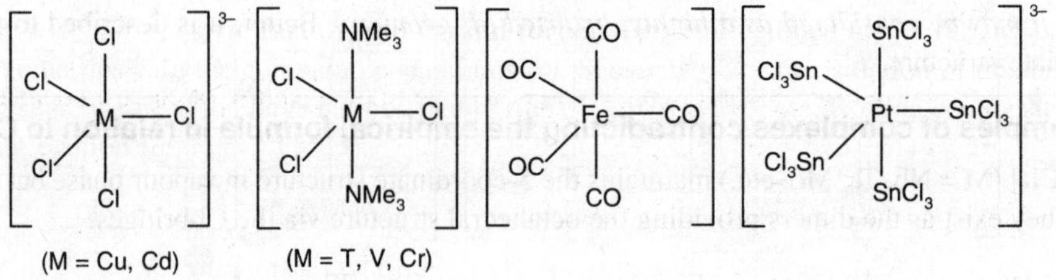

(M = Cu, Cd)   (M = T, V, Cr)

Fig. 2.11.7.6 Trigonal bipyramidal geometries of some representative complexes having the 5 monodentate ligands.

C. Site preference by the ligands in TBP

● For the nontransition central elements, the more electronegative substituents go to the axial directions (*see* Bent's rule and VSEPR, Vol. 2 of Fundamental Concepts of Inorganic Chemistry). It also happens so in many cases of the TBP complexes of transition metals. *However, the d^n configurations may have some important roles in this regard.* In the case of P (in phosphoranes), the Me–groups go to the equatorial position but for the d^8–systems, the Me–groups go to the axial positions **allowing the better π–acceptor ligands to occupy the equatorial sites.**

● The π–acceptor ligands will preferably occupy the equatorial sites. In terms of π–acceptance, the preference of the ligands to occupy the equatorial site can be arranged as follows:

$$NO^+ \rangle CN^- \rangle SnCl_3^- \rangle PR_3 \rangle C_2H_4 \rangle CH_3^-$$

Thus very often, the non–π bonding ligands like CH_3^- occupy the axial positions allowing the π–acceptor ligands to occupy the equatorial sites.

● In AX$_5$ (TBP structure), the equatorial bonds are shorter than the axial bonds (*cf.* Bent's rule and VSEPR, Vol. 2 of Fundamental Concepts of Inorganic Chemistry). It happens so for the *s*– and *p*–block central elements. For the complexes of transition elements, the axial bond length may be longer or shorter than the equatorial bond length depending on the conditions. *It depends on the relative importance of the π–bonding (metal → ligand) by the ligands occupying the equatorial positions.*

If the π–bonding is efficient then the equatorial bond length becomes shorter. If the π–bounding is less efficient then to minimise the ligand-ligand repulsion in the equatorial plane, the bond length in the equatorial plane may be elongated. It is illustrated for [Fe(CO)$_5$] and [Mn(CO)$_5$]$^-$. Both are the d^8– systems but **in the anionic complex, the metal → ligand π–bonding in the equatorial plane is more important to contract the bond length.**

	M—L$_{ax}$ (pm)	M—L$_{eq}$ (pm)
[Fe(CO)$_5$]	181	183
[Mn(CO)$_5$]$^-$	182	180
[CuCl$_5$]$^{3-}$	229	239

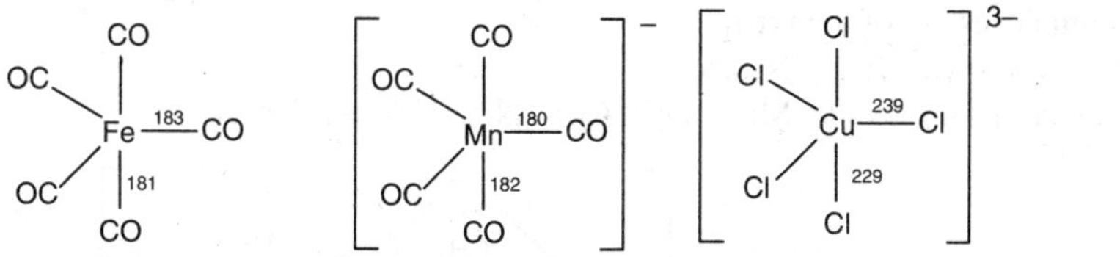

(equatorial bond is longer)   (axial bond is longer)   (equatorial bond is unusually longer)

(Reversed)   **(Normal)**   **(Exception)**

Fig. 2.11.7.7 Equatorial and axial bond lengths (in pm) of some representative TBP complexes.

Note: In d^0–systems, the TBP produces the longer axial bonds than the equatorial bonds. It has been explained by using the concept of nonequivalent hybrid orbitals (*i.e.* sp^2 + pd, resulting from s + p_x + p_y and p_z + d_{z^2}) to participate in bonding. In d^n–systems, the situation is different. If the d_{z^2} orbital does not carry any electron density (as in **low-spin d^8 system**, *see* **Fig. 2.11.7.1**), *then the axial ligand experiences less repulsion and the M–L_{ax} bond becomes shorter compared to the M–L_{eq} bond that experiences more repulsion*. However, the strong π–bonding in the equatorial plane can reverse the situation (*cf.* [Fe(CO)$_5$] and [Mn(CO)$_5$]$^-$).

In **[CuCl$_5$]$^{3-}$** (d^9–system), the d_{z^2} orbital possesses one electron while the other d–orbitals are completely filled in (*cf.* Fig. 2.11.7.1). Cl$^-$ cannot act as a π-acceptor ligand and thus the π–bonding phenomenon to contract the equatorial bond length does not arise here. Obviously, the d_{z^2} orbital bearing only one electron produces less repulsion in the axial directions compared to the other filled d-orbitals lying in the equatorial plane. It makes the Cu–Cl$_{ax}$ bond shorter than the Cu–Cl$_{eq}$ bond. In d^{10} systems, it is expected that all the d–orbitals are equivalent to cause the repulsion and the situation is similar to that of d^0 and d^5 (high-spin state) and in absence of π-bonding, the equatorial bond should be shorter or at least the equatorial and axial bonds should be comparable in length. In fact, in [CdCl$_5$]$^{3-}$ found in [Co(NH$_3$)$_6$][CdCl$_5$], the equatorial and axial bonds are comparable (within 1% range). It may be noted that the case of [ZnCl$_5$]$^{3-}$ cannot be considered for comparison as it exists as [ZnCl$_4$]Cl^{3-} with the cation [Co(NH$_3$)$_6$]$^{3+}$.

Thus for the transition metal complexes of TBP structure, possibility of π–bonding in the equatorial plane by the π–acceptor ligands and electron density in the d–orbitals causing repulsion in the approach of the ligands are to be considered to explain the relative difference in the bond lengths of axial and equatorial bonds.

D. Ligand imposed trigonal bipyramidal geometry: Tripodal umbrella like tetradentate ligands

The **tripodal ligands** represented by X($\backsim$Y)$_3$ where X = N, P or As; Y = NR$_2$, PR$_2$, AsR$_2$, etc., the connecting segment ($\backsim$) = —CH$_2$CH$_2$—, —CH$_2$CH$_2$CH$_2$—, o—C$_6$H$_4$ (o–phenylene), *i.e.*

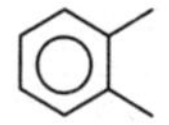

(*cf.* diars) enforce to provide the TBP structure of the metal complexes. The representative examples of such tripodal ligands are:

N(CH$_2$CH$_2$NH$_2$)$_3$ (**tren**),   N(CH$_2$CH$_2$NMe$_2$)$_3$ (**Me$_6$tren**)
i.e. tris(2–aminoethyl)amine   *i.e.* tris(2–dimethylaminoethyl)amine
P(CH$_2$CH$_2$CH$_2$AsMe$_2$), N(CH$_2$CH$_2$AsPh$_2$)$_3$, N(CH$_2$CH$_2$PPh$_2$)$_3$, etc.

The tetradentate umbrella-like tripodal ligands X($\backsim$Y)$_3$ *coordinate symmetrically to occupy three equatorial positions and one axial position of a TBP in which the remaining axial position is occupied by a unidentate ligand (L).*

The TBP complexes are of the type:

[M(Ligand)L]$^+$ (L = Cl$^-$, Br$^-$, CN$^-$, etc.)

(M^{2+} = bivalent metal ions like Mn^{2+}, Fe^{2+}, Co^{2+}, Ni^{2+}, Cu^{2+} and Zn^{2+}).

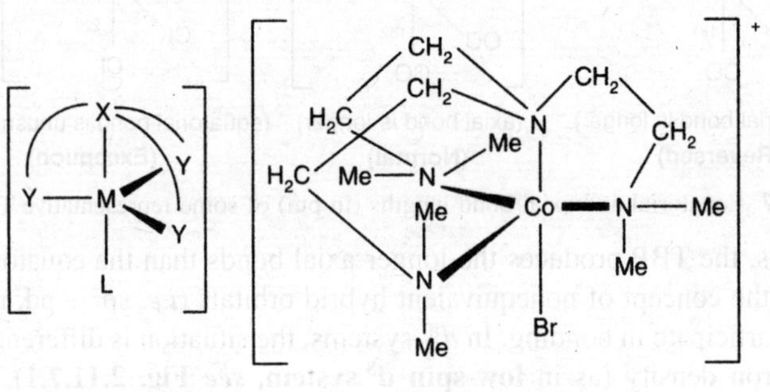

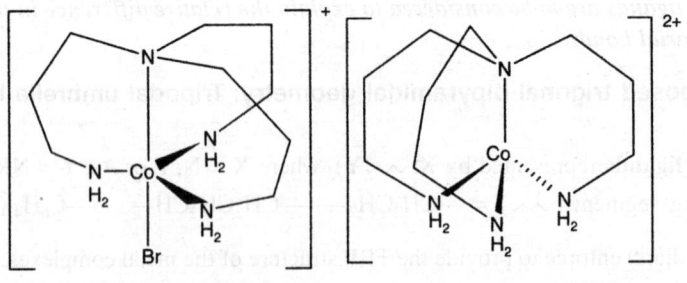

[MX($\sim\sim$Y)$_3$L]
(charge not shown)
(a)

[Co(Me$_6$tren)Br]$^+$
(b)

Fig. 2.11.7.8 Illustration of the **forced trigonal bipyramidal configuration** of the 5-coordinate complexes by the tripodal umbrella like ligands (*cf.* Fig. 4.5.2.9, Sec. 1.7.2).

When the connecting segment ($\sim\sim$) is —C$_2$—, the tripodal ligands cannot span comfortably the four coordination sites of a square planar or a tetrahedral geometry. It cannot fold to coordinate the four coordination sites of an octahedron. **Such tripodal ligands are ideal for providing the TBP geometry because of the steric requirements of the ligand itself.** If the connecting segment ($\sim\sim$) is elongated from —C$_2$— to —C$_3$—, then the ligand becomes more flexible and *it may act then as a* **facultative ligand**. In fact, the ligand having the arm X(—C$_3$—Y)$_3$ can occupy the four coordination sites of a trigonal bipyramidal or a tetrahedral geometry. In fact, N(CH$_2$CH$_2$CH$_2$NH$_2$)$_3$, *i.e.* *tris*(3-aminopropyl)amine can form both the tetrahedral and trigonal bipyramidal complex.

[Co{N(CH$_2$CH$_2$CH$_2$NH$_2$)$_3$}]Br$_2$; [Co{N(CH$_2$CH$_2$CH$_2$NH$_2$)$_3$}Br]Br

(tetrahedral cation)   **(trigonal bipyramidal complex)**

Trigonal bipyradimal geometry   **Tetrahedral geometry**

Fig. 2.11.7.9 Illustration of the facultative behaviour of N(CH$_2$CH$_2$CH$_2$NH$_2$) – a tripodal ligand.

E. Examples of square pyramidal (SP) complexes

In the SP geometry, the metal center may sit at the plane of basal ligands or somewhat above the basal plane. **In most of the cases, the metal center lies slightly above the basal plane** (*cf.* VSEPR, theory).

● **[M(diars)₂X]⁺ (X = Br⁻, Cl⁻); (M = Ni, Pd, Pt):** Formation of this type of square pyramidal complex can be rationalised by the addition of a weakly bound ligand (X^-) to the strongly π-bonded square planar complex $[M(diars)_2]^{2+}$ of d^8 systems.

$$[M(diars)_2]^{2+} + X^- \rightarrow [M(diars)_2X]^+$$

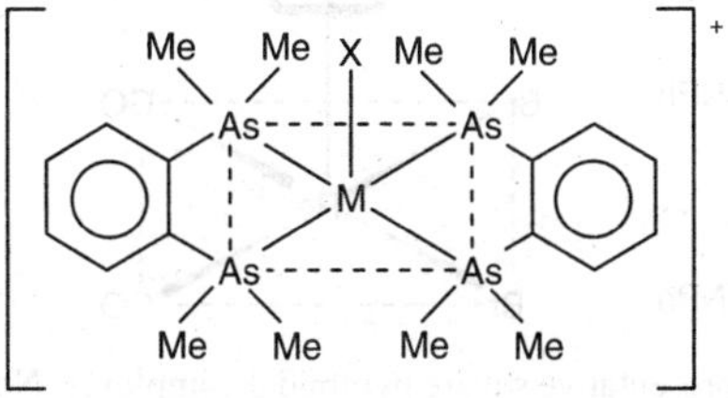

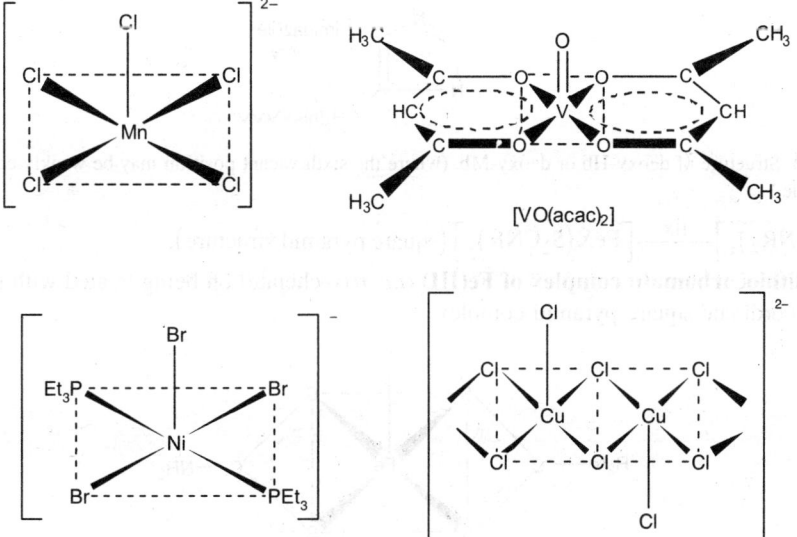

Fig. 2.11.7.10 Structure of $[M(diars)_2X]^+$.

● **[VO(acac)₂]:** The electronic spectra of $[VO(acac)_2]$ (more correctly, $[V(acac)_2O]$) differ in different solvents depending on the coordinating power of the solvents. In solution, they produce the octahedral complexes, $[VO(acac)_2(S)]$ (S = solvent) through the coordination by the solvent (S) at the vacant coordination site. *In fact, this is a common feature for the square pyramidal complexes. The vacant coordination site is very often occupied by a ligand to provide the octahedral geometry.*

● **Other examples** are: $[NiBr_3(PEt_3)_2]^-$, $[Cu(bpy)_2X]^+$ (X = halogen), $[Co(CNPh)_5]$, $[InCl_5]^{2-}$, $[MnCl_5]^{2-}$, $[Sb(Ph)_5]$, $[ReBr_2(CO)_2(\eta^5-C_5H_5)]$, $[Mo(CO)_2(\eta^5-C_5H_5)Cl(PR_3)]$, $[Mo(CO)_3(\eta^5-C_5H_5)Cl]$, etc. In the last three examples, if η^5-Cp is argued to occupy 5 coordination sites then such complexes may be considered as the 9-coordinate complexes (*cf.* Sec. 2.5.2).

● **[Cu₂Cl₆]²⁻** indicates the square pyramidal coordination around Cu(II) via the Cl–bridges between the adjacent anions.

Fig. 2.11.7.11 (*Contd.*)

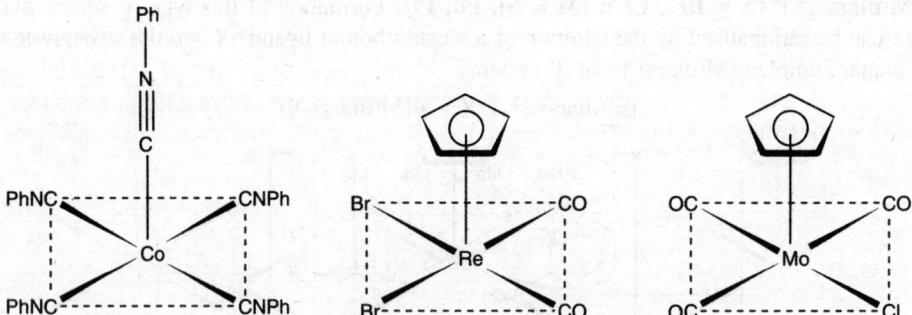

Fig. 2.11.7.11 Structures of some representative square pyramidal complexes. **Note:** In general, the metal center slightly lies above the basal plane.

● **Deoxy-hemoglobin (Hb) or myoglobin (Mb):** It approximately adopts the square pyramidal structure in which 'Fe' sits near to the plane of the porphyrin ring and the imidazole-N of a histidine moiety coordinates the axial position. On oxygenation, the sixth vacant site is occupied by O_2 to provide the octahedral geometry.

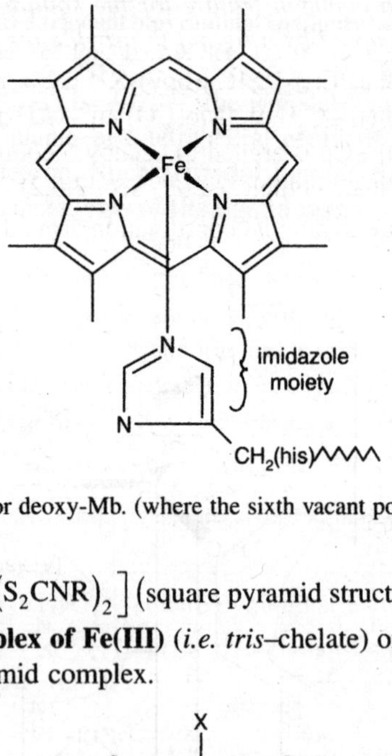

Fig. 2.11.7.12 Structure of deoxy-Hb or deoxy-Mb. (where the sixth vacant position may be weakly coordinated by a water molecule.)

● $\left[Fe(S_2CNR_2)_3 \right] \xrightarrow{HX} \left[FeX(S_2CNR)_2 \right]$ (square pyramid structure).

Dialkyldithiocarbamato complex of Fe(III) (*i.e. tris*–chelate) on being treated with HX produces the said 5-coordinate square pyramid complex.

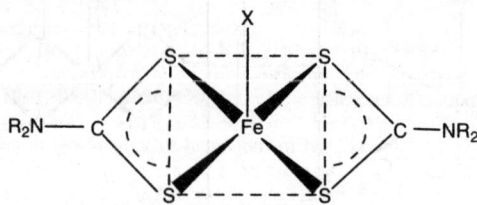

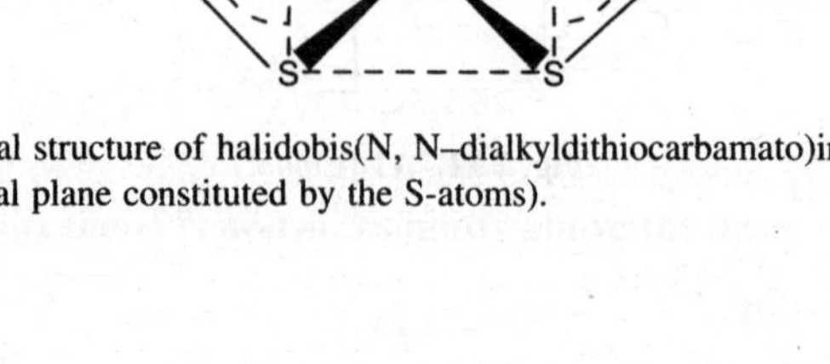

Fig. 2.11.7.13 Square pyramidal structure of halidobis(N, N–dialkyldithiocarbamato)iron(III). (Actually the Fe-center lies about 62 pm above the basal plane constituted by the S-atoms).

• $[:SbF_5]^{2-}$ and $[:SbCl_5]^{2-}$ are square pyramidal because of the presence of *a **stereochemically** active lone pair* (*cf.* Geometry of $[XeF_5]^-$; VSEPR theory in Vol. 2 of Fundamental Concepts of Inorganic Chemistry).

Stereochemically active/inactive lone pair (*cf.* Sec. 10.8.5, Vol. 2 of Fundamental Concepts of Inorganic Chemistry)

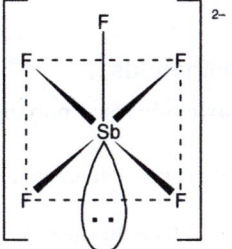

For the bulky ligands like Cl^-, Br^-, the lone pair is generally kept **stereochemically inactive** by housing it in an inner spherical *s*-orbital to elongate the bond length. It happens so in $[TeX_6]^{2-}$, $[SbX_6]^{3-}$(X = Br, Cl) to minimise the ligand-ligand repulsion. For the **smaller ligands like F^-, O^{2-}**, the lone pair is **stereochemically active** as expected from VSEPR. Thus $[SbF_6]^{3-}$, $[Sb(C_2O_4)_3]^{3-}$ are **pentagonal bipyramidal** with the active lone pair while $[SbCl_6]^{3-}$, $[SbBr_6]^{3-}$ are **octahedral.**

Fig. 2.11.7.14 Square pyramidal structure of $[SbF_5]^{2-}$ in terms of VSEPR theory.

• **Bond length parameters:** VSEPR and hybridisation scheme for the square pyramidal complexes of the nontransition elements suggest that the metal centre resides slightly above the basal plane, and the **basal bond is longer than the axial bond** (*cf.* Sec. 10.8.3, Vol. 2 of Fundamental Concepts of Inorganic Chemistry). In the square pyramidal complexes of transition metals, the situation is different as in the trigonal bipyramical systems.

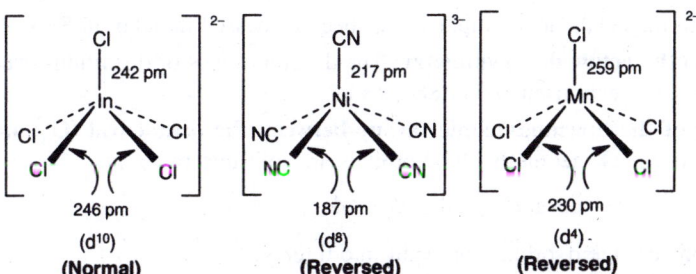

Fig. 2.11.7.15 Bond length parameters in some representative square pyramidal complexes.

The vacant $d_{x^2-y^2}$ orbital for the d^8 (low spin) and d^4 systems (*cf.* Fig. 2.11.7.1) causes a lesser repulsion (compared to the filled $d_{x^2-y^2}$ orbital as in the d^{10} system) towards the basal ligands to contract the basal bond.

F. Isomerism in five coordinate complexes

These are discussed in Secs. 2.5.2, 2.8. The geometrical and optical isomerism are expected for the 5–coordinate complexes, but they are isolated in some rare cases only. *This is because of the very low activation energy in the interconversion of one configuration into the other.*

2.11.8 Examples of the Six Coordinate Complexes (Coordination Number Six)

The six coordination number is probably the most common in coordination chemistry. Among the three possible structures—planar hexagon, octahedral and trigonal prism, the octahedron is the most common one and the trigonal prism structure is reported only in some rare cases. No complex having the hexagonal planar structure is yet reported.

Some metal centers like Cr(III) (d^3), Co(III) (d^6), and Pt(IV) (d^6) almost exclusively form the octahedral complexes. At the time of Werner, six coordinate complexes of these metal centers were only investigated. Werner established their octahedral structure by considering their possible strereoisomerism. Werner's this methodology to exclude the possibilities of hexagonal planar and trigonal prism has been discussed in detail in Sec. 1.4.5.

Stereochemical nonrigidity in the coordination compounds

- Stereochemical nonrigidity can lead to two different phenomena: **fluxionality** and **polytopal isomerism** (*cf.* Sec. 10.9, Vol. 2).
- It is common for C.N. = 5 and ≥ 7 because the interconversion among the possible geometries having very much comparable energies is kinetically favourable.
- Square planar/tetrahedral interconversion of some Ni(II) complexes has been illustrated.
- Octahedral complexes are **generally stereochemically rigid.** However, **twisting mechanism** can lead to the stereochemical nonrigidity in the *bis-* and *tris-* chelates.
- Octahedral complexes **[MH$_2$(PR$_3$)$_4$]** (M = Fe, Ru) can show the sterochemical nonrigidity at temperature $> 50°$ (indicated by one PMR signal). However, at lower temperature, the *cis-* and *trans-* isomers can be detected separately.

A. Characteristics of the octahedral complexes: There is an endless number of examples of the octahedral complexes. These may be *high-spin* and *low-spin* for the d^{4-7} configurations.

The **isomerism** in the octahedral complexes has been discussed in detail in Secs. 2.5.3 and 2.8.

B. Distortions in the octahedral symmetry: The different types of distortions occurring in the octahedral geometry have been discussed in Sec. 2.11.2.

C. Comparison of the thermodynamic favour between the octahedral (O$_h$) and trigonal prism (TP) geometry: In the O$_h$ geometry, the d-orbitals are split into two sets:

$$\textbf{Energy: } d_{xy}, d_{yz}, d_{zx} < d_{x^2-y^2}, d_{z^2}$$

In the TP geometry, the d–orbitals are split into three sets:

$$\textbf{Energy: } d_{z^2} < d_{xy}, d_{x^2-y^2} < d_{xz}, d_{yz}, \textit{(cf. Sec. 3.5.10)}$$

In terms of *cfse*, all the configurations except d^0, d^5 (high-spin) and d^{10} (*i.e.* the cases having no *cfse* for the both geometries) are preferred for the octahedral geometry over the trigonal prism geometry. However, for the d^1 and $d^{6,7}$ (high-spin) configurations, in moving from the octahedral to the trigonal prismatic geometry, the loss of *cfse* is not so high.

- *Thus, in terms of cfse, the trigonal prismatic geometry may arise for d^0, d^1, d^5 (high spin), d^6 (high spin), d^7 (high-spin) and d^{10} configurations as in such cases there is no strong preference for the octaheral geometry.*
- *For d^6 (low-spin), the loss of cfse is maximum in moving from the O_h to the TP geometry.*
- The **relative instability** of the trigonal prismatic structure compared to that of the octahedral structure arises due to the following factors.

(i) **lower *cfse*** (in general) in the TP structure; (ii) **enhanced ligand-ligand steric repusion** in the TP structure and it is important to consider for the bulky and anionic ligands and small size of the metal center.

● *The small bite distance of the relatively inflexible didentate ligands, steric requirements of the polydentate ligands and bonding interaction among the different segments/moieties of the coordinated ligands can only favour the trigonal prismatic coordination.*

● The **bite distance** in somewhat inflexible ligands like **1, 2-dithiolene** and **tropolonato anion** is too small to coordinate two *cis*-positions of an octahedron but this distance is suitable for the trigonal prismatic geometry. Because of the same ground, $\eta^2-NO_3^-$ favours the TP structure.

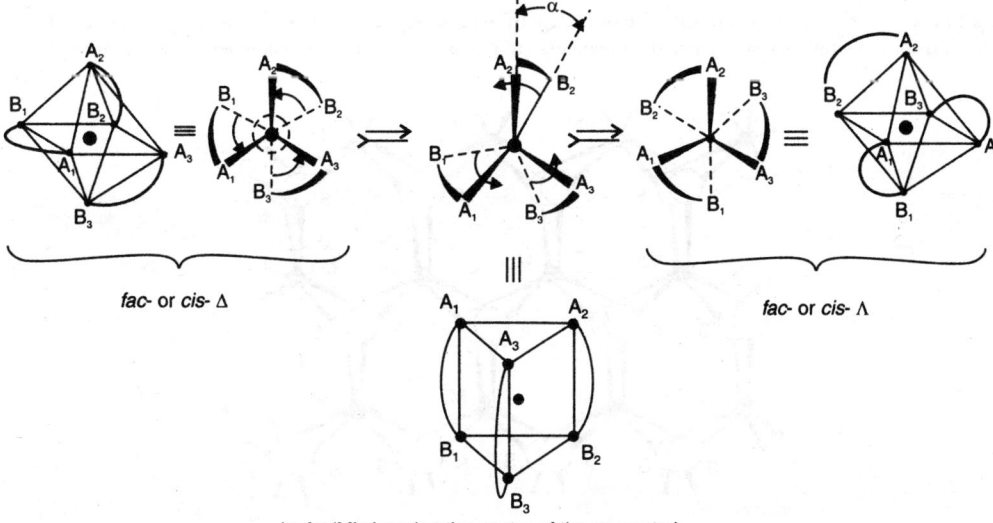

● It will be discussed later that by rotating a triangular octahedral face by 60° relative to the opposite triangular face, a trigonal prismatic structure may be obtained from an octahedral structure without any *bond rupture* (*cf.* **Fig. 2.11.2.5 and Bailer twist mechanism**).

D. Interconversion between the octahedral (O_h) and trigonal-prismatic (TP) geometries (*cf.* Fig. 2.11.2.5): *Bailer twist* can convert the change and it is illustrated for three didentate ligands (A – B) supposed to give the 6 C.N.

$$\left[M(A-B)_3\right](O_h, \alpha = 60°) \rightleftharpoons \left[M(A-B)_3\right](TP, \alpha = 0°), \text{(Bailer twist)}$$

In fact, Bailer twist mechanism can lead to racemization in the tris-didentate complexes like $[M(A-A)_3]$ or $[M(A-B)_3]$. This aspect will be discussed in detail in connection with the mechanism of racemisation of tris-chelates. The Bailer-twist is illustrated in Fig. 2.11.8.1.

Fig. 2.11.8.1 Bailer twist (*i.e.* trigonal twist) and twist angle (α) for recemisation of a *tris*–chelate. Illustrated for the $[M(A - B)_3]$ type chelates; rotation along the C_3–axis passing through the octahedral face bearing the three ligating sites A's. (if A = B, the intermediate is a trigonal prism having the D_{3h} symmetry) (*cf.* Fig. 2.11.2.5a; and Figs. 10.9.3.1-2 in Vol. 2 of Fundamental Concepts of Inorganic Chemistry).

In terms of Bailer twist, a trigonal prism is obtained from an octahedron by twisting one trigonal face relative to the other. In terms of the *twist angle* (α), **$\alpha = 0$ for a perfect TP and $\alpha = 60°$ for a pertect O_h.** Thus, α can very continously in the range 0°–60° giving rise to all the possible geometries. In fact, **an octahedron is basically a** *trigonal antiprism* **(TAP, $\alpha = 60°$).**

The *tris-*(dithiolato) metal complexes, *i.e.* [M(L–L)$_3$]$^{n-}$, very often adopt the trigonal prismatic structure but depending on the nature of the metal centre and 1,2-dithiolene ligand, α (twist angle) has been found to vary. It is illustrated below.

[V(L–L)$_3$] ($\alpha = 0°$); [Nb(L–L)$_3$]$^-$ ($\alpha \approx 0.5°$); [Ta(L–L)$_3$]$^-$ ($\alpha \approx 16°$); [V(L–L)$_3$]$^{2-}$ ($\alpha \approx 17°$); [Zr(L–L)$_3$]$^{2-}$ ($\alpha \approx 19.5°$); [Mo(L–L)$_3$] ($\alpha \approx 0°$); [Mo(L–L)$_3$]$^{2-}$ ($\alpha = 14°$); [W(L–L)$_3$]$^{2-}$ ($\alpha = 14°$); [Re(L–L)$_3$] ($\alpha = 0°$); [Tc(L–L)$_3$] ($\alpha = 4.5°$); [Fe(L–L)$_3$]$^{2-}$($\alpha \approx 25°$)

E. Examples of trigonal prismatic (TP) coordination: Formation of a trigonal prismatic structure by the unidentate and didentate ligands is illustrated in Fig. 2.11.8.2.

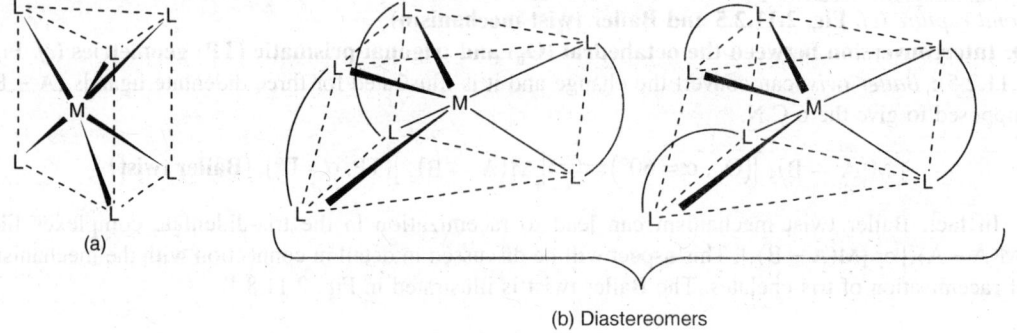

(a)

(b) Diastereomers

Fig. 2.11.8.2 (a) Trigonal prismatic structure of (ML$_6$). (b) Chelation by the didentate ligands in the [M(L – L)$_3$] type trigonal prismatic complexes. (Only two diastereomers are possible in such cases. These are shown above.)

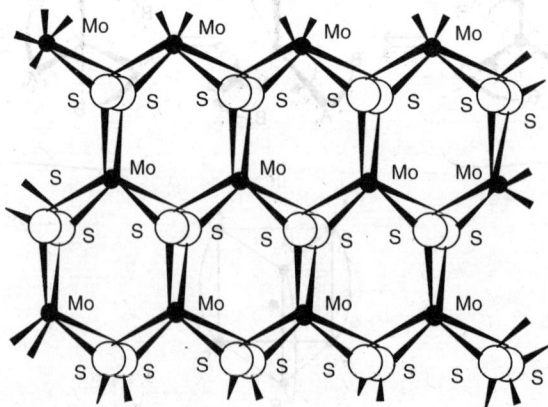

Fig. 2.11.8.3 Structural representation of the MoS$_2$ polymer having the trigonal prismatic coordination around the Mo–center (denoted by ●) by S–atoms. (denoted by O)

● In the crystal lattices of **ThI$_2$, MoS$_2$, WS$_2$, ReS$_2$, NbS$_2$** etc. the trigonal prismatic coordination is known. The structure of MoS$_2$ can be explained by considering the super imposed layer of S–atoms and placing the Mo–atoms at the trigonal pyramidal holes produced by the S–atoms. Thus it possesses a layer structure (*cf.* CdI_2 structure) and it can be *used as a lubricant.* WS_2, ReS_2, NbS_2, etc. possess the similar structure.

- In 1965, Werner's proposition (*i.e.* **6 coordinate complexes are always octahedral**) was **invalidated with the discovery of the trigonal prismatic complex, $[Re(S_2C_2Ph_2)_3]$. This was the first report of TP complex.** Later on, by using the ligand $R_2C_2S_2$, other trigonal prismatic complexes $[M(S_2C_2R_2)_3]^{n-}$ (M = Mo, W, V, Zr, Nb, etc.), have been characterised. **$[Re\{S_3(CH_2)_3CMe\}_2]$** is also an example of TP–structure.

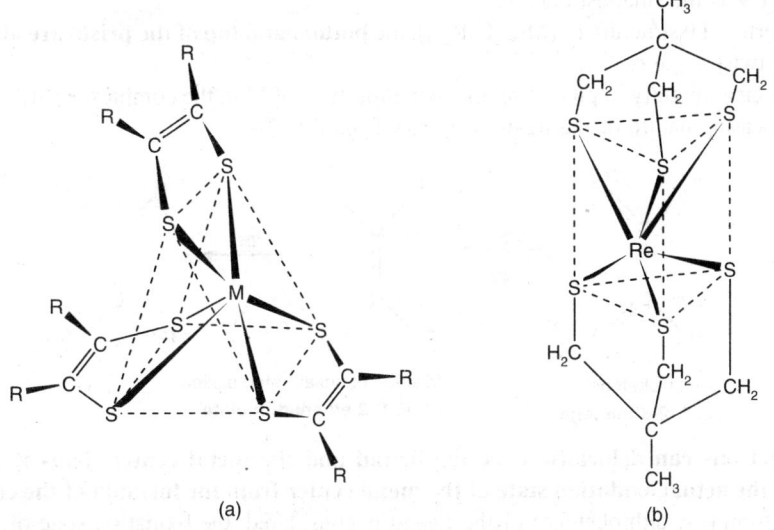

(a)   (b)

Fig. 2.11.8.4 Structural representation of the trigonal prismatic complexes. (a) $[M(S_2C_2R_2)_3]$ (M = Re, Mo, V), (b) $[Re\{S_3(CH_2)_3CMe\}_2]$.

F. Characteristic features of the neutral trigonal prismatic $[M(S_2C_2R_2)_3]$ complexes ($\alpha = 0$): The important characteristic features of the trigonal prismatic complexes $[M(S_2C_2R_2)_3]$ (M = Re, Mo, V) are outlined below.

(i) The quadrilateral sides of the trigonal prism are almost equal (*i.e.* **almost a perfect square**). It indicates that the *intra–* and *interligand* S—S distances are very much comparable (~304 pm when R = Ph, M = Re). But this is not an essential condition. Generally, *it is expected that in a trigonal prismatic structure, the quadrangular faces are rectangular.*

(ii) Here, it may be mentioned that the small **S—S the bite distance** of 1,2-dithiolene is suitable for the trigonal prismatic structure. **In fact, the bite distance of somewhat inflexible 1,2-ditholene is too short to span the two *cis*-vertices of an octahedron.**

(iii) The observed S—S distance (ca. 304 pm) in the compex is shorter by about 60 pm than the sum of their van der Waals radii. *This shrinkage in S—S distance indicates a fairly strong bonding interaction in the S—S linkage of the TP structure.* **This additional bonding interaction is the main driving force to stabilise the TP structure and this makes the intra- and interligand S—S distances comparable.**

(iv) To maintain the S—S bonding interaction in the TP structure of the complex, the S-atoms are pulled towards each other. It reduces the *bite angle* to about 80° (*cf.* 90° in a perfect octahedral structure).

(v) *Twist of a S_3–triangle by 60° will give the octahedral structure but it will lead to rupture the S—S bonding interaction.* **This S—S bonding interaction prevents the movement towards the octahedral structure from the TP structure.**

Note: It may be pointed out that in the TP structure of MoS_2, the tetragonal face is almost perfectly square and the S—S distance (about 305 pm) is also very short to support the bonding interaction in the S—S linkage.

(vi) For the complexes of different metals like Mo, Re and V, the M—S distance is remarkably similar. For example, the M—S distance varies in the range 232.5 to 234 pm (for R = Ph) though the radius of V is the smallest one.

(vii) In the perfect TP structure of $[M(S_2C_2R_2)_3]$, the **bottom and top of the prism are almost eclipsed** (*i.e.* the twist angle $\alpha \approx 0$).

(viii) There is an ambiguity in predicting the oxidation state of M in the complexes $[M(S_2C_2R_2)_3]$ as the charge bearing nature of the ligand can vary from 0 to 2–.

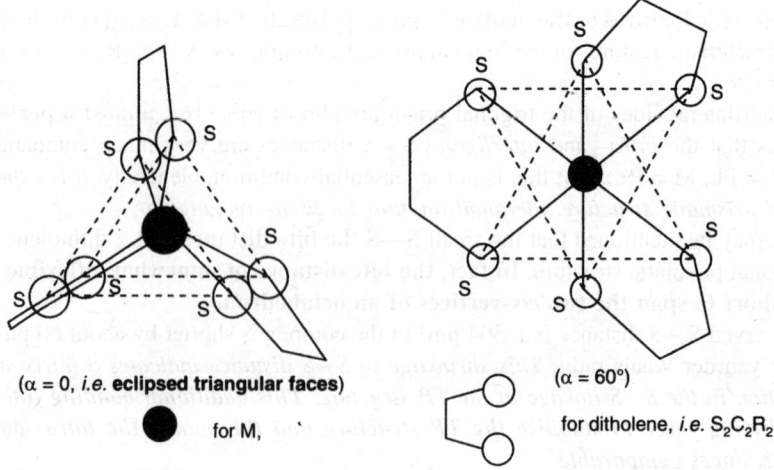

(dithioketone)
(*i.e.* **1, 2-dithiolene**)

(dianion of unsaturated dithiol)
(*i.e.* **1, 2-ethenedithiolate**)

The electrons can delocalise over the ligand and the metal center. Thus it is difficult to predict the actual oxidation state of the metal center from the formula of the complex. If the neutral form (*i.e.* dithioketone) of the ligand is considered, the oxidation state of M becomes 0 (zero) and if the dianionic form of the ligand is considered in the complex, then the oxidation state becomes +6.

($\alpha = 0$, *i.e.* **eclipsed triangular faces**)
for M,

($\alpha = 60°$)
for ditholene, *i.e.* $S_2C_2R_2$

Fig. 2.11.8.5 (a) Trigonal prismatic structure of $[M(S_2C_2R_2)_3]$ where S- - - -S bonding interactions (not shown) stabilises the structure. (b) Octahedral structure of $[M(S_2C_2R_2)_3]$ where S- - - -S bonding interaction is absent. (R–groups are not shown in both cases.)

(ix) The neutral complexes, $[M(S_2C_2R_2)_3]$ can be reduced to produce the anionic complex.

$$\left[M\left(S_2C_2R_2\right)_3\right]\xrightarrow{ne}\left[M\left(S_2C_2R_2\right)_3\right]^{n-}$$
$$\underset{(\alpha=0°)}{\qquad} \underset{(\alpha>0°)}{\qquad\qquad\qquad}$$

$$(n = 1, 2, 3)$$

The reduced complex $[M(S_2C_2R_2)_3]^{n-}$ still maintains the TP structure with a distortion towards the octahedral structure ($\alpha = 60°$). The additional electrons probabily occupy the antibonding orbitals of the S—S linkage to destroy the S—S bonding interaction. *It destabilises the TP structure and favours the O_h structure.* (*cf*. Fig. 2.11.8.5).

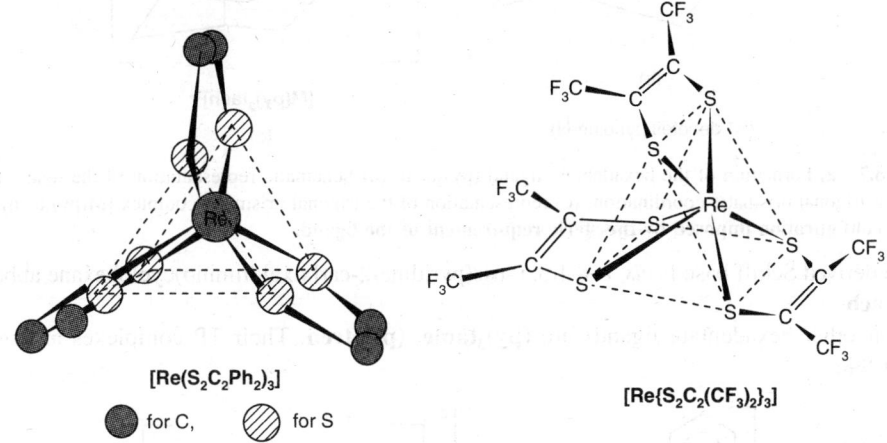

[Re(S₂C₂Ph₂)₃]

● for C, ◐ for S

[Re{S₂C₂(CF₃)₂}₃]

Fig. 2.11.8.6 Structural representation of trigonal prismatic [Re(S₂C₂Ph₂)₃] (the Ph-groups are not shown) and [Re{S₂C₂(CF₃)₂}₃].

● **[Re{S₃(CH₂)₃CMe}₂]** is another example of TP structure having the S–ligands (*cf*. Fig. 2.11.8.4).
● **[W(CH₃)₆]** and **[Zr(CH₃)₆]²⁻** also adopt the TP structure.
● **[Co(η²–NO₃)₃]** (*cf*. Sec. 2.11.11) adopts the trigonal prismatic structure because the **small bite distance in η²–NO₃⁻** is not suitable for the octahedral geometry.

G. Ligand imposed TP structure

● Some rigid *hexadentate ligands* are ideal for providing the trigonal prismatic structure because of the steric requirements of the ligands. One such typical ligand is the Schiff based obtianed from the condensation of pyridine-2-carboxaldehyde and *cis, cis*–1,3,5–triaminocyclohexane (tach).

(tach)

(py)₃tach

(a)

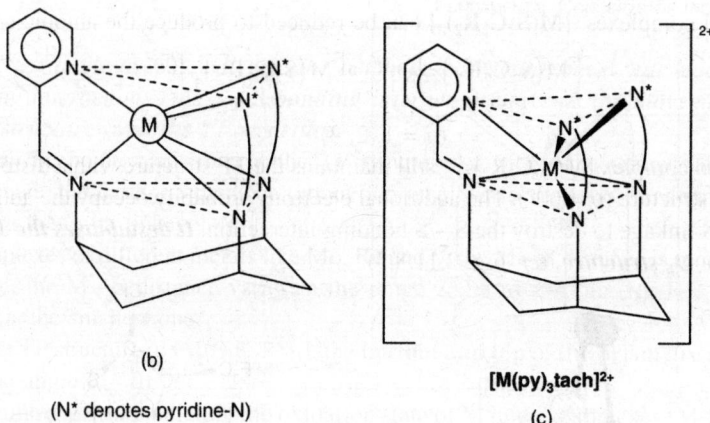

(b)

(N* denotes pyridine-N)

[M(py)₃tach]²⁺

(c)

Fig. 2.11.8.7 (a) Formation of the hexadentale ligand (py)₃tach. (b) Schematic representation of the ligand (py)₃tach suitable for trigonal prismatic coordination; (c) representation of the trigonal prismatic complex [M(py)₃tach]—**a case of forced configuration imposed by the steric requirement of the ligand.**

Thus the derived Schiff base is *cis, cis*–1,3,5–tris(pyridine-2-carboxaldiimino)cyclohexane abbreviated as **(py)₃tach**.

● Similar other hexadentate ligands are **(py)₃tame, (py)₃tren**. Their TP complexes are shown in Fig. 2.11.8.8.

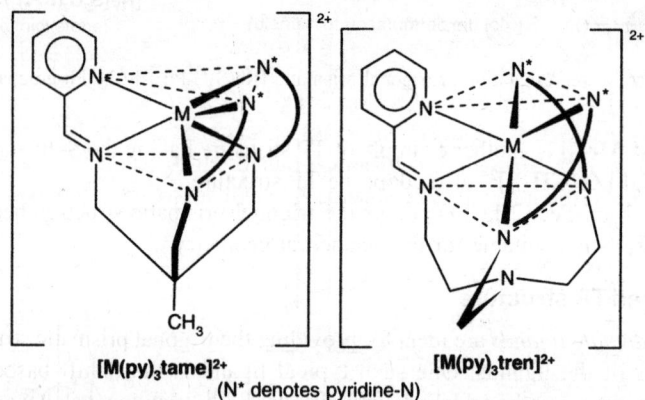

[M(py)₃tame]²⁺

[M(py)₃tren]²⁺

(N* denotes pyridine-N)

Fig. 2.11.8.8 Trigonal prismatic configuration of [M(py)₃tame]²⁺ and [M(py)₃tren]²⁺ – **cases of forced configuration imposed by the ligand.**

The ligand rigidity in such cases is the main condition to favour the TP geometry. **The tendency to provide the TP geometry decreases with the increasing of flexibility of the ligand.** The tendency to favour the TP structure changes as follows for the following hexadentate Schiff base ligands.

$$(py)_3 tach > (py)_3 tame > (py)_3 tren$$

Note: (py)₃tren can provide C.N. 7, but the M—N distance by the bridging N is too long to contribute to C.N. However, very weak interaction has been noted for high-spin $Mn^{2+}(d^5)$ and $Co^{2+}(d^7)$ complexes for which *cfse* is not of an important consideration. It is discussed in detail in the next Section for C.N. 7.

2.11.9 Examples of Seven Coordinate Complexes

Seven coordinate complexes are known in some limited cases. The disfavour to provide the 7 C.N. arises from the following facts.

(i) less *cfse*, *i.e.* loss of *cfse* in attaining 7 C.N. instead of C.N. 6 present in an octahedral geometry.

(ii) increased ligand-ligand repulsion, specially when the ligands are relatively bulkier and anionic and the smaller metal ion.

(iii) weaker metal-ligand bonding interaction.

Favourable conditions for higher coordination number > 6

The higher C.N. like 7, 8, etc. is generally found for the 2nd and 3rd transition series metal ions which are relatively larger in size. It is due to the following facts:

(i) *larger size* of the metal center can accommodate the larger number of ligands better.

(ii) *availability of the larger number of orbitals which* are energetically comparable to allow the higher coordination number.

(iii) hard donor sites (*i.e. small but electronegative donor site*) like F, N, O, are ideal for providing the higher coordination; chelating ligands with a *small bite*.

(**Note:** Polarisable and soft donor sites are ideal for the lower C.N.; for such soft donor sites, the metal-ligand π-bonding can compensate the effect of lesser number metal-ligand σ–bonding interaction).

(iv) in terms of the *electroneutrality principle*, for the higher C.N. generally provided by the σ–donor hard ligands, the metal center should have a high formal oxidation state; in the case low oxidation states, formation of a larger number L → M σ–dative bonds will create an unfavourable situation.

A. **Possible geometries for coordination number seven.**
For C.N. 7, the common geometries are: **monocapped octahedron** (C_{3v}); **pentagonal bipyramid** (D_{5h}) **and monocapped trigonal prism** (C_{2v}); **tetragonal base–trigonal base** (C_s polyhedron).

The C_s polyhedron actually represents an intermediate geometry between the pentagonal bipyramid and the capped trigonal prism.

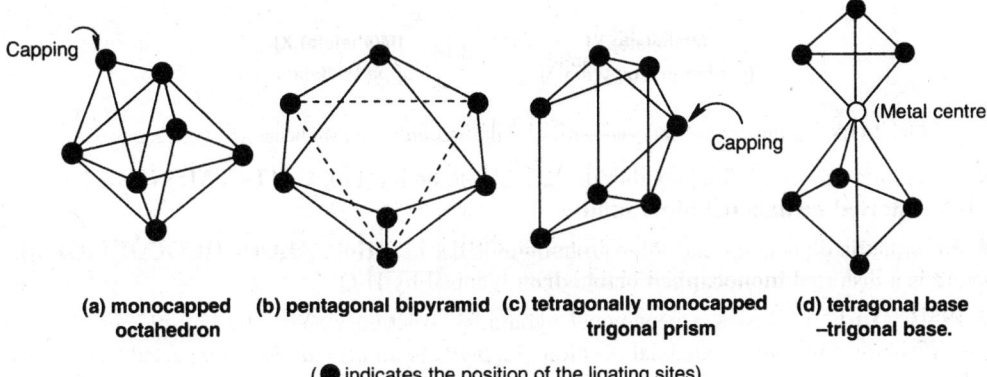

Capping   Capping   (Metal centre)

| (a) monocapped | (b) pentagonal bipyramid | (c) tetragonally monocapped | (d) tetragonal base |
| octahedron | | trigonal prism | –trigonal base. |

(● indicates the position of the ligating sites)

Fig. 2.11.9.1 Geometries for 7 coordination number (*cf.* Fig. 2.11.2.7).

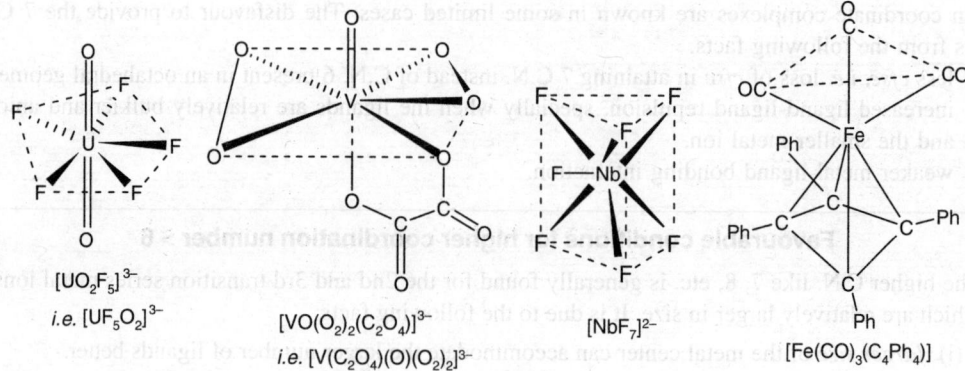

Fig. 2.11.9.2 Some representative examples of 7-coordinate complexes.

● **Different 7 coordinate geometries of comparable energy:** *All the polyhedra describing the C.N. 7 are very much comparable in energy and it leads to the stereochemical nonrigidity.* In fact, it happens so for C.N. ≥ 7 (in general). The actual geometry to be adopted in a particular case can hardly be predicted as it depends on many factors like electronic configuration of the metal ion, size of the metal center, ligand basicity, etc. The geometries actually found are very often distorted and *these distorted structures can be explained in terms of different regular geometries.* This is illustrated in the following examples having the formula **[M(chelate)₃X]** where the chelate stands for a didentate ligand and X is a monodentate ligand.

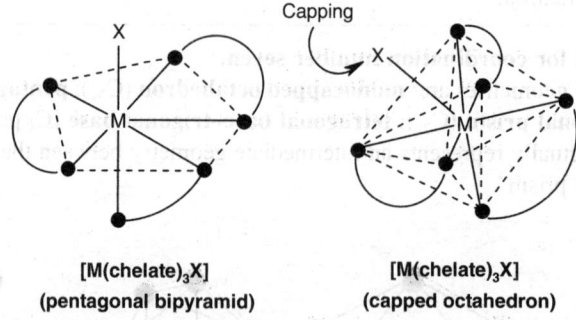

| [M(chelate)₃X] | [M(chelate)₃X] |
| (pentagonal bipyramid) | (capped octahedron) |

Fig. 2.11.9.3 Different possible geometries for the 7-coordinate [M(chelate)₃X] type complexes.

● For chloridotris(N, N–dimethyldithiocarabamato)titanium(IV), *i.e.* **[TiCl(Me₂NCS₂)₃]**, the structure is **a distorted pentagonal bipyramid**;

● For aquatris(diphenylpropanedionato)holmium(III), *i.e.* **[Ho(C₆H₅COCHCOC₆H₅)₃(OH₂)]**, the structure is a distorted **monocapped octahedron** (capped by H₂O).

● **[Sb(C₂O₄)₃]³⁻** possesses a pentagonal pyramidal structure which arises by placing the *stereochemically active lone pair* in an axial position of a *pentagonal bipyramid* (as expected from VSEPR). The Sb centre lies 35 pm below the pentagonal plane and the axial Sb—O bond is shorter than the equatorial one by about 20 pm (*cf. VSEPR, Vol. 2 of Fundamental Concepts of Inorganic Chemistry*).

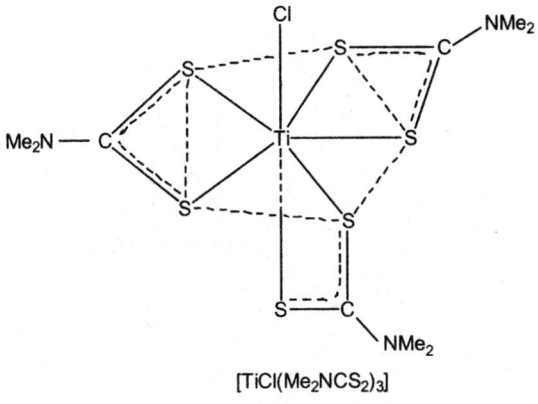

[TiCl(Me₂NCS₂)₃]
(Distorted pentagonal bipyramid)

Aquatris(diphenylpropanedionato)holmium(III)
(Capped octahedron)

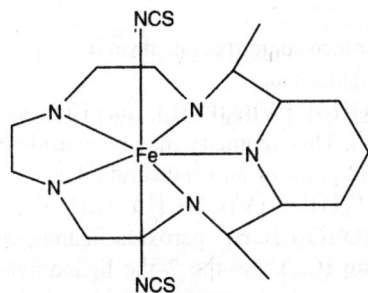

Pentagonal bipyramidal

2, 13- dimethyl- 3,6,9,12,18- pentaazabicyclo [12.3.1]-octadeca-1(18), 2, 12, 14, 16-pentane complex of Fe(II) with two NCS⁻ axial ligands. (H-atoms not shown)

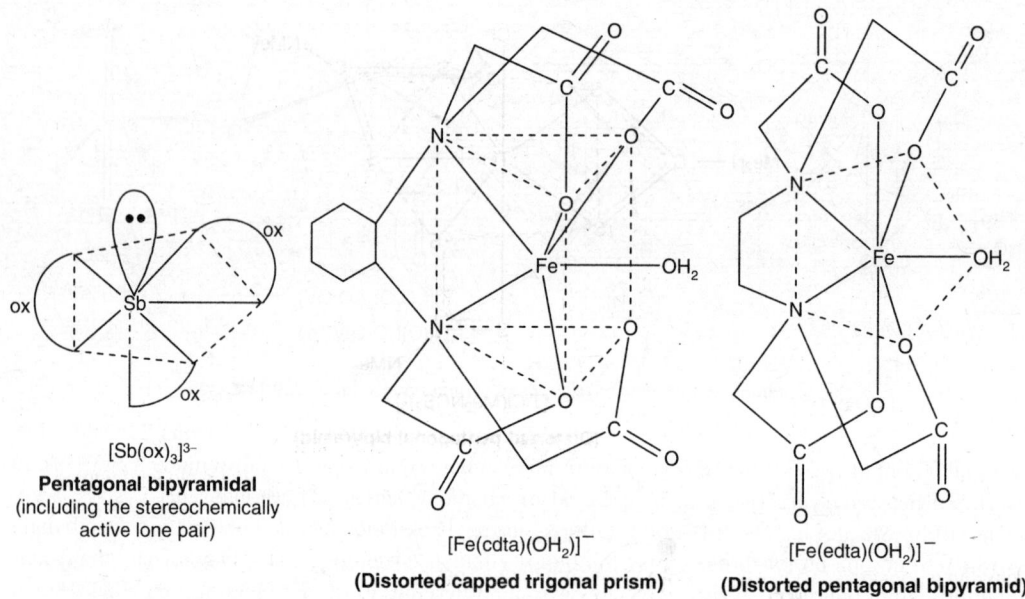

[Sb(ox)₃]³⁻

Pentagonal bipyramidal
(including the stereochemically
active lone pair)

[Fe(cdta)(OH₂)]⁻
(Distorted capped trigonal prism)

[Fe(edta)(OH₂)]⁻
(Distorted pentagonal bipyramid)

Fig. 2.11.9.4 Schematic representation of some representative 7–coordinate complexes having the chelating ligands.

For aqua(1, 2–diaminocyclohexane–N, N′–tetraacetato)ferrate(III), *i.e.* **[Fe(cdta)(OH₂)]⁻**, the structure is a **distorted monocapped trigonal prism** in which 4O sites constitute a capped quadrilateral face and H_2O is placed as a cap on this face. Interestingly, **[Fe(edta)(OH₂)]⁻** is having a **distorted pentagonal bipyramidal structure** in which the pentagonal basal plane contains H_2O, 2N and 2O.

Note: [Co(edta)]⁻ is six coordinate while **[Mn(edta)(OH₂)]²⁻**, **[Fe(edta)(OH₂)]⁻** are seven coordinate. For Mn^{2+} (d^5), Fe^{3+}(d^5), the 7-coordinate systems are possible because of two reasons: (i) no loss of *cfse* with respect to the octahedral geometry, (ii) larger size (*cf.* crystal field radius for d^5 configuration) of Mn^{2+}, Fe^{3+} compared to that of Co^{2+}; in fact, edta cannot span well around the larger cation Fe^{3+} in its octahedral geometry.

B. Examples

(i) **Monocapped octahedron** or face centerd octahedron (C_{3v}), *i.e.* 7th ligand/ligating site coordinates through the center of a triangular face:
[Ho(C_6H_5COCHCOC₆H₅)₃(OH₂)], [WBr₃(CO)₄], (one CO caps)

(ii) **Pentagonal bipyramid** (D_{5h}): This geometry may be considered to be developed by placing the 7th ligating site along the mid-point of an octahedral edge.
[UF₇]³⁻, [UF₅O₂]³⁻, [NbF₆O]³⁻, [V(CN)₇]⁴⁻, [ZrF₇]³⁻, [HfF₇]³⁻, [TiCl(S_2CNMe₂)₃],
[Fe(edta)(OH₂)]⁻, [V(C_2O_4)(O)(O₂)₂]³⁻ (η^2–peroxido ligand), etc.

(iii) **Monocapped trigonal prism** (C_{2v}), *i.e.* the 7–the ligand/ligating site through the center of a rectangular or quadrilateral face of a trigonal prism.
[NiF₇]²⁻, [NbF₇]²⁻, [TaF₇]²⁻, [Fe(cdta)(OH₂)]⁻, [TiF₅(O₂)]³⁻ (η^2–peroxido ligand)

(iv) **Tetragonal base-trigonal base structure,** *i.e.* a geometry describing the intermediate structure of two idealised geometries — pentagonal bipyramid (D_{5h}) and capped trigonal prism (C_{2v}). This intermediate structure is attained in [Fe(CO)₃(C_4Ph_4)].

C. Ligand imposed 7–coordination number

● A pentadentate macrocyclic ligand (L) (shown in Fig. 2.11.9.5) provides five binding sites in a plane. By placing two *trans*-axial ligand (X) (*e.g.* Cl⁻, Br⁻, NCS⁻, N₃⁻), then a pentagonal bipyramidal geometry in [MX₂(L)] can be attained.

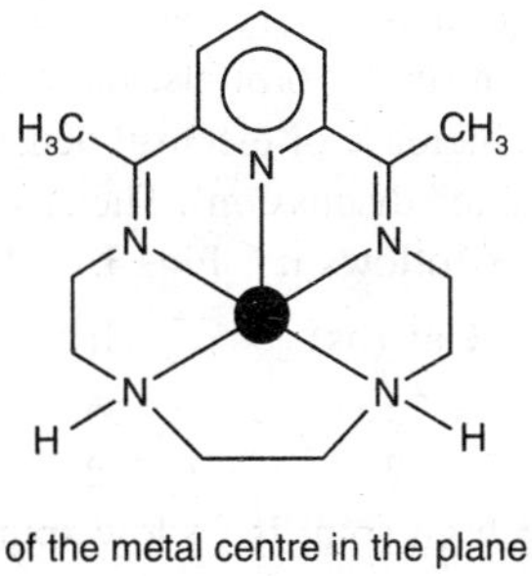

(● indicates position of the metal centre in the plane of the macrocyclic ligand).

Fig. 2.11.9.5 Macrocyclic ligand (L) providing 5 donor sites (5N) in the basal plane. The ligand allows coordination to two unidentate ligands in the axial directions to provide the C.N. total 7. (The *trans*-axial ligands are not shown.)

● The Schiff-base ligand **(py)₃tren** obtained through the condensation of N(CH₂CH₂NH₂)₃ (*i.e.* tren) and pyridine-2-carboxaldehyde is ideal for providing the 7 coordinate geometry (which is approximately a **face-centered octahedron** or a **face-centered trigonal antiprism**) of the complex [M(py)₃tren)]²⁺ (M = Mn, Fe, Co, Ni, Cu, Zn).

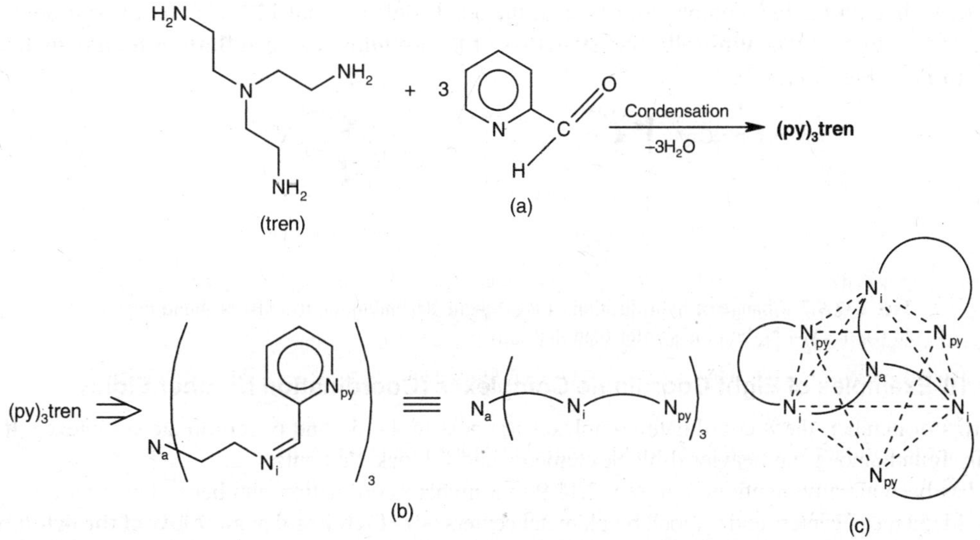

Fig. 2.11.9.6 (a) Synthesis of the polydentate Schiff base ligand **(py)₃tren** (having 7 donor sites); (b) Schematic representation of the ligand (py)₃tren; (c) Schematic representation of the structure of **[M{(py)₃tren}]²⁺** where 3N$_i$ and 3N$_{py}$ ligating sites occupy the octahedral sites and the N$_a$-site (acting as the 7ᵗʰ ligating site) coordinates through the center of a trigonal face. (M–center is not shown).

The ligand (py)₃tren is (C₅ H₄NCH = NCH₂CH₂)₃N that possesses three types of N–donor sites. **3 immine–N** (*i.e.* N$_i$), **3 pyridine N** (*i.e.* N$_{py}$) and **one ammine–N** (*i.e.* N$_a$). The ligand (py)₃tren acting as a **hexadentate ligand** by using $3N_i$ and $3N_{py}$ donor sites can provide an **octahedral geometry**. The 7th binding site (*i.e.* N_a) resides at the center of a trigonal face of the octahedron. This N–site (N_a) is projected towards the t_{2g} orbitals of the octahedron (*cf.* other 6 N sites are projected towards the e_g orbitals). This N-site (N_a) can coordinate to give the 7-coordinate geometry, *i.e.* **capped octahedron or face centerd octahedron**. This 7th bond along the t_{2g} orbital is also described as an **antibond** and the nature of this bond is different from the remaining six 'M—N' bonds formed along the e_g orbitals.

With the increase of electron density in the t_{2g} orbitals, the repulsion increases to elongate the 7th M—N bond (*i.e.* M—N_a bond) while the increase of electron density in the e_g level elongates the other six M—N bonds (*cf.* Sec. 3.14.9 for a detailed discussion). The electron distribution in the d-orbitals for different $[M\{(py)_3tren\}]^{2+}$ complexes is as follows (*cf.* Fig. 3.14.9.1):

	Mn^{2+}(hs)	Fe^{2+} (ls)	Co^{2+}(hs)	Ni^{2+}	Cu^{2+}	Zn^{2+}
t_{2g}:	3	6	5	6	6	6
e_g:	2	0	2	2	3	4

(hs $\Rightarrow$ high spin; ls $\Rightarrow$ low spin)

The **capped M—N_a bond** is longer than the other remaining 6 bonds for the above mentioned complexes. The repulsion causes the maximum bond length for the M—N_a bond for the t_{2g}^6 configuration while it is minimum for the t_{2g}^3 configuration. For the t_{2g}^6 configuration, the repulsion pushes away the N–donor site from the octahedral face (*i.e.* **M—N_a maximum bond length**) while for the t_{2g}^3 configuration the N-donor site is closest to the metal center (*i.e.* **M—N_a minimum bond length**) (*cf.* Fig. 3.14.9.1). It causes a gradual change in the bond angle around N_a in moving from the t_{2g}^3 to t_{2g}^6 configuration. For the Mn^{2+} complex (*i.e.* $t_{2g}^3 e_g^2$), it is about 112° (*i.e.* approximately sp^3 hybridisation of the ammine nitrogen) while for the Fe^{2+} complex (*i.e.* $t_{2g}^6 e_g^0$), the bond angle is about 120° (*i.e.* sp^2 hybridisation of the ammine nitrogen). The **umbrella-like structure of the ammine nitrogen flattens in moving from Mn^{2+} to Fe^{2+}** (Fig. 2.11.9.7).

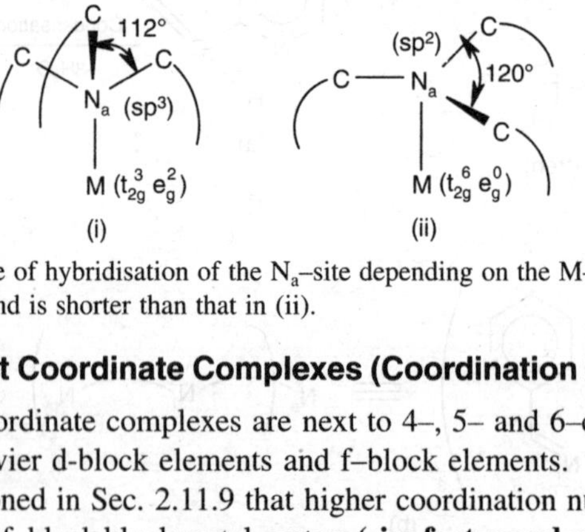

Fig. 2.11.9.7 Change of hybridisation of the N_a–site depending on the M—N_a bond length. In (i), the M—N_a bond is shorter than that in (ii).

2.11.10 Examples of Eight Coordinate Complexes (Coordination Number Eight)

In terms of number, the 8 coordinate complexes are next to 4–, 5– and 6–coordinate complexes. It is mainly found among the heavier d-block elements and f–block elements.

It has been already mentioned in Sec. 2.11.9 that higher coordination number is favoured for:

    4d, 5d metal centers and f–block block metal centers (**size factor and availability of the orbitals**); **small and hard donor sites** (*e.g.* F⁻, CN⁻); chelate ligands with a **small bite** (*e.g.* bpy); high formal oxidation states of the metal center.

A. Geometries for the C.N. 8

The possible geometries are:

● *Cubic, square antiprism* (**D_{4d}**) and *dodecahedron* (**D_{2d}**): These are **mutually interconvertible** and they are of comparable energy. **It leads to the stereochemical nonrigidity for C.N. 8.**

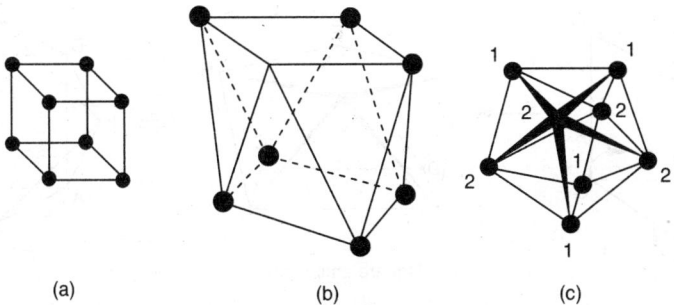

Fig. 2.11.10.1 Some common geometries for 8-coordinate complexes. (a) cube, (b) square antiprism, (c) dodecahedron (having two different types of positions for coordination denoted by 1 and 2) (● denotes the position of the ligating sites)

Other possible geometries are:
- *hexagonal bipyramid* (D_{6h}), *bicapped trigonal prism* (D_{3h}) and *bicapped trigonal antiprism* (D_{3d}).

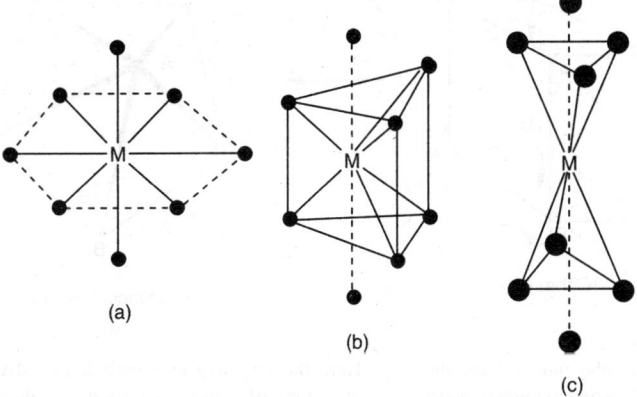

Fig. 2.11.10.2 Octacoordination in (a) hexagonal bipyramid (D_{6h}), (b) bicapped (through the trigonal faces) trigonal prism (D_{3h}) and (c) bicapped trigonal antiprism (D_{3d}) (● denotes the positions of ligating sites).

Note: These three geometries are relatively rarely found in the actinide and lanthanide complexes. Hexagonal bipyramid very often describes the complexes of dioxidocations OMO^{n+}, *e.g.* $O{=}U{=}O^{2+}$ where $O{=}M{=}O$ lies along the axial direction.

- **Stereochemical preference for C.N. 8:** We shall first discuss the structural features of the cubic, square antiprism and dodecahedron geometries. *Though in the ionic solids, cubic coordination (i.e. CsCl type structure) is well known, in the complexes, the 8 coordination number is mainly provided by the square antiprism and dodecahedron geometries*. In fact, the ligand-ligand repulsion disfavours the cubic geometry. The **ligand-ligand repulsion** changes as follows:

cubic >> dodecahedron > square antiprism

Thus in terms of ligand-ligand repulsion, **the square antiprism geometry is favoured** (but slightly) over the dodecahedron geometry.

- **Interconvertible geometries (*i.e.* stereochemical nonrigidity):** The square antiprism geometry can be obtained from a cubic geometry by rotating one face of the cube by 45° about the C_4 axis relative to the opposite face. *It reduces the ligand-ligand repulsion keeping the M—L bond distance unchanged.*

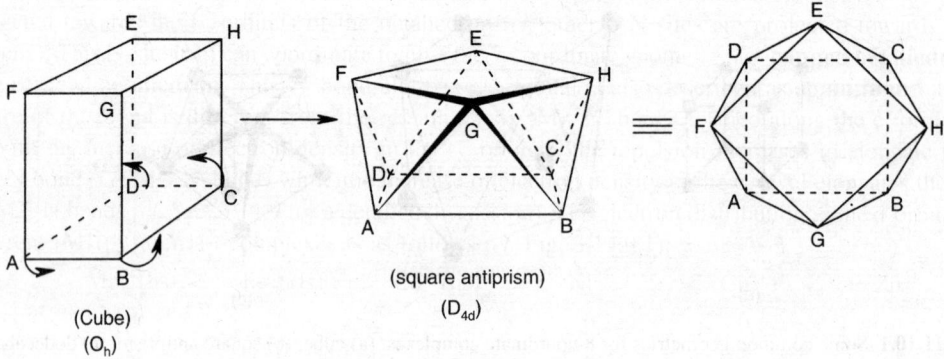

(a) Conversion of a cube into a square antiprism. (Here the face ABCD is rotated by 45° relative to the opposite face EFGH).

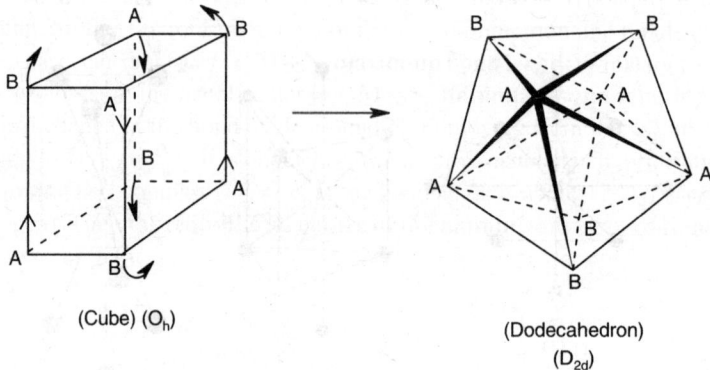

(b) Conversion of a cube into a dodecahedron. (Here the tetrahedron constituted by 4A is flattened by the displacement of its 4 vertices; the tetrahedron constituted by 4B is elongated by the displacement of its vertices.

Fig. 2.11.10.3 (a) Conversion of a cube into a square antiprism. (b) Conversion of a cube into a dodecahedron.

Why not cubic complexes for 8 C.N.?: Orbital restriction and ligand-ligand repulsion

Considering 9 orbitals (*i.e.* 1 ns + 3 np + 5 $(n-1)d$), we can have maximum 7–orbitals for the bonding interaction in a cubic geometry. The $(n-1)d_{z^2}$ and $(n-1)d_{x^2-y^2}$ orbitals are not available for this purpose. These two orbitals are projected towards the faces of the cube while the ligands are approaching along the corners of the cube. The nonavailability of these two d–orbitals in the metal-ligand interaction can also be realised by considering the cube to be constituted by two interlocked tetrahedra. *However, for the f–block elements, i.e. lanthanides and actinides, availability of the $(n-2)f$ orbitals removes the **orbital restriction** applicable for the d–block elements.*

Even when sufficient orbitals are available (as in the lanthanides and actinides) for the cubic coordination, this geometry is not attained in the 8 coordinate complexes. **Because, in the cubic geometry, the ligand-ligand repulsion is maximum** compared to that in other possible 8-coordinate polyhedra like square antiprism, dodecahedron.

The attainment of the dodecahedron from a cubic geometry may be visualised *by considering the starting cube to be constituted by two **interpenetrating tetrahedra**.* If one tetrahedron is elongated and the other tetrahedron is flattened (*i.e.* compressed) then the cube produces the dodecahedron geometry. Thus a system of *two **interlocking tetrahedra** (i.e.* cube) is converted into a system of *two **interlocking trapezoids** (i.e.* dodecahedron). Alternatively, a D_{2d} dodecahedron may be considered as a ***tetra-capped tetrahedron.***

The distortions leading to the generation of square antiprism (D_{4d}) and dodecahedron (D_{2d}) from a cube are illustrated in Fig. 2.11.10.3.

Importance of the bite distance: Square antiprism $\rightleftharpoons$ Dodecahedron;(polytopism)

Though the square antiprism is marginally more stable than the dodecahedron, but in terms of the metal — ligand bonding interactions, both the geometries earn the comparable stabilisation. The **comparable energies of these two stereoisomers give the origin of stereochemical nonrigidity (*i.e. fluxionality* and *polytopism*).** This is why, it may be difficult to predict the actual structure. This is illustrated for the $[M(didentate)_4]^{n+}$ type complexes showing C.N. 8. The actual geometry to be adopted by this type of complexes depends on the ***ligand bite distance*** (distance between the donor atoms of a chelate ring). *The bite distance mainly depends on the size of the chelate ring.* **If the bite distance gradually increases, the geometry gradually changes from the dodecahedron to square antiprism.** *In other words, the smaller bite distance favours the* D_{2d} *dodecahedron geometry while the larger bite distance favours the square antiprism (cf. smaller bite distance favours the trigonal prism over the octahedron).* This prediction has been verified in $[M(didentate)_4]^{n+}$ type complexes (*cf.* Fig. 2.11.10.5).

It is also illustrated in the following examples:

● **[Zr(acac)₄] (Square antiprism**; 6–membered chelate ring, *i.e.* **bite distance is larger.)**

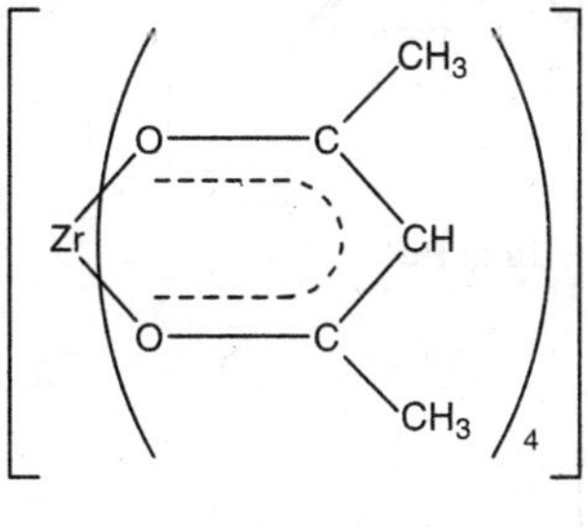

(square antiprism)

● **[Zr(ox)₄]⁴⁻(Dodecahedron;** 5–membered chelate ring, *i.e.* **bite distance is smaller.)**

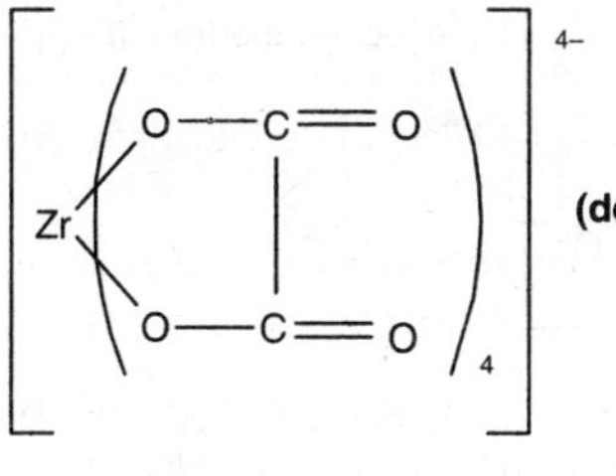

(dodecahedron)

● **Hybridisation concept consideration:** The square antiprism structure may be explained by considering **sp^3d^4 hybridisation** while the dodecahedral structure may arise from the sp^3d^4 or p^3d^5 hybridisation. *Participation of a large number of d–orbitals needs a high formal positive charge on the metal center and low d–electron density* (like d^0, d^1, d^2 configuration) (*see* Vol. 2). *Thus the concept of VBT explains the need of high oxidation state and* d^{0-2} *systems for the 8–coordinate complex formation.* In fact, the 8-coordinate complexes are well known for Zr(IV) (d^0), Mo(IV) (d^2), Mo(V) (d^1), Re(V) (d^2), Re(VI) (d^1) centers.

● **CFT concept consideration** (*cf.* Fig. 3.5.9.1): *In terms of CFT, both the square antiprism and dodecahedron give one d–orbital of very low energy that can accommodate maximum 2–electrons* (*under the low spin condition*). For d^n ($n > 2$), it needs the placement of additional electrons in the energetically unfavourable d–orbitals. *This is why, in terms of* **CFT**, *only the* d^{0-2} *configurations are favoured for the square antiprism and dodecahedron geometries.* This prediction is supported from diamagnetic character of $[Mo(CN)_8]^{4-}$ (a d^2 system that remains paired in the lowest available d–orbital).

B. Examples of complexes having the cubic, square antiprism and dodecahedron structures

● *Cubic geometry* (**distorted**): LnO_2 (Ln = Ce, Pr), $[UF_8]^{3-}$ (in $Na_3[UF_8]$), $[PaF_8]^{3-}$ (in $Na_3[PaF_8]$), $[U(NCS)_8]^{4-}$ (in $(Et_4N)_4[U(NCS)_8]$). High spin Mn(II) attains a cubic coordination with the [2, 2, 2]–cryptand.

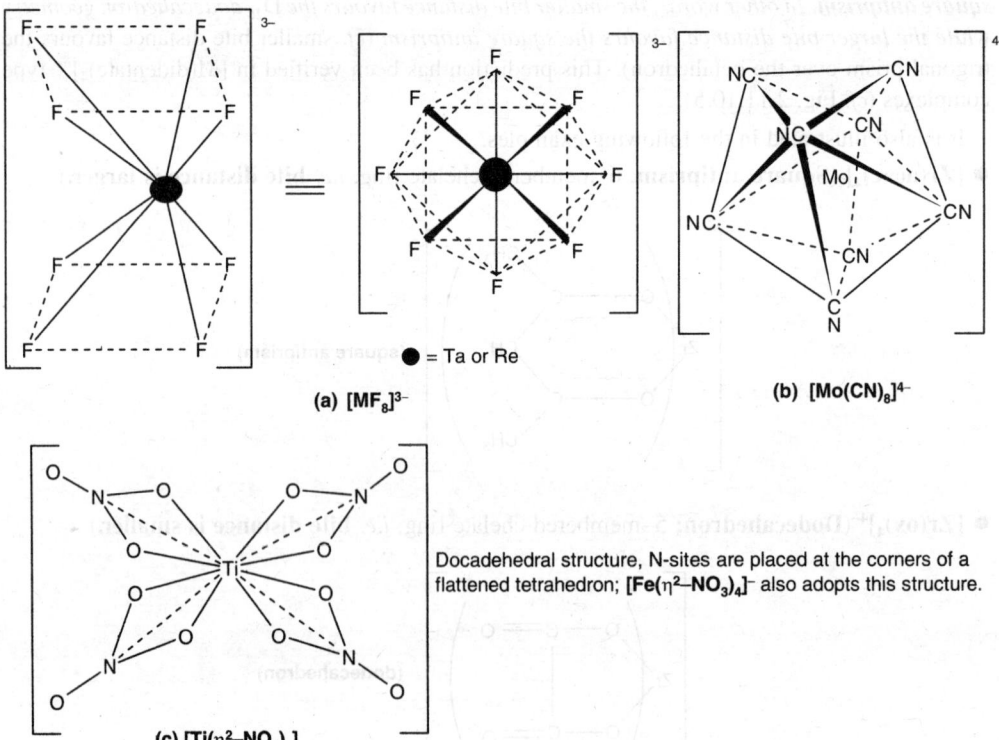

● = Ta or Re

(a) $[MF_8]^{3-}$

(b) $[Mo(CN)_8]^{4-}$

Docadehedral structure, N-sites are placed at the corners of a flattened tetrahedron; $[Fe(\eta^2-NO_3)_4]^-$ also adopts this structure.

(c) $[Ti(\eta^2-NO_3)_4]$

Fig. 2.11.10.4 (a) Square antiprismatic structure of $[ReF_8]^{3-}$; $[TaF_8]^{3-}$. (b) Dodecahedral structure of $[Mo(CN)_8]^{4-}$ (Mo—L) bonds are not shown. (c) Dodecahedral structure of $[Ti(\eta^2-NO_3)_4]$.

Again it may be mentioned that the D_{2d} and D_{4d} symmetries are favoured more than the cubic symmetry in terms of ligand-ligand repulsion.

● **Square antiprism** (D_{4d}): $[Yb(NH_3)_8]^{3+}$, $[M(acac)_4]$ (M = Zr, Th, Ce), $[W(CN)_8]^{3-}$, $[Mo(CN)_8]^{3-}$, $[TaF_8]^{3-}$, $[ReF_8]^{3-}$, $[U(NCS)_8]^{4-}$ [in $Cs_4[U(NCS)_8]$], $[Eu(acac)_3(phen)]$.

Note: $[U(NCS)_8]^{4-}$ can attain both the cubic and square antiprismatic structure in solid state depending on the nature of counter ions, *i.e.* Et_4N^+ and Cs^+ respectively. However, the anion attains the square antiprism geometry in solution.

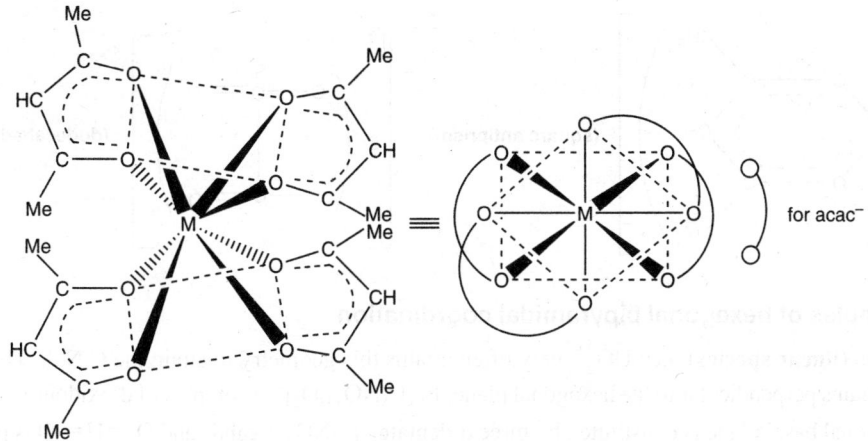

(a) [M(acac)₄] (square antiprism)

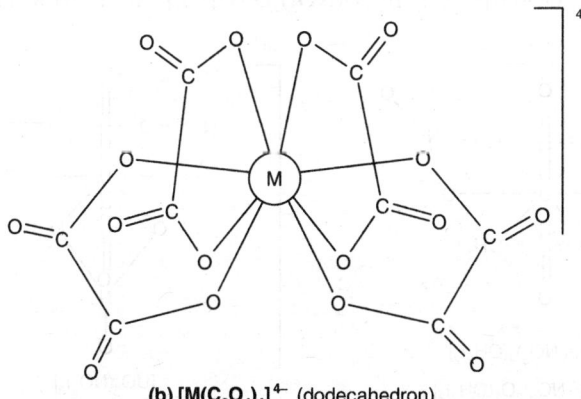

(b) [M(C₂O₄)₄]⁴⁻ (dodecahedron)

Fig. 2.11.10.5 (a) Square antiprismatic structure of $[M(acac)_4]$ (M = Th, U, Zr, etc.; (b) Dodecahedral structure of $[M(ox)_4]^{4-}$(M = Zr, Th, etc.).

● **Dodecahedron** (D_{2d}): $[Ho(tropolonate)_4]^-$, $[Ti(\eta^2-NO_3)_4]$, $[Mo(CN)_8]^{4-}$, $[Mo(CN)_8]^{3-}$, $[Zr(C_2O_4)_4]^{4-}$ etc.

Note: With some didentate ligands having a **small bite, dodecahedral geometry is attained for the 3d–series metal ions**, *e.g.* $[Ti(NO_3)_4]$, $[Mn(NO_3)_4]^{2-}$, $[Cr(O_2)_4]^{3-}$, $[Fe(bpy)_4]^{2+}$, $[Co(NO_3)_4]^{2-}$, $[Fe(NO_3)_4]^-$, etc.

For the 3d series metal ions, the C.N. 8 (in dodecahedron) is only possible when the *bite distance* in the chelate ring (as in bpy, $\eta^2-O_2^{2-}$, $\eta^2-NO_3^-$) is small. *This* **small bite distance** *favours the situation for the smaller cations to provide the C.N.* 8 (dodecahedral geometry). *(**cf. The smaller bite distance*** *favours the dodecahedral geometry while the larger bite distance favours the square antiprism geometry)*

● **Dodecahedron vs. square antiprism:** That the **square antiprism and dodecahedron geometries differ marginally in stability** is supported from the following examples:

● $[Mo(CN)_8]^{4-}$ is dodecahedral in solid state but in solution both the square antiprismatic and dodecahedral structures are noted. $[Zr(acac)_4]$ (**square antiprism**, relatively larger bite distance) vs. $[Zr(ox)_4]^{4+}$ (**dodecahedron**, relatively smaller bite distance)

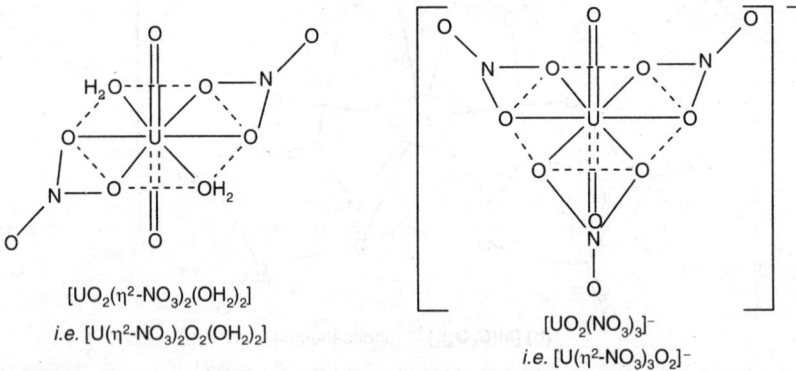

C. Examples of hexagonal bipyramidal coordination

Uranyl ion (**linear species**), *i.e.* UO_2^{2+} very often attains this geometry to attain the C.N. 8. The UO_2^{2+} group remains perpendicular to the hexagonal plane. In $[U(NO_3)_3O_2]^-$ (a complex of dioxidouranium(VI)), the hexagonal basal plane is constituted by three didentate η^2-NO_3^- ligands and $O=U=O$ is projected along the axial direction keeping 'U' at the center of the hexagonal plane. A similar situation is attained in $[U(acetate)_3O_2]^-$, $[U(CO_3)_3O_2]^{4-}$, etc. In $[U(NO_3)_2O_2(OH_2)_2]$, in the basal plane, water molecules coordinate.

$[UO_2(\eta^2\text{-}NO_3)_2(OH_2)_2]$

i.e. $[U(\eta^2\text{-}NO_3)_2O_2(OH_2)_2]$

$[UO_2(NO_3)_3]^-$

i.e. $[U(\eta^2\text{-}NO_3)_3O_2]^-$

Fig. 2.11.10.6 8–coordinate complexes of UO_2^{2+} having the hexagonal bipyramidal structure.

D. Capped trigonal prism

In $[AmCl_2(OH_2)_6]^+$, $2Cl^-$ ligands cap on the trigonal faces of a trigonal prism of water ligands.

2.11.11 Coordination Number Higher than Eight

These higher coordination numbers like 9, 10, 12, etc. are generally shown by *the lighter f–block elements*. Some examples are discussed below.

C.N. 9 (tricapped trigonal prism): $[ReH_9]^{2-}$, $[TeH_9]^{2-}$, $[Nd(OH_2)_9]^{3+}$ (in general $[Ln(OH_2)_9]^{3+}$), etc.

The metal ion is placed at the center of a trigonal prism in which there is an additional ligand over each tetragonal or quadrilateral face. This gives C.N. 9 (= **6 vertices + 3 tetragonal faces**) generating the tricapped trigonal prism.

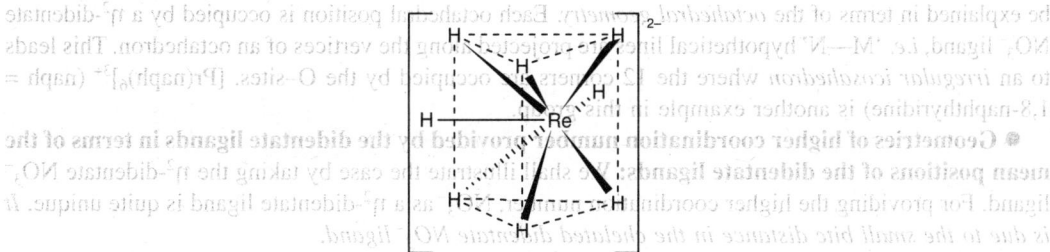

Fig. 2.11.11.1 Tricapped trigonal prismatic structure of $[ReH_9]^{2-}$, (9th H not shown; it resides at the back side).

C.N. 9 (monocapped square antiprism): In this geometry (rarely found), the ninth ligand is placed on a square face of a square antiprism. In solid state, $[ThF_8]^{4-}$ attains the capped square antiprism by the capped F–acting as a bridging ligand. Cl–bridged $[LaCl(OH_2)_7]_2^{4+}$ also attains this structure. $[La(NH_3)_9]^{3+}$ possesses a capped square antiprism structure.

C.N. 10 (double trigonal bipyramid): $[Ce(\eta^2\text{-}NO_3)_5]^{2-}$ in which the NO_3^- ions as the η^2-didentate ligands are arranged about a trigonal bipyramidal geometry, *i.e.* each η^2-didentate NO_3^- is occupying an apex of a trigonal bipyramidal geometry. This leads to 'Ce—N' imaginary lines along the vertices of the trigonal bipyramidal geometry.

C.N. 10 (bicapped square autiprism): Uranium(IV) acetate polymer; $K_4[Th(ox)_4(OH_2)_2]\cdot2H_2O$; $[Ce(\eta^2\text{-}CO_3)_5]^{6-}$.

C.N. 10 (bicapped dodecahedron): $[Ln(bpy)_2(\eta^2\text{-}NO_3)_3]$, $[Ln(NO_3)_5]^{2-}$ (Ln = Ce, Eu).

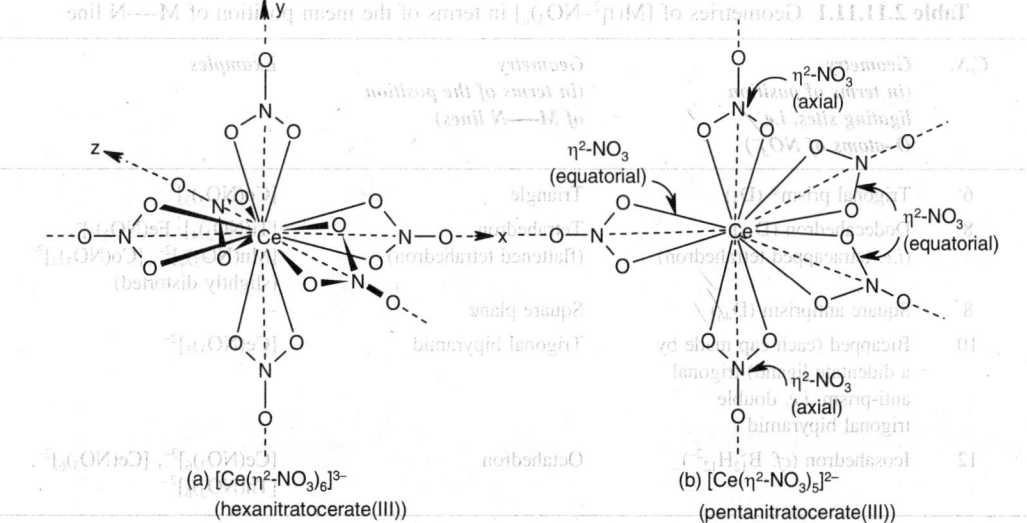

(a) $[Ce(\eta^2\text{-}NO_3)_6]^{3-}$
(hexanitratocerate(III))

(b) $[Ce(\eta^2\text{-}NO_3)_5]^{2-}$
(pentanitratocerate(III))

Fig. 2.11.11.2 Icosahedral structure of $[Ce(NO_3)_6]^{3-}$ having 12 C.N. (The structure can be described in terms of the octahedral structure). **Double trigonal bipyramidal structure** of $[Ce(NO_3)_5]^{2-}$. (This structure can be explained in terms of the trigonal bipyramidal geometry).

C.N. 10 in $[Ce(\eta^2\text{-}NO_3)_4(OPPh_3)_2]$ can be explained in terms of the octahedral geometry. Four η^2-didentate NO_3^- ligands occupy the four basal positions, *i.e.* Ce—N imaginary lines are projected along the corners of the square basal plane. The axial positions are occupied by the phosphine oxygens, *i.e.* $OPPh_3$ acts as a unidentate ligand. Thus the C.N. 10 is attained.

C.N. 12 (icosahedral arrangement): $[Ce(NO_3)_6]^{3-}$, $[Ce(NO_3)_6]^{4-}$, $[Th(NO_3)_6]^{2-}$, etc. The structure can be explained in terms of the *octahedral geometry*. Each octahedral position is occupied by a η^2-didentate NO_3^- ligand, *i.e.* 'M—N' hypothetical lines are projected along the vertices of an octahedron. This leads to an *irregular icosahedron* where the 12 corners are occupied by the O–sites. $[Pr(naph)_6]^{3+}$ (naph = 1,8-naphthyridine) is another example in this group.

● **Geometries of higher coordination number provided by the didentate ligands in terms of the mean positions of the didentate ligands:** We shall illustrate the case by taking the η^2-didentate NO_3^- ligand. For providing the higher coordination number, NO_3^- as a η^2-didentate ligand is quite unique. *It is due to the small bite distance in the chelated didentate NO_3^- ligand.*

$$- - - - M - \!\!-\!\! - N - O - - - - \qquad \nearrow O \quad \text{Bite-distance}$$

(chelating $\eta^2\text{-}NO_3^-$)

The chelated NO_3^- occupies two adjacent coordination sites. The *average position* of the chelated NO_3^- may be reasonably considered to lie along the M—N imaginary line. **Very often, the coordinated O–atoms are twisted about the M—N line to minimise the repulsion.** In terms of the average position (*i.e.* position of the M—N line) of the didentate NO_3^- ligands, the geometries of higher coordination number (say x) can be explained by some simpler geometries satisfying $x/2$ coordination number. It is illustrated in the Table 2.11.11.1.

Table 2.11.11.1 Geometries of $[M(\eta^2\text{-}NO_3)_n]$ in terms of the mean position of M----N line

C.N.	Geometry (in terms of position ligating sites, i.e. O–atoms of NO_3^-)	Geometry (in terms of the position of M—N lines)	Examples
6	Trigonal prism* (D_3)	Triangle	$[Co(NO_3)_3]$
8	Dodecahedron (D_{2d}) (i.e. tetracapped tetrahedron)	Tetrahedron (flattened tetrahedron)	$[Ti(NO_3)_4]$, $Fe(NO_3)_4]^-$, $[Mn(NO_3)_4]^{2-}$, $[Co(NO_3)_4]^{2-}$ (slightly distorted)
8	Square antiprism (D_{4d})	Square plane	–
10	Bicapped (each cap made by a didentate ligand) trigonal anti-prism, i.e. double trigonal bipyramid	Trigonal bipyramid	$[Ce(NO_3)_5]^{2-}$
12	Icosahedron (cf. $B_{12}H_{12}^{2-}$)	Octahedron	$[Ce(NO_3)_6]^{3-}$, $[Ce(NO_3)_6]^{2-}$, $[Th(NO_3)_6]^{2-}$

* Because of the small bite angle, it produces the **trigonal prism instead of the octahedron. In fact, with the decrease of bite angle, $[M(didentate)_3]$ tends to move from the octahedral geometry towards the trigonal prismatic geometry** (cf. Sec. 2.11.8).

For the other η^2-**didentate ligands like** CO_3^{2-}, **acac⁻**, the geometries can be explained by considering the mean positions of such didentate ligands, *i.e.* positions of the M——C lines. However, **the bite distance in** $M(\eta^2$-**acac) is larger than that in** $M(\eta^2$-$CO_3)$ or in M(naph).

EXERCISE 2

A. General Questions

1. What do you mean by isomerism? Compare the different types of isomerism in organic compounds and coordination compounds.
2. What are the different types of structural isomerism in coordination compounds.
3. Illustrate the ionisation isomerism, hydration isomerism, coordination isomerism, and ligand isomerism with examples.
6. Discuss the different factors to control the relative stabilities of the linkage isomers.
7. Illustrate the symbiotic and antisymbiotic effect in connection with the relative stabilities of linkage isomers.
8. Discuss the competitive and noncompetitive π-bonding interaction in determining the relative stabilities of the linkage isomers.
9. How do the steric factors determine the relative stabilities of the linkage isomers?
10. Discuss the conformational or polytopal isomerism in 4-coordination complexes with suitable examples.
11. Find the total number of stereoisomers in the following cases (octahedral complexes).
    (i) $[MA_4B_2]$, (ii) $[MA_3B_3]$, (iii) $[MA_3BCD]$, (iv) $[MA_4BC]$, (v) $[MA_2B_2C_2]$, (vi) $[MA_2B_2CD]$, (vii) $[MA_2B_2(CC)]$, (viii) $[M(AB)_3]$, (ix) $[MABCDE_2]$, (x) $[MABCD(EE)]$; (xi) $[M(AB)_2X_2]$, (xii) $[M(AA)B_2CD]$, (xiii) $[M(AB)_2CD]$, (xiv) $[M(AA)_2BC]$, (xv) $[M(AA)_2(BB)]$, (xvi) $[M(AAA)_2]$, (xvii) $[M(AAAA)B_2]$
12. Explain the origin of optical rotation.
13. Illustrate the chirality symbols R/S and C/A (*cf.* Sec. 1.15.5). What do you mean by absolute configuration in the optically active compounds? How do you define and determine the Δ and Λ configurations for the optically active octahedral complexes like $[M(AA)_3]$, $[M(AB)_3]$, *cis*–$[M(AA)_2B_2]$.
14. How do you define the δ– and λ–conformations in the M(en) chelate ring? Give the isomers in $[M(en)(NH_3)_4]$ (octahedral) and $[M(en)_2]$ (planar) complexes in terms of the δ– and λ–comformations of the chelate rings.
15. Give the optical isomers in $[M(en)_3]$, $[M(pn)_3]$ and $[M(ala)_3]$ in terms of the δ– and λ– conformations of the chelate rings. Comment on the stability order of these isomers.
16. What do you mean by stereoselectivity in the optically active compounds having the chelate rings?

17. What do you mean by ORD and CD curves? Give their applications to determine the absolute configurations of the optically active octahedral complexes.
18. Illustrate the optical isomerism in the tetrahedral, square planar and square pyramidal complexes.
19. Discuss the optical isomerism in the mononuclear and dinuclear octahedral complexes.
20. Discuss the optical activity arising from the chelate ring conformations.
21. Discuss the different factors to control the preferences towards the tetrahedral, square planar and octahedral geometries.
22. Sometimes, the steric requirements of ligands may enforce some specific geometries of coordination compounds—illustrate with examples for TBP, square planar, trigonal prismatic and pentagonal bipyramidal geometries.
23. C.N. 2 for the transition metals cannot be simply considered as the result of *sp*–hybridisation– explain.
24. C.N. 6 can be attained by both the octahedral and trigonal prismatic geometries but the octahedral complexes are more abundant—explain.
25. Discuss the methods of resolution of the optically active coordination compounds.
26. Writes notes on:
    (i) Symbiotic and antisymbiotic effect; (ii) Berry's pseudorotation; (iii) Bailer's twist mechanism; (iv) trigonal and tetragonal distortion in octahedral complexes, (v) rhombic distortion in octahedral complexes, (vi) chelate ring conformations and stereoselectivity; (vii) polytopism and fluxional molecules, (viii) stereochemical nonrigidity in coordination compounds; (ix) plane polarised light, (x) Cotton effect in ORD and CD curves, (xi) stereochemical nonrigidity of the 8-coordinate complexes of cubic geometry, square antiprismatic geometry and dodecahedral geometry; (xii) capped octahedral and capped trigonal prismatic complexes; (xiii) square antiprismatic and dodecahedral complexes, (xiv) 9, 10 and 12 coordinate complexes, (xv) R/S and C/A chirality symbols. (xvi) Stereoisomerism in $[Co(dien)X_2Y]^+$, $[Co(dien)_2]^{3+}$ and $[CoCl_2(trien)]^+$, (xvii) Optical isomers of $[Co(en)_3]^{3+}$ by considering the chelate ring conformations, (xviii) octahedron *vs.* trigonal prism, (xix) square antiprism *vs.* dodecahedron, (xx) tetrahedron *vs.* square planar, (xxi) Complexes of coordination number 1, 3, 7, 8, 9, 10, 12, (xxii) Ligand imposed C.N. 5, 4 and 7.

B. Justify the following statements

1. There are various types of isomerisms in coordination compounds.
2. Ionisation isomerism may be studied by measuring the molar conductivities.
3. Ionisation isomerism may lead to different types of colours.
4. The concept of ionisation isomerism is meaningful only for the kinetically inert centers.
5. Green solution of commercial $CrCl_3 \cdot 6H_2O$ on standing becomes gradually violet coloured.
6. Ionisation and hydration isomerism of the coordination compounds are basically of the same type.
7. *tren* and *trien* are the isomeric ligands but they prefer different types of geometries in coordination compounds.
8. α–ala and β–ala are the isomeric ligands but they form the complexes of different stabilities.
   ● tren and trien are the isomeric ligands but they favour different coordination geometries.

9. $\left[Co(NH_3)_5(OH_2) \right]^{3+} \xrightarrow{\text{NaNO}_2/\text{HCl}} $ Red compound $\xrightarrow{\text{Heating}}$ Yellow compound

10. The linkage isomers produced by the ambidentate ligands like SCN^- or NO_2^- can be identified by *ir*–studies.

11. $\left[Pd(SCN)_4\right]^{2-}$ $\xrightarrow[\text{(low temp.)}]{\text{bpy}}$ Orange yellow compound $\xrightarrow{\text{Heating}}$ Light yellow compound

12. Thermodynamic stabilities of the linkage isomers follow the sequences:

    (a) $[M(SCN)_4]^{2-} \rangle [M(NCS)_4]^{2-}$, (M = Pd, Pt, Cd, Hg).

    (b) $[M(NCS)(NH_3)_5]^{2+} \rangle [M(NH_3)_5(SCN)]^{2+}$, (M = Co, Rh)

    (c) $[Co(SCN)_4]^{2-} \langle [Co(NCS)_4]^{2-}$

    (d) $[Co(CN)_5(SCN)]^{3-} \rangle [Co(CN)_5(NCS)]^{3-}$

    (e) $cis-[M(NH_3)_2(SCN)_2] \rangle cis-[M(NCS)_2(NH_3)_2]$, (M = Pd, Pt)

    (f) $cis-[M(NCS)_2(PR_3)_2] \rangle cis-[M(PR_3)_2(SCN)_2]$, (M = Pd, Pt)

    (g) $[Cr(OH_2)_5(SCN)]^{2+} \langle [Cr(NCS)(OH_2)_5]^{2+}$

    (h) $[Fe(NCSe)_4]^{2-} \rangle [Fe(SeCN)_4]^{2-}$

    (i) $[Fe(CO)_2(C_5H_5)(SeCN)] \rangle [Fe(CO)_2(C_5H_5)(NCSe)]$

    (j) $trans-[Pd(SCN)_2(PR_3)_2] \rangle trans-[Pd(NCS)_2(PR_3)_2]$

    (k) $[M(NCS)_2(bpy)] \rangle [M(SCN)_2(bpy)]$ (M = Pd, Pt)

    (l)

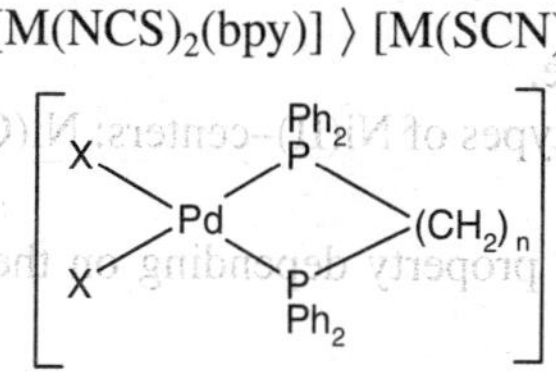

    (i) n = 2, X preferably binds through the S–end

    (ii) n = 3, X preferably binds through the N–end.

    (m)

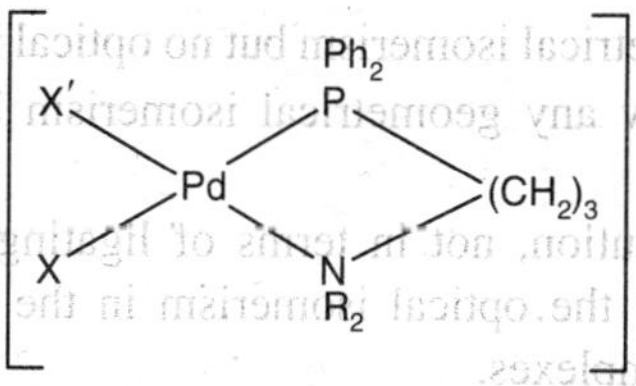

    X' preferably binds through the S–end while
    X preferably binds through the N–end

    (n)

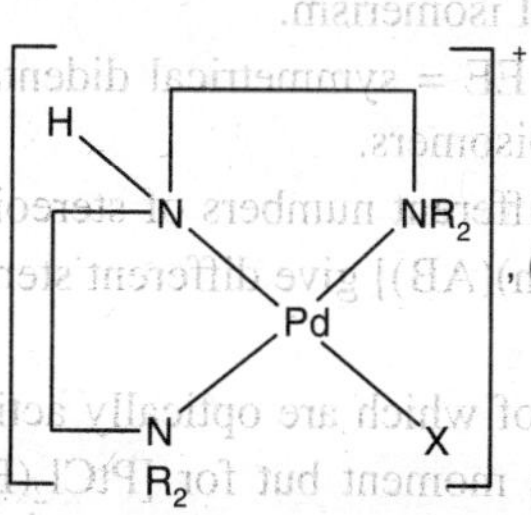

    (i) R = H, X binds preferably through the Se–end,
    (ii) R = Et, X binds preferably through the N–end.

13. Dithiooxalate and thiosulfate can produce the linkage isomers.

14. XCN (X = S, Se, O) can produce linkage isomers but the steric requirements of these isomers are different.

15. $\left[Co(CN)_5 (thiosulphate) \right]^{4-}$ (red) $\xrightarrow{H^+}$ liberation of H_2S

$\left[Co(CN)_5 (thiosulphate) \right]^{4-}$ (yellow) $\xrightarrow{H^+}$ liberation of S.

16. The nature of linkage isomers produced by the flexidentate ligands like H_4edta, Hhis, H_2cys etc. depends on pH.

17. The ligating properties of a peptide linkage depends on pH.

18. (i) $K_3\left[Cr(CN)_6 \right] + FeSO_4 \xrightarrow{Cold\ condition}$ Brick red solid

$$\downarrow \text{Heating}$$

Dark green solid.

(ii) $\begin{array}{c} K_4\left[Fe(CN)_6 \right] + FeCl_3 \\ K_3\left[Fe(CN)_6 \right] + FeCl_2 \end{array}$ $\begin{array}{c} \searrow \\ \nearrow \\ \text{Heating} \end{array}$ Prussian blue

(iii) Turnbul blue on standing isomerises to prussian blue.

19. In the polymeric structure of $Ni(CN)_2 \cdot NH_3$, there are two types of Ni(II)–centers: $Ni(C)_4$ (planar) and $Ni(N_6)$ (octahedral).

20. $[NiX_2(PR_3)_2]$ (X = halide) shows an anomalous magnetic property depending on the degree of steric crowding.

21. $[NiX_2(PR_3)_2]$ can show the conformational isomerism.

22. $[NiL_2]$ (LH = salicylaldimine) shows an anomalous magnetic property that depends on the degree of steric crowding on the N–site.

23. $[M(A-B)_2]$ type planar complexes can show the geometrical isomerism but no optical isomerism.

24. $[M(A-B)_2]$ type tetrahedral complexes cannot show any geometrical isomerism but optical isomerism.

25. Unsymmetrical didentate ligands (in terms of substitution, not in terms of ligating sites; *e.g.* substituted glyoximes, β–diketones etc.) can produce the optical isomerism in the tetrahedral complexes and geometric isomersim in the planar complexes.

26. $[Mo(CO)_2(\eta^5-C_5H_5)Cl(PR_3)]$ and $[ReBr_2(CO)_2(\eta^5-C_5H_5)]$ can lead to the geometric isomers. The former can also develop the optical isomerism but the later cannot.

27. $[MA_2B_2C_2]$ type octahedral complexes can show the optical isomerism.

28. $[MABCDE_2]$ and $[MABCD(EE)_2]$ (E = unidentate ligand, EE = symmetrical didentate ligand) type octahedral complexes give different numbers of stereoisomers.

29. $[M(AA)_3]$ and $[M(AB)_3]$ type octahedral complexes give different numbers of stereoisomers.
    - $[M(dien)ABC]$, $[M(tren)XY]$, $[M(tren)(AB)]$ and $[M(trien)(AB)]$ give different stereoisomers.
    - $[M(dien)_2]$ give different stereoismers.
    - $[M(OH_2)_2(trien)]$ give different stereoisomers and some of which are optically active.

30. For $[PtCl_2(PR_3)_2]$, the *trans*–isomer shows 0 (zero) dipole moment but for $[PtCl_2(Et_2S)_2]$, the *trans*–isomer shows a nonzero dipole moment.

31. Intensity of the electronic spectra of the *cis–trans* isomers of octahedral complexes may differ.

32. *cis–trans* isomers of the octahedral complexes may differ in optical activity.

33. *cis–trans* isomers can be distinguished by chemical methods (*e.g.* Grinberg's method, Kurnakov's method).

34. Cis - $\left[PtCl_2 (NH_3)_2 \right] \xrightarrow{S_2O_3^{2-}} \left[Pt(S_2O_3)_2 \right]^{2-}$

$\xrightarrow{tu} \left[Pt(tu)_4 \right]^{2+}$

$trans$ - $\left[PtCl_2 (NH_3)_2 \right] \xrightarrow[\text{(aqueous solution)}]{S_2O_3^{2-}} trans$ - $\left[Pt(NH_3)_2 (OH_2)(S_2O_3) \right]$

$\xrightarrow{tu} trans$ - $\left[Pt(NH_3)_2 (tu)_2 \right]^{2+}$

35. Plane of the plane polarised light is rotated when it interacts with an optically active substance.

36. The designations *dextrorotatory* and *laevorotatory* in terms of the sign of the angle of optical rotation cannot tell us the absolute configurations of the optically active compounds.

37. The *d/l* designation is determined by the sign of optical rotation that depends on the wavelength of the light, but the designation by Δ/Λ or D/L does not require the knowledge of the sign of optical rotation.

38. (a) By considering the nature of Cotton effect (in ORD and CD curves), the absolute configurations of an optically active compound can be determined.

    (b) In terms of the conformations of the individual chelate ring, [M(en)(NH_3)_4] and [M(en)_2] (planar) complexes can produce the stereoisomers that may be optically active.

39. By considering the conformations of the individual chelate rings, [M(en)_3] should produce 8–stereoisomers (*i.e.* 4 pairs of enantiomers) of different thermodynamic stabilities.

40. Because of stereoselectivity, all the predicted 8-stereoisomers of [M(en)_3] are not obtained.

41. In [M(en)_3], the Δ–$\lambda\lambda\lambda$ isomer (*i.e. lel*–from) is more stable than the Δ–$\delta\delta\delta$ isomer (*i.e. ob*–form).

42. For Λ-[M(en)_3], the δ–conformation of a chelate ring is more stable than the λ–conformation. But in solution, the Λ–$\delta\delta\lambda$ isomer is more abundant than the Λ–$\delta\delta\delta$ isomer.

43. ORD (optical rotatory dispersion) and CD (circular dichorism) curves are important to find out the absolute configurations of the optically active compounds.

44. In M(pn) chelate ring, for R(–)–pn, λ–chelate ring is more stable than the δ–chelate ring while for S(+) – pn, the reverse is true.

45. For [M(pn)_3], the stable isomers are Λ–$\delta\delta\delta$, Λ–$\lambda\lambda\lambda$, Δ–$\delta\delta\delta$ and Δ–$\lambda\lambda\lambda$.

46. In the electronic spectra of [Co(en)_3]^{3+}, two peaks are noticed (without splitting); in the CD curve, for the lower energy band of electronic spectrum, there are two Cotton effects but for the higher energy band there is only one Cotton effect.

47. It is very difficult to resolve the optical isomers of tetrahedral complexes.

48. In some special cases, optical isomers may be obtained for the square planar complexes.

49. There are some optically active octahedral complexes having no organic ligands.

50. [Co(pn)_3]^{3+} can generate several optically active isomers.

51. Ligand chirality can introduce the optical activity in both the octahedral and planar complexes.

52. For coordination number 4, the T_d, D_{2d} and D_{4h} symmetries can lead to polytopism.

53. In the equilibrium: $T_d \rightleftharpoons D_{2d} \rightleftharpoons D_{4h}$, position of the equilibrium is determined by the *cfse* factor and steric factor.

54. In five coordinate systems, the stereochemical nonrigidity may arise.

56. The stereochemical nonrigidity described by the equilibrium, $T_d \rightleftharpoons D_{4h}$ is a case of polytopism but not a case of fluxional character.

57. The Berry–pseudorotation leading to: TBP (I) $\rightleftharpoons$ SP $\rightleftharpoons$ TBP (II) is a case of fluxional behaviour of the TBP geometries.

58. For the 6 coordinate complexes, octahedral geometry is abundant but trigonal prismatic geometry is found only in some limited cases.

59. The octahedral geometry may experience different types of distortions to reduce the symmetry.

60. The 7–coordinate complexes are stereochemically nonrigid.

61. Coordination number 2 is shown by the metal centers like Ag(I), Au(I), Hg(II), etc. characterised by the d^{10} configuration.

62. Among Cu(I), Ag(I) and Au(I), the heaviest congener, *i.e.* Au(I) is the most suitable candidate for showing the coordination number 2.

63. The simple *sp*-hybridisation concept is insufficient for explaining the development of coordination number 2 in transition metal complexes.

64. Some bulky and anionic ligands can provide the C.N. 2 to the metal centers of d^n configuration ($n \neq 10$).

65. In terms of step-wise stability constants, there is a huge drop in K_3 for the $Hg^{2+} - CN^-$ system.

66. In $Ag^+ - NH_3$ system, the sequence $K_2 \rangle K_1$ goes against the statistical prediction.

67. $[Ag(NH_3)_2]^+$ is more stable than $[Ag(en)]^+$. This denies the chelate effect.

68. In $\left[Ag \begin{array}{c} H_2 \\ N \\ \diagdown \\ N \\ H_2 \end{array} (CH_2)_n \right]$, (n = 2, 3, 4), the stability increases with the increase of n. This is apparently unusual.

69. The tendency to adopt C.N. 2 changes in the sequence Hg(II) $\rangle\rangle$ Cd(II), Zn(II). In fact, for Zn(II) and Cd(II), C.N. 2 is not practically known.

● Among Cu(I), Ag(I) and Au(I), Au(I) is the most suitable centre for C.N.2.

70. In the three coordinate complexes, even for the same ligand, symmetrical geometry is not obtained.

71. In the anion of K[Cu(CN)$_2$], C.N. 2 is not maintained in solid state.

72. For $K_2[Ni(CN)_3]$ or $Cs[CuCl_3]$, in the anionic part, the apparent C.N. 3 is not observed in solid state.

73. Bulky, anionic character and weak field strength of the ligands favour the T_d geometry in the 4–coordinate complexes.

74. Under the comparable conditions, Co(II) is a better candidate to adopt a tetrahedral geometry than Ni(II).

75. The D_{2d} or T_d symmetry of $[CuCl_4]^{2-}$ depends on the mode of crystal packing.

76. Tetrahedral complexes are, in general, high-spin but in some rare cases, low spin tetrahedral complexes have been reported.

77. Formation of $[CoCl_4]^{2-}$ in presence of conc. HCl is well known but $[NiCl_4]^{2-}$ can only be produced in noncoordinating solvents.

78. The structural properties of the complexes simply represented by $[M(acac)_2]$ (M = Ni, Co) are widely different in solid state for Co(II) and Ni(II).

79. The magnetic properties and colours of *Lifschitz salts* depend on the nature of solvent.

80. The d^8–systems can form the square planar complexes easily.

81. Some ligands can enforce the planar geometry for the four coordinate complexes.

82. The 5-coordinate TBP and SP geometries are very often easily intercovertible.

83. [CoCl$_2$(dien)] does not really exist as a 5–coordinate complex.

84. K$^+$ cannot stabilise [Ni(CN)$_5$]$^{3-}$ but the bulky cations like [Cr(en)$_3$]$^{3+}$, [Cr(pn)$_3$]$^{3+}$, etc. can stabilise the anion.

85. In [Cr(en)$_3$][Ni(CN)$_5$] · 1.5 H$_2$O, the anion can exist in both TBP and SP geometries.

86. [CoCl(dppe)$_2$]$^+$ can give two isomers of different colours.

87. [Ni(Et$_4$dien)X$_2$] does not exist as a 5–coordinate complex.

88. In [Fe(CO)$_5$], the equatorial bond is longer, while in [Mn(CO)$_5$]$^+$ the axial bond is longer. (Both are having the TBP structures).

89. In PCl$_5$, the axial bond is longer but in [CuCl$_5$]$^{3+}$, the equatorial bond is longer.

90. In [CuCl$_5$]$^{3-}$, the axial and equatorial bonds differ more than those in [CdCl$_5$]$^{3-}$. (Both are assumed to have the TBP geometry).

91. In the TBP structure of P–compounds, Me–groups are preferably positioned in the equatorial directions while in the transition metal complexes, Me–groups occupy the axial directions keeping the equatorial positions reserved for the π–bonding ligands.

92. The tripodal ligands very often can induce the TBP geometry in presence of some suitable monodentate ligands.

93. N(CH$_2$CH$_2$NH$_2$)$_3$ is more suitable than N(CH$_2$CH$_2$CH$_2$NH$_2$)$_3$ for providing the TBP geometry. In fact, N(CH$_2$CH$_2$CH$_2$NH$_2$)$_3$ can act as a **facultative ligand** to provide both the tetrahedral and trigonal bipyramidal geometry.

94. [M(diars)$_2$]$^{2+}$ of d^8–systems can easily adopt the square pyramidal geometry in presence of suitable monodentate ligands like halides.

95. The colour of [VO(acac)$_2$] depends on the nature of solvents.

96. Through distortion, an octahedron can be converted into a trigonal prism.

97. In terms of C.F.T, the d^0, d^1, d^5 (hs), d^7(hs) and d^{10} configurations may adopt the trigonal prismatic (TP) structure if other factors favour the TP structure.

98. [M(S$_2$C$_2$R$_2$)$_3$] can adopt both the trigonal prismatic (TP) and octahedral(O$_h$) structure:

$$O_h \rightleftharpoons TP \xrightarrow{\text{Reduction}} O_h.$$

i.e. the TP structure is favoured over the O$_h$ structure but on reduction the O$_h$ structure is favoured. (S$_2$C$_2$R$_2$ denotes dithiolene)

99. There are many factors in the complex [M(S$_2$C$_2$R$_2$)$_3$] to favour the trigonal prismatic structure.

100. There is an ambiguity in assigning the oxidation state of the metal center in [M(S$_2$C$_2$R$_2$)$_3$] type complexes.

101. MoS$_2$ can act as a lubricant and it can be explained from its structure.

102. The rigid hexadentate ligands like py$_3$tach, py$_3$tren, etc. can enforce the trigonal prismatic structure.

103. The ligand py$_3$tren can act as a **facultative ligand** to provide the both 6 (trigonal prismatic)- and 7–coordinate (capped octahedral) complexes.

104. Higher coordination number is favoured under the conditions: heavier congeners of the d–block elements and f–block elements; small and hard ligands; and high oxidation state of the metal center.

105. The C.N. 6 is favoured over the C.N. 7 (in general).

106. The possible geometries for C.N. 7 are very often stereochemically nonrigid.

107. py$_3$tren can produce the face-centerd octahedral geometry (*i.e.* C.N. 7) of the complex [M(py$_3$tren)]$^{2+}$ (M = Mn and Fe): The 7th 'M – L' bond differs significantly for the Fe–complex (low spin) and Mn–complex (high-spin).

● State of hybridisation of the ammine-N differs in different metal complexes.

108. The geometries: cube, square antiprism and dodecahedron for C.N. 8 are interconvertible.

109. For C.N. 8, cubic geometry is disfavoured compared to the other geometries like square antiprism, and dodecahedron.

110. For the d–block elements, cubic geometry cannot attained from the concept of orbitals restriction.

111. $[M(ox)_4]^{4-}$ adopts the dodecahedral structure while $[M(acac)_4]$ adopts the square antiprism structure (M = Zr, Th, U).
    ● Both dodecahedral and square antiprismatic structures of $[Mo(CN)_8]^{4-}$ are known.

112. For the square antiprism or dodecahedron, d^0, d^1 and d^2 configurations are only favoured.

113. The geometries of $[Ce(NO_3)_6]^{3-}$ (C.N. 12), and $[Ce(NO_3)_5]^{2-}$ (C.N. 10) can be explained by considering some simple geometries describing C.N. = 6 and 5 respectively.

114. $[Co(edta)]^-$ is a six coordinate complex but $[Mn(edta)(OH_2)]^{2-}$ and $[Fe(edta)(OH_2)]^{2-}$ are the 7–coordinate complexes.

115. $[Co(X)_2(trien)]^+$ (X = CN) can produce two linkage isomers that may remain in equilibrium.

116. NO_3^- as a didentate chelating ligand is suitable for providing the higher coordination number even for the first transition series metal centers.

117. Hexagonal bipyramidal (C.N. 8) geometries are common for UO_2^{2+}.

118. $[Sb(ox)_3]^{3-}$ is not octahedral but pentagonal pyramidal.

119. Low spin tetrahedral complexes can be generated under some special conditions.

120. 1, 2-dithiolates specially favour the square planar geometry.

121. High spin square planar complexes of Ni(II) are rarely known.

122. In $[Cu_2Cl_6]^{2-}$, Cu(II) maintains a square pyramidal geometry.

123. Octatedron and trigonal prism describe a common geometry but they differ in twist angle.
    ● Octahedron· is basically a trigonal antiprism.

124. Different geometries of $[U(NCS)_8]^{4-}$ are known.

125. Stabilities of the square antiprism and dodecahedron geometries differ marginally.

CHAPTER 3

Bonding Theories of Metal Complexes

3.1 INTRODUCTION

Different theories like **electrostatic theory** (developed by Kossel, Magnus, Van Arekel, Garrick), **Sidgwick's effective atomic number rule** (EAN), **Pauling's valence bond theory** (VBT), **crystal field theory** (CFT), **adjusted crystal field theory** (ACFT) **or ligand field theory** (LFT) providing some allowance of covalence in CFT, and **moleculer orbital theory** (MOT) have been proposed for understanding of bonding in coordination compounds.

3.2 ELECTROSTATIC THEORY OF BONDING IN COORDINATION COMPOUNDS

This theory considers the electrostatic attraction between the positively charged metal ion and the negatively charged ligands or negative ends of polar molecules acting as the ligands. Thus complexing power of a particular metal ion depends on its **charge density** called **ionic potential** (= charge/radius). In fact, ionic potential is a good parameter to measure the complexing power of metal ions (*cf.* Table 3.2.1).

Table 3.2.1 Ionic potential values of some metal ions and their complexing power.

Metal ion	Ionic potential*	Complexing power
Cs$^+$	0.55	
Rb$^+$	0.60	
K$^+$	0.66	very poor
Na$^+$	0.86	
Li$^+$	1.11	
Ba^{2+}	1.34	
Cu$^+$	1.35 (T_d)	
Pb^{2+}	1.50	
Ca^{2+}	1.75	Large number of complexes are known
Hg^{2+}	1.82 (T_d)	
Cd^{2+}	1.83	

(Contd.)

Table 3.2.1 (*Contd.*)

Metal ion	Ionic potential*	Complexing power
Mn^{2+}	2.06	
Fe^{2+}	2.17	
Co^{2+}	2.25	Large number of complexes are known.
Zn^{2+}	2.27	
Cu^{2+}	2.30	
Mg^{2+}	2.32	
Ni^{2+}	2.41	
Rh^{3+}	3.70	
Fe^{3+}	3.80	
Ga^{3+}	3.94	Very High (*i.e.* Complexing power most
Cr^{3+}	3.95	pronounced)
Co^{3+}	4.00	
Al^{3+}	4.41	
Sn^{4+}	4.81	
Be^{2+}	4.88 (T_d)	
Pt^{4+}	5.20	

* Ionic potential is given by q/r where q denotes the number of units of positive charge on the metal ion and r gives the radius of the metal ion in A° for 'high-spin' 6 coordination number (octahedral), if not mentioned.

Merits and Demerits of the Theory

- **Complexation by neutral ligands:** For a particular metal ion, the *anionic ligands are expected to form more stable complexes compared to the neutral ligands* like CO, C_2H_4, PR_3, etc. For the anionic ligand, the electrostatic force of attraction is of much more importance than the ion-dipole interaction that prevails for the neutral ligands. However, this prediction is not experimentally always verified. In fact, the neutral ligands like CO, C_2H_4, PR_3, etc. can form very stable complexes with the metal ions of low oxidation states.

- **Relative complexing power:** *The electrostatic theory can explain the relative complexing power of the nontransition metal ions but it fails very often for the transition metal ions.* In terms of the q/r value, Al^{3+} complexes should be more stable than the complexes formed by Co^{3+}, Fe^{3+}, Cr^{3+}, etc. but in reality it does not happen so.

- **Magnetic properties:** For the transition metal complexes, electrostatic theory cannot explain the observed magnetic properties. The illustrative examples are: $[Fe(OH_2)_6]^{2+}$ is paramagnetic while $[Fe(CN)_6]^{4-}$ is diamagnetic; $[Ni(CN)_4]^{2-}$ is diamagnetic while $[NiCl_4]^{2-}$, $[Ni(NH_3)_6]^{2+}$ are paramagnetic; $[CoF_6]^{3-}$ is paramagnetic while $[Co(NH_3)_6]^{3+}$, $[Co(en)_3]^{3+}$ are diamagnetic.

- **Colour of the complexes:** Electrostatic theory fails to explain the colour of the complexes, specially transition metal complexes.

- **Introduction of covalency:** For the metal ions of high q/r values, polarization (*cf.* Fajan's rule) is expected to introduce the covalency in the metal-ligand interaction. In such cases, the concept of simple electrostatic attraction does not work.

3.3 SIDGWICK'S EFFECTIVE ATOMIC NUMBER RULE: THE 18-ELECTRON RULE
(cf. Secs. 3.17.5-6 for explanation and Sec. 12.8.5, Vol. 2 of Fundamental Concepts of Inorganic Chemistry)

3.3.1 Illustration of the EAN Rule

According to this rule, the ligands donate the electrons to the central metal ion through the **covalent-coordinate bonds** and the total number of electrons on the central metal ion including those gained from the ligands is the atomic number of the nearest noble gas. This is referred to as the **effective atomic number (EAN) rule.** For the transition metal complexes, this corresponds to the filling of $(n-1)d$, ns and np orbitals of the central metal. **These 9 (= 5 + 1 + 3) valency orbitals can accommodate 18 valency electrons. This is why, the EAN rule is also described as the 18-electron rule.**

Table 3.3.1.1 Illustration of the EAN rule in different metal complexes.

Complex	Electrons on the metal centre	Electrons donated by the ligands	Total No. of valence shell	EAN	Next Noble gas electrons
$[Co(CN)_6]^{3-}$	24 (as Co^{3+})	6×2	$6 + 12 = 18$	$24 + 12 = 36$	^{36}Kr
$[Cd(NH_3)_4]^{2+}$	46 (as Cd^{2+})	4×2	$10 + 8 = 18$	$46 + 8 = 54$	^{54}Xe
$[Cr(CO)_6]$	24	6×2	$6 + 12 = 18$	$24 + 12 = 36$	^{36}Kr
$[Mn(CO)_6]^+$	24 (as Mn^+)	6×2	$6 + 12 = 18$	$24 + 12 = 36$	^{36}Kr
$[Mn(CH_3)(CO)_5]$	24 (as Mn^+)	6×2	$6 + 12 = 18$	$24 + 12 = 36$	^{36}Kr
$[Mn(CO)_5Cl]$	24 (as Mn^+)	6×2	$6 + 12 = 18$	$24 + 12 = 36$	^{36}Kr
$[Fe(CO)_5]$	26	5×2	$8 + 10 = 18$	$26 + 10 = 36$	^{36}Kr
$[Co(CO)_4Cl]$	26 (as Co^+)	5×2	$8 + 10 = 18$	$26 + 10 = 36$	^{36}Kr
$[Co(CO)_3(NO)]$	27	$3 \times 2 + 3 \times 1$	$9 + 9 = 18$	$27 + 9 = 36$	^{36}Kr
$[Mn(CO)_3(C_2H_4)]^+$	24 (as Mn^+)	6×2	$6 + 12 = 18$	$24 + 12 = 36$	^{36}Kr
$[Cr(\eta^6 - C_6H_6)_2]$	24	2×6	$6 + 12 = 18$	$24 + 12 = 36$	^{36}Kr
$[Fe(\eta^5 - C_5H_5)_2]$	26	2×5	$8 + 10 = 18$	$26 + 10 = 36$	^{36}Kr
	24 (as Fe^{2+})	2×6 (as $C_5H_5^-$)	$6 + 12 = 18$	$24 + 12 = 36$	^{36}Kr
$[Fe(CO)_2(\eta^1 - Cp)(\eta^5 - Cp)]$	24 (as Fe^{2+})	$2 \times 2 + 6 + 2 \times 1$	$6 + 12 = 18$	$24 + 12 = 36$	^{36}Kr
$[Co(CO)_3(\eta^3 - C_3H_5)]$	26 (as Co^+)	$3 \times 2 + 4 \times 1$	$8 + 10 = 18$	$26 + 10 = 36$	^{36}Kr
$[Fe(CN)_6]^{4-}$	24 (as Fe^{2+})	6×2	$6 + 12 = 18$	$24 + 12 = 36$	^{36}Kr
$[PtCl_2(NH_3)_4]^{2+}$	74 (as Pt^{4+})	6×2	$6 + 12 = 18$	$74 + 12 = 86$	^{86}Rn
Complexes disobeying the EAN Rule					
$[Fe(CN)_6]^{3-}$	23 (as Fe^{3+})	6×2	$5 + 12 = 17$	$23 + 12 = 35$	^{36}Kr
$[Pt(\eta^2 - C_2H_4)Cl_3]^-$	76 (as Pt^{2+})	$3 \times 2 + 1 \times 2$	$8 + 8 = 16$	$76 + 8 = 84$	^{86}Rn
$[Pt(NH_3)_4]^{2+}$	76 (as Pt^{2+})	4×2	$8 + 8 = 16$	$76 + 8 = 84$	^{86}Rn
$[PdCl_4]^{2-}$	44 (as Pd^{2+})	4×2	$8 + 8 = 16$	$44 + 8 = 52$	^{54}Xe
$[Ag(NH_3)_2]^{2+}$	46 (as Ag^+)	2×2	$10 + 4 = 14$	$46 + 4 = 50$	^{54}Xe
$[Cu(NH_3)_4]^{2+}$	27 (as Cu^{2+})	4×2	$9 + 8 = 17$	$27 + 8 = 35$	^{36}Kr
$[CuCl_2]^-$	28 (as Cu^+)	2×2	$10 + 4 = 14$	$28 + 4 = 32$	^{36}Kr

In different transition metal complexes (cf. Table 3.3.1.1), the tendency to attain the next noble gas configuration as predicted from the EAN Rule is well documented. Some illustrative examples are given below.

(i) **[Co(NH₃)₆]³⁺:** Co^{3+}:   24e,   (*cf.* Atomic No. of Co = 27)

$6NH_3$: $6 \times 2e$

Total = 36e   (*cf.* Atomic No. of Kr = 36).

$Co^{3+}(d^6)$ + 6 ligand pairs = 18 valence electrons.

(ii) **[Fe(CN)₆]⁴⁻:** Fe^{2+}:   24e   (*cf.* Atomic No. of Fe = 26)

$6CN^-$: $6 \times 2e$

Total = 36e   (*cf.* Atomic No. of Kr = 36)

(iii) **[Pt(NH₃)₆]⁴⁺:** Pt^{4+}:   74e   (*cf.* Atomic No. of Pt = 78)

$6NH_3$: $6 \times 2e$

Total = 86e   (*cf.* Atomic No. of Rn = 86)

3.3.2 Exception to the EAN Rule (18e Rule) and Application of the 16e Rule

Some examples disobeying the EAN rule have been given in Table 3.3.1.1. Such cases are illustrated here.

● When a particular metal centre adopts **different coordination numbers**, the EAN rule is not maintained. This is illustrated in the following examples.

Ni(II) (d⁸-system): $[Ni(CN)_4]^{2-}$ (EAN = 34), $[Ni(NH_3)_6]^{2+}$ (EAN = 38)

Fe(III) (d⁵-system): $[Fe(CN)_6]^{3-}$ (EAN = 35), $[FeCl_4]^-$ (EAN = 31)

Co(II) (d⁷-system): $[CoCl_4]^{2-}$ (EAN = 33), $[Co(OH_2)_6]^{2+}$ (EAN = 37).

● When a particular metal centre forms the stable complexes **in different oxidation states**, for the same coordination number, the EAN rule is not maintained. Fe(II) gives stable $[Fe(CN)_6]^{4-}$ complex (EAN = 36) and Fe(III) also gives stable $[Fe(CN)_6]^{3-}$ complex (EAN = 35). Obviously, $[Fe(CN)_6]^{4-}$ supports the EAN rule but $[Fe(CN)_6]^{3-}$ violates the EAN rule. Similarly, EAN rule is maintained in $[Cr(CO)_6]$ (EAN = 24 + 12 = 36) but not in $[Cr(NH_3)_6]^{3+}$ (EAN = 21 +12 = 33), and both are stable.

● **16e-Rule (cf. Sec. 3.17.6):** The 18e rule is not also applicable for the metals (*e.g.* Rh, Ir, Pd and Pt) residing at the right bottom corner of the d-block elements. Rh(I), Ir(I), Pd(0), Pt(0), Pd(II) and Pt(II) very often prefer the **16e rule** instead of the 18e rule. For such systems, EAN is less than 2 compared to the atomic number of the nearest inert gas. The examples are given

Pt(II) or Pd(II) (d⁸-system) $[PdCl_4]^{2-}$, $[Pt(CN)_4]^{2-}$, $[PtCl_4]^{2-}$, $[Pt(C_2H_4)Cl_3]^-$;
(8 + 4 ligand bond pairs = 16e)   For Pt(II) complexes, EAN = 76 + 4 × 2 = 84
                                         (*cf.* Atomic No. of Rn = 86)

Ir(I) (d⁸-system) $[Ir(CO)Cl(PPh_3)_2]$
8 + 4 ligand bond pairs   EAN = 76 + 4 × 2 = 84
= 16e   (*cf.* Atomic No. of Rn = 86)

Pt(0) (d¹⁰-system) $[Pt(PPh_3)_3]$, EAN = 78 + 3 × 2 = 84
(10 + 3 ligand bond pairs = 16e   (*cf.* At. No. of Rn = 86)

The examples given above follow the 16e rule instead of the 18e rule.

The **16e rule for the square planar complexes** can be explained in terms of CFT and MOT (Sec. 3.17.6).

3.3.3 18-Electron Rule in Organometallics

The 18-electron rule is also important to understand the coordination number of stable organometallic coordination compounds. This is illustrated for d^6 (*e.g.* Fe^{2+}, Co^{3+}, Cr^0, Mn^+, V^-), d^8 (*e.g.* Fe^0, Mn^-, Co^+, etc.) and d^{10} (Zn^{2+}, Ni^0, Cu^+, Co^-, Fe^{2-}, etc.) systems.

- **d^6-system:** 6 + 6 ligand bond pairs = 18e.

    [ML$_6$] (octahedral)

    Examples: $[V(CO)_6]^-$, $[Cr(CO)_6]$, $[Mn(CO)_6]^+$, $[Mn(CO)_5CH_3]$, $[Cr(CO)_3(\eta^6-C_6H_6)]$, $[Cr(CO)_3(\eta^5-C_5H_5)]^-$, etc. $[Mn(CO)_3(\eta^5-C_5H_5)]$. For the 6e-donor ligands like $\eta^6-C_6H_6$, $\eta^5-C_5H_5^-$, geometries of the complexes are different.

- **d^8-system:** 8 + 5 ligand bond pairs = 18e.

    [ML$_5$] (trigonal bipyramidal or square pyramidal).

    Examples: $[Fe(CO)_5]$, $[Mn(CO)_5]^-$, $[Mn(CO)_4NO]$, $[Co(CO)_2(\eta^5-C_5H_5)]$, etc. For the $\eta^5-C_5H_5^-$ ligand (6e donor), geometry of the complex is different.

- **d^{10}-system:** 10 + 4 ligand bond pairs = 18e.

    [ML$_4$] (tetrahedral geometry).

    Examples: $[Ni(CO)_4]$, $[Co(CO)_4]^-$, $[Fe(CO)_4]^{2-}$, $[Fe(CO)_2(NO)_2]$, $[Co(CO)_3NO]$, etc.

In calculating the EAN, the ligands like CO, PR_3, CN^-, NO^+, H^-, C_2H_4, etc. are considered as the **2e-donor ligands**. For the ligands $C_5H_5^-$ (*i.e.* Cp$^-$), C_6H_6, C_4H_4 or $C_4H_4^{2-}$ (cyclobutadiene), C_4H_6 (1,3-butadiene), $C_3H_5^-$ (allyl group), C_7H_8 (cycloheptatriene), $C_7H_7^+$ (tropylium ion), C_8H_8 (cyclooctatetraene), etc. the donor properties depend on their coordinating behaviour (*cf.* Chapter 9). This is illustrated below.

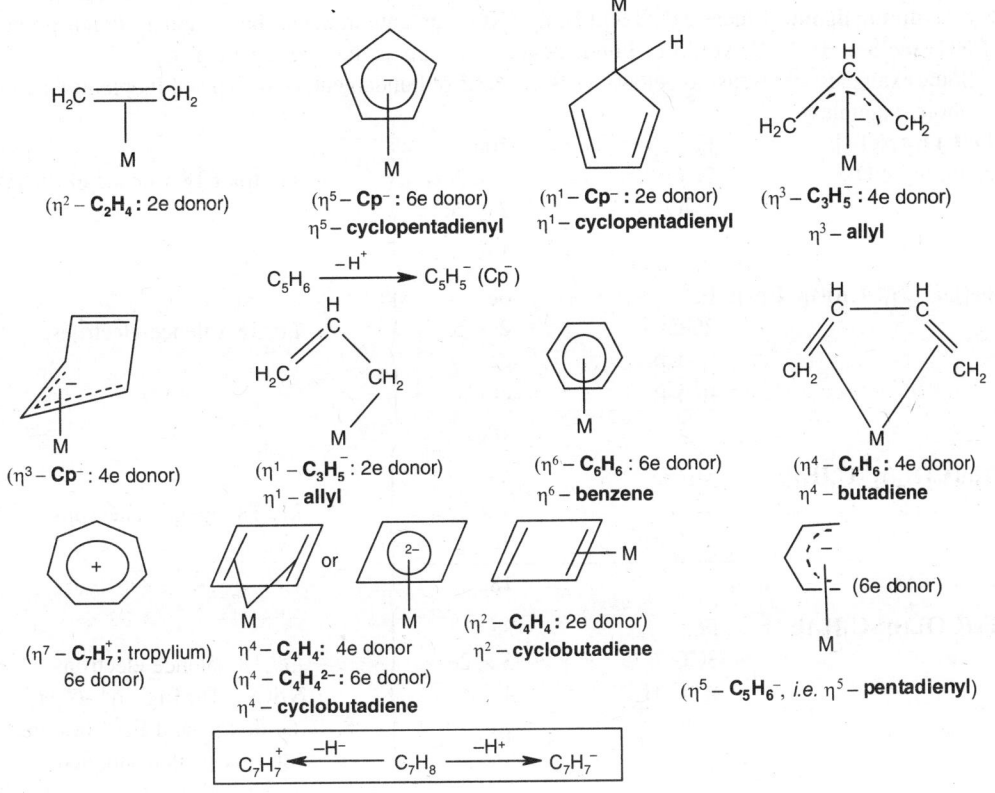

3.3.3 18-Electron Rule in Organometallics

The 18-electron rule is also important to understand the coordination number of stable organometallic coordination compounds. This is illustrated for d^6, e.g. Co^{3+}, Cr^0, Mn^+, V^-, Ru^{4+}, Fe^0, Mn^-, Co^+, etc.) and d^{10} (Zn^{2+}, Ni^0, Cu^+, Co^{-}, Fe^{2-}, etc.) systems.

- d^6-system: $6 + 6$ ligand bond pairs = 18e.
  ([ML_6] (octahedral))

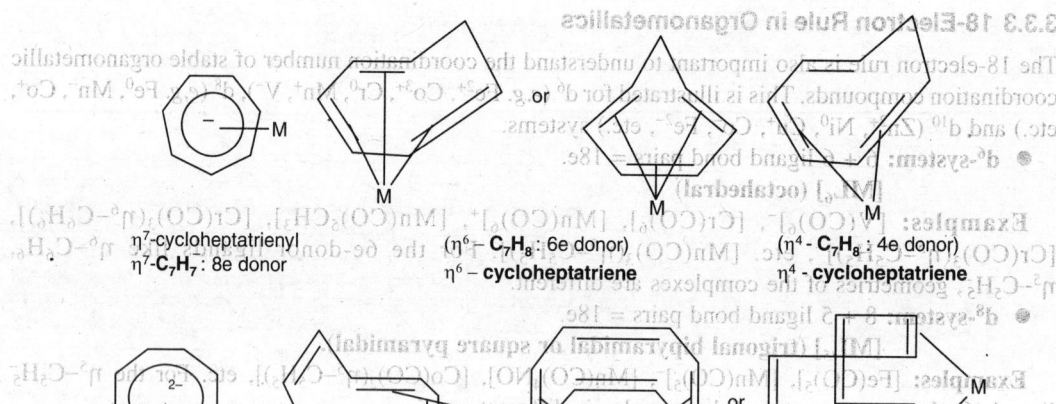

η^7-cycloheptatrienyl
η^7-**C$_7$H$_7$** : 8e donor

($\eta^6 - $**C$_7H_8$** : 6e donor)
$\eta^6 - $**cycloheptatriene**

($\eta^4 - $**C$_7H_8$** : 4e donor)
$\eta^4 - $**cycloheptatriene**

10e donor

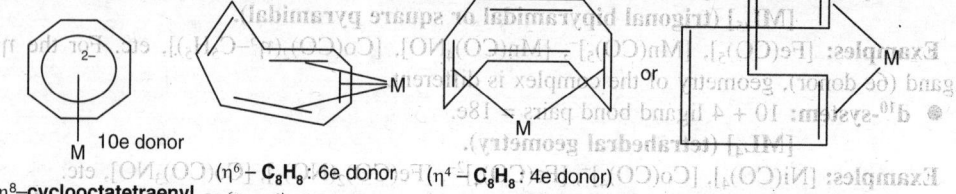

$\eta^8 - $**cyclooctatetraenyl**
$\eta^8 - $**cot^{2-}**

($\eta^6 - $**C$_8H_8$** : 6e donor)
$\eta^6 - $**cot**

($\eta^4 - $**C$_8H_8$** : 4e donor)
$\eta^4 - $**cyclooctatetraene**, *i.e.* $\eta^4 - $**cot or 1,2,5,6 $- \eta^4 - $cot**

NO as a ligand: The **electron-donor property of NO** (an odd electron molecule) is quite interesting. If it is considered as NO^+ then the metal centre is supposed to be reduced by one electron, *i.e.* NO acts as a **3e-donor ligand.** If it acts as NO^-, then the metal centre is oxidized by one electron, *i.e.* NO act as a **1e-donor ligand.** Linear MNO and bent MNO segments indicate the 3e and 1e donor properties of NO (*see* Sec. 16.1.2D, Vol. 3 of Fundamental Concepts of Inorganic Chemistry).

Some examples are discussed below (*see* **Sec. 12.8.5** of Fundamental Concepts of Inorganic Chemistry, for more examples).

[Fe(CO)$_2$(NO)$_2$]: Fe^{2-}:   10e
(assuming NO^+)   2CO: $2 \times 2e$ $\Big\}$ *i.e.* total 18 valence electrons.
                 $2NO^+$: $2 \times 2e$
                      _____
                      18e

[Fe(CO)$_2$(η^1-Cp)(η^5-Cp)]: Fe^{2+}:   6e
                        2CO: $2 \times 2e$ $\Big\}$ *i.e.* 18 valence electrons.
                        η^5-Cp$^-$:   6e
                        η^1-Cp$^-$:   2e
                        _____
                        18e

[Mn(CO)$_4$(η^3-C$_3$H$_5$)]: Mn^+:   6e
                    4CO: $4 \times 2e$ $\Big\}$ *i.e.* 18 valence electrons
                    η^3-C$_3$H$_5^-$:   4e
                    _____
                    18e

[Fe(CO)$_3$(η^4-C$_4$H$_4$)]:   Fe: 8e
                  3CO: $3 \times 2e$ $\Big\}$ *i.e.* 18 valence electrons
                  η^4-C$_4$H$_4$:   4e   **Note:** Taking η^4-C$_4$H$_4^{2-}$
                      _____   (6e donor) and Fe^{2+} (6e), the
                      18e   18e rule is also satisfied.

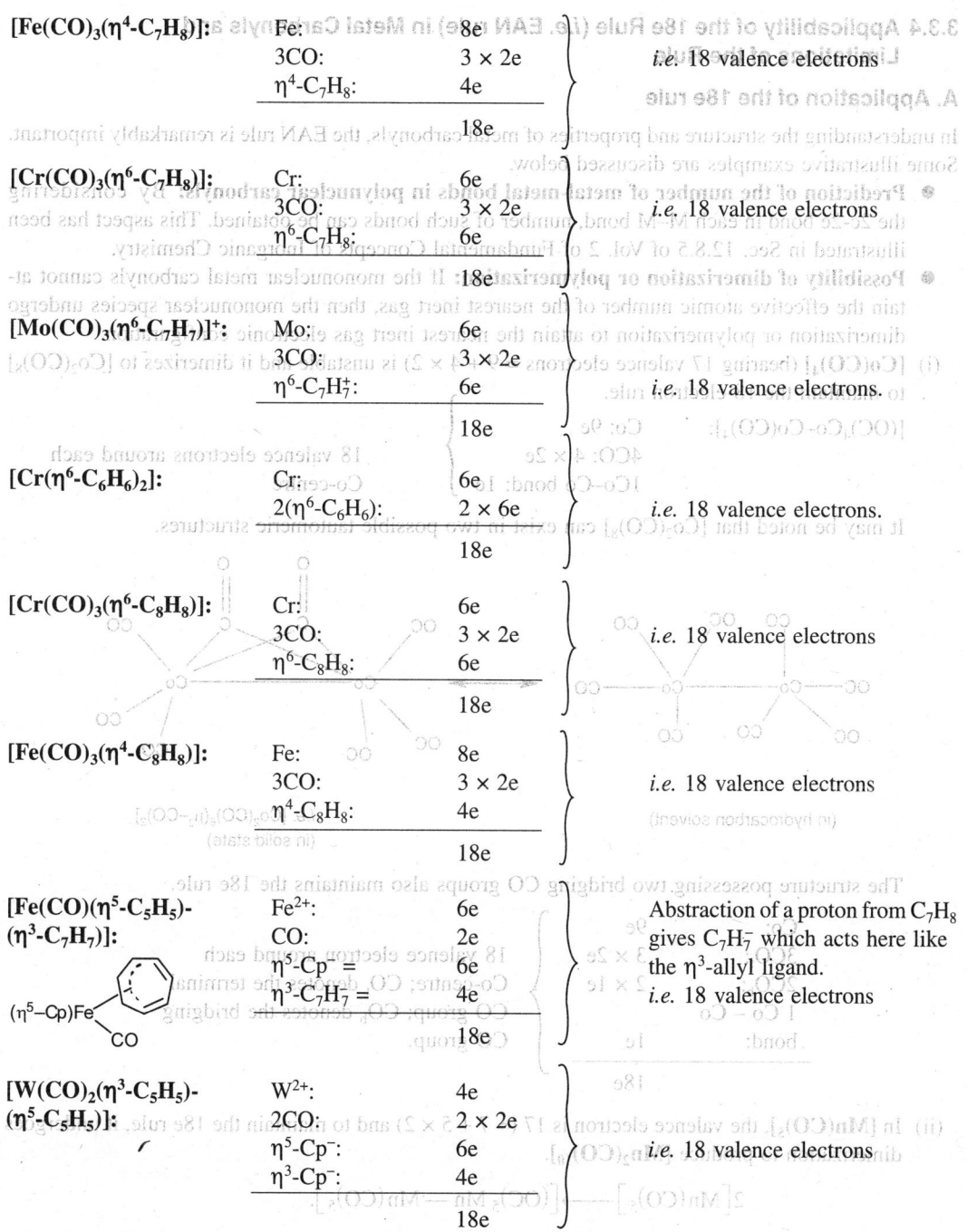

$[Fe(CO)_3(\eta^4-C_7H_8)]$:

Fe:	8e	
3CO:	$3 \times 2e$	*i.e.* 18 valence electrons
$\eta^4-C_7H_8$:	4e	
	18e	

$[Cr(CO)_3(\eta^6-C_7H_8)]$:

Cr:	6e	
3CO:	$3 \times 2e$	*i.e.* 18 valence electrons
$\eta^6-C_7H_8$:	6e	
	18e	

$[Mo(CO)_3(\eta^6-C_7H_7)]^+$:

Mo:	6e	
3CO:	$3 \times 2e$	
$\eta^6-C_7H_7^+$:	6e	*i.e.* 18 valence electrons.
	18e	

$[Cr(\eta^6-C_6H_6)_2]$:

Cr:	6e	
$2(\eta^6-C_6H_6)$:	$2 \times 6e$	*i.e.* 18 valence electrons.
	18e	

$[Cr(CO)_3(\eta^6-C_8H_8)]$:

Cr:	6e	
3CO:	$3 \times 2e$	*i.e.* 18 valence electrons
$\eta^6-C_8H_8$:	6e	
	18e	

$[Fe(CO)_3(\eta^4-C_8H_8)]$:

Fe:	8e	
3CO:	$3 \times 2e$	*i.e.* 18 valence electrons
$\eta^4-C_8H_8$:	4e	
	18e	

$[Fe(CO)(\eta^5-C_5H_5)-(\eta^3-C_7H_7)]$:

Fe^{2+}:	6e	Abstraction of a proton from C_7H_8
CO:	2e	gives $C_7H_7^-$ which acts here like
$\eta^5-Cp^- =$	6e	the η^3-allyl ligand.
$\eta^3-C_7H_7^- =$	4e	*i.e.* 18 valence electrons
	18e	

$[W(CO)_2(\eta^3-C_5H_5)-(\eta^5-C_5H_5)]$:

W^{2+}:	4e	
2CO:	$2 \times 2e$	
η^5-Cp^-:	6e	*i.e.* 18 valence electrons
η^3-Cp^-:	4e	
	18e	

(**Note:** If both the Cp⁻-ligands are considered to be as η^5 in the above compound, then the 18e rule will be disobeyed and **the compound will be unstable.**)

3.3.4 Applicability of the 18e Rule (*i.e.* EAN rule) in Metal Carbonyls and Limitations of the Rule

A. Application of the 18e rule

In understanding the structure and properties of metal carbonyls, the EAN rule is remarkably important. Some illustrative examples are discussed below.

- **Prediction of the number of metal-metal bonds in polynuclear carbonyls:** By considering the 2c-2e bond in each M–M bond, number of such bonds can be obtained. This aspect has been illustrated in Sec. 12.8.5 of Vol. 2 of Fundamental Concepts of Inorganic Chemistry.
- **Possibility of dimerization or polymerization:** If the mononuclear metal carbonyls cannot attain the effective atomic number of the nearest inert gas, then the mononuclear species undergo dimerization or polymerization to attain the nearest inert gas electronic configuration.

(i) $[Co(CO)_4]$ (bearing 17 valence electrons $= 9 + 4 \times 2$) is unstable and it dimerizes to $[Co_2(CO)_8]$ to maintain the 18-electron rule.

$[(OC)_4Co–Co(CO)_4]$:   Co: 9e

                                     4CO: 4×2e   18 valence electrons around each

                                     1Co–Co bond: 1e   Co-centre

It may be noted that $[Co_2(CO)_8]$ can exist in two possible tautomeric structures.

(in hydrocarbon solvent)   *i.e.* $[Co_2(CO)_6(\mu_2–CO)_2]$ (in solid state)

The structure possessing two bridging CO groups also maintains the 18e rule.

Co: 9e

3CO$_t$: 3×2e   18 valence electron around each

2CO$_b$: 2×1e   Co-centre; CO$_t$ denotes the terminal

1 Co – Co   CO group; CO$_b$ denotes the bridging

bond: 1e   CO group.

            18e

(ii) In $[Mn(CO)_5]$, the valence electron is 17 $(= 7 + 5 \times 2)$ and to maintain the 18e rule, it undergoes dimerization to produce $[Mn_2(CO)_{10}]$.

$$2[Mn(CO)_5] \longrightarrow [(OC)_5 Mn — Mn(CO)_5].$$

(iii) According to the EAN rule, $[V(CO)_6]$ should undergo dimerization leading to a V–V bond giving rise to $[V_2(CO)_{12}]$. But in fact, it does not happen. Probably, in the dimer, the **steric hindrance does not favour the situation**. However, it satisfies the rule in another way discussed below.

- **Unstable carbonyls for the early members of transition series:** For the early members of transition series, the number of d-electrons is few (say 2, 3, etc.). To attain the 18e configuration, it needs a larger number of CO groups (*i.e.* requirement of higher coordination number) which is disfavoured due to the steric grounds. This is why, the carbonyls formed by the early members of the series, is relatively unstable. For example, $[Ti(CO)_6]$, $[V(CO)_6]$ are unstable.

- **Change of oxidation state:** Sometimes, to maintain the 18e rule, the metal centre may experience the oxidation state change. For example, $[V(CO)_6]$ (17 valence electrons = $5 + 6 \times 2$) is not stable but $[V(CO)_6]^-$ (satisfying 18 rule) is stable. In the electron count, V^- is considered to provide 6 valence electrons. Similarly, $[Mn(CO)_5]$ (17 electron), changes easily to $[Mn(CO)_5]^-$ or $[Mn^{+1}(CO)_5Cl]$ to satisfy the 18e-rule.

- **Mechanistic paths of ligand substitution reactions in carbonyls and organometallics:** This can be understood in terms of 18e rule. This aspect has been discussed in Secs. 9.4.3 and 10.2 in detail.

- **Polymerization by losing CO groups and producing the M—M bonds:** Sometimes, the mononuclear carbonyls obeying the 18e rule may also undergo polymerization through the releasing of CO groups and producing the M—M bonds. In such polynuclear carbonyls, both bridging and nonbridging CO groups may be present. It is illustrated for $[Fe(CO)_5]$ (18e = $8 + 5 \times 2$).

$$2\left[Fe(CO)_5\right] \longrightarrow \left[Fe_2(CO)_9\right] + CO;$$

$$3\left[Fe(CO)_5\right] \longrightarrow \left[Fe_3(CO)_{12}\right] + 3CO$$

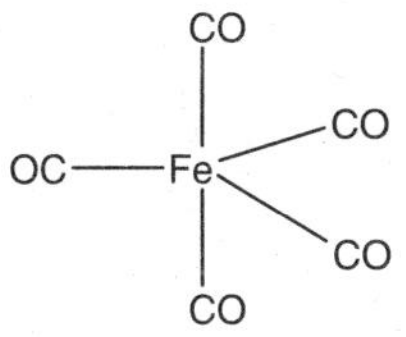

[Fe(CO)₅]
18e (= 8e + 5 × 2e)

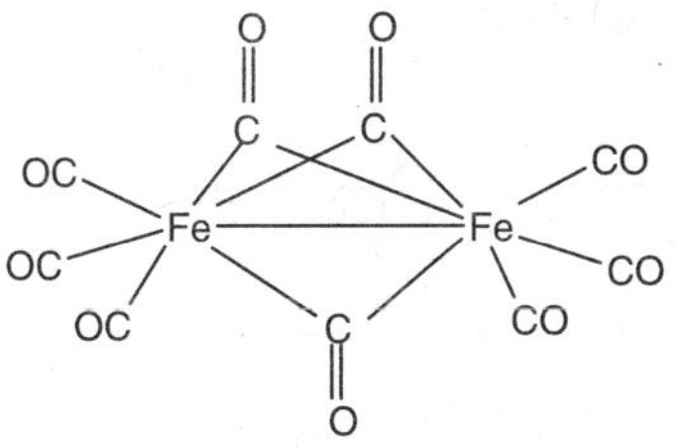

[Fe₂(CO)₆(μ₂–CO)₃]

Fe	:	8e
3CO_t	:	3 × 2e
3CO_b	:	3 × 1e
1Fe–Fe	:	1e
		18e

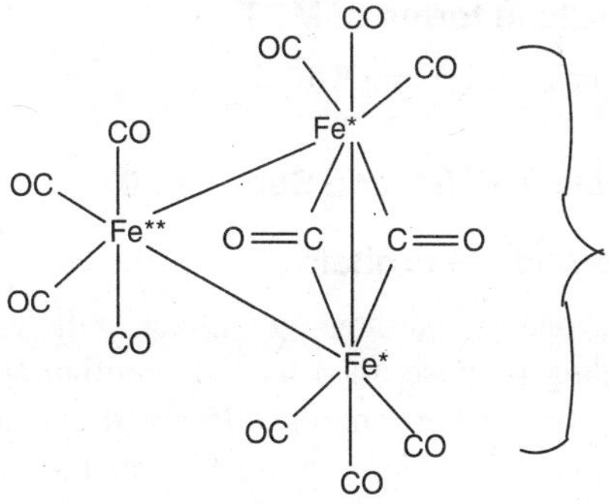

[Fe₃(CO)₁₀(μ₂–CO)₂]

For Fe**

Fe	=	8e
4CO_t	=	4 × 2e
2Fe–Fe	=	2 × 1e
		18e

For Fe*

Fe	=	8e
3CO_t	=	3 × 2e
2CO_b	=	2 × 1e
2Fe–Fe	=	2 × 1e
		18e

● **Some other applications of 18e Rule:** This aspect has been illustrated in Sec. 12.8.5 (Vol. 2 of Fundamental Concepts of Inorganic Chemistry).

B. Limitations of the EAN Rule without the structural information

When the structural information is not available, for the polynuclear carbonyls, then EAN rule cannot predict the number of bridging CO (*i.e.* μ_2-CO) groups. It is illustrated for [Fe$_3$(CO)$_{12}$] having 3Fe–Fe bonds. Without considering the nature of the CO groups, we have:

$$\text{No. of electrons per Fe} = \frac{3 \times 8 + 12 \times 2 + 3 \times 2}{3} = 18$$

However, the actual structure is: [Fe$_3$(CO)$_{10}$(μ_2–CO)$_2$] which also satisfies the 18e Rule.

Here it may be noted that for the heavier congeners of iron, the corresponding trinuclear carbonyls, *i.e.* [Ru$_3$(CO)$_{12}$] and [Os$_3$(CO)$_{12}$] possess the structure having no bridging CO group. **It is believed that for the heavier congeners (which are larger in size), the metal-metal bonds are too long to allow the bridge formation by CO.** A similar situation prevails for [M$_4$(CO)$_{12}$] (M = Co, Rh, Ir) where for M = Co, Rh there are two μ_2–CO groups while for M = Ir, there is no μ_2–CO group. However, in each case, there are 6M–M bonds along the edges of a tetrahedron and in each case, the EAN rule is satisfied.

The structural difference among the species [M$_3$(CO)$_{12}$] (M = Fe, Ru, Os) or [M$_4$(CO)$_{12}$] (M = Co, Rh, Ir) cannot be predicted by EAN rule. This aspect has been illustrated in Sec. 9.4.3.

In this regard, [Co$_2$(CO)$_8$] is also relevant. In solid state, it predominantly exists with two μ_2–CO groups but in solution it may remain in equilibrium with different structural isomers. There is an isomer having no bridging CO group (*i.e.* all CO groups as the terminal groups). In fact, in terms of electron count, **both the following structures are equivalent.**

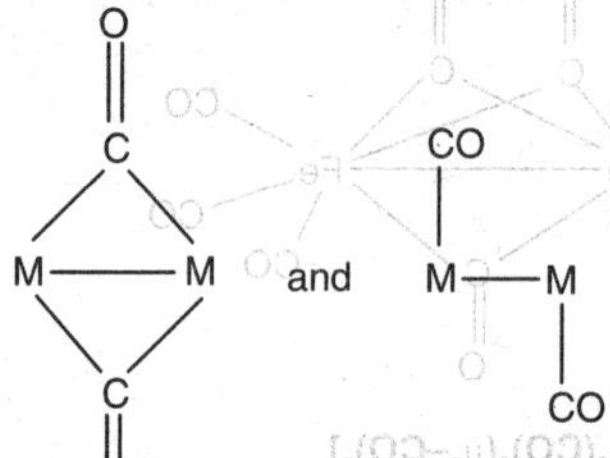

> **Explanation of the 18e and 16e rule in terms of MOT**
>
> This aspect has been discussed in Secs. 3.17.5 and 3.17.6

3.3.5 Interpretation of the 18e and 16e Rule (*cf.* Table 3.17.5.1 and Sec. 3.17.6)

A. Interpretation in terms of relative energy of the valence orbitals

For the transition metals, the nine orbitals *i.e.* five $(n-1)d$, one ns and three np will act as the valence orbitals. If these nine orbitals are available for the bonding purpose, then the 18e configuration is attained. If we consider the energy change trends for the $(n-1)d$, ns and np orbitals, then with the increase of effective nuclear charge (increasing trend along the series), the energy of the $(n-1)d$ orbitals falls more sharply compared to the ns and np orbitals (Fig. 3.3.5.1). For the 1st transition series, upto Co, all the nine orbitals being of fairly similar energy can participate in bonding to satisfy the 18e rule. In the case of Ni, the 18e rule is attained automatically even when the 3d orbitals do not participate in bonding. In other words, the 18e electron rule cannot ascertain the participation of the *d*-orbitals in bonding for Ni. **Beyond Ni, the 3d-orbitals are so stabilized that they are not expected to participate in bonding to satisfy the 18e rule.** Thus, it is reasonable to expect that for Cu, Zn and other main group metals for which the 3*d* orbitals are highly stabilized, the 18e rule will not be satisfied.

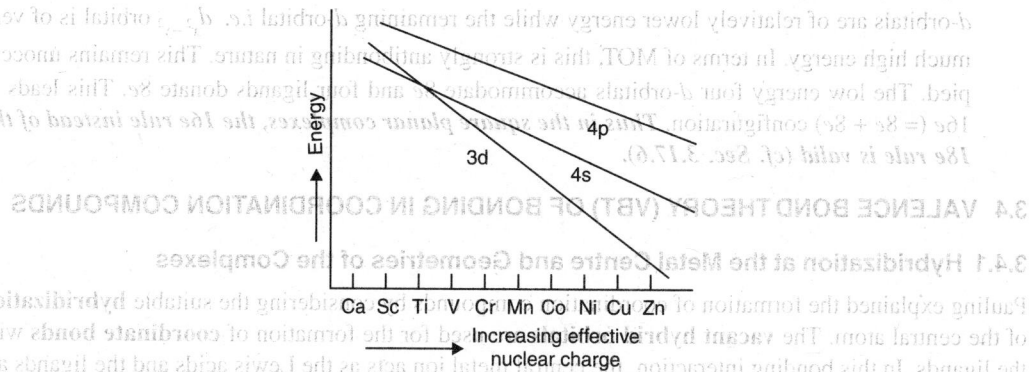

Fig. 3.3.5.1 Schematic representation of the variation trends of energy of 3*d*, 4*s*, and 4*p* orbitals along the first transition series.

● **For the beginners of the series, due to the steric factors, sufficient number of ligands may not be accommodated to satisfy the 18e rule** (*cf.* instability of [Ti(CO)₆], [V(CO)₆], etc.).

● Here it may be noted that with the increase of positive oxidation state, the energy of the *d*-orbitals falls more sharply with the increase of atomic number.

● **In general it can be concluded that the metals (in low or negative oxidation states) in the middle part of the series are the best candidates to satisfy the 18e rule.**

Explanation of the **18e rule and 16e rule in terms of MOT** has been given in Secs. 3.17.5 and 3.17.6.

The basis of 18e and 16e rule can be also rationalized in terms of CFT. This aspect is discussed below.

B. Interpretation of the 18e and 16e rule in terms of CFT

This aspect has been illustrated for different geometries.

 (i) **Octahedral geometry for the *d*⁶ system** (*cf.* Sec. 3.17.5): Six ligands donate 12 electrons. Crystal field splitting of the *d*-orbitals stabilizes three *d*-orbitals (d_{xy}, d_{yz}, d_{zx} *i.e.* t_{2g} set) and destabilizes two *d*-orbitals ($d_{x^2-y^2}$, d_{z^2} *i.e.* e_g set). Because of the high energy of these two *d*-orbitals, they remain unoccupied. In terms of MOT, these two *d*-orbitals are antibonding in character. The t_{2g}-set is filled in to accommodate 6-electrons. It leads to $(6 + 12)e$ *i.e.* 18e configuration. This is why, [ML₆] satisfies the 18*e* rule for the *d*⁶ configuration. This simple explanation can be questioned because many complexes are known where the electrons are present in the e_g-set.

 (ii) **Tetrahedral geometry for the *d*¹⁰ system:** Crystal field splitting stabilizes the *e*-set (*i.e.* $d_{x^2-y^2}$ and d_{z^2}) and destabilizes the t_2-set. But the energy separation between the t_2- and *e*-set is not too high. Consequently, all these five *d*-orbitals may be occupied to accommodate 10 electrons. Four ligands donate 8-electrons. This leads to $(10 + 8)e$ *i.e.* 18e configuration for the *d*¹⁰ system.

 (iii) **Trigonal bipyramidal geometry for the *d*⁸ system** (*cf.* Sec. 3.17.6): In terms of crystal field splitting, out of the five *d*-orbitals, four *d*-orbitals are of relatively lower energy while the remaining

d-orbital *i.e.* d_{z^2} is of much higher energy. In terms of MOT, the d_{z^2} orbitals is antibonding in character. The four low energy d-orbitals accommodate $8e$ and the five ligands donate $10e$. This leads to $18e$ configuration.

(iv) **Square planar geometry for the d^8 system** (*cf.* Sec. 3.17.6): Out of the five d-orbitals, the four d-orbitals are of relatively lower energy while the remaining d-orbital *i.e.* $d_{x^2-y^2}$ orbital is of very much high energy. In terms of MOT, this is strongly antibonding in nature. This remains unoccupied. The low energy four d-orbitals accommodate $8e$ and four ligands donate $8e$. This leads to $16e$ ($= 8e + 8e$) configuration. *Thus in the square planar complexes, the 16e rule instead of the 18e rule is valid (cf. Sec. 3.17.6).*

3.4 VALENCE BOND THEORY (VBT) OF BONDING IN COORDINATION COMPOUNDS

3.4.1 Hybridization at the Metal Centre and Geometries of the Complexes

Pauling explained the formation of coordination compounds by considering the suitable **hybridization** of the central atom. The **vacant hybrid orbitals** are used for the formation of **coordinate bonds** with the ligands. In this bonding interaction, the central metal ion acts as the Lewis acids and the ligands act as the Lewis bases (*i.e.* electron pair donors). The important aspects of VBT are outlined as follows:

(i) **Geometry of the complexes:** The mode of hybridization of the central atom determines the geometry of the complexes. For the hybridization, the ns, np and nd or $(n-1)d$ orbitals are available. Out of the five d-orbitals, generally, d_{z^2} and $d_{x^2-y^2}$ participate in the hybridization. For the tetrahedral complexes, d_{xy}, d_{yz} and d_{zx} orbitals may participate in the d^3s hybridization.

(ii) **Coordination number:** The metal centre provides the vacant hybrid orbitals for accepting the electron pairs from the ligands. Thus the coordination number is given by the number of vacant hybrid orbitals available for receiving the σ-bonding electron pairs from the ligands.

(iii) **Pi-bonding orbitals:** Out of the five d-orbitals, generally, only the d_{z^2} and $d_{x^2-y^2}$ orbitals participate in σ-bonding and d_{xy}, d_{yz} and d_{xz} may participate in the π-bonding interaction. To reduce the amount of negative charge accumulated on the metal centre through the σ-donation of electron pairs by the ligands, the metal $\rightarrow$ ligand π-bonding becomes important when the ligand possesses the suitable vacant orbitals to receive back the electron.

(iv) **Outer and inner orbital complexes:** For the **8- and 18-electron ions**, *e.g.* Be^{2+}, Mg^{2+}, Ca^{2+}, Al^{3+}, Ga^{3+}, Cu^+, Ag^+, Zn^{2+}, Cd^{2+}, Hg^{2+}, etc. ns, np and nd orbitals are available for the hybridization. Thus, if the nd orbitals participate in the hybridization for σ-bonding, then such complexes are referred to as **the outer orbital complexes.** Here it may be pointed out that **energy of the nd orbitals is relatively higher compared to that of the ns and np orbitals.** Consequently, in such cases, very often, the hybridization tends to get confined within the ns and np orbitals.

For the **transition metal ions**, both the $(n-1)d$ and nd orbitals are available. For the **octahedral complexes** of the d^4, d^5, d^6 and d^7 systems, both the $(n-1)d$ and nd orbitals may participate in the hybridization depending on the condition. For the **square planar complexes** of the d^8 and d^9 systems, again both the possibilities may also arise.

In such cases, if the $(n-1)d$ orbitals participate in the hybridization, then the complexes are described as the **inner orbital complexes;** if the nd orbitals participate in the hybridization, then the complexes are described as the **outer orbital complexes.**

For the **inner orbital complexes**, the hybridizations are denoted by d^2sp^3 **(for the octahedral complexes)** and dsp^2 **(for the square planar complexes).** The d-electrons are accommodated in the remaining d-orbitals (i.e. d_{xy}, d_{yz}, d_{xz} for the octahedral geometry; d_{xy}, d_{yz}, d_{zx} and d_{z^2} orbitals for the square planar geometry) excluded from the hybridization. Sometimes, it may be required to excite the d-electrons to some higher energy levels to provide the required d-orbitals for the hybridization. Thus, **rearrangement of electrons** present in the $(n-1)d$ levels may be required to allow the d^2sp^3, i.e. $(n-1)d^2nsnp^3$ or dsp^2, i.e. $(n-1)dnsnp^2$ hybridization. This will cause pairing of electrons in the $(n-1)d$ level. Such inner orbital type complexes are also described as the **low-spin complexes** in modern terminology. These were mistakenly described as the **covalent complexes** or **hypoligated complexes** in old literature.

Table 3.4.1.1 Some representative complexes along with their hybridization and geometries.

Types of hybridization	Geometry	Examples
sp; $(5s, 5p)$ (cf. Sec. 2.11.4)*	Linear	$[Ag(NH_3)_2]^+$, $[Ag(CN)_2]^-$
sp^2; $(6s, 6p)$ (cf. Sec. 2.11.5)*	Trigonal planar	$[HgI_3]^-$
sp^3; $(4s, 4p)$	Tetrahedral	$[Ni(CO)_4]$, $\{Zn(NH_3)_4]^{2+}$, $[CoCl_4]^{2-}$, $[ZnCl_4]^{2-}$, $[MX_4]^{2-}$ (M = Cu, Mn, Ni; X = Cl, Br, I), $[FeCl_4]^-$
d^3s; $(3d_{xy}, 3d_{yz}, 3d_{zx}, 4s)$	Tetrahedral	$[MnO_4]^-$
dsp^2; $(3d_{x^2-y^2}, 4s, 4p_x, 4p_y)$	Square planar	$[Ni(CN)_4]^{2-}$, $[Ni(bigH)_2]^{2+}$, $[Cu(bigH)_2]^{2+}$, $[Cu(NH_3)_4]^{2+}$, etc.
dsp^2 ($4d_{x^2-y^2}$ for Pd^{II}, $5d_{x^2-y^2}$ for Pt^{II})	Square planar	$[PdCl_4]^{2-}$, $[PtCl_4]^{2-}$, $[Pt(NH_3)_4]^{2+}$, etc.
dsp^3; $(3d_{z^2}, 4s, 4p)$	Trigonal bipyramidal	$[CuCl_5]^{3-}$, $[Ni(CN)_5]^{3-}$, $[Fe(CO)_5]$, etc.
dsp^3; $(3d_{x^2-y^2}, 4s, 4p)$	Square pyramidal	$[VO(acac)_2]$, $[Ni(CN)_5]^{3-}$
d^2sp^3; $(3d_{x^2-y^2}, 3d_{z^2}, 4s, 4p)$	Octahedral (inner orbital)	$[Ti(OH_2)_6]^{3+}$, $[Cr(NH_3)_6]^{3+}$, $[Mn(CN)_6]^{3-}$, $[Fe(CN)_6]^{3-}$, $[Fe(CN)_6]^{4-}$, $[NiF_6]^{4-}$, $[Co(NH_3)_6]^{3+}$, etc.
sp^3d^2; $(4s, 4p, 4d_{x^2-y^2}, 4d_{z^2})$	Octahedral (outer orbital)	$[FeF_6]^{3-}$, $[Fe(OH_2)_6]^{2+}$, $[CoF_6]^{3-}$, $[CuF_6]^{3-}$, $[Ni(NH_3)_6]^{2+}$; $[Cu(OH_2)_6]^{2+}$, $[Zn(NH_3)_6]^{2+}$, $[Zn(OH_2)_6]^{2+}$, $[Al(OH_2)_6]^{3+}$, etc.
d^2sp^3; $(4d_{xz}, 4d_{yz}, 5s, 5p)$	Trigonal prism	$[Mo(S_2C_2Ph_2)_3]$, etc.

* See Secs. 2.11.4 and 2.11.5 to understand the role of d-orbitals in such cases.

It may be noted that for the d^1, d^2, d^3 electronic configurations, the d^2sp^3 hybridization (required for the octahedral geometry) can occur without disturbing any electronic configuration or arrangement in the $(n-1)d$ level.

When the required d-orbitals for the hybridizations are not available from the $(n-1)d$ level, the nd level instead of the $(n-1)d$ level provides the required d-orbitals for the hybridization. **In such cases, the electronic configuration or arrangement of the electrons in the $(n-1)d$ level remains unchanged.** The hybridizations are represented by sp^3d^2, i.e. $nsnp^3nd^2$ (for the octahedral complexes) and sp^2d, i.e. $nsnp^2nd$ (for the square planar complexes). Such complexes are described as the **outer-orbital type complexes**. These are described as the **high-spin** complexes in modern terminology. These were mistakenly described as the **ionic or hyperligated complexes.**

Outer orbital complexes keeping the electronic configuration in the $(n-1)d$ level unchanged show the same magnetic behaviour of the free ions (*i.e.* uncomplexed). In other words, the metal centres of such outer orbital complexes possess the same electronic configurations of the corresponding free metal ions. This is why, such complexes were mistakenly considered as the **ionic complexes.** The metal centres of inner orbital complexes causing a rearrangement of electronic configuration in the $(n-1)d$ level possess fewer number of unpaired electrons compared to those present in the corresponding free ions. This is why, such inner orbital complexes were described as the **covalent complexes.**

- **Outer orbital complexes:** No rearrangement of electronic configuration in the underlying electron shells. It involves the hybridization like $(n)s$; $(n)p^3$, *i.e.* sp^3 **(tetrahedral);** $(n)s(n)p^2(n)d$, *i.e.* sp^2d **(square planar) and** $(n)s(n)p^3(n)d^2$, *i.e.* sp^3d^2 **(octahedral).**
- **Inner orbital complexes:** Depending on the conditions, sometimes, it may be required to rearrange or pair the electrons in the underlying shell. The hybridization are: $(n-1)d(n)s(n)p^2$, *i.e.* dsp^2 **(square planar);** $(n-1)d^2(n)s(n)p^3$, *i.e.* d^2sp^3 **(octahedral).**
  For the $(n-1)d^{0-3}$ **electronic configurations,** no rearrangement or pairing of electrons in the $(n-1)d$ level is required for the inner orbital octahedral complexes. For the $(n-1)d^{4-7}$ **electronic configurations,** it requires rearrangement of electrons for the formation of inner orbital octahedral complexes.
- $(n-1)d^{4-7}$ **systems:** Depending on the conditions, both the inner and outer orbital octahedral complexes may be formed. For the d^4, d^5 and d^6 systems, pairing of the electrons in the d_{xy}, d_{yz} and d_{xz} orbitals can provide the required vacant d-orbitals for the formation of inner orbital type octahedral complexes. ***Both pairing and excitation*** of the seventh $(n-1)d$ electron are required for the formation of inner orbital octahedral complexes of the d^7-system. Six d-electrons can be placed in the d_{xy}, d_{yz}, and d_{xz} orbitals through pairing and the 7th d-electron is to be promoted to some higher energy level to provide the $d_{x^2-y^2}$ and d_{z^2} orbitals for the hybridization in the inner orbital octahedral complexes.
- $(n-1)d^{0-3}$ **systems:** Such systems can always produce the **inner orbital octahedral complexes.**
- $(n-1)d^{8-10}$ **systems:** Such systems generally produce the **outer orbital octahedral complexes.**
- d^9 **system:** For the inner orbital square planar complex formation, the 9th d-electron is to be promoted to some higher energy level to provide the vacant $d_{x^2-y^2}$ orbital for the hybridization.

3.4.2 Octahedral Complexes in Terms of VBT

(i) $(n-1)d^{0-3}$ **systems:** The required two d-orbitals, *i.e.* $d_{x^2-y^2}$ and d_{z^2} for the hybridization are always available from the $(n-1)$ shell without rearranging the electrons in the $(n-1)d$ level. *Such systems always form the octahedral complexes through the d^2sp^3 hybridization (i.e. inner orbital octahedral complexes).*

It may be noted that for the d^1, d^2, d^3 electronic configurations, d^2sp^3 hybridization (required for the octahedral geometry) can occur (without disturbing any electronic configuration or arrangement in the $(n-1)d$ level).

When the required d-orbitals are not available from the $(n-1)d$ level, the $(n)d$-level instead of the $(n-1)d$-level provides the required d-orbitals for the hybridization. In such cases, the electronic configuration or arrangement of the electrons in the $(n-1)d$ level remains unchanged.

The hybridization required for the octahedral complexes is d^2sp^3, *i.e.* $(n-1)d^2(n)s(n)p^3$ type for the square planar complexes. Such complexes are described as the outer-orbital type complexes. These are described as the high-spin complexes in modern terminology. These were mistakenly described as the ionic or hyperligated complexes.

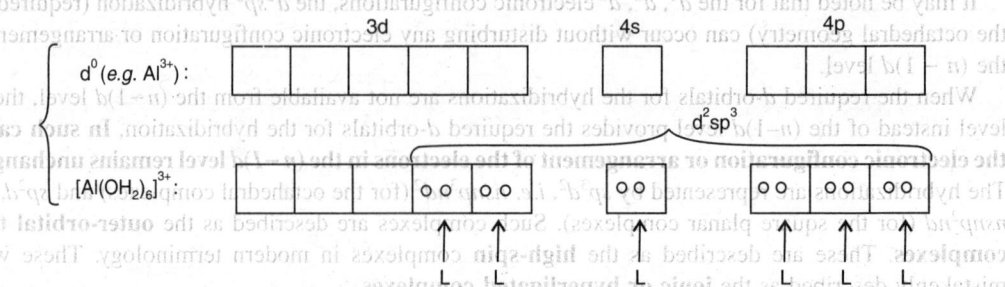

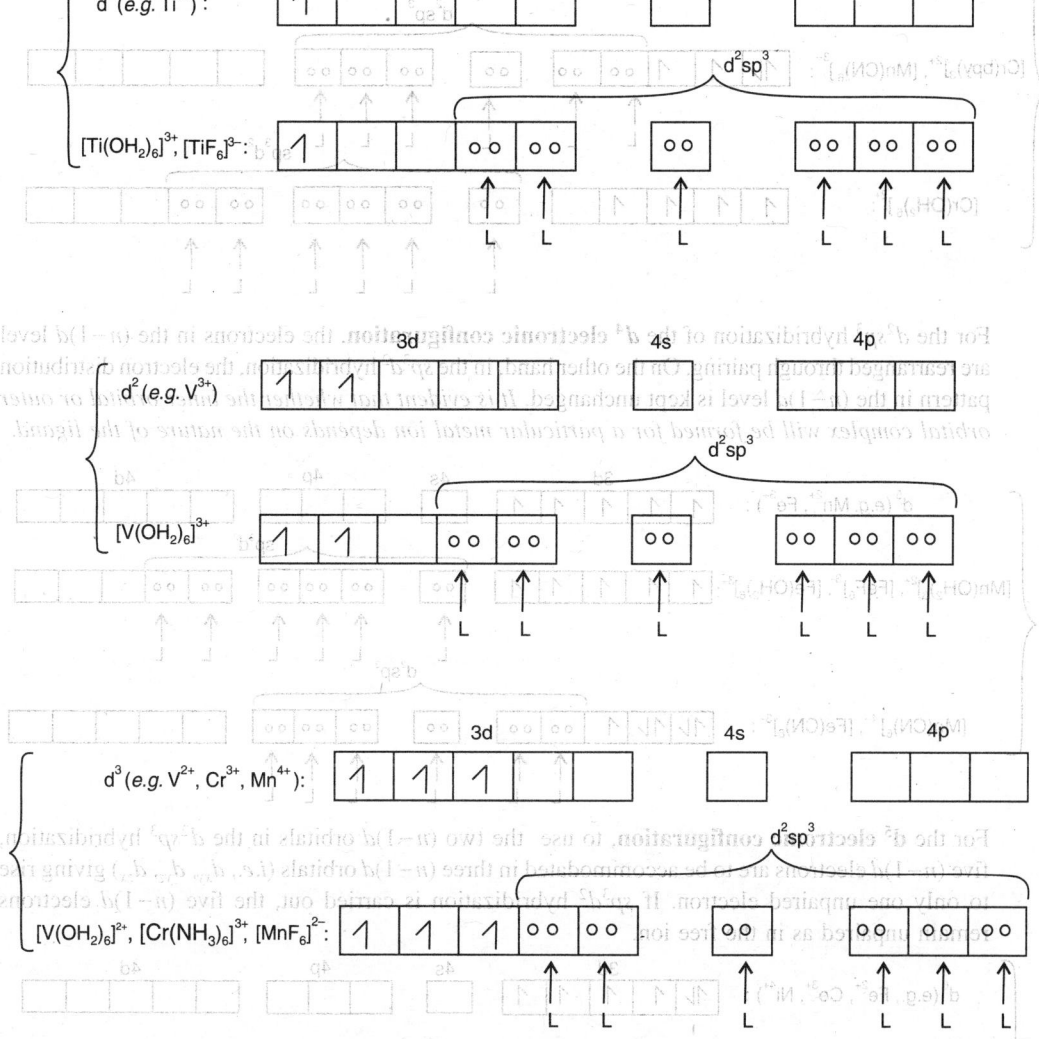

The electron pairs donated by the ligands (L) in the vacant hybrid orbitals are denoted by (oo).

(ii) **d^{4-6} systems:** To utilize the $(n-1)d_{x^2-y^2}$ and $(n-1)d_{z^2}$ in the hybridization scheme d^2sp^3 (*i.e.* **inner orbital** complex formation), it requires rearrangement of the electrons in the $(n-1)d$ level. If the electronic configuration in $(n-1)d$ level is kept undisturbed, then the required $d_{x^2-y^2}$ and d_{z^2} orbitals for the hybridization purpose may be obtained from the n-shell. It will lead to sp^3d^2 hybridization (*i.e.* outer orbital complex formation). **Thus, depending on the conditions, the d^{4-6} systems can generate either the inner orbital or the outer orbital octahedral complexes.** These are illustrated in the following examples.

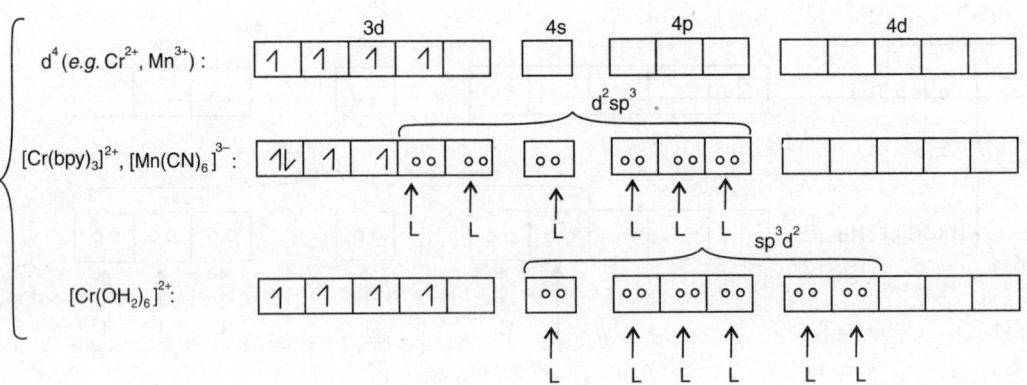

For the d^2sp^3 hybridization of the **d^4 electronic configuration**, the electrons in the $(n-1)d$ level are rearranged through pairing. On the other hand, in the sp^3d^2 hybridization, the electron distribution pattern in the $(n-1)d$ level is kept unchanged. *It is evident that whether the inner orbital or outer orbital complex will be formed for a particular metal ion depends on the nature of the ligand.*

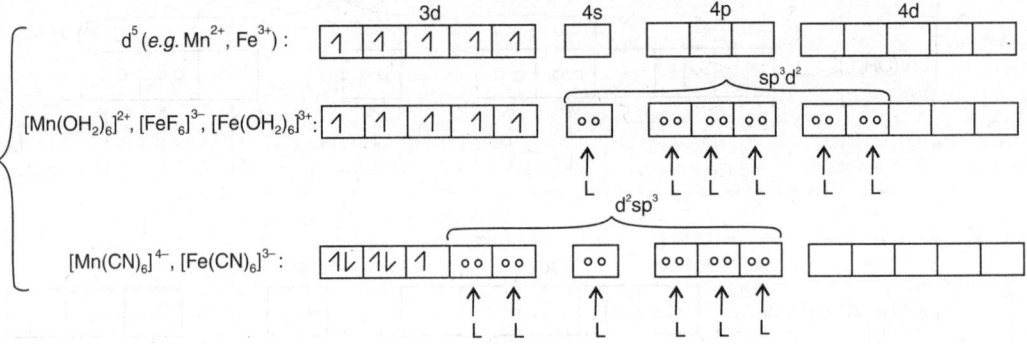

For the **d^5 electronic configuration**, to use the two $(n-1)d$ orbitals in the d^2sp^3 hybridization, five $(n-1)d$ electrons are to be accommodated in three $(n-1)d$ orbitals (*i.e.*, d_{xy}, d_{yz}, d_{zx}) giving rise to only one unpaired electron. If sp^3d^2 hybridization is carried out, the five $(n-1)d$ electrons remain unpaired as in the free ion.

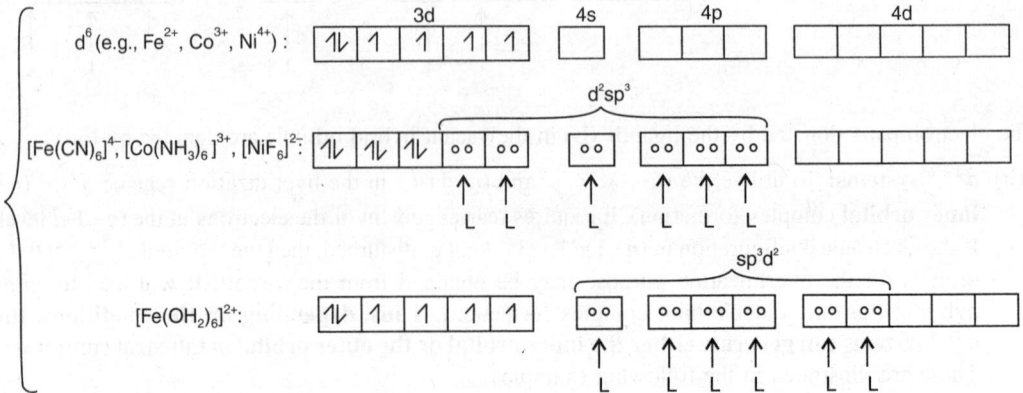

For the d^6 **electronic configuration**, if the d^2sp^3 hybridization is carried out, there will be no unpaired electron, *i.e.* six $(n-1)d$ electrons are paired up in three d-orbitals (*i.e.* d_{xy}, d_{yz}, d_{xz}). On the other hand, if the sp^3d^2 hybridization is carried out then four unpaired electrons will be present as in the free ions.

(iii) **d^7-system:** To execute the d^2sp^3 hybridization, there are only three $(n-1)d$ orbitals (*i.e.*, d_{xy}, d_{yz}, d_{xz}) available to accommodate the electrons. These three $(n-1)d$-orbitals can accommodate maximum six electrons and then **the seventh $(n-1)d$ electron is to be promoted to some higher energy level.** Pauling proposed that for the inner orbital complexes like $[NiF_6]^{3-}$, $[NiCl_2(diars)_2]^+$ where diars = *ortho*-$C_6H_4(AsMe_2)_2$, the seventh 3d electron is promoted to the 5s orbital. Without reorganizing the electron distribution in the $(n-d)$ level, the d^7-system can adopt the sp^3d^2 hybridization. *Thus, for the d^7-system, both the possibilities may exist depending on the conditions.*

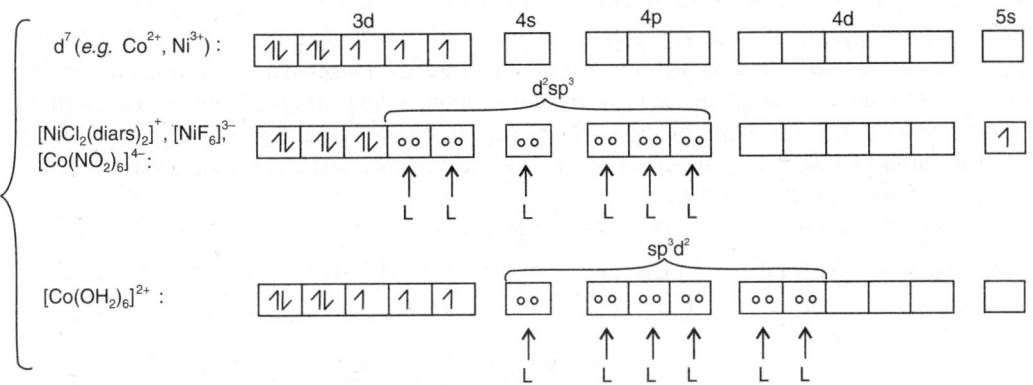

(iii) **d^{8-10} system:** In such cases, through the rearrangement of $(n-1)d$ electrons in the $(n-1)d$ level and excitation of $(x-6)$ ($x = 8, 9, 10$) electrons to some higher energy levels, two vacant $(n-1)d$ orbitals **cannot be provided for the hybridization** to produce the σ-bonds because the process becomes energetically unfavourable. *Thus such, systems fail to produce the inner orbital octahedral complexes and they generally produce the outer orbital octahedral complexes.*

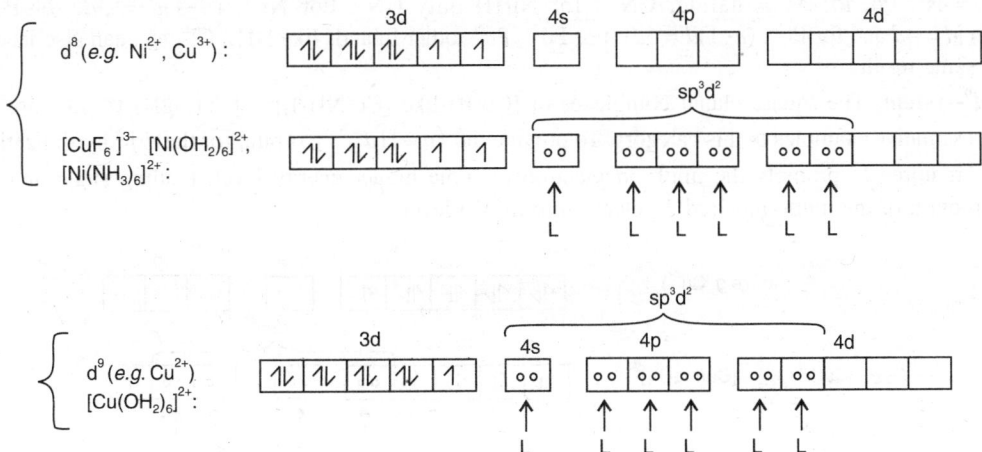

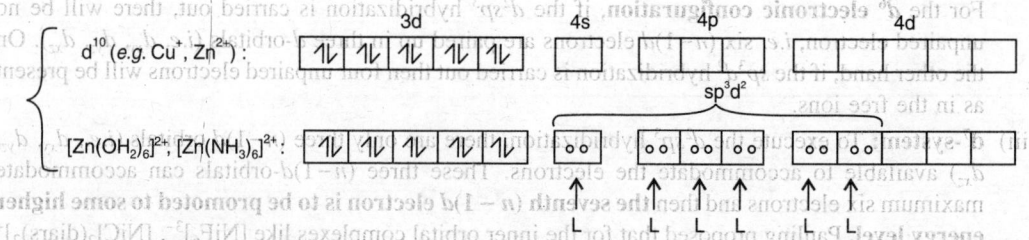

3.4.3 Square Planar Complexes in Terms of VBT

If the square planar geometry is supposed to lie in the xy-plane then the required orbitals for hybridization are: $d_{x^2-y^2}$, s, p_x and p_y. Theoretically, the d^{0-4} systems can adopt the dsp^2 hybridization (*i.e.* inner orbital type complex) without any change of electronic configuration in the $(n-1)d$ level; the d^{5-8} systems can adopt the dsp^2 hybridization through the rearrangement of electrons in the $(n-1)d$ level; the d^9 system can attain the dsp^2 hybridization through the pairing of eight $(n-1)d$ electrons in the $(n-1)d$ level and promotion of the ninth $(n-1)d$ electron to some higher energy level; the d^{5-10} systems can adopt the sp^2d hybridisation (*i.e. outer orbital type complex*) keeping the electron distribution pattern in the $(n-1)d$ level unchanged.

In reality, the square planar geometry is well characterized for the d^8 and d^9 systems.

d^8-system (*e.g.* Ni^{2+}, Pd^{2+}, Pt^{2+}, Cu^{3+}, etc.): To provide the $(n-1)d_{x^2-y^2}$ orbital in the required dsp^2 hybridization, the eight $(n-1)d$ electrons are to be paired up in the remaining four $(n-1)d$ orbitals.

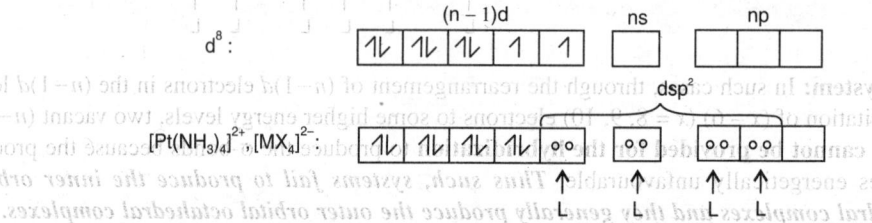

(M = Ni, Pd, Pt; X$^-$ = halide, CN$^-$; for Ni(II) only CN$^-$: For Ni^{2+}, $(n-1)d = 3d$; for Pd^{2+}, $(n-1)d = 4d$ and for Pt^{2+}, $(n-1)d = 5d$). For Pd^{2+}, Pt^{2+}, other ligands like NH$_3$, Cl$^-$, etc. can also lead to the same result.

d^9-system: The square planar complexes of [Cu(II)] like [Cu(NH$_3$)$_4$]$^{2+}$, [Cu(bigH)$_2$]$^{2+}$, etc. are the representative examples in this category. To provide the $(n-1)d_{x^2-y^2}$ orbital for the dsp^2 hybridization, it is required to promote the ninth $3d$ electron to some higher energy level. Pauling suggested the promotion of the ninth unpaired $3d$ electron to the $4p$-level.

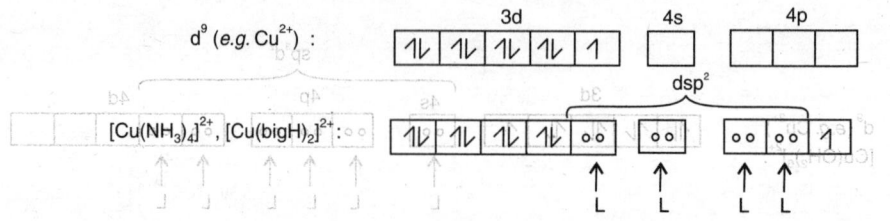

(*cf.* Inner orbital octahedral complex formation by the d^7 system where the seventh $(n-1)d$ electron is also required to be promoted to some higher energy level).

3.4.4 Tetrahedral Complexes in Terms of VBT

For the tetrahedral complexes, the most common mode of hybridization is $(n)s(n)p^3$, *i.e.* sp^3 ($s + p_x + p_y + p_z$). Thus it should not depend on the nature of electronic configuration in the $(n-1)d$ level, *i.e.* all system represented by $(n-1)d^{0-10}$ can adopt the tetrahedral geometry keeping the electron distribution pattern in the $(n-1)d$ level unchanged.

In reality, the tetrahedral complexes are well documented for the d^0 (*e.g.* $[Be(OH_2)_4]^{2+}$), d^1 (*e.g.* $[TiCl_4]^-$), d^5 (*e.g.* $[FeCl_4]^-$, $[MnCl_4]^{2-}$), d^7 (*e.g.* $[CoCl_4]^{2-}$); d^8 (*e.g.* $[NiCl_4]^{2-}$), d^9 (*e.g.* $[CuCl_4]^{2-}$), d^{10} (*e.g.* $[ZnCl_4]^{2-}$, $[Ni(CO)_4]$) systems.

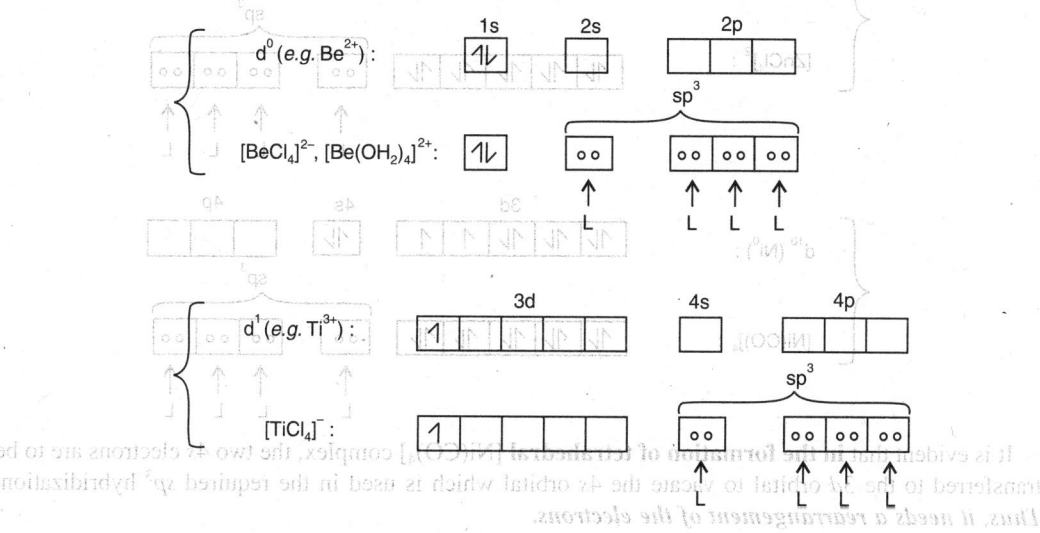

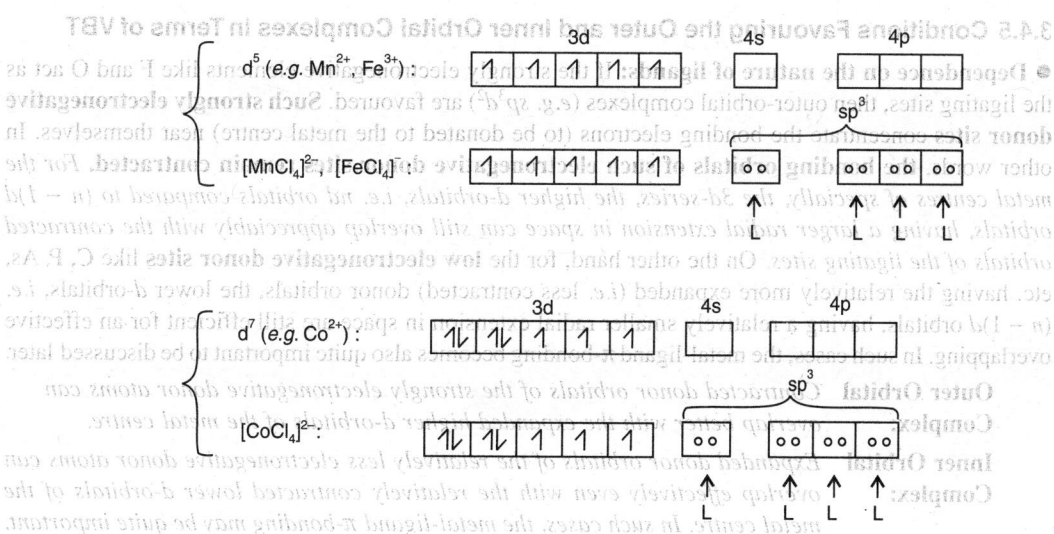

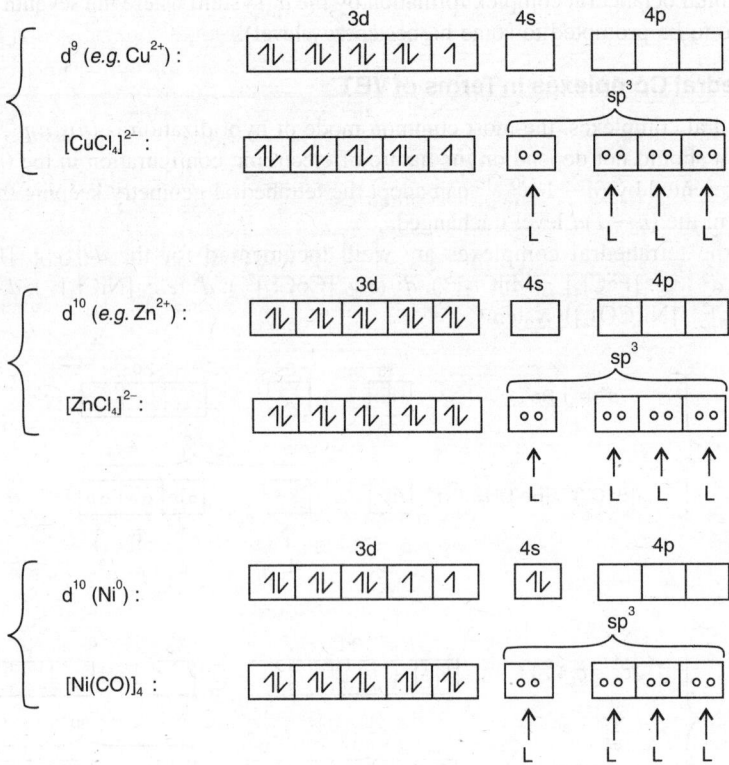

It is evident that **in the formation of tetrahedral** $[Ni(CO)_4]$ complex, the two $4s$ electrons are to be transferred to the $3d$ orbital to vacate the $4s$ orbital which is used in the required sp^3 hybridization. *Thus, it needs a rearrangement of the electrons.*

3.4.5 Conditions Favouring the Outer and Inner Orbital Complexes in Terms of VBT

● **Dependence on the nature of ligands:** If the strongly electronegative elements like F and O act as the ligating sites, then outer-orbital complexes (*e.g.* sp^3d^2) are favoured. **Such strongly electronegative donor sites** concentrate the bonding electrons (to be donated to the metal centre) near themselves. In other words, **the bonding orbitals of such electronegative donor sites remain contracted.** *For the metal centres of specially, the 3d-series, the higher d-orbitals, i.e. nd orbitals compared to $(n-1)d$ orbitals, having a larger radial extension in space can still overlap appreciably with the contracted orbitals of the ligating sites.* On the other hand, for the **low electronegative donor sites** like C, P, As, etc. having the relatively more expanded (*i.e.* less contracted) donor orbitals, the lower d-orbitals, *i.e.* $(n-1)d$ orbitals, having a relatively smaller radial extension in space are still efficient for an effective overlapping. In such cases, the metal-ligand π-bonding becomes also quite important to be discussed later.

Outer Orbital Complex:	*Contracted donor orbitals of the strongly electronegative donor atoms can overlap better with the expanded higher d-orbitals of the metal centre.*
Inner Orbital Complex:	*Expanded donor orbitals of the relatively less electronegative donor atoms can overlap effectively even with the relatively contracted lower d-orbitals of the metal centre. In such cases, the metal-ligand π-bonding may be quite important.*

● **Dependence on the periodic position of the metal ions:** Generally, the heavier congeners, *i.e.* the members of the 2nd and 3rd transition series, prefer the inner orbital complex.

For the heavier congeners, compared to the lower *d*-orbitals, *i.e.* $(n-1)d$ orbitals, the higher *d*-orbitals, *i.e.* *nd* orbitals are too diffused (*cf.* **relativistic expansion of *d*-orbitals; see Vol. 1 of Fundamental Concepts of Inorganic Chemistry**) to overlap effectively with the donor orbitals of the ligands. This is why, the heavier congeners generally prefer the $(n-1)d$ orbitals for bonding, *i.e.* inner-orbital complexes (*e.g.* $[Fe(OH_2)_6]^{3+}$ (outer orbital complex) *vs.* $[Ru(OH_2)_6]^{3+}$ (inner orbital complex).

● **Dependence on the magnitude of positive charge on the metal centre:** Increased positive charge on the metal centre prefers the participation of $(n-1)d$ orbitals compared to the *nd* orbitals. For the higher positive charge on the metal centre, the *nd* orbitals are too diffused compared to the $(n-1)d$ orbitals to overlap with the ligand orbitals. These are illustrated by the examples: $[Co(NH_3)_6]^{2+}$ is an outer orbital complex while $[Co(NH_3)_6]^{3+}$ is an inner orbital complex.

$[MnF_6]^{3-}$, $[FeF_6]^{3-}$, $[CoF_6]^{3-}$ are the outer orbital complexes (sp^3d^2) while $[NiF_6]^{2-}$, $[PtF_6]^{2-}$, etc. are the inner orbital complexes (d^2sp^3).

3.4.5 Inner Orbital Complex Formation at the Cost of Excitation of One Unpaired Electron

● **Octahedral complexes of the d^7-system** (*e.g.* **Co^{II}, Ni^{III}**): It has been already discussed that in such cases, for the formation of inner orbital complexes (*i.e.* d^2sp^3 hybridization), the 7th $(n-1)d$ electron is to be promoted to some higher energy level like the 5s-orbital.

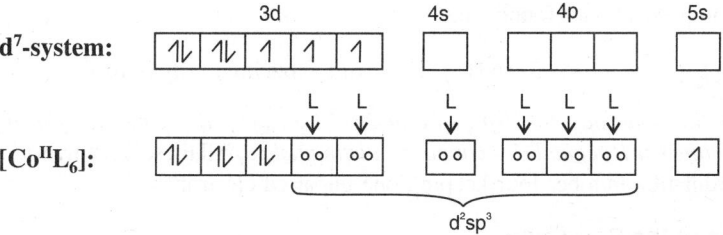

For the excitation of the 7th electron to the higher energy level, it will require some energy. This disfavour is compensated by the stronger bonding interaction provided by the $(n-1)d$ orbitals (compared to the *nd* orbitals). ***The excited electron at the higher energy level can be lost easily to favour the oxidation of Co(II) to Co(III).*** In fact, many Co(II) complexes are prone to oxidation to Co(III). However, this prediction cannot be generalized.

$$[Co^{II}L_6] - e \longrightarrow [Co^{III}L_6], \text{ (favourable condition and } \textbf{experimentally realised).}$$

● **Octahedral complexes of the d^8-system** (*e.g.* Ni^{II}, Cu^{III}): To execute the d^2sp^3 hybridization, two $(n-1)d$ electrons are to be promoted to some higher energy level (say 5s orbital).

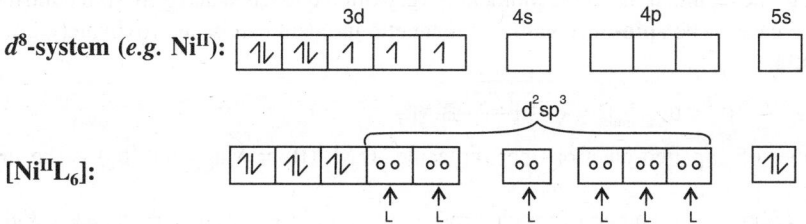

For the excitation of two $(n-1)d$ electrons to the higher energy level, it will require a large amount of energy. This high disfavour cannot be compensated by the favoured bonding interaction provided by the inner orbital complex formation (*i.e.* d^2sp^3 hybridization). This is why, such systems avoid the d^2sp^3 hybridization and adopt the sp^3d^2 hybridization which does not require any excitation energy for the promotion of $(n-1)d$ electrons to the higher energy levels.

● **Square planar complexes of the d^9-systems** (*e.g.* Cu^{II}): If the dsp^2 hybridization, *i.e.* utilization of the $(n-1)d_{x^2-y^2}$ orbital, is considered, then the ninth $(n-1)d$ electron is to be promoted to some higher energy level (say np orbital).

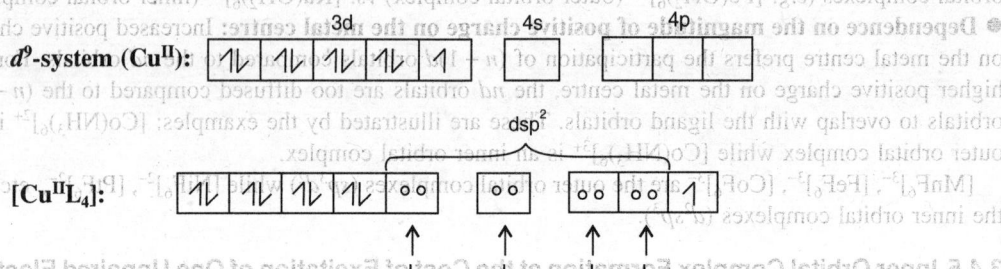

If it is argued that the excited electron will be lost easily, then it will be reasonable to expect that the square planar complexes of Cu(II) will be prone to oxidation to Cu(III) (*cf.* ease of oxidation of the octahedral Co^{II}-complexes having the d^7 configuration). However, experimentally this prediction for the Cu(II) square planar complexes is not found true.

$$\left[Cu\left(NH_3\right)_4 \right]^{2+} - e \longrightarrow \left[Cu\left(NH_3\right)_4 \right]^{3+}; \text{ (not experimentally realised)}$$

Note: *In terms of magnetic properties, the dsp^2, sp^2d and sp^3 (tetrahedral geometry) hybridization schemes for the four coordinate Cu(II) complexes cannot be distinguished.* All these three hybridization schemes maintain the coordination number four keeping one unpaired electron.

3.4.6 Magnetic Criterion of the Bond Type

It is established that the substances bearing the unpaired electrons are **paramagnetic** (*i.e.* attracted in an external magnetic field) while the substances bearing the paired electrons only are **diamagnetic** (*i.e.* repelled in an external magnetic field). An electron (which is equivalent to a moving electric charge) produces an electric current and consequently an electron is associated with a magnetic moment that can interact with an external magnetic field. The magnetic moments of two electrons of opposite spin cancel each other. **Thus the paired electrons do not possess any magnetic moment.**

Paramagnetism originates from the both **orbital angular momentum** (*i.e.* orbital motion) and **spin angular momentum** (*i.e.* spinning motion). However, for the complexes of the 1st transition series, orbital contribution to the resultant magnetic moment is very often quenched and only spin contribution exists in most of the cases. The **spin-only value** of magnetic moment can be approximately given by **Bose-Stoner formula.**

$$\mu_s = \mu_{spin-only} = \sqrt{n(n+2)} \text{ B.M.,}$$

Where n = number of unpaired electrons per molecule, B.M. (Bohr Magneton, μ_β) is the unit of magnetic moment:

1B.M. = $1\mu_\beta$ = $eh/4\pi mc$, e = charge of an electron, m = mass of electron, h = Planck's constant, c = velocity of light.

$$1\mu_\beta = 9.274 \times 10^{-24} \text{ J T}^{-1} \text{ (Joules per Tesla)}$$

Thus from the number of unpaired electrons, $\mu_{\text{spin-only}}$ value (in B.M. unit) can be calculated.

$n = 1$	2	3	4	5
$\mu_s = \sqrt{3} = 1.73$	$\sqrt{8} = 2.83$	$\sqrt{15} = 3.87$	$\sqrt{24} = 4.90$	$\sqrt{30} = 5.92$

Table 3.4.6.1 Number of unpaired electrons in different hybridizations for the different d^n systems.

d^n	No. of unpaired electrons				
	Free ion	sp^3d^2	d^2sp^3	sp^3	dsp^2
d^1	1	1	1	1	1
d^2	2	2	2	2	2
d^3	3	3	3	3	3
d^4	4	4	2	4	4
d^5	5	5	1	5	3
d^6	4	4	0	4	2
d^7	3	3	1*	3	1
d^8	2	2	2* or 0*	2	0
d^9	1	1	1* or 3*	1	1*

* Remaining in the higher energy levels. For the d^8 system, in the d^2sp^3 hybridization, the two electrons from the $(n-1)d$ level are to be promoted to some higher energy level. These promoted electrons may remain in a paired condition (*i.e.* $n = 0$) or an unpaired ($n = 2$) condition. Similarly, for the d^9 system, three electrons are to be promoted and these may remain in a paired (*i.e.* $n = 1$) or an unpaired condition (*i.e.* $n = 3$). However, in reality, the inner orbital octahedral complexes for the d^8 and d^9 configurations are not known.

Table 3.4.6.2 Magnetic moments of some complexes of first transition series and their structures.

$(3d^n)$ n	*Complexes*	*Geometry*	*Hybridization*	*Unpaired electrons*	μ_s *(B.M.)*	μ_{obs} *(B.M.)*
1	$K_3[TiF_6]$	Octahedral	d^2sp^3	1	1.73	1.70
	$[V(acac)_2O]$	Square pyramidal	dsp^3	1	1.73	1.70
	$Cs_2[VCl_6]$	Octahedral	d^2sp^3	1	1.73	1.8
2	$[V(acac)_3]$	Octahedral	d^2sp^3	2	2.83	2.80
	$[V(OH_2)_6]^{3+}$ in $(NH_4)_2SO_4$, $V_2(SO_4)_3$, $24H_2O$	Octahedral	d^2sp^3	2	2.83	2.70
3	$[Cr(bigH)_3]Cl_3$ $[Cr(NH_3)_6]Cl_3$	Octahedral	d^2sp^3	3	3.87	3.8–3.9
4	$[Cr(OH_2)_6]SO_4$	Octahedral	sp^3d^2	4	4.90	5.0
	$K_3[Mn(CN)_6]$	Octahedral	d^2sp^3	2	2.83	3.2*
5	$Na_3[FeF_6]$	Octahedral	sp^3d^2	5	5.92	5.85
	$K_3[Fe(CN)_6]$	Octahedral	d^2sp^3	1	1.73	2.25*
	$[Mn(NCS)_6]^{4-}$	Octahedral	sp^3d^2	5	5.92	6.1*

(Contd.)

Table 3.4.6.2 (*Contd.*)

6	$(Et_4N)_2[FeCl_4]$	Tetrahedral	sp^3	4	4.90	5.40*
	$[Co(NH_3)_6]Cl_3$	Octahedral	d^2sp^3	0	0	~0
	$K_4[Fe(CN)_6]$	Octahedral	d^2sp^3	0	0	~0
	$[Fe(OH_2)_6]^{2+}$ (in Mohr's salt)	Octahedral	sp^3d^2	4	4.90	5.5*
7	$Cs_2[CoCl_4]$	Tetrahedral	sp^3	3	3.87	4.60*
	$Hg[Co(SCN)_4]$	Tetrahedral	sp^3	3	3.87	4.4*
	$[Co(bigH)_2]SO_4$	Square planar	dsp^2	1	1.73	2.49*
	$[Co(OH_2)_6]SO_4$ in $(NH_4)_2SO_4$, $CoSO_4$, $6H_2O$	Octahedral	sp^3d^2	3	3.87	5.10*
	$K_2Ba[Co(NO_2)_6]$	Octahedral	d^2sp^3	1	1.73	1.85
8	$(Et_4N)_2[NiCl_4]$	Tetrahedral	sp^3	2	2.83	3.89*
	$[Ni(bigH)_2]Cl_2$	Square planar	dsp^2	0	0	0
	$K_2[Ni(CN)_4]$					
	$[Ni(NH_3)_6]Cl_2$	Octahedral	sp^3d^2	2	2.83	3.32*
9	$[Cu(bigH)_2]Cl_2$	Square planar	dsp^2	1	1.73	1.80
	$Cs_2[CuCl_4]$	Tetrahedral	sp^3	1	1.73	2.0*

* Values significantly different from the spin only value, can be rationalized by considering the orbital contribution and spin-orbit coupling phenomena (see Chapter 8 for explanation).

The observed magnetic moments of some representative complexes of the 1st transition series are given in Table 3.4.6.2. In most of the cases, from the μ_{spin} values (assuming $\mu_{obs} \approx \mu_{spin}$), the calculated number of unpaired electrons (by using the **Bose-Stoner equation**) can explain the stereochemistry and nature of hybridization. But when the different hybridization schemes lead to the existence of the same number of unpaired electrons, the magnetic moments (μ_{obs}) cannot discriminate the possibilities. For example, the μ_{obs} value is almost the same for both the square planar (dsp^2) and tetrahedral complexes of the d^9 systems (*e.g.* Cu^{2+}).

For the **Co(II) and Ni(II) complexes**, the magnetic moment ($\mu_{obs} \rangle \mu_s$) cannot be simply explained by considering the μ_s value. It needs the consideration of orbital contribution and spin-orbit coupling. These aspects will be discussed later in Chapter 8.

3.4.7 Electroneutrality Principle: Compensation of Negative Charge Accumulated on the Metal Centre in Terms of VBT

Coordination bond formation by the ligands is expected to increase the accumulation of negative charge on the metal centre. Even if the bonding pair is equally shared then very often, the metal centre is to exist in some negative oxidation states. But such a situation is highly unlikely for the metal ions. According to Pauling, there are two pathways to avoid such an unfavourable situation.

(i) **Partial ionic character** in the metal-ligand bond to reduce the negative charge on the metal centre.

(ii) **Formation of double bond** leading to the donation of unused d-electron from the metal centre to the suitable vacant orbital of the ligand (*i.e.* metal $\rightarrow$ ligand π-bonding).

The possibility of **π-back bonding** does not arise in all complexes. It occurs only in the complexes where the ligand possesses some suitable vacant orbitals. However, the charge neutralization through the *introduction of partial ionic character* can occur in all complexes.

According to Pauling, the above mentioned two operations continue until the resultant charge on the metal centre becomes close to zero. *The resultant charges are mainly possessed by the most electropositive and electronegative elements to provide the maximum electrostatic stability.*

● **(A) Charge neutralization through the introduction of partial ionic character:** The per cent ionic character in the metal-ligand bond can be obtained from the knowledge of electronegativity difference. It is illustrated for **$[Be(OH_2)_4]^{2+}$.**

$$\chi(Be) = 1.5, \chi(O) = 3.5, \Delta\chi = 2.0, \text{ (Pauling's electronegativity values)}$$

$$\% \text{ of ionic character} = 16(\chi_A \sim \chi_B) + 3.5(\chi_A - \chi_B)^2$$
$$= 16 \times 2 + 3.5 \times 2^2 = 46$$

(*cf.* Modified Pauling Eqn. to calculate the % of ionic character in A–B bond; Chapter 10 of Fundamental Concepts of Inorganic Chemistry, Vol. 2).

It indicates about **54% covalent character** in the 'Be—O' bond. Thus in $[Be(OH_2)_4]^{2+}$, 4 oxygen atoms actually donate $4 \times 0.54 = 2.16$ valency electrons to the Be^{2+} centre. Consequently, the resultant charge on Be-centre can be obtained as follows.

$$\text{The resultant charge on Be} = 2 - 2.16 = -0.16$$

To calculate the resultant charge on oxygen, we are to consider the both Be—O and O—H bonds. In the Be—O bond (O is the donor site; and covalency = 54%), O-atom gains 0.54 unit of positive charge. By taking, $\chi(H) = 2.1$, $\chi(O) = 3.5$, *i.e.* $\Delta\chi = 1.4$, we get:

$$\% \text{ of ionic character in O–H bond} = 16\Delta\chi + 3.5(\Delta\chi)^2$$
$$= 22.4 + 6.86$$
$$= 29.26$$

Taking 29% (approximate value) ionic character in the O–H bond, each O-atom from two H-atoms (*i.e.* two O–H bonds) gains 2×0.29 unit of negative charge. Thus, the resultant charge on each O-atom is obtained as follows:

$$\text{The resultant charge on each O-atom} = +0.54 - 0.58$$
$$= -0.04$$

The resultant charge distribution on different centres of **$[Be(OH_2)_4]^{2+}$** is denoted as follows:

Be:	-0.16	
4O:	$4 \times -0.04 = -0.16$	Charge distribution
8H:	$8 \times 0.29 = +2.32$	in **$[Be(OH_2)_4]^{2+}$**
	Total $= +2.0$	

Similar calculations on the following species give the charge distribution as follows:

$[Be(OH_2)_6]^{2+}$			**$[Al(OH_2)_6]^{3+}$**, $\chi(Al) = 1.5$		
Be:	$2 - 6 \times 0.54 = -1.24$		Al:	$3 - 6 \times 0.54 = -0.24$	
6O:	$6 \times -0.04 = -0.24$		6O:	$6 \times -0.04 = -0.24$	
12 H:	$12 \times 0.29 = +3.48$		12 H:	$12 \times 0.29 = +3.48$	
Net	$= +2.0$		Net	$= +3.0$	

Calculation of charge distribution in **$[Be(NH_3)_4]^{2+}$**:

$$\chi(Be) = 1.5, \chi(N) = 3.0, \text{ i.e. } \Delta\chi = 1.5$$

$\%$ of ionic character in Be—N bond $= 16 \times 1.5 + 3.5 \times (1.5)^2 \approx 32$

$\%$ of covalent character in Be—N bond $\approx 68\%$

$$\chi(N) = 3.0, \chi(H) = 2.1, \text{ i.e., } \Delta\chi = 0.9$$

% of ionic character in N—H bond = $16 \times 0.9 + 3.5 \times (0.9)^2 \approx 17\%$

$$\left.\begin{array}{ll} & \text{Charge} \\ \text{Be:} & 2 - 4 \times 0.68 = -0.72 \\ \text{4N:} & 4\,(0.68 - 3 \times 0.17) = +0.68 \\ \text{12H:} & 12 \times 0.17 = +2.04 \\ \hline & \text{Net} = +2 \end{array}\right\} \begin{array}{l}\text{Charge distribution} \\ \text{in} \left[\text{Be}\left(\text{NH}_3\right)_4\right]^{2+}\end{array}$$

Taking $\chi(\text{Al}) = 1.5$, $\chi(\text{N}) = 3.0$, $\chi(\text{H}) = 2.1$ charge distribution in $[\text{Al(NH}_3)_6]^{3+}$ can be shown as follows:

$$\left.\begin{array}{ll} & \text{Charge} \\ \text{Al:} & 3 - 6 \times 0.68 = -1.08 \\ \text{6N:} & 6\,(0.68 - 3 \times 0.17) = +1.02 \\ \text{18H:} & 18 \times 0.17 = +3.06 \\ \hline & \text{Net} = +3 \end{array}\right\} \begin{array}{l}\text{Charge distribution} \\ \text{in} \left[\text{Al}\left(\text{NH}_3\right)_6\right]^{3+}\end{array}$$

Now let us consider the following Be(II) complexes:

$$[\text{Be(OH}_2)_4]^{2+} \qquad [\text{Be(OH}_2)_6]^{2+} \qquad [\text{Be(NH}_3)_4]^{2+}$$

Charge on Be: -0.16 -1.24 -0.72

In $[\text{Be(OH}_2)_4]^{2+}$, the 2+ charge of Be^{2+} is just neutralized while in the other two complexes, accumulation of the negative charge on the Be-centre is fairly high. Accumulation of the negative charge on the metal-centre destabilizes the complexes. Thus, the expected **stability order** should run as follows:

$$[\text{Be(OH}_2)_4]^{2+} \gg\!\!\!\rangle [\text{Be(NH}_3)_4]^{2+} \gg\!\!\!\rangle [\text{Be(OH}_2)_6]^{2+}$$

Because of the higher electronegativity of oxygen than that of nitrogen, Be-centre gains more electron from the Be—N bond compared to that from the Be—O bond. This allows the accumulation of more negative charge on the metal centre in $[\text{Be(NH}_3)_4]^{2+}$ compared to that in $[\text{Be(OH}_2)_4]^{2+}$. Compared to $[\text{Be(OH}_2)_4]^{2+}$, in $[\text{Be(OH}_2)_6]^{2+}$ accumulation of the negative charge on the metal centre is more because the metal centre accumulates the electron from the six Be—O bonds while in $[\text{Be(OH}_2)_4]^{2+}$, this accumulation of negative charge occurs from the four Be—O bonds.

(b) Relative stabilities of $[\text{M(OH}_2)_6]^{3+}$ and $[\text{M(NH}_3)_6]^{3+}$ (M = Al, Fe): On the same ground of argument, we can explain the **higher stability of $[\text{Al(OH}_2)_6]^{3+}$ compared to that of $[\text{Al(NH}_3)_6]^{3+}$.** Nitrogen is less electronegative than oxygen. Consequently, the Al-centre shares more electron from the Al—N bond compared to that from the Al—O bond. This is why, in $[\text{Al(NH}_3)_6]^{3+}$, the accumulation of negative charge on the metal centre is more. In the same line of argument, we can explain the following

$$[\text{Fe(OH}_2)_6]^{3+} \gg\!\!\!\rangle [\text{Fe(NH}_3)_6]^{3+}, \text{ (stability order)}$$

(c) Relative stabilities of $[\text{M(OH}_2)_x]^{2+}$ (x = 4, 6; M = Be, Mg): Here it may be pointed out that charge calculation explains the **unstable character of $[\text{Be(OH}_2)_6]^{2+}$** and stable character of $[\text{Be(OH}_2)_4]^{2+}$. However, Mg^{2+} forms the **stable complex, $[\text{Mg(OH}_2)_6]^{2+}$.** Be^{2+} is more electronegative than Mg^{2+}. **This prevents the excessive accumulation of negative charge on the metal centre** from the six 'Mg—O' bonds in $[\text{Mg(OH}_2)_6]^{2+}$.

Here it is worth mentioning that **in terms of ionic radius**, the smaller Be^{2+} ion cannot accommodate the six ligands because of the **steric factor.** However, the larger Mg^{2+} ion does not experience this steric crowding to accommodate the six water molecules in its coordination sphere.

- **(B) Charge neutralization on the metal centre through the metal $\rightarrow$ ligand π-bonding:** The electroneutrality principle in terms of only partial ionic character in the metal-ligand linkage does not work always. *In the complexes where the metal centre is in the low oxidation state and the* *ligating sites are less electronegative, it is apparently expected that there will be a huge accumulation of negative charge on the metal centre*. In such cases, the metal → ligand (*e.g.* CO, CN^-, PR_3, etc.) π-back bonding introduces the partial double bond character in the metal-ligand linkage. The ligand → metal σ-bonding interaction enhances the negative charge on the metal centre but the metal → ligand π–back bonding removes the electron density from the metal centre. These are illustrated below:

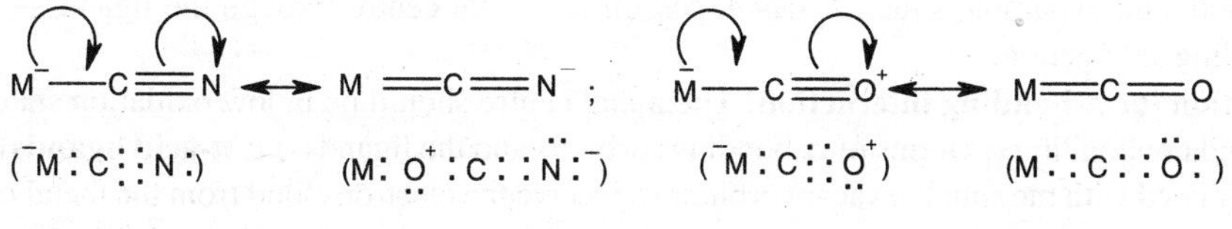

$(^-M : C :: N :)$ $(M : \overset{+}{O} : C :: N :^-)$ $(^-M : C :: \overset{+}{O})$ $(M :: C :: O :)$

Case of $[Fe(CN)_6]^{4-}$: Thus in $[Fe(CN)_6]^{4-}$, formation of three Fe=C linkage minimizes the accumulation of negative charge on the Fe-centre.

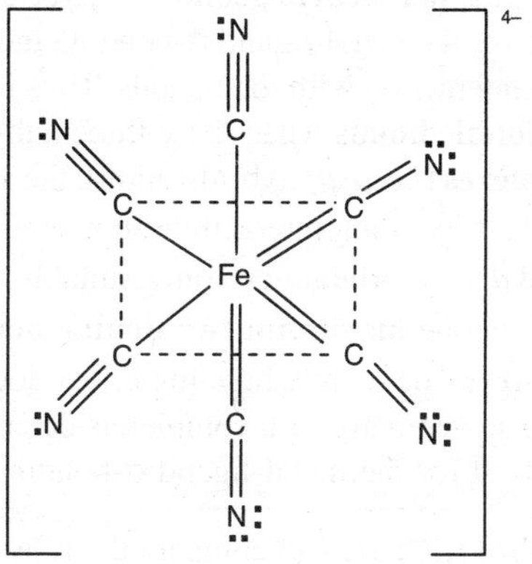

By considering the six 'Fe–CN' single bonds (without any ionic character), there may be an accumulation of 4 units of negative charge on the Fe centre. Consideration of the existence of three 'Fe=CN' bonds in $[Fe(CN)_6]^{4-}$ indicates *the presence of only one unit of negative charge on the metal centre.* Thus this type of double bond formation prevents the accumulation of excessive amount of negative charge on the metal centre.

In $[Fe(CN)_6]^{4-}$, three single 'Fe—C' bonds and **three 'Fe=C' double bonds** may be present simultaneously. Variation in the position of these bonds can produce the **different resonating structures.** Without considering any ionic character in the *iron-carbon bond*, the metal centre is to bear one unit of negative charge and each nitrogen bears –1/2 unit of charge (average value over the all possible resonating structures). It is reasonable to consider that in addition to this mechanism of charge neutralization on the metal centre, the *mechanism of charge neutralization through the introduction of partial ionic character will also operate simultaneously.* Both these mechanisms operating simultaneously may lead to reduce the negative charge on the metal centre further and the resultant charge on the metal centre may be even positive. Considering, ~17% ionic character in the iron-carbon bond, it can be shown that in $[Fe(CN)_6]^{4-}$, the resultant charge on the metal centre becomes 0 (zero) and –2/3 unit charge on each CN. If 25% ionic character is considered then the effective charge on Fe and CN becomes to +0.50 and – 0.75 respectively.

Case of [Ni(CO)$_4$)]: In Ni(CO)$_4$, existence of two 'Ni=C' bonds along with two 'Ni—C' bonds can maintain 0 (zero) charge on Ni-centre without considering any ionic character in the metal-ligand bond.

Characteristic features of the double bonding effect (*i.e.* metal → ligand π-bond)
(*cf.* Fig. 10.7.8.1, Vol. 2 of Fundamental Concepts of Inorganic Chemistry)

(i) **Removal of negative charge from the metal centre:** This argument made by Pauling can remove the unfavourable situation developed on the metal centre through the ligand → metal σ-bonding interaction.

(ii) **Condition for π-bonding interaction:** The metal centre should be in low oxidation state with the filled nonbonding (in terms of σ-bonding) orbitals and the ligands (*i.e.* **π-acid ligands**) must be associated with the suitable vacant orbitals to receive the electron cloud from the metal centre.

(iii) **Synergistic Interaction:** The ligand → metal σ-bonding and metal → ligand π-bonding work in a synergistic fashion.

(iv) **Number of double bonds:** It depends on the nature of hybridization and stereochemistry of the metal complex. In d^2sp^3 (*i.e.* **octahedral geometry**) hybridization, the d-orbitals which are not utilized in hybridization for the metal-ligand σ-bonding interactions, *i.e.* d_{xy}, d_{yz} and d_{xz}, are available for the π-bonding interaction with the ligands. Thus for the d^2sp^3 hybridization, **there may be maximum three double bonds** when these three unhybrid d-orbitals remain filled in. In **the square planar complexes** (*i.e.* dsp^2 hybridization), the same set of d-orbitals (*i.e.* d_{xy}, d_{yz} and d_{zx}) can participate in the π-bonding interaction. In a tetrahedral geometry (*i.e.* sp^3 and d^3s hybridization), the d_{z^2} and $d_{x^2-y^2}$ orbitals are only suitable for such π-bonding. Thus, in a tetrahedral complex, there may be **maximum two double bonds at the same time**.

(v) **No effect on stereochemistry:** The double bonding effect does not produce any effect on the geometry of the complex. The geometry of a complex is determined by the spatial orientation of the hybrid orbitals involved for the metal-ligand σ-bonding.

Case of [Fe(NH$_3$)$_6$]$^{3+}$ vs. [Fe(bpy)$_3$]$^{3+}$: We can compare the **relative stabilities of [Fe(NH$_3$)$_6$]$^{3+}$ and [Fe(bpy)$_3$]$^{3+}$** where both the complexes use the N-donor sites. [Fe(NH$_3$)$_6$]$^{3+}$ where no π-back bonding is possible does not exist and in fact in this complex, the metal centre is to bear a high amount of negative charge from the σ-donation by six sp^3–N sites. In [Fe(bpy)$_3$]$^{3+}$, the accumulated negative charge on the metal centre can be pushed back into the vacant π*–MO of the ligand, bpy. This removes the unfavourable situation. Here it may be also mentioned that compared to [Fe(NH$_3$)$_6$]$^{3+}$, in [Fe(bpy)$_3$]$^{3+}$, the extent of σ-donation by the ligand is less because the sp^2–N of bpy is more electronegative than the sp^3–N of NH$_3$.

There are many experimental evidences to support the fact that in many complexes, the double bonding effect (*i.e.* metal → ligand π-bonding) exists. These aspects will be discussed in detail in Sec. 9.4 and Chapter 9.

3.4.8 Limitations of VBT in Explaining the Bonding in Coordination Complexes

Pauling's VBT (developed in 1931–32) **qualitatively** explained the structure and magnetic properties of many complexes. In fact, this theory rendered a great service for nearly two decades in understanding the properties of complexes. However, with the development of CFT, it lost its importance. The major limitations of VBT are mentioned below.

(i) **Conditions leading to outer and inner orbital complexes:** It was attempted to explain the fact in terms of electronegativity of the donor sites of the ligands, relative bonding properties of the

$(n-1)d$ and nd orbitals. But no quantitative explanation to predict the nature of the complex (*i.e.* outer or inner orbital complex) was available from VBT. In fact, when the magnetic properties are known, then it may be possible in many cases (but not in all cases) to identify the outer or inner orbital complexes.

(ii) **No prediction regarding the preference of geometry of the complexes:** Ni(II) can produce octahedral (*e.g.* [Ni(NH$_3$)$_6$]$^{2+}$), tetrahedral (*e.g.* [NiCl$_4$]$^{2-}$), and square planar (*e.g.* [Ni(CN)$_4$]$^{2-}$) complexes depending on the nature of the ligands. Interestingly, the heavier congeners (*e.g.* Pd^{II}, Pt^{II}) of Ni(II) always produce the square planar complexes. The different preferences for Ni(II), Pd(II) and Pt(II) to attain the stable geometries cannot be explained by VBT. The difference in the geometries of [Cu(NH$_3$)$_4$]$^{2+}$ and [Zn(NH$_3$)$_4$]$^{2+}$ cannot be explained by VBT. Co(II) forms both the tetrahedral (*e.g.* [CoCl$_4$]$^{2-}$) and octahedral (*e.g.* [Co(NO$_2$)$_6$]$^{4-}$) complexes. Fe(III) can produce both the tetrahedral (*e.g.* [FeCl$_4$]$^-$) and octahedral (*e.g.* [Fe(CN)$_6$]$^{3-}$) complexes but Co(III), Cr(III) form only the octahedral complexes. VBT cannot explain the relative stabilities for different geometries in metal complexes.

(iii) **No explanation for the characteristic colours of the complexes:** Very often, the complexes of transition metal ions are characterized by their colours (*i.e.* absorption spectra). But no convincing explanation is available from VBT.

(iv) **No explanation for distortion in the Cu(II) complexes:** Cu(II) complexes are always tetragonally distorted even when all the ligands are equivalent. VBT fails to offer any explanation behind this observation.

(v) **Lability and inertness of the metal complexes:** VBT was applied to explain these aspects but no quantitative and even semiquantitative explanation was available from VBT (see Chapter 5).

(vi) **Excitation of the $(n-1)d$ electron to form the inner orbital complexes:** It has been pointed out that for Co(II) (d^7-system), the d^2sp^3 hybridization needs the promotion of the 7th $(n-1)d$ electron to some higher energy level. This is why, the excited electron is expected to be lost easily. In fact, the complexes like [Co(NH$_3$)$_6$]$^{2+}$, [Co(NO$_2$)$_6$]$^{4-}$ are easily oxidized to their corresponding Co(III) complexes. However, this concept cannot be generalized. For example, the dsp^2 hybridization of Cu(II) (d^9-system) requires the promotion of ninth $3d$ electron into a higher orbital like $4p$. Thus, it may be argued that the square planar complexes of Cu(II), *e.g.* [Cu(NH$_3$)$_4$]$^{2+}$, [Cu(bigH)$_2$]$^{2+}$ should be easily oxidized to their corresponding Cu(III) complexes. But this prediction is not experimentally supported. Moreover, the ESR (electron spin resonance) data do not support the presence of the unpaired electron in the $4p$ orbital. In fact, the ESR data indicate the unpaired electron in the $3d$-orbital. *The experimental findings of the square planar complexes of Cu(II) can be explained by considering the sp^2d hybridization keeping the unpaired electron in the 3d level.* For Ni(III) (d^7-system), the octahedral inner orbital complex (d^2sp^3) suggests the promotion of the 7th $3d$ orbital to some higher energy orbital ($5s$). But the Ni(III) octahedral complexes (inner orbital complex, supported by the magnetic data) are not susceptible to oxidation.

Thus the properties of the square planar complexes of Cu(II), the octahedral complexes of Ni(III) cannot be explained in terms of VBT involving the promotion of the unpaired electron to some higher level.

(vii) **Magnetic criterion to distinguish between the inner and outer orbital complexes:** For the d^n-systems ($n = 4, 5, 6, 7$), where the both possibilities exist for the inner and outer orbital complexes, the magnetic data can be successfully applied to identify the outer and inner orbital complexes. But this idea fails when the observed magnetic moment possesses the orbital contribution, and spin-orbit coupling contribution. **In fact, the magnetic criterion to identify the** nature of hybridization works well when the observed magnetic moment (μ_{obs}) is only due to the spin-only value (μ_{spin}).

The said magnetic criterion also fails when both the possibilities lead to the same number of unpaired electron. For example, both the dsp^2 and sp^2d hybridizations of Cu(II) (d^9-system) lead to one unpaired electron. Similarly, if two different geometries produce the same number of unpaired electrons, then the concept of magnetic criterion to identify the geometry also fails. For example, both the tetrahedral (sp^3) and square planar (dsp^2 or sp^2d) geometries of the four coordinate complexes of Cu(II) give one unpaired electron.

(viii) **Misleading classification of the ionic (*i.e.* outer orbital) and covalent (*i.e.* inner orbital) complexes:** The complexes which show the same number of unpaired electrons present in the free ions were considered as the ionic complexes (*i.e.* outer orbital complex). Similarly, the inner orbital complexes where the number of unpaired electron is less than that present in the free ions were called the covalent complexes. The **so-called ionic complex,** [Fe(acac)$_3$] (sp^3d^2 hybridization, supported by the magnetic moment value) possess the properties of covalent compounds like *high volatility, high solubility in organic solvents.*

In the aqua-complexes, $[M(OH_2)_6]^{n+}$, the **Brönsted acidity** (*i.e.* protonic character of the hydrogen atom of the coordinated water) is expected to increase with the increase of the covalent character of the 'M—O' bond for a particular value of ionic charge of the complex. Let us consider the following aqua complexes.

$$\left[(H_2O)_5M-O\begin{array}{c}H\\\\H\end{array} \right]^{3+} \rightleftharpoons [(H_2O)_5M(OH)]^{2+} + H^+, K_a$$

Acidity: $\left[Cr(OH_2)_6\right]^{3+} < \left[Fe(OH_2)_6\right]^{3+} < \left[Co(OH_2)_6\right]^{3+}$

pK_a:	3.8	2.2	0.7
Hybridization:	d^2sp^3	sp^3d^2	d^2sp^3
Bond type:	covalent	ionic	covalent

Among the above mentioned aqua-complexes, $[Fe(OH_2)_6]^{3+}$ is the only ionic complex and consequently it should be the weakest acid. But, the experimental value does not support this prediction. $[Co(OH_2)_6]^{3+}$ is the strongest acid and $[Fe(OH_2)_6]^{3+}$ is a stronger acid than $[Cr(OH_2)_6]^{3+}$. The **water exchange rate (k_{ex})** in the ionic complex (*i.e.* aqua complex) is expected to be faster than the water exchange rate of the covalent complexes. But this prediction is not experimentally supported in many cases as in the above mentioned aqua complexes. $[Fe(OH_2)_6]^{3+}$ and $[Co(OH_2)_6]^{3+}$ exchange with H_2O^* almost instantaneously while $[Cr(OH_2)_6]^{3+}$ exchanges very slowly.

(ix) **No explanation for the spin isomerism:** VBT theory finds no explanation for the spin isomerism (*i.e.* isomerism between outer and inner orbital complexes) found in many complexes like the Fe(III) complexes. This phenomenon leads to the anomalous magnetic properties.

(x) **Magnetic properties:** The observed magnetic moments cannot always be explained by considering the spin-only value. The phenomena of orbital contribution, spin-orbit coupling, spin–isomerism, etc. may be important in many cases to interpret the observed magnetic moment.

Co(II) (d^7-system) gives the same number of unpaired electrons in both the tetrahedral (sp^3) and outer-orbital octahedral complexes (sp^3d^2) where the expected μ_{spin} value is 3.88 B.M. But the tetrahedral complexes of Co(II) show the magnetic moment in the range of 4.2–4.8 B.M. while the outer-orbital (*i.e.* high spin) octahedral complexes of Co(II) give the magnetic moment in the range 4.8–5.2 B.M. Similarly, for the octahedral complexes of Ni(II), μ_{obs} (= 2.9–3.5 B.M.) differs significantly from the μ_{spin} value (= 2.83 B.M.). This significant deviation from the μ_{spin} value cannot be explained by considering the Pauling's VBT.

(xi) **Relative stabilities of the complexes among the congeners for different oxidation states:** It is illustrated for the Gr 11 elements (*i.e.* Cu, Ag, Au). The Cu(II) complexes are stable while the Ag(II)-complexes and Au(II)-complexes are unstable. The Cu(III) complexes are unstable and oxidizing while the Au(III) complexes are relatively more stable and less oxidizing. Variation of these properties among the congeners does not find any explanation in VBT.

(xii) **Nonexistence of many complexes:** Many complexes like $[Co(CN)_6]^{4-}$, $[Ni(CN)_6]^{4-}$, $[Cu(NH_3)_6]^{2+}$, $[Pt(NH_3)_6]^{2+}$, etc. do not exist. The reason behind the nonexistence of such complexes is not available in VBT.

3.5 CRYSTAL FIELD AND LIGAND FIELD THEORIES: SPLITTING OF THE METAL ORBITALS IN DIFFERENT CRYSTAL FIELDS OF LIGANDS

The **Crystal Field Theory (CFT)** was first proposed by Bethe (in 1929) by considering the *electrostatic interaction* between the metal and the ligands to explain the colour and magnetic properties of the *crystalline solids of metal salts*. It was further modified by Van Vleck (in 1935) to account for the magnetic properties of transition metal complexes. However, before 1950, the theory was mainly confined in solid state physics. However, in 1950's, the theory was popularized by many workers (L.E. Orgel was one of pioneers) in understanding the problems of coordination chemistry.

The original crystal field theory (CFT) was modified in the adjusted crystal field theory (ACFT) or the ligand field theory (LFT) **through the incorporation of some metal-ligand orbital overlap.** However, the basic approach of the electrostatic model of CFT is retained in LFT. In fact, in a practical sense, we do not make any difference between CFT and LFT. LFT may be considered to bridge between the crude CFT and more sophisticated MOT (molecular orbital theory).

Note: CFT was first introduced to explain the properties of **crystalline substances.** This is why the theory was named so.

3.5.1 Basic Features of CFT and LFT

(i) **Ligands as the point-like charges:** In the CFT model, the approaching ligands are considered as the *point-like negative charges* (that may be the negative charge on the ligand or the negative end of the ligand dipole).

(ii) **Wave mechanical identity of the atomic orbitals of the metal centre:** *Though the ligands are considered as the point-like negative charges, the metal electrons are supposed to maintain their wave mechanical identity. In other words, the metal centre is supposed to bear the appropriate atomic orbitals to interact with the electrostatic field provided by the point-like negative charges, i.e. the ligands.*

(iii) **Electrostatic perturbation of the metal orbital electrons by the ligands:** The metal-ligand interactions are basically electrostatic and these may be of ion-ion or ion-dipole type. For the transition metal complexes, the electrostatic effect of the ligands on the energies of the *d*-orbitals, *i.e. d*-electrons, is the most important aspect of CFT. The intensity and symmetry of the electrostatic field of the ligands (considered as the point-like charges) are important to interact with the atomic orbitals of metal.

(iv) **No covalent interaction:** In true CFT, no covalent interaction between the ligand and metal is considered. *It is due to the fact that the ligands are treated as the point like charges and their* *orbitals are of no consideration.* Thus, in this concept, the metal electrons are considered as the nonbonding electrons. However, in the modified CFT described as LFT, the metal-ligand overlap is considered to some extent.

(v) **Effect on the energy and degeneracy of the metal orbitals:** The question of degeneracy of the metal orbitals in the presence of the ligands depends on the symmetry of the electrostatic field of the ligands. The extent of energy change of the metal orbitals depends on the intensity of the electrostatic field of the ligands. These aspects are discussed in detail in Sec. 3.5.2.

3.5.2 Effect of the Electrostatic Field of the Ligands on the Energies of the Metal Orbitals

(A) **Effect on the metal s-orbital:** The *s*-orbital is spherically symmetrical and consequently this orbital is equally affected by the electrostatic field of the ligands in all directions. *Thus, in the presence of the electrostatic field of the ligands, energy of the s-orbital is simply changed.*

(B) **Effect of the spherical electrostatic field of the ligands:** When the electrostatic field is spherically symmetrical, *the metal p, d or f-sets of orbitals are simply changed in energy but the degeneracy in each set of the orbitals is retained* as in the case of free metal ions (*i.e.* in the absence of any external field). Thus, the three *p*-orbitals or five *d*-orbitals or seven *f*-orbitals remain degenerate in the presence of a spherically symmetrical ligand field. Only their energies are affected depending on the intensity of the field.

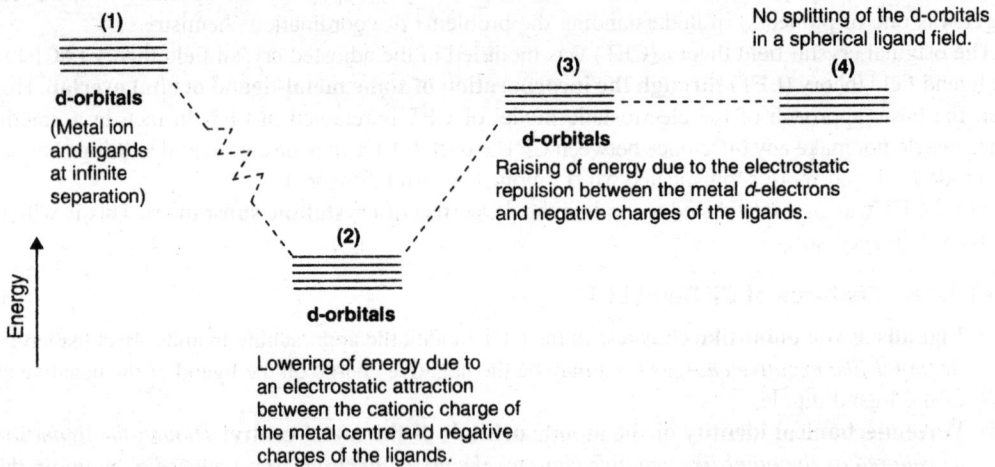

Fig. 3.5.2.1 Effect of the spherical ligand field of negative charges on the metal *d*-orbitals. See Figs. 3.5.4.2, 4, 7 for the significance of (1), (2), 3($\equiv$3' + 3") and (4).

The metal centre is supposed to sit at the centre of a hollow sphere whose *radius may be considered as the metal-ligand distance.* The negative charges of the ligands may be considered to spread uniformly over the sphere. The **electrostatic attraction** between the cationic charge of the metal and ligand negative charges spreading over the sphere will lower the energy of the system. But, energy of the system will be somewhat raised by the **repulsion** of the metal electrons with the ligand negative charges. Because of the spherical symmetry of the ligand field, degeneracy in each set of the orbitals (*e.g. p* or *d* or *f* set of orbitals) is retained. This is illustrated in Fig. 3.5.2.1 for the five-fold degenrate *d*-orbitals.

(C) Effect of the unsymmetrical electrostatic field of the ligands: Because of the directional character of the *p*, *d* or *f* sets of the orbitals, they are no longer degenerate in the presence of an unsymmetrical field of the ligands. The nature of loss of degeneracy of the metal orbitals depends on both the symmetry of the ligand field and symmetry of the orbitals (in a particular set) experiencing the ligand field. These are illustrated in the following sections.

Generally for the transition metals, the $(n-1)p$ and $(n-1)d$ orbitals are affected by the ligand field. But the $(n-1)p$ orbitals are generally fully occupied and consequently they will not have any net effect on the energy of the system through their splitting. Splitting of the $(n-1)d$ orbitals (which generally remain partially filled in) is quite important. For the lanthanides, $(n-2)f$, *i.e.* **the 4f orbitals are deeply seated,** *i.e.* **well shielded by the outer electrons.** Consequently, the electrostatic field of the ligands on the 4f orbitals is not important. For the actinides, the **5f orbitals which are relatively more exposed** (*cf.* **relativistic expansion of the f-orbitals;** Vol. 1 of Fundamental Concepts of Inorganic Chemistry) to the ligands experience the splitting in presence of an unsymmetrical ligand field.

3.5.3 Splitting of the Metal *p* – and *d* – Orbitals in a Linear Ligand Field

If the ligands are considered to approach along the $\pm z$ directions to give the linear geometry of the complex, then the metal orbitals having the lobes along the $\pm z$ directions will be more affected compared to the other orbitals lying in other directions. Thus, the metal orbitals having the *z*-component in their wave function will experience a stronger field of negative charge from the ligands and such orbitals will be repelled more. By considering the directional characters, it is evident that the p_z and d_{z^2} orbitals will be repelled most, followed by the d_{xz} and d_{yz} orbitals. **Thus the three-fold degeneracy of the *p*-orbitals and five-fold degeneracy of the *d*-orbitals will be destroyed.** The orbitals being repelled by the ligand field will be energetically raised and the orbitals which are not repelled (or relatively less repelled) will be energetically lowered in a way so that the **Barycentre Rule (*i.e.* principle of conservation of energy)** is maintained. According to the barycentre rule, the centre of gravity of the set of orbitals

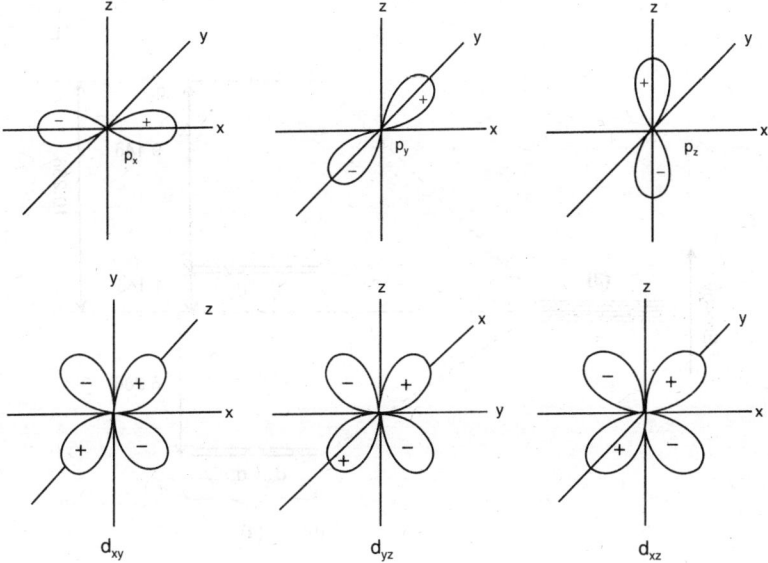

(Contd.)

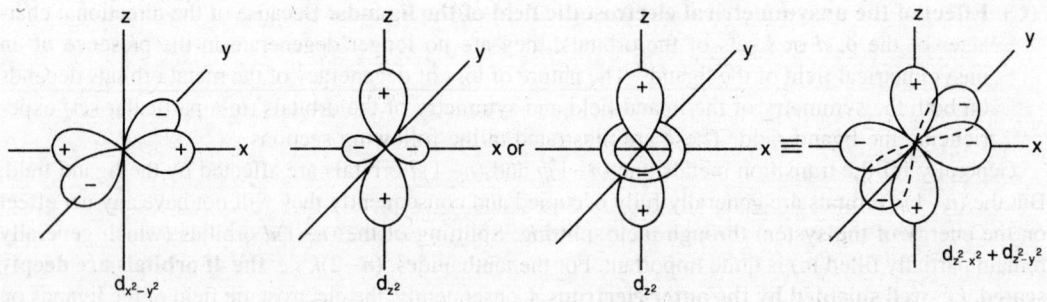

Fig. 3.5.3.1 Shapes of the *p* and *d*-orbitals (*see* Fig. 4.1.6.1, Vol. 1 of Fundamental Concepts of Inorganic Chemistry).

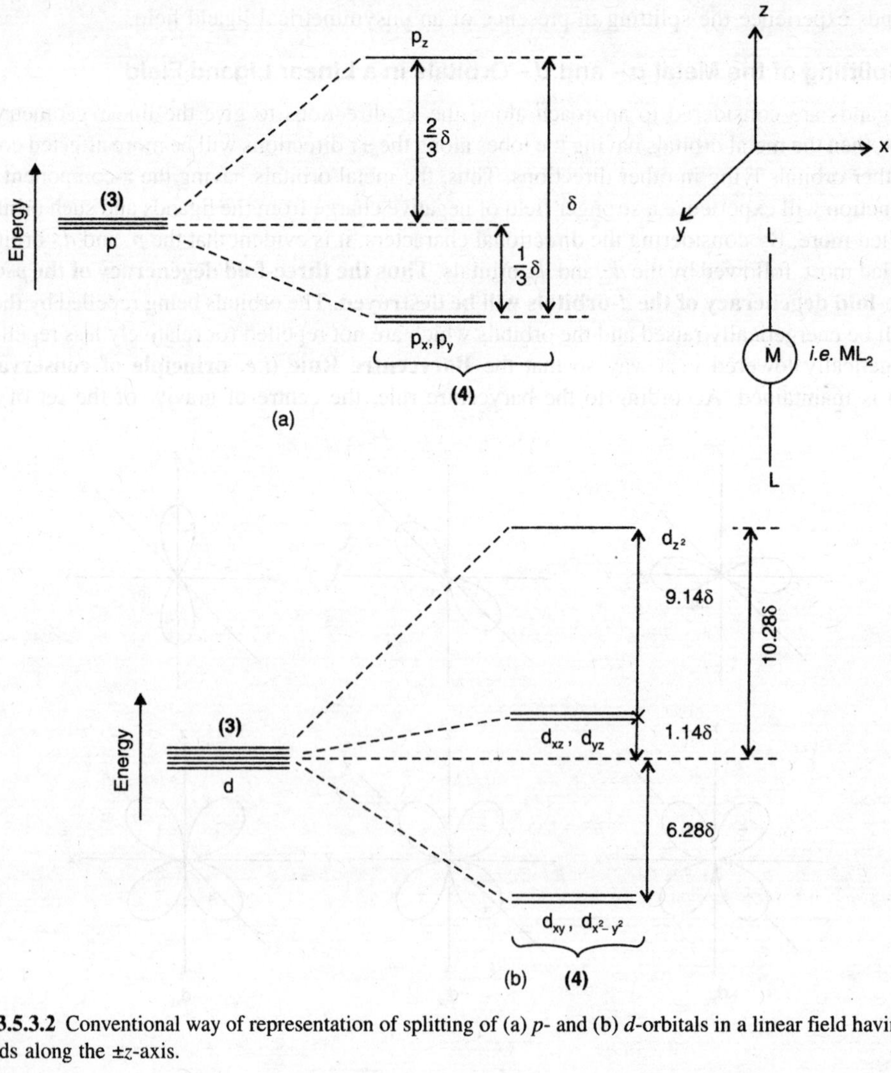

Fig. 3.5.3.2 Conventional way of representation of splitting of (a) *p*- and (b) *d*-orbitals in a linear field having the ligands along the ±*z*-axis.

experiencing the splitting will remain unchanged, *i.e.* **centre of gravity will be the same before and after splitting of the orbitals.** In the present case, the orbitals having no z-component in their wave functions (*i.e.* p_x and p_y; d_{xy} and $d_{x^2-y^2}$) will be relatively stabilized (*i.e.* energetically lowered) to maintain the barycentre rule.

Note: Actual energy change of the orbitals is not shown in Fig. 3.5.3.2 (*cf.* Figs. 3.5.4.1, 2, 7 for the significance of (**3**)).

3.5.4 Splitting of the Orbitals in an Octahedral Ligand Field

(A) *p*-orbitals: The three *p*-orbitals (*i.e.* p_x, p_y and p_z) are degenerate in a field-free condition. Each *p*-orbital projects its two lobes on either side of the nucleus. Thus, if the metal centre is supposed to sit at the origin of the Cartesian coordinate system, then the metal p_x orbital projects its one lobe along the $+x$ direction while the other equivalent lobe is projected along the $-x$ direction. The similar situation arises for the p_y and p_z orbitals.

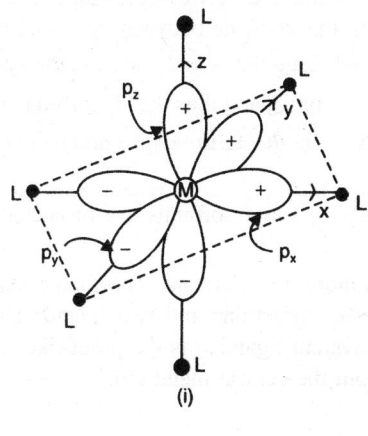

(i)

Fig. 3.5.4.1 (i) Three *p*-orbitals in an **octahedral geometry**. If two axial ligands from the $\pm z$ directions are removed, then the p_x and p_y orbitals are positioned in a square planar geometry (in xy-plane).

$$\left[\mathbf{ML_6}\right](\mathbf{O_h}) \xrightarrow[\text{($\pm z$ directions)}]{-2L} \left[\mathbf{ML_4}\right](\mathbf{D_{4h}}, \text{square planar})$$

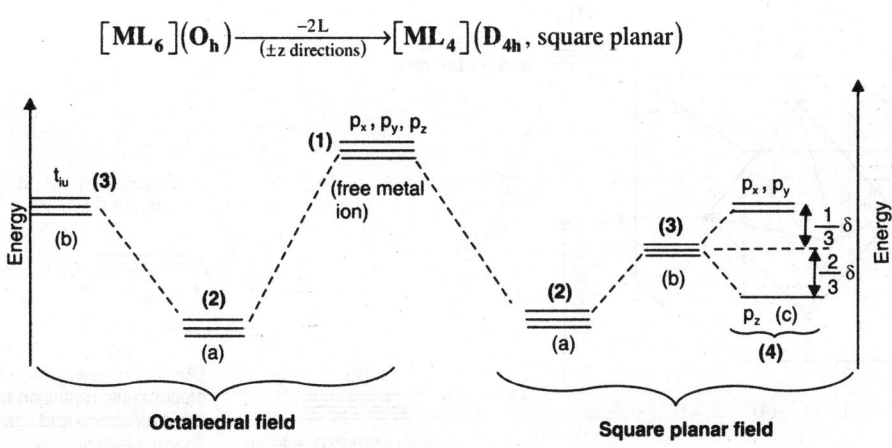

Fig. 3.5.4.1 (ii) Energy of the *p*-orbitals in presence of an **octahedral field and a square planar field** (qualitative representation). The position (a), *i.e.* (**2**) represents the lowering of energy due to the electrostatic attraction between the cationic metal centre and negative charge of the ligands. The position (b), *i.e.* (**3**) represents the raising of energy due to the repulsion between the metal *p*-electrons and negative charges of the ligands. (c), *i.e.* (**4**) Splitting of the *p*-orbitals in the presence of ligand field. (*See* Fig. 3.5.4.7 for the significance of (**1**), (**2**), **3** (≡**3′** + **3″**), (**4**)).

In the octahedral ligand field around the metal centre lying at the origin of the Cartesian coordinate system, the six ligands (as the point-like negative charges) reside along the $\pm x$, $\pm y$ and $\pm z$ directions at an equal metal-ligand distance. **Thus, each of the p-orbitals equally faces two ligands along the coordinate axes.** Consequently, fate of the three p-orbitals remains unchanged, *i.e.* **degeneracy of the p-orbitals will remain unchanged in an octahedral ligand field**. Here, simply, the electrostatic repulsion between the metal electrons and ligand charges will raise the energy of the three-fold degenerate p-orbitals illustrated in Fig. 3.5.4.1. However, there will be a net decrease in energy of the orbitals due to the electrostatic attraction between the cationic charge of the metal centre and ligand negative charges.

Note: In a square planar geometry, considering the ligands to reside in the xy-plane, the p_x and p_y orbitals will be destabilized more; and the p_z orbital will be relatively stabilized. **Thus, degeneracy the three-fold p-orbitals will be destroyed in a square planar geometry.**

(B) d-orbitals: The five d-orbitals are degenerate in a field-free condition or in a spherically symmetrical field. The five d-orbitals are not all alike. There are two types of d-orbitals:

d_{xy}, d_{yz} **and** d_{xz}**:** The four lobes of each orbital are concentrated between the respective coordinate axes, *e.g.* the four lobes of d_{xy} orbital are concentrated in the xy-plane between the x and y axes.

$d_{x^2-y^2}$ **and** d_{z^2} **:** The four lobes of $d_{x^2-y^2}$ are projected along the x- and y-axes in the xy-plane. The d_{z^2} orbital is actually a combination of $d_{z^2-x^2}$ and $d_{z^2-y^2}$ orbitals and thus the d_{z^2} orbital is equivalent to $d_{2z^2-x^2-y^2}$. The d_{z^2} orbital bears *two equivalent lobes along the $\pm z$ directions and a concentric lobe around the nucleus in the xy-plane.*

Thus qualitatively it is realized that the lobes of $d_{x^2-y^2}$ and d_{z^2} orbitals are projected along the coordinate axes.

In an octahedral geometry, the six ligands approach along the Cartesian coordinate axes, *i.e.* two ligands along the $\pm x$ directions, two ligands along the $\pm y$ directions and two ligands along the $\pm z$ directions. Thus in an octahedral geometry, the six equivalent ligands (as the point-like charges) are placed on the coordinate axes and at equal distances from the central metal ion.

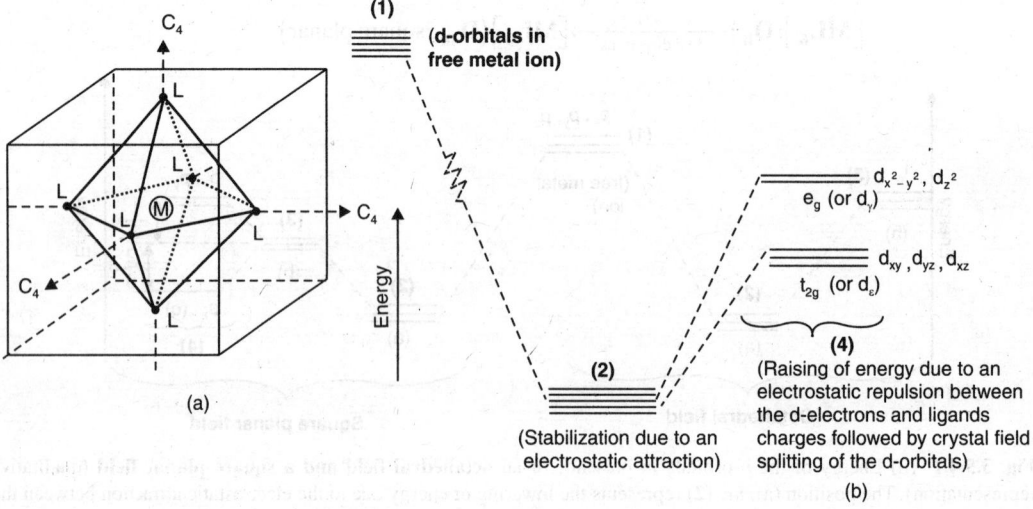

Fig. 3.5.4.2 (a) Octahedral crystal-field generated by placing 6 ligands (L) at the centres of 6 faces of a cube keeping the metal centre (M) at the centre of the cube (*cf.* Fig. 3.5.2.1, 3.5.4.4, 7). (b) Splitting of the d-orbitals in an octahedral field.

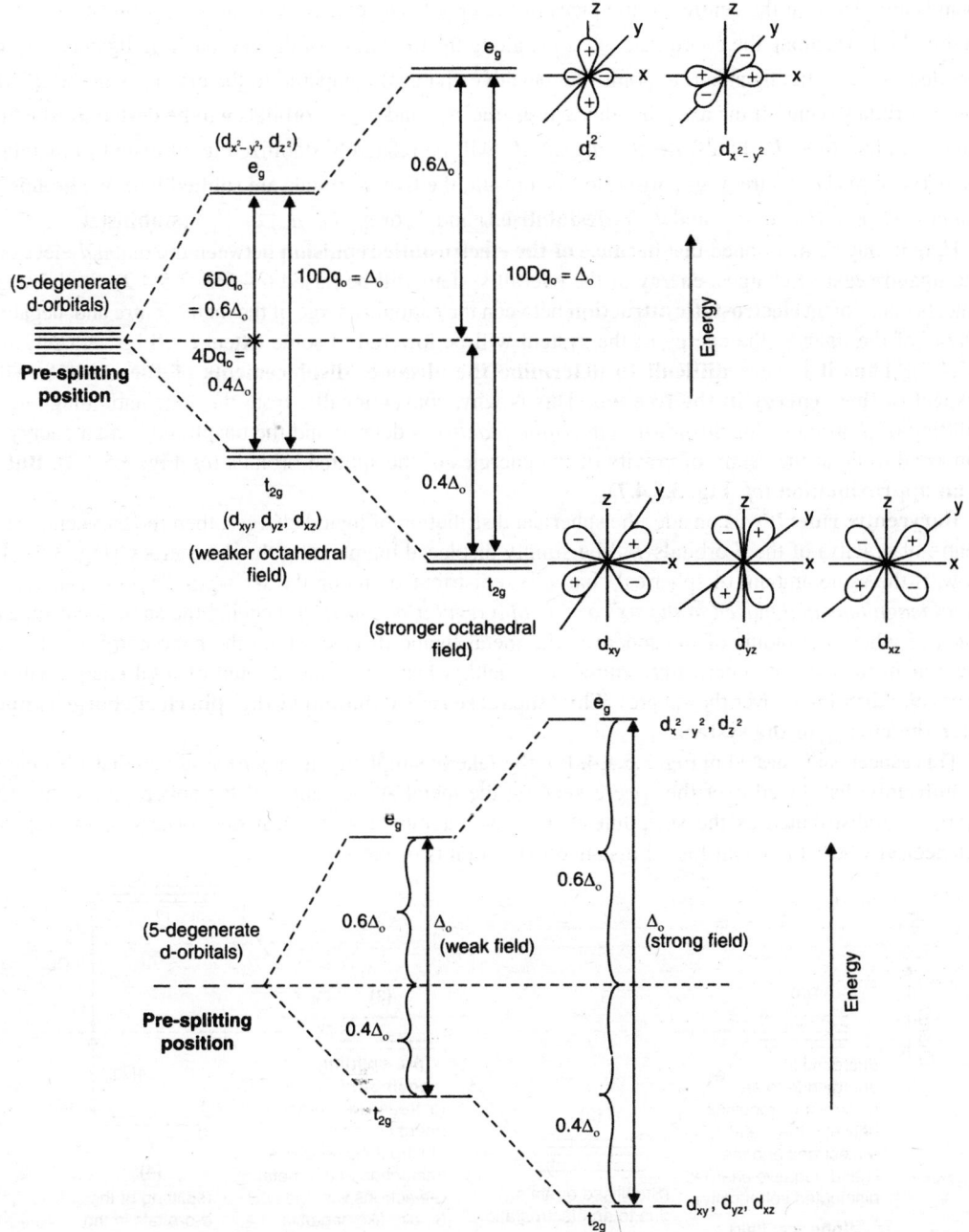

Fig. 3.5.4.3 Conventional representation of splitting of the *d*-orbitals in an octahedral field (**without considering the absolute energy of the *d*-orbitals in the field**, *cf.* Fig. 3.5.4.7).

The octahedral geometry may be considered to be generated **by placing the metal at the centre of a cube where the Cartesian axes act as the four-fold axes (C_4 axes) of the octahedron (*i.e.* the** ligands are placed at the centres of the faces of the cube). The electrons in the $d_{x^2-y^2}$ and d_{z^2} orbitals having the lobes along the coordinate axes, *i.e.* along the directions of the approaching ligands, will be repelled more by the ligands (*i.e.* point-like negative charges) compared to the electrons in the d_{xy}, d_{yz} and d_{xz} orbitals lying off the axes. In other words, the d_{z^2} and $d_{x^2-y^2}$ orbitals will be destabilized while the remaining three d-orbitals, *i.e.* d_{xy}, d_{yz} and d_{xz} will be relatively stabilized to an extent to maintain the barycentre (*i.e.* centre of gravity) rule. As a result, the five d-orbitals are splitted into two groups of orbitals: e_g or d_γ (*i.e.* $d_{x^2-y^2}$ and d_{z^2}) (**destabilised**) and t_{2g} or d_ε (*i.e.* d_{xy}, d_{yz} d_{xz}) (**stabilised**).

Here it may be mentioned that **because of the electrostatic repulsion** between the metal d-electrons and ligand negative charges, energy of the overall system will be raised (*cf.* Fig. 3.5.4.2). At the same time, because of an **electrostatic attraction** between the cationic charge of the metal centre and negative charge of the ligands, the energy of the system will be lowered to some extent (*cf.* Figs. 3.5.2.1, and 3.5.4.7). **Thus it is very difficult to determine the absolute displacements of the orbitals with respect to their energy in the free ion.** This is why, conventionally, from the schematic diagram of splitting of d-orbitals, this *unknown / uncertain quantity* is deleted and the unsplitted orbital energy is supposed to lie at the centre of gravity of the energies of the splitted orbitals (*cf.* Fig. 3.5.4.3). **But it is an approximation (*cf.* Fig. 3.5.4.7).**

Barycentre rule: If we consider the spherical distribution of ligand charges, then the barycentre (*i.e.* centre of gravity) of the d-orbitals will be simply displaced maintaining the degeneracy (Fig. 3.5.2.1). Now, if the same amount of spherical charge is redistributed among the *six separate point charges of equal magnitude and placed at the six corners of a regular octahedron* (keeping the same metal-ligand distance where the radius of the sphere is the metal-ligand distance) then the barycentre will be the same as in the case of spherically symmetrical field (where the same amount of total charge will be uniformly distributed over the sphere). **Thus the mere redistribution of the spherical charge cannot alter the energy of the system.**

This aspect is illustrated in Fig. 3.5.4.4. For the sake of simplicity, it may be assumed that $12e$ charge is uniformly distributed over the sphere keeping the metal at the centre of the sphere. Now this $12e$ charge is redistributed as the six point charges of magnitude $2e$ each at the corners of the regular octahedron where the metal-ligand distance is the radius of the sphere.

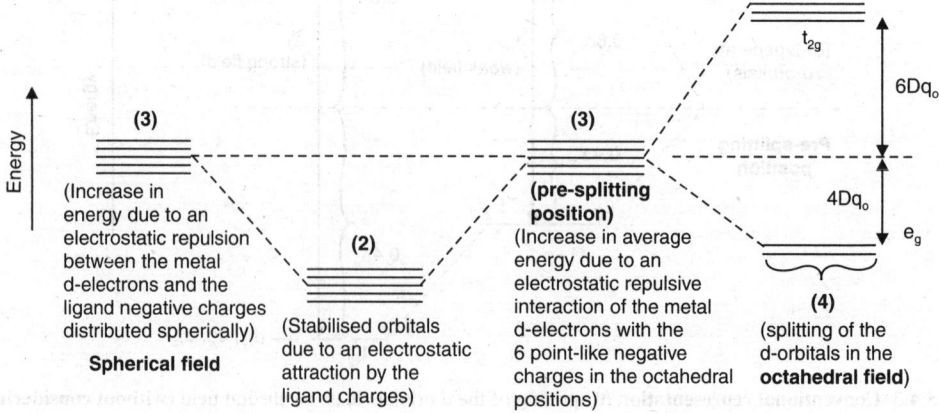

Fig. 3.5.4.4 (*cf.* Figs. 3.5.2.1, 3.5.4.2, 7) Positions of the barycentre of the d-orbitals in a spherical field (of total charge $12e$) and in an octahedral field (of total charge $6 \times 2e$, at six corners of the octahedron) keeping the same metal-ligand distance in both the geometries. (Position of the d-orbitals (**1**) in free ion is not shown here).

Since the mere splitting of the d-orbitals into two sets of orbitals in the octahedral field cannot change the barycentre (*i.e.* centre of gravity of the d-orbitals), then it must satisfy the following condition.

Destabilization caused by the e_g set (*i.e.* $d_{x^2-y^2}$ and d_{z^2}) = stabilization caused by the t_{2g} set

(i.e. d_{xy}, d_{yz}, d_{xz})

This destabilization or stabilization is measured with respect to the **pre-splitting barycentre.** If the e_g set is displaced upward by x (say) from the pre-splitting barycentre then the t_{2g} set will move downward by y (say) so that the following condition is maintained.

$$10E = 4(E + x) + 6(E - y), \ i.e. \ 2x = 3y.$$

Here, E denotes the energy of each electron **prior to splitting.** It leads to:

$$4x = 6y.$$

i.e. total increase of energy of the four electrons in the $d_{x^2-y^2}$ and d_{z^2} orbitals = total decrease of energy of the six electrons in the d_{xy}, d_{yz} and d_{xz} orbitals.

Taking, $x + y = \Delta_o = 10Dq_o$, we can write:

$$x + \frac{2x}{3} = \Delta_o; \ \text{or} \ \frac{5x}{3} = \Delta_o; \ \text{or} \ x = \frac{3}{5}\Delta_o = 6Dq_o \ \text{and} \ y = \frac{2}{5}\Delta_o = 4Dq_o.$$

The right-hand subscript 'o' stands for the octahedral splitting. **Thus each t_{2g} electron is stabilized by $4Dq_o$ while each e_g electron is destabilized (relatively) by $6Dq_o$. In other words, energy of a t_{2g} electron is $-4Dq_o$ while energy of an e_g electron is $+6Dq_o$** (measured with respect to the energy of the electron in the pre-splitting condition).

- **Significance of the term Dq:** The term Dq is the product of two terms D and q required in the quantitative treatment of CFT to describe the Hamiltonian operator of the system. Thus D and q are the **potential energy terms** and in the octahedral crystal field, these are given by:

$$D = \frac{35ze}{4a^5}, \ q = \frac{2e\overline{r^4}}{105}, \ i.e. \ Dq = \frac{ze^2\overline{r^4}}{6a^5}$$

where z = charge on the ligand; e = charge of an electron, a = metal-ligand distance; $\overline{r^4}$ = mean fourth power radius of the d-electron (*i.e.* distance of the d-electron from the nucleus). If the **point dipole** (having μ dipole moment) ligands are considered then, Dq is given by:

$$Dq = 5\mu\overline{r^4}/6a^6$$

Dq is defined so that the energy difference between the t_{2g} and e_g set is $10Dq$ (for the octahedral system, it is represented as $10Dq_o$). It leads to:

$$E_{(e_g)} - E_{(t_{2g})} = 10Dq_o;$$

and $4E_{(e_g)} + 6E_{(t_{2g})} = 0 \ \left(i.e. \ \text{barycentre rule}\right).$

The above two relations give: $E_{(e_g)} = 6Dq_o$ and $E_{(t_{2g})} = -4Dq_o$

These energy values are with respect to that of the pre-splitting barycentre.

- $10Dq_o$ values are experimentally determined from the electronic spectra. These aspects will be discussed later.
- **Energy of the t_{2g} and e_g electrons:** If E denotes the energy of an electron in the **pre-splitting condition** then we can write:

$$\text{energy of a } t_{2g} \text{ electron} = E - 4Dq_o$$
$$\text{energy of an } e_g \text{ electron} = E + 6Dq_o$$

● **Significance of the t_{2g} and e_g symbol:** The symbol t_{2g} set indicates the *three-fold degeneracy of the orbital set, i.e. triplet orbital degeneracy.* The symbol, e_g indicates the *two-fold degeneracy* of the *orbital set* (*cf.* the symbol *e* stands for a two-fold degeneracy).

The subscript '*g*' denotes the '*gerade*' (meaning even). The symbol '*g*' denotes the presence of centre of inversion (*i.e.* centre of symmetry). In the octahedral complex having the 6 identical ligands, the metal centre lies at the centre of inversion and **with respect to this centre of inversion,** the ***d*-orbitals** maintain the same sign of the wave functions in the opposite lobes (*i.e.* the opposite lobes bear the same sign) (*cf.* Fig. 3.5.4.5).

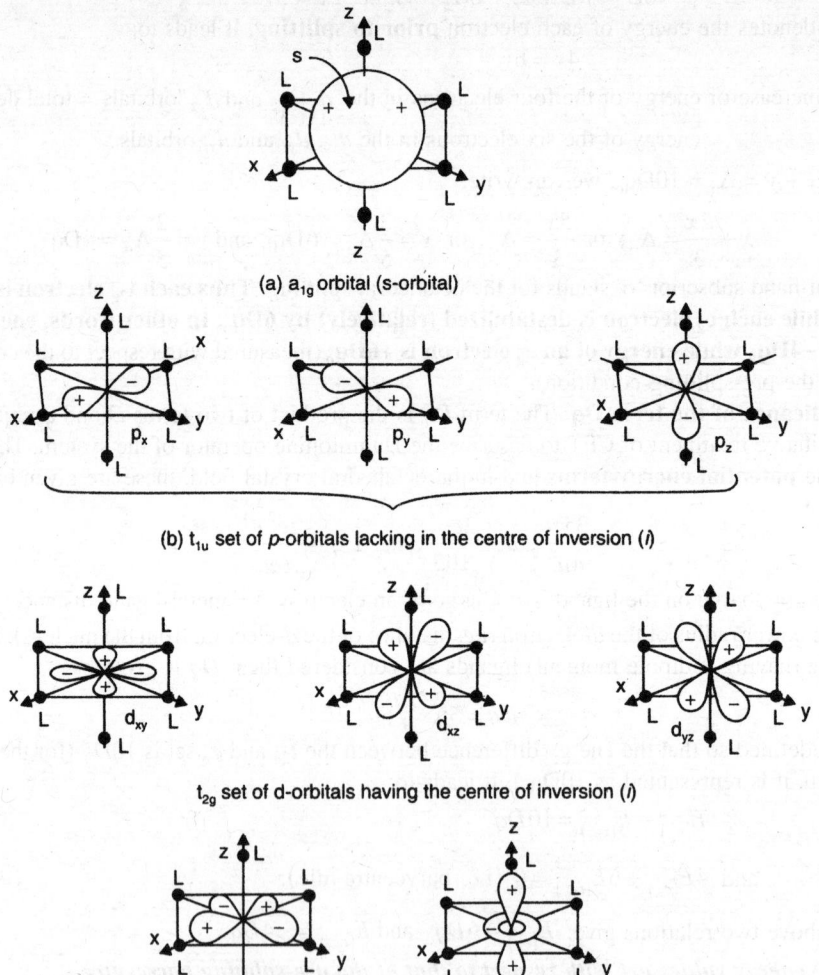

(a) a_{1g} orbital (s-orbital)

(b) t_{1u} set of *p*-orbitals lacking in the centre of inversion (*i*)

t_{2g} set of d-orbitals having the centre of inversion (*i*)

(c) e_g set of d-orbitals having the centre of inversion (*i*)

Fig. 3.5.4.5 The a_{1g}, t_{1u}, t_{2g} and e_g symmetry metal orbitals in an **octahedral system.**

Here it may be pointed out that for **the *p*-orbitals,** their signs of the lobes change on inversion in the octahedral field. This is why, the set of *p*-orbitals (degenerate in the octahedral field) is described as *ungerade* (*i.e.* uneven) and the set is described as t_{1u} (*u* stands for ungerade). The *s*-orbital in an octahedral field is described as a_{1g}.

Thus with respect to the centre of inversion of the octahedron, the *d*-orbitals and *s*-orbitals are **centrosymmetric** and the *p*-orbitals are **not centrosymmetric**.

$$s\text{-orbital} \xrightarrow{\;O_h\text{-field}\;} a_{1g}; \quad p\text{-orbitals} \xrightarrow{\;O_h\text{-field}\;} t_{1u} \text{ set,}$$

$$d\text{-orbital} \xrightarrow{\;O_h\text{-field}\;} t_{2g} \text{ set and } e_g \text{ set.}$$

- **Equivalent character of the orbitals in each set, *i.e.* t_{2g} set and e_g set:** The three *d*-orbitals, *i.e.* d_{xy}, d_{yz} and d_{xz} constituting the t_{2g} set are evidently equivalent. These are equivalently situated with respect to the ligands. Moreover, these orbitals can be mutually interchanged by simple symmetry operation (C_3 operation; C_3 axis passes through the centres of opposite trigonal faces; $4C_3$ axes prevai in O_h) or interchanging the labeling of the Cartesian coordinate system ($x \to y \to z \to x$) (Fig. 3.5.4.6a).

$$d_{xy} \xrightarrow[(i.e.\ C_3\text{-operation})]{x \to y \to z \to x} d_{yz}; \quad d_{yz} \xrightarrow[(i.e.\ C_3\text{-operation})]{x \to y \to z \to x} d_{xz}; \quad d_{xz} \xrightarrow[(i.e.\ C_3\text{-operation})]{x \to y \to z \to x} d_{xy}.$$

The symmetry operation or relabeling of the Cartesian coordinate system cannot change the energy of the orbitals. **In fact, the degenerate orbitals can be interchanged through symmetry operation.**

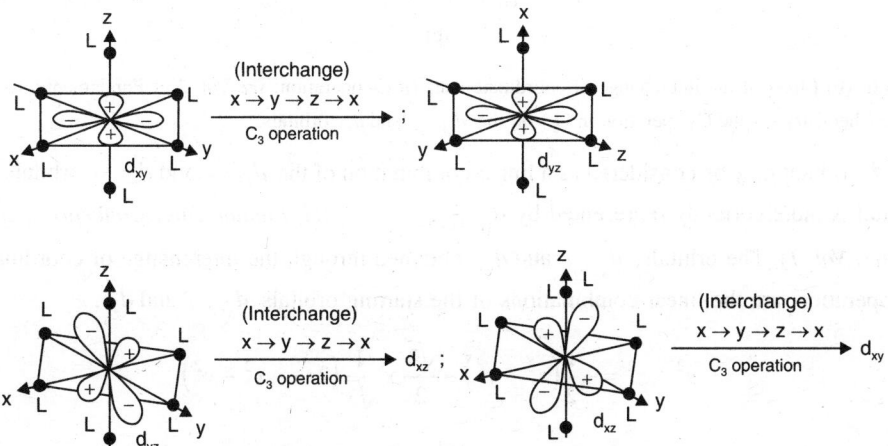

Fig. 3.5.4.6 (a) Effect of the C_3 operation or interchange of the coordinate axes on the d_{xy}, d_{yz} and d_{xz} orbitals.

Are the two *d*-orbitals in the e_g set equivalent? The two *d*-orbitals, *i.e.* $d_{x^2-y^2}$ and d_{z^2} constituting the e_g set are **not apparently equivalent** and they cannot be mutually interconverted through symmetry operation. However, both the orbitals possess their maximum amplitudes along the coordinate axes, *i.e.* both the orbitals project their lobes along the directions of the ligands. But, *qualitatively, it is difficult to understand that the orbitals $d_{x^2-y^2}$ and d_{z^2} (different in shapes) are equally affected by the ligands in an octahedral field.* However, the detail quantitative treatment can prove that these two orbitals are equally affected. This quantitative treatment lies beyond the scope of the book. However, **the equivalence can be demonstrated by the symmetry arguments.**

The symmetry operation (C_3) on $d_{x^2-y^2}$ and d_{z^2} cannot interchange the orbitals but produce new orbitals $d_{y^2-z^2}$ and d_{x^2} respectively (Fig. 3.5.4.6b).

Thus the interchange $x \to y \to z \to x$ in Cartesian coordinate system gives the following conversions.

$$d_{x^2-y^2} \xrightarrow[\substack{(x \longrightarrow y \longrightarrow z \longrightarrow x)}]{\text{Interchange of coordinate axes}} d_{y^2-z^2} \; ; \; d_{z^2} \xrightarrow[\substack{(x \longrightarrow y \longrightarrow z \longrightarrow x)}]{\text{Interchange of coordinate axes}} d_{x^2}$$

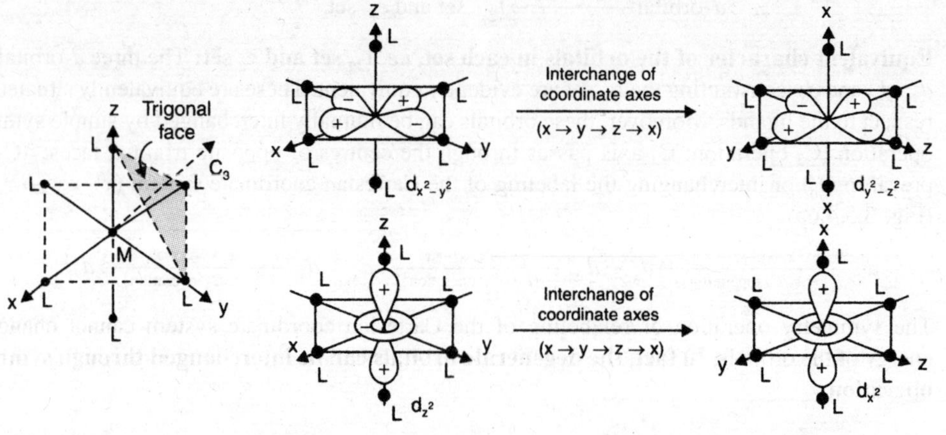

(b)

Fig 3.5.4.6 (b) Effect of the interchange of coordinate axes ($\equiv C_3$-operation; *see* Vol. 2 of Fundamental Concepts of Inorganic Chemistry for the C_3 operation in O_h) on the $d_{x^2-y^2}$ and d_{z^2} orbitals.

The d_{z^2} orbital may be considered as a linear combination of the $d_{z^2-x^2}$ and $d_{z^2-y^2}$ orbitals; and the d_{z^2} orbital is more correctly represented by $d_{(1/\sqrt{3})(2z^2-x^2-y^2)}$ (*cf. Fundamental Concepts of Inorganic Chemistry, Vol. 1*). The orbitals, $d_{y^2-z^2}$ and d_{x^2} obtained through the interchange of coordinate axes, i.e. C_3-operation, are the linear combinations of the starting orbitals $d_{x^2-y^2}$ and d_{z^2}.

$$y^2 - z^2 = -\frac{1}{2}(x^2 - y^2) - \frac{\sqrt{3}}{2} \times \frac{1}{\sqrt{3}}(2z^2 - x^2 - y^2)$$

i.e.

$$d_{y^2-z^2} = -\frac{1}{2}d_{x^2-y^2} - \frac{\sqrt{3}}{2}d_{z^2}$$

The d_{x^2} orbital is actually $d_{\frac{1}{\sqrt{3}}(2x^2-y^2-z^2)}$ (*cf. d_{z^2} orbital*)

$$\frac{1}{\sqrt{3}}(2x^2 - y^2 - z^2) = \frac{\sqrt{3}}{2}(x^2 - y^2) - \frac{1}{2} \times \frac{1}{\sqrt{3}}(2z^2 - x^2 - y^2)$$

i.e.

$$d_{x^2} = \frac{\sqrt{3}}{2}d_{x^2-y^2} - \frac{1}{2}d_{z^2}$$

Thus, the C_3-symmetry operation on the $d_{x^2-y^2}$ and d_{z^2} orbitals produce the new orbitals, *i.e.* $d_{y^2-z^2}$ and d_{x^2} which are the linear combinations of the original orbitals. **It is possible if the new orbitals generated through the symmetry operation are degenerate and also the starting orbitals are degenerate and energy of each orbital is the same.** It is also quite reasonable as relabeling of the coordinate axes would not change the energies of the orbitals.

- **Total energy of the system:** Figures 3.5.4.2-4 represent only the change of energy of d-orbitals in an octahedral crystal field. To have the idea regarding the **total energy change**, we are to consider the following interactions (*cf.* Fig 3.5.4.7).

(i) electrostatic attraction between the cationic charge of the central metal ion and the point-like negative charges of the ligands; (ii) **destabilization** of the metal electrons in the orbitals other than the d-orbitals through the **electrostatic repulsion by the negative charges of the ligands;** (iii) destabilization of the metal d-electrons by the **electrostatic repulsion of the negative charges of the ligands;** (iv) crystal field splitting of the d-orbitals (assuming no splitting of other sets of degenerate orbitals).

The stabilization energy through the electrostatic attraction is more than the destabilization energies of the d-electrons and other electrons. *This is why, the coordination cluster gains a net stabilization energy. In fact, relative energies* (commonly called crystal field stabilization energy, *cfse*) *of the electrons in the t_{2g} and e_g set is only one of the many contributing factors to determine the overall energy of the system.* Figure 3.5.4.7 qualitatively describes the effect of different contributing factors to determine the total energy. Similar situations will arise also for other geometries.

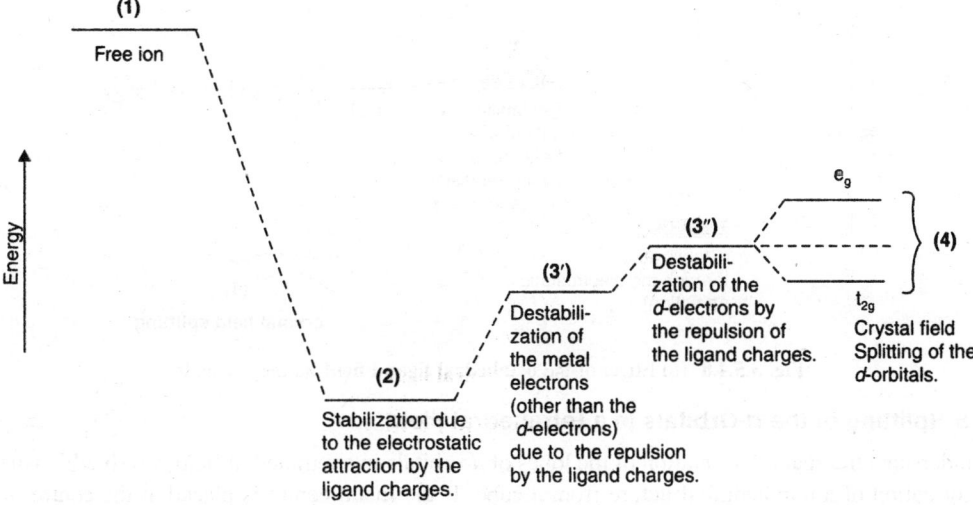

Fig. 3.5.4.7 Qualitative representation of the total energy change of a coordination cluster in an **octahedral crystal field**.

(C) Crystal field splitting of the f-orbitals in an octahedral field: The equivalent orbitals f_{x^3}, f_{y^3} and f_{z^3} having the maximum amplitudes along the Cartesian x, y and z-axes respectively (*see* Fig. 4.1.6.4, Vol. 1 of Fundamental Concepts of Inorganic Chemistry) are destabilized maximum. While the orbital f_{xyz} experiencing the minimum repulsion from the ligand charges is the most stable one. The equivalent orbitals: $f_{x(y^2-z^2)}$, $f_{y(z^2-x^2)}$ and $f_{z(x^2-y^2)}$ are of intermediate stability. The splitting pattern (*cf.* Fig. 3.5.4.8) of the f-orbitals is evident from the orientation of their lobes in the octahedral ligand field.

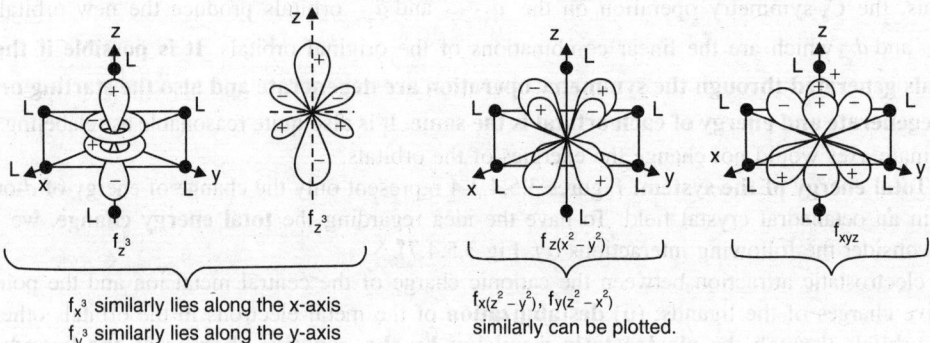

f_x^3 similarly lies along the x-axis
f_y^3 similarly lies along the y-axis

$f_{x(z^2-y^2)}$, $f_{y(z^2-x^2)}$
similarly can be plotted.

Fig. 3.5.4.8 (a) Representation of the lobes of f-orbitals in an **octahedral ligand field.**

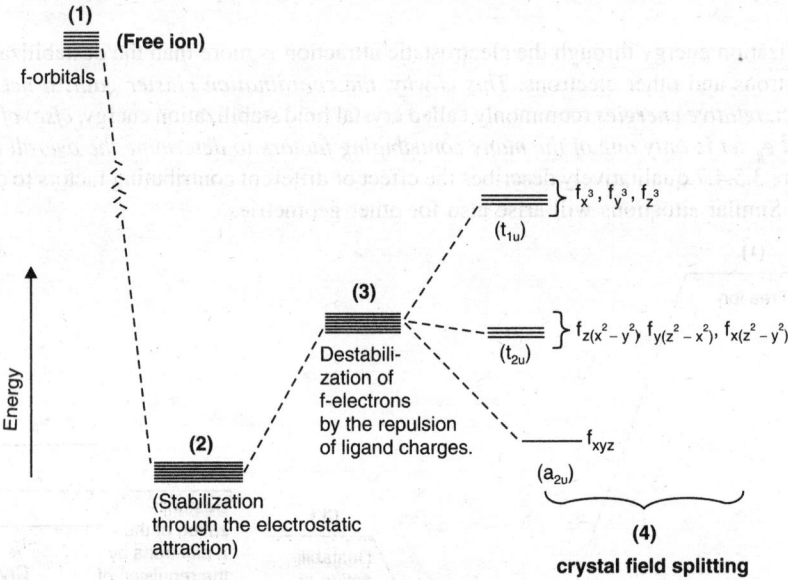

Fig. 3.5.4.8 (b) Effect of the **octahedral ligand field** on the f-orbitals.

3.5.5 Splitting of the d-Orbitals in a Tetrahedral Field

To understand the spatial orientation of the lobes of d-orbitals in a tetrahedral field, we should visualize the formation of a tetrahedral structure from a cube. If the metal centre is placed at the centre of the cube, then the alternate corners are the vertices of a regular tetrahedron (*cf.* Fig. 3.5.5.1). *Thus the C_4 axes of cube or octahedron lie along the C_2 axes of the tetrahedron.* **These three C_2-axes are passing through the centres of the opposite faces of the cube and these are the Cartesian coordinate axes.** In other words, the C_2-axes bisecting the six tetrahedral edges are the coordinate axes.

In this tetrahedral crystal field (developed from the cubic field by placing the ligands at the alternate corners of the cube), the d_{xy}, d_{yz} and d_{xz} orbitals project their lobes to intersect the edges of the cube **at a distance equal to half a cube edge away from the nearest ligand position.** On the other hand, the $d_{x^2-y^2}$ and d_{z^2} orbitals project their lobes along the centres of the cube faces, *i.e.* their lobes lie **half a**

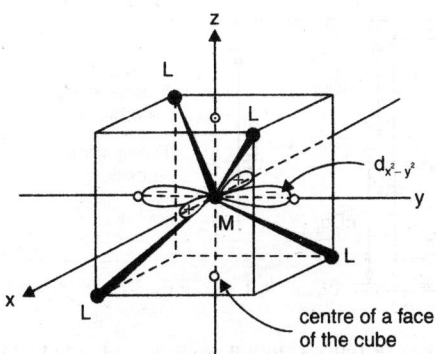

Fig. 3.5.5.1 Development of a **tetrahedral geometry** by placing the ligands (L) at the alternate corners of a cube and keeping the metal centre at the centre of the cube and position of the Cartesian coordinate axes.

Note: From the relationship of Cartesian coordinate axes, orientation of the d-orbitals in the cube can be understood. Here position of the $d_{x^2-y^2}$ orbital is shown.

face diagonal away from the ligand positions. If a be the edge length of the cube, then the face diagonal is $\sqrt{2}a$. Thus the ligands are at $\dfrac{1}{\sqrt{2}}a$ distance away from the lobes of the $d_{x^2-y^2}$ and d_{z^2} orbitals while the distance between the ligand and the lobes of the d_{xy}, d_{yz} and d_{xz} orbitals is $\dfrac{1}{2}a\left(<\dfrac{1}{\sqrt{2}}a\right)$.

Thus in the tetrahedral field, none of the five d-orbitals is projected directly towards the ligands but the d_{xy}, d_{yz} and d_{xz} orbitals can come closer to the ligands compared to the other two d-orbitals $d_{x^2-y^2}$ and d_{z^2}. **Thus, the set consisting of d_{xy}, d_{yz} and d_{xz} is relatively destabilized while other set consisting of $d_{x^2-y^2}$ and d_{z^2} is stabilized with respect to the barycentre.** This relative stabilization and destabilization of the splitted d-orbitals maintain the *barycentre rule*. The barycentre of the unsplitted d-orbitals lies at the position raised by a hypothetical spherical field of these four ligand charges where the radius of the sphere represents the metal-ligand distance of the tetrahedral complex.

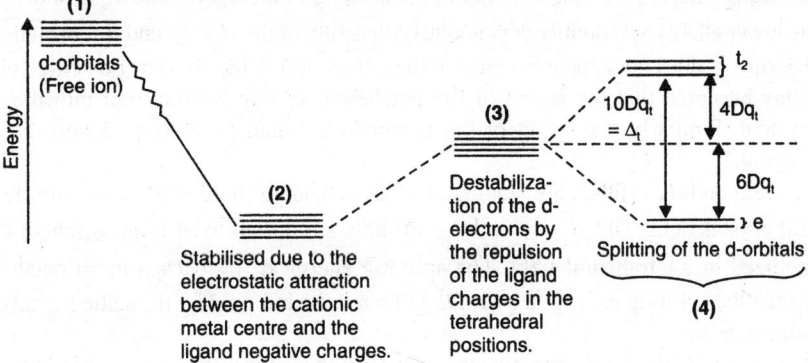

Fig. 3.5.5.2 (a) Effect of the **tetrahedral ligand field** on the d-orbitals. *See* Figs. 3.5.2.1; 3.5.4.2, 4, 7 for the significance of (**1**), (**2**), (**3**), (**4**).

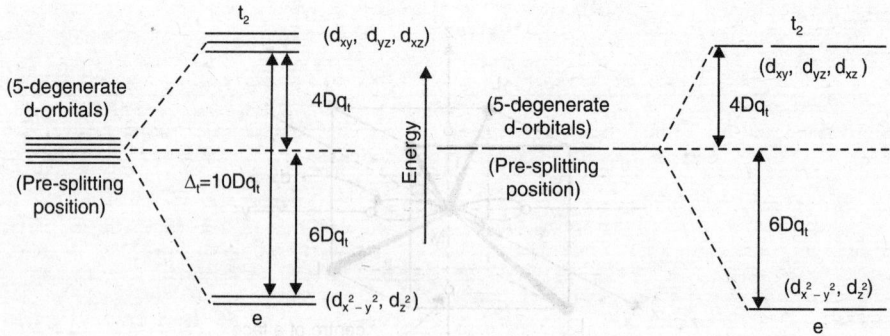

Fig. 3.5.5.2 (b) **Conventional way of representation** of crystal field splitting of the *d*-orbitals in a tetrahedral ligand field.

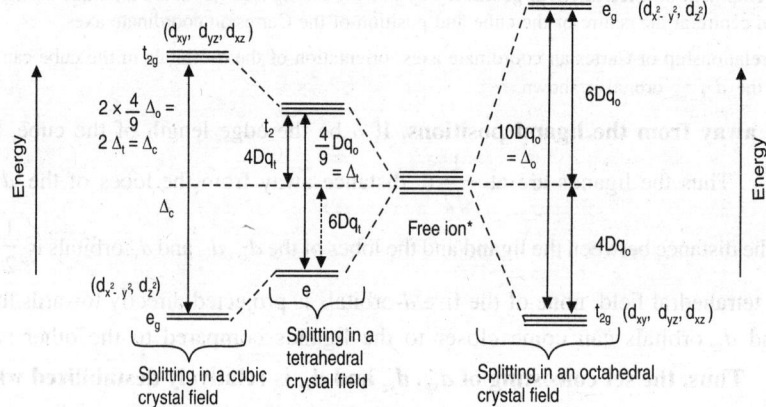

Fig. 3.5.5.3 Comparison of the **conventional crystal field splitting pattern** of the *d*-orbitals in an octahedral and a tetrahedral field (splitting in a cubic field is included for comparison).* It actually represents the energy of the d-orbitals in the electrostatic field of the ligand charges assumed to be spherically distributed at the metal-ligand distance (*cf.* Fig. 3.5.4.4, 7).

The energetically higher set (triply degenerate) consisting of the d_{xy}, d_{yz} and d_{xz} orbitals is designated as t_2 while the lower energy set (doubly degenerate) consisting of the $d_{x^2-y^2}$ and d_{z^2} orbitals is described as e. The subscript '*g*' does not appear because in the tetrahedral field– **there is no centre of symmetry.**

Note: It may be noted that the t_{2g} set of the octahedral system is converted into the t_2 set of the tetrahedral system. Similarly, the e_g set of the octahedral system is converted into the e set of the tetrahedral system.

Thus the d_{xy}, d_{yz} and d_{xz} orbitals are stabilized in an octahedral field while these are destabilized in the tetrahedral crystal field. The $d_{x^2-y^2}$ and d_{z^2} orbitals are destabilized in an octahedral field while these are stabilized in a tetrahedral field. The splitting energy Δ_o (= $10Dq_o$) in an octahedral field is much higher than the splitting energy Δ_t (= $10Dq_t$) of a tetrahedral field by the same ligands. This is due to the following reason:

In the octahedral field 6 ligands directly approach the lobes of orbitals of e_g set but in the tetrahedral field none of the 4 ligands directly approaches the lobes of the ligands. These aspects are discussed below in detail.

● **Comparison between Δ_t and Δ_o:** The energy difference (Δ_t or $10Dq_t$) between the t_2 and e set in the tetrahedral crystal field is much less than the Δ_o or $10Dq_o$ value (energy difference between the t_{2g} and e_g set). In other words, *crystal field splitting in an octahedral field is more than that in a tetrahedral field* for the same ligands coordinating to a particular metal centre.

Let us consider that the same ligands are present in both the octahedral and tetrahedral geometries for the same metal ion *at the same metal-ligand distance.* Under this condition, the electrostatic effect of 4 ligands (*i.e.* 4 point charges) of the tetrahedral field will be 2/3rd of that produced by the 6 ligands of the octahedral field. This prediction will operate if the orbital lobes of the metal centre interact in the same way in both the geometries. But in the octahedral geometry, the ligands directly face the orbital lobes while in the tetrahedral geometry, none of the ligands interact directly with the orbital lobes. This is why, Δ_t is less than $\dfrac{2}{3}\Delta_o$. In fact for the indirect interaction between the ligands and d-orbital lobes, it requires an additional factor 2/3 to compare with Δ_o. The origin of this additional factor 2/3 is not discussed here.

$$\Delta_t = \left(10Dq_t\right) = \frac{2}{3} \times \frac{2}{3}\Delta_0 = \frac{4}{9}\Delta_0 \left(= 10Dq_0\right).$$

The above relation holds good under the conditions: same metal ion, **same ligands and same metal-ligand distance.**

(**Note:** More rigorous mathematical treatment is needed to prove the quantitative relationship, $\Delta_o = \dfrac{4}{9}\Delta_t$. Interested readers may consult the book, "Introduction to Ligand Fields" by B.N. Figgis. This treatment lies beyond the scope of this book.)

Based on the above results, it may be concluded, in general, that the crystal field splitting power of a tetrahedral field is approximately half of that of the octahedral field. This is supported experimentally in many cases. If we consider the electronic spectra of the tetrahedral and octahedral chlorido complexes of V(IV) (d^1 system), we get the following results:

$\left\{\begin{array}{l}\text{[VCl}_4\text{] (tetrahedral)}, \, c^1t_2^0 : \text{ Absorption band at about 8000 cm}^{-1} \text{ for the electronic transition,} \\ e \rightarrow t_2, \, i.e. \, \Delta_t \, (= 10Dq_t) \approx 8000 \text{ cm}^{-1}. \\ \text{[VCl}_6\text{]}^{2-} \text{ (octahedral)}, \, t_{2g}^1 e_g^0 : \text{ Absorption band at about, 15,500 cm}^{-1} \text{ for the electronic transition,} \\ t_{2g} \rightarrow e_g, \, i.e. \, \Delta_o \, (= 10Dq_o) \approx 15{,}500 \text{ cm}^{-1} \approx 2\Delta_t\end{array}\right.$

$\left\{\begin{array}{ll}\text{[CoCl}_4\text{]}^{2-} \text{ (tetrahedral)}, \, e^4t_2^3: & \Delta_t \approx 3120 \text{ cm}^{-1} \\ \text{[CoCl}_6\text{]}^{4-} \text{ (in solid CoCl}_2\text{)}, \, t_{2g}^5 e_g^2 \text{ (high spin)}: & \Delta_o \approx 6900 \text{ cm}^{-1}\end{array}\right.$

$\left\{\begin{array}{ll}\text{[Co(NCS)}_4\text{]}^{2-} \text{ (tetrahedral)}, \, e^4t_2^3: & D_t \approx 4.9 \times 10^3 \text{ cm}^{-1} \\ \text{[Co(NH}_3\text{)}_6\text{]}^{2+} \text{ (Octahedral)} \, t_{2g}^5 e_g^2 \text{ (high spin)}: & \Delta_o \approx 10.1 \times 10^3 \text{ cm}^{-1}\end{array}\right.$

These results support the fact that Δ_t is approximately half of Δ_o.

● **Comparison among Δ_{cubic}, Δ_o and Δ_t:** Δ_{cubic} denotes the crystal field splitting in a cubic field (*i.e.* 8 ligands as the point-like charges at the 8 corners of a cube keeping the metal centre at the centre of the cube). An octahedral field is generated by placing 6 ligands on the point charges at the centres of 6 faces of a cube keeping the metal centre at the centre of the cube. A tetrahedral crystal field is generated by placing the 4 ligands at the alternate corners of a cube. **Thus a cubic field is generated by two interlocking tetrahedral fields.** Thus we can reasonably write (for the same metal ion, same ligand and same metal-ligand distance):

$$\Delta_{cubic} = \Delta_c = 2\Delta_t = 2 \times \frac{4}{9}\Delta_o = \frac{8}{9}\Delta_o.$$

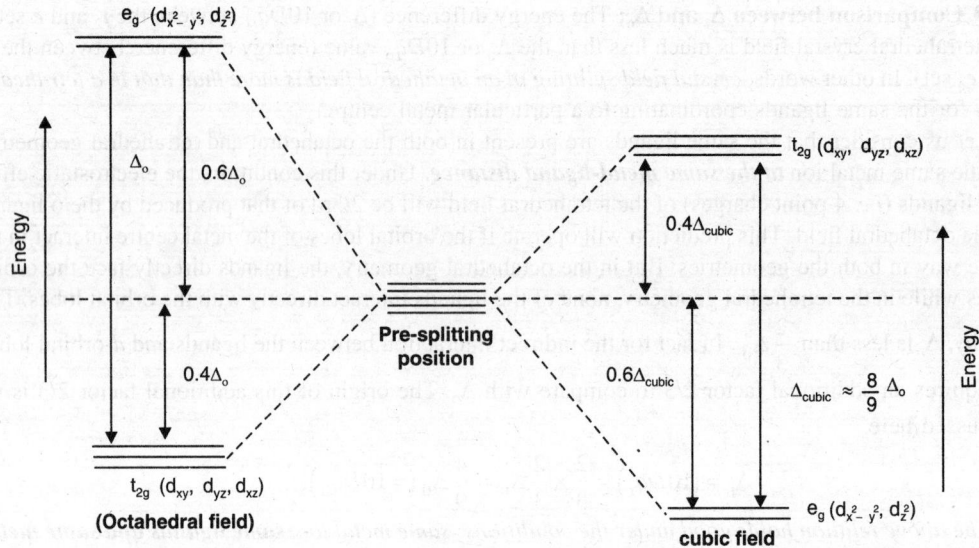

Fig. 3.5.5.4 Comparison of splitting pattern of the *d*-orbitals in octahedral and cubic fields (*cf.* Fig. 3.5.5.3).

It has been pointed out that two interlocking tetrahedrons constitute a cube. **This is why, the pattern of splitting of the *d*-orbitals in a cubic field will be the same as in a tetrahedral field** but the magnitude of Δ_{cubic} will be twice as large as Δ_t. The cubic field is centrosymmetric (*cf.* tetrahedral field is lacking in a centre of symmetry). This is why, the suffix *g* is required to represent the two sets of orbitals, *i.e.* **destabilized set** t_{2g} (= d_{xy}, d_{yz}, d_{xz}) and **stabilized set** e_g (= $d_{x^2-y^2}, d_{z^2}$). It may be noted that in the octahedral field, the t_{2g} set is stabilized and the e_g set is destabilized, *i.e.* opposite pattern of splitting with respect to that of cubic splitting.

● **Comparison of crystal field splitting of *f*-orbitals in the octahedral and tetrahedral field:**
   These are shown in Fig. 3.5.5.5.

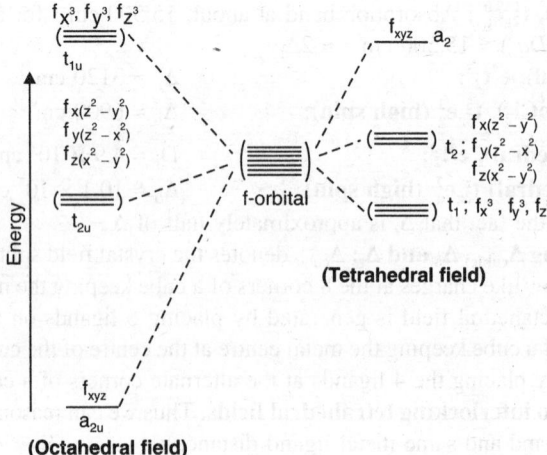

Fig. 3.5.5.5 Comparison of crystal field splitting of *f*-orbitals in an **octahedral field** and in a tetrahedral field. (*cf.* Fig. 3.5.4.8a for orientation of lobes of *f*-orbitals).

3.5.6 Crystal Field Splitting of the *d*-orbitals in a Square Planar Crystal Field (*cf.* Sec. 3.13)

There are **conceptually different approaches** to understand the splitting of *d*-orbitals in a square planar crystal field. We can start with the square planar arrangement of the ligands along the ±*x* and ±*y* directions in the *xy*-plane. Through the **compression of a tetrahedral geometry,** the square planar geometry can be attained (*i.e.* $T_d \longrightarrow D_{2d} \longrightarrow D_{4h}$; *see* Fig. 2.11.2.2). In terms of the other concept, to understand the crystal field splitting of the *d*-orbitals in a square planar crystal field, **we can start with the octahedral crystal field.**

(A) Octahedral ligand field → Square planar ligand field $\left(i.e. O_h \longrightarrow D_{4h} \right)$: If the two *trans*–axial ligands (say, along the ±*z* directions) of an octahedral geometry are taken to infinity, then the square planar crystal field is developed where the four ligands lie in the ±*x* and ±*y* directions. Removal of the ligands along the ±*z* directions from an octahedral geometry, **stabilizes (relatively) the orbitals having the z-components.** Because, such orbitals will experience a less electrostatic repulsion from the ligands compared to the other orbitals.

In the square planar geometry (in the *xy*-plane), the ligands directly face the lobes of the $d_{x^2-y^2}$ orbital. Thus, **the $d_{x^2-y^2}$ orbital will be destabilized maximum.** The orbital d_{xy} lying in the *xy*-plane and projecting the lobes between the adjacent ligands will be also destabilized next to the $d_{x^2-y^2}$ orbital. There is no electrostatic repulsive ligand field along the ±*z* directions and consequently, the d_{z^2} **orbital is stabilized maximum though it has a small cocentric lobe in the xy-plane.** The lobes of the d_{xz} and d_{yz} orbitals are not in the immediate vicinity of the ligands but they bear the *x*– and *y*– component respectively. Both of them also possesses the *z*-component. The presence of *z*-component stabilizes the orbitals but the presence of *x*– and *y*– components will destabilize these orbitals to some extent. **In fact, the d_{xz} and d_{yz} orbitals are stabilized to some extent compared to the starting case of the octahedral geometry.**

- **Relative positions of the d_{xz}, d_{yz} and d_{z^2} orbitals:** In fact, in the square planar geometry, the orbitals bearing the *z*-component, *i.e.* the d_{z^2} orbital (which was an e_g orbital in the starting octahedral complex) and d_{xz}, d_{yz} orbitals (which were the t_{2g} orbitals in the starting octahedral complex), are stabilized and the extent of stabilization depends on the ligand field strength. In fact, in some cases, the d_{z^2} orbital is the most stable one while in some cases, the d_{xz}, d_{yz} orbitals are the most stable ones. **This is why, it may be difficult to predict whether the set consisting of the d_{xz}, d_{yz} orbitals or the d_{z^2} orbital represents the lowest energy level.** It happens so because in these orbitals, besides the *z*-component, there are some *x*, *y* components. It is evident for the d_{xz} and d_{yz} orbitals but the d_{z^2} orbital also bears the *x*, *y* component. In the d_{z^2} orbital, the component in the *xy*-plane also experiences the ligand field repulsion. Existence of the component in the *xy*-plane for the d_{z^2} orbital is evident from the following representation.

$$d_{z^2} \equiv d_{z^2-x^2} + d_{z^2-y^2} \quad i.e. \quad d_{z^2} \equiv d_{2z^2-x^2-y^2}$$

In fact, the d_{z^2} orbital possesses a concentric lobe around the nucleus in the *xy*-plane besides the two major lobes in the ±*z* directions. *This concentric lobe in the xy plane experiences a repulsive field from the square planar ligand field spreading in the xy-plane.* Similarly, the *x*-component of the d_{xz} orbital and the *y*- component of the d_{yz} orbital will experience the repulsive field to some extent from the square planar ligand field. **This repulsive effect is less than the stabilising effect due to the presence of the z-component.**

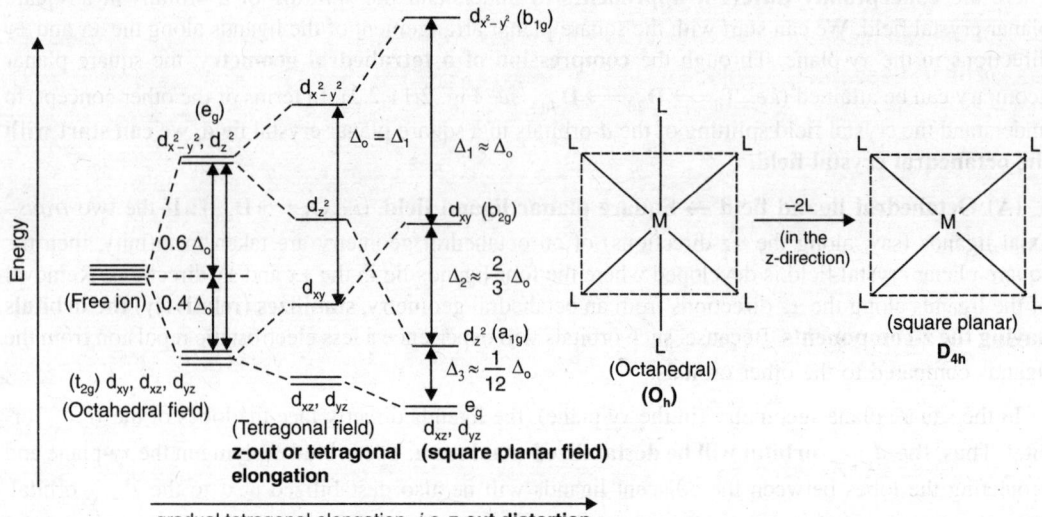

Fig. 3.5.6.1 Qualitative and schematic representation of splitting of the d-orbitals when the *trans*- axial ligands at the $\pm z$ directions of an octahedral complex are removed.

The splitting of d-orbitals in the square planar crystal field is illustrated in Fig. 3.5.6.1 by considering the gradual removal of two *trans*-axial ligands ($\pm z$ directions) of an octahedral ligand field.

(**Note:** Here the g suffix is maintained because of the presence of centre of symmetry in the square planar complex. **There is some freedom in the designation of b_{1g} and b_{2g} symbols.** Some authors use b_{1g} for the $d_{x^2-y^2}$ orbital while some authors use b_{2g} for the same orbital. The same thing happens for the d_{xy} orbital.)

- **Crystal field splitting parameters:** In Fig. 3.5.6.1, the crystal field splitting parameters Δ_1, Δ_2 and Δ_3 in a square planar complex can be correlated with the corresponding hypothetical octahedral crystal field splitting parameter Δ_o. Semi-quantitative calculations on the various square planar complexes of Co(II), Ni(II) and Cu(II) furnish the following relations (*cf.* Sec. 3.13).

$$\Delta_1 \approx \Delta_o, \Delta_2 \approx \frac{2}{3}\Delta_o, \Delta_3 \approx \frac{1}{12}\Delta_o, \ i.e. \ \Delta_{sp} = \Delta_1 + \Delta_2 + \Delta_3 \approx 1.75\Delta_o$$

Here, *it may be noted that the above relations are not always found true for all square planar complexes.* For example, in the Pd(II) and Pt(II) square planar complexes, $\Delta_1 + \Delta_2 + \Delta_3 \approx 1.3\Delta_o$ has been found; in $[Ni(CN)_4]^{2-}$ the results: $\Delta_1 \approx \Delta_o, \Delta_2 \approx \frac{2}{5}\Delta_o, \Delta_3 \approx \frac{1}{38}\Delta_o, \ i.e. \ \Delta_{sp} \approx 1.42\ \Delta_o$ have been noticed.

- $E_{d_{x^2-y^2}} \sim E_{d_{xy}} = \Delta_1 = \Delta_o$ (**in a square planar crystal field**): It has been already shown in Fig. 3.5.6.1 that at any stage of formation of the square planar complex from an octahedral complex through the removal of the *trans*– axial ligands (lying in the $\pm z$ directions), the energy difference between the $d_{x^2-y^2}$ and d_{xy} orbitals remains constant which is Δ_o. These two orbitals lie in the xy-plane. The lobes of the $d_{x^2-y^2}$ orbital are projected towards the ligands while the lobes of the d_{xy} orbital are projected between the adjacent ligands. In the starting octahedral complex, the energy difference between these orbitals is Δ_o (*i.e.* $10\ Dq_o$). **Whatever may be the positions of the two ligands in the ±z directions, the effect of these two ligands on the $d_{x^2-y^2}$ and d_{xy} orbitals lying in the xy-plane will be the same.** This is why, the initial energy difference (= Δ_o) between the $d_{x^2-y^2}$ and d_{xy} orbitals will be maintained at any stage of the tetragonal distortion leading to a limiting situation as in the square planar geometry.

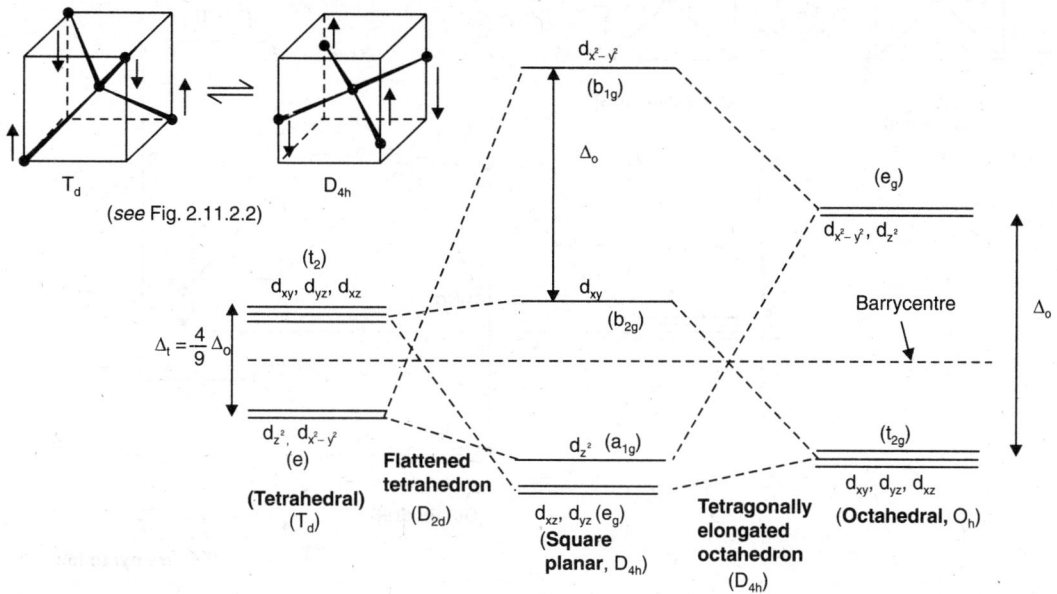

Fig. 3.5.6.2 Relationship of crystal-field splitting in three geometries tetrahedron, square planar and octahedron.

(B) Relationship between the crystal field splitting of the d-orbitals in the tetrahedral and square planar geometry (Fig. 3.5.6.2): It has been already discussed the development of a square planar geometry from an octahedral geometry through the gradual removal of the *trans*-axial ligands lying along the ±z-directions. *Thus the process moves through the tetragonal elongations (i.e. z-out distortion).* From a tetrahedral geometry, a square planar geometry can be obtained through compression, *i.e. the process moves through the flattened tetrahedron (i.e.* $\mathbf{T_d} \longrightarrow \mathbf{D_{2d}} \longrightarrow \mathbf{D_{4h}}$, Fig. 2.11.2.2). These are illustrated in Fig. 3.5.6.2. If a tetrahedron is compressed or flattened leading to place the four ligands in the xy-plane, then the orbitals bearing the x,y-components will be destabilized more while the orbitals bearing the z-component will be stabilized more. This flattening will destabilize maximum the $d_{x^2-y^2}$ orbital which faces the ligands directly.

3.5.7 Crystal Field Splitting of the d-orbitals in a Square Pyramidal Geometry (*cf.* Sec. 8.25.1)

In a square pyramidal geometry, four lobes of the $d_{x^2-y^2}$ orbital directly face the ligands and this is the most destabilized orbital. The d_{z^2} orbital directly faces one ligand and it is also destabilized next to the

$d_{x^2-y^2}$ orbital. Out of the d_{xy}, d_{yz} and d_{xz} orbitals, the d_{xy} orbital lying in the xy plane (*i.e.* plane of the ligand charges) will be relatively more destabilized compared to the d_{xz} and d_{yz} orbitals. The crystal field splitting of d-orbitals in a square pyramidal geometry can be understood **in two ways**: removal of one axial ligand from the octahedral geometry; and addition of one ligand in the axial direction to the square planar geometry.

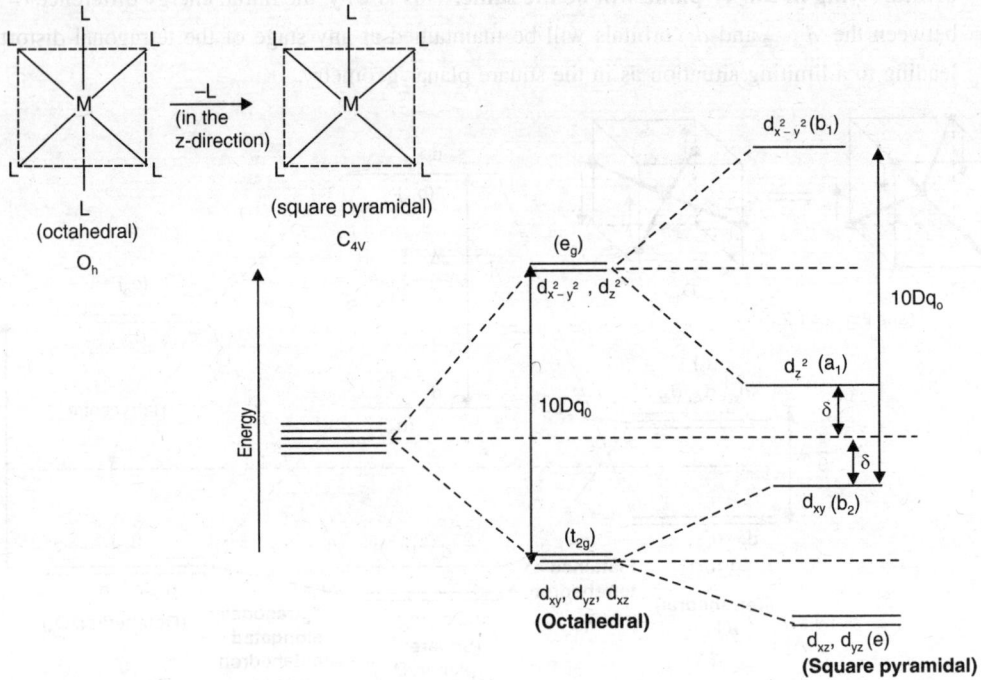

Fig. 3.5.7.1 Comparison (qualitatively) of crystal field splitting of the d-orbitals in an octahedral and a square pyramidal ligand field (*cf.* Effect of removal of one ligand in the z-direction from an octahedral geometry).

 (A) Generation of a square pyramidal geometry from the octahedral geometry: If one axial ligand is gradually removed then ultimately it will produce the square pyramidal geometry. Removal of one axial ligand from the z-direction will reduce the repulsion by the ligand in the z-direction. **In an octahedral geometry, two ligands along the z-axis cause the repulsion while in a square pyramidal geometry, only one ligand causes the repulsion along the z-axis.** Thus, with respect to the octahedral system, the orbitals bearing the z-component will be relatively more stabilized while the other orbitals will be destabilized accordingly to conserve the barycentre. This is qualitatively illustrated in Fig. 3.5.7.1.

(Note: ● Tetragonal elongation, *i.e.* z-out distortion also gives the same pattern of splitting the d-orbitals as in a square pyramidal geometry.

 ● Strong π-bonding in the basal plane and placement of the metal centre above the basal plane in the square pyramidal geometry may lead to the following energy order:

$$d_{x^2-y^2} > d_{z^2} > d_{xz}, d_{yz} > d_{xy} \ (cf. \text{ Sec. 8.25.1, Fig. 8.24.5.3})$$

(B) Generation of a square planar pyramidal geometry from the square planar geometry: If one ligand is added along the z-axis to a square planar geometry (lying in the xy-plane), we get the square pyramidal geometry. Thus with respect to the square planar geometry, the orbitals having the z-component are destabilized more and to maintain the barycentre rule, other orbitals will be stabilized accordingly. This is qualitatively illustrated in Fig. 3.5.7.2.

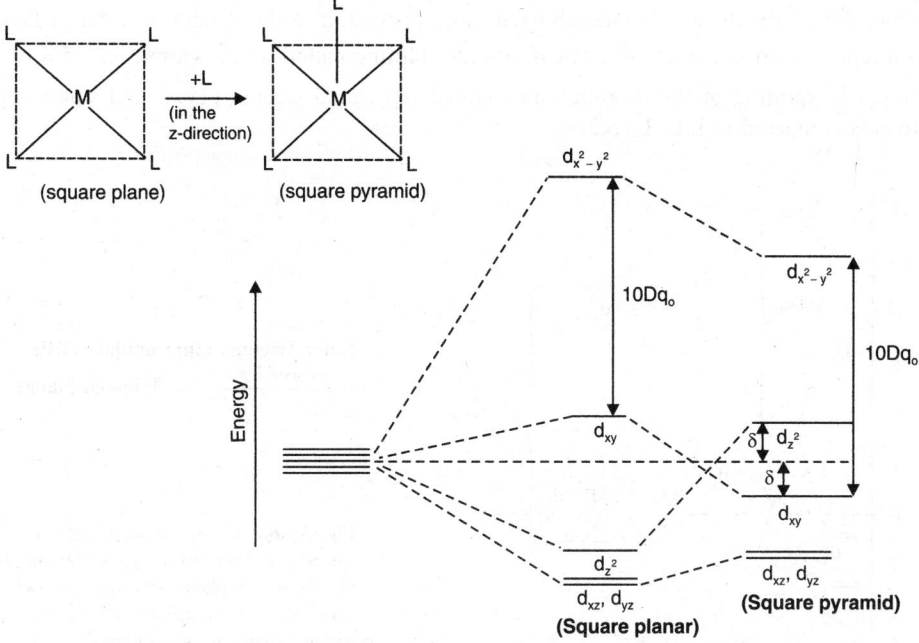

Fig. 3.5.7.2 Comparison (qualitatively) of crystal field splitting of the d-orbitals in a square planar complex and a square pyramidal complex (*cf.* Effect of adding one ligand in the z-direction to a square planar complex).

3.5.8 Crystal Field Splitting of the d-orbitals in a Trigonal Bipyramidal Geomtry

If the equatorial plane (*i.e.* basal plane) lies in the xy-plane, the axial ligands lie in the $\pm z$-directions. In the equatorial plane, the $d_{x^2-y^2}$ and d_{xy} orbitals do not directly face all the equatorial ligands but

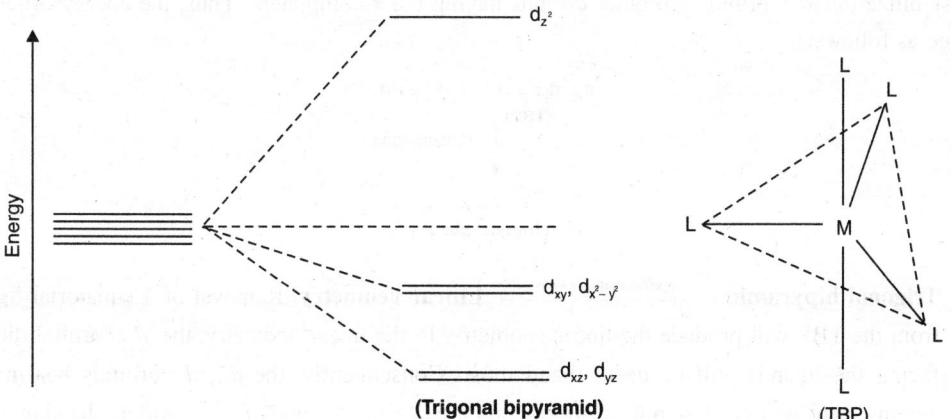

Fig. 3.5.8.1 Crystal field splitting of the d-orbitals in a trigonal bipyramidal field (axial ligands along the z-axis).

they are equally affected by the equatorial ligands. Thus the d_{z^2} orbital experiences the maximum repulsion from the direct facing ligands and this orbital is destabilized maximum. The $d_{x^2-y^2}$ and d_{xy} orbitals lying in the equatorial plane also experience the repulsion to some extent from the equatorial ligands and these orbitals are also destabilized more compared to the d_{xz} and d_{yz} orbitals facing the minimum repulsion from the ligands. The d-orbitals splitting pattern is shown in Fig. 3.5.8.1.

Crystal field splitting of the d-orbitals in trigonal bipyramid, square planar and square pyramid geometries is compared in Fig. 3.5.8.2.

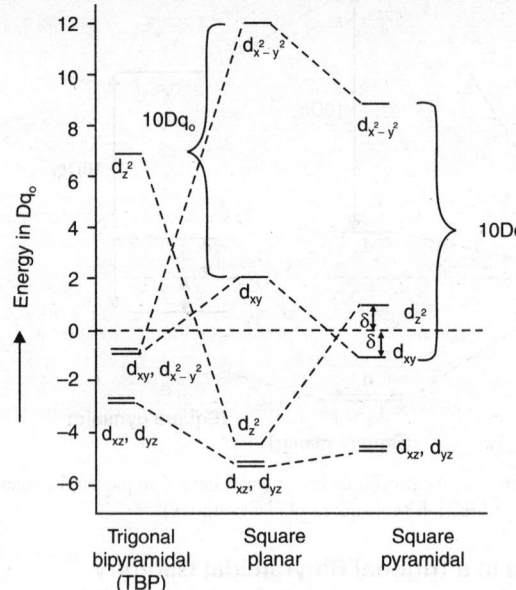

Note: Trigonal bipyramidal (TBP)
$$\xrightarrow{-2 \text{ axial ligands}} \textbf{Trigonal planar}$$

Fig. 3.5.8.2 Comparison of crystal field splitting of d-orbitals in trigonal bipyramidal (TBP), square planar and square pyramidal geometries. (Square plane, pyramid base and trigonal plane lie in xy-plane).

● **Trigonal bipyramid** $\longrightarrow$ **Trigonal planar:** Removal of two axial ligands from the TBP geometry will stabilize the d_{z^2} orbital and other orbitals having the z-component. Thus, the energy order will change as follows:

$$d_{xz}, d_{yz} < d_{x^2-y^2}, d_{xy} << d_{z^2}$$
$$\textbf{(TBP)}$$
$$\downarrow -2 \text{ axial ligands}$$
$$d_{x^2-y^2}, d_{xy} > d_{xz}, d_{yz} > d_{z^2}$$
$$\textbf{(Trigonal planar)}$$

● **Trigonal bipyramid** $\xrightarrow{-3 \text{ equatorial ligands}}$ **Linear geometry**: Removal of 3 equatorial ligands from the TBP will produce the linear geometry. In the linear geometry, the d_{z^2} orbital directly facing the ligands will be destabilized most. Consequently, the d_{xz}, d_{yz} orbitals bearing the z-component will be destabilized more compared to the d_{xy} and $d_{x^2-y^2}$ orbitals lacking in the z-component.

$$d_{xz}, d_{yz} < d_{x^2-y^2}, d_{xy} << d_{z^2}$$
(TBP)
|
−3 equatorial ligands
↓

$$d_{z^2} > d_{xz}, d_{yz} > d_{xy}, d_{x^2-y^2}$$
(Linear geometry)

3.5.9 Crystal Field Splitting of the d-orbitals in 8-Coordinate Structures– Cubic, Square Antiprismatic (D_{4d}) and Dodecahedral (D_{2d}).

These are shown in Fig. 3.5.9.1.

Cubic:
t_{2g} (d_{xy}, d_{yz}, d_{xz})
e_g $(d_{x^2-y^2}, d_{z^2})$

Square antiprismatic (D_{4d}):
e_3 (d_{xz}, d_{yz})
e_2 $(d_{x^2-y^2}, d_{xy})$
a_1 (d_{z^2})

Dodecahedral (D_{2d}):
b_2 (d_{xy})
e (d_{xz}, d_{yz})
a_1 (d_{z^2})
b_1 $(d_{x^2-y^2})$

Cubic   **Square antiprismatic (D_{4d})**   **Dodecahedral (D_{2d})**

Fig. 3.5.9.1 Qualitative representation of splitting of the *d*-orbitals in eight-coordinate complexes of different geometries like cubic, square antiprismatic and dodecahedral.

3.5.10 Crystal Field Splitting of the d-orbitals in a Trigonal Prismatic Stereochemistry

Only very few complexes, *e.g.* $[M(S_2C_2R_2)_3]$ (M = Re, V, Mo, W) having this structure have been characterized. In this geometry, the lobes of the d_{z^2} orbital are not affected much by the ligands. The main lobes of this orbital are passing through the centres of two triangular faces of the prism. The lobes of the d_{xz} and d_{yz} orbitals are affected most. The lobes of the d_{xy} and $d_{x^2-y^2}$ orbitals lie in a parallel plane between the two triangular faces and these two orbitals are affected least. These are understandable from the Fig. 3.5.10.1. The energy order of the orbitals will be:

$$d_{xz}, d_{yz} \rangle d_{z^2} \rangle d_{xy}, d_{x^2-y^2}$$

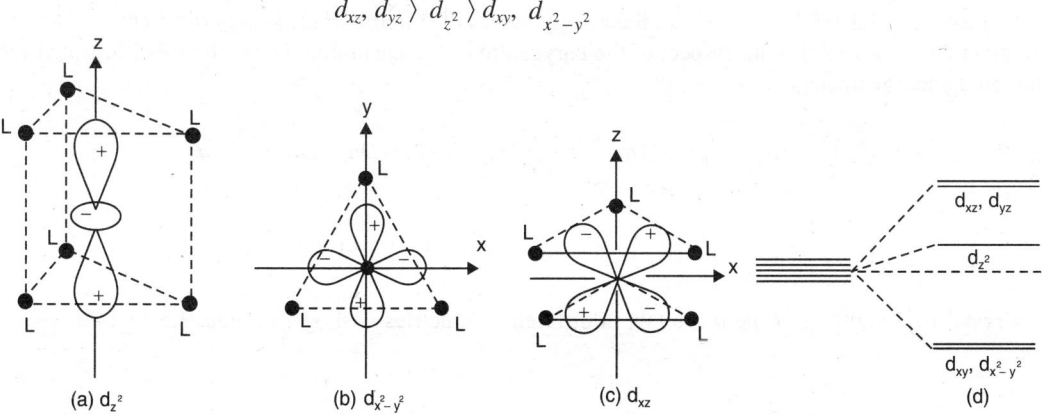

(a) d_{z^2}   **(b) $d_{x^2-y^2}$**   **(c) d_{xz}**   **(d)**

Fig. 3.5.10.1 Spatial orientation (a–c) and splitting (d) of the *d*-orbitals with respect to the ligands in a trigonal prismatic coordination. In (b), out of the six ligands, only three ligands are shown. The $d_{x^2-y^2}$ orbital lies between the two trigonal planes.

3.5.11 Energy of the d-orbitals (*i.e.* d-orbital Energy) in Crystal Fields of Different Symmetries

Table 3.5.11.1   *d*-Orbital energy values (in terms of Dq_o) in crystal fields of different geometries.

Coordination number	Structure (*i.e.* Geometry)	(Energy value in Dq_o with respect to the barycentre of the unsplitted d-orbitals)					
		$d_{x^2-y^2}$	d_{z^2}	d_{xy}	d_{yz}	d_{xz}	Largest splitting
2	Linear (a)	−6.28	10.28	−6.28	1.14	1.14	9.14
3	Trigonal planar (b)	5.46	−3.21	5.46	−3.86	−3.86	8.67
4	Tetrahedral	−2.67	−2.67	1.78	1.78	1.78	4.45
4	Square planar (b)	12.28	−4.28	2.28	−5.14	−5.14	10.00
5	Trigonal bipyramidal (c)	−0.82	7.07	−0.82	−2.72	−2.72	7.90
5	Square pyramidal (c)	9.14	0.86	−0.86	−4.57	−4.57	8.28
6	Octahedral	6.0	6.0	−4.0	−4.0	−4.0	10.00
6	Tigonal prism	−5.84	0.96	−5.84	5.36	5.36	6.80
7	Pentagonal bipyramidal (c)	2.82	4.93	2.82	−5.28	−5.28	8.10
8	Square antiprism	−0.89	−5.34	−0.89	3.56	3.56	4.45
8	Cubic	−5.34	−5.34	3.56	3.56	3.56	8.90

(a) Ligands along the $\pm z$ axis.

(b) Bonds in the *xy*-plane.

(c) Equatorial ligands in the *xy*-plane.

In Table 3.5.11.1 (*cf.* Fig. 3.5.11.2), the energy values of different *d*-orbitals in different geometries are given in terms of Dq_o with respect to the barycentre of the unsplitted *d*-orbitals. It is illustrated for the tetrahedral geometry.

$$E\left(d_{z^2}\right) = E\left(d_{x^2-y^2}\right) = -6Dq_t = -\frac{4}{9}\left(6Dq_o\right) = -2.67Dq_o, \ Dq_t = \frac{4}{9}Dq_o$$

$$E\left(d_{xy}\right) = E\left(d_{yz}\right) = E\left(d_{zx}\right) = +4Dq_t = +\frac{4}{9}\left(4Dq_o\right) = +1.8Dq_o$$

Crystal field splitting of the *d*-orbitals in different geometries is shown in Figs. 3.5.11.1–3.

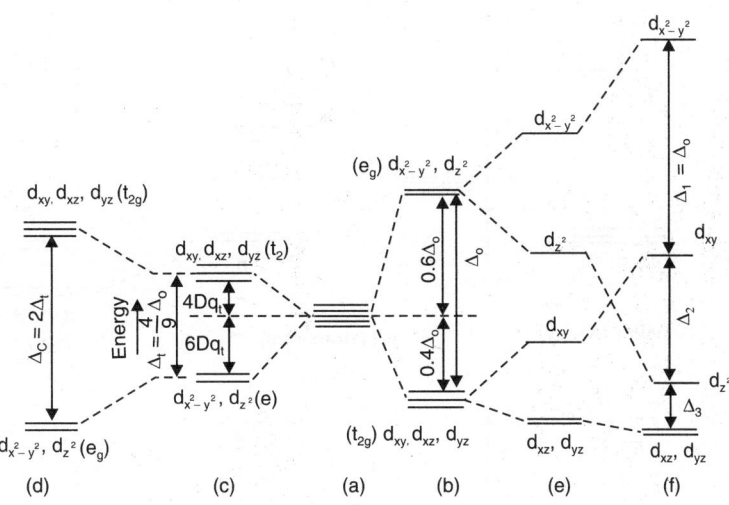

Fig. 3.5.11.1 Crystal field splitting of d-orbitals in different geometries. (a) pre-splitting condition of five degenerate *d*-orbitals. (b) Octahedral field (O_h); (c) Tetrahedral field (T_d), (d) Cubic field; (e) Tetragonal elongation (*z*-out distortion) and square pyramid; (f) Square planar field.

(**Note:** In a square planar geometry, the d_{z^2} orbital may lie below the d_{xz}, d_{yz} orbitals in presence of strong field ligands).

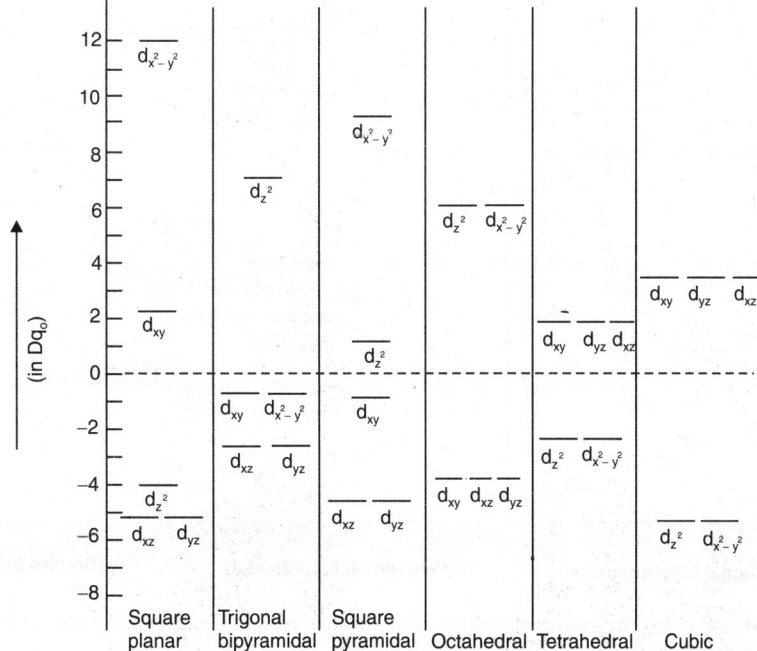

Fig. 3.5.11.2 Relative energies (in Dq_o) of different d-orbitals in different crystal field geometries.

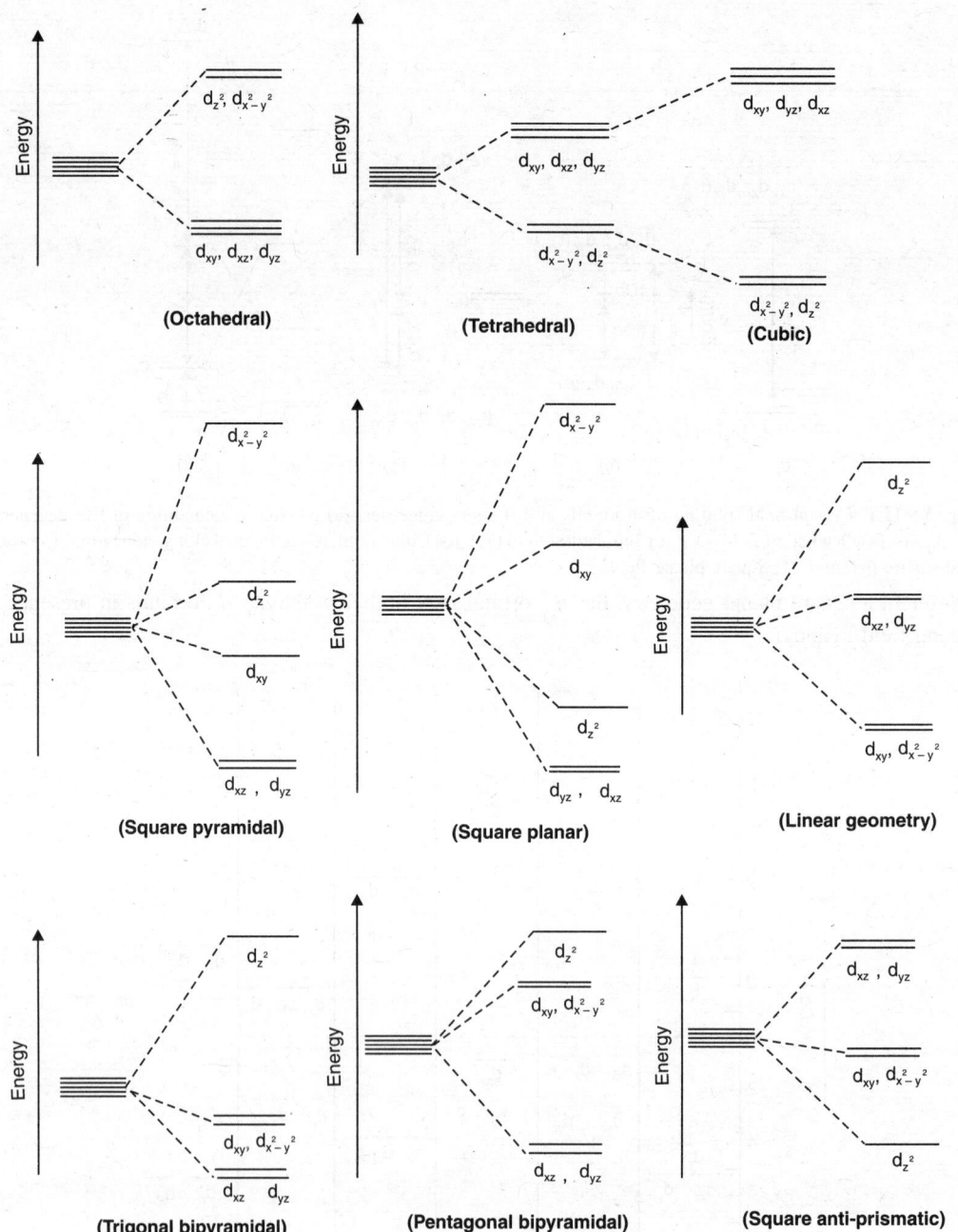

Fig. 3.5.11.3 Qualitative way (conventional way) of representation of splitting of the d-orbitals in different systems. (Energy values not in scale).

(**Note:** Equatorial or basal plane in the xy-plane while the axial ligand (s) in the z-direction)

3.6 FACTORS AFFECTING THE CRYSTAL FIELD SPLITTING PARAMETER (Δ)

The crystal field splitting parameter (Δ) is generally experimentally obtained from the analysis of electronic spectra. Let us illustrate the case with $[Ti(OH_2)_6]^{3+}$ having the d^1 configuration. The absorption peak arises from the promotion of the t_{2g} electron to the e_g level, *i.e.* the electronic transition is:

$$t_{2g}^1 e_g^0 \xrightarrow{+10Dq_o} t_{2g}^0 e_g^1$$

The maximum of the absorption peak lies at 20,300 cm^{-1} (= 243 kJ mol^{-1}). Thus, Δ_o or $10Dq_o$ of the complex, $[Ti(OH_2)_6]^{3+}$ is 20,300 cm^{-1}.

(**Note:** 1000 cm^{-1} = 1kK (Kilo Kayser) = 11.97 kJ mol^{-1}. E = hν = $\dfrac{hc}{\lambda}$ per molecule; EN_o = energy per mole).

● **Stereochemistry:** The magnitude of crystal field splitting parameter strongly depends on the nature of the geometry. This is illustrated below (for the same metal ion, same ligand and same metal-ligand distance):

$$\Delta_{sp} (\approx 1.75\Delta_o) \rangle \Delta_o \rangle \Delta_{cubic} \left(\approx \frac{2 \times 4}{9}\Delta_o \right) \rangle \Delta_t \left(\approx \frac{4}{9}\Delta_o \right).$$

● **Metal factor and ligand factor:** In a particular stereochemistry, the magnitude of crystal field splitting (*i.e.* $10Dq$) depends on the nature of the metal centre and the nature of the ligand. According to Jorgensen, $10Dq_o$ for an octahedral complex can be expressed as the product of two **independent factors** f and g arising from the ligand and metal respectively, *i.e.*

$$\Delta_o \text{ or } 10Dq_o = f.g$$

The *ligand factor* (*f*) describes the ligand field strength which is expressed with respect to the reference of aqua ligands (taking $f_{aq} = 1$). The *metal factor* (*g*) depends on the nature of the metal centre and the g-value may vary in the range 8kK to 36kK (1kK = 1000 cm^{-1}).

Table 3.6.1 Some f (ligand factor) and g (metal factor) values in octahedral systems.

Ligand	f	Metal centre	g (in kK)
Br$^-$	0.72	Mn(II)	8.0
SCN$^-$	0.73	Ni(II)	8.7
Cl$^-$	0.78	Co(II)	9.0
N$_3^-$	0.83		
F$^-$	0.90		
ox^2 (*i.e.* C$_2$O$_4^{2-}$)	0.99	Fe(III)	14.0
H$_2$O	1.0		
NCS$^-$	1.02	Cr(III)	17.4
py	1.23		
NH$_3$	1.25	Co(III)	18.2
en	1.28	Ru(II)	20.0
bpy	1.33	Rh(III)	27.0
CN$^-$	1.70	Pt(IV)	36.0

3.6.1 Effect of the Nature of Metal Centre on Δ in a Particular Stereochemistry

(i) **Oxidation state of the metal centre:** With the increase of positive oxidation state of the metal centre, the electrostatic attraction between the ligand centre and metal centre increases. The ionic size of the metal centre also decreases with the increase of the positive oxidation state. *Thus the increased electrostatic metal-ligand attraction and decreased ionic radius of the metal centre shortens the metal-ligand distance with the increase of positive charge of the metal centre. The shorter metal-ligand distance causes a better electrostatic effect to split the metal d-orbitals.*

The magnitude of Δ is **highly sensitive to the metal-ligand interacting distance** (a) as follows:

$$\Delta_0 \propto a^{-x}, \, x = 5 - 6$$

Thus **for the higher oxidation state**, a decreases and consequently the splitting becomes higher. In fact, in moving from +2 state to +3 state, the crystal field splitting increases by $\sim 50\%$.

$\Delta(M^{2+}): \Delta(M^{3+}): \Delta(M^{4+}) \approx$ **1: 1.6: 1.9** (ratio of the g values, *i.e.* Δ values).

This is illustrated in the following examples.

Complex	Δ_o (kK)	Complex	Δ_o (kk)	Complex	Δ_o (kK)
$[Fe(OH_2)_6]^{2+}$	10.4	$\{$ $[Co(NH_3)_6]^{2+}$	10.20	$\{$ $[Co(OH_2)_6]^{2+}$	9.20
$[Fe(OH_2)_6]^{3+}$	14.0	$[Co(NH_3)_6]^{3+}$	22.90	$[Co(OH_2)_6]^{3+}$	20.76

(ii) **Position of the metal centre in the periodic group:** In the d-block elements, in moving down in a group, the *effective nuclear charge* on the metal centre gradually increases because of the gradual accumulation of the low screening d– and f– electrons. In addition to this, the *higher spatial distribution* of the larger d-orbitals of the heavier congeners enhances the interaction with the point-like ligand charges. These two factors, *i.e. greater effective nuclear charge and greater spatial distribution of the d-orbitals of the heavier congeners are jointly responsible for their higher splitting power.*

For a particular oxidation state, the splitting power increases by about 50% in moving from the 1st transition series to the 2nd transition series and again by about 30% in moving from the 2nd transition series to the 3rd transition series. It indicates that the crystal field splitting power varies as:

$$3d \lll 4d \lll 5d.$$

Experimental findings indicate that the g-value (*i.e.* crystal field splitting power) for the $3d^n$, $4d^n$ and $5d^n$ metal ions varies in the ratio **1: 1.45: 1.75**. This is illustrated in the following examples.

Complex:	$[Co(NH_3)_6]^{3+}$	$[Rh(NH_3)_6]^{3+}$	$[Ir(NH_3)_6]^{3+}$	$[Co(OH_2)_6]^{3+}$	$[Rh(OH_2)_6]^{3+}$
$10Dq_o$ (kK)	22.90	34.10	41.20	20.76	27.20

Here it may be noted that because of the higher crystal field splitting power and lower pairing energies of the heavier congeners, the tendency of low-spin complex formation increases for the heavier congeners.

(iii) **Exchange energy (See. 3.7):** The higher value of Δ_o prevents the exchange phenomenon between the t_{2g} and e_g set. Thus to gain the stabilisation through the exchange energy, Δ_o is reduced (*cf.* d^5 system of Mn^{2+}).

(iv) **Metal ions of a particular period:** The trend of Δ is controlled by many factors like *crystal ionic radii trend, electronegativity trend, effective nuclear charge trend, exchange energy, etc.* These parameters do not vary smoothly in a period of the d-block elements (*cf.* Sec. 3.14.8). It is illustrated for a particular oxidation state of the 3d metal ions. For the bivalent metal ions of 1st transition series, the splitting power does not remarkably change and the change does not follow any regular trend. It is illustrated below.

$$Mn^{2+} (d^5) \langle Ni^{2+} (d^8) \langle Co^{2+} (d^7) \langle Fe^{2+} (d^6) \langle V^{2+} (d^3)$$

$$g = \quad 8.0 \quad\quad 8.7 \quad\quad 9.0 \quad\quad - \quad\quad 12.0$$

Note: Higher Δ value disfavours the **exchange phenomenon**. To gain the stabilisation through the exchange process, Δ is reduced as in the d^5 system (*cf.* Sec. 3.7).

● **Spectrochemical series of metal ions (based on the g-values):**

Mn(II) $\langle$ Ni(II) $\langle$ Co(II) $\langle$ Fe(II) $\langle$ V(II) $\langle$ Fe(III) $\langle$ Cr(III) ~ V(III) $\langle$ Co(III) $\langle$ Mn(IV) $\langle$ Mo(III) $\langle$ Rh(III) ~ Ru(III) $\langle$ Pd(IV) $\langle$ Ir(III) $\langle$ Re(IV) $\langle$ Pt(IV)

3.6.2 Effect of the Nature of the Ligands on Crystal Field Splitting Power (*i.e.* measured by Δ): Ligand Field Strength

The ligand field strength measured by crystal field splitting power in a particular geometry is obtained from the analysis of electronic spectra of the metal complexes. If the ligands are arranged in order of their crystal field splitting power, the series obtained is described as the **Spectrochemical Series** (named so because the series is constructed from the spectral data of the metal complexes). Ideally, to compare the ligand field strength (*i.e.* crystal field splitting power) of the ligands, it is required to study the spectral data of their metal complexes of a particular geometry with a particular metal ion.

But it is not possible to maintain this ideal condition because all the ligands may not form the same type of complex with a particular metal ion. In reality, the series has been developed by *overlapping the different sequences determined for a limited number of ligands from the spectroscopic studies of their complexes (same geometry) with a single metal ion.* Each overlapping sequence constitutes a portion of the series. For example, spectral studies of the complexes, $[Co^{III}(NH_3)_5L]$ helped us to construct the spectral series for various types of monodentate ligands (L). For such complexes, the *spectral shift* depends on the relative ligand field strength of the ligands (L).

In terms of f (*i.e.* ligand field factor measuring the crystal field splitting power of the ligand; *cf.* Table 3.6.1), the ligands are placed in the following sequence called **spectrochemical series** (or Fajans-Tsuchida series). For the ambidentate ligands, the ligating site is underlined.

$I^- \langle Br^- \langle \underline{S}CN^- \langle Cl^- \langle N_3^- \langle F^- \langle Me_2SO \langle$ urea ($\underline{O}$), $OH^- \langle RCO_2^- \langle ox^{2-}, O^{2-} \langle H_2O \langle NCS^- \langle NC^- \langle edta^{4-} \sim gly^- \langle py \langle NH_3 \langle en \langle H_2\underline{N}OH \langle bpy \langle phen \langle \underline{N}O_2^- \langle CH_3^- \langle C_6H_5^- \langle \underline{C}N^- \langle CO$

Table 3.6.2.1 $10Dq_o$ or Δ_o values (in kK) in some octahedral complexes to illustrate the variation of ligand field strength.

		Ligands					
d^n	M^{n+}	$6F^-$	$6Cl^-$	$6H_2O$	$6NH_3$	$3en$	$6CN^-$
d^3	Cr^{3+}	15.10	13.70	17.40	21.50	22.30	26.60
d^5	Mn^{2+}	7.80	7.50	8.50	–	10.10	30.00
d^6	Co^{3+}	13.10	–	20.76	22.90	23.16	33.50
d^8	Ni^{2+}	7.30	7.50	8.50	10.80	11.90	–

The above mentioned series is approximately valid irrespective of the nature of metal centre. The ligands positioned at the right end of the series are called the **strong field ligands,** *i.e.* relatively higher value of f) to cause more splitting of the metal orbitals while the ligands positioned at the left end of the series are called the **weak field ligands** (*i.e.* relatively lower value of f) to cause less splitting of the orbitals.

Explanation of the spectrochemical series in terms of CFT and its limitations

● **Anionic ligand** *vs.* **neutral ligands:** In terms of the basic postulates of CFT, it is very difficult to explain the full spectrochemical series. Because, the anionic ligands are expected to interact electrostatically more strongly with the metal orbitals compared to the neutral dipolar ligands. But this expectation is not maintained in the series. In fact, many anionic ligands (*e.g.* Br$^-$, Cl$^-$, F$^-$, etc.) are placed at the extreme left end of the series, *i.e.* these are the weak field ligands. On the other hand, many neutral dipolar ligands (*e.g.* bpy, phen, CO, etc.) are found as the strong-field ligands.

● **H$_2$O vs. OH$^-$:** It is quite interesting to note that H$_2$O (neutral) is a stronger field ligand than the anionic OH$^-$ ligand.

● **NH$_3$ vs. H$_2$O:** Similarly, NH$_3$ is a stronger field ligand than H$_2$O, though their dipole moment sequence is reversed (*i.e.* $\mu_{NH_3} < \mu_{H_2O}$).

To rationalize the complete series, metal-ligand covalent interaction is to be considered. These aspects will be discussed later in detail.

● **Electrostatic interaction and spectrochemical series:** In spite of the above mentioned limitations of CFT to explain the complete spectrochemical series, it may be mentioned that the ligand field strength can be sensed to some extent in terms of the electrostatic interaction, the basis of CFT. **It is expected that smaller donor atoms and less electronegative donor atoms** (which can donate more negative charge to split the metal orbitals) **will cause more splitting of the metal orbitals.** Smaller size of the donor atom will increase the electric field strength of the point-like ligand and the metal-ligand interacting distance becomes shorter $\left(cf.\ \Delta \propto \dfrac{1}{a^x},\ x = 5 - 6 \right)$. These will cause more splitting of the metal orbitals.

Spectrochemical series in terms of the donor atom:

Ligand field strength: $\quad \underset{\text{decreasing size}}{\underrightarrow{I < Br < Cl < S < F}} \qquad\qquad \underset{\text{decreasing electronegativity}}{\underrightarrow{O < N < C}}$

This helps us to understand the higher ligand field strength of CN$^-$ (κC) than that of NC$^-$ (κN); higher ligand field strength of NH$_3$ than that of H$_2$O. *The less electronegative donor atom can donate more electronic charge to cause a better electrostatic effect to split the metal orbitals.*

The size factor helps us to understand the position of halides in the spectrochemical series. Electrical field intensity (measured by the charge/radius ratio) of the point-like ligands increases with the decrease in the size of the donor atom. *Higher electrical field intensity of the ligands will cause more splitting of the metal orbitals through the better electrostatic effect.* Moreover, for the smaller donor atoms, the metal-ligand interacting distance becomes shorter to cause the more crystal field splitting ($\propto \dfrac{1}{a^x},\ x = 5 - 6,\ a$ = metal-ligand distance).

● **Position of the π-acid ligands in terms of the electrostatic interaction:** The π-acceptor ligands (*e.g.* CO, CN$^-$, etc.) will increase the positive charge on the metal centre. This will shorten the metal-ligand interacting distance (a) to increase the crystal field splitting power (*cf.* $\Delta \propto \dfrac{1}{a^x},\ x = 5 - 6$). *Thus,* *even the strong field character of the π-acid ligands can be partly understood in terms of the electro-static model.* However, this can be better rationalized in terms of the MOT to be discussed later.

For the better understanding the position of different ligands in the spectrochemical series, we shall consider the MOT at the appropriate place.

3.7 PAIRING ENERGY AND FACTORS CONTROLLING THE PAIRING ENERGY (*cf.* Sec. 8.24.1)

The pairing energy is required to overcome the **electrostatic repulsion** to place two electrons in the same orbital. It also leads to the **loss of exchange energy** due to the reduction in the number of unpaired electrons with parallel spins. Thus the two important factors to determine the pairing energy are:

electrostatic repulsion and loss of exchange energy

(i) **Electrostatic repulsion:** It is determined by the size of the orbital where the electron pairing occurs and the degree of covalence in the metal-ligand bond that can spread the electron cloud in a larger space. The electrostatic repulsion part of the pairing energy can be rationalised in terms of the **nephelauxetic effect** (Sec. 3.16C).

With the increase in the size of the d-orbitals, the electrostatic repulsion decreases in pairing the electrons. This is why, P varies in the sequence: $5d \langle 4d \langle 3d$. In fact, the more diffuse $5d$ orbital can more comfortably accommodate two electrons (*i.e.,* two units of negative charge) than the smaller $3d$ orbital. *Because of this fact, it is easier to produce the low-spin complexes for the heavier congeners.*

For a particular metal centre, with the increase of positive oxidation state, the nd-orbital **will be gradually contracted to increase the electrostatic repulsion to disfavour the pairing.** The electrostatic repulsion increases as: $Fe^{3+} \rangle Fe^{2+}$; $Co^{3+} \rangle Co^{2+}$, etc. Here it may be pointed out that in the higher oxidation state, the less electron density will partly reduce the electron-electron repulsion. However, the effect of contraction in size of the d-orbitals is more predominant.

For a particular d^n electronic configuration, the pairing energy goes to increase with the increase of positive oxidation state. In fact, with the increase of positive oxidation state of the metal centre, the d-orbital experiences contraction to increase the electrostatic repulsion. This is supported by the fact:

$$\left. \begin{array}{l} d^4 \text{ system: } P_{es}(Cr^{2+}) \langle P_{es}(Mn^{3+}) \\ d^5 \text{ system: } P_{es}(Mn^{2+}) \langle P_{es}(Fe^{3+}) \\ d^6 \text{ system: } P_{es}(Fe^{2+}) \langle P_{es}(Co^{3+}) \end{array} \right\} \quad (cf. \text{ Table 3.7.1})$$

Here P_{es} denotes the contribution of electrostatic repulsive factor to the total pairing energy (P).

Important factors controlling the pairing energy

- **Pairing energy decreases for the heavier congeners.**
- **Pairing energy increases with the increase of positive oxidation state of the metal centre.**
- **Pairing energy decreases with the decreases of electron density for a particular oxidation state.**
- **Pairing energy increases with the increase of loss of exchange energy.**
- **Covalency in the metal-ligand bond and nephelauxatic effect in the complex reduce the pairing energy.**
- **Pi-acid ligands can reduce the pairing energy.**

In the complex, because of the possibility of the overlapping interaction between the metal d-orbital and ligand orbital, a larger space is available compared to that in free ion to accommodate the electrons.

In fact, this overlapping interaction leads to the **nephelauxetic effect** (electron cloud expansion) and it reduces the electrostatic repulsion and consequently the pairing energy compared to that of the free ion. P value is decreased by about 15 to 30% in the complexes due to this effect. Obviously, the higher covalence will reduce the pairing energy more. **Because of this fact, it is easier to produce the low-spin complexes (*i.e.* 10Dq$_o$ $>$ P)** where the covalence is high.

The **powerful π-acid ligands** can withdraw the d-electron cloud to reduce the electron-electron repulsion which consequently lowers the pairing energy.

(Note: The π-acid ligands are the **strong field ligands** and they can also reduce the pairing energy. Both these factors favour the low-spin complex formation).

(ii) **Loss of exchange energy:** Exchange energy (E_{ex}) is the driving force of Hund's rule of maximum spin multiplicity. Exchange energy of a particular electronic configuration is proportional to the number of pairs of electrons with the parallel spins.

$$E_{ex} \propto {}^x C_2 = \frac{x(x-1)}{2} \text{ where } x \text{ gives the number of electrons with the same spin } \left(i.e. +\frac{1}{2} \text{ or } -\frac{1}{2}\right).$$

$$i.e. \ E_{ex} = \frac{K_{ex}x(x-1)}{2}, \ K_{ex} = \text{proportionality constant.}$$

Obviously, with the pairing of electrons, the number of electrons with a particular spin is decreased. *It will lead to the loss of exchange energy.* It is illustrated for the d^6 system assuming that the energy difference (10Dq$_o$ or Δ_o) will not prevent the spin exchange between the t_{2g} and e_g levels. Strictly speaking, the spin exchange can only occur among the electrons of same energy. **In fact, the higher Δ_o value reduces the exchange energy.**

High-spin ($t_{2g}^4 e_g^2$) state: There are 5 electrons with $+\frac{1}{2}$ spin and there is one electron with $-\frac{1}{2}$ spin.

$$E_{ex} = \frac{K_{ex}5(5-1)}{2} + \frac{K_{ex}(1)(1-1)}{2} = 10K_{ex}$$

Low-spin ($t_{2g}^6 e_g^0$) state: There are 3 electrons with $+\frac{1}{2}$ spin and there are three electrons with $-\frac{1}{2}$ spin.

$$E_{ex} = \frac{K_{ex}3(3-1)}{2} + \frac{K_{ex}3(3-1)}{2} = 6K_{ex}$$

Thus, the loss of exchange energy is $4K_{ex}$ in going from the high-spin state to the low-spin state.

The loss of exchange energy (ΔE_{ex}) and gain of cfse in moving from the high-spin state to the low-spin state can be expressed as follows:

Gain of cfse *(ignoring the pairing energy)*		*Loss of exchange energy* *(Assuming the exchange between the t_{2g} and e_g set in* *both the h.s. and l.s. states)*	
d^4:	$1 \times 10Dq_o$	$\Delta E_{ex} = K_{ex} \, {}^4C_2 - K_{ex} \, ({}^3C_2 + {}^1C_2)$	$= 6K_{ex} - 3K_{ex} = 3K_{ex}$
d^5:	$2 \times 10Dq_o$	$\Delta E_{ex} = K_{ex} \, {}^5C_2 - K_{ex} \, ({}^3C_2 + {}^2C_2)$	$= 10K_{ex} - 4K_{ex} = 6K_{ex}$
d^6:	$2 \times 10Dq_o$	$\Delta E_{ex} = K_{ex} \, ({}^5C_2 + {}^1C_2) - K_{ex} \, (2 \times {}^3C_2)$	$= 10K_{ex} - 6K_{ex} = 4K_{ex}$
d^7:	$1 \times 10Dq_o$	$\Delta E_{ex} = K_{ex} \, ({}^5C_2 + {}^2C_2) - K_{ex} \, ({}^4C_2 + {}^3C_2)$	$= 11K_{ex} - 9K_{ex} = 2K_{ex}$

Thus in moving from h.s. to l.s., per gain of $10Dq_o$ cfse, the loss of exchange energy is: $3K_{ex}$ for d^4 and d^5, $2K_{ex}$ for d^6 and d^7 systems. Thus in terms of the loss of pairing energy, **the ease of spin pairing follows the sequence**: d^6, $d^7 \rangle d^4$, d^5.

Obviously, the loss of exchange energy is very high for the d^4 and d^5 configurations and it is generally relatively more difficult to produce the low-spin complexes of the d^4 (*e.g.* Cr^{2+}, Mn^{3+}) and d^5-systems (*e.g.* Fe^{3+}, Mn^{2+}).

The above mentioned predictions are supported by the fact that most of Co(III) (d^6-system) complexes involve the spin pairing (*i.e.* low-spin complexes). On the other hand, **only the strong field ligands like CN^- can induce the spin pairing in Fe(III)(d^5), Mn(II)(d^5), Cr(II)(d^4), Mn(III)(d^4) systems**.

(**Note:** *Strictly, the electrons of the same energy can only exchange.* Here for the sake of simplicity, it may be assumed that in the *high spin state, the energy difference between the t_{2g} and e_g levels is not too high to prevent the exchange process* between the t_{2g} and e_g levels but in the low spin state, the high energy difference (*cf.* in general, $\Delta_{l.s.} > \Delta_{h.s.}$) may not allow the spin exchange process between the t_{2g} and e_g levels. For the sake of simplicity, in the above calculation, it has been assumed that the exchange process will occur between the t_{2g} and e_g set in both the h.s. and l.s. states. If, the energy difference between the t_{2g} and e_g set in the l.s. state prevents the electron exchange between these two levels, then the calculation is different for the d^7 system only. It is as follows (assuming the exchange between the t_{2g} and e_g set in the h.s. state but not in the l.s. state):

$$\Delta E_{ex} = K_{ex}\left({}^5C_2 + {}^2C_2\right) - K_{ex}\left(2 \times {}^3C_2 + {}^1C_2\right) = 11K_{ex} - 6K_{ex} = 5K_{ex}.$$

It indicates the highest loss of exchange energy per gain of $10Dq_o$ cfse in the d^7 system).

Table 3.7.1 Calculated pairing energies (P) for some gaseous 3d metal ions.

		P (cm^{-1}) for gaseous M^{n+}		
d^n	Ion	Coulombic (P_{es})	Exchange (P_{ex})	Total (P)
d^4	Cr^{2+}	5,950	14,470	20,420
	Mn^{3+}	7,350	17,870	25,220
d^5	Mn^{2+}	7,610	16,210	23,820
	Fe^{3+}	10,050	19,830	29,880
d^6	Fe^{2+}	7,460	11,690	19,150
	Co^{3+}	9,450	14,170	23,620
d^7	Co^{2+}	8,400	12,400	20,800

Note: Here P denotes the energy required for pairing two electrons. For the different d^n-configurations, the number of pairing depends on n. It is illustrated for octahedral systems:

	High spin (h.s.)	Low spin (l.s.)	No. of additional pairing (in going from h.s. to l.s.)	Required pairing energy
d^4 (*e.g.* Cr^{2+}, Mn^{3+}):	$t_{2g}^3 e_g^1$	$t_{2g}^4 e_g^0$	1	1P
d^5 (*e.g.* Mn^{2+}, Fe^{3+}):	$t_{2g}^3 e_g^2$	$t_{2g}^5 e_g^0$	2	2P
d^6 (*e.g.* Fe^{2+}, Co^{3+}):	$t_{2g}^4 e_g^2$	$t_{2g}^6 e_g^0$	2	2P
d^7 (*e.g.* Co^{2+}):	$t_{2g}^5 e_g^2$	$t_{2g}^6 e_g^1$	1	1P

3.8 CFT AND OCTAHEDRAL COMPLEXES

3.8.1 Crystal Field Stabilization Energy (cfse) in the Octahedral Systems: High-spin and Low-spin Complexes

(A) Calculation of cfse and contribution of P in calculating cfse: cfse indicates the **additional stabilization** earned through the splitting of d-orbitals by the ligand field. If, already, the paired electrons are present in the pre-splitting condition, then we should consider the number of additional pairing required. This is illustrated for the d^4 and d^7 systems.

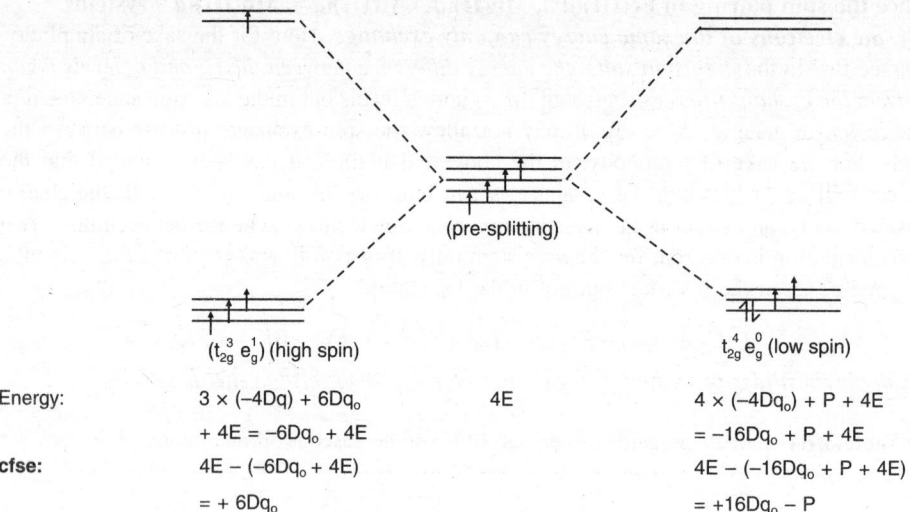

Energy:	$3 \times (-4Dq) + 6Dq_0$	4E	$4 \times (-4Dq_0) + P + 4E$
	$+ 4E = -6Dq_0 + 4E$		$= -16Dq_0 + P + 4E$
cfse:	$4E - (-6Dq_0 + 4E)$		$4E - (-16Dq_0 + P + 4E)$
	$= + 6Dq_0$		$= +16Dq_0 - P$

Fig. 3.8.1.1 Calculation of cfse for the high-spin and low-spin octahedral complexes of the d^4 system. E = energy of an electron in the degenerate pre-splitting d-orbitals. P = **pairing energy** (*i.e.* required energy to pair two electrons).

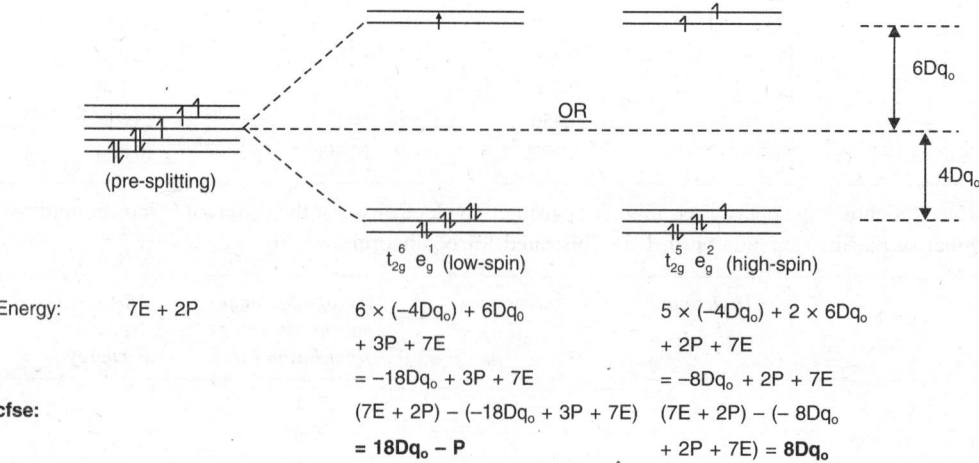

Energy:	7E + 2P	$6 \times (-4Dq_0) + 6Dq_0$	$5 \times (-4Dq_0) + 2 \times 6Dq_0$
		$+ 3P + 7E$	$+ 2P + 7E$
		$= -18Dq_0 + 3P + 7E$	$= -8Dq_0 + 2P + 7E$
cfse:		$(7E + 2P) - (-18Dq_0 + 3P + 7E)$	$(7E + 2P) - (- 8Dq_0$
		$= 18Dq_0 - P$	$+ 2P + 7E) = 8Dq_0$

Fig. 3.8.1.2 Calculation of cfse for the high-spin and low-spin complexes of the d^7 system. E = energy of an electron in the degenerate pre-splitting d-orbitals. P = pairing energy.

Thus, for the d^6, d^7, d^8, d^9 and d^{10} systems, the number of pairing already present in pre-splitting condition is: 1, 2, 3, 4 and 5. These are to be considered to calculate the additional pairing required in redistribution of electrons after splitting of the d-orbitals.

(**Note:** ● Many authors show as: cfse $(t_{2g}^6 e_g^1) = 18Dq_o - 3P$; cfse $(t_{2g}^5 e_g^2) = 8Dq_o - 2P$)

Here, the number of *pairing* present already in the pre-splitting condition has been ignored **but it is not justified.**

● Many authors, do not consider the pairing energy in expressing the cfse, *e.g.* cfse for $t_{2g}^6 e_g^1 \approx$ $18Dq_o$ or $-18Dq_o$. **Crystal field energy** may be taken as $-18Dq_o$ but **cfse should not be denoted by $-18Dq_o$.** Stabilisation indicates its energy as $-18Dq_o$ with respect to the barycentre.

● Many authors express the cfse as the energy of the system after splitting as follows: $t_{2g}^6 e_g^1$: $-18Dq_o$ + pairing energy but **the sign convention is not justified.**

Different Conventions to express the cfse

● The energy of an electron in the t_{2g} level (of octahedral system) is $-4Dq_o$ with respect to that of the barycentre and similarly, the energy of an electron in the e_g level is $+6Dq_o$ with respect to that of the barycentre. The $-$ sign (for the t_{2g} level) signifies the stabilization. Thus it indicates that a t_{2g} electron is stabilized by $4Dq_o$ while an e_g electron is destabilized by $6Dq_o$ (measured with respect to that of the barycentre). Thus for the $t_{2g}^3 e_g^1$ configuration we can write:

$t_{2g}^3 e_g^1$: $-3 \times 4Dq_o + 6Dq_o = -6Dq_o$ (**crystal field energy** with respect to that of the barycentre)

It means that the $t_{2g}^3 e_g^1$ configuration is stabilized by $6Dq_o$ (with respect to the pre-splitting condition) due to crystal field splitting. **Thus it is reasonable to say that cfse of the system is $6Dq_o$.** However, many authors express the cfse of the system as $-6Dq_o$, *i.e. crystal field energy of the system is treated as the cfse*. **It is not justified.**

Thus the convention to be used to express the cfse depends on the choice of the readers.

● Pairing energy (P) is to reduce the cfse and this is why it is added with a negative sign with the **cfse** (*e.g.* $t_{2g}^4 e_g^0$: cfse $= 4 \times 4Dq_o - P$). However, in the alternative convention, **cfse** of the $t_{2g}^4 e_g^0$ system is given by: $-4 \times 4Dq_o + P$. **It is not justified.**

● Many authors do not include the pairing energy in expressing the cfse. **It is an approximation.**

● If the pairing energy (P) is to be included in the cfse, then the additional pairings, required compared to that of the pre-splitted condition are to be considered. **Obviously, cfse of a high spin complex (where no additional pairing is required compared to that in free ion) does not contain any pairing energy term and cfse of a low spin complex (d⁴, d⁵, d⁶ and d⁷ systems) contains the additonal pairing energy term.**

But some authors consider the total number of pairings present in the splitted condition. For example, in $t_{2g}^6 e_g^0$ configuration, though there are two additional pairings (*i.e.* 2P), some authors consider three pairings (*i.e.* 3P) in expressing the cfse. **However, it is not justified.**

(B) High-spin and Low-spin complexes: Very often, in presence of a crystal field of ligands, the d-orbitals are splitted. In the splitted d-orbitals, the d-electrons will be accommodated. The lower energy levels will be first occupied (*cf.* auf-bau principle), *i.e. the electrons will occupy the orbitals in terms of energy order.* But in deciding the electron distribution, the pairing energy and energy difference between the successive energy levels are the important factors to be considered. This is illustrated in the octahedral systems for the different d^n-electronic configurations.

Case-I: d^n (n = 1, 2, 3, 8, 9, 10) in the octahedral system

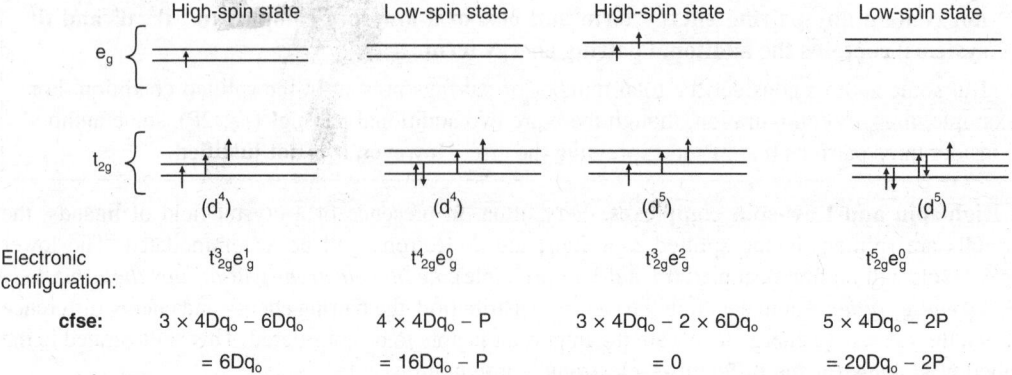

Fig. 3.8.1.3 Unique ground state electronic configurations for the d^1, d^2, d^3, d^8, d^9 and d^{10} systems in the octahedral field.

For the d^n (n = 1, 2, 3, 8, 9, 10) electronic configurations, in each case there is only one possible way of electron distribution in the t_{2g} and e_g orbitals of an octahedral system. The cfse can be calculated by considering the stabilisation of one electron in the t_{2g} level by $4Dq_o$ while destabilisation for one electron in the e_g level by $6Dq_o$. For pairing of two electrons in a particular d-orbital, it requires the pairing energy (P). Thus for one additional pairing, the system is destabilised by 1P. In the above mentioned cases, there is no such additional pairing in redistributing the electrons in the splitted orbitals. **Thus the cfse does not contain the pairing energy term.**

Case-II: d^n (n = 4, 5, 6, 7) in the octahedral systems

In these cases, there are two possible ways to distribute the d-electrons in the t_{2g} and e_g level for each d^n system. There will be two opposing factors. The crystal field splitting energy, *i.e.* $10Dq_o$ (or Δ_o) will try to place the electrons into the more stable t_{2g} level as many as possible while the pairing energy (P) required to pair two electrons in the same orbital will try to avoid the pairing as far as possible. In a particular system, whether pairing will occur or not will depend on the relative values of $10Dq_o$ and P. Both $10Dq_o$ and P vary from system to system.

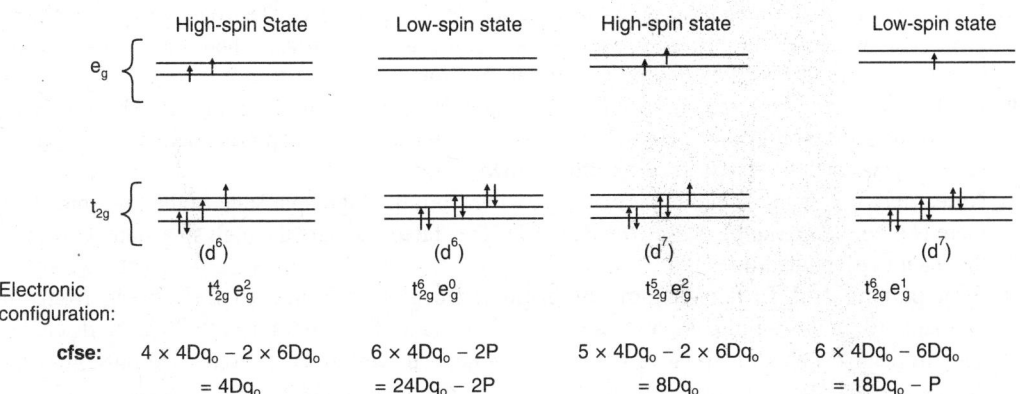

Fig. 3.8.1.4 High- and low-spin states for the d^4, d^5, d^6 and d^7 systems in an octahedral geometry.

Note: ● For the d^4, d^5, d^6 and d^7 systems, in the pre-splitted condition, the number of pairings are 0, 0, 1 and 2 respectively. In their high-spin complexes, no additional pairing is required. In their low-spin complexes, the number of additional pairings required are 1, 2, 2 and 1 respectively and these are considered in calculating their cfse values.

● Though the d-orbitals are not degenerate in the high spin complexes, the spin arrangement is the same as in the free ion where the d-orbitals are degenerate.

Table 3.8.1.1 Crystal field stabilization energy (cfse) of octahedral complexes.

d^n	Weak Field, i.e. high-spin complexes		Strong Field, i.e. low-spin complexes	
	Electronic configuration	cfse	Electronic configuration	cfse
d^1	$t_{2g}^1 e_g^0$	$4Dq_o$	$t_{2g}^1 e_g^0$	$4Dq_o$
d^2	$t_{2g}^2 e_g^0$	$2 \times 4Dq_o = 8Dq_o$	$t_{2g}^2 e_g^0$	$2 \times 4Dq_o = 8Dq_o$
d^3	$t_{2g}^3 e_g^0$	$3 \times 4Dq_o = 12Dq_o$	$t_{2g}^3 e_g^0$	$3 \times 4Dq_o = 12Dq_o$
d^4	$t_{2g}^3 e_g^1$	$3 \times 4Dq_o - 6Dq_o = 6Dq_o$	$t_{2g}^4 e_g^0$	$4 \times 4Dq_o - P = 16Dq_o - P$
d^5	$t_{2g}^3 e_g^2$	$3 \times 4Dq_o - 2 \times 6Dq_o = 0$	$t_{2g}^5 e_g^0$	$5 \times 4Dq_o - 2P = 20Dq_o - 2P$
d^6	$t_{2g}^4 e_g^2$	$4 \times 4Dq_o - 2 \times 6Dq_o = 4Dq_o$	$t_{2g}^6 e_g^0$	$6 \times 4Dq_o - 2P = 24Dq_o - 2P$
d^7	$t_{2g}^5 e_g^2$	$5 \times 4Dq_o - 2 \times 6Dq_o = 8Dq_o$	$t_{2g}^6 e_g^1$	$6 \times 4Dq_o - 6Dq_o - P = 18Dq_o - P$
d^8	$t_{2g}^6 e_g^2$	$6 \times 4Dq_o - 2 \times 6Dq_o = 12Dq_o$	$t_{2g}^6 e_g^2$	$6 \times 4Dq_o - 2 \times 6Dq_o = 12Dq_o$
d^9	$t_{2g}^6 e_g^3$	$6 \times 4Dq_o - 3 \times 6Dq_o = 6Dq_o$	$t_{2g}^6 e_g^3$	$6 \times 4Dq_o - 3 \times 6Dq_o = 6Dq_o$
d^{10}	$t_{2g}^6 e_g^4$	$6 \times 4Dq_o - 4 \times 6Dq_o = 0$	$t_{2g}^6 e_g^4$	$6 \times 4Dq_o - 4 \times 6Dq_o = 0$

For d^n (n = 1, 2, 3, 8, 9, 10) configurations, only one possibility.

(C) Condition for pairing: $10Dq_o > P$.

● Thus, if $10Dq_o < P$, then pairing of electrons in the t_{2g} level will be avoided as far as possible. This gives the **high-spin or spin-free complex** where the number of unpaired electrons is maximum. This number of unpaired electrons is identical with the number of unpaired electrons in the d-orbitals of the free ion (*cf.* outer-orbital or ionic complex in VBT).

- If, $10Dq_o \rangle P$, then the system will place the maximum number of electrons possible in the t_{2g} level. This will give the **low-spin complex or spin-paired complex** where the number of unpaired electrons is less than that present in the free ion.

- If $10Dq_o \approx P$, then both the high-spin and low-spin complexes may remain in equilibrium giving rise to the **spin isomerism**. The $10Dq_o$ value ($\approx P$) leading to the spin isomerism is called the **critical crystal field strength** (*i.e.* **critical $10Dq_o$**)

  Note: Here it is worth mentioning that in moving from the high spin state to the low spin state, there is a *loss of exchange energy (cf.* Sec. 3.7). This factor favours the high spin state. However, this factor is included in P.

- **Spin pairing and Tanabe-Sugano diagram:** Spin pairing is needed for moving from the high-spin to the low-spin complexes and it happens under the condition $\Delta_o > P$ (to be discussed later in detail). For a quantitative comparison among the different d^n configurations, from the Tanabe-Sugano diagram, we get the following Dq_o/B values at which the cross-over regions (*i.e.* h.s. $\rightleftharpoons$ l.s.) occur.

  $$d^4: 2.7(Dq_o/B) \; ; \; d^5: 2.8(Dq_o/B)$$
  $$d^6: 2.0(Dq_o/B) \; ; \; d^7: 2.2(Dq_o/B)$$

For the first transition series, the Racah parameter B can be taken approximately constant. ***Thus it appears that spin-pairing is easiest for the d^6-systems and it is most difficult for the d^5 and d^4 systems.*** Thus the ease of spin pairing leading to the low-spin complexes run as: $3d^6 \rangle 3d^7 \rangle 3d^4 \rangle 3d^5$. This sequence, *i.e.* ease of spin pairing is also maintained for the gaseous ions (where *nephelauxetic effect* is absent).

- **Ease of spin pairing $d^6 > d^7 > d^4 > d^5$:** In view of the fact mentioned above, experimentally it is established that there are very few Co(III) (d^6) complexes (*e.g.* $[CoF_6]^{3-}$, $[CoF_3(OH_2)_3]$ etc.) which are spin free while a large number of spin paired (*i.e.* low-spin) complexes of Co(III) even with the moderately strong field ligands are reported. In this regard, we must mention the fact that in the low-spin complex, there is a large cfse (= $24Dq_o - 2P$). On the other hand, for Fe(III) (d^5) and Mn(III) (d^4) systems, only strong field ligands like CN^- can induce the spin pairing. Here it is interesting to mention that $[CoF_6]^{2-}$ (*i.e.* Co^{IV}, d^5 system) is a low-spin complex where the very high value of Δ_o (*cf.* the effect of oxidation state) can favour the spin pairing.

- **Pi-acid ligands favouring the spin pairing:** Here it is worth mentioning that the π-acid ligands can withdraw the metal d-electrons to enhance Δ_o and to reduce the electron-electron repulsion. It reduces the pairing energy and makes Δ_o high to favour the low-spin complex formation.

- ***The relative values of $10Dq_o$ and P will determine whether the t_{2g} set will be occupied in preference to the e_g set.*** If $10Dq_o \rangle P$, then the cfse of the low-spin state will be more than that of the corresponding high-spin state. On the other hand, under the condition, $10Dq_o \langle P$, the cfse of the high-spin state will be more than that of the low-spin state. It is illustrated for the d^6 system.

	cfse	*difference in cfse*
High-spin state ($t_{2g}^4 e_g^2$):	$4Dq_o$	$20Dq_o - 2P$
Low-spin state ($t_{2g}^6 e_g^0$):	$24Dq_o - 2P$	

(i) If $10Dq_o \rangle P$, *i.e.* $P = 10Dq_o - x$ (say), then we have:

$24Dq_o - 2(10Dq_o - x) = 4Dq_o + 2x \rangle 4Dq_o$

i.e. **cfse of the low-spin state is higher than that of the high-spin state.**

(ii) If, $P \rangle 10Dq_o$, *i.e.* $P = 10Dq_o + x$ (say), then we have:

$24Dq_o - 2(10Dq_o + x) = 4Dq_o - 2x \langle 4Dq_o$

i.e. **cfse of the high-spin state is higher than that of the low-spin state.**

For a particular metal ion, the strong field ligand (i.e. $10Dq_o$ is high) favours the low-spin complex formation while the weak-field ligand (i.e. relatively smaller $10Dq_o$ value) favours the high-spin complex formation. This prediction has been experimentally verified in many cases (*cf.* Table 3.8.1.2).

Table 3.8.1.2 Low-spin and high-spin octahedral complexes depending on the $10Dq_o$ and P (mean pairing energy) values (1kK = 1000 cm^{-1}).

d^n	Complex	P (kK)	$10Dq_o$ (kK)	Spin State	
				Predicted	Observed
d^4	$[Cr(OH_2)_6]^{2+}$	23.5	12.6	h.s.	h.s.
	$[Mn(OH_2)_6]^{3+}$	28.0	21.0	h.s.	h.s.
d^5	$[Mn(OH_2)_6]^{2+}$	25.5	8.5	h.s.	h.s.
	$[Fe(OH_2)_6]^{3+}$	30.0	14.0	h.s.	h.s.
	$[Fe(CN)_6]^{3-}$	30.0	35.0	l.s.	l.s.
d^6	$[Fe(OH_2)_6]^{2+}$	17.6	10.5	h.s.	h.s.
	$[Fe(CN)_6]^{4-}$	17.6	32.2	l.s.	l.s.
	$[Co(OH_2)_6]^{3+}$	~ 20.0 -	~ 20.0	h.s. $\rightleftharpoons$	h.s. $\rightleftharpoons$
		21.0		l.s.	l.s.
d^6	$[CoF_6]^{3-}$	21.0	13.1	h.s.	h.s.
	$[Co(NH_3)_6]^{3+}$	21.0	22.9	l.s.	l.s.
	$[Co(CN)_6]^{3-}$	21.0	33.5	l.s.	l.s.
d^7	$[Co(OH_2)_6]^{2+}$	19.5	9.2	h.s.	h.s.

Source: Coordination Chemistry by D. Banerjea, p. 207, Tata McGraw-Hill Publishing Company Ltd., N. Delhi, 1993.

Note: (i) This mean pairing energy (P) used in this Table differs slightly from the calculated values given in Table 3.7.1.1. The given pairing energy (P) gives the energy required for pairing a pair of electrons.

(ii) l.s. stands for low-spin and h.s. stands for high-spin.

(D) Conditions favouring the low-spin complex formation: The condition, $\Delta_o > P$ is easily attained when *the crystal field splitting is high* and *pairing energy is less*.

(i) *High positive charge on the metal centre* and the strong field ligand produce the high Δ_o value to favour the low-spin complex formation. Here it is worth mentioning that the higher positive charge on the metal centre slightly enhances the pairing energy (Sec. 3.7) to disfavour the low spin complex formation. However, the relative effect on Δ_o is more than that on P. Thus, in general, the high oxidation state favours the low spin complex formation.

(ii) *Heavier congeners* give higher the Δ_o value and the lower pairing energy to favour the low-spin state.

(iii) The *pairing energy* runs as: $d^5 > d^4 > d^6$ for a particular oxidation state. Thus the spin pairing is easiest for the d^6 system (*e.g.* Co(III)) while it is most difficult for the d^5 system (*e.g.* Fe(III)). In fact, most of the Co(III) complexes are the low-spin complexes.

(iv) *Pi-acid ligands* can favour formation of the low-spin complexes in two ways: strong field character and favouring electron pairing by reducing the electron-electron repulsion through the metal $\rightarrow$ ligand π-back bonding.

3.8.2 Critical Crystal Field Strength (*i.e.* $\Delta_o \approx$ P): Spin-state Equilibrium: Temperature Controlled Spin Isomerism (*cf.* Sec. 8.24, Chapter 8 for details)

We have already mentioned the following conditions:

Low-spin (l.s.) complex: $\Delta_o \rangle$ P

High-spin (h.s.) complex: $\Delta_o \langle$ P $\left.\begin{array}{l} \\ \\ \end{array}\right\}$ for the d^4, d^5, d^6 and d^7 systems.

Under the condition of critical value of Δ_o (= $10Dq_o$), *i.e.* $\Delta_o \approx$ P, both the h.s. and l.s. complex will have the same cfse. This will lead to the *spin-state equilibrium, i.e.*

$$\text{h.s.} \rightleftharpoons \text{l.s., (for } d^n, n = 4, 5, 6, 7)$$

The critical value of Δ_o is described as the *high-spin low-spin crossover point.* If the crystal field strength lies close to the crossover region, then the spin state equilibrium is attained and it leads to anomalous magnetic moments (*cf.* Sec. 8.24).

If the energy difference between the two spin states is of the order of $k_B T$ (*i.e.* **thermal energy**) then *position of the equilibrium will depend on the temperature.* It is illustrated in Fig. 3.8.2.1.

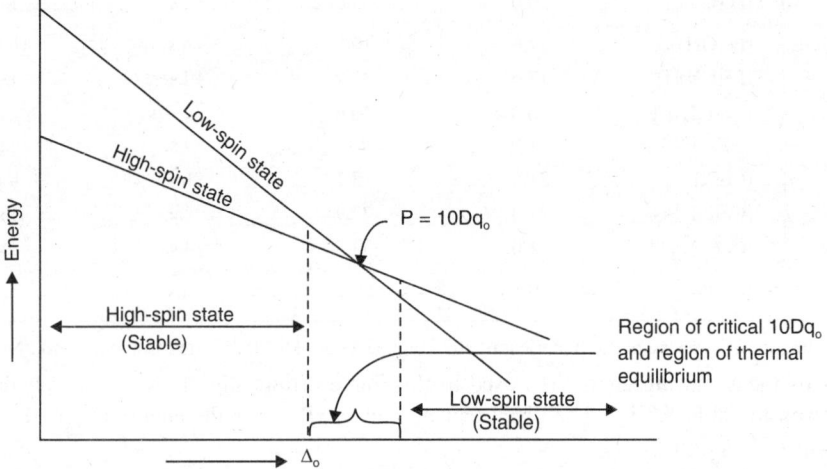

Fig. 3.8.2.1 Qualitative representation of the variation of energy of the high-spin and low-spin state with Δ_o for the d^4, d^6 and d^7 systems; qualitative representation of the critical Δ_o value for the change-over from the high-spin to the low-spin state of the octahedral complexes of the d^4, d^6 and d^7 systems.

		Energy of the system	
	Pre-splitting state	*high-spin state*	*low-spin state*
d^4:	$4E$	$-\dfrac{3}{5}\Delta_o + 4E$	$-\dfrac{8}{5}\Delta_o + P + 4E$
d^5:	$5E$	$0 \times \Delta_o + 5E$	$-2\Delta_o + 2P + 5E$
d^6:	$6E + P$	$-\dfrac{2}{5}\Delta_o + P + 6E$	$-\dfrac{12}{5}\Delta_o + 3P + 6E$
d^7:	$7E + 2P$	$-\dfrac{4}{5}\Delta_o + 2P + 7E$	$-\dfrac{9}{5}\Delta_o + 3P + 7E$

(E = energy of an electron for the degenerate d-orbitals in pre-splitting conditon; P = pairing energy required for pairing of a pair of electrons.)

The slopes of the two lines in Fig. 3.8.2.1 are different. As the Δ_o value increases, energies of both states (*i.e.* h.s. and l.s.) will decrease but energy of the l.s. will decrease more steeply. This can be rationalized by considering the energy of the two spin states.

It is evident that the energy of the low-spin state decreases more sharply compared to the corresponding high-spin state with the increase of ligand field strength. For the d^4 system, for the plot of energy *vs.* Δ_o, slope of the h.s. state is $-3/5$ while for the l.s. state it is $-8/5$. Similarly, for the plot of energy *vs.* Δ_o, the slopes are as follows:

	slope (h.s.)	*slope (l.s.)*
d^4 :	$-3/5$	$-8/5$
d^5 :	0 (*i.e.* parallel to the Δ_o axis)	-2
d^6 :	$-2/5$	$-12/5$
d^7 :	$-4/5$	$-9/5$

Thus we can conclude (*cf.* Fig. 3.8.2.1):

below the critical crystal field strength, the high-spin state is the ground state while above the critical crystal field strength, the low-spin state is the ground state.

Conversion of a h.s. state to a l.s. state leads to the transfer of one electron (for the d^4 and d^7 systems) or two electrons (for the d^5 and d^6 systems) from the e_g orbitals which directly face the ligands to the

t_{2g} level and it will shrink the metal-ligand bond distance (r). **Thus Δ_o value for the low-spin state is**

relatively higher $\left(cf.\ \Delta_o \propto \dfrac{1}{r^x},\ x = 5-6 \right)$. Thus the l.s. state will have its minimum energy for the

smaller r value compared to that for the h.s. state. The potential energy curves for the h.s. and l.s. states are shown in Fig. 3.8.2.2. For the crossover region, the minimum energies for the two curves are the same.

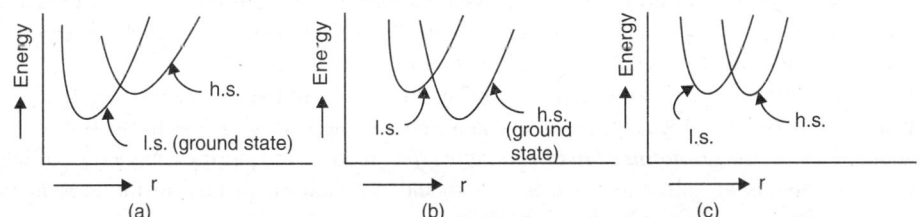

Fig. **3.8.2.2** Potential energy curves for (a) low-spin ground state, (b) high-spin ground state and (c) cross-over situation.

● **Effect of temperature on the h.s. $\rightleftharpoons$ l.s. process:** Origin of the thermally controlled spin crossover can be rationalized in terms of the potential wells of the two states. The metal-ligand bond distance in the low-spin state is relatively shorter than that in the high-spin state. Crossover point leading to the thermally controlled spin isomerism arises when the difference (ΔE_o) in **zero-point energies** (Fig. 3.8.2.3) of the two states is in the order of $k_B T$ (*i.e.* thermal energy), *i.e.* $\Delta E_o \approx k_B T$. This condition is attained for $10Dq_o \approx$ P.

● **Effect of pressure on the h.s. $\rightleftharpoons$ l.s. process:** In the region of crossover point, the position of equilibrium may also *depend on the external pressure* that will try to reduce the volume. If the e_g electrons (directly facing the approaching ligands) are shifted to the t_{2g} level. (*i.e.* change from the high-spin state to the low-spin state), then molar volume of the complex will be reduced. *Thus the external pressure will derive the equilibrium from the high-spin state to the low-spin state.*

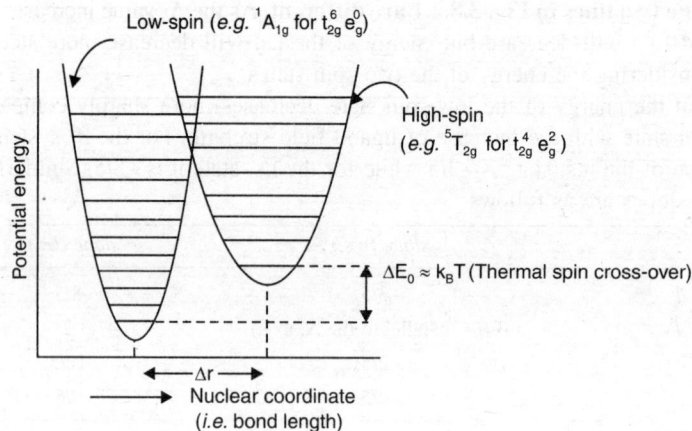

Fig. 3.8.2.3 Potential energy curve for the high-spin and low-spin curves at crossover point. (**Note:** Thermal spin cross-over under the condition $\Delta E_0 \approx k_B T$).

● **Examples experiencing h.s. $\rightleftharpoons$ l.s. (cf. Sec. 8.24):**

(i) *Tris*–**dialkyldithiocarbamatoiron(III)**, Fe $\left(\begin{array}{c} S \\ \\ S \end{array} \!\!\! \right) C\!\!-\!\!NR_2 \Big)_3$, *i.e.* [Fe(S$_2CNR_2$)$_3$] shows the spin-

state equilibrium in benzene solution in the monomeric form. The magnetic moment (at room temperature) varies in the range $\mu = 5.80 - 3.0$ B.M. depending on the nature of the alkyl group (R). At low temperature, the magnetic moment tends to 2.1 B.M. corresponding to $t_{2g}^5 e_g^0$ (*i.e.* low-spin state). With the increase of temperature, the observed magnetic moment gradually increases. The tris(dithiocarbamato)iron(III) complexes prefer to adopt the low-spin state at higher pressure because the transformation into the low-spin state reduces the molar volume.

Note: It has been already pointed out that the pairing energy (P) in the d^5-systems (*e.g.* Fe^{III}) is very high because of the loss of a huge amount of exchange energy on pairing the electrons. But in presence of the S-donor ligands, the ***pronounced nephelauxetic effect*** reduces the pairing energy (*cf.* participation of the vacant d-orbitals of sulfur in the metal $\rightarrow$ ligand π-bonding). In fact, in the present Fe(III)-complex, the nephelauxetic effect helps to satisfy the condition, $\Delta_o \approx P$.

(ii) **[Fe(phen)$_2$X$_2$] (X = NCS, NCSe)** also shows the spin-state isomerism. The observed magnetic moment increases with the increase of temperature.

(iii) **[Co(terpy)$_2$]Cl$_2$, 3.5H$_2$O** also shows the spin isomerism.

(iv) **Bis-(2,6-pyridinedialdehydrazone)cobalt(II) iodide** shows the spin isomerism.

Different aspects of spin-state isomerism have been discussed in detail in Sec. 8.24.

(**Note:** Many paramagnetic tetrahedral complexes of Ni(II) are converted into the diamagnetic square planar complexes when pressure is increased. This actually represents the **conformation isomerism.**

tetrahedral–Ni(II) $\rightleftharpoons$ square planar–Ni(II).

The above equilibrium is shifted to the right side with the increase of pressure. The paramagnetic tetrahedral complexes may be considered as the h.s. complexes and the diamagnetic square planar complexes may be considered as the l.s. complexes. The complex, bis(N,N-diethylamino-troponeiminato) nickel(II) remains in a such pressure dependent equilibrium.

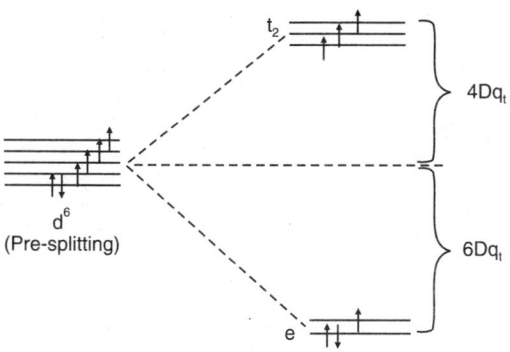

This type of **conformation isomerism** in Ni(II) complexes leads to an anomalous magnetic behaviour. The examples are discussed in Chapters 2 (Sec. 2.3) and 8.

3.9 CFT AND THE TETRAHEDRAL COMPLEXES

The tetrahedral crystal field splits the d-orbitals into two sets: t_2 (d_{xy}, d_{yz} and d_{zx}) and e $\left(d_{x^2-y^2}, d_{z^2}\right)$. The t_2-set is of higher energy while the e-set is of lower energy.

The crystal field splitting (Δ_t) in the tetrahedral complexes is generally too low (cf. $\Delta_t \approx \dfrac{4}{9}\Delta_0$) to cause the spin pairing (i.e. $P \gg \Delta_t$). **This is why, the tetrahedral complexes are generally the high-spin complexes.**

$$t_2 \uparrow\downarrow\ \uparrow\ \uparrow$$

$$4Dq_t$$

$$d^6$$
(Pre-splitting)

$$6Dq_t$$

$$e \uparrow\downarrow\ \uparrow\downarrow$$

Fig. 3.9.1 Illustration of calculation of the cfse for the high-spin d^6 configuration in a tetrahedral complex. E = energy of an electron in the pre-splitted condition, P = pairing energy.

Energy: $6E + P$ $3 \times (-6Dq_t) + P + 3(+4Dq_t) + 6E = -6Dq_t + P + 6E$

cfse: $(6E + P) - (-6Dq_t + P + 6E) = 6Dq_t$

Table 3.9.1 cfse for the d^n-systems in the high-spin tetrahedral complexes ($Dq_t \approx \dfrac{4}{9} Dq_0$).

d^n	Electronic configuration	cfse
d^1	$e^1 t_2^0$	$6Dq_t$
d^2	$e^2 t_2^0$	$2 \times 6Dq_t = 12Dq_t$
d^3	$e^2 t_2^1$	$2 \times 6Dq_t - 1 \times 4Dq_t = 8Dq_t$
d^4	$e^2 t_2^2$	$2 \times 6Dq_t - 2 \times 4Dq_t = 4Dq_t$
d^5	$e^2 t_2^3$	$2 \times 6Dq_t - 3 \times 4Dq_t = 0$
d^6	$e^3 t_2^3$	$3 \times 6Dq_t - 3 \times 4Dq_t = 6Dq_t$
d^7	$e^4 t_2^3$	$4 \times 6Dq_t - 3 \times 4Dq_t = 12Dq_t$
d^8	$e^4 t_2^4$	$4 \times 6Dq_t - 4 \times 4Dq_t = 8Dq_t$
d^9	$e^4 t_2^5$	$4 \times 6Dq_t - 5 \times 4Dq_t = 4Dq_t$
d^{10}	$e^4 t_2^6$	$4 \times 6Dq_t - 6 \times 4Dq_t = 0$

The cfse can be calculated as usual (*cf.* Table 3.9.1 and Fig. 3.9.2).

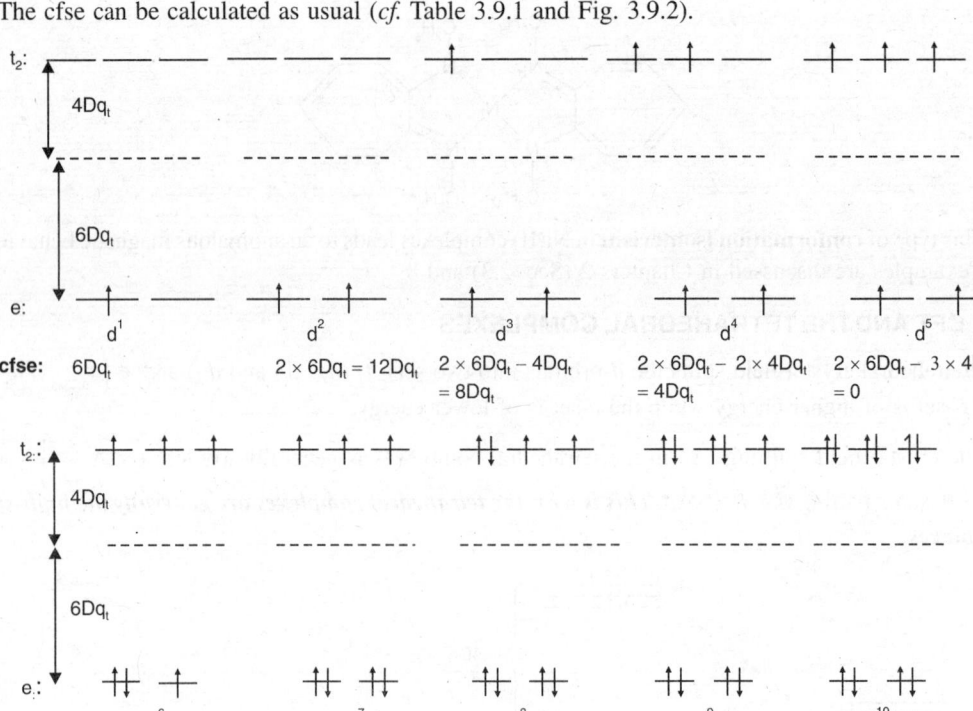

Fig. 3.9.2 Calculations of cfse in the high-spin tetrahedral complexes for the different d^n– configurations.

Note: In redistribution of the electrons in the e and t_2 levels of high spin tetrahedral complexes, *no additional electron pairing (with respect to that in the free ion) is required.* **This is why, cfse of the high spin tetrahedral complexes does not contain the pairing energy term.** It happens also so for the high spin octahedral complexes.

● **Examples of tetrahedral complexes:**

Complex	Electronic configuration	cfse
$[FeCl_4]^-$	$e^2 t_2^3$	$2 \times 6Dq_t - 3 \times 4Dq_t = 0$
$[CoCl_4]^{2-}$	$e^4 t_2^3$	$4 \times 6Dq_t - 3 \times 4Dq_t = 12Dq_t$
$[ZnCl_4]^{2-}$	$e^4 t_2^6$	$4 \times 6Dq_t - 6 \times 4Dq_t = 0$

The cfse of tetrahedral complexes (which are high-spin) is generally less and such tetrahedral complexes predominantly occur for the d^0, d^1, d^2, d^5, d^7, d^{10} systems for which the cfse of the corresponding octahedral complexes is not remarkably high.

In fact, **the high cfse favours the octahedral and square planar geometries while the less ligand-ligand repulsion in the tetrahedral complexes favours the tetrahedral complexes.** This aspect has been discussed in detail in Secs. 2.3, 2.11 and Ch. 8.

Possibility of the low-spin tetrahedral complexes: The most favourable condition arises for the d^4 system ($e^4 t_2^0$) having the cfse $24Dq_t - 2P$ (*cf.* for the octahedral system, the most favourable situation for the low-spin complex arises for the d^6 system having $t_{2g}^6 e_g^0$ with the cfse $24Dq_o - 2P$).

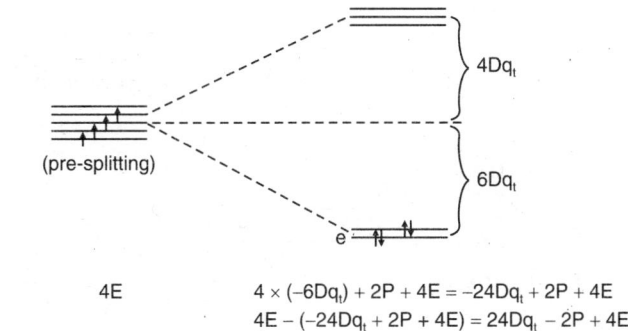

Energy:	4E	$4 \times (-6Dq_t) + 2P + 4E = -24Dq_t + 2P + 4E$
cfse:		$4E - (-24Dq_t + 2P + 4E) = 24Dq_t - 2P + 4E$

Fig. 3.9.3 Calculations of cfse for the low-spin tetrahedral complexes of the d^4-system. E = energy of an electron in the degenerate pre-splitting condition. P = pairing energy.

In fact, for the bulky anionic ligands, Cr(II) (d^4) has been found to produce the **distorted low-spin tetrahedral complex,** *e.g.* **[Cr{N(SiMe₃)₂}₃(NO)].** Similarly, **1-norbornyl anion** has been found to produce a low-spin tetrahedral complex with Co(IV) (d^5). These are discussed in Chapter 2.

The **bulky anionic ligands** produce the more steric repulsion (*i.e.* ligand-ligand repulsion) in other possible geometries like square-planar or octahedral (*cf.* bond angle 109° vs. 90°).

3.10 CFT AND TRIGONAL BIPYRAMIDAL AND SQUARE PYRAMIDAL COMPLEXES

The splitting patterns of the d-orbitals in these two geometries are shown in Fig. 3.10.1.

(A) Trigonal bipyramidal (TBP) geometry: The energy difference (δ_1) between the two stabilized doublet sets (*i.e.* d_{xz}, d_{yz} and d_{xy}, $d_{x^2-y^2}$) cannot exceed the pairing energy (P) but the energy difference (δ_2) between d_{z^2} and the (d_{xy}, $d_{x^2-y^2}$) set may exceed the pairing energy (P) depending on the ligand field strength.

$$\begin{cases} \delta_2 \gg \delta_1, \quad P \gg \delta_1; \quad \delta_2 > P \text{ (low-spin)}, \quad \delta_2 < P \text{ (high-spin)}, \\ \delta_2 \approx P \text{ (spin state equilibrium } i.e. \text{ spin isomerism).} \end{cases}$$

- For the d^1, d^2, d^3, d^4 systems, *always the high-spin TBP complexes* are produced.
- For the d^5, d^6, d^7 and d^8 systems, depending on the relative values of δ_2 and P, *high-spin or low-spin or spin-isomerism* conditions may arise.
- For the d^9 and d^{10} systems, *only one possibility exists.*

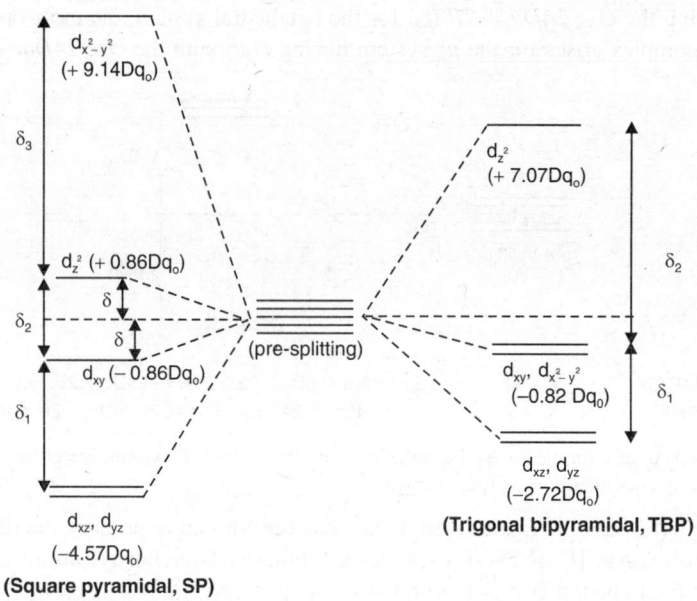

Fig. 3.10.1 Splitting pattern (not in scale) of the d-orbitals in the square pyramidal and trigonal bipyramidal geometries.

[**Note:** For the **square pyramidal geometry, in presence of the strong π-bonding ligands,** the energy order of the orbitals is: $d_{x^2-y^2} \rangle d_{z^2} \rangle d_{xz}, d_{yz} \rangle d_{xy}$ $(cf.$ Sec. 8.25.1$)$]

The two possibilities of electron distribution in the d^6 system of the TBP geometry are given below:

$$\left. \begin{array}{l} \textbf{high-spin:} \quad (d_{xz}, d_{yz})^3, (d_{xy}, d_{x^2-y^2})^2, (d_{z^2})^1 \\[4pt] \textbf{low-spin:} \quad (d_{xz}, d_{yz})^4, (d_{xy}, d_{x^2-y^2})^2, (d_{z^2})^0 \end{array} \right\} (cf.\ \text{Sec. 8.25})$$

Though theoretically for the d^5, d^6, d^7 and d^8 systems, both high-spin and low-spin complexes are possible, **in reality, most of the trigonal bipyramidal complexes are high-spin in nature.** However, for Ni(II) (*i.e.* d^8 system), the low-spin trigonal bipyramidal complexes have been found in some rare cases.

(B) Square pyramidal (SP) geometry $(cf.$ Fig. 3.10.1$)$: The energy differences are: $\delta_1 \rangle \delta_2; \delta_3 \rangle\rangle \delta_1, \delta_2$.

$$P \rangle \delta_1, \delta_2; \delta_3 \rangle P \text{ (low-spin)}, \delta_3 \langle P \text{ (high-spin)}, \delta_3 \approx P \text{ (spin isomerism)}.$$

● Thus, the d^1, d^2, d^3 and d^4 systems will always produce the **high-spin square pyramidal complexes** while for the d^5, d^6, d^7 and d^8 systems, the **high-spin and low-spin complexes** may be formed depending on the conditions.

(C) [Ni(CN)$_5$]$^{3-}$: Considering the low-spin TBP and SP complexes $(cf.$ CN$^-$ strong field ligand), the cfse values may be calculated as follows:

Low spin TBP: $(d_{xz}, d_{yz})^4 (d_{xy}, d_{x^2-y^2})^4$; cfse $= 4 \times (2.72Dq_o) + 4 \times (0.82Dq_o) - (4P - 3P)$

$$= 14.16Dq_o - P$$

Low spin SP: $(d_{xz}, d_{yz})^4 (d_{xy})^2 (d_{z^2})^2$; cfse $= 4 \times (4.57Dq_o) + 2 \times (0.86Dq_o) - 2 \times (0.86Dq_o) -$

(4P − 3P)

$$= 18.28Dq_o - P$$

In terms of the cfse and also the π-bonding effect, the SP structure of $[Ni(CN)_5]^{3-}$ is relatively more stable. But in reality, the stabilities of these two geometries differ marginally and $[Ni(CN)_5]^{3-}$ has been characterized in both forms (*cf.* Sec. 2.11.7). Now let us compare the bond lengths in these two geometries.

Low-spin TBP $[Ni(CN)_5]^{3-}$: *The axial bond is slightly shorter than the average equatorial bond.* The electronic configuration of the low-spin TBP (d^8-system) indicates that the d_{z^2} orbital (projecting towards the axial ligands) is vacant. On the other hand, the filled d_{xy} and $d_{x^2-y^2}$ orbitals are concentrated in the equatorial plane. *It makes the axial bond relatively shorter.* Here it may be mentioned that the π-acceptor property of the CN^- ligands will try to contract the equatorial bond through the π-bonding effect. These aspects have been discussed in detail in Sec. 2.11.7.

Low-spin SP $[Ni(CN)_5]^{3-}$: *The axial bond is significantly longer than the average equatorial bonds.* Here the $d_{x^2-y^2}$ orbital directly facing the equatorial ligands is vacant while the d_{z^2} orbital facing directly the axial ligand is filled in. *This makes the equatorial bond shorter than the axial bond.*

(D) TBP $[CuCl_5]^{3-}$: The electronic configuration is:

$$(d_{xz}, d_{yz})^4 (d_{xy}, d_{x^2-y^2})^4 (d_{z^2})^1$$

The d_{z^2} orbital projecting towards the axial ligands bears only one electron. The orbitals lying in the plane of the equatorial ligands are completely filled in. It makes the axial bond relatively shorter. Here it may be mentioned that Cl^- is not a π-bonding (π-acceptance) ligand to contract the equatorial bonds through the π-bonding.

Properties of the 5 coordinate complexes (*i.e.* TBP and SP) have been discussed and compared in Sec. 2.11.7.

3.11 TETRAGONAL DISTORTION IN THE OCTAHEDRAL SYMMETRY: JAHN-TELLER DISTORTION: OCTAHEDRAL AND TETRAHEDRAL COMPLEXES

3.11.1 Tetragonally Distorted Octahedrons and Crystal Field Splitting of the d-Orbitals

In the octahedral system, if the two *trans*– ligands lying along the z-axis are compressed or elongated compared to the other four ligands lying in the xy-plane, then we get the **tetragonally distorted octahedrons.** These are very often described as the **tetragonal structures** (*cf.* Fig. 2.11.2.6). These tetragonal structure may be of two types discussed below.

Cubic Symmetry: $x = y = z$; **Tetragonal Symmetry:** $x = y \neq z$; where x, y and z denote the metal-ligand bond lengths along the respective axis. Thus an octahedron maintains the cubic symmetry.

(A) **Tetragonally elongated structure (*i.e.* z-out distortion):** In order to understand the splitting pattern of the d-orbitals in this tetragonal system, it is better to start with the octahedral system in which the *trans*-axial ligands lying along the z-axis are being taken away. Consequently, the d_{z^2} orbital starts to experience the less repulsion compared to the $d_{x^2-y^2}$ orbital. Thus degeneracy in the e_g set of the octahedral geometry is lifted and the d_{z^2} orbital is stabilized relatively more. To conserve barycentre of the unsplit e_g level, the $d_{x^2-y^2}$ orbital will be destabilized to the same extent. Thus the e_g set is split and this splitting energy is denoted by δ_1 (say).

Because of the removal of crystal/ligand field along the z-direction, the orbitals (*i.e.* d_{xz}, d_{yz}) having the z-component in the t_{2g} set of the octahedral system will be also relatively stabilized more compared to the d_{xy} orbital. Thus, the t_{2g} set will also be split and the splitting will also conserve the barycentre of the unsplit t_{2g}-set. If splitting of the t_{2g} level is measured by δ_2, then the doublet set consisting of d_{xz} and d_{yz} will be stabilized by $\frac{1}{3}\delta_2$ while d_{xy} will be destabilized by $\frac{2}{3}\delta_2$ with respect to the barycentre of the unsplit t_{2g} set.

It may be noted that the d_{z^2} orbital faces the ligands directly in the octahedral system while the d_{xz} and d_{yz} orbitals are away from the ligands. ***Thus effect of the removal of the ligands along the z-direction influences the splitting of the e_g level more than that for the t_{2g} level, i.e.***

$$\Delta_o, P \rangle\rangle \delta_1 \rangle\rangle \delta_2$$

In such cases, δ_1 is generally less than the pairing energy (P). Thus the situation does not allow the spin pairing. *The limiting situation of z-out distortion leads to the square planar geometry.*

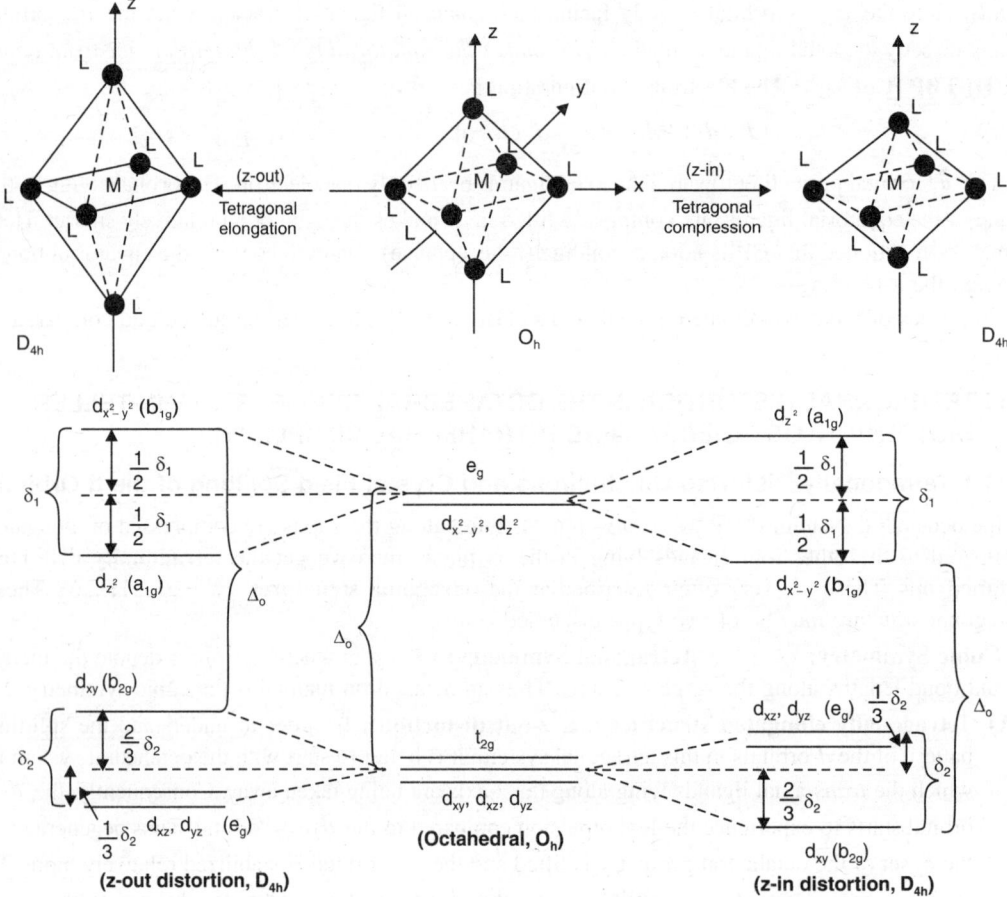

Fig. 3.11.1.1 Splitting of the t_{2g} and e_g orbitals of an octahedral system due to the z-in (*i.e.* tetragonal compression) and z-out (*i.e.* tetragonal elongation) distortion. (Δ_o, P $\rangle\rangle$ δ_1 $\rangle\rangle$ δ_2).

(B) Tetragonally compressed structure (*i.e. z-in distortion*): In the octahedral geometry, if the *trans*–axial metal-ligand bonds along the $\pm$ z-directions are compressed (*i.e.* shortened) with respect to the remaining four equatorial bonds, then the situation arises. *Obviously, in this tetragonal distortion, energy of the orbitals possessing the z-component will be increased relatively.* Consequently, the degeneracy in both the octahedral e_g and t_{2g} sets will be lost. Thus, the d_{z^2} orbital will be destabilized and to conserve the barycentre of the e_g set, the $d_{x^2-y^2}$ orbital will be stabilized to the same extent. This splitting of the e_g set may be represented by δ_1. Similarly, the octahedral t_{2g} will be also split and the doublet set consisting of d_{xz} and d_{yz} will be destabilized and the d_{xy} orbital will be stabilized in a way to conserve the barycentre of the t_{2g} set (*cf.* Fig. 3.11.1).

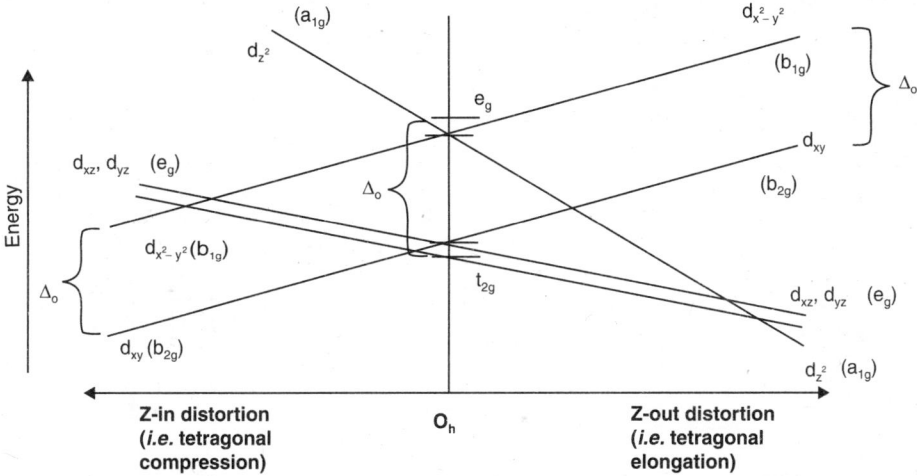

Fig. 3.11.1.2 Splitting of the t_{2g} and e_g orbitals of an octahedral system due to the z-in (*i.e.* tetragonal compression) and z-out (*i.e.* tetragonal elongation) distortion.

(C) Conditions of the tetragonal distortion

(i) The complexes like *trans*–[ML_4X_2] where X is weaker (in terms of the ligand field strength) than L experience *tetragonal elongation*. Here the metal-ligand interaction along the axial direction is relatively weaker. The examples are: *trans*–[$CoF_2(NH_3)_4$]$^+$, *trans*–[$Co(en)_2F_2$]$^+$, etc.

(ii) The stronger metal-ligand interaction along the axial direction will produce the condition of tetragonal compression in the complexes like *trans*–[ML_4X_2], where X is a stronger field ligand. A complex like *trans*–[$M(NH_3)_2(ox)_2$], can experience the z-in distortion. Oxo-vanadium(IV) complexes are the examples of tetragonal compression. In [$VO(OH_2)_5$]SO_4, the vanadyl oxygen-vanadium bond length (160 pm) is shorter than the vanadium aqua-oxygen bond ($\sim$ 230 pm) lying in the equatorial plane.

(iii) It has been already mentioned that if the axial and equatorial ligands are different (in terms of the crystal field strength) then the tetragonal distortion may occur. *Thus the mixed ligand complexes are always distorted.* But there are cases where tetragonal distortion can occur even in presence of six identical ligands of the octahedral complexes. These are governed by the **Jahn-Teller theorem** and the distortion is described as the **Jahn-Teller distortion.** This aspect will be discussed separately (Sec. 3.11.3). The following configurations experience the J.T. distortion in the octahedral systems: d^1, d^2, d^4 (both h.s. and l.s.), d^5 (l.s.), d^6 (h.s.), d^7 (both h.s. and l.s.) and d^9. Fig. 3.11.1.2 indicates that the relative energies of the orbitals depend on the extent of distortion.

3.11.2 Distortion (z-out and z-in) in the Tetrahedral Complexes and Splitting of the d-Orbitals

The splitting of t_2 and e sets of the tetrahedrons due to the distortion is shown in Fig. 3.11.2.1.

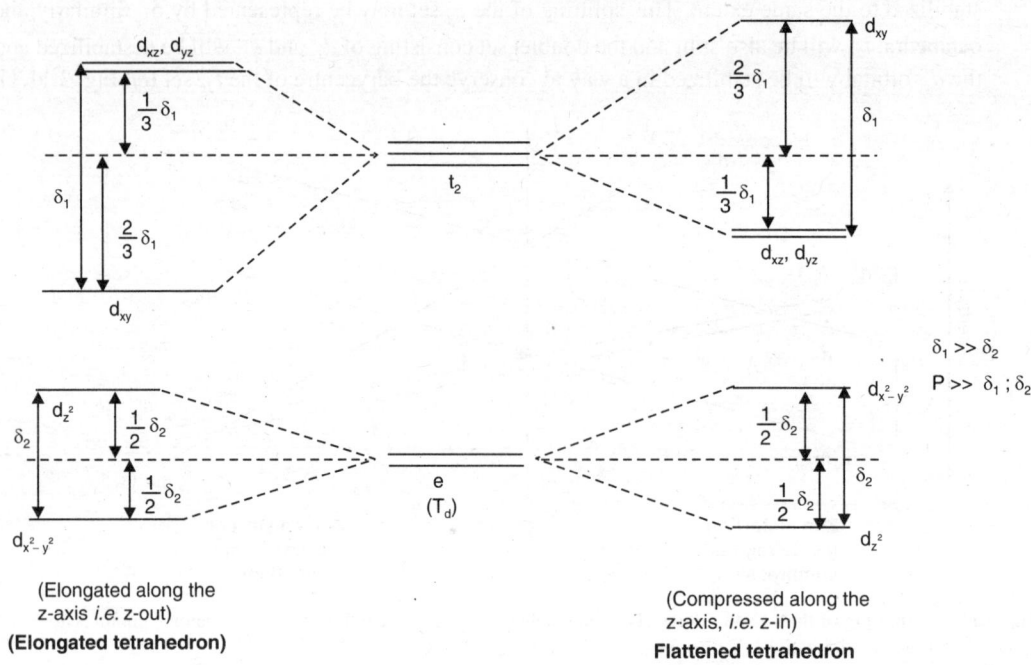

Fig. 3.11.2.1 Splitting of the e and t_2 levels of a regular tetrahedron (T_d) during distortion.

- Compression along the z-axis will flatten the tetrahedron and it will tend to place the ligands in the xy-plane. Evidently, *the limiting situation of this flattening will lead to the square-planar geometry* (*i.e.* $\mathbf{T_d} \longrightarrow \mathbf{D_{2d}} \longrightarrow \mathbf{D_{4h}}$; *see* Fig. 2.11.2.2). Thus, compression along the z-axis will destabilize (relatively) the $d_{x^2-y^2}$ and d_{xy} orbitals more and stabilise the d_{z^2}, d_{xz}, d_{yz} orbitals having the z-component (*cf.* the d-orbital splitting in a square planar geometry; Fig. 3.5.6.2).

- Elongation along the z-axis will destabilise the orbitals d_{z^2}, d_{xz}, d_{yz} having the z-component and stabilize the d_{xy} and $d_{x^2-y^2}$ orbitals more. To maintain the barycentre rule, the orbitals will be placed accordingly.

3.11.3 Jahn-Teller Theorem and Jahn-Teller Distortion in the Octahedral and Tetrahedral Complexes

The generalized statement of the theorem was made by Jahn and Teller (in 1937) as:

- *Any nonlinear molecule in a degenerate electronic state will undergo distortion to remove the degeneracy and to lower the energy.*

● *According to theorem, if there is a centre of symmetry in the undistorted system, then the centre of symmetry will also be maintained after distortion.*

The theorem tells the condition of distortion but it cannot tell the magnitude or direction of the distortion.

The **driving force** behind the J.T. distortion is: **removal of degeneracy of the energy levels will give an additional stabilization.** Thus the J.T. distortion will occur if only the splitted energy levels can yield an additional stabilization through the distortion.

(A) J.T. distortion in the octahedral systems

Now let us apply the J.T. theorem **in the octahedral systems** where the degenerate d-orbitals are splitted into the t_{2g} (3-fold degenerate) and e_g (2-fold degenerate) sets to earn the cfse. Further removal of the degeneracy of the t_{2g} and e_g sets can yield an **additional stabilization in some cases.**

In the octahedral systems, the d^3 ($^4A_{2g}$), high-spin d^5 ($^6A_{1g}$) and low-spin d^6 ($^1A_{1g}$) and d^8 ($^3A_{2g}$) configurations possess the **nondegenerate ground states** while the other configurations, *i.e.* d^1, d^2, d^4 (both h.s. and l.s.), d^5 (l.s.), d^6 (h.s.), d^7 (both h.s. and l.s.) and d^9 are characterized by the **electronic degeneracy** in the ground states. **These configurations experiencing the electronic degeneracy at the ground state are subjected to the J.T. distortion.** The *electronic degeneracy* will prevail when the assignment of electrons in the degenerate orbitals (*i.e.* t_{2g} or e_g) can be done in more than one way. This is illustrated below.

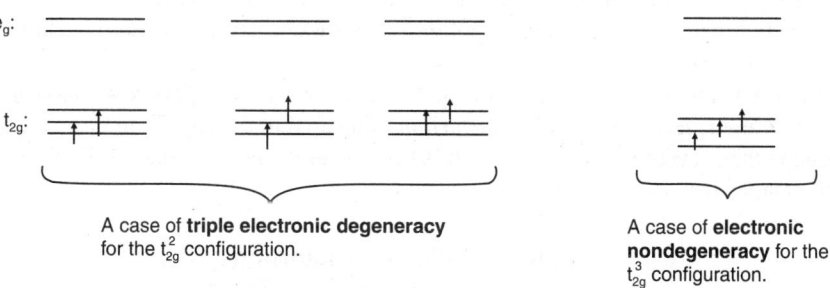

A case of **triple electronic degeneracy** for the t_{2g}^2 configuration.

A case of **electronic nondegeneracy** for the t_{2g}^3 configuration.

According to the J.T. theorem when applied to an octahedral system, it gives the following condition the under which the J.T. distortion will occur.

*If the t_{2g} and/or e_g set of an octahedral system are/is unsymmetrically filled in then the the system will experience the **J.T. distortion.*** Thus we can write:

(i) t_{2g}^n $(n = 1, 2, 4, 5)$; e_g^n $(n = 1, 3)$: J.T. distortion will occur.

(ii) t_{2g}^n $(n = 3, 6)$; e_g^n $(n = 2, 4)$: No J.T. distortion.

(B) J.T. distortion in the tetrahedral systems

Similarly, we can conclude for the **J.T. distortion in tetrahedral systems:**

(i) e^n $(n = 1, 3)$, t_2^n $(n = 1, 2, 4, 5)$: J.T. distortion

(ii) e^n $(n = 2, 4)$, t_2^n $(n = 3, 6)$: No J.T. distortion

Table 3.11.3.1 Jahn-Teller distortion in ground states.

d^n	Octahedral Complex		Tetrahedral Complex
	high-spin	Low-spin	
d^1	$t_{2g}^1 e_g^0$, (yes)	–	$e^1 t_2^0$, (yes)
d^2	$t_{2g}^2 e_g^0$, (yes)	–	$e^2 t_2^0$, (no)
d^3	$t_{2g}^3 e_g^0$, (no)	–	$e^2 t_2^1$, (yes)
d^4	$t_{2g}^3 e_g^1$, (yes)	$t_{2g}^4 e_g^0$, (yes)	$e^2 t_2^2$, (yes)
d^5	$t_{2g}^3 e_g^2$, (no)	$t_{2g}^5 e_g^0$, (yes)	$e^2 t_2^3$, (no)
d^6	$t_{2g}^4 e_g^2$, (yes)	$t_{2g}^6 e_g^0$, (no)	$e^3 t_2^3$, (yes)
d^7	$t_{2g}^5 e_g^2$, (yes)	$t_{2g}^6 e_g^1$, (yes)	$e^4 t_2^3$, (no)
d^8	$t_{2g}^6 e_g^2$, (no)	–	$e^4 t_2^4$, (yes)
d^9	$t_{2g}^6 e_g^3$, (yes)	–	$e^4 t_2^5$, (yes)

Note:

- **J.T. distortion in O_h vs. T_d:** The J.T. distortion is of *higher magnitude in the octahedral complexes* compared to that in the tetrahedral complexes.
- **Origin of J.T. distortion – e_g vs. t_{2g} levels:** In the octahedral system, if the origin of J.T. distortion lies in the e_g level, then it is more pronounced compared to the cases where the origin of J.T. distortion lies in the t_{2g} level. **The e_g-orbitals are more sensitive to the approaching ligands** (*cf.* Fig. 3.11.1.1).
- **Origin of J.T. distortion – e vs. t_2 levels:** In tetrahedral systems, *the unsymmetrical filling of electron in the t_2 level causes more distortion* compared to the case of unsymmetrical filling of electron in the e level (*cf.* Fig. 3.11.2.1). It is because of the fact that the t_2-orbitals experience the ligand charges better (*cf.* Fig. 3.5.5.1).

Ni(II) vs. Cu(II): J.T. Distortion

Ni(II) (d^8) can form the perfectly octahedral complexes, but because of the inherent J.T. distortion, Cu(II) (d^9) can never form the perfectly octahedral complexes. Both Ni(II) and Cu(II) will form the distorted tetrahedral complexes.

(C) Calculation of additional stabilisation due to the J.T. distortion in the octahedral complexes

It has been already pointed out that the d^1, d^2, d^4 (both h.s and l.s.), d^5 (l.s.), d^6 (h.s.), d^7 (both h.s. and l.s.) and d^9 configurations *cannot form the perfectly octahedral complexes even in the presence of six identical ligands because of the inherent J.T. distortion.* **The distortion gives an additional stabilization with respect to that of the perfectly octahedral complex.** This additional stabilization can be calculated in terms of δ_1 and δ_2 (*cf.* Fig. 3.11.1.1, $\delta_1 \rangle \delta_2$; $P \rangle\rangle \delta_1, \delta_2$). These are given in Table 3.11.3.2.

Calculation of additional stabilization for the $t_{2g}^2 e_g^0$ configuration through the J.T. distortion is illustrated below (*cf.* Fig. 3.11.1.1)

$$t_{2g}^2 e_g^0 \xrightarrow{z-out} (d_{xz}, d_{yz})^2 (d_{xy})^0;$$

Table 3.11.3.2 Additional stabilization earned through the J.T. distortion in the octahedral systems.

Configuration and ground-state term for undistorted O_h	Additional stabilization with respect to that of O_h due to the J.T. distortion (cf. Fig. 3.11.1.1)	
	z-out	z-in
d^1 i.e. $t_{2g}^1 e_g^0$, $(^2T_{2g})$	$\frac{1}{3}\delta_2$	$\frac{2}{3}\delta_2$
d^2 i.e. $t_{2g}^2 e_g^0$, $(^3T_{1g})$	$2 \times \frac{1}{3}\delta_2 = \frac{2}{3}\delta_2$	$\frac{2}{3}\delta_2 - \frac{1}{3}\delta_2 = \frac{1}{3}\delta_2$
d^4 i.e. $t_{2g}^3 e_g^1$, (^5E_g), h.s.	$\frac{1}{2}\delta_1$	$\frac{1}{2}\delta_1$
d^4 i.e. $t_{2g}^4 e_g^0$, $(^3T_{1g})$, l.s.	$3 \times \frac{1}{3}\delta_2 - \frac{2}{3}\delta_2 = \frac{1}{3}\delta_2$	$2 \times \frac{2}{3}\delta_2 - 2 \times \frac{1}{3}\delta_2 = \frac{2}{3}\delta_2$
d^5 i.e. $t_{2g}^5 e_g^0$, $(^2T_{2g})$, l.s.	$4 \times \frac{1}{3}\delta_2 - \frac{2}{3}\delta_2 = \frac{2}{3}\delta_2$	$2 \times \frac{2}{3}\delta_2 - 3 \times \frac{1}{3}\delta_2 = \frac{1}{3}\delta_2$
d^6 i.e. $t_{2g}^4 e_g^2$, $(^5T_{2g})$, h.s.	$\frac{1}{3}\delta_2$	$\frac{2}{3}\delta_2$
d^7 i.e. $t_{2g}^5 e_g^2$, $(^4T_{1g})$, h.s.	$\frac{2}{3}\delta_2$	$\frac{1}{3}\delta_2$
d^7 i.e. $t_{2g}^6 e_g^1$, (^2E_g), l.s.	$\frac{1}{2}\delta_1$	$\frac{1}{2}\delta_1$
d^9 i.e. $t_{2g}^6 e_g^3$, (^2E_g)	$2 \times \frac{1}{2}\delta_1 - \frac{1}{2}\delta_1 = \frac{1}{2}\delta_1$	$2 \times \frac{1}{2}\delta_1 - \frac{1}{2}\delta_1 = \frac{1}{2}\delta_1$

It leads to an additional stabilization with respect to that of the perfectly octahedral system:

$$2 \times \left(\frac{1}{3}\delta_2\right) = \frac{2}{3}\delta_2$$

$$t_{2g}^2 e_g^0 \xrightarrow{\text{z-in}} (d_{xy})^1 (d_{xy}, d_{yz})^1 ; \quad P \gg \delta_2$$

i.e. additional stabilization earned is given by:

$$1 \times \left(\frac{2}{3}\delta_2\right) - \frac{1}{3}\delta_2 = \frac{1}{3}\delta_2 .$$

It is evident that for the d^2 configuration, the z-out (i.e. tetragonal elongation) distortion earns the more stabilization than that for the z-in (i.e. tetragonal compression) distortion. Similarly, the predictions (cf. Table 3.11.3.2) are:

z-in: d^1, d^4 (l.s.), d^6 (h.s.)

z-out: d^2, d^5 (l.s.), d^7 (h.s.)

z-in and z-out (equally probable): d^4 (h.s.), d^7 (l.s.), d^9

However, these predictions are not always experimentally verified. For example, for the Cu(II)-complexes (d^9-system), it is evident that both the z-out and z-in distortion give the same amount of additional stabilization (by $\frac{1}{2}\delta_1$) (i.e. both z-out and z-in equally probable) **but in reality, most of the Cu(II) complexes experience the z-out distortion.**

For the e_g^1 and e_g^3 configurations, both the z-out and z-in distortions will give the same amount of additional cfse, but the **preference for the z-out distortion** has been explained by some authors as follows:

The $d_{x^2-y^2}$ orbital experiences directly the repulsive action from four ligands while the d_{z^2} orbital experiences the similar repulsive action from two ligands. Thus placement of electrons in the $d_{x^2-y^2}$ orbital causes a more repulsion. This is why, to minimise the repulsion with the ligands, the e_g-electrons are preferably housed in the d_{z^2} orbital giving rise to the z-out distortion.

However, the above argument can be called in question because in terms of the octahedral crystal field, both the $d_{x^2-y^2}$ and d_{z^2} orbitals are equivalent. In solid crystals, the **crystal packing forces** may favour the z-out distortion compared to the z-in distortion.

● It can be realized *why the symmetrically filled t_{2g}^3, t_{2g}^6, e_g^2, e_g^4 sets do not experience the J.T. distortion.* It is illustrated for the t_{2g}^3 system.

$$\left(d_{xy}\right)^1 \left(d_{xz}, d_{yz}\right)^2 \xleftarrow{\ z\text{-in}\ } t_{2g}^3 \xrightarrow{\ z\text{-out}\ } \left(d_{xz}, d_{yz}\right)^2 \left(d_{xy}\right)^1$$

Additional $1 \times \dfrac{2}{3}\delta_2 - 2 \times \dfrac{1}{3}\delta_2$ $\left(^4A_{2g}\right)$ $2 \times \dfrac{1}{3}\delta_2 - \dfrac{2}{3}\delta_2$
stabilisation
energy: $= 0$ $= 0$

Similarly, it can be shown that the splitting of the t_{2g}^6, e_g^2 and e_g^4 configurations cannot earn any additional stabilization energy through the J.T. distortion.

(D) J.T. distortion in the tetrahedral complexes and the gain of additional cfse

The tetrahedral complexes of the d^3, d^4, d^8 and d^9 configurations experience the J.T. distortion due to the unsymmetrical filling of electrons in the t_2-level.

Theoretically, the e^1 and e^3 electronic configurations can cause the distortions in the tetrahedral complexes according to the J.T. theorem. But this distortion is of less significance compared to that originated from the unsymmetrical filling in the t_2 orbitals. *Here it is worth mentioning that overall, the J.T. distortion in the tetrahedral complexes is less important than that in the octahedral complexes.*

Table 3.11.3.3 Gain of additional cfse due to the J.T. distortion in the tetrahedral complexes.

Ground state electronic configuration in the tetrahedral complex	*Additional cfse with respect to that of the tetrahedron (cf. Fig. 3.11.2.1)*	
	z-out (i.e. elongated)	*z-in (i.e. flattened)*
d^3 i.e. $e^2 t_2^1$:	$\dfrac{2}{3}\delta_1$	$\dfrac{1}{3}\delta_1$
d^4 i.e. $e^2 t_2^2$:	$\dfrac{2}{3}\delta_1 - \dfrac{1}{3}\delta_1 = \dfrac{1}{3}\delta_1$	$2 \times \dfrac{1}{3}\delta_1 = \dfrac{2}{3}\delta_1$
d^8 i.e. $e^4 t_2^4$:	$2 \times \dfrac{2}{3}\delta_1 - 2 \times \dfrac{1}{3}\delta_1 = \dfrac{2}{3}\delta_1$	$3 \times \dfrac{1}{3}\delta_1 - \dfrac{2}{3}\delta_1 = \dfrac{1}{3}\delta_1$
d^9 i.e. $e^4 t_2^5$:	$2 \times \dfrac{2}{3}\delta_1 - 3 \times \dfrac{1}{3}\delta_1 = \dfrac{1}{3}\delta_1$	$4 \times \dfrac{1}{3}\delta_1 - \dfrac{2}{3}\delta_1 = \dfrac{2}{3}\delta_1$

It is evident that the d^3 and d^8 systems gain more stabilization in the z-out (*i.e.* elongation) distortion while the d^4 and d^9 systems gain more stabilization in the z-in (*i.e.* flattening) distortion. These predictions have been experimentally verified.

[CuCl$_4$]$^{2-}$ (d^9 system): Flattened tetrahedral structure (predicted and experimentally supported).

- The distortion and gain of additional cfse $\left(=\dfrac{1}{2}\delta_2\right)$ due to the unsymmetric filling of electrons in the e-level *i.e.* e^1 and e^3, are not much important. *In other words, if the origin of J.T. distortion lies in the e-set of the tetrahedral systems, then the distortion is not so important.*

Note: ● The J.T. theorem is only applicable for the octahedral and tetrahedral complexes where all the ligands are identical. **Thus the J.T. Theorem is applicable only for the [ML$_6$] and [ML$_4$] type complexes.** Thus, the theorem is not applicable for the mixed ligand complexes. In such mixed ligand complexes, because of the difference in the ligand field strength in different directions, the degeneracy of the orbitals is already lifted.

- **J.T. distortion can occur in both the ground and excited states.** The effect of J.T. distortion in the excited state is well understood in the electronic spectra of [Ti(OH$_2$)$_6$]$^{3+}$, [TiCl$_6$]$^{3-}$ of the d^1 system. These will be discussed later.
- For the **octahedral complexes of the d^8 configuration** ($t_{2g}^6 e_g^2$), the J.T. distortion is not expected but it can occur only under the condition δ_1 (splitting energy in e_g level) $> P$ (pairing energy) (*cf.* Sec. 3.12)
- For the **octahedral complexes**, if the **origin of J.T. distortion lies in the e$_g$ set** (*i.e.* e_g set is unsymmetrically filled in as in e_g^1, e_g^3) for which the orbital lobes are directly projected towards the ligands, the distortion is more severe than the cases where the origin of J.T. distortion lies in the t_{2g} set. It can be also understood in terms of:

$$\delta_1 \text{ (splitting energy of } e_g \text{ set)} \rangle\rangle \delta_2 \text{ (spliiting energy of } t_{2g} \text{ set)}.$$

- For the **tetrahedral complexes**, the J.T. distortion is significant, if the **origin lies in the t$_2$-set.**

3.11.4 Examples of the Octahedral and Tetrahedral Complexes Experiencing the J.T. Distortion: Effect of the J.T. Distortion on Different Properties (*cf.* Sec. 7.14 for spectral properties)

(A) Octahedral Complexes:

d^1, Octahedral: [Ti(OH$_2$)$_6$]$^{3+}$, [TiCl$_6$]$^{3-}$ (in TiCl$_3$ crystal))

d^4 (high-spin), octahedral: CrF$_2$ (solid), MnF$_3$ (solid), KCrF$_3$ (solid), K$_2$[MnF$_5$(OH$_2$)], Mn$_2$O$_3$ (solid)

d^7 (low-spin), Octahedral: NaNiO$_2$ (crystal), Co(II) complexes.

d^9, octahedral: [CuCl$_2$] (solid), [CuF$_2$] (solid), Na$_2$[CuF$_4$] (solid), [Cu(OH$_2$)$_6$]$^{2+}$, etc.

d^6 (high spin), octahedral: [Fe(OH$_2$)$_6$]$^{2+}$, [CoF$_6$]$^{3-}$

- **(a) Ti(III) (d^1-system)** (*cf.* Sec. 7.14.1 for details): [Ti(OH$_2$)$_6$]$^{3+}$ (t_{2g}^1) experiences the J.T. distortion in the ground state but the distortion due to splitting of the t_{2g} level is insignificant compared to the distortion caused in the excited state ($t_{2g}^0 e_g^1$) due to splitting of the e_g-level. **In other words, the excited state is more sensitive to the J.T. distortion.**

In the ground state of d^1-system, the *z-in distortion* gives more stabilization compared to the z-out distortion (*cf.* Table 3.11.3.2). This effect is reflected in the electronic absorption spectrum of [Ti(OH$_2$)$_6$]$^{3+}$. In the absence of distortion, we can expect one peak due to the transition.

$$t_{2g}^1 \xrightarrow{\ h\nu\ } e_g^1$$

In the presence of z-in distortion, the possible transitions are (*cf.* Fig. 3.11.4.1):

$$d_{xy} \xrightarrow{\nu_1} (d_{xz}, d_{yz}); \ d_{xy} \xrightarrow{\nu} d_{x^2-y^2}; \ d_{xy} \xrightarrow{\nu_2} d_{z^2}$$

The transition, $d_{xy} \xrightarrow{\nu_1} (d_{xz}, d_{yz})$ will not be observed in the visible spectrum because of the very small energy difference. The transition will occur in the i.r. region. **The other two peaks will overlap to produce a broad peak which is due to the superposition of two peaks** (at ~ 21,000 cm^{-1} and ~ 19,500 cm^{-1}). The peak at ~ 19,500 cm^{-1} appears **as a shoulder** in the broad peak. Average of these two peaks at about 20,300 cm^{-1} represents the peak position of the hypothetical ideal octahedral $[Ti(OH_2)_6]^{3+}$ complex.

$$d_{xy} \xrightarrow[(\equiv 10Dq_0)]{\nu} d_{x^2-y^2} \ (\sim 19,500 \text{ cm}^{-1}); \ d_{xy} \xrightarrow{\nu_2} d_{z^2} \ (\sim 21,000 \text{ cm}^{-1}).$$

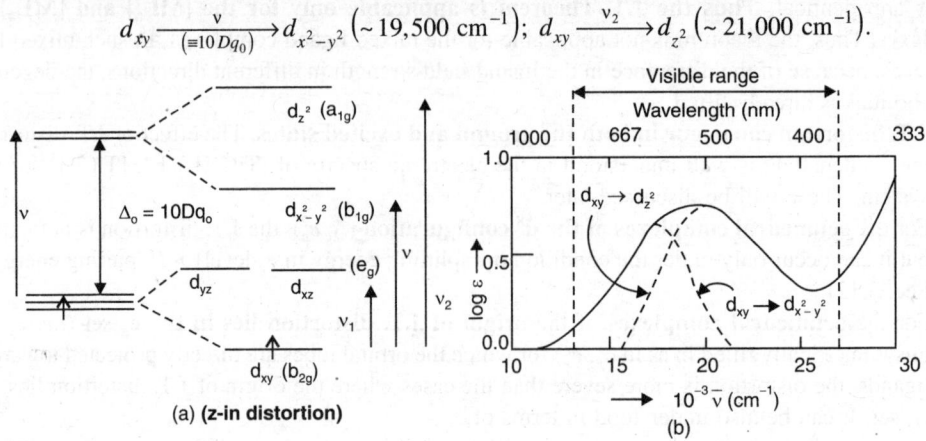

Fig. 3.11.4.1 (a) Splitting of the d-orbitals in $[Ti(OH_2)_6]^{3+}$, $[TiCl_6]^{3-}$ (b) The absorption spectrum of $[Ti(OH_2)_6]^{3+}$.

The electronic absorption spectrum of $[TiCl_6]^{3-}$ in $M_3 [TiCl_6]$ ($M_3 = Rb_3$, Cs_2K, Rb_2Na), shows the peak broadening phenomenon due to overlapping of two peaks υ and υ_2. These are the evidences in favour of the J.T. distortion (z-in).

● **(b) CrF$_2$ (solid), KCrF$_3$ (solid), MnF$_3$ (solid), K$_2$[MnF$_5$(OH$_2$)] — high-spin d^4-systems** (*cf.* Sec. 7.14.7 for details): In all these cases (d^4, high-spin), the two M – F *trans*– axial bonds are relatively longer compared to the remaining equatorial M–F bonds. This happens due to the z-**out distortion** of the d^4 system ($t_{2g}^3 e_g^1$) where the origin of the J.T. distortion lies at the e_g-level.

● **(c) Mn$_3$O$_4$ ($\equiv$ MnO.Mn$_2$O$_3$) crystal:** It also shows the z-out J.T. distortion due to Mn$_2$O$_3$ having the Mn(III) (d^4) centres in high-spin state.

For the d^4 (high-spin) system, the ground state electron distribution is:

$$\left(d_{xz}, d_{yz}\right)^2 < d_{xy}^1 < d_{z^2}^1 < d_{x^2-y^2}^0$$

The $d_{x^2-y^2}$ orbital is unoccupied and consequently, it does not experience the electrostatic repulsion from the approaching 4 equatorial ligand while the partially filled in d_{z^2} orbital experiences more electrostatic repulsion from the approaching two axial ligands. **It makes the four equatorial bonds shorter and two axial bonds longer** (*i.e.* z-out distortion).

● **(d) Spectral features of [Mn(dmso)$_6$]$^{3+}$:** Here dmso (dimethyl sulfoxide) coordinates through the O-donor sites. The spectral features are:

three bands (instead of one band as expected for the perfectly octahedral system) in the electronic absorption spectrum.

This can be explained by considering the **z-out distortion.**

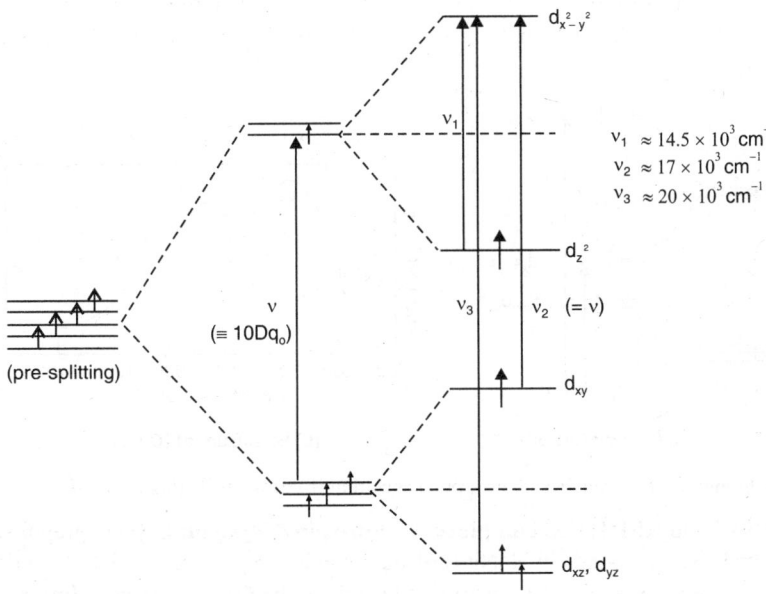

$$v_1 \approx 14.5 \times 10^3 \text{ cm}^{-1}$$
$$v_2 \approx 17 \times 10^3 \text{ cm}^{-1}$$
$$v_3 \approx 20 \times 10^3 \text{ cm}^{-1}$$

Fig. 3.11.4.2 Electronic transitions in $[Mn(dmso)_6]^{3+}$.

The characteristic features of the ir-spectra are:

$$\underbrace{v_{S-O} = 915 \text{ cm}^{-1}, \ v_{S-O} = 960 \text{ cm}^{-1},}_{\text{Complex } [Mn\,(dmso)_6]^{3+}} \qquad \underbrace{v_{S-O} = 1055 \text{ cm}^{-1}}_{\text{Free dmso}}$$

The electron distribution in Mn(III) is:

$$\left(d_{xz}, d_{yz}\right)^2 < d^1_{xy} < d^1_{z^2} < d^0_{x^2-y^2}$$

The vacant $d_{x^2-y^2}$ orbital experiences the minimum repulsion from the approaching four equatorial ligands while the d_{z^2} orbital experiences the relatively more repulsion from the approaching axial ligands. It makes the equatorial bonds stronger than the axial bonds *i.e.* **the Mn—O (axial) bond is weaker than the Mn—O (equatorial) bond.** It makes the 'S—O' bonds coordinated to the metal centre (through the O-end) at the equatorial plane weaker (*i.e.* corresponding to 915 cm^{-1}) and the relatively stronger S—O bonds coordinated to the metal centre in the axial directions correspond to $v_{S-O} = 960$ cm^{-1}. It is also supported by the fact that the intensity of the 915 cm^{-1} band is almost double the intensity of the band at 960 cm^{-1} (*cf.* the ratio of the number of equatorial and axial Mn—O bonds is 4:2 *i.e.* 2:1).

- **(e) Electronic spectra of $[CoF_6]^{3-}$, $[Fe(OH_2)_6]^{2+}$ – high-spin d^6-system:** In the ground state ($t^4_{2g} e^2_g$), the weak J.T. distortion arises from the t^4_{2g} level which is unsymmetrically filled in. In fact, the bond length parameters are not so sensitive to this weak J.T. distortion. *However, the excited state ($t^3_{2g} e^3_g$) is more sensitive to the J.T. distortion.* A similar situation has been already illustrated for the Ti(III) (d^1)-complexes. The splitting of e_g level of the first excited state causes the peak broadening phenomenon in the electronic absorption spectra of $[Fe(OH_2)_6]^{2+}$ and $[CoF_6]^{3-}$.

In the perfectly octahedral $[CoF_6]^{3-}$ (high-spin) complex, there should be a single peak ($^5T_{2g} \rightarrow {}^5E_g$) but due to the z-in J.T. distortion in the excited state, the peak splits into two components at ~ 700 nm and ~ 900 nm (*see* Sec. 7.14.8, v_1 lies in the *ir*-range). This splitting supports the splitting of the e_g level into two energy levels.

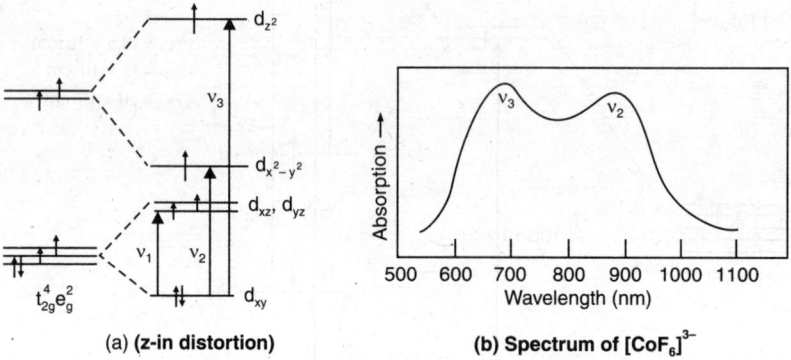

(a) **(z-in distortion)**   (b) **Spectrum of $[CoF_6]^{3-}$**

Fig. 3.11.4.3 Splitting (a) of d-orbitals and absorption spectrum (b) of $[CoF_6]^{3-}$ (high-spin, d^6) (*See* Sec. 7.14.8).

● **(f) Co(II) (l.s.) and Ni(III) (l.s) complexes — low-spin d^7 system:** Crystalographic data indicate two elongated 'Ni—O' bonds in $NaNiO_2$. If the ligands are very strong field ligands, then the e_g level will experience a severe z-out distortion leading to the *four or five coordinate intermediate*. The *nonexistence of* $[Co(CN)_6]^{4-}$ can be explained by considering this argument.

$$\left[Co(CN)_6\right]^{4-} \xrightarrow[\substack{+solvent \\ (weaker\ field\ ligand)}]{-CN^-} \left[Co^{II}(CN)_5(solvent)\right]^{3-}$$

$$\substack{(Hypothetical) \\ t_{2g}^6 e_g^1}$$

The strong field ligand CN^- generates *a severe z-out distortion* through the splitting of the e_g orbitals of the hypothetical $[Co(CN)_6]^{4-}$ complex. This practically leads to the loss of an axial CN^- ligand and this vacant site is very often occupied by a solvent molecule (generally a weak field ligand). This expulsion of one axial CN^- ligand and entry of a weak field ligand (generally solvent) reduce the splitting of the e_g level to such an extent to retain the other axial CN^- ligand. By using the relatively weaker field ligands (but sufficiently strong to allow the formation of low-spin complexes), the six-coordinate Co(II) complexes can be isolated. *In such six-coordinate complexes, the z-out distortion is not so severe to allow the expulsion of the axial ligand but the axial bonds are relatively elongated.* The examples are:

$[Co(NO_2)_6]^{4-}$ (z-out distortion; *cf.* NO_2^- is a weaker field ligand compared to CN^-).

$[Co(py)_2(salen)]$, H_2salen = bis(salicylidene)ethylenediamine.

● **(g) Cu(II)-complexes (d^9-system)** (*cf.* Sec. 7.14.2 for details): A large number of Cu(II) complexes are known to experience the J.T. distortion because of the unsymmetrical filling of electrons in the e_g-level, *i.e.* e_g^3. Both the z-out and z-in distortions will bring about the same amount of additional cfse but in reality most of the Cu(II) complexes experience the z-out distortion. The probable explanation for the **preference of z-out** distortion has been discussed earlier in Sec. 3.11.3. The electron distribution is:

$$\left(d_{xz}, d_{yz}\right)^4 < d_{xy}^2 < d_{z^2}^2 < d_{x^2-y^2}^1$$

The singly occupied $d_{x^2-y^2}$ orbital experiences relatively a less repulsion from the four equatorial ligands while the doubly occupied d_{z^2} orbital experiences a more repulsion from the axial ligands. It makes the equatorial bonds shorter and the axial bonds longer.

Table 3.11.4.1 Metal-ligand bond length (pm) in some compounds of Cu(II)-compounds.

Compound	4 short Cu—L bonds (pm)	2 long Cu—L bonds (pm)
$CuCl_2$	230 (4 Cu—Cl bonds)	295 (2 Cu—Cl bonds)
$CuCl_2.2H_2O$	229 (2 Cu—Cl bonds) ⎤ 196 (2 Cu—O bonds) ⎦	294 (2 Cu—Cl bonds)
CuF_2	193 (4 Cu—F bonds)	227 (2 Cu—F bonds)
$CuF_2.2H_2O$	190 (2 Cu—F bonds) ⎤ 194 (2 Cu—O bonds) ⎦	247 (2 Cu—F bonds)
$CuCl_2.2py$	202 (2 Cu—N bonds) ⎤ 228 (2 Cu—Cl bonds) ⎦	305 (2 Cu—Cl bonds)
Na_2CuF_4	191 (4 Cu—F bonds)	237 (2 Cu—F bonds)
$[Cu(NH_3)_6]^{2+}$ (in liquid NH_3)	207 (4 Cu—N bonds)	262 (2 Cu—N bonds)
$[Cu(OH_2)_6]^{2+}$	~195 (4 Cu—O bonds)	238 (2 Cu—O bonds)

The crystal structures of $CuSO_4.5H_2O$ and $CuCl_2.2H_2O$ strongly indicate the z-out distortion (Fig. 3.11.4.4)

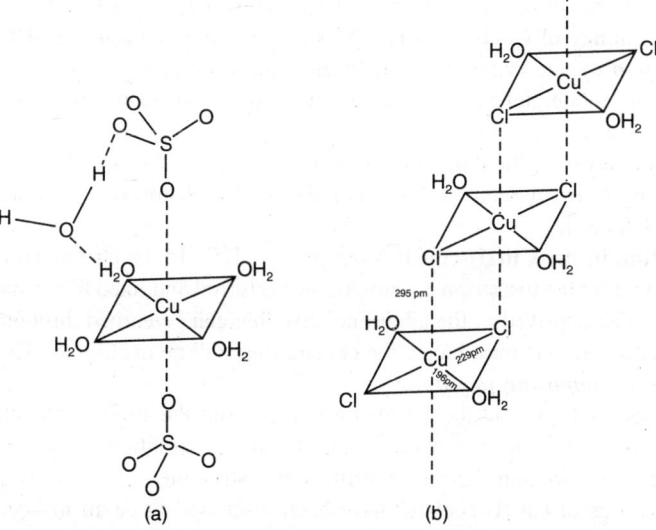

(a) (b)

Fig. 3.11.4.4 Crystal structure of $CuSO_4.5H_2O$ (a) and $CuCl_2.2H_2O$ (b).

The existence of tetragonally distorted structure of Cu(II)-compounds is a strong evidence in favour of CFT. In fact, VBT can explain either the perfectly square planar (dsp^2 or sp^2d) or perfectly octahedral (sp^3d^2) Cu(II) complexes through hybridization. *But no hybridization scheme is available to explain the tetragonally distorted structures.*

Chemistry of the Cu(II)-complexes is largely controlled by the J.T. distortion.

● **(h) Non-existence of $[Cu(NH_3)_6]^{2+}$ in aqueous system:** In the d^9-system, origin of the J.T. distortion lies at the e_g-level and NH_3 is a fairly strong field ligand to produce the splitting of e_g-orbitals significantly. This significant z-out distortion makes the axial 'Cu—N' bonds of $[Cu(NH_3)_6]^{2+}$ sufficiently weak to be replaced by the solvent. Here it is worth mentioning that in $[Cu(OH_2)_6]^{2+}$, the extent of splitting of the e_g-orbitals is relatively less compared to the case of $[Cu(NH_3)_6]^{2+}$ because H_2O is a much weaker field ligand.

       $[Cu(NH_3)_6]^{2+}$: axial Cu—N = 262 pm 〉 equatorial Cu—N = 207 pm

                   (difference by 55 pm)

       $[Cu(OH_2)_6]^{2+}$: axial Cu—O = 238 pm 〉 equatorial Cu—O = 196 pm

                   (difference by 42 pm)

The severe bond weakening in the axial direction of $[Cu(NH_3)_6]^{2+}$ leads to the **loss of the axial NH_3 ligands** in aqueous solvents.

$$\left[Cu(NH_3)_6\right]^{2+} \xrightarrow[-2NH_3]{H_2O} \left[Cu(NH_3)_4(OH_2)_2\right]^{2+}$$

$[Cu(NH_3)_4(OH_2)_2]^{2+}$ is actually square planar (very often represented as **$[Cu(NH_3)_4]^{2+}$**) because the two *trans*–axial aqua ligands are weakly held.

$$cf. \left[Co(CN)_6\right]^{4-} \xrightarrow[-CN^-]{H_2O} \left[Co(CN)_5(OH_2)\right]^{3-}$$

● **(i) Nonexistence of $[CuCl_6]^{4-}$:** The nonexistence of $[CuCl_6]^{4-}$ can also be explained in the same way. In fact, it produces $[CuCl_4]^{2-}$ having a flattened tetrahedral structure close to square planar. Here, besides the J.T. effect, the steric crowding caused by the **bulky and anionic Cl⁻ ligands** in $[CuCl_6]^{4-}$ plays a crucial role to convert $[CuCl_6]^{4-}$ to $[CuCl_4]^{2-}$ (flattened tetrahedral).

● **(j) Inherent tendency of Cu(II) to form the square planar complexes:** The *limiting situation of z-out distortion* of the octahedral complexes is the formation of square planar complexes. In fact, even for the moderately strong field ligands, the z-out distortion is so severe that it can produce the square planar complexes.

● **(k) Stability constants of the Cu(II)-complexes:** To explain the position of Cu(II) in the Irving-Williams stability order, it needs the consideration of J.T. distortion effect. This aspect has been discussed in Chapter 4.

● **(l) J.T. distortion in the Cu(II)-chelates** (*cf.* Sec. 3.11.5): In the chelate complexes, the chelate ring will try to restrict the distortion because of the preferred and fixed bite distance. Consequently, the complex will be deprived of the additional cfse that can be earned through the J.T. distortion. But, if the J.T. distortion is introduced, the chelate ring will be in strain to destabilize the system. Thus there are *two opposing factors*:

    (i) Chelate effect will prevent the J.T. distortion to avoid the strain in the chelate ring.

    (ii) J.T. distortion will try to distort the system to earn an additional cfse.

Because of these two opposing factors, stabilities and structures of the Cu(II)-chelates are quite interesting. The stabilities of Cu(II)-chelates have been discussed in detail in Sec. 4.5.2 with special reference to the Cu(II)-en system (*cf.* Fig. 4.5.2.11), Cu(nta)–amino acids systems (*i.e.* ternary complexes).

However, both the **distorted** and **undistorted Cu(II)-chelates** are known (*cf.* Sec. 3.11.5).

Distorted Chelates: $[Cu(en)_3]^{2+}$ (two axial Cu—N bonds are relatively longer), $[Cu(bpy)(hfa)_2]$ where bpy occupies two *cis*– equatorial positions and hfa (= hexafluoroacetylacetonate) spans the *cis*–axial-equatorial positions (axial Cu—O = 229.6 pm 〉 equatorial Cu—O = 196.7 pm) (*cf.* Fig. 3.11.4.5).

These unequal bond lengths partially reduces the stability of *the acetylacetonate chelate ring experiencing the resonance* (*cf.* Sec. 4.5.2).

Here the additional cfse earned though the J.T. distortion is the driving force to distort the structure.

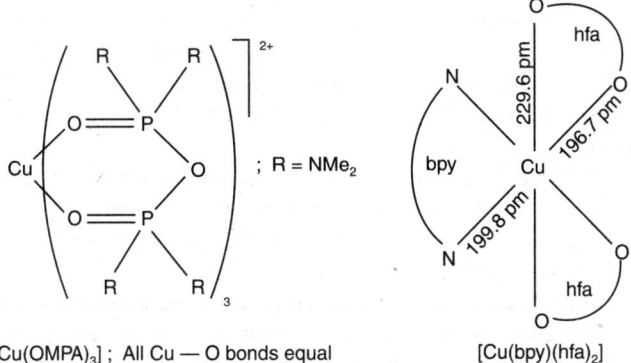

[Cu(OMPA)$_3$] ; All Cu — O bonds equal   [Cu(bpy)(hfa)$_2$]

Fig. 3.11.4.5 Structural parameters of [Cu(OMPA)$_3$]$^{2+}$ (undistorted chelate) and [Cu(bpy)(hfa)$_2$] (*cf.* Sec. 4.5.2 and Fig. 4.5.2.15).

Undistorted Chelates: [Cu(OMPA)$_3$]$^{2+}$ where OMPA = octamethylpyrophosphoramide acting as a didentate chelating ligand (6 membered ring) using the P $=$ O linkages as the donor sites.

Here the additional cfse earned through the J.T. distortion is less than the destabilization caused by the strain introduced by the J.T. distortion. *This is why, the system avoids the J.T. distortion.*

- **(m) Lability of the Cu(II)-complexes:** The high lability of Cu(II) complexes is also a consequence (in part) of the J.T. distortion. This aspect has been discussed in Sec. 5.13 and Chapter 5.
- **(n) Spectral properties of the Cu(II)-complexes:** Splitting of the energy levels by the J.T. distortion shows a marked effect on the nature of spectra of the Cu(II)-complexes. These are discussed in Secs. 7.11-12, Chapter 7.

(B) Distorted tetrahedral complexes due to the J.T. distortion

In the tetrahedral complexes, the J.T. distortion originated from the t_2-level is quite important. It has been already shown that in terms of the additional cfse to be earned through the J.T. distortion, the d^3 ($e^2t_2^1$) and d^8 ($e^4t_2^4$) systems will prefer the **elongated tetrahedral structure** while the d^4 ($e^2t_2^2$) and d^9 ($e^4t_2^5$) systems will prefer the **flattened tetrahedral structure** (*cf.* Table 3.11.3.3).

- **[CuCl$_4$]$^{2-}$:** It adopts the **flattened tetrahedral structures** as expected for the d^9-system (*cf.* Sec. 2.11.6)
- **NiCr$_2$O$_4$:** In this spinel structure, Ni(II) (d^8) adopts the **elongated tetrahedral structure** as expected.

3.11.5 Symmetrical Structure in Spite of the J.T. Distortion: Static and Dynamic J.T. Distortion

- **Symmetrical structure imposed by the chelate rings:** It has been already mentioned (Sec. 3.11.4) that in some chelate complexes, *the chelate ring may prevent the J.T. distortion to avoid the strain in the chelate ring because the strain will destabilize the system.* [Cu^{II}(OMPA)$_3$]$^{2+}$ (OMPA = octamethylpyrophosphoramide) can avoid the J.T. distortion because the chelate rings oppose the distortion to avoid the strain.

● **Static and dynamic J.T. distortion in terms of the crystallographic and spectroscopic data:** In many cases, both the **crystallographic** and **spectral data** are in conformity with the J.T. distortion. Such examples have been discussed in Sec. 3.11.4. *These are the examples of static J.T. distortion. There are some examples where the crystallographic data at room temperature indicate the symmetrical structure (i.e. no distortion) but the spectroscopic data are in favour of the distortion.* In such cases, if the crystallographic data are recorded **at very low temperature,** the distortion is revealed. Such complexes are described to show the **dynamic J.T. distortion.**

● **Static J.T. distortion:** It is characterized by the different bond length parameters (*i.e.* crystallographic data) at room temperature.

● **Dynamic J.T. distortion in terms of the time scale of observation:** It is characterized by the *symmetric structure* (*i.e.* no distortion) revealed from the room temperature crystallographic data but the *distorted structure* revealed from the spectroscopic data and the crystallographic data at very low temperature.

● **Dynamic J.T. distortion in terms of rapid interconversion among the three possible forms of the tetragonal distortions through the tunneling and or crossing the energy barrier:** To explain the dynamic J.T. distortion, it has been suggested that the tetragonal distortion can occur along the x, y and z axes giving rise to three equivalent distorted structures with an equal probability. These three equivalent structures are mutually interconvertible through the **oscillation** (*i.e.* molecular vibration). If the activation energy for these mutual interconversions is not high then it can be obtained from the thermal energy ($k_B T$). If the oscillation is fast enough compared to the time scale of a particular physical method used to observe the structure, **then the method will simply display the time averaged symmetrical structure.** When the energy barrier is not too high, the interconversion can also occur through the **tunneling**. If the measurement is done at very low temperature at which the oscillation rate becomes sufficiently slow to freeze out a particular distorted structure then the frozen out distorted structure can be experimentally observed. In other words, at low temperature, the oscillation rate becomes significantly slower than the time scale of the physical method to observe the structure and then the physical method can observe the single distorted structure frozen-out.

Generally, the time scale of the spectroscopic method is much smaller than the time scale of the crystallographic method. This is why, when the characterization / measurement is done by the spectroscopic methods which are fast enough compared to the oscillation rate, the distorted structure is observed.

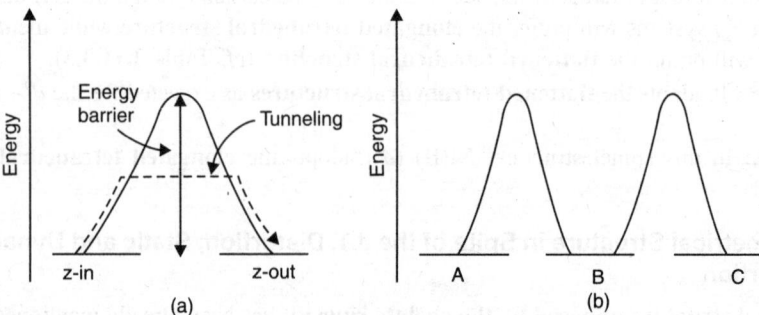

Fig. 3.11.5.1 Dynamic Jahn Teller distortion. (a) Interconversion between the z-in and z-out distorted forms of equal energy. (b) Interconversion among the three tetragonally distorted forms (A, B, C of equal energy and experiencing the distortions along the x, y and z-axes respectively).

(**Note:** In such cases, perfectly octahedral *i.e.* undistorted system is of little higher energy).

● **Dynamic J.T. distortion in terms of rapid interconversion between the z-in and z-out distortions:**
Dynamically, the J.T. Distortion can be explained in a different way. If the z-out and z-in distortions are of the same energy (as in the cases e_g^1, e_g^3) and the activation energy or the energy barrier for their interconversion is low (*i.e.* comparable to the thermal energy k_BT), then a rapid interconversion between the two distorted forms will occur through the tunneling or/and crossing the barrier.

z-out distorted form	$\rightleftharpoons$	z-in distorted form; (rapid)
(2 long axial bonds,		(2 contracted axial bonds,
4 smaller equatorial bonds)		4 longer equatorial bonds)

Under the condition, the time averaged structure will appear as an undistorted octahedron. If the activation energy for the interconversion is higher than the thermal energy k_BT, then a particular distorted structure is frozen out to display the static J.T. distortion.

Note: Fluxionality, *i.e.* stereochemical non-rigidity may be compared with the dynamic J.T. distortion.

Examples of dynamic J.T. distortion

● **[Mn(acac)$_3$]** ($t_{2g}^3 e_g^1$): The crystallographic data (at room temperature) indicate no distortion but the spectroscopic data indicate the distortion as expected from the J.T. theorem.

● **Salts containing the [Cu(NO$_2$)$_6$]$^{4-}$ anion:** A series of compounds containing the anion shows the full range of static and dynamic J.T. distortion. In these crystalline compounds, both the **crystal packing force and additional cfse from the J.T. distortion** are important.

K$_2$Ca[Cu(NO$_2$)$_6$], K$_2$Ba[Cu(NO$_2$)$_6$], K$_2$Sr[Cu(NO$_2$)$_6$]: distorted structure (z-out) (room temperature crystallographic data)

M$_2$Ba[Cu(NO$_2$)$_6$] (M = Rb, Cs): distorted structure (z-in) (room temperature crystallographic data).

K$_2$Pb[Cu(NO$_2$)$_6$]: no distortion (room temperature crystallographic data).

The variation in the nature of the counter cations influences the **crystal packing forces**. This is why, the distortion or no distortion in the counter anion depends on the nature of the cation.

3.12 A SPECIAL CASE OF JAHN-TELLER DISTORTION IN THE d^8-CONFIGURATION

The d^8-system is expected to form the perfectly octahedral complex ($t_{2g}^6 e_g^2$) and it should not experience any J.T. distortion under the condition, δ_1 (splitting energy of the e_g level) and δ_2 (splitting energy of the t_{2g} level) $\langle\langle$ P (pairing energy) (*cf.* Fig. 3.11.1.1). In fact, under the condition, $\delta_1, \delta_2 \langle\langle$ P, there will be no gain of additional cfse through the distortion. **Consequently, the d^8-system should form the perfectly octahedral complexes without any distortion.** This expectation is supported by the fact: [Ni(OH$_2$)$_6$]$^{2+}$, [Ni(NH$_3$)$_6$]$^{2+}$, [Ni(en)$_3$]$^{2+}$, etc. are the examples of perfectly octahedral complexes. The magnitudes of Crystal Field Splitting parameters δ_1 and δ_2 (*i.e.* splitting energy) depend on the nature of both the ligand and metal centre (*cf.* Sec. 3.6, factors controlling the crystal field splitting parameter Δ). It has been already mentioned that δ_1 is larger than δ_2 but both δ_1 and δ_2 are significantly smaller than P (pairing energy). *If the splitting can induce the condition, $\delta_1 \rangle$ P then the distortion (i.e. z-out distortion) can earn the additional cfse* (*cf.* Fig. 3.12.1).

It is evident (*cf.* Fig. 3.12.1) that for the condition, $\delta_1 \langle$ P, there is no gain of additional cfse through the z-out distortion. *But under the condition, $\delta_1 \rangle P$, there is a gain of additional cfse* (with respect to that of the undistorted octahedron) and it is given by: $2 \times \left(\frac{1}{2}\delta_1\right) - P = \delta_1 - P$.

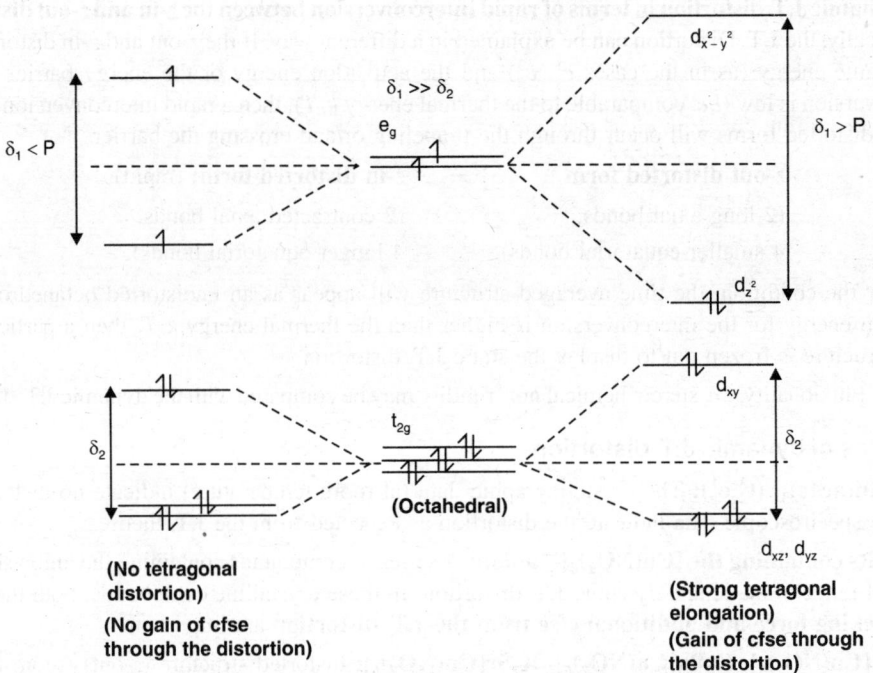

Fig. 3.12.1 Possibility of z-out distortion in the d^8-system.

- **Limiting situation of the z-out distortion:** The condition, $\delta_1 \rangle P$ leading to the spin pairing in the d_{z^2} orbital through the splitting of the e_g level actually gives the **square planar geometry** (*cf.* Figs. 3.5.6.1-2; Crystal Field Splitting by square planar geometry), **a limiting situation of the z-out distortion.**

- **Ni(II) ($3d^8$-system):** For Ni(II), the condition of spin pairing (*i.e.* $\delta_1 \rangle P$) through the splitting of the e_g orbitals can only be attained for the very strong field ligands like CN⁻.

  Thus, with the CN⁻ ligand, Ni(II) forms the square planar complex [Ni(CN)₄]²⁻. If we consider the **hypothetical [Ni(CN)₆]⁴⁻**, then the two *trans*–axial CN⁻ ligands lying along the ±z directions will experience a severe repulsive action from the filled d_{z^2}-orbital. This will virtually place these two *trans*–axial ligands at the non-bonding distances *i.e.,*

$$\left[Ni(CN)_6\right]^{4-} \xrightarrow[\text{(from the } \pm z \text{ directions)}]{-2CN^-} \left[Ni(CN)_4\right]^{2-}$$
  (Hypothetical)

For the weak field ligands (*e.g.* H₂O) and even the moderately strong field ligands (*e.g.* NH₃, en), the condition, $\delta_1 \rangle P$ is not attained for Ni(II). Consequently, they form the perfectly octahedral complexes with Ni(II).

The other *well known square planar complexes of Ni(II)* are:

[Ni(dmgH)₂], [Ni(bigH)₂]Cl₂.

Here it is worth mentioning that sometimes, the very weak field ligands (*e.g.* S-donor sites having the vacant d-orbitals to delocalise the metal d-electrons through the metal-ligand covalent interaction) can produce the diamagnetic square planar complex with Ni(II). In fact, such ligands can show the **pronounced nephelauxetic effect** that substantially reduces the *pairing energy* (P).

Under these conditions, even the moderate crystal field splitting can satisfy the required condition: $\delta_1 \rangle$ P *i.e.* **the condition of imposed J.T. distortion.**

● **Pd(II) ($4d^8$), Pt($5d^8$):** These are the heavier congeners of Ni(II). The crystal field splitting power runs in the sequences: Pt(II) $\rangle$ Pd(II) $\rangle$ Ni(II)

The pairing energy (P) decreases for the heavier congeners and the pairing energy sequence runs as: Ni(II) $\rangle$ Pd(II) $\rangle$ Pt(II)

By considering the sequence of both crystal field splitting power (δ_1) and pairing energy (P), the ease of satisfying the condition, $\delta_1 \rangle$ P runs in the sequence:

$$\text{Pt(II)} \rangle \text{Pd(II)} \rangle \text{Ni(II)}$$

This is the sequence of the ease of square planar complex formation.

Thus, for Ni(II), the condition, $\delta_1 \rangle$ P is attained only for the very strong field ligands but for the heavier congeners, the condition is attained easily even for the much weaker field ligands (*e.g.* Cl⁻, NH_3, etc.). *In fact, because of this situation, Pt(II) and Pd(II) generally form the square planar complexes with almost all types of ligands.*

● **Cu(III) ($3d^8$), Au(III) ($5d^8$):** Though copper is a member of 3d series, because of its higher charge in Cu^{3+}, its crystal field splitting power is sufficiently high to induce the condition, $\delta_1 \rangle$ P even for the so called weak field ligands. For Au(III), the crystal field splitting power is very high (*cf.* high charge and position in the 5d-series) and the condition $\delta_1 \rangle$ P is easily attained. In fact, Cu(III) and Au(III) generally form the square planar complexes like $[AuCl_4]^-$.

---

d^8-system: *No inherent J.T. distortion, but the J.T. distortion can be imposed under the conditions of strong crystal field splitting and reduced pairing energy. It leads to the limiting situation giving rise to the square planar geometry through the expulsion of the two trans-axial ligands from the hypothetical octahedral geometry.*

3.13 CRYSTAL FIELD THEORY OF THE SQUARE PLANAR GEOMETRY

● **Limiting situation of the z-out distortion (*i.e.* tetragonal elongation) in the octahedral system** (*cf.* Sec. 3.5.6): It has been already mentioned that the complete removal of the two *trans*–axial ligands along the $\pm z$-directions gives the square planar geometry in the xy-plane. Removal of repulsion from the ligands along the z-direction stabilizes relatively the d-orbitals bearing the z-component, *i.e.* d_{z^2}, d_{xz} and d_{yz}. The energy order (*cf.* Figs. 3.5.6.1, 3.11.1.1) is:

$$d_{xz}, d_{yz} \, (e_g) \, \langle \, d_{z^2} \, (a_{1g}) \, \langle \, d_{xy} \, (b_{2g}) \, \langle \, d_{x^2-y^2} \, (b_{1g})$$

In some special cases, d_{z^2} may come below d_{xz} and d_{yz} (*i.e.* e_g set).

The total splitting energy Δ_{sp} is given by (*cf.* Figs. 3.5.6.1-2):

$$\Delta_{sp} = E_{d_{x^2-y^2}} - E_{d_{yz}, \, d_{zx}}$$

$$= \left(E_{d_{x^2-y^2}} - E_{d_{xy}} \right) + \left(E_{d_{xy}} - d_{z^2} \right) + \left(E_{d_{z^2}} - E_{d_{xz}, \, d_{yz}} \right)$$

$$= \Delta_1 + \Delta_2 + \Delta_3 = 10Dq_0 + \Delta_2 + \Delta_3$$

The energy difference between $d_{x^2-y^2}$ and d_{xy} gives the measure of $10Dq_0$ (*i.e.* Δ_0) of the corresponding hypothetical octahedral complex. The approximate relations of Δ_1, Δ_2 and Δ_3 for the square planar complexes of Co(II) (d^7), Ni(II) (d^8) and Cu(II) (d^9) are as follows:

$$\Delta_1 \approx \Delta_0, \Delta_2 \approx \frac{2}{3}\Delta_0 \text{ and } \Delta_3 \approx \frac{1}{12}\Delta_0$$

i.e.
$$\Delta_{sp} = \Delta_1 + \Delta_2 + \Delta_3 \approx 1.75\Delta_0$$

The above mentioned approximate relations are not always satisfied. It is illustrated below for $[Ni(CN)_4]^{2-}$:

$$[Ni(CN)_4]^{2-}: \Delta_1 \approx \Delta_0, \Delta_2 \approx \frac{2}{5}\Delta_0, \Delta_3 \approx \frac{1}{38}\Delta_0$$

i.e.
$$\Delta_{sp} \approx 1.42\Delta_0$$

For the square planar complexes of Pd(II) and Pt(II), the following relationship has been found.

$$\Delta_{sp} = \Delta_1 + \Delta_2 + \Delta_3 \approx 1.3\Delta_0$$

● The square planar complexes are generally formed by the d^8, d^9 *and* d^7 (strong field) systems where the *z-out J.T. distortion can originate from the e_g level* of the corresponding hypothetical octahedral complexes.

● d^8-system (*e.g.* Ni(II), Pd(II), Pt(II), etc.): The electronic configuration is:

$$(d_{xz}, d_{yz})^4 \langle d_{z^2}^2 \langle d_{xy}^2 \langle d_{x^2-y^2}^0$$

With the increase of crystal field splitting, the $d_{x^2-y^2}$ orbital becomes gradually more destabilized and the orbitals bearing the z-component *i.e.* d_{xz}, d_{yz}, d_{z^2} will be more stabilized. *As the $d_{x^2-y^2}$ orbital does not carry any electron, with the increase of crystal field splitting, the square planar complex becomes more and more stable.*

The driving force for the formation of square planar complexes in the d^8 system through the **imposed J.T. distortion** has been discussed in Sec. 3.12. The square planar complexes of the d^8-systems are always spin-paired (*i.e.* low-spin) and **such complexes are diamagnetic.**

● d^9-system (*e.g.* Cu(II)): The electronic configuration is:

$$(d_{xz}, d_{yz})^4 \langle d_{z^2}^2 \langle d_{xy}^2 \langle d_{x^2-y^2}^1$$

Here it may be mentioned that in the corresponding hypothetical octahedral complex ($t_{2g}^6 e_g^3$), the **inherent J.T. distortion** exists to earn the additional cfse ($= \frac{1}{2}\delta_1$, δ_1 = splitting energy of the e_g level, *i.e.* energy difference between the $d_{x^2-y^2}$ and d_{z^2} orbitals). Thus the d^9 system is expected to experience the z-out distortion, *i.e.* two *trans*–axial bonds are elongated. When δ_1 increases significantly, the two *trans*–axial ligands are lost from the z-direction giving rise to the square planar geometry.

The square planar complexes of d^9-system are paramagnetic due to the presence of an unpaired electron in the $d_{x^2-y^2}$ orbital. Here we may recall the VBT which requires the excitation of the unpaired electron from the $d_{x^2-y^2}$ orbital to make it available for the dsp^2 hybridization of the square planar complex. But CFT does not require any such excitation.

Relative stabilities of Cu(II) and Au(II): For copper, +2 is the most stable oxidation state which gains the high cfse through the tetragonal elongation (and the limiting situation is square planar). On the other hand, for its heavier congener gold, +1 and +3 are the most stable oxidation states. For the +2 oxidation state, the ninth electron is to be placed in the $d_{x^2-y^2}$ orbital which is more destabilized in Au^{2+}

compared to that in Cu^{2+} because, the crystal field splitting power of Au^{2+} is about 80% greater than that for Cu^{2+}. This is why, it is easier to ionize Au^{2+} to Au^{3+} compared to the case of Cu^{2+}. In fact, this case favours the **disproportionation of Au^{2+} to Au^{3+} and Au^{+}**. In Au^{3+} (d^8-system) there is a huge gain of cfse in the square planar geometry and no electron is required to be placed in the highly destabilised orbital $d_{x^2-y^2}$. This is the driving force behind the disproportionation of Au^{2+}.

- **Low-spin d^7-system (e.g. Ni(III), Co(II)):** The corresponding low-spin octahedral complex (*i.e.* $t_{2g}^6 e_g^1$) experiences the J.T. distortion through the splitting of e_g level. If the crystal field the splitting (*i.e.* δ_1) in the e_g-set is sufficiently high to attain the limiting situation of z-out distortion, then the square planar complexes will be formed. Here it may be mentioned that in the d^7-system, the d_{z^2} orbital bears only one electron while in the d^9 and d^8 systems, the d_{z^2} orbital bears two electrons. ***Thus, in the d^7-system, the ligands along the z-direction experience relatively a less repulsion from the d_{z^2} orbital compared to that in the d^9 and d^8 systems. Thus, it is relatively more difficult to produce the square planar complex for the d^7-system.*** It is illustrated below.

d^8-system: $\left[Ni(CN)_6 \right]^{4-} \xrightarrow{-2CN^-} \left[Ni(CN)_4 \right]^{2-}$

d^7-system: $\left[Co(CN)_6 \right]^{4-} \xrightarrow[+S]{-CN^-} \left[Co(CN)_5 S \right]^{3-}$, S = Solvent

The well known square planar complexes of Co(II) are:

$$[Co(diars)_2](ClO_4)_2, \; [Co(salen)], \; [Co(bigH)_2]SO_4, \; [Co(acacen)]$$

diars = *o*-phenylenebis(dimethylarsane), salenH$_2$ = N,N'-ethylenebis(salicylidieneimine), bigH = protonated biguanide, acacen = N,N'-ethylenebis(acetylacetoneimine).

In [Co(salen)] and [Co(acacen)], the quadridentate (N, N, O, O) Schiff base ligand property fits into the basal plane and such complexes may bind two unidentate ligands (*e.g.* solvent) along the axial directions. These *trans*– axial bonds are elongated.

Here it may be mentioned that the low-spin d^7-systems are very **much prone to oxidation**, giving rise to the d^6-system that can earn a huge amount of cfse (= $24Dq_0 - 2P$) in the low-spin octahedral complexes.

Note: For the **low-spin d^4 (*i.e.* t_{2g}^4) systems**, the square planar complexes may be theoretically produced through the splitting of the t_{2g} level which is unsymmetrically filled in. In such cases, the splitting δ_2 (= splitting energy of the t_{2g} level) should be greater than P and that would lead to the electronic configuration: $(d_{xz}, d_{yz})^4 d_{xy}^0$.

But such a situation can hardly be attained. **This is why, we get [Mn(CN)$_6$]$^{3-}$ (low-spin, d^4) rather than the square planar [Mn(CN)$_4$]$^-$.** Here it is needless to say that [Mn(CN)$_6$]$^{3-}$ is a distorted octahedral complex due to the J.T. distortion from the t_{2g} level.

3.14 APPLICATIONS OF CFT: EVIDENCES IN FAVOUR OF CFT

3.14.1 Magnetic Properties of the Coordination Compounds in Terms of CFT

CFT can predict the electron distribution pattern of the metal electrons in the splitted orbitals. From this electronic configuration, the number of unpaired electrons can be calculated and from this electronic configuration, the **spin-only magnetic moment** can be estimated (*cf.* Sec. 3.4.6).

The number of unpaired electrons for a particular d^n–configuration depends on the high-spin or low-spin state of the complex. By the relative magnitudes of Δ (*i.e.* crystal field splitting parameter) and pairing energy (P), it is also possible to predict the spin state (including the spin isomerism) of the complex.

The possibility of **orbital contribution and spin-orbit coupling interaction** to determine the resultant magnetic moment can also be understood from the knowledge of CFT.

The magnetic properties of the complexes have been discussed in detail in Chapter 8.

3.14.2 Spectral Properties of the Coordination Compounds in Terms of CFT

These aspects have been discussed in detail in Chapter 7.

3.14.3 Kinetic Properties of the Coordination Compounds in Terms of CFT

The crystal field activation energy (CFAE) is an important contributing factor to determine the overall activation energy. In fact, CFT is quite important to understand the lability and inertness of coordination compounds. These aspects have been discussed in detail is Chapter 5.

Thermodynamic properties of the metal complexes and CFT

By considering the crystal field splitting parameter, it can be shown that for most of the coordination compounds, the cfse may lie in the range 60-500 kJ mol^{-1}. This stabilisation energy is quite significant to determine the different thermodyanimic properties like **metal-ligand stability constants** (*cf.* Chapter 4), **hydration energy, lattice energy,** etc. These aspects are discussed separately.

3.14.4 Contribution of cfse to the Metal-ligand Bond Energy

Metal-ligand bond energy can be calculated theoretically by considering the electrostatic interaction, polarisation effect, and the contribution of cfse. Thus the calculated value nicely agrees with the experimental value (*cf.* Table 3.14.4.1).

Table 3.14.4.1 Contribution of cfse to the bond energy of some representative metal-ligand bonds.

Complex	Energy per metal-ligand bond, (kJ mol^{-1})			
	Calculated value without cfse	Contribution of cfse	Calculated value with the cfse contribution	Experimental value
$[AlF_6]^{3-}$	890	0	890	978
$[K(OH_2)_6]^+$	54	0	54	67
$[Zn(NH_3)_4]^{2+}$	361	0	361	374
$[Fe(OH_2)_6]^{3+}$	458	0 (h.s.)	458	487
$[Fe(OH_2)_6]^{2+}$	210	$\frac{1}{6}$ (4 Dq_o), (h.s.)	219	243.5
$[Cr(OH_2)_6]^{3+}$	466	$\frac{1}{6}$ (12 Dq_o)	504	512
$[Co(NH_3)_6]^{3+}$	492	$\frac{1}{6}$ (24 Dq_o – 2P), (l.s.)	525	562

Table 3.14.4.2 Contribution of cfse in the stability of the octahedral $[ML_6]$ and tetrahedral $[ML_4]$ complexes.

$[d^n]$	cfse for $[ML_6]$		cfse for $[ML_4]$
	High-spin	*Low-spin*	*tetrahedral*
d^0	0	0	0
d^1	$4\,Dq_o$	$4\,Dq_o$	$6\,Dq_t \equiv 2.67\,Dq_o$
d^2	$8\,Dq_o$	$8\,Dq_o$	$12\,Dq_t \equiv 5.33\,Dq_o$
d^3	$12\,Dq_o$	$12\,Dq_o$	$8\,Dq_t \equiv 3.55\,Dq_o$
d^4	$6\,Dq_o$	$16\,Dq_o - P$	$4\,Dq_t \equiv 1.78\,Dq_o$
d^5	0	$20\,Dq_o - 2P$	0
d^6	$4\,Dq_o$	$24\,Dq_o - 2P$	$6\,Dq_t \equiv 2.67\,Dq_o$
d^7	$8\,Dq_o$	$18\,Dq_o - P$	$12\,Dq_t \equiv 5.33\,Dq_o$
d^8	$12\,Dq_o$	$12\,Dq_o$	$8\,Dq_t \equiv 3.55\,Dq_o$
d^9	$6\,Dq_o$	$6\,Dq_o$	$4\,Dq_t \equiv 1.78\,Dq_o$
d^{10}	0	0	0

The contribution of cfse to the overall stability of the complexes is qualitatively shown in Fig. 3.14.4.1. For the high-spin octahedral and tetrahedral complexes, **there are double humped curves while for the low-spin octahedral complexes there is a single humped curve (ingnoring the P).**

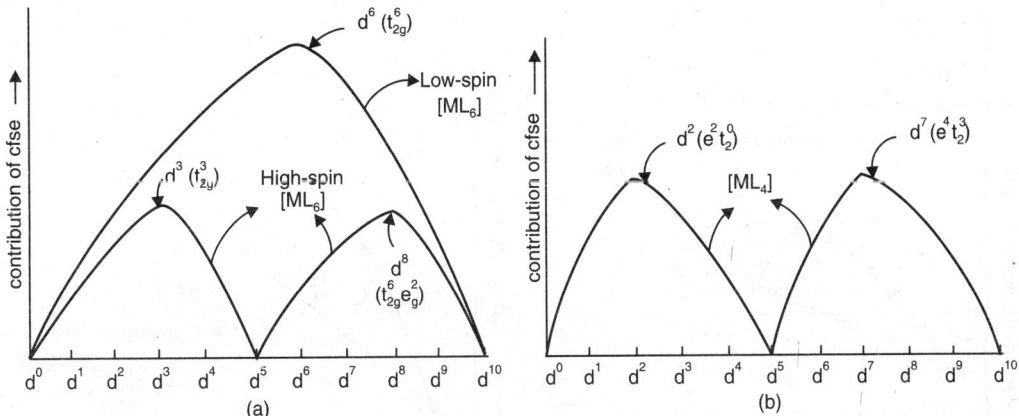

Fig. 3.14.4.1 Qualitative representation of contribution of cfse to the **stability of the octahedral $[ML_6]$ and tetrahedral $[ML_4]$ complexes.**

3.14.5 Contribution of cfse to the Metal-Ligand Step-wise Stability Constants

These aspects have been discussed in detail in Chapter 4. In explaining the natural stability order, *i.e.* the Irving-Williams stability order, the cfse plays an important role (*cf.* Sec. 4.5.1).

3.14.6 Contribution of cfse to the Hydration Energies of Metal Ions

The *heat of hydration, i.e. enthalpy of hydration* of metal ions may be considered as the *enthalpy of formation of their hexaaqua complexes.*

$$M^{n+}(g) + H_2O(\text{excess}) \rightleftharpoons \left[M(OH_2)_6\right]^{n+}$$

The enthalpy values (ΔH) obtained at different finite concentrations can be extrapolated upto the infinite dilution. *These values at infinite dilution are taken for comparison.*

If we move from Mg^{2+} to Ba^{2+} (in Gr. 2 *i.e.* IIA), there is a gradual increase of ionic radius radius and this is why the hydration energy decreases gradually from Mg^{2+} to Ba^{2+}. Thus, in terms of electrostatic model, **the hydration energy and ionic radius follow the opposite sequences.** Similarly, in moving from Ca^{2+} to Zn^{2+} along the 1st transition series, the hydration energy should increase gradually with the gradual expected decrease of ionic radius. But the actual values make a **double humped curve** in which the values for Ca^{2+} (d^0), Mn^{2+} (d^5) and Zn^{2+} (d^{10}) systems lie on a more or less straight line (with a positive slope) as expected from the pure electrostatic interaction model. In the double humped curve, one maximum lies in the region of d^3 (*i.e.* V^{2+}), d^4 (*i.e.* Cr^{2+}) systems while the other maximum lies in the region of d^8 (*i.e.* Ni^{2+}), d^9 (*i.e.* Cu^{2+}) systems.

By considering the weak field nature of H_2O, the aqua complexes of the said metal ions are expected to be the high-spin octahedral complexes. Thus, except the d^0, d^5 (high-spin) and d^{10} systems, in all cases there is a finite contribution of cfse. If the Jahn-Teller stabilisation is ignored for the d^4 (*i.e.* $t_{2g}^3 e_g^1$) and d^9 ($t_{2g}^6 e_g^3$) systems, then the maxima of cfse exist at the d^3 (= 12 Dq_0) and d^8 (= 12 Dq_0) systems along the series. Thus the deviation of the experimental curve from the curve constructed by only Ca^{2+} (d^0), Mn^{2+} (d^5) and Zn^{2+} (d^{10}) is due to the contribution of cfse. *If from the estimated value, the contribution of cfse is substracted, then all the metal ions will be placed on a single smooth curve (almost a straight line) with an upward slope.*

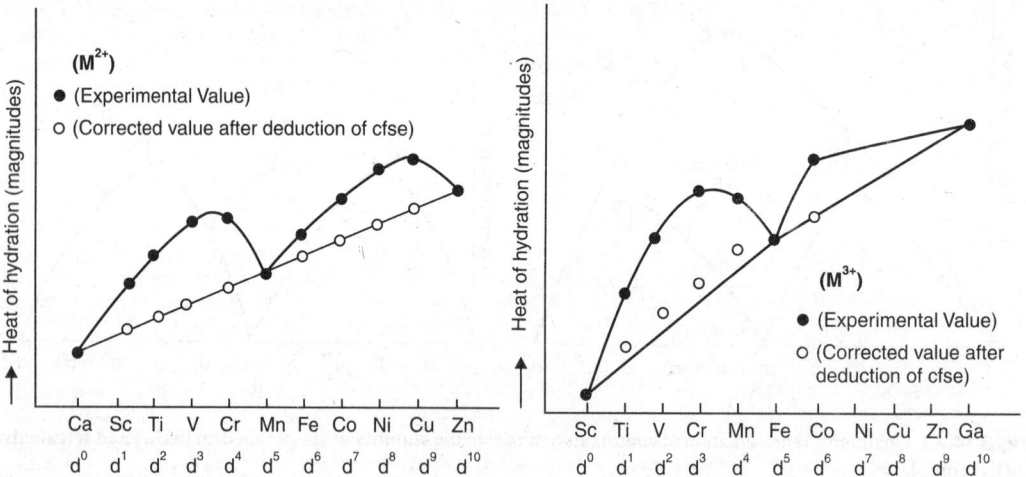

Fig. 3.14.6.1 Qualitative representation of the variation of heat of hydration (**magnitudes**) of bi- and trivalent metal ions of the 1st transition series.

If variation of enthalpies of hydration of trivalent metal ions of the first transition series (*i.e.* Sc^{3+} to Ga^{3+}, in which data for Ni^{3+}, Cu^{3+} and Zn^{3+} are not available), then again a double humped curve is obtained. The two maxima lie in the regions of Cr^{3+} (d^3), Mn^{3+} (d^4) and Co^{3+} (d^6). $[Co(OH_2)_6]^{3+}$ remains in a low-spin $\rightleftharpoons$ high-spin equilibrium. It may be pointed out that the $[Mn(OH_2)_6]^{3+}$ and $[Fe(OH_2)_6]^{3+}$ are the high spin complexes. The nature of the curve can be explained by considering the contribution of cfse as in the case of bivalent metal ions of the 1st transition series.

Contribution of cfse to the hydration energy

The $10Dq_o$ value for $[Cr(OH_2)_6]^{2+}$ is 13,900 cm^{-1}. The calculated hydration energy by using the ionic radius of gaseous Cr^{2+} is -1830 kJ mol^{-1}. What would be the expected heat of hydration if experimentally determined?

The electronic configuration of h.s. $[Cr(OH_2)_6]^{2+}$ is $t_{2g}^3 e_g^1$ and the cfse $6Dq_o = \dfrac{6}{10} \times 13,900$ cm^{-1}

$= 8340$ cm^{-1} $= 8340 \times 100$ m^{-1} $= 8.34 \times 10^5$ m^{-1}

We have: $E = h\nu = \dfrac{hc}{\lambda}$

i.e. cfse $= 6.26 \times 10^{-34}$ J s $\times 2.99 \times 10^8$ m s^{-1} $\times 8.34 \times 10^5$ m^{-1}

$\qquad = 156.10 \times 10^{-21}$ J

$\qquad = \dfrac{156.10 \times N_0 \times 10^{-21}}{1000}$ kJ mol^{-1}

$\qquad = \dfrac{156.1 \times 6.02 \times 10^{23} \times 10^{-21}}{1000}$ kJ mol^{-1}

$\qquad = 93.97$ kJ mol^{-1}.

Simply, the relation: 1 kk (kilo kayser) = 1000 cm^{-1} = 10^5 m^{-1} = 11.97 kJ mol^{-1} can be used.

Thus we can write:

experimental heat of hydration = Calculated value $-$ cfse

$\qquad\qquad = -1830 - 93.97$

$\qquad\qquad = -1923.97$ kJ mol^{-1}

3.14.7 Contribution of cfse to the Lattice Energy

According to the Born-Lande equation, the lattice energy is inversely proportional to the internuclear separation. Here for Ca^{2+} to Zn^{2+}, in a particular type crystal, the lattice energy should increase smoothly from Ca^{2+} to Zn^{2+}. But experimentally, we get a **double humped curve** (Fig. 3.14.7.1) for MX_2 (X^-= halide) having maxima at the regions $d^3 - d^4$ and $d^8 - d^9$ and minima at d^0, d^5 and d^{10} centres. In fact, $Ca^{2+}(d^0)$, Mn^{2+} (d^5) and Zn^{2+} (d^{10}) lie on a more or less straight line with an upward slope as expected from the Born-Lande equation. The halides (MX_2) having the cfse in the high-spin state (*cf.* halides provide weak field ligand strength) are deviated from the straight line curve of d^0, d^5 and d^{10} systems having no cfse.

By considering the additional cfse from the J.T. distortion, existence of maxima at the regions of $d^3 - d^4$ and $d^8 - d^9$ can be explained. If from the observed lattice energy values, the contribution of cfse is subtracted then all the halides of M^{2+} ions of the series $Ca^{2+} - Zn^{2+}$ lie on the same straight line.

3.14.8 Variation of the Octahedral Crystal Ionic Radii for the Transition Metal Ions

Along the transition metal series, the ionic radius is expected to decrease **with the increase of effective nuclear charge**. The effective nuclear charge experienced by the electrons at the periphery (*i.e.* outermost shell) gradually increases because of the gradual accumulation of the electrons in the $(n-1)d$ orbitals having the poor screening or shielding power. ***But actually the crystal radii do not change in a smooth way***. When the M—O distance (in oxide lattice) is plotted against the d^n–configuration along the Ca^{2+} —Zn^{2+} series, it produces a **double humped curve** for the oxides of M^{2+} ions of the 1st transition series. The curve shows two minima at V^{2+} (d^3) and Ni^{2+} (d^8). The M—O bond length is the sum of ionic radii of M^{2+} and O^{2-}, *i.e.* $d_{M-O} = r_{M^{2+}} + r_{O^{2-}}$ where $r_{O^{2-}}$ may be considered to be the same for all the metal-oxides. Thus, the variation trend of M—O distance reflects the variation trend of $r_{M^{2+}}$ (Fig. 3.14.8.1).

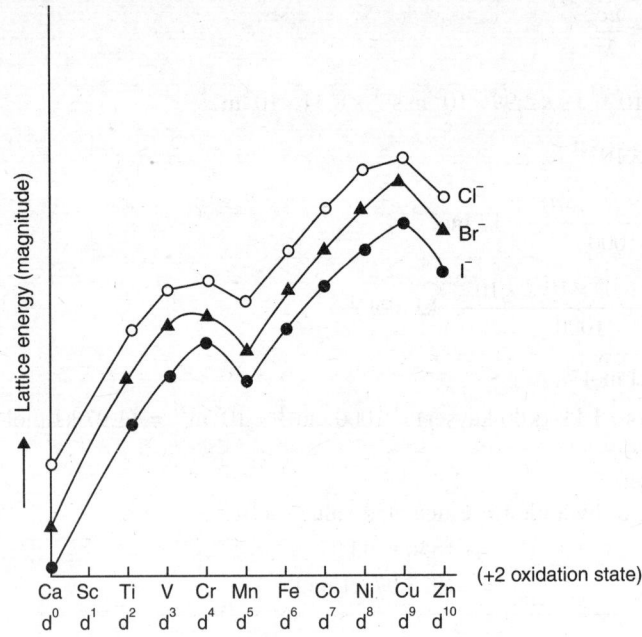

Fig. 3.14.7.1 Variation of the lattice energies (**magnitudes**) of the dihalides MX_2 (X = Cl, Br, I) of the 1st transition series.

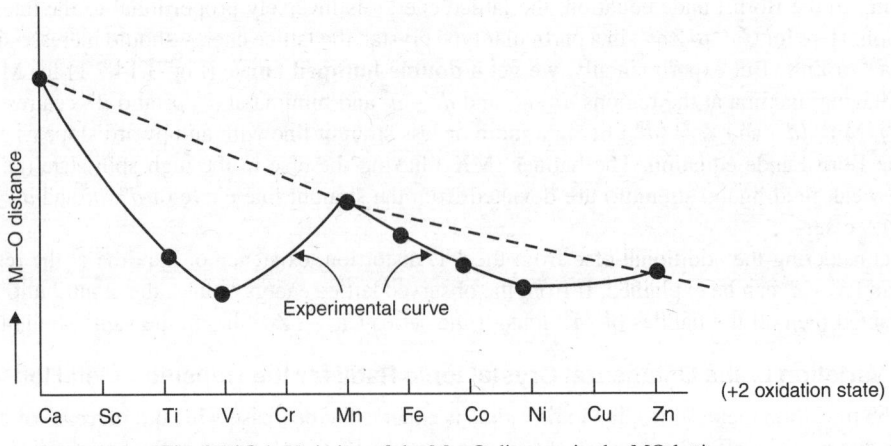

Fig. 3.14.8.1 Variation of the M—O distance in the MO lattice.

The observation can be explained by applying CFT for the metal oxides having the weak field ligand to provide the high-spin octahedral 'MO_6' stereochemistry in their oxide crystal.

In terms of CFT, in the octahedral crystal field, the e_g orbitals (*i.e.* $d_{x^2-y^2}$ and d_{z^2}) project their lobes directly to the approaching ligands (*i.e.* oxide–O in the MO crystal). Thus, the approaching ligating sites experience a direct repulsive action from the e_g–electrons. On the other hand, the ligating sites do not experience any such direct repulsive action from the t_{2g}–electrons.

The electronic configurations in the high-spin metal oxides (MO) are:

$$Sc^{2+}\left(t_{2g}^1 e_g^0\right), Ti^{2+}\left(t_{2g}^2 e_g^0\right), V^{2+}\left(t_{2g}^3 e_g^0\right), Cr^{2+}\left(t_{2g}^3 e_g^1\right), Mn^{2+}\left(t_{2g}^3 e_g^2\right),$$

$$Fe^{2+}\left(t_{2g}^4 e_g^2\right), Co^{2+}\left(t_{2g}^5 e_g^2\right), Ni^{2+}\left(t_{2g}^6 e_g^2\right), Cu^{2+}\left(t_{2g}^6 e_g^3\right), Zn^{2+}\left(t_{2g}^6 e_g^4\right).$$

Upto $VO\left(t_{2g}^3 e_g^0\right)$, the M—O distance decreases as expected from the increase of effective nuclear charge. After VO, the electron is gradually accumulated in the e_g–level and these e_g–electrons repel the donor sites (*i.e.* O) and consequently the M—O distance increases upto MnO ($t_{2g}^3 e_g^2$). After Mn^{2+}, *i.e.* MnO, the M—O distance decreases upto NiO ($t_{2g}^6 e_g^2$) as expected from the periodic trend. After NiO, again electron density starts to increase in the e_g–set upto ZnO ($t_{2g}^6 e_g^4$). This is why, after NiO, the M—O distance again starts to increase. ***Thus in the double humped curve, the 'V—O' and 'Ni—O' distances represent the two minima.***

This ***double humped nature*** is also maintained for the ***crystallographic radii*** (or metal-ligand distances) of the M^{2+} ions of the 1st transition series under the condition of weak octahedral crystal field giving rise to the high-spin state (*cf.* Table 3.14.8.1).

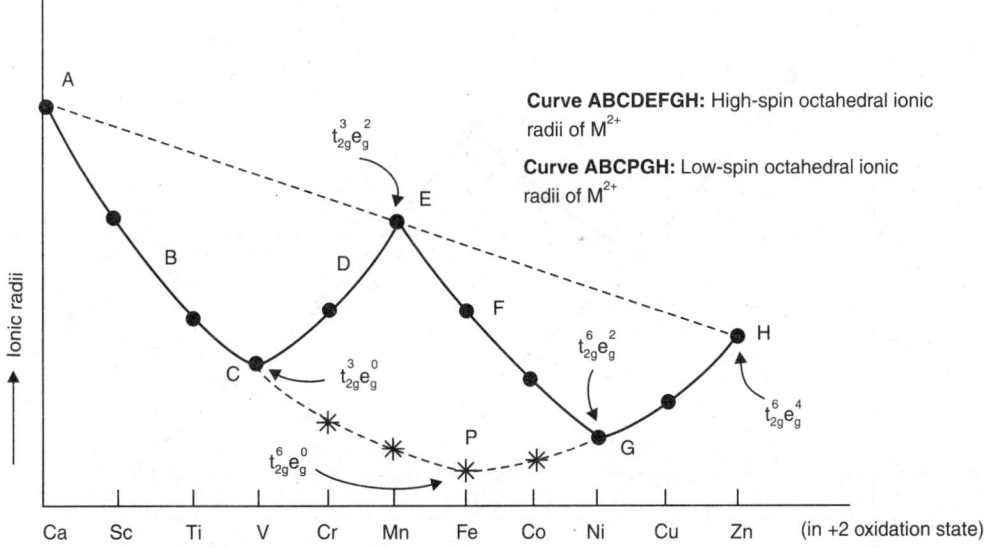

Fig. 3.14.8.2 Qualitative representation of variation of the high-spin and low-spin octahedral crystallographic ionic radii of the M^{2+} ions along the series Ca^{2+}—Zn^{2+}.

Table 3.14.8.1 Crystallographic octahedral ionic radii (pm) for 1st transition series metal ions.

	Ca	Sc	Ti	V	Cr	Mn	Fe	Co	Ni	Cu	Zn	Ga
M^{2+} (high-spin):	115	–	100	93	94	97	92	88.5	83	87	88	–
M^{2+} (low-spin):						81	75	79				
M^{3+} (high-spin):		88.5	81	78	75.5	78.5	78.5	75	74	–	–	76
M^{3+} (low-spin):						72	69	68.5	70	–		

If the crystal field strength can produce the condition of low-spin octahedral configuration for the M^{2+} ions of the 1st transition series then, the ionic radius will decrease smoothly from Ca^{2+} (d^0) to Fe^{2+} ($t_{2g}^6 e_g^0$) as expected from the periodic trend, then it will increase upto Zn^{2+} ($t_{2g}^6 e_g^4$) with the gradual accumulation of electron in the e_g level. These e_g–electrons will repel the approach of the donor sites. **Thus for the high-spin octahedral radii of M^{2+} ions, we get two minima at V^{2+} (d^3, i.e. $t_{2g}^3 e_g^0$) and Ni^{2+} (d^8, i.e. $t_{2g}^6 e_g^2$) leading to a typical double humped curve** but, for the low-spin octahedral radii there is a single minimum at Fe^{2+} (d^6, i.e. $t_{2g}^6 e_g^0$) (cf. Fig. 3.14.8.2).

If the high-spin octahedral ionic radii of M^{3+} of the 1st transition series are plotted against d^n then we shall get a double humped curve having two minima at Cr^{3+} ($t_{2g}^3 e_g^0$) and Cu^{3+} ($t_{2g}^6 e_g^2$) (cf. Fig. 3.14.8.3). For the low-spin octahedral radii, there will be a single minimum at Co^{3+} ($t_{2g}^6 e_g^0$). These can be explained by considering the effect of both the effective nuclear charge and e_g–electron density causing a direct repulsion to the donor sites.

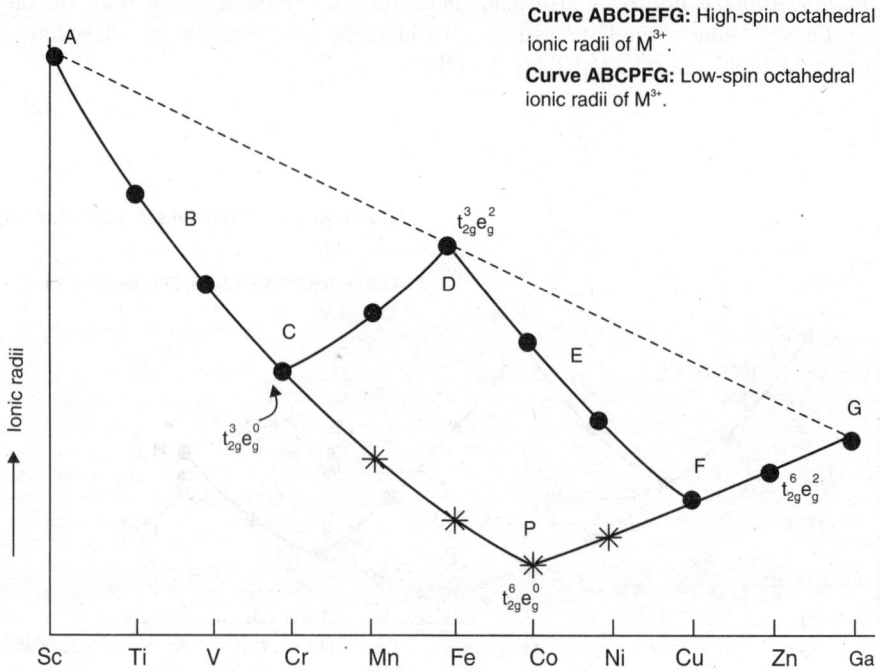

Curve ABCDEFG: High-spin octahedral ionic radii of M^{3+}.

Curve ABCPFG: Low-spin octahedral ionic radii of M^{3+}.

Fig. 3.14.8.3 Qualitative representation of the variation of the high-spin and low-spin octahedral crystallographic ionic radii of M^{3+} along the series Sc^{3+} — Ga^{3+}.

Note: Magnitudes of hydration and lattice energy curves in terms of the ionic radii curve

It is evident that the curves for the variation of hydration energy or lattice energy with the d^n–configurations are **the inverted (qualitative sense) curves for the variation of ionic radii** with d^n. For the high-spin states, the curve for the ionic radius shows the minimima in the regions of d^3 (*i.e.* t_{2g}^3) and d^8 ($t_{2g}^6 e_g^2$) and for the curves of hydration energy or lattice energy, the maxima occur at the same regions. It is quite expected because the smaller ionic radius will produce the higher hydration energy and lattice energy.

3.14.9 Variation of the Metal-Ligand Bond Distance in the Capped Octahedral System

The bond length dependence on the d^n configuration for the seven coordinate complex with the ligand **(py)₃tren**, *i.e.* $N(CH_2CH_2N = CHC_5H_4N)_3$ is quite interesting. The structural features of the seven coordinate complexes with the ligand (py)₃tren have been discussed in detail in Sec. 2.11.9 (*cf.* Fig. 2.11.9.6). Six N–donor sites (3 pyridine–N + 3 imine–N) of the ligand occupy the octahedral sites and the seventh N-donor site (tertiary amine–N) coordinates at the centre of a trigonal face of the octahedron. Thus this seventh donor site directly approaches towards the lobes of the t_{2g} orbitals. The important features of this complex are (*cf.* Fig. 2.11.9.6):

six N–donor sites approach directly towards the lobes of the e_g–orbitals as usual in the octahedral complexes, while the seventh N-donor site residing at the centre of a trigonal face approaches directly towards the t_{2g} orbital.

The six octahedral M — N bonds (along the vertices of a regular octahedron) will be elongated with the increase of electron density in the e_g orbitals as usual. *On the other hand, the seventh M — N bond projected along the centre of an octahedral face will be elongated with the increase of the electron density in the t_{2g} level.* For the complexes with this ligand, (py)₃tren along the series, Mn^{2+} — Zn^{2+}, Fe^{2+} forms a low-spin complex ($t_{2g}^6 e_g^0$). The electron distribution patterns in the t_{2g} and e_g orbitals are shown below for the series.

M^{2+}:	Mn^{2+}	Fe^{2+}	Co^{2+}	Ni^{2+}	Cu^{2+}	Zn^{2+}
Electron distribution:	$t_{2g}^3 e_g^2$	$t_{2g}^6 e_g^0$	$t_{2g}^5 e_g^2$	$t_{2g}^6 e_g^2$	$t_{2g}^6 e_g^3$	$t_{2g}^6 e_g^4$
	High-spin	low-spin	High-spin			

Because of the *effect of the effective nuclear charge*, the M—N bond length should decrease along the series, Mn^2 — Zn^{2+}. But this decreasing trend is not monotonic. These are discussed here.

- **Octahedral M — N bonds:** In moving from Mn^{2+} to Fe^{2+}, there will be a sharp decrease in the M — N bond length because of the removal of electrons from the e_g level to the t_{2g} level for the formation of the low-spin complex of Fe(II). Then for Co^{2+}, two electrons are placed in the e_g–level. This increased electron density in the e_g–level will elongate the metal-ligand bond length. Thus, there will be a sharp hike in the M — N bond length in moving from Fe^{2+} to Co^{2+}. For the movement from Co^{2+} to Ni^{2+} there is no change of electron density in the e_g–level. This is why, the decrease of the M — N bond length from Co — N to Ni — N will be as usual predicted from the effective nuclear charge. After Ni^{2+}, electrons will be gradually placed in the e_g–level and consequently the bond length will increase in the sequence: Ni — N ⟨ Cu—N ⟨ Zn — N.

- **Axial M—N bond, *i.e.* 7ᵗʰ M — N bond** (*cf.* R.M. Kirchner et at, *Coord. Chem. Rev.*, **77**, 89-163, 1987): This bond is projected toward the t_{2g} orbitals. This bond length increases with the increase of electron density in the t_{2g} level. In moving from Mn^{2+} to Fe^{2+}, the t_{2g}–electron density is increased because of the formation of the low-spin complex for Fe^{2+} and this will make the Fe—N bond longer than the Mn—N bond. In moving from Fe^{2+} to Co^{2+}, the electron density is reduced in the t_{2g} level. This will make the Co—N bond shorter than the Fe—N bond. Moving from Co^{2+} to Ni^{2+} leads to an increase of electron density in the t_{2g} level and this will make the Ni—N bond longer than the Co—N bond. After Ni^{2+}, the t_{2g} electron density will remain unchanged upto Zn^{2+}. Thus the M—N bond will decrease as usual (*i.e.* periodic trend) in the sequence Ni—N 〉 Cu—N 〉 Zn—N.

- **Change of hybridisation state of the amine-N:** Depending on the bond length of the axial M — N bond (*i.e.* 7th M — N bond), hybridisation state of the amine-N changes. This aspect has been already illustrated in Fig. 2.11.9.7.

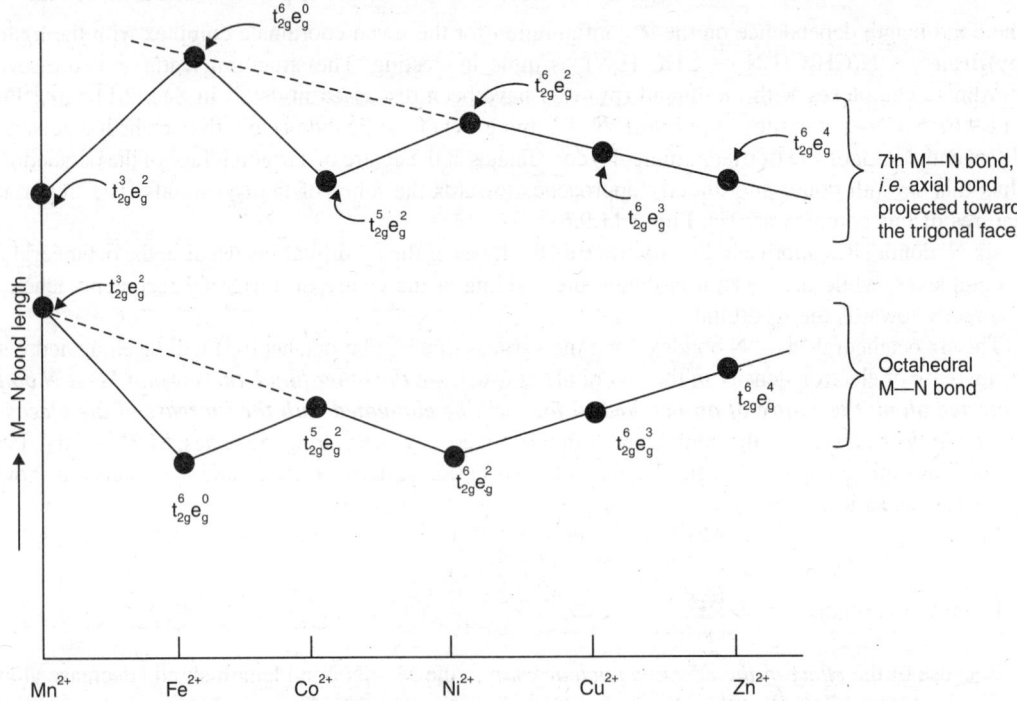

Fig. 3.14.9.1 Qualitative representation of the effect of t_{2g} and e_g–electron density on the metal-ligand bond length in the 7–coordinate complex with the ligand (py)$_3$tren, $N(CH_2CH_2N = CHC_5 H_4N)_3$ (*cf.* Figs. 2.11.9.6, 7).

3.14.10 Variation of Tetrahedral Ionic Radii for the Transition Metal Ions

In the tetrahedral complexes, the approaching ligands experience relatively more repulsion from the t_2-electrons compared to the e-electrons. Thus with the increase of electron density in the t_2–level, the tetrahedral crystal radius should increase.

The effect of effective nuclear charge expects the gradual decrease of the ionic radii along the series, $Ca^{2+} — Zn^{2+}$. The radius decreases upto d^2 (*i.e.* $e^2t_2^0$) (Fig. 3.14.10.1) as expected from the effect of effective nuclear charge. After d^2, the electrons are gradually added to the t_2-level upto d^5 (*i.e.* $e^2t_2^3$) and consequently, the ionic radius increases in the sequence, $d^2 〈 d^3 〈 d^4 〈 d^5$. After d^5, the electrons are added in the e-level upto d^7 (*i.e.* $e^4t_2^3$) and the radius decreases upto d^7 as expected from the periodic trend. After d^7, the electrons are gradually added in the t_2-level and the enhanced repulsion towards the ligand will increase the radius in the sequence, $d^7 \langle d^8 \langle d^9 \langle d^{10}$ (Fig. 3.14.10.1).

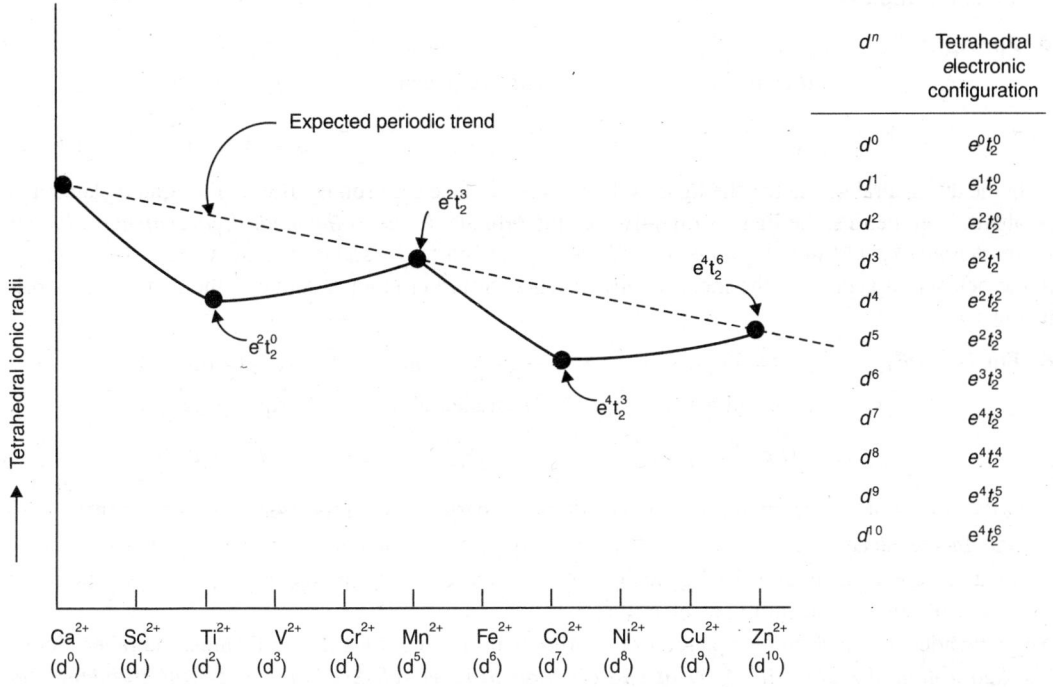

d^n	Tetrahedral electronic configuration
d^0	$e^0 t_2^0$
d^1	$e^1 t_2^0$
d^2	$e^2 t_2^0$
d^3	$e^2 t_2^1$
d^4	$e^2 t_2^2$
d^5	$e^2 t_2^3$
d^6	$e^3 t_2^3$
d^7	$e^4 t_2^3$
d^8	$e^4 t_2^4$
d^9	$e^4 t_2^5$
d^{10}	$e^4 t_2^6$

Fig. 3.14.10.1 Qualitative representation (not in scale) of the expected variation of the tetrahedral ionic radii along the series $Ca^{2+}-Zn^{2+}$ showing the effect of electron density in the e and t_2 level.

3.14.11 Effect of Electron Occupancy in Different Orbitals on the Ionic Radii in other Geometries

A. Square Planar Geometry: The order of energy of the d–orbitals are $\left(d_{xz}, d_{yz}\right) < d_{z^2} < d_{xy} < d_{x^2-y^2}$. The ligands approach directly to the lobes of $d_{x^2-y^2}$ orbital. The d_{xy} orbital projects the lobes in the xy-plane between the ligands. ***Thus, the metal-ligand bond distance is very much sensitive to the electron density in the $d_{x^2-y^2}$ orbital and it is also slightly affected by the d_{xy}-electron density.*** If the $d_{x^2-y^2}$ orbital is vacant (as ***in the d^8-system***), then the ligands do not experience any repulsive force from the $d_{x^2-y^2}$ orbital and the metal-ligand bond distance is shortened. But in the ***d^9-system***, the $d_{x^2-y^2}$ orbital is half filled in and consequently, the metal-ligand bond is relatively elongated more than the case of the d^8 system.

Thus, effect of the ***effective nuclear charge*** should make the Cu(II)—L bond smaller than the Ni(II)—L bond in the square planar geometry, but the ***effect of orbital occupancy*** will make the Cu(II)—L bond longer than the Ni(II)—L bond.

In the ***d^{10}-system,*** the $d_{x^2-y^2}$ orbital is full-filled and a approaching ligand will experience a severe repulsion from the $d_{x^2-y^2}$ orbital and the metal-ligand bond will be elongated. If the d_{xy} electron density is decreased, the metal-ligand bond will be slightly shortened (*cf.* ***d^8 vs. d^7 systems***).

- **M—L distance (Square:**
  $$d^2_{x^2-y^2}d^2_{xy} \quad \rangle \quad d^1_{x^2-y^2}d^2_{xy} \quad \rangle \quad d^0_{x^2-y^2}d^2_{xy} \quad \rangle \quad d^0_{x^2-y^2}d^1_{xy};$$
  planar complex)
  $$(d^{10}) \qquad\qquad (d^9) \qquad\qquad (d^8) \qquad\qquad (d^7)$$

- **For Ag$^+$ (d^{10}):** r_{Ag^+} (= 129 pm) $\rangle$ r_{Ag^+} (= 116 pm) $\rangle$ r_{Ag^+} (= 114)

  (octahedral)   (square planar)   (tetrahedral)

  $$t^6_{2g}e^4_g \left(i.e.\ d^2_{x^2-y^2}\ d^2_{xy}\right) \qquad\qquad\qquad \left(e^4t^6_2\right)$$

In the tetrahedral geometry, the ligands do not directly face the orbitals. But, in the square planar and octahedral geometries, the ligands directly face the orbitals. *It makes the metal-ligand distance longer in the octahedral and square planar complexes.* Compared to the square planar (C.N. = 4) geometry, in the octahedral geometry, the metal-ligand distance is further elongated due to the increased coordination number.

- **For Ni^{2+} (d^8):** $r_{Ni^{2+}}$ (= 83 pm) $\rangle$ $r_{Ni^{2+}}$ (= 69 pm) $\rangle$ $r_{Ni^{2+}}$ (= 63 pm)

  (octahedral)   (tetrahedral)   (square planar)

  $$t^6_{2g}e^2_g \left(i.e.\ d^1_{x^2-y^2}\ d^1_{z^2}\right) \qquad e^4t^4_2 \qquad \left(d^0_{x^2-y^2}\ d^2_{xy}\right)$$

In the square planar geometry, the ligands do not experience any repulsion from the vacant $d_{x^2-y^2}$ orbital. In the octahedral geometry, the ligands experience repulsion from the e_g–electrons. In the tetrahedral geometry, though the ligands do not face the orbitals directly but the ligands experience some sort of repulsion from the t^4_2–electrons.

Note: Besides the electron occupancy in the orbitals projecting towards the ligands, the other factors like *ligand-ligand repulsion, effect of coordination number (cf. Sec. 11.4, Vol. 2 of Fundamental Concepts of Inorganic Chemistry) also contribute to determine the ionic radius.* Higher C.N. gives the higher radius. Higher ligand-ligand repulsion gives the higher radius.

B. Square Pyramidal Geometry: The basal ligands directly face the lobes of $d_{x^2-y^2}$ orbital and the axial ligand faces one lobe of the d_{z^2} orbital. Thus, the metal-ligand bond length increases in the basal plane with the increase of electron occupancy in the $d_{x^2-y^2}$ orbital. The axial bond length increases with the increase of electron occupancy in the d_{z^2} orbital. The possibility of π–bonding with the basal ligands will shrink the basal metal-ligand bonds. These aspects have been illustrated in Secs. 2.11.7 and 3.10.

The energy order is:

$$\left(d_{xz}, d_{yz}\right) < d_{xy} < d_{z^2} < d_{x^2-y^2}\ (cf.\ Fig.\ 3.5.7.1)$$

C. Trigonal Bipyramidal Geometry: The axial ligands directly face the d_{z^2} orbital and in the equatorial plane, the lobes of the $d_{x^2-y^2}$ and d_{xy} orbitals are projected but the equatorial ligands do not directly face these two orbibals. The axial bond length increases with the increase of the electron occupancy in the d_{z^2}–orbital. Besides the electron occupancy in the different orbitals, many other factors like nonequivalent hybrid orbitals, VSEPR, π-bonding in the equatorial plane also contribute to determine the metal-ligand bond distance. These aspects have been illustrated in Secs. 2.11.7 and 3.10.

The energy order is:

$$\left(d_{xz}, d_{yz}\right) < \left(d_{xy}, d_{x^2-y^2}\right) < d_{z^2}\ (cf.\ Fig.\ 3.5.8.1)$$

3.14.12 Preferred Stereochemistry: Octahedral versus Tetrahedral (O_h vs. T_d)

Different factors operate in different directions to favour or disfavour a particular crystal geometry. These are discussed below.

(i) **Bond energy:** In the octahedral geometry, the larger number of bond formation (*i.e.* metal-ligand interaction) compared to that in a tetrahedral geometry, will release the larger amount of energy. *Thus the octahedral geometry is favoured over the tetrahedral geometry, in general, in terms of the number of metal-ligand interaction.*

(ii) **Steric hidrance:** In accommodating the six ligands in an octahedral geometry, the ligand-ligand repulsion will occur at $90°$ while in a tetrahedral geometry, the four ligands are placed around the metal centre and the ligand-ligand repulsion occurs at $109°$. *Thus, in terms of steric crowding, the tetrahedral geometry is preferred over the octahedral geometry. Specially, if the ligands are bulky and anionic, then the reduced ligand-ligand repulsion strongly favours the tetrahedral geometry.*

(iii) **Cationic charge:** The higher cationic charge of the metal ion is expected to earn the higher stabilisation energy through the larger number of stronger metal-ligand interaction. *In fact, this is why, compared to the bivalent metal ions, the trivalent metal ions favour the octahedral geometry more than the tetrahedral geometry.*

(iv) **cfse:** Here it should be remembered that cfse makes only a small contribution (*ca.* 10%) in the total metal-ligand bond energy of a complex. *But, when all other factors are comparable, the cfse contribution may sit on the driver's seat to determine the stereochemical preference.*

In general, the cfse of an octahedral geometry is more than that of a tetrahedral geometry. Thus, if an octahedral geometry is converted into a tetrahedral geometry, there will be a loss of cfse. This loss of cfse is described as the ***octahedral site selection energy (OSSE)*** or the ***octahedral structure preference energy (OSPE)*** (*cf. Table 3.14.12.1 and Fig. 3.14.12.1*)

$$\text{OSSE or OSPE} = \text{cfse (oct.)} - \text{cfse (tet.)}$$

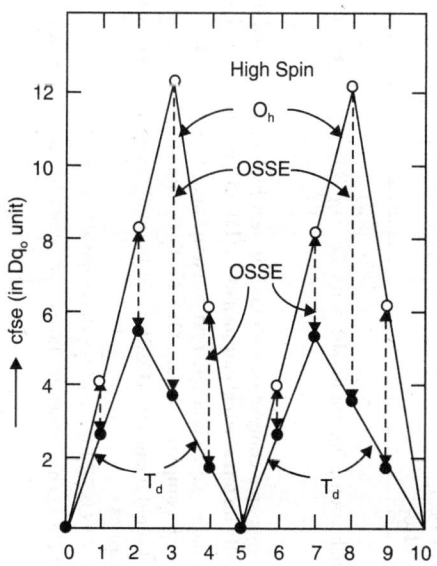

Fig. 3.14.12.1 Comparison of the cfse for the d^n ions in the octahedral (high spin) and tetrahedral (high-spin) fields. (i) OSSE = 0 for the d^0, d^5 and d^{10} systems; (ii) OSSE : small for the d^1, d^2, d^6 and d^7 systems; (iii) OSSE: high for the d^4, d^9 systems; (iv) OSSE: very high for the d^3 and d^8 systems; OSSE: d^3, $d^8 \rangle\rangle d^4$, $d^9 \rangle d^2$, $d^7 \rangle d^1$, $d^6 \rangle d^0$, d^5, $d^{10} = 0$

Ease of T_d complex formation: d^0, d^5, $d^{10} \rangle d^1$, $d^6 \rangle d^2$, $d^7 \rangle d^4$, $d^9 \rangle d^5$, d^8.

(A) Conditions favouring the tetrahedral geometry: *If the OSSE is not too large, then the steric relaxation in moving from the octahedral geometry to tetrahedal geometry will favour the tetrahedral*

Table 3.14.12.1 cfse values for the different d^n configurations in the octahedral and tetrahedral crystal fields and octahedral site selection energy (OSSE).

d^n:	d^0	d^1	d^2	d^3	d^4	d^5	d^6	d^7	d^8	d^9	d^{10}
High-spin											
● *Octahedral:* (in terms of Dq_o)	0	$4Dq_o$	$8Dq_o$	$12Dq_o$	$6Dq_o$	0	$4Dq_o$	$8Dq_o$	$12Dq_o$	$6Dq_o$	0
● *Tetrahedral:* (in terms of Dq_t)	0	$6Dq_t$	$12Dq_t$	$8Dq_t$	$4Dq_t$	0	$6Dq_t$	$12Dq_t$	$8Dq_t$	$4Dq_t$	0
(in terms of Dq_o):	0	$2.67Dq_o$	$5.33Dq_o$	$3.55Dq_o$	$1.78Dq_o$	0	$2.67Dq_o$	$5.33Dq_o$	$3.55Dq_o$	$1.78Dq_o$	0
● OSSE (in terms of Dq_o): $= \text{cfse}_{o_h} - \text{cfse}_{T_d}$	0	$1.33Dq_o$	$2.67Dq_o$	$8.45Dq_o$	$4.22Dq_o$	0	$1.33Dq_o$	$2.67Dq_o$	$8.45Dq_o$	$4.22Dq_o$	0
Low-spin Octahedron and High-spin tetrahedron											
● *Octahedral:* (in terms of Dq_o)	0	$4Dq_o$	$8Dq_o$	$12Dq_o$	$16Dq_o - P$	$20Dq_o - 2P$	$24Dq_o - 2P$	$18Dq_o - P$	$12Dq_o$	$6Dq_o$	0
● OSSE (in terms of Dq_o): $= \text{cfse}_{o_h} - \text{cfse}_{T_d}$	0	$1.33Dq_o$	$2.67Dq_o$	$8.45Dq_o$	$14.22Dq_o - P$	$20Dq_o - 2P$	$21.33Dq_o - 2P$	$12.67Dq_o - P$	$8.45Dq_o$	$4.22Dq_o$	0

Dq_o (for octahedron); Dq_t (for tetrahedron), $D_{q_t} = \dfrac{4}{9}Dq_o$; very often, in calculating the OSSE, the Pairing energy (P) is ignored.

Note:

- For the high-spin complexes, the OSSE is very high for the d^3 and d^8 systems *i.e.* cfse strongly favours the octahedral system over the tetrahedral system. **Practically no tetrahedral complex of Cr(III) (d^3) is known. Ni(II) (d^8) also predominantly gives the octahedral complexes with the weak field ligands.**
- For the high-spin complexes, next to the d^3 and d^8 systems, **cfse favours the octahedral complexes for the d^4 and d^9 configurations more over the tetrahedral complexes,** *i.e.* the OSSE is fairly large for the d^4 and d^9 systems. The J.T. distortion further gives an additional cfse for the d^4 and d^9 systems.
- For weak-field ligands, the OSSE is fairly small for the d^1, d^2, d^6 and d^7 systems. Thus such d^n configurations may form the tetrahedral complexes for the steric reasons.
- There is no OSSE for the d^0, d^5 (h.s.) and d^{10} systems.
- Strong field ligands generally favour the octahedral complexes because of the high OSSE.

geometry over the octahedral geometry. It is further favoured for the bulky and anionic ligands that cause more steric repulsion in the octahedral geometry.

For the d^0, d^5 **(h.s.) and d^{10} systems,** OSSE = 0, *i.e.* there is no loss of cfse in the conversion of the octahedral to the tetrahedral germetry. For d^1, d^2, d^6 **(h.s.), and d^7 (h.s.),** OSSE is not very high (*cf.* Fig. 3.14.12.1). Such electronic configurations may favour the tetrahedral geometry when the ligands are having the properties: *weak field ligands, bulky and anionic.*

(B) Conditions favouring the octahedral geometry: When the OSSE is very high as in the cases d^3, d^4, d^8, d^9, etc., the octahedral geometry is preferred over the tetrahedral geometry. Obviously, the strong field nature of the ligand will favour the octahedral geometry more. If the ligands are sufficiently strong to produce the low-spin octahedral geometry then the OSSE will be very high to favour the octahedral geometry strongly. Here, it should be mentioned that the tetrahedral complexes are always the high spin complexes because of the low splitting in the tetrahedral crystal field.

It is illustrated by the following examples.

Tetrahedral: $[VCl_4]$ (d^1), $[VCl_4]^-$ (d^2), $[FeCl_4]^-$ (d^5), $[FeCl_4]^{2-}$ (d^6), $[CoCl_4]^{2-}$ (d^7)
$\quad\quad\quad$ Cl^- (bulky ligand, anionic ligand, weak field ligand).

Octahedral: $[M(CN)_6]^{x-}$, where M = V(III) (d^2), Cr(III) (d^3), Mn(III) (d^4), Fe(II) (d^6), Fe(III)
$\quad\quad\quad$ (d^5), Co(III) (d^6)
$\quad\quad\quad$ CN^- (a strong field ligand to induce the spin pairing in the octahedral geometry).

The above predictions are supported by the following examples.

(i) **d^0–system (OSSE = 0):** Tetrahedral complexes like $[AlCl_4]^-$, $[VOCl_3]$, $[MnO_4]^-$, $[CrO_4]^{2-}$ are well documented.

(ii) **d^5–system:** If the ligand is sufficiently weak and the octahedral complexes are high-spin, then **OSSE becomes zero.** Then the tetrahedral complexes are preferred specially when the ligands are bulky and anionic (*e.g.* Cl^-). The well known examples are: $[FeCl_4]^-$, $[MnCl_4]^{2-}$, etc.

(iii) **d^{10}–system (OSSE = 0):** A large number of tetrahedral complexes are known. The steric relaxation (90° *vs* 109°) is the driving force for the tetrahedral geometry compared to the other geometries like octahedral, square planar, etc. It may be noted that cfse is zero for all geometries. The representative examples are:
$\quad$ $[MX_4]^{2-}$ (X = halogen), $[M(NH_3)_4]^{2+}$, $[M(CN)_4]^{2-}$, (M = Zn, Cd, Hg), $[Ni(CO)_4]$, $[Cu(CN)_4]^{3-}$

(iv) **d^1 and d^2–systems:** For the weak field ligands, the OSSE is not very high. Thus the anionic and bulky ligands lead to the tetrahedral complexes. The representative examples are:
$\quad$ $[TiCl_4]^-$ (d^1), $[VCl_4]$ (d^1), $[VCl_4]^-$ (d^2)

(v) **d^6 and d^7–systems:** For the weak field ligands, the OSSE is not very high and the anionic bulky ligands will prefer the tetrahedral geometry, *e.g.*
$$[FeCl_4]^{2-}, [CoX_4]^{2-} (X = Cl, Br, CNS)$$

Note: In presence of the strong field ligands, the d^6–system will produce the low-spin octahedral complex and the d^7–system will produce the strongly distorted (tetragonally elongated) octahedral complex, highly prone to oxidation to the d^6–system.

Octahedral Complexes perferred over the Tetrahedral Complexes

● **d^3, d^4, d^8, d^9:** For such d^n–configurations, the OSSE is fairly high and the octahedral complexes are preferred over the tetrahedral complexes even for the weak field ligands.

● **Strong field and nonbulky ligands:** OSSE for the strong field ligands forming the low-spin octahedral complexes is very high to favour the octahedral geometry. For the nonbulky ligands, steric crowding in the octahedral geometry is not so high.

3.14.13 Stereochemistry Preferred by Co(II) (d^7), Ni(II) (d^8) and Cu(II) (d^9) Complexes

Square planar geometry (*cf.* Sec. 3.13): **The strong field ligands** and conditions leading to the crystal field splitting more than the pairing energy (*i.e.* $\Delta \gg P$) will prefer the square planar geometry over the octahedral and tetrahedral geometries for the d^7, d^8 and d^9 systems. The **ease of square planar geometry formation** runs in the sequence (Sec. 3.13): $d^9 > d^8 \gg d^7$

Octahedral *vs.* tetrahedral: For the **weak field, bulky and anionic ligands**, among the metal centres, Co(II), Ni(II) and Cu(II), *cobalt(II) can readily form the tetrahedral complexes.* In fact, it is difficult to have the true tetrahedral complexes of Cu(II) and Ni(II). This is illustrated by the following fact that if concentrated HCl solution is added to an aqueous solution (light pink) of Co(II), *i.e.* $[Co(OH_2)_6]^{2+}$, it becomes blue coloured due to the formation the tetrahedral species.

$$\left[Co(OH_2)_6\right]^{2+} \text{(light pink)} \xrightarrow{\text{conc. HCl}} \underbrace{\left[CoCl_3(OH_2)\right]^-, \left[CoCl_4\right]^{2-}}_{\text{deep blue, tetrahedral}}$$
$$\text{(Octahedral)}$$

In the case of $[Ni(OH_2)_6]^{2+}$, **even in presence of 12 M HCl,** only two Cl^- ligands can substitute the coordinated water molecules.

$$\left[Ni(OH_2)_6\right]^{2+} \xrightarrow{\text{12 M HCl}} \left[NiCl_2(OH_2)_4\right]$$
$$\text{(Octahedral)} \qquad\qquad\qquad \text{(Octahedral)}$$

This example illustrates the strong tendency of Ni(II) to maintain the octahedral geometry. Tetrahedral $[NiCl_4]^{2-}$ can be isolated with the large cations like $N(C_2H_5)_4^+$ or $As(C_6H_5)_4^+$ from the non-aqueous media. Highly distorted tetrahedral structure of $[CuCl_4]^{2-}$ has been discussed in Sec. 2.11.

With the bulky and anionic ligands, the **steric factors** may favour the tetrahedral geometry for Ni(II) · $[NiX_2L_2]$ (L = PR$_3$) gives the four coordinate complexes maintaining an equilibrium between the tetrahedral and square planar structure. This aspect has been discussed in Sec. 2.3.

Here it is worth mentioning that $[NiCl_4]^{2-}$ is a distorted tetrahedron because of the **inherent J.T. distortion ($e^4 t_2^4$).**

---

cfse and cyanido-complex of Ni(II)

	$[Ni(CN)_4]^{2-}$	$[Ni(CN)_6]^{4-}$	$[Ni(CN)_4]^{2-}$	$[Ni(CN)_5]^{3-}$	$[Ni(CN)_5]^{3-}$
	(Square planar)	(Octahedral)	(Tetrahedral)	(TBP)	(Square pyramidal)
cfse:	$24.56 Dq_0 - P$	$12 Dq_0$	$3.56 Dq_0$	$14.16 Dq_0 - P$	$18.28 Dq_0 - P$

The cfse values are calculated in terms of $10Dq_0 (= \Delta_0)$ by using the relations given in Table 3.5.11.1. It is obvious, that the *cfse is maximum for the square planar geometry while it is minimum for the tetrahedral structure.* In other words, among the possible geometries, the square planar geometry is the most stable one and the **tetrahedral geometry is the least stable one.** In terms of the cfse, the square planar, TBP (trigonal bipyramidal) and square pyramidal (SP) geometries are more preferred over the hypothetical octahedral structure. In fact, the **octahedral $[Ni(CN)_6]^{4-}$ complex does not exist** under any condition. However, under certain conditions (*i.e.* in presence of excess CN^-), the most stable square planar structure shows a feeble tendency to add an additional CN^- ligand to produce TBP and/or SP. These two 5-coordinate complexes possess more or less comparable cfse. In fact **both the TBP and SP structures of $[Ni(CN)_5]^{3-}$** have been isolated and characterised (*cf.* Secs. 2.11.7, 3.10 and Fig. 2.11.7.4 for details).

[Ni(acac)$_2$] was originally considered to be tetradedral (*i.e.* C.N. = 4) but it is now established that it attains the octahedral structure through polymerisation (*cf.* Sec. 2.11.6, Fig. 2.11.6.6). **[Co(acac)$_2$]** can exist both as the monomers (tetrahedral structure) and tetramers (octahedral coordination) (*cf.* Fig. 2.11.6.5). The **Lifschitz salts** (Fig. 2.11.6.7) are the special types of complexes of Ni(II). These are structurally quite interesting. They can remain in the following equilibrium.

$$\text{Square planar} \rightleftharpoons \text{Distorted octahedral} \rightleftharpoons \text{Octahedral}$$

These are discussed in detail in Sec. 2.11.6 and Chapter 8.

All these support the fact that Ni(II) is reluctant to adopt the tetrahedral structure.

3.14.14 Site Selection of the Cations in Spinel and Inverse Spinel and OSSE (*cf.* Sec. 11.11.1, Vol. 2 of Fundamental Concepts of Inorganic Chemistry)

The mixed oxides, $A^{II}B_2^{III}O_4$ are named after the naturally occurring one **spinel, $Mg^{II}Al_2^{III}O_4$**. In spinel, oxides (O^{2-}) are distributed in a **cubic close packed** (CCP) **array.** Each oxide ion provides two tetrahedral (T_d) holes and one octahedral (O_h) hole. Thus, per formula, we have eight T_d holes and four O_h holes. *Generally, the trivalent cations are placed in the octahedral holes while the bivalent cations prefer the tetrahedral holes.* This distribution pattern gives a **better metal-ligand electrostatic inter-action.** But in the inverse spinel, half of the trivalent cations (*i.e.* B) are placed in the T_d holes and the bivalent cations are placed in the O_h holes. The structural features of the normal and inverse spinels are shown below.

- **Normal Spinel:** Bivalent cations (*i.e.* A^{II}) occupy 1/8 of the T_d holes and the trivalent cations (*i.e.* B^{III}) occupy 1/2 of the O_h-holes. It is: $A^{tet}B_2^{oct}O_4$
- **Inverse Spinel:** Half of the trivalent cations (*i.e.* B^{III}) occupy 1/8th of the T_d holes; and the other half of the trivalent cations and the bivalent cations (*i.e.* A^{II}) jointly occupy 1/2 of the octahedral holes (O_h-holes). It is: $B^{III}(A^{II}B^{III})O_4$, *i.e.* $B^{tet}(AB)^{oct}O_4$.

Different factors to control the site selection preference of the cations: These are discussed below:

- **Electrostatic interaction and lattice energy:** The higher cationic charge is expected to earn the higher lattice energy through the larger number of metal-ligand electrostatic attraction. This is why, the trivalent cation is preferably placed in the octahedral holes (C.N. = 6) than in the tetrahedral holes (C.N. = 4). *This is the main driving force for the normal spinel structure.*
- **Size factor:** The cations of higher charge are generally smaller in size. The tetrahedral holes are smaller in size. This size factor may prompt to place half of the higher valent cations in the tetrahedral holes. *This will lead to the inverse spinel structure* provided in shifting the higher valent cations from the O_h holes to the T_d holes, the loss of lattice energy and cfse are not too large.
- **cfse factor:** Sometimes, there may be a gain of cfse in attaining the *inverse spinel structure. This happens for the some transition metal ions.*

Examples

Normal spinels: $M^{II}Al_2^{III}O_4$ (M = Mg, Mn, Zn);
$Fe^{II}Cr_2^{III}O_4$; $Co^{II}Co_2^{III}O_4$; $Mn^{II}Mn_2^{III}O_4$; etc.

Inverse spinels: $Fe^{III}(Co^{II}Fe^{III})O_4$; $Fe^{III}(Fe^{II}Fe^{III})O_4$; $Fe^{III}(Ni^{II}Fe^{III})O_4$; $Fe^{III}(Cu^{II}Fe^{III})O_4$;
$Al^{III}(Ni^{II}Al^{III})O_4$; etc.

- **cfse and OSSE:** Preference for the inverse spinel structure can be explained in terms of CFT. The oxide ions are the weak field ligands and they generally produce the high-spin complexes.

If Fe^{III} (h.s., d^5), $Al^{III}(d^0)$, $Ga^{III}(d^{10})$ are shifted from the octahedral site to the tetrahedral site, there will be no loss of cfse (*i.e.* OSSE = 0). If the O_h–hole vacated by such trivalent cations (d^0, d^5 or d^{10}) are occupied by the bivalent cations of the d^6 (*e.g.* Fe^{II}), d^7 (*e.g.* Co^{II}), d^8 (*e.g.* Ni^{II}) or d^9 (*e.g.* Cu^{II}) configurations, there will be a gain of cfse. **In such cases, the inverse spinel structures are favoured over the normal spinel structures.**

If the trivalent cations are of the d^3 (*e.g.* Cr^{III}), d^4 (*e.g.* Mn^{III}) and d^6 (l.s.; Co^{III}) configurations, they will enjoy a high amount of cfse in the octahedral holes. Such trivalent cations will lose a fairly large amount of cfse (*i.e.* OSSE very high) when shifted to the T_d holes from the O_h holes. **Such trivalent cations will strongly favour the normal spinel structure.**

Table 3.14.14.1 cfse values of the M^{2+} and M^{3+} cations in the O_h and T_d holes in terms of $Dq_o(M^{2+})$ values.

cfse values calculated by using the relations: $\left[Dq_t = \dfrac{4}{9}Dq_o, Dq(M^{3+}) \approx 1.5Dq(M^{2+}) \right]$.

Configuration	cfse (in terms of Dq_o of M^{2+})			
	T_d–sites		O_h–sites	
	M^{2+}	M^{3+}	M^{2+}	M^{3+}
d^0, d^5 (h.s.)	0	0	0	0
d^1, d^6 (h.s.)	$6 \times \dfrac{4}{9} = \dfrac{8}{3}$	$\dfrac{8}{3} \times \dfrac{3}{2} = 4$	4	$4 \times \dfrac{3}{2} = 6$
d^2, d^7 (h.s.)	$12 \times \dfrac{4}{9} = \dfrac{16}{3}$	$\dfrac{16}{3} \times \dfrac{3}{2} = 8$	8	$8 \times \dfrac{3}{2} = 12$
d^3, d^8	$8 \times \dfrac{4}{9} = \dfrac{32}{9}$	$\dfrac{32}{9} \times \dfrac{3}{2} = \dfrac{16}{3}$	12	$12 \times \dfrac{3}{2} = 18$
d^4 (h.s.), d^9	$4 \times \dfrac{4}{9} = \dfrac{16}{9}$	$\dfrac{16}{9} \times \dfrac{3}{2} = \dfrac{8}{3}$	6	$6 \times \dfrac{3}{2} = 9$
d^5 (h.s.) d^{10}	0	0	0	0

Table 3.14.14.2 Octahedral site selection energy (OSSE) in terms of $Dq_o(M^{2+})$ (*cf.* Tables 3.14.14.1 and 3.14.12.1).

Configuration (h.s.)	OSSE (for M^{2+})	OSSE (for M^{3+})
d^1, d^6	$4 - \dfrac{8}{3} = \dfrac{4}{3}$	$6 - 4 = 2$
d^2, d^7	$8 - \dfrac{16}{3} = \dfrac{8}{3}$	$12 - 8 = 4$
d^3, d^8	$12 - \dfrac{32}{9} = \dfrac{76}{9}$	$18 - \dfrac{16}{3} = \dfrac{38}{3}$
d^4, d^9	$6 - \dfrac{16}{9} = \dfrac{38}{9}$	$9 - \dfrac{8}{3} = \dfrac{19}{3}$

Table 3.14.14.3 cfse [in terms of Dq_o of M^{2+}, $Dq_t \approx \dfrac{4}{9}Dq_o$, $Dq_o(M^{3+}) \approx 1.5\, Dq_o(M^{2+})$] of some mixed oxides in both the normal and inverse spinel forms (assuming the high spin states of M^{II} and M^{III}).

Mixed Oxides:	Mn_3O_4	Fe_3O_4	Co_3O_4
cfse for the Normal Spinel:	$Mn^{II}Mn_2^{III}O_4$	$Fe^{II}Fe_2^{III}O_4$	$Co^{II}Co_2^{III}O_4$
	$d^5,\ d^4$	$d^6,\ d^5$	$d^7,\ d^6$
	$0 + 2 \times 9 = 18$	$\dfrac{8}{3} + 0 = 2.7$	$\dfrac{16}{3} + 2 \times 6 = 17.3$
cfse for the Inverse Spinel:	$Mn^{III}(Mn^{II}M^{III})O_4$	$Fe^{III}(Fe^{II}Fe^{III})O_4$	$Co^{III}(Co^{II}Co^{III})O_4$
	$d^4,\ d^5,\ d^4$	$d^5,\ d^6,\ d^5$	$d^6,\ d^7,\ d^6$
	$\dfrac{8}{3} + 0 + 9 = 11.7$	$0 + 4 + 0 = 4$	$4 + 8 + 6 = 18$
Observed Structure:	Normal	Inverse	Normal

• **Mn_3O_4 and Fe_3O_4:** It is evident that for **Mn_3O_4**, cfse in the normal spinel structure is more than that in the inverse spinel structure. This is why, it adopts the normal spinel structure. In terms of cfse, the inverse spinel structure of **Fe_3O_4** is favoured over the normal spinel structure.

Crystal field stabilisation energy (cf. Sec. 11.11.1, Vol. 2 of Fundamental Concepts of Inorganic Chemistry)

Normal spinel   **Inverse spinel**

$$A_{tet}^{II}\left(B_2^{III}\right)_{oct}O_4 \longrightarrow B_{tet}^{III}\left(A^{II}B^{III}\right)_{oct}O_4$$

Gain of cfse = cfse of $\{(A^{II})_{oct} + (B^{III})_{oct} + (B^{III})_{tet} - 2(B^{III})_{oct} - (A^{II})_{tet}\}$

= cfse of $\{(A^{II})_{oct} + (B^{III})_{tet} - (B^{III})_{oct} - (A^{II})_{tet}\}$

For, d^0 and d^5 (h.s.) B^{III},

gain of cfse in the transformation of normal to inverse spinel = cfse$\left\{(A^{II})_{oct} - (A^{II})_{tet}\right\}$

• **Co_3O_4:** For **Co_3O_4**, if Co(III) (d^6) is supposed to be in the high-spin state in the octahedral hole, then cfse is more or less the same in both the normal and inverse spinel structure. Thus, it should remain in an equilibrium mixture of both forms. *However, Co_3O_4 exists in the normal spinel form.* This can be explained by considering the *critical crystal field splitting* ($10Dq_o$) for Co(III) by the oxide ligand fields. In fact, $[Co(OH_2)_6]^{3+}$ remains in a **h.s. $\rightleftharpoons$ l.s. equilibrium.** Thus we can consider the existence of Co(III) in low-spin state to some extent in the octahedral hole. **The high cfse for the low-spin d^6** $(t_{2g}^6 e_g^0)$ **configuration in the octahedral hole strongly favours the normal structure,** i.e. OSSE is very high for the low-spin d^6 (t_{2g}^6) configuration.

	$Co^{II}Co_2^{III}O_4$ (Normal)	$Co^{III}(Co^{II}Co^{III})O_4$ (Inverse)
cfse	$e^4 t_2^3,\ t_{2g}^6$	$e^3 t_2^3,\ t_{2g}^5 e_g^2,\ t_{2g}^6$
	$\dfrac{16}{3}Dq_o(Co^{II}) + 2 \times 36Dq_o(Co^{II})$	$4Dq_o(Co^{II}) + 8Dq_o(Co^{II}) + 36Dq_o(Co^{II})$
	$- 2 \times 2\, P(Co^{III})$	$- 2P(Co^{III})$
	$= \dfrac{232}{3}Dq_o(Co^{II}) - 4P(Co^{III})$	$= 48Dq_o(Co^{II}) - 2P(Co^{III})$

By considering low-spin Co^{III} (t_{2g}^6) in octahedral hole, it is evident that the normal spinel sructure gains more cfse compared to that of the inverse spinel structure. **This is the driving force for the spinel structure of Co_3O_4.**

3.14.15 Preference between the Trigonal Bipyramidal (TBP) Structure and the Octahedral Structure in Terms of CFT

In terms of cfse, generally the octahedral structure is preferred over the TBP structure. However, calculation indicates that for the d^9 configuration, the loss of cfse in moving from the octahedral to the TBP structure is relatively small and it is zero for the d^{10} configuration.

3.14.16 Effect of cfse on the Reduction Potential (E^0) of the Redox Couple $M^{(n+1)+}/M^{n+}$ Couple (*cf.* Sec. 4.9.1 and Sec. 16.4.4, Vol. 3 of Fundamental Concepts of Inorganic Chemistry)

(A) Co^{III}/Co^{II} couple: This aspect is illustrated for the redox couple Co^{III}/Co^{II}

Redox Couple	E^0(V)
$\left[Co(OH_2)_6\right]^{3+} + e \rightleftharpoons \left[Co(OH_2)_6\right]^{2+}$ (h.s.$\rightleftharpoons$l.s.) (h.s.)	+1.84
$\left[Co(edta)\right]^- + e \rightleftharpoons \left[Co(edta)\right]^{2-}$ (l.s.) (h.s.)	+0.60
$\left[Co(ox)_3\right]^{3-} + e \rightleftharpoons \left[Co(ox)_3\right]^{4-}$ (l.s.) (h.s.)	+0.57
$\left[Co(bpy)_3\right]^{3+} + e \rightleftharpoons \left[Co(bpy)_3\right]^{2+}$ (l.s.) (h.s.)	+0.31
$\left[Co(phen)_3\right]^{3+} + e \rightleftharpoons \left[Co(phen)_3\right]^{2+}$ (l.s.) (h.s.)	+0.42
$\left[Co(NH_3)_6\right]^{3+} + e \rightleftharpoons \left[Co(NH_3)_6\right]^{2+}$ (l.s.) (h.s.)	+0.10
$\left[Co(en)_3\right]^{3+} + e \rightleftharpoons \left[Co(en)_3\right]^{2+}$ (l.s.) (h.s.)	−0.26
$\left[Co(CN)_6\right]^{3-} + e \rightleftharpoons \left[Co(CN)_5(OH_2)\right]^{3-} + CN^-$ (l.s.) (l.s.)	−0.83

- **The decrease of reduction potential (E^0) means the increase of relative stability of the Co(III)–complex compared to that of the Co(II)–complex.** Thus, the decrease of E^0 value indicates the ease of oxidation of Co(II) to Co(III).
- The striking observation is:
  The order of decreasing E^0 value (*i.e.* increasing stability of the Co^{III}–state compared to that of the Co^{II}–state) is roughly the order of increasing ligand field strength.
- **This means that with the increase of ligand field strength, the Co(III)–state becomes more stable than the Co(II)–state.**

- The cfse is only one of the contributing factors to determine the E^0 value. The other factors like the **entropy effect** (mainly determined by the relative extent of electrorestriction on the surrounding solvent by the oxidised state and reduced state), chelate effect, π–bonding effect etc. are also important to determine the E^0 value. **This is why, the decreasing trend of the E^0 value of the Co^{III}/Co^{II} couple does not exactly run with the increasing ligand field strength.**

E^0 **(for ligands):** $H_2O \rangle$ edta$^{4-} \rangle$ ox$^{2-} \rangle$ phen $\rangle$ NH$_3 \rangle$ en $\rangle$ CN$^-$

Ligand field strength: CN$^- \rangle$ phen $\rangle$ en $\rangle$ edta$^{4-} \rangle$ NH$_3 \rangle$ ox$^{2-} \rangle$ H$_2$O

It indicates that en is a weaker field ligand than phen, but en stabilises the Co^{III} state (compared to the Co^{II}–state) **more than phen, i.e.** $\dfrac{K_3}{K_2}(en) \rangle \dfrac{K_3}{K_2}(phen)$. In fact, the E° value is determined by both the **enthalpic** and the **entropic factor.** The enthalpic contribution is controlled by the factors like cfse, metal-ligand π-bonding, etc. while the entropic contribution is controlled by many other factors like the relative electrorestriction over the solvent by the oxidant and reductant of the couple, etc.

- **Role of cfse:** In most of the cases, cfse stabilises the low-spin Co(III) complex more than the high-spin complex of Co(II). The net gain of cfse can be calculated as follows:

low-spin $Co^{III}(t_{2g}^6 e_g^0)$: $24Dq_0$ (Co^{III}) $- 2P(Co^{III}) \approx 36Dq_0(Co^{II}) - 2P(Co^{III})$

high-spin $Co^{II}(t_{2g}^5 e_g^2)$: $8Dq_0$ (Co^{II})

Thus the gain of cfse for attaining Co(III) (l.s.) from Co(II) (h.s.) $\approx 28Dq_0(Co^{II}) - 2P(Co^{III})$.

This is the driving force for the ease of oxidation of Co(II) to Co(III). Obviously, with the increase of ligand field strength (*i.e.* Dq_0), oxidation of Co(II) to Co(III) becomes favoured. This makes the E^0 value more negative, in general, for the stronger field ligand. In other words, in presence of the weak field ligand, the Co(III) state becomes less stable compared to the Co(II) state. On the other hand, **in presence of the strong field ligand, Co(III) becomes more stable than Co(II).** This is illustrated by the following facts:

$Co(III)$-aqua complex $+ H_2O \longrightarrow Co(II)$-aqua complex $+ O_2$, **(effect of the weak field ligand)**

$Co(II)$-en complex $+ O_2 \longrightarrow Co(III)$-en complex, **(effect of the strong field ligand)**

$Co(aq)^{3+}$ complex oxidises water to release O_2. While in presence of the strong field ligands like NH$_3$, en, etc., **air (*i.e.* O_2 present in air) can oxidise Co(II) to Co(III).**

- **en vs. phen:** Gain of cfse for h.s Co(II) to l.s. Co(III) $\approx 28Dq_0(Co^{II}) - 2P(Co^{III})$.

For the en and phen ligands, it becomes *ca.* 1410 kJ mol^{-1} and *ca.* 1500 kJ mol^{-1} respectively. It indicates that for the phen ligand, E^0 value of the Co^{III}/Co^{II} couple should be more negative than that for the en ligand. But, in reality, the reverse is true. It indicates that besides the cfse gain, other factors are also important.

In fact, phen is a π–acid ligand but en is a σ–donor ligand. The π–bonding (metal $\rightarrow$ ligand) is favoured for the lower oxidation states of the metal centre. *Thus, phen stabilises the Co(II)–complex more than the Co(III)–complex by dint of the metal-ligand π–bonding interaction while cfse stabilises the Co(III) complex more than the Co(II) complex.* By considering the both π–bonding effect and cfse effect, en stabilises Co(III) more relative to Co(II) than phen. This explains: E^0 **(phen)** $\rangle$ E^0**(en)**

- **CN$^-$ vs phen:** Both CN$^-$ and phen are the π–acid ligands but CN$^-$ is a stronger field ligand than phen. In fact, CN$^-$ gives the low-spin complex for both Co(II) and Co(III) while phen gives the low-spin complex for Co(III) but the high-spin complex for Co(II). For the oxidation of Co(III) to Co(II), there is a gain of cfse.

Gain of cfse (for phen): $24Dq_0(Co^{III}) - 2P (Co^{III}) - 8Dq_0 (Co^{II}) \approx 1500$ kJ mol^{-1}

Gain of cfse (for CN$^-$): $24Dq_0(Co^{III}) - 2P(Co^{III}) - \{18Dq_0 (Co^{II}) - P(Co^{II})\} \approx 1680$ kJ mol^{-1}

Thus in terms of the gain of cfse, in presence of CN^-, the E^0 value is expected to be more negative than the case of phen ligand. Higher positive charge on Co(III) can accommodate the anionic ligands better than Co(II). *For the oxidation of Co(II) to Co(III), there is a large gain of cfse.*

Both phen and CN^- are the π–acid ligands and the lower oxidation state (*i.e.* Co^{II}) is expected to be stabilised more than the higher oxidation state (*i.e.* Co^{III}) due to the metal $\rightarrow$ ligand π–bonding effect. *But the higher oxidation state (i.e. Co^{III}) is stabilised better by the anionic CN^- ligands than the neutral phen ligand.* In fact, compared to the neutral π-acid ligand, the anionic π-acid ligand CN^- can stabilise Co(III) better.

Moreover, the **entropic factor** (determined mainly by the **degree of electrorestriction over the solvent molecules** by the reactant and product) favours the oxidation of Co(II) to Co(III) for the CN^- ligand (*cf.* Sec. 4.9.1).

Charge annihilation: $\left[\text{Co(CN)}_5\right]^{3-} + CN^- \rightleftharpoons \left[\text{Co(CN)}_6\right]^{3-} + e$, ΔS = **positive**;

Charge creation: $\left[\text{Co(phen)}_3\right]^{2+} \rightleftharpoons \left[\text{Co(phen)}_3\right]^{3+} + e$, ΔS = **negative**

Thus compared to phen, the more negative E^0 value for CN^- arises due to the following reasons:

(i)   more gain of cfse for the conversion of Co(II) to Co(III) for the CN^- ligand (compared to the phen ligand) (*cf.* 1680 kJ mol^{-1} *vs.* 1500 kJ mol^{-1}).

(ii) better stabilisation of the Co(III) state by by anionic CN^- ligand (compared to phen).

(iii) For the oxidation of Co(II) to Co(III), **entropic favour** runs as: $L = CN^- \rangle L = \text{phen}$.

The high negative value of E^0 for the CN^- ligand allows the reduction of H_2O (to H_2) by $[\text{Co(CN)}_5(\text{OH}_2)]^{3-} + CN^-$ giving rise to $[\text{Co(CN)}_6]^{3-}$

$$\left[\text{Co(CN)}_5\left(\text{OH}_2\right)\right]^{3-} + CN^- + H_2O \xrightarrow{\text{slow}} \left[\text{Co(CN)}_6\right]^{3-} + H_2(\uparrow)$$

(B) Fe^{III}/Fe^{II} couple: Variation of the reduction potential of the couple Fe(III)/Fe(II) with the variation of the nature of ligand:

Couple	E^0(V)
$[\text{Fe(OH}_2)_6]^{3+}/[\text{Fe(OH}_2)_6]^{2+}$	0.76
$[\text{FeF}_6]^{3-}/[\text{FeF}_6]^{4-}$	0.4
$[\text{Fe(CN)}_6]^{3-}/[\text{Fe(CN)}_6]^{4-}$	0.36
$[\text{Fe(phen)}_3]^{3+}/[\text{Fe(phen)}_3]^{2+}$	1.14

The reasons behind the variation of these E^0 values have been discussed in Sec. 4.9.1 and Sec. 16.4.4, Vol. 3 of Fundamental Concepts of Inorganic Chemistry.

3.15 MERITS AND DEMERITS OF CFT

3.15.1 Merits of CFT

(i)   **Spectral properties:** The spectral properties (specially the ligand field bands) of the coordination compounds can be explained by CFT. This aspect has been discussed in detail in Chapter 7.

(ii) **Magnetic properties:** The theory can interpert the magnetic properties of the complexes (*cf.* Chapter 8). It can interpret the temperature dependent paramagnetism (TIP), specially at the cross-over region.

(iii) **Stereochemical preference:** It can explain the stereochemical preference of the different metal centres.

(iv) **Distortion:** It can explain the distortion in the complexes.

(v) **Thermodynamic properties:** The theory can explain the different thermodynamic properties of complexes like the Irving-Williams stability order, hydration energies, lattice energy, crystal radii, etc.

(vi) **Kinetic properties:** The theory can explain the kinetic properties of the complexes in terms of crystal field activation energy (CFAE) (Chapter 5).

$$CFT \xleftarrow{\quad LFT \quad} MOT$$

Ligand field theory (LFT), *i.e.* adjusted crystal field theory (ACFT) invokes the basic assumptions of the **purely electrostatic CFT model** with some covalent interaction between the metal and ligand. Thus LFT deviates to some extent from the ideal CFT but does not ignore the basic concepts of CFT. **Consequently, LFT is a blend of two extreme concepts: CFT and MOT.** In other words, LFT bridges between these two extreme concepts—electrostatic model and covalent model. The pattern splitting of the *d*–orbitals proposed in the original CFT, remains unchanged in the modified theories LFT and MOT. *Here it is worth mentioning that when the metal centres are in their normal oxidation states, the overlap between the metal and ligand orbitals is not too large.* Thus, LFT is ideal for most of the complexes. Here it may be mentioned that the important conclusions derived from LFT and MOT are the same in most of the cases. In literature, if not mentioned strictly, the terms LFT and CFT are equivalently used.

3.15.2 Demerits of CFT

(i) **Theoretical drawback:** CFT considers the ligands as the **point-like charges** while the metal centre is considered to be associated with its electrons in its orbitals. Thus in the metal-ligand interaction, the metal centres are considered to bear the *wave mechanical term* while the ligands are considered as the point-like charges. In fact, the electrons on the ligands are also associated with different orbitals. Thus, CFT invokes the self-contradictory ideas.

(ii) **Purely electrostatic model:** The CFT assumes the purely electrostatic interaction between the metal centre and ligand. But this idea can never be strictly true. The metal centre is associated with its electrons in different orbitals and the ligands are also associated with their electrons in different orbitals. Thus, the orbital overlap, *i.e.* mixing of the metal and ligand orbitals must occur at least to some extent. This ovarlaping interaction is experimentally supported by many facts, *e.g. nephelauxetic effect, reduction of interelectronic repulsion (i.e. Racah parameter decreases on complexation), hyperfine and super-hyperfine structure of the esr signal as in $[IrCl_6]^{2-}$,* etc.

(iii) **Charge transfer band:** The purely electrostatic model fails to explain the origin of the intense charge transfer bands.

(iv) **Intensities of the ligand field bands due to d—d transition:** CFT cannot explain these as the *d—d* transitions are the forbidden transitions.

(v) **Antiferromagnetic exchange interaction:** The pure electrostatic model (without any overlap between the metal and ligand orbital) cannot explain this magnetic property.

(vi) **Spectrochemical series:** In terms of the pure electrostatic model, the anionic ligands are expected to be the stronger field ligands than the neutral ligands. But many anionic ligands like halides are the weak field ligands while many neutral ligands (*e.g.* CO, bpy, phen, etc.) are the strong field ligands. H_2O is a stronger field ligand than OH^-. The dipole moments of H_2O and NH_3 are in the reversed order of their crystal field strength order. *It fact, interpretation of the spectrochemical series requires the concept of overlap between the metal and ligand orbitals.*

(vii) **π–Bonding:** The π–bonding interaction is important to explain the ligand field strength of many ligands (i.e. position of the ligands in the spectrochemical series), stability of many complexes, kinetic properties of many complexes. But, the π–bonding interaction is not considered in CFT.

3.16 EXPERIMENTAL EVIDENCES IN FAVOUR OF METAL-LIGAND OVERLAP IN THE COORDINATION COMPOUNDS

(A) Electron spin resonance (ESR) spectrum of the coordination compounds (see Chapter 12) with special reference to that of $[IrCl_6]^{2-}$

When there is an unpaired electron, application of an external magnetic field will split the two degenerate spin states $\left(M_s = +\frac{1}{2}, -\frac{1}{2} \right)$. The transition from the lower energy state $\left(M_s = -\frac{1}{2} \right)$ to the higher energy state $\left(M_s = +\frac{1}{2} \right)$ will produce a **single absorption band** (*cf.* Fig. 3.16.2a). But, many transition metal complexes (even bearing only one unpaired electron) show the ESR spectra with hyperfine and superhyperfine structure. This hyperfine and superhyperfine structure can be explained by considering the overlap between the metal d–orbital and ligand orbital. *The nuclear magnetic moments of the ligand nuclei can affect the energy states of the metal electron through the metal-ligand covalent interaction.*

ESR Spectrum of $[IrCl_6]^{2-}$ ($K_2[IrCl_6]$ in $K_2[PtCl_6]$): Hyperfine or Superhyperfine?

- When an unpaired electron is not subjected to interact with the other unpaired electrons or magnetic nuclei, it will show a single absorption peak.
- When the electronic spin is coupled with *its nuclear spin,* then the peak splits to give the **hyperfine structure.** This peak splitting giving rise to the hyperfine structure may also occur when the electronic spin of the metal is coupled with the nuclear spin of the ligand.
- The unpaired electron in t_{2g}^5 of $[IrCl_6]^{2-}$ may interact with the nuclear spin of Ir and Cl. *Some authors believe that the nuclear magnetic moments of the Ir–isotopes are very small and coupling the electronic spin with the nuclear spin of Ir may be neglected.* They believe that the hyperfine structure of the ESR peak occurs due to the coupling of the electronic spin with the nuclear spins of Cl ($I = 3/2$). **This hyperfine structure of the ESR signal of $[IrCl_6]^{2-}$ strongly supports the overlap of the ligand orbital with the metal orbital.** The components (n) are given by: $n = 2I_{total} + 1 = 2 \times 6 \times 3/2 + 1 = 19$.

  However, instead of 19 components, 13 components have been experimentally observed. To explain this difference, it has been proposed that the nuclear moments of the two Cl–isotopes (*i.e.* ^{35}Cl, ^{37}Cl) are not quite equal and some of the lines are overlapped. Detailed calculation indicates that the Ir–electron spends about 30% of its time on the ligands (*i.e.* 5% on each Cl–atom). It gives the delocalisation factor (k) 0.7.

- The **superhyperfine structure** of the ESR signal is explained by the coupling of the electronic spin with the nuclear spins of both Ir and ligand Cl. This aspect has been described in detail in the text.

The ESR spectrum of $[IrCl_6]^{2-}$ shows a complex pattern **(superhyperfine structure caused by the ligands).** This can be explained by considering the splitting by the chloride ligands. The overlap between the Ir–orbital (bearing the unpaired electron) and the ligand (*i.e.* Cl^-) orbital allows the unpaired

Ir(IV)–electron (t_{2g}^5) to spend some time on the chloride ligands. The extent of splitting by the nuclear magnetic moments of the Cl⁻ ligand depends on the fraction of the Ir(IV)–electron spending time in the ligand orbitals.

Note: The actual nature of the splitting of ESR spectrum of $[IrCl_6]^{2-}$ anion depends on the nature of crystal. ***Depending on the nature of crystal packing, it may maintain the perfectly cubic or octahedral symmetry or tetragonal distortion.*** In the case of cubic symmetry, the ESR spectrum does not depend on the direction of the field axis. In the case of tetragonal symmetry (*i.e.* J.T. distortion), all the three axes are not equivalent and the nature of ESR spectrum depends on the choice of axis of the applied magnetic field. These are discussed in detail below.

(a) Details of the ESR spectrum of low-spin hexachloridoiridate(IV) ion (*i.e.* [IrCl₆]²⁻) of Na₂[IrCl₆] · 6H₂O crystallised in the diamagnetic host lattice of Na₂[PtCl₆] · 6H₂O

The electronic configuration of Ir^{IV} is t_{2g}^5. It indicates the J.T. distortion in the octahedral structure. The nuclear spin quantum number (I) is 3/2 for each of the nuclei ^{193}Ir, ^{35}Cl and ^{37}Cl. If the Ir(IV)–unpaired electron is expected to be localised in the d–orbital of Ir(IV) to experience the nuclear spin (= 3/2) of ^{193}Ir, then it should give four $\left[2 \times 1 \times 3/2 + 1 = 4\right]$ hyperfine lines in the ESR spectrum. The complex places two Cl⁻ ligands along each of the Cartesian axes. If all these Cartesian axes are considered to be equivalent, then the two Cl⁻ ligands along any axis will bring about *superhyperfine splitting* and each of the 4 hyperfine lines will be further split into seven $\left[2 \times 2 \times 3/2 + 1 = 7\right]$ lines. Thus if the magnetic field is applied along one Cl — Ir — Cl axis, then superhyperfine splitting will generate (7 × 4 =) 28 lines. *This superhyperfine structure (cf. Fig. 3.16.2) actually observed* (when the magnetic field is parallel to the x or y axis) supports the overlap interaction between the Ir(IV)–d-orbital and ligand (Cl⁻) orbital. When the magnetic field is along the z–axis, there are only 4 hyperfine lines caused by the ^{193}Ir nucleus alone, *i.e.* for this field direction no superhyperfine structure is noticed. The superhyperfine

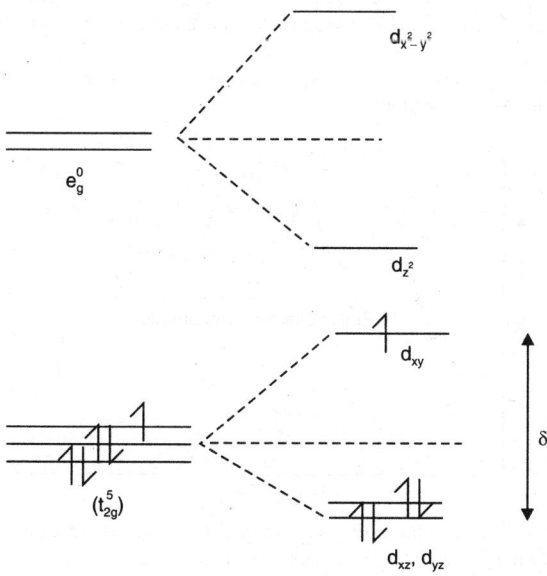

Fig. 3.16.1 J.T. distortion (z-out) in low-spin [IrCl₆]²⁻ (t_{2g}^5 system).

structure (when magnetic field is along the x or y axis) and hyperfine structure (when the applied magnetic is in the z–direction, *cf.* Fig. 3.16.3) can be explained by considering the **static z–out J.T. distortion** (Fig. 3.16.1).

The z–out distortion gives the additional efse $\left(=\dfrac{2}{3}\delta\right)$. *In this static tetragonally elongated complex (Fig. 3.16.1), the two Cl⁻ ligand along the ±z directions lie far away from the Ir–nucleus and they fail to interact through the orbital overlap.* On the other hand, the Cl⁻ ligands in the equatorial plane (*i.e.* along the ±x and ±y directions) are at closer distances to interact with the Ir(IV)–unpaired electron.

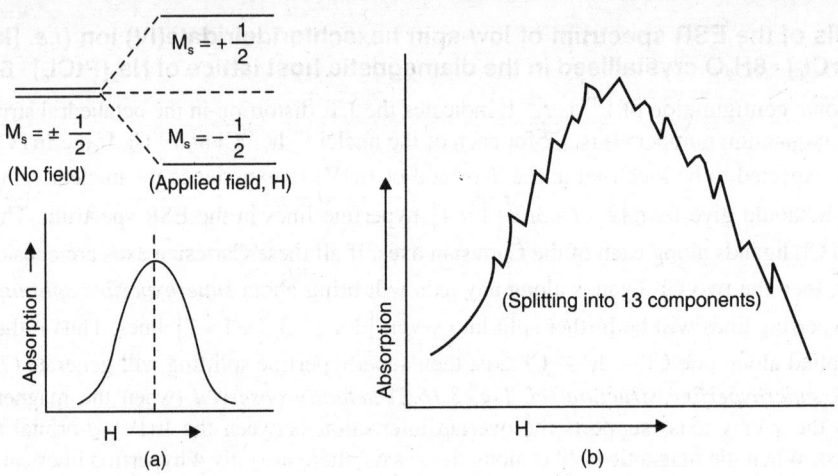

Fig. 3.16.2 (a) Splitting of two degenerate spin states of an unpaired electron on application of a magnetic field (H) causing a single absorption peak without any splitting peak, (*see* Chapter 12). (b) Qualitative representation of the ESR spectrum of $[IrCl_6]^{2-}$ present in $Na_2[IrCl_6] \cdot 6H_2O$ crystallised in the diamagnetic host crystal $Na_2[PtCl_6] \cdot 6H_2O$ with the applied field parallel to x– or y–axis.
(**Note:** Hyperfine splitting by ¹⁹³Ir followed by superhyperfine splitting by the chloride ligands. Tetragonal elongation, *i.e.* D_{4h} symmetry due to static z–out distortion).

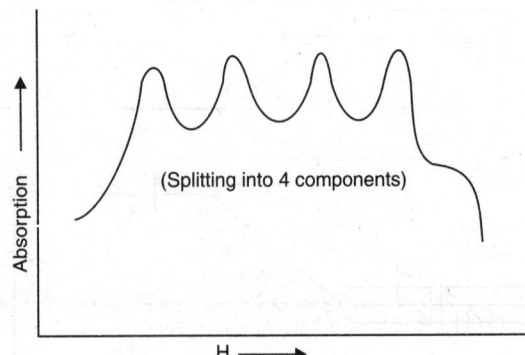

Fig. 3.16.3 Qualitative representation of the ESR spectrum of $[IrCl_6]^{2-}$ present in $Na_2[IrCl_6] \cdot 6H_2O$ crystallised in the diamagnetic host lattice $Na_2[PtCl_6] \cdot 6H_2O$ with the applied field along the z-axis.
Note: Only 4 hyperfine splittings due to ¹⁹³Ir and no superhyperfine splitting by the chloride ligands lying in the ±z directions. z–out distortion due to the static J.T. distortion.

When the ESR spectrum of $[IrCl_6]^{2-}$ is recorded with the applied field parallel to the x or y axis, 13 lines instead of 28 lines are obtained (*cf.* Fig. 3.16.2). It is due to the fact that some lines have overlapped. The ESR spectrum (*cf.* Figs. 3.16.2-3) of $Na_2[IrCl_6]$. $6H_2O$ crystallied in the **diamagnetic host lattice of $Na_2[PtCl_6] \cdot 6H_2O$** indicates the **static z–out J.T. distortion** in the anion.

(b) Details of the ESR spectrum of $(NH_4)_2[IrCl_6]$ crystallised in the diamaganetic host lattice of $(NH_4)_2[PtCl_6]$ (*cf.* Fig. 3.16.4)

(i) It has been already pointed out that the pattern of splitting of the ESR spectrum (*cf.* Fig. 3.16.2-3) of $[IrCl_6]^{2-}$ present in $Na_2[IrCl_6]$ indicates that the system lacks in cubic symmetry and three axes are not equivalent due to the **static z-out distortion**. But $[IrCl_6]^{2-}$ in $(NH_4)_2[IrCl_6]$ maintains the cubic symmetry (*i.e.* **dynamic J.T. distortion** or crystal packing force effect) and the ESR spectrum is the same along any of the three axes (*cf.* Fig. 3.16.4). In other words, the Cl^- ligands lying along any Cartesian axis can cause the superhyperfine splitting of the ESR signal.

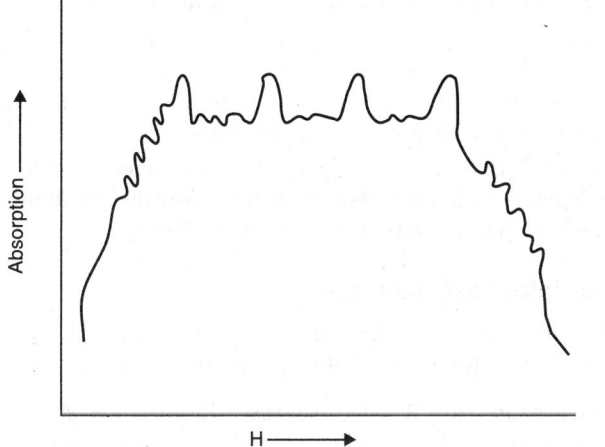

Note:
- The 4 hyperfine lines due to the nuclear spin of ^{193}Ir experience the superhyperfine splitting by the chloride ligands.
- Cubic symmetry due to the **dynamic J.T. distortion** (Ir^{IV}, t_{2g}^5) is maintained.

Fig. 3.16.4 Qualitative representation of the ESR spectrum of $[IrCl_6]^{2-}$ present in $(NH_4)_2[IrCl_6]$ crystallised in the diamagnetic host crystal $(NH_4)_2[PtCl_6]$ with the applied field parallel to x or y or z–axis.

(B) **NMR spectrum of the complexes:** Shift of the resonance frequency may be considered as an indirect evidence in favour of the metal-ligand orbital overlap. Comparison of the **PMR signals** of the ring proton (*i.e.* H–attached to the ring C) in the **paramagnetic complex [V(acac)₃]** and the **diamagnetic complex [Al(acac)₃]** indicates that the electronic environments of the ring proton are different in two cases. In [V(acac)₃], the t_{2g}^2 electrons are not really confined within the t_{2g} level. They can partly move into the pi-electron system of the acetylacetonate ligand and eventually into the $1s$ orbitals of the H–atoms. In the case of diamagnetic [Al(acac)₃] (d^0–system), there is no way to change the electronic environment of the ring proton. This explains the PMR signals for the said two comparable *tris*–(acetylacetonato) complexes at different resonance frequencies.

In the $[MF_6]^{2-}$ type octahedral complexes, the electrostatic character is expected to be maximum in the M—F bonds (*cf.* hard F^- is highly nonpolarisable). But, in reality, the ^{19}F—NMR spectra indicate ca. $2 - 6\%$ delocalisation of the metal d–electron into the ligand orbitals.

(C) **Nephelauxetic effect:** To interpret the electronic spectra (due to d—d absorption bands) of the transition metal complexes, the energy separations between the Russel-Saunders states are found smaller in the complexes compared to those in the free ions. These energy separations depend on the extent of repulsion among the d-electrons. The extent of ***inter-electron repulsion*** is measured by the parameter called **Racah parameter (B)**. The Racah parameters are found to be much smaller in the complexes compared to those in their free ions. For example, in $[NiX_4]^{2-}$ (X = Cl, Br, I), the B value is approximately reduced by about 30% compared to the value found in the free Ni^{2+} ion. **This reduction in the interelectron repulsion parameter (B) is an indication of the wider separation among the mutually repelling d-electrons.** In other words, it indicates the **expansion of d-electron cloud** causing a less repulsion among the d-electrons. The expansion of d-electron cloud is due to the metal d-electron delocalisation into the metal-ligand bond produced through the overlap of the metal and ligand orbitals.

The effect of ligands to expand the d-electron cloud to reduce the interelectorn repulsion is described as the nephelauxetic effect (*cf.* Greek word nephelauxetic meaning *cloud-expanding*).

The ligands with greater ability to delocalise the metal electrons will show the higher nephelauxetic effect, *i.e.* the lower B value indicates the greater delocalisation of the metal d-electron into the ligand orbital. The extent of this delocalisation is expressed in terms of the β-value as defined below:

$$\beta \text{ (Nephelauxetic constant)} = \frac{B_{complex}}{B_{gaseous}},$$

B = interelectronic repulsion parameter called the Racah parameter.

The lower value of β (*i.e.* higher nephelauxetic effect) indicates the higher delocalisation of the metal electrons into the ligand orbital. **This β-value depends on the extent of covalency in the metal-ligand bond and the nephelauxetic effect depends on the nature of both the metal and ligand.**

B value (in cm^{-1}) for gaseous ions

M^{2+}:	$V^{2+}(d^3)$	$Cr^{2+}(d^4)$	$Mn^{2+}(d^5)$	$Fe^{2+}(d^6)$	$Co^{2+}(d^7)$	$Ni^{2+}(d^8)$	$Cu^{2+}(d^9)$
B (in cm^{-1}):	765 $\langle$	830 $\langle$	960 $\langle$	1060 $\langle$	1120 $\langle$	1080 $\langle$	1240

- With the increase of the **number of d-electrons,** the B value increases. Besides this, the **spatial distribution** of the $3d$-orbitals also gradually decreases from V to Cu (*cf.* **contraction of d-orbitals** along the periods). It also causes an increasing trend of interelectronic repulsion.

M^{2+}:	$Fe^{2+}(d^6)$	$Ru^{2+}(d^6)$		$Ni^{2+}(d^8)$	$Pd^{2+}(d^8)$
B (in cm^{-1}):	1060	$\rangle$	620	1080 $\rangle$	830

- For the **heavier congeners,** the B value decreases because of the **larger spatial distribution** of the d-orbitals for the heavier congeners.

M^{n+}:	$Cr^{2+}(d^4)$	$Cr^{3+}(d^3)$
B (in cm^{-1}):	830 $\langle$	1030

- With the increase of **positive oxidation state,** the d-orbital is gradually **contracted** to increase the interelectronic repulsion.

- The interelectronic repulsion parameter B bears a good correlation with the electrostatic repulsion part of the pairing energy P (*cf.* Sec. 3.7).

The higher covalent bond forming property of the ligands will show the higher nephelauxetic effect. The ligands can be placed in the order of decreasing β-value and the series is called the **nephelauxetic series of the ligands** (with respect to a particular metal centre).

$$\beta = \dfrac{B_{complex}}{B_{gaseous}} \left\{ \begin{array}{l} \overrightarrow{\text{decreasing trend of } \beta\text{-value}} \\[2pt] F^- > H_2O > Urea > NH_3 > C_2O_4^{2-} \sim en \sim NCS^- > Cl^- \sim CN^- > Br^- > I^- \\[6pt] \overrightarrow{\text{Increasing tendency to introduce the covalence into the metal-ligand bond}} \end{array} \right.$$

It is evident that the *nephelauxetic series* is different from the *spectrochemical series*.

- From the definition, β can never be greater than unity (*i.e.* $\beta \le 1$). The value $\beta = 1$ indicates 100% ionic or electrostatic interaction in the metal-ligand bond. For the hard, *i.e.* **nonpolarisable ligands** like F^-, β value is close to unity while for the more covalently bonding ligands (*i.e.* **polarisable ligands**) like I^-, S^{2-}, the β value drops to about 0.3.

- The extent of covalency (*i.e.* nephelauxetic effect) in the metal-ligand bonds depends on the nature of both the metal and ligands (*cf.* **Fajans' Rule**, Vol. 2 of Fundamental Concepts of Inorganic Chemistry). Higher polarisability of the ligands and higher polarising power of the metal centre will introduce the higher covalence into the metal-ligand bond (*i.e.* lower β–value). In terms of the covalent bond forming power of the metal ions, a **nephelauxetic series of the metal ions** (for a particular ligand) can be constructed.

The approximate nephelauxetic series of the metal ions (with a particular ligand) is given below.

$$\overrightarrow{\text{increasing trend to introduce the covalency into the metal-ligand bond}}$$
$$Mn(II) < V(II) < Ni(II) < Mo(III) < Cr(III) < Fe(III) < Rh(III) < Co(III) < Pt(IV) < Mn(IV)$$

According to Jorgensen, the nephelauxetic effect is proportional to the product of the nephelauxetic parameters of the ligands (h) and metals (k), *i.e.*

$$\text{nephelauxetic effect} \propto hk.$$

$$\beta = \dfrac{B_{complex}}{B_{gaseous}} = 1 - hk.$$

Ligands:	F^-	H_2O	Urea	NH_3	en	ox^{2-}	Cl^-	CN^-	Br^-	N_3^-	I^-
h:	0.8	1.0	1.2	1.4	1.5	1.5	2.0	2.1	2.3	2.4	2.7

Metal Centres:	Mn(II)	V(II)	Ni(II)	Mo(III)	Cr(III)	Co(II)	Fe(III)	Rh(III)	Co(III)	Pt(IV)
k:	0.07	0.08	0.12	0.15	0.20	0.24	0.24	0.30	0.35	0.5

(D) **Spectral intensity:** The d–d transitions (*i.e.* ligand field bands) are forbidden. Thus in terms of 100% electrostatic interaction in the metal-ligand bond, we cannot explain the intensity of the d–d transitions. The tetrahedral complexes (lacking in centre of symmetry) give the more intense ligand field bands compared to the centrosymmetric octahedral complexes. All these aspects find an explanation if we consider the overlap between the metal d–orbitals and ligand orbitals. This aspect will be discussed in detail in Chapter 7.

(E) **Charge transfer band in metal complexes:** The existence of intense charge transfer bands in different metal complexes can only be explained by considering the overlap between the metal and ligand orbitals. This aspect will be discussed in detail in Chapter 7.

(F) **Intensity stealing phenomenon:** The intensity of a d–d transition lying close to a charge transfer band is increased. This is called the *intensity stealing phenomenon*. It indicates the mixing between the d–d transition and the charge-transfer transition (*cf.* Chapter 7).

(G) **Antiferromagnetic interaction through the superexchange mechanism:** This phenomenon is observed in many metal oxides (*e.g.* MnO, FeO, CoO, NiO, etc.) and in many ligand bridged polynuclear complexes. To explain this observation, we need the consideration of overlap between the metal and ligand orbitals (*cf.* Chapter 8).

(H) **Orbital reduction factor (k) and magnetic properties of the complexes:** This factor is required to explain the effect of covalency on the magnetic properties of complexes (*cf.* Chapter 8).

(I) **Spin-orbit coupling constant:** The reduction of spin-orbit coupling constant in a complex compared to that in the free ions is an indication of covalence in the metal-ligand bond (cf. Chapter 8).

3.17 MOLECULAR ORBITAL APPROACH OF BONDING IN COORDINATION COMPOUNDS

The MOT of bonding in coordination compounds considers: symmetry permitted overlap between the atomic orbitals of metal centre and suitable **ligand group orbitals** (LGOs). These LGOs actually represent the **terminal atom symmetry orbitals (TASOs)** (*cf.* Chapter 9, Vol. 2 of Fundamental Concepts of Inorganic Chemistry). In the resultant MOs, electrons are placed in terms of their energy as usual.

3.17.1 MOT of σ–Bonding in the Octahedral Complexes

Let us illustrate the case with the metal ions of 1st transition series. The filled 3s and 3p orbitals of the metal centres are of very low energy (*i.e.* **deeply seated**). They are radially so contracted that they **cannot participate in the overlap with the ligand orbitals.** The available metal orbitals to overlap with the ligand orbitals are given below:

five 3d orbitals + one 4s orbital + three 4p orbitial, *i.e.* total 9 atomic orbitals are available for the purpose of bonding with the ligand orbitals.

Bonding properties of the 9 atomic orbitals of metal in the octahedral system: The three Cartesian axes (*i.e.* x, y, an z axes) may be considered as the $3C_4$ axes, *i.e.* ligands are in the $\pm x$, $\pm y$ and $\pm z$ directions. The following six orbitals can participate in σ–type bonding interactions with the ligands at the corners of the octahedron (*i.e.* along the three Cartesian axes).

$$3d_{x^2-y^2}, 3d_{z^2}, 4s \text{ (spherically symmetrical)}, 4p_x, 4p_y \text{ and } 4p_z$$

The remaining three valence orbitals, *i.e.* $3d_{xy}$, $3d_{yz}$, and $3d_{xz}$ (projecting their lobes between the Cartesian coordinates) are not suitable for the σ–bonding.

Symmetry properties of the 6 σ–bonding atomic orbitals in an octahedral system: These are as follows:

$$4s: a_{1g} \text{ (spherically symmetrical)}$$

$$4p_x, 4p_y \text{ and } 4p_z: t_{1u} \text{ (a set of three equivalent orbitals)}$$

$$3d_{x^2-y^2} \text{ and } 3d_{z^2} : e_g \text{ (a pair of equivalent orbitals)}.$$

The above classifications are done by using the concept of Group Theory. Here we can rationalise the classifications by the method of pictorial representation.

Formation of the σ–LGOs with proper symmetries: We are to consider the σ–bonding orbitals of the ligands. For most of the ligands, these are composed of s and p atomic orbitals (*e.g.* sp^3 for H_2O, NH_3; p for halides, etc.). These obitals (available for the σ–type interaction with the metal) approaching from the directions $+x$, $-x$, $+y$, $-y$, $+z$ and $-z$ are described as σ_x, σ_{-x}, σ_y, σ_{-y}, σ_z, and σ_{-z} respectively. From these ligand orbitals, through the linear combinations, different *ligand group orbitals* (*LGOs*) can be constructed to match with the metal orbitals (*cf.* Fig. 3.17.1.1).

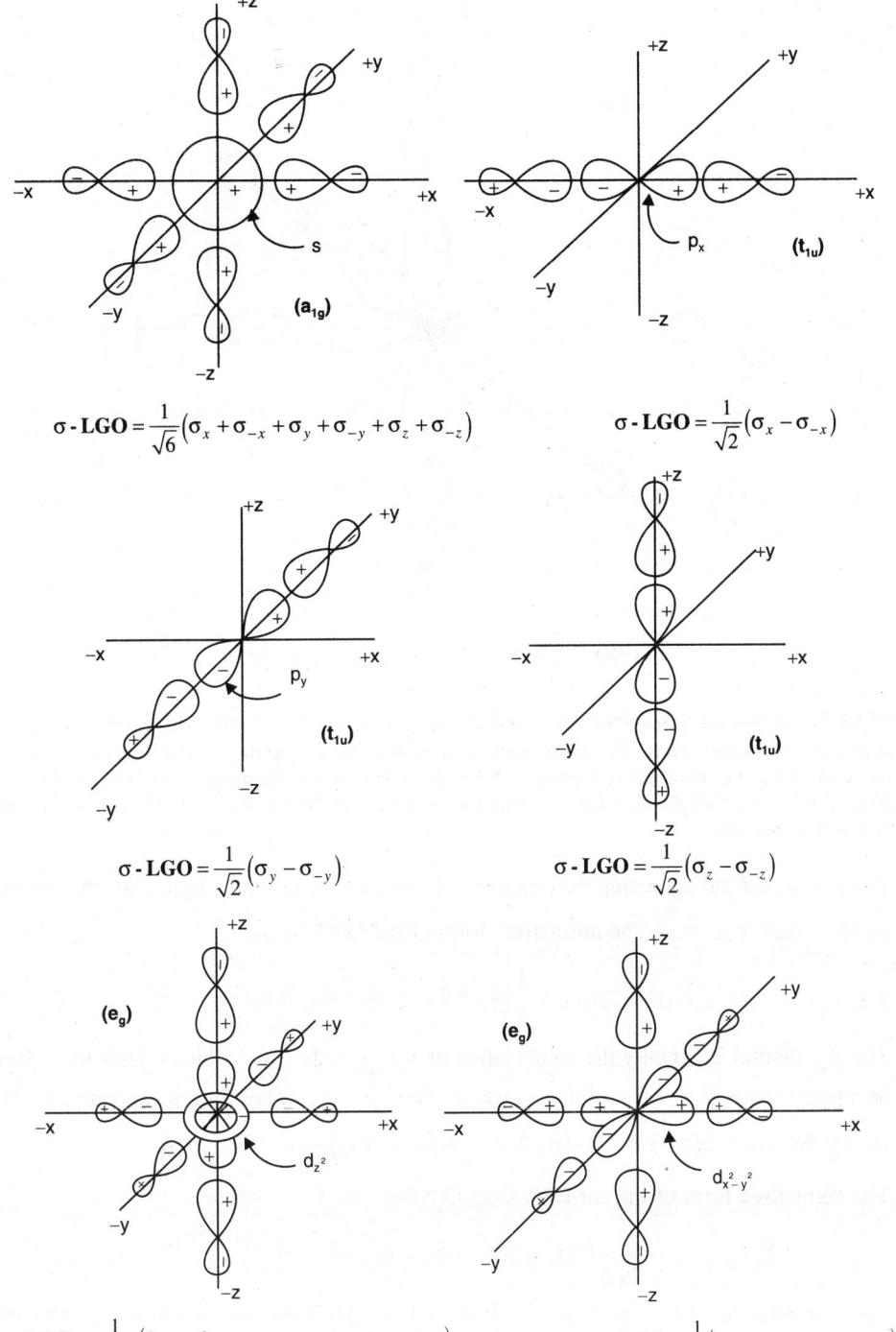

$$\sigma\text{-LGO} = \frac{1}{\sqrt{6}}\left(\sigma_x + \sigma_{-x} + \sigma_y + \sigma_{-y} + \sigma_z + \sigma_{-z}\right)$$

$$\sigma\text{-LGO} = \frac{1}{\sqrt{2}}\left(\sigma_x - \sigma_{-x}\right)$$

$$\sigma\text{-LGO} = \frac{1}{\sqrt{2}}\left(\sigma_y - \sigma_{-y}\right)$$

$$\sigma\text{-LGO} = \frac{1}{\sqrt{2}}\left(\sigma_z - \sigma_{-z}\right)$$

$$\sigma\text{-LGO} = \frac{1}{2\sqrt{3}}\left(2\sigma_z + 2\sigma_{-z} - \sigma_x - \sigma_{-x} - \sigma_y - \sigma_{-y}\right)$$

$$\sigma\text{-LGO} = \frac{1}{2}\left(\sigma_x + \sigma_{-x} - \sigma_y - \sigma_{-y}\right)$$

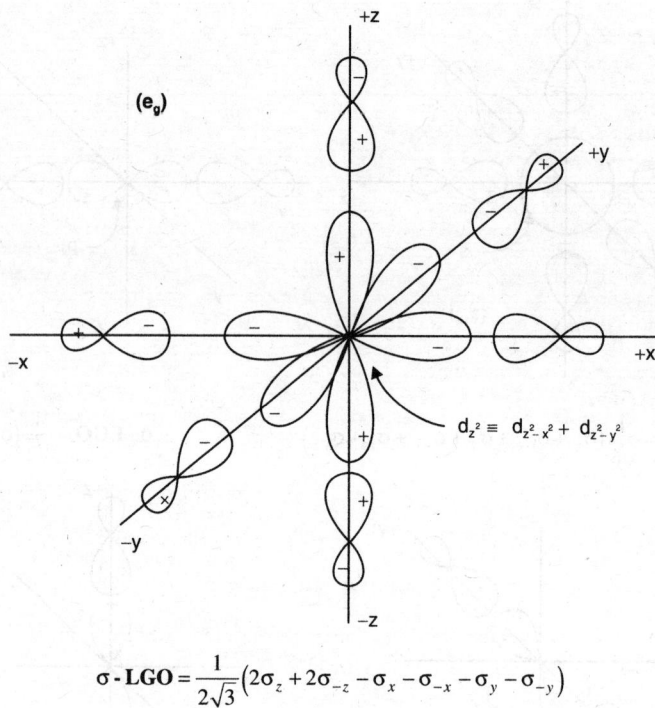

$$\sigma\text{-}\mathbf{LGO} = \frac{1}{2\sqrt{3}}\left(2\sigma_z + 2\sigma_{-z} - \sigma_x - \sigma_{-x} - \sigma_y - \sigma_{-y}\right)$$

Fig. 3.17.1.1 Sigma bonding metal orbitals $s(a_{1g})$, $p(t_{1u})$, $d_{x^2-y^2}\left(e_g\right)$ and $d_{z^2}\left(e_g\right)$ and their corresponding σ–LGOs in an octahedral system (LGOs). (**Note:** The linear combination method to construct the σ–LGOs is to be carried out by the inspection method. The ligand orbital projecting the lobe with +sign towards the metal centre is considered to make a positive (+) combination while the ligand orbital projecting the lobe with −sign makes a negative (−) combination. This convention is followed here).

(i) The $d_{x^2-y^2}$ **orbital** is bearing the opposite pairs of lobes with + or − sign. Thus the combination is: $(\sigma_x + \sigma_{-x} - \sigma_y - \sigma_{-y})$. The normalised form of the LGO is:

$$\Sigma_{x^2-y^2(\sigma-LGO)} = \frac{1}{2}\left(\sigma_x + \sigma_{-x} - \sigma_y - \sigma_{-y}\right);\ e_g$$

(ii) The d_{z^2} **orbital** is actually the combination of $d_{z^2-x^2}$ and $d_{z^2-y^2}$ orbitals. Thus the orbital may be represented as $d_{2z^2-x^2-y^2}$ or $d_{3z^2-r^2}$ where $r^2 = x^2 + y^2 + z^2$. Thus the corresponding LGO needs the combination: $2\left(\sigma_z + \sigma_{-z}\right) - \left(\sigma_x + \sigma_{-x}\right) - \left(\sigma_y + \sigma_{-y}\right)$

The normalised form of the corresponding LGO is:

$$\Sigma_{z^2(\sigma-LGO)} = \frac{1}{2\sqrt{3}}\left(2\sigma_z + 2\sigma_{-z} - \sigma_x - \sigma_x - \sigma_y - \sigma_{-y}\right);\ e_g$$

(iii) The **s–orbital** has the same sign in all directions. Thus the corresponding combination is:

$$\left(\sigma_x + \sigma_{-x}\right) + \left(\sigma_y + \sigma_{-y}\right) + \left(\sigma_z + \sigma_{-z}\right)$$

The normalised form of the LGO is:

$$\sum\nolimits_{s(\sigma-LGO)} = \frac{1}{\sqrt{6}}\left(\sigma_x + \sigma_{-x} + \sigma_y + \sigma_{-y} + \sigma_z + \sigma_{-z}\right); a_{1g}$$

(iv) For the **p–orbitals**, the opposite lobes are bearing the opposite sign, *i.e.* if one lobe bears the + sign, the opposite lobe bears the −sign. Thus the required combinations are:

$$\sigma_x - \sigma_{-x}, \sigma_y - \sigma_{-y}, \sigma_z - \sigma_{-z}$$

The normalised forms of the LGOs are:

$$\left.\begin{aligned}
\sum\nolimits_{x(\sigma-LGO)} &= \frac{1}{\sqrt{2}}\left(\sigma_x - \sigma_{-x}\right)\\
\sum\nolimits_{y(\sigma-LGO)} &= \frac{1}{\sqrt{2}}\left(\sigma_y - \sigma_{-y}\right)\\
\sum\nolimits_{z(\sigma-LGO)} &= \frac{1}{\sqrt{2}}\left(\sigma_z - \sigma_{-z}\right)
\end{aligned}\right\} t_{1u}$$

- **Formation of the MOs:** The σ–LGOs, *i.e.* $\sum_s, \sum_x, \sum_y, \sum_z, \sum_{x^2-y^2}$ and $\sum_{z^2}$ will combine with the corresponding metal orbitals to produce the six bonding molecular orbitals (σ–BMOs) and six antibonding molecular orbitals (σ*–ABMOs). The combinations are:

$$4s + \sum\nolimits_s : \sigma_{a_{1g}} \text{ and } \sigma^*_{a_{1g}} ; \quad 4p_x + \sum\nolimits_x : \sigma_{t_{1u}(x)} \text{ and } \sigma^*_{t_{1u}(x)}$$

$$4p_y + \sum\nolimits_y : \sigma_{t_{1u}(y)} \text{ and } \sigma^*_{t_{1u}(y)} ; \quad 4p_z + \sum\nolimits_z : \sigma_{t_{1u}(z)} \text{ and } \sigma^*_{t_{1u}(z)}$$

$$3d_z + \sum\nolimits_{z^2} : \sigma_{e_g(z^2)} \text{ and } \sigma^*_{e_g(z^2)} ; \quad 3d_{x^2-y^2} + \sum\nolimits_{x^2-y^2} : \sigma_{e_g(x^2-y^2)} \text{ and } \sigma^*_{e_g(x^2-y^2)}.$$

- **Energy order of the MOs:** The actual energy order of the MOs experiences some uncertainty. But the approximate order can be predicted with some reasoning. For the transition metal ions, the energy order of the atomic orbitals is: $(n-1)d \langle ns \langle np$, *i.e.* $3d \langle 4s \langle 4p$. For most of the ligands like NH_3, H_2O and F^-, the ligand σ–orbitals are lower in energy compared to the metal valence orbitals.

(i) Compared to the 4s (*i.e.* a_{1g}) and 4p (*i.e.* t_{1u}) orbitals, the 3d orbitals (*i.e.* e_g in the present case) are large and diffuse. **This is why, the 4s and 4p orbitals can make the overlap better than the 3d orbitals.** This is why, the BMOs of a_{1g} and t_{1u} symmetry are more stabilised compared to the BMOs of e_g symmetry. Consequently, the ABMOs of a^*_{1g} and t^*_{1u} symmetry are destabilised more compared to the ABMOs of e^*_g symmetry.

(ii) The remaining three metal orbitals, *i.e.* d_{xy}, d_{yz} and d_{zx} (of t_{2g} symmetry) remain as the nonbonding ones.

(iii) The 6 BMOs $(3t_{1u} + 2e_g + 1a_{1g})$ are relatively energetically closer to the ligand orbitals, *i.e.* the BMOs are more enriched with the character of the ligand orbitals. *In other words, the electrons placed in the BMOs are attracted more towards the ligands to introduce polarity in the metal-ligand bond.*

(iv) The 6 ABMOs are relatively energetically closer to the metal orbitals, *i.e.* they are more enriched with the metal orbital character.

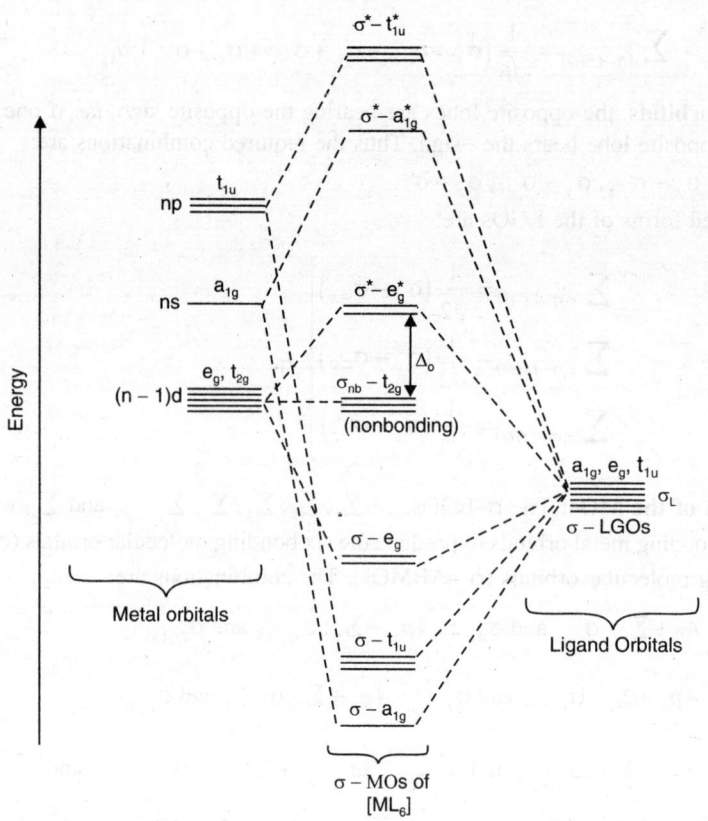

Fig. 3.17.1.2 MO energy level diagram of the metal-ligand σ–bonding in an octahedral complex.

- **Relation between the CFT and MOT:** In CFT, the metal *d*–orbitals are splitted into the t_{2g} and e_g sets. *These are the pure metal d-orbitals in terms of CFT.* The central part of the molecular orbital diagram consists of NBMO- t_{2g} (which is purely metal orbital) and e_g^* (enriched with the character of metal orbital). *This central part actually represents with the idea of CFT.* Qualitatively, both the theories lead to the same result. The energy difference between t_{2g} and e_g^* ($\equiv e_g$ in terms of CFT) is called $10Dq_0$ or Δ_0.

- **Electronic configuration of the octahedral metal complexes:** Here, the possibility of pairing in the t_{2g} level leading to the low-spin complex depends on the relative values of Δ_0 (difference between t_{2g} and e_g^*) and P (pairing energy). Some representative examples are given below.

$$\left[\text{Ti}(\text{OH}_2)_6\right]^{3+} (13e = 6 \times 2e + 1e): a_{1g}^2\, t_{1u}^6\, e_g^4\, t_{2g}^1$$

$$\left[\text{Co}(\text{NH}_3)_6\right]^{3+} (18e = 6 \times 2e + 6e): a_{1g}^2\, t_{1u}^6 e_g^4\, t_{2g}^6,\ 10Dq_0 > P$$

$$\left[\text{CoF}_6\right]^{3-} (18e = 6 \times 2e + 6e): a_{1g}^2 t_{1u}^6 e_g^4\, t_{2g}^4\, e_g^{*2},\ 10Dq_0 < P.$$

The reasons behind the conditions, $10Dq_o \rangle P$ for $[Co(NH_3)_6]^{3+}$ and $10Dq_o \langle P$ for $[CoF_6]^{3-}$ will be discussed in detail later in terms of MOT.

3.17.2 Molecular Orbital Theory of π–Bonding in the Octahedral Complexes

Without the group theoretical treatment, here we shall proceed to develop the π–bonding scheme just by considering the pictorial representation of the orbitals.

The orbitals on the metal centre which can participate in π–bonding are:

$$t_{1u} \ (i.e. \ p_x, \ p_y, \ p_z) \ \text{and} \ t_{2g} \ (i.e. \ d_{xy}, \ d_{yz}, \ d_{xz})$$

The t_{1u} set, *i.e.* p–orbitals can also participate in σ–bonding. The σ–bonds are stronger than the π–bonds. *This is why, the metal p–orbitals prefer to form the stronger σ–bonds and they do not participate in the π–bonding process.*

For the ligands, the orbitals which are perpendicular to the internuclear axis can participate in π–bonding. In such π–bonding, the ligands can act as the π–donors as well as the π-acceptor ligands.

For the π–bonding purpones, the ligands can provide the following orbitals (*cf.* Fig. 10.7.8.1 Vol. 2 of Fundamental Concepts of Inorganic Chemistry)

(i) p_π–orbitals perpendicular to the M—L σ–bond;
(ii) d_π–orbitals as in R_3P, R_3As, R_2S, etc.
(iii) Suitable molecular orbitals like π^* in the ployatomic ligands *e.g.* CO, CN⁻ etc.

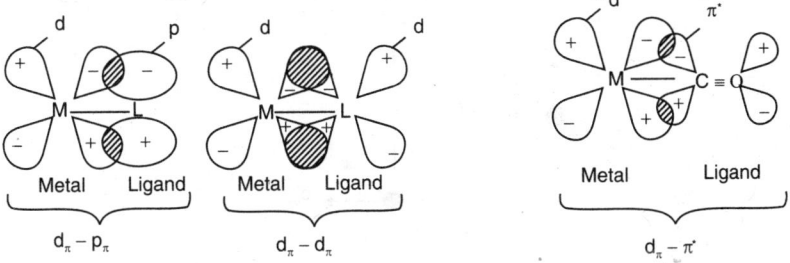

Fig. 3.17.2.1 Different types of metal-ligand π–bonding. (*cf.* Fig. 10.7.8.1, Vol. 2 of Fundamental Concepts of Inorganic Chemistry).

Table 3.17.2.1 Pi(π)–bonding in the metal-ligand bonds.

	Type	*Examples of ligands*
(a)	Ligand → metal $(p_\pi \to d_\pi)$	X⁻ (halide), RO⁻, O^{2-} ⁻OH, R_2N^-, etc.
(b)	Metal → Ligand $(d_\pi \to d_\pi)$ From the filled metal d–orbital to the vacant d–orbital of the ligand	R_3P, R_3As, R_2S
(c)	Metal → ligand $(d_\pi \to \pi^*)$ From the filled metal d–orbital to the vacant π^* of the ligand	CO, CN⁻, bpy, phen, py, alkene, N_2, NO_2^-, etc.
(d)	Metal → ligand $(d_\pi \to \sigma^*)$ From the filled metal d–orbital to empty σ^*–MO of the ligand	H_2, alkane, R_3P

Metal centre may use the d-p hybrid orbitals instead of the pure d-orbitals. R_3P may use the vacant d-orbital or vacant σ^*-MO to receive electron from the metal.

Different types of π-bonding between the metal d-orbital and ligand orbitals are shown in Fig. 3.17.2.1.

We shall first pay attention to the ligand p-orbitals. The six ligands can provide 12 p_π-orbitals. These may be represented by a set of two arrows attached to each of the donor site of the ligands of the octahedron (Fig. 3.17.2.2). The arrows actually denote the directions of Cartesian axes.

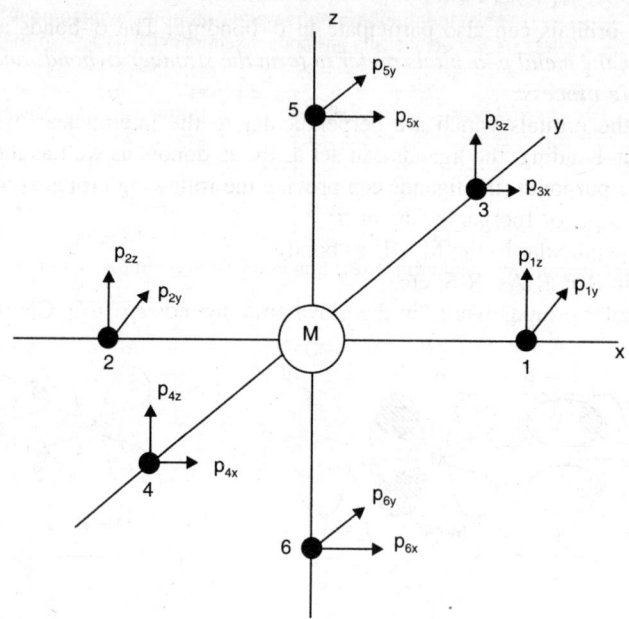

Fig. 3.17.2.2 Representation of the p_π orbitals of the ligands in an octahedral complex. The ligands (●) are denoted by the numbers 1, 2, 3, 4, 5 and 6.

The 12 p_π-orbitals of the ligands are given below:

L_1 (+x direction): $p_{\pi(1z)}$, $p_{\pi(1y)}$

L_2 (−x direction): $p_{\pi(2z)}$, $p_{\pi(2y)}$

L_3 (+y direction): $p_{\pi(3x)}$, $p_{\pi(3z)}$

L_4 (−y direction): $p_{\pi(4x)}$, $p_{\pi(4z)}$

L_5 (+z direction): $p_{\pi(5x)}$, $p_{\pi(5y)}$

L_6 (−z direction): $p_{\pi(6x)}$, $p_{\pi(6y)}$

From the given 12 p_π-orbitals of the ligands, we can have 4 sets of LGOs: t_{1g}, t_{2g}, t_{1u} and t_{2u}. (For details, it needs the knowledge of *Group Theoretical Treatment*). There are no metal t_{1g} and t_{2u} orbitals and consequently the corressponding LGOs will remain as the nonbonding ones. The t_{1u} set will weaken the σ-bonding as the t_{1u} set (*i.e.* metal p-orbitals) is already engaged in the σ-bonding. This is why, the t_{1u} set will also remain as a nonbonding one. Generation of the two LGO sets, *i.e.* t_{1u} and t_{2g} through the combination of the ligand p_π-orbitals is shown in Table 3.17.2.1.

Table 3.17.2.1 Generation of the LGOs for π–bonding from the p_π–orbitals of the ligands in an octahedral complex (*cf.* Figs. 3.17.2.1, 3).

π–LGO set	Corresponding metal orbitals	Combination of the ligand p_π–orbitals
t_{1u}	$\begin{cases} p_x \\ p_y \\ p_z \end{cases}$	$\begin{cases} \Sigma_{\pi_x} = \dfrac{1}{2}\left(p_{3x} + p_{4x} + p_{5x} + p_{6x}\right) \\[2mm] \Sigma_{\pi_y} = \dfrac{1}{2}\left(p_{1y} + p_{2y} + p_{5y} + p_{6y}\right) \\[2mm] \Sigma_{\pi_z} = \dfrac{1}{2}\left(p_{1z} + p_{2z} + p_{3z} + p_{4z}\right) \end{cases}$
t_{2g}	$\begin{cases} d_{xy} \\ d_{yz} \\ d_{xz} \end{cases}$	$\begin{cases} \Sigma_{\pi_{xy}} = \dfrac{1}{2}\left(p_{1y} - p_{2y} + p_{3x} - p_{4x}\right) \\[2mm] \Sigma_{\pi_{yz}} = \dfrac{1}{2}\left(p_{3z} - p_{4z} + p_{5y} - p_{6y}\right) \\[2mm] \Sigma_{\pi_{xz}} = \dfrac{1}{2}\left(p_{1z} - p_{2z} + p_{5x} - p_{6x}\right) \end{cases}$

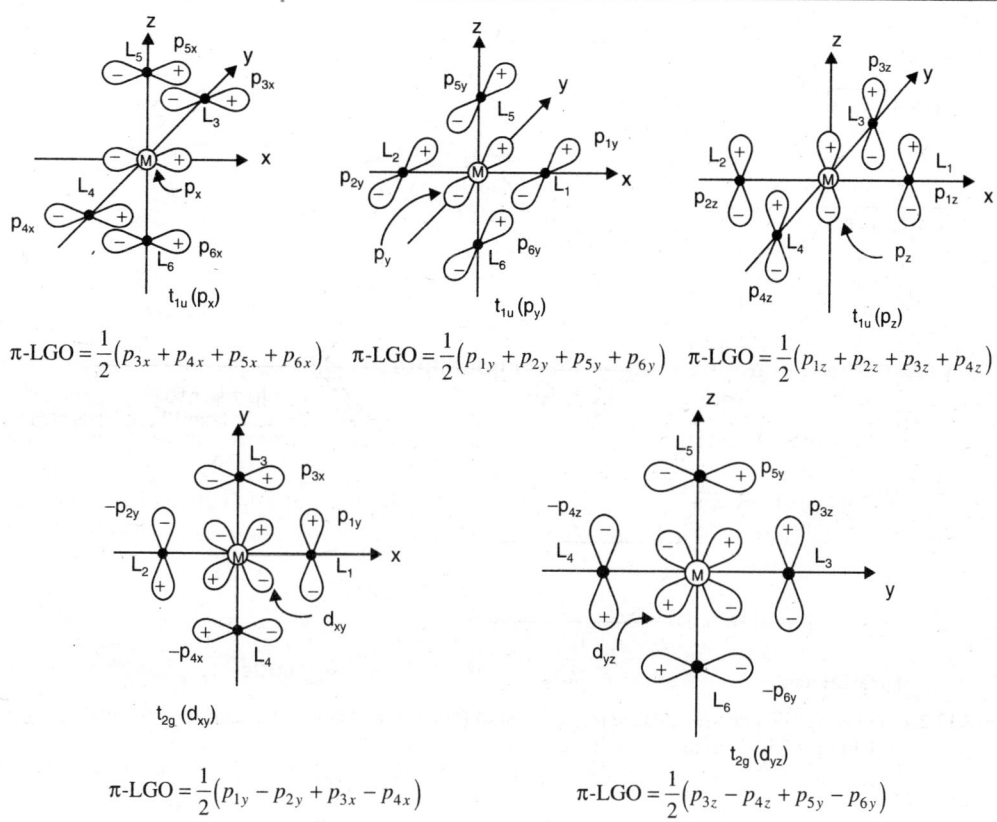

$\pi\text{-LGO} = \dfrac{1}{2}\left(p_{3x} + p_{4x} + p_{5x} + p_{6x}\right)$ $\pi\text{-LGO} = \dfrac{1}{2}\left(p_{1y} + p_{2y} + p_{5y} + p_{6y}\right)$ $\pi\text{-LGO} = \dfrac{1}{2}\left(p_{1z} + p_{2z} + p_{3z} + p_{4z}\right)$

$t_{1u}(p_x)$ $t_{1u}(p_y)$ $t_{1u}(p_z)$

$t_{2g}(d_{xy})$ $t_{2g}(d_{yz})$

$\pi\text{-LGO} = \dfrac{1}{2}\left(p_{1y} - p_{2y} + p_{3x} - p_{4x}\right)$ $\pi\text{-LGO} = \dfrac{1}{2}\left(p_{3z} - p_{4z} + p_{5y} - p_{6y}\right)$

(Contd.)

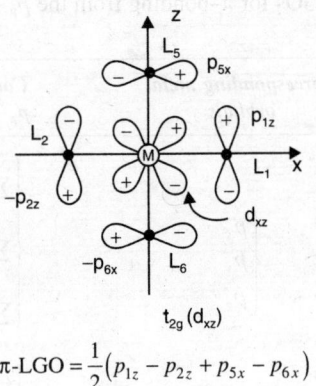

$$\pi\text{-LGO} = \frac{1}{2}\left(p_{1z} - p_{2z} + p_{5x} - p_{6x}\right)$$

Fig. 3.17.2.3 Pictorial representation of the formation of π–LGOs matching with the metal t_{1u} (*i.e.* p_x, p_y and p_z) and t_{2g} (*i.e.* d_{xy}, d_{yz} and d_{zx}) orbitals of an octahedral system.

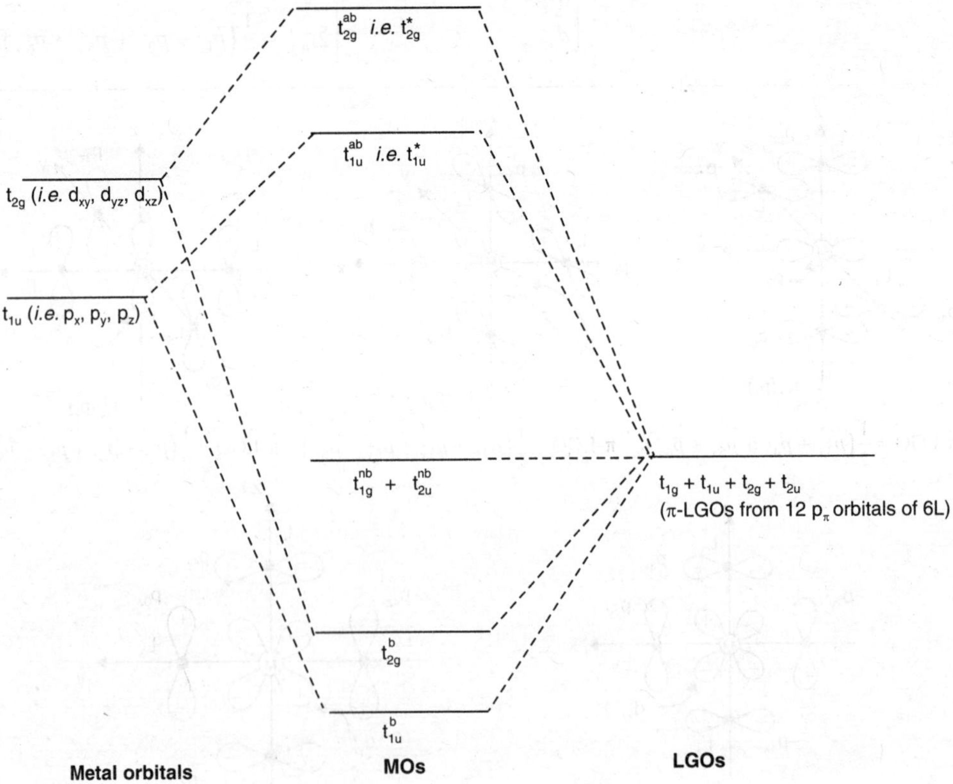

Fig. 3.17.2.4 MO energy diagram (qualitative representation) for the π–bonding in the octahedral complexes (assuming the t_{1u} set of metal to participate in the π–bonding).
(**Note:** Generally, t_{1u} (π-LGO) remains as nonbonding because the metal t_{1u} set is involved in the stronger σ-bonding interaction.)

The π–bonding in octahedral complexes is shown in Fig. 3.17.2.4.

The MO energy level diagram of an octahedral complex with the both σ and π bonding is shown in Fig. 3.17.2.5 where the π–bonding with the t_{1u}–set of π–LGO is ignored, *i.e.* t_{1u} (π–LGO) remains as nonbonding.

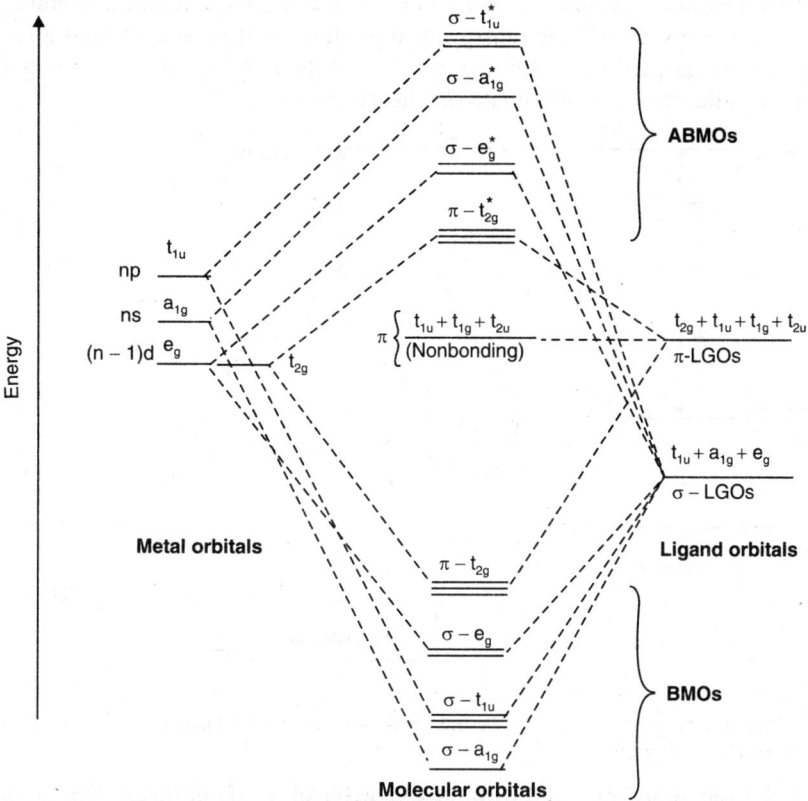

Fig. 3.17.2.5 MO energy level diagram of an octahedral complex involving both the σ–and π–bonding simultaneously. (**Note:** The π–LGOs of t_{1u} are kept as nonbonding because the corresponding metal orbitals preferably participate in σ–bonding. There may be some uncertainty in the relative positions of BMOs as well as the ABMOs.)

d_π–orbitals on the ligands: In PR_3, AsR_3, R_2S, etc., the vacant $d\pi$–orbitals (*i.e.* d_{xy}, d_{yz} and d_{xz}) can participate in π–bonding with the t_{2g} set of the metal centre in the octahedral complexes.

π^* MOs on the polyatomic ligands: In CO, CN^-, C_2H_4, etc. the vacant π^*–MOs can participate in π–bonding. In $[Cr(CO)_6]$, the t_{2g} set (*i.e.* d_{xy}, d_{yz} and d_{xz}) of the metal centre can participate in π–bonding with the vacant π^*–MOs of the carbonyls.

3.17.3 Effect of π–Bonding on the Value of $10Dq_o$ (*i.e.* Δ_o) of Octahedral Complexes

The π–LGO (t_{2g}) orbitals interact with the t_{2g}–set (that remains nonbonding in the metal-ligand σ–bonding interaction only) of the metal centre to produce the t_{2g} bonding π–MO and t_{2g}^* antibonding π–MO. In absence of such π–bonding interaction, the energy difference between t_{2g} (non-bonding) of the metal centre and $\sigma_{e_g^*}^*$, *i.e.* $\sigma - e_g^*$ gives the measure of $10Dq_o$ (*cf.* Fig. 3.17.1.2). But in presence of the π–bonding interaction, the energy difference between $\sigma - e_g^*$ and $\pi - t_{2g}^*$ or $\sigma - e_g^*$ and $\pi - t_{2g}$ determines the $10Dq_o$ value depending on the conditions.

The magnitude of $10Dq_0$ depends on the following factors:

(i) Whether π–LGO (t_{2g}) is of higher or lower energy compared to that of the metal t_{2g} orbital.

(ii) Whether the ligand π–orbitals are filled or empty.

(A) Case I: Filled ligand π–obitals (*i.e.* π–LGO t_{2g}) of lower energy than the metal t_{2g} orbitals.

If the ligand π–orbitals are of lower energy, then positions of the resulting MOs appear as in Fig. 3.17.3.1. The π–t_{2g}^* is enriched more with the character of the metal t_{2g} orbitals while π–t_{2g} (BMO) is enriched more with the character of π–LGO (*i.e.* ligand orbitals).

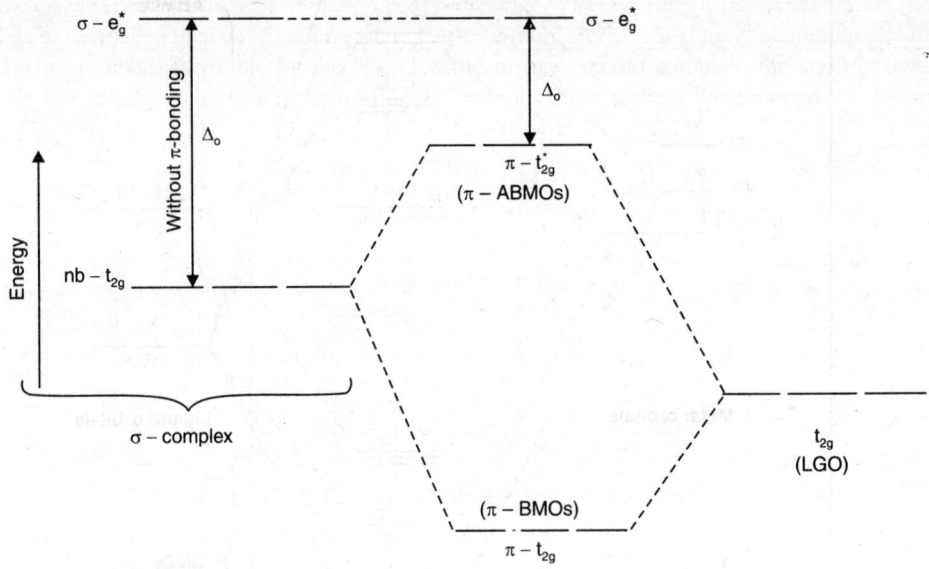

Fig. 3.17.3.1 Effect of π–bonding on Δ_0 when the ligands possess the filled orbitals (*i.e.* π-LGO t_{2g}) of lower energy. (compared to the metal t_{2g}–orbital).

The π–t_{2g}^*(relatively **more enriched with the character of t_{2g}–set of metal**) lies closer to the σ–e_g^* orbital. The energy difference between σ–e_g^* and π–t_{2g}^* (ABMO) gives the new $10Dq_0$ value. **Thus $10Dq_0$ value decreases compared to that found in absence of such π-bonding.**

● Energy difference between σ–e_g^* and $nb - t_{2g}$ of metal = $10Dq_0$ (in absence of π–bonding)

● Energy difference between σ–e_g^* and π–t_{2g}^* = $10Dq_0$ (in presence of both π–and σ–bonding).

● Here it may be noted that both σ–e_g^* and π–t_{2g}^* are more (relatively) enriched with the character of metal orbitals.

If π–LGO (t_{2g}) orbitals are filled in, then the π–t_{2g} (BMO) accommodates the electrons and the electrons of metal t_{2g}–orbital go to the π^*–t_{2g} (ABMO) orbital.

Thus in such cases, the π–bonding reduces the $100q_0$ value and such ligands are described as **weak field ligands.** The examples of such π–**donor ligands** are halides, OH⁻, H_2O, etc. **With the increase of π-donor properties of the ligands, $10Dq_0$ value decreases.**

(i) **Relative positions of –OH and H_2O in the spectrochemical series:** The $:\overset{..}{O}H^-$ is a better π–donor ligand than $H_2\overset{..}{O}$. It makes OH⁻ a weaker field ligand than H_2O (*cf.* Fig. 3.17.4.2).

(ii) **Relative positions of the halides in the spectrochemical series:** All the halides are the π–donor ligands. The filled atomic orbitals to construct the π–LGO (t_{2g}) are $2p$, $3p$, $4p$, and $5p$ for F⁻, Cl⁻, Br⁻ and I⁻ respectively. Their energy order is:

$$\pi\text{--LGO } (t_{2g})\text{: } F^- \langle \text{ Cl}^- \langle \text{ Br}^- \langle \text{ I}^-$$

The energy difference between t_{2g} (metal) and t_{2g}-LGO (F$^-$) is maximum while the energy difference is minimum for t_{2g}-LGO (I$^-$) for a particular metal centre (*cf.* Fig. 3.17.3.2). Thus the t_{2g}-LGO of I$^-$ overlaps most efficiently with the metal t_{2g} orbital while the t_{2g}-LGO of F$^-$ overlaps with metal t_{2g} orbital most inefficiently. This makes the $\pi\text{--}t_{2g}^*$ MO closest to the $\sigma\text{--}e_g^*$ MO for I$^-$ and $10Dq_o$ becomes minimum while the energy difference between the $\sigma\text{--}e_g^*$ MO and the $\pi\text{--}t_{2g}^*$ MO becomes maximum for F$^-$, *i.e.* $10Dq_o$ becomes maximum for F$^-$, *i.e.* $10Dq_o$ becomes maximum for F$^-$ among the halides. It explains the crystal field splitting power of the halides.

$$F^- \rangle \text{ Cl}^- \rangle \text{ Br}^- \rangle \text{ I}^-$$

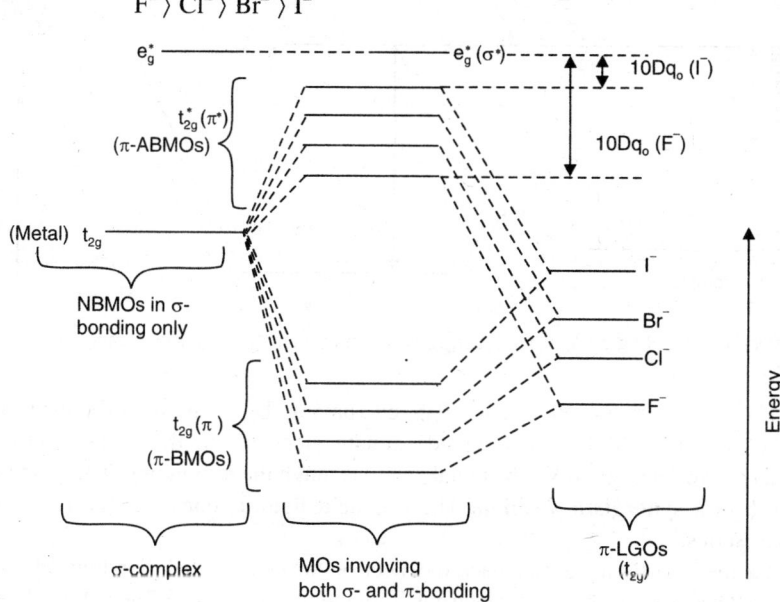

Fig. 3.17.3.2 Relative effects of π–bonding by the halide ligands to determine the $10Dq_o$ value of an octahedral complex.

Here it may be noted that in terms of electronegativity, I$^-$ is the **best π–donor** while F$^-$ is the **poorest π–donor**. It explains their crystal field splitting power in terms of MOT. In other words, the **π-donor property of a ligand reduces the positive charge on the metal centre.** Consequently, it makes the $10Dq$ value smaller. In terms of pure CFT, among the halides (treated as the point-like charges), F$^-$ having the highest charge density (measured by charge/radius ratio) shows the highest crystal field splitting power.

Note: Except F$^-$, all other halides possess the **empty d–orbitals** that may participate in metal → ligand π–back bonding, *i.e.* **π–acceptance**. But these ligand d–orbitals do not effectively participate in the π–bonding. This is why, they appear as the weak field ligands. In fact, if the metal → ligand ($d_\pi \rightarrow d_\pi$) bonding were efficient, then the halides would appear as the strong field ligands.

(B) Case II: Empty ligand π–orbitals (*i.e.* π–LGO–t_{2g}) of higher energy than the metal t_{2g} orbitals

As the π–LGO (t_{2g}) orbitals are of higher energy than the metal t_{2g} orbitals, the bonding $\pi\text{--}t_{2g}$ MO becomes enriched with the character of the metal t_{2g} orbital and the antibonding $\pi\text{--}t_{2g}^*$ MO becomes enriched with the character of the ligand orbitals. Thus the bonding $\pi\text{--}t_{2g}$ MO is of lower energy than the starting metal t_{2g} orbital (*i.e.* NBMOs in the σ–bonding scheme). The energy difference between the π–t_{2g} MO (enriched with the metal t_{2g}–character) and the σ–e_g^* MO (that remains unaffected in the π–bonding interaction) gives the new $10Dq_0$ which is larger than the $10Dq_0$ value in absence of π–bonding. Thus such π–bonding interaction enhances the $10Dq_0$ value (Fig. 3.17.3.3). **It makes the ligands strong field ligands.**

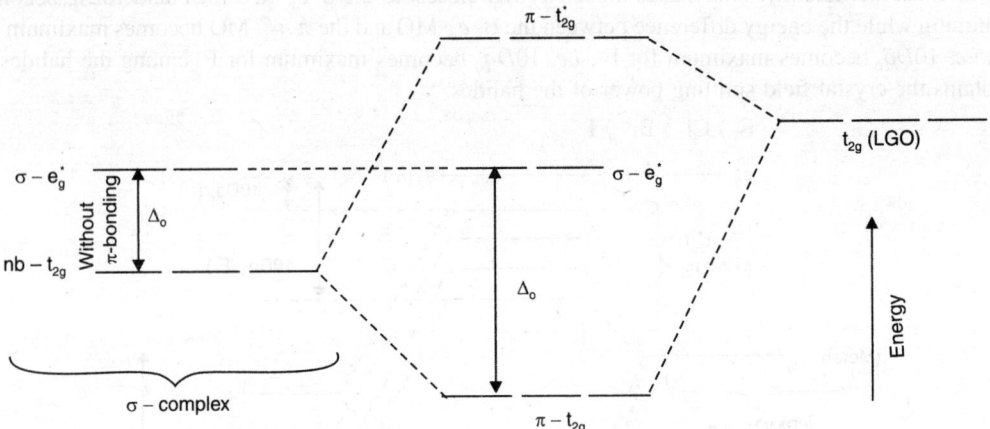

Fig. 3.17.3.3 Effect of π–bonding on Δ_0 when the ligands possess vacant π–orbitals of higher energy (compared to the metal t_{2g}–orbital).

Examples of such ligands are: PR_3, AsR_3, R_2S, etc. having the vacant $d\pi$–orbitals; CO, CN^-, C_2H_4, etc. having the vacant π^*–MOs. Such ligands donate electrons to the metal centre through the σ–bonding and receives electron from the metal centre through the π–bonding. This σ–donation and π–acceptance work in a **synergistic fashion.** These **π–acid ligands** construct the right portion of the spectrochemical series.

Note: H_2 can use the σ^*–MO to receive back the electron through π–back bonding. PR_3 and AsR_3 may also use the σ^*–MOs to receive back the electrons from the metal centre. Some authors have argued that for PR_3, the hybrid orbitals arising from the mixing of the σ^*–MO and $3d$ orbitals are used to receive back the metal electron through the π–bonding.

What is $10Dq_0$?

Ligands having only the σ-donor properties (e.g. NH_3): In the absence of π–bonding, the energy difference between the σ–e_g^* (ABMO) and nonbonding t_{2g} orbital (NBMO) of the metal gives the $10Dq_0$ value.

Ligands with the π-donor and σ-donor properties (e.g. ^-OH): When, the π–LGO (t_{2g}) is of lower energy than the metal t_{2g} orbital, then the $\pi^*(t_{2g})$ ABMO *becomes more enriched with the metal t_{2g} character* and $\pi(t_{2g})$ BMO becomes more enriched with ligand orbital character. In such cases, the energy difference between σ–e_g^* and $\pi^*(t_{2g})$ gives the $10Dq_0$value. It reduces the $10Dq_0$ value. It happens so when the ligand uses the filled orbitals for π–donation.

Ligands with the π-acceptor and σ-donor properties (e.g. CO): When the π–LGO (t_{2g}) is of higher energy than the metal t_{2g}–orbital, the π–t_{2g} BMO becomes more enriched with the character of the metal t_{2g} orbital. In such cases, the energy difference between the σ–e_g^* MO and π–t_{2g} MO gives the $10Dq_0$ value. It enhances the $10Dq_0$ value. It happens so for the π–acceptor ligands, *i.e.* π–acid ligands.

(C) Ligand field strength and π–bonding property of the ligands

● It has been shown that the **π–donor ligands show the lower 10Dq_o** value (energy difference between the σ–e_g^* MO and π–t_{2g}^* MO which is more enriched with the metal t_{2g}–character).

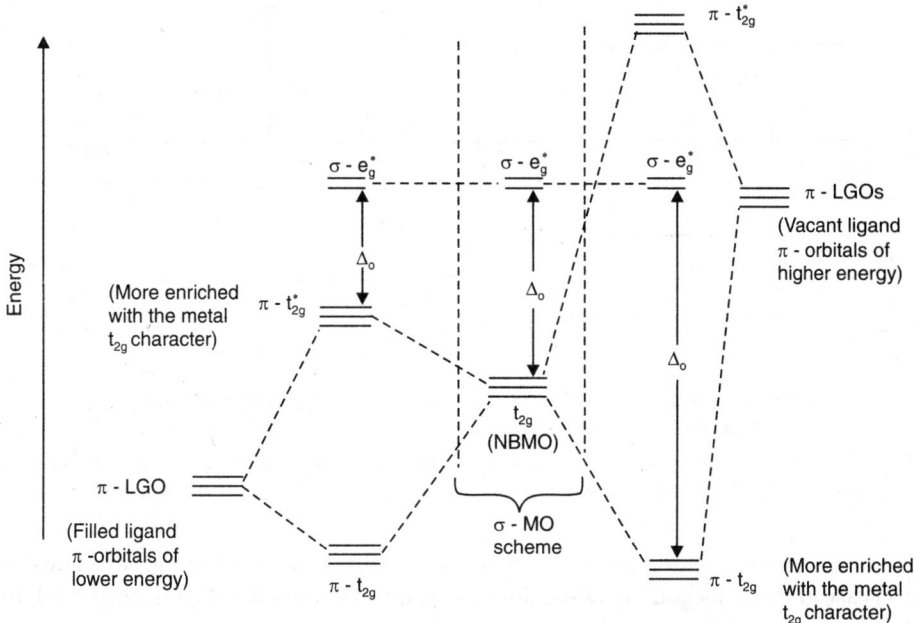

Fig. 3.17.3.4 Effect of metal-ligand π–bonding on Δ_0 (*i.e.* 10Dq_o) in an octahedral system.

● The **π-acceptor ligands show the higher 10Dq_o** value (energy diference between σ–e_g^* and π-t_{2g} (BMO) which is more enriched with the metal t_{2g}-character).

● Qualitatively, it can be said that the **π-donor property of the ligands will reduce the positive charge on the metal centre** and it will consequently reduce the crystal field splitting power, *i.e.* weak field ligand character. Withdrawal of electron cloud from the metal centre through the **π-acceptor property of the ligand will increase the positive charge** on the metal centre and it will increase the crystal field splitting power, *i.e.* strong field ligand character.

● Thus the **π-donor ligands constitute the left portion** of the spectrochemical series while the **π-acceptor ligands constitute the right portion** of the spectrochemical series. Purely **good σ-donor** ligands (having no π-bonding properties) constitute the **middle portion of the series.**

3.17.4 Spectrochemical Series in the Light of MOT

The different factors to control the magnitude of 10Dq_o for a particular metal centre depend on the bonding properties of the ligands. These are discussed below.

(i) **σ–bonding properties of the ligands:** With the increase of σ–basicity of the ligands, the six bonding MOs are stabilised more and correspondingly 6 antibonding MOs are destabilised more (*cf.* Fig. 3.17.1.2, 3.17.2.5 MO energy diagram for only the σ–bonding in ML$_6$). These are summarised in Fig. 3.17.4.1.

With the increase of σ–basicity of the ligands, the ABMOs are destabilised more and consequently the energy difference between σ–e_g^* and nonbonding–t_{2g} (metal) increases, *i.e.* 10Dq_o increases.

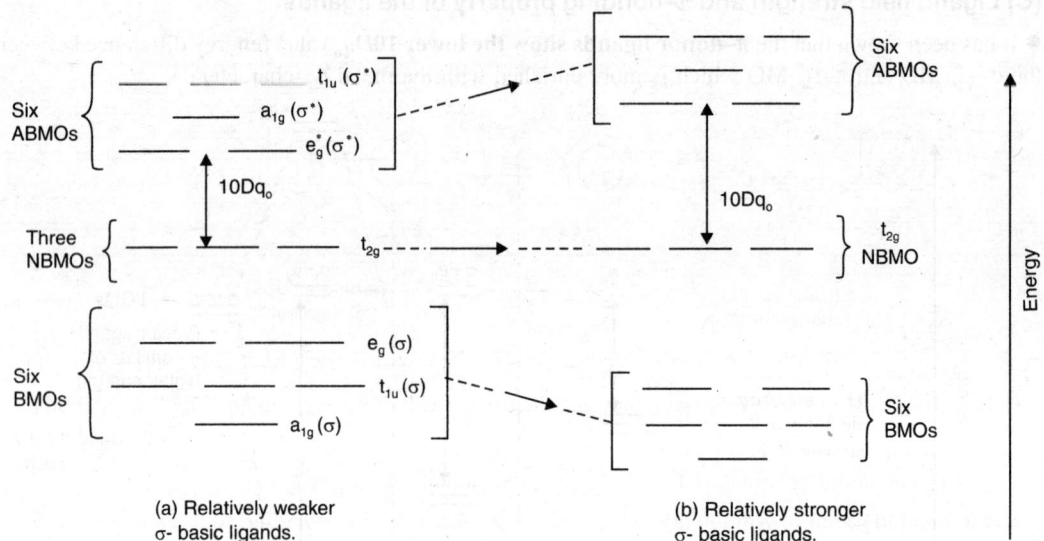

Fig. 3.17.4.1 (*cf.* Fig. 3.17.1.2) Schematic representation of more stabilisation of the σ–BMOs (bonding molecular orbitals) and more destabilisation of the σ–ABMOs (antibonding molecular orbitals) with the increase of σ–basicity of the ligands.

The more electronegative donor site will act as the poorer σ–donor ligands. **Thus with the decrease of electronegativity of the donor site, the crystal field splitting power will increase.** This is experimentally verified.

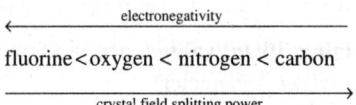

(ii) **π–donor properties of the ligands:** For the π–donor ligands, the π–LGOs are of lower energy compared to the metal t_{2g}–orbitals. Then π–bonding interactaction reduces the energy difference (= $10Dq_o$) between π–t_{2g}^* (ABMOs **enriched with the metal t_{2g}-orbitals**) and σ–e_g^* and consequently the ligand field strength decreases (*cf.* Fig. 3.17.3.1). In other words, the π–donor property of the ligands **reduces the positive charge on the metal centre** and consequently the $10Dq_o$ value decreases.

(iii) **π–Acceptor properties of the ligands:** In such cases, the vacant π–LGOs are of higher energy compared to the metal t_{2g}–set. It makes the energy difference (= $10Dq_o$) between π–t_{2g} (BMOs **enriched with the metal t_{2g} orbitals**) and σ–e_g^* larger (*cf.* Fig. 3.17.3.3), *i.e.* crystal field splitting power increases. In other words, the metal → ligand π–bonding **increases the positive charge on the metal centre** and consequently, the crystal field splitting power increases.

● Ligands with the weak σ–donor property and good π–donor property: **very weak field ligands** to constitute the **extreme left portion** of the spectrochemical series.

● Ligands with the good σ–donor property and weak π–donor property or no π-bonding property: **moderately strong field ligands** to constitute the **middle portion** of the spectrochemical series.

● Ligands with good σ–donor property and strong π–acceptor property (working in a synergistic fashion): **very strong field ligands** to constitute the **extreme right portion** of the spectrochemical series.

Weak σ donor	Good σ donor	Good σ donor
+ good π-donor	+ no π-bonding property	+ good π-acceptor

Now let us consider some specific cases.

(a) **F⁻ ⟩ Cl⁻ ⟩ Br⁻ ⟩ 1⁻:** This has been already explained (*cf.* Fig. 3.17.3.2).

(b) **H₂O ⟩ OH⁻:** In terms of CFT, it is very difficult to explain the higher crystal field splitting power of H₂O compared to that of anionic OH⁻. In terms of MOT, this can be explained. In H₂O̤ , out of the two lone pairs, one is involved in σ–donation while the other lone pair is available for π–donation. In :O̤H⁻ , out of the three lone pairs, one lone pair is involved in σ–donation while the other two lone pairs are involved in π–donation. **Thus, OH⁻ is a better π–donor ligand the H₂O.** This makes ⁻OH a weaker field ligand. *Moreover, in ⁻OH, the lone pairs engaged in π–donation bear more p–character (i.e. less s–character) compared to* the lone pair of H₂O where the lone pair engaged in π–donation is basically residing in a *sp³* hybrid orbital. Thus, the ⁻OH ion can make a better π–donation in terms of the electronegativity of the orbital housing the lone pair. In other words, **the energy of the π–donor orbital of H₂O is lower than that of the π–donor orbital of OH⁻** (*cf.* more s–character lowers the energy more).

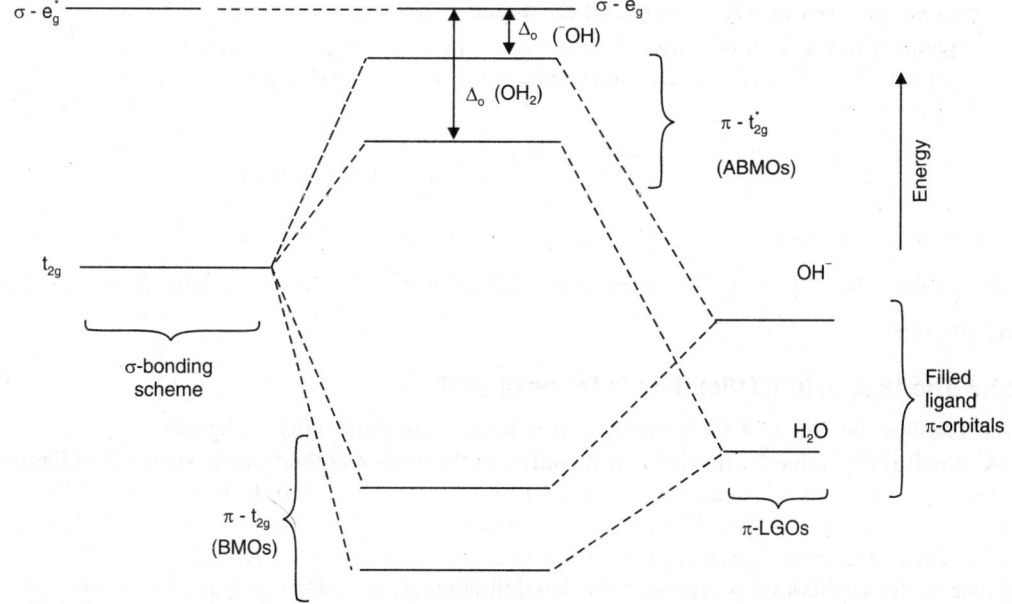

Fig. 3.17.4.2 Relative crystal field splitting powers of the π–donor ligands OH⁻ and H₂O in terms of MOT.

(c) **py ⟨ bpy ~ phen:** It appears so because of the poorer π–donor properties and better π–acceptor properties (by using the vacant π*–MO) of bpy and phen compared to those of py.

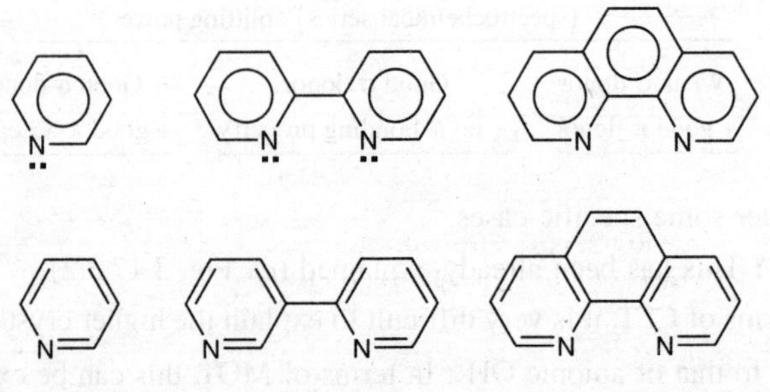

(py)   (bpy)   (phen)

In bpy and phen, the π–electron clouds are delocalised over an extended carbon skeleton compared to that in py. **Thus for bpy and phen, both the HOMO and LUMO (*i.e.* π^*–MO) are of lower energy.** This lower energy of π–HOMO, makes bpy and phen the poorer π–donor ligands and the lower energy of π–LUMO makes them the better π–acceptors.

(d) **$NH_3 \rangle OH_2$:** In terms of dipole moment, the reverse sequence is expected. :NH_3 can act as a σ–donor ligand only while $H_2\ddot{O}$ can act as a σ–donor ligand as well as a π–donor ligand because of the additional lone pair. :NH_3 can in no way act as a π–donor ligand. *The additional π–donor property of H_2O makes it a weaker field ligand.* Moreover, NH_3 is a better a σ–donor ligand than H_2O because of the higher electronegativity of O–site compared to that of the N–site. *This better σ–donor property of NH_3 also makes the ligand a stronger one* (*cf.* Fig. 3.17.4.1).

(e) **Ligands with the N–donor sites:** If the s–character on the lone pair of nitrogen increases then the σ–basicity of the N–site decreases and it reduces the ligand field strength. This explains the ligand field strength series.

$$\underset{sp-N}{NCS^-} < \underset{sp^2-N}{py} \sim \underset{sp^3-N}{NH_3} \text{ (ligand field strength)}$$

For $NCS^-\left(i.e.\ M\!-\!\ddot{N}\!=\!C\!=\!S\right)$, there is also a possibility of π–donation and it makes NCS^- further weaker. The very strong field character of bpy, phen and NO_2^- arises because of their good π–acceptor property.

3.17.5 The 18–Electron (18e) Rule in Terms of MOT

Let us consider the Fig. 3.17.5.1 for energy level diagrams of different types ligands.

(A) Small $10Dq_0$ value for the σ-donor ligands: For **the weak and moderately strong field ligands** (*cf.* Fig. 3.17.5.1a, σ–donor ligands) (*i.e.* small $10Dq_0$ value), there are 6 σ–BMOs and 3 NBMOs (*i.e.* metal t_{2g} set) and 6 σ^*–ABMOs. The 6 BMOs can accommodate 12e and the 3 NBMOs can accommodate 6e. Thus 18 electrons can be accommodated in these 9 MOs. The $6e$ in NBMOs neither contribute anything in the stabilisation process nor the destabilisation process. *Thus at best 18 electrons can be accommodated.* The 12 electrons in the BMOs contribute in the stabilisation process. If the number of electrons is less than 12, then the BMOs will remain vacant and stability of the system will decrease. If the number of electron is greater than 18, then the excess electrons are to be placed in the ABMOs and again the stability will decrease. **Thus the most favourable situation arises for the 12 – 18 valence electrons when the ligands are moderately strong.** If the $10Dq_0$ value is not very high, then placement of a few electrons in the ABMOs will not significantly reduce the stability because in such cases the ABMOs are not severely destabilised. **This is why for the very weak field ligands, the valence electron count may lie in the range 12-22e.**

The above predictions are supported by the examples given in Table 3.17.5.1.

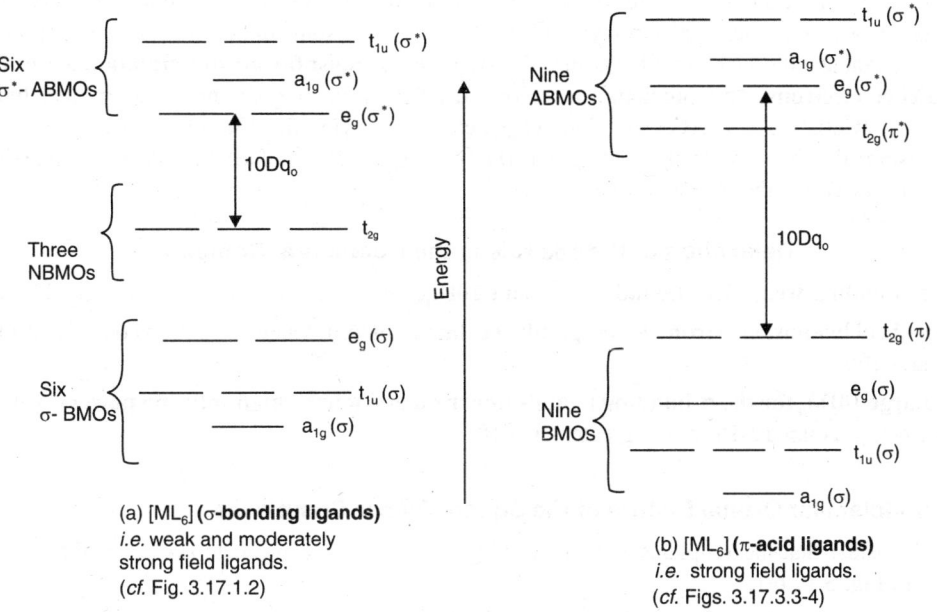

(a) $[ML_6]$ (σ-**bonding ligands**)
i.e. weak and moderately
strong field ligands.
(cf. Fig. 3.17.1.2)

(b) $[ML_6]$ (π-**acid ligands**)
i.e. strong field ligands.
(cf. Figs. 3.17.3.3-4)

Fig. 3.17.5.1 The central sections of the energy level diagrams of $[ML_6]$ complexes with weak field ligand (a) and strong field ligands (b) where there may be an uncertainty in the relative positions of the 9 ABMOs.

Table 3.17.5.1 Valence electron count for some octahedral complexes having very weak and fairly strong field ligands (**all σ–bonding ligands**)

Complex	Valence electron count	Comment	
$[TiF_6]^{2-}$	12		
$[Ti(OH_2)_6]^{3+}$	13(= 12 + 1)		
$[VCl_6]^{2-}$	13(= 12 + 1)	$10Dq_o$ not very high	
$[Fe(OH_2)_6]^{3+}$	17(= 12 + 5)		
$[Fe(OH_2)_6]^{2+}$	18(= 12 + 6)		(electron count may lie
$[Co(OH_2)_6]^{2+}$	19(= 12 + 7)		in the range 12-22)
$[Ni(NH_3)_6]^{2+}$	20(= 12 + 8)	$10Dq_o$ not very high	
$[Zn(NH_3)_6]^{2+}$	22(= 12 + 10)		
$[ZrF_6]^{2-}$	12		
$[PtF_6]^-$	17(= 12 + 5)	$10Dq_o$ very high	
$[PtF_6]^{2-}$	18(= 12 + 6)	(for the transition metals of the 2nd and 3rd series)	
$[WCl_6]^-$	13(= 12 + 1)	**in such cases, the electron count cannot exceed 18**	

Conclusion: The 18e rule is not strictly followed for the weak field and moderately strong field ligands where $10Dq_o$ is not very high.

(B) High $10Dq_o$ value for the π-acceptor ligands: For the **strong field ligands** (*cf.* Fig. 3.17.5.1b, π–acid ligands), there are 9 BMOs to accommodate the 18 electrons. If the valence electron count exceeds 18, then the additional electrons are to be placed in the ABMOs to deststabilise the system. On the other hand, if the valence electron count is less than 18, then some of the strongly bonding MOs will remain vacant and the system will lack in stability. **Thus the most favourable situation arises for the 18 valence electrons.** This prediction is supported by the following examples (all π–acid ligands): $[Cr(CO)_6]$, $[Co(CN)_6]^{3-}$, $[Fe(CN)_6]^{4-}$, $[V(CO)_6]^-$, etc. (all obeying the 18e rule).

Conclusion: For the very strong field ligands (*i.e.* very high $10Dq_o$ value), the 18e rule is more or less strictly followed for the octahedral complexes.

Relaxation of the 18e rule in the Octahedral Complexes

- σ–**Bonding weak field ligands** (*i.e.* **small $10Dq_o$**): Possible valence electron count **12-22.**

- π–**Acid ligands,** *i.e.* **strong field ligands** (*i.e.* **large $10Dq_o$**): Valence electron count **18 (almost strictly).**

- **Large $10Dq_o$ for the σ-bonding ligands (for the metal ions of high splitting power):** Valence electron count **12-18;** never greater than 18.

3.17.6 Molecular Orbital Picture of the Square Planar Complexes

Let us consider that the xy–plane is the molecular plane and the coordinate system of the complex is shown in Fig. 3.17.6.1.

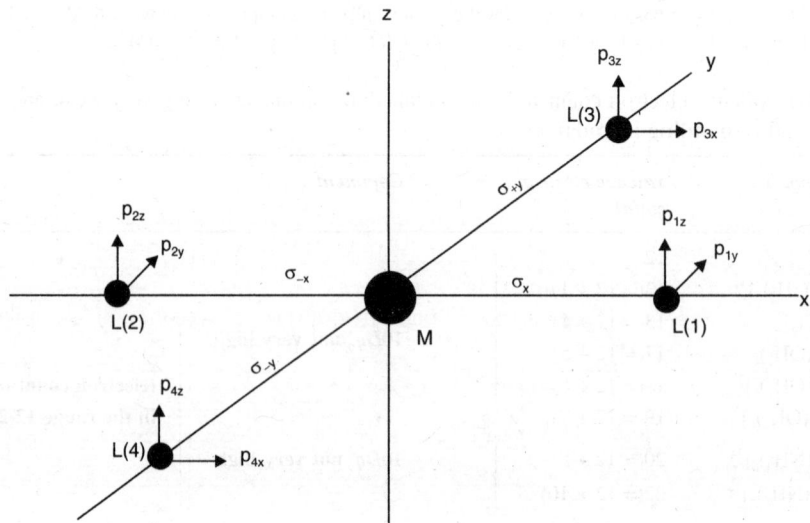

Fig. 3.17.6.1 Coordinate system of a square planar complex $[ML_4]$.

In a square planar $[ML_4]$ (D_{4h} symmetry) complex, ligands lying in the xy–plane, the metal centre can provide the following orbitals for the σ–type interaction with the ligands.

Atomic orbitals: $\underset{a_{1g}}{\underbrace{s}}$ $\qquad$ $\underset{e_u}{\underbrace{p_x, p_y}}$ $\qquad$ $\underset{a_{1g}}{\underbrace{d_{z^2}}}$ $\qquad$ $\underset{b_{1g}}{\underbrace{d_{x^2-y^2}}}$

Symmetry:

The $p_z(a_{2u})$, $d_{xz}(e_g)$, $d_{yz}(e_g)$, and $d_{xy}(b_{2g})$ orbitals cannot make any net σ–overlap in the xy–plane and these remains as the nonbonding orbitals in the σ–only system.

The d_{z^2} orbital $\left(\equiv d_{z^2-x^2} + d_{z^2-y^2}\right)$ bears a collar in the xy–plane but its main lobe is directed along the z–axis. **This is why, the d_{z^2} orbital weakly interacts with the ligand σ–orbitals.** The s–orbital is spherically symmetrical and it can interact with the σ–orbitals of the ligands. The lobes of p_x, p_y and $d_{x^2-y^2}$ orbitals are directly projected towards the ligands and these orbitals can strongly interact with the ligand σ–orbitals.

Formation of the σ–LGOs with proper symmetries: The ligand orbitals available for the σ–type interaction with the metal orbitals may be designated as $σ_x$, $σ_{-x}$, $σ_y$ and $σ_{-y}$ approaching from the $+x$, $-x$, $+y$ and $-y$ directions respectively. Formation of the σ–LGOs through the combination of $σ_x$, $σ_{-x}$, $σ_y$ and $σ_{-y}$ is explained in Fig. 3.17.6.2.

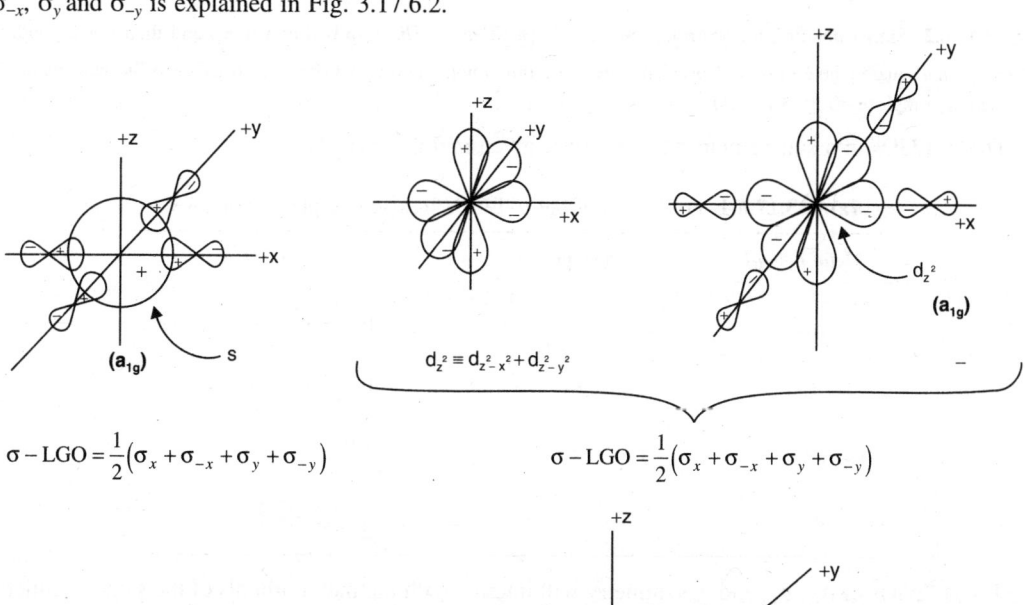

$$\sigma - LGO = \frac{1}{2}\left(\sigma_x + \sigma_{-x} + \sigma_y + \sigma_{-y}\right)$$

$$d_{z^2} \equiv d_{z^2-x^2} + d_{z^2-y^2}$$

$$\sigma - LGO = \frac{1}{2}\left(\sigma_x + \sigma_{-x} + \sigma_y + \sigma_{-y}\right)$$

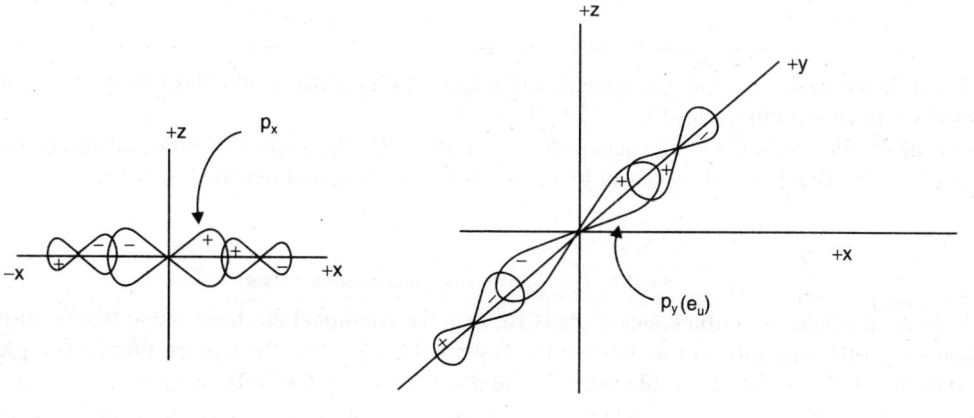

$$\sigma - LGO = \frac{1}{\sqrt{2}}\left(\sigma_x - \sigma_{-x}\right)$$

$$\sigma - LGO = \frac{1}{\sqrt{2}}\left(\sigma_y - \sigma_{-y}\right)$$

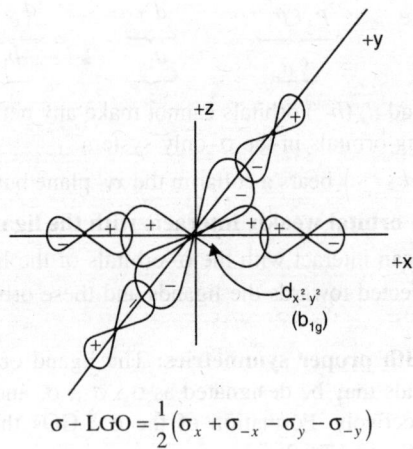

$$\sigma - LGO = \frac{1}{2}\left(\sigma_x + \sigma_{-x} - \sigma_y - \sigma_{-y}\right)$$

Fig. 3.17.6.2 Sigma bonding metal orbitals $s(a_{1g})$, $d_{z^2}(a_{1g})$, $d_{x^2-y^2}(b_{1g})$, $p_x(e_u)$ and $p_y(e_u)$ and their corresponding σ–LGOs in a square planar system. **Note:** The linear combination method to construct the σ-LGOs is the same as in the octahedral complexes (Fig. 3.17.1.1).

The σ–LGOs in a square planar complex are given in Table 3.17.6.1.

Table 3.17.6.1 Formation of the σ–LGOs in a square planar complex.

Metal orbital	Symmetry	σ–LGO
s, d_{z^2}	a_{1g}	$\frac{1}{2}\left(\sigma_x + \sigma_{-x} + \sigma_y + \sigma_{-y}\right)$
$d_{x^2-y^2}$	b_{1g}	$\frac{1}{2}\left(\sigma_x + \sigma_{-x} - \sigma_y - \sigma_{-y}\right)$
p_x	e_u	$\frac{1}{\sqrt{2}}\left(\sigma_x - \sigma_{-x}\right)$
p_y	e_u	$\frac{1}{\sqrt{2}}\left(\sigma_y - \sigma_{-y}\right)$

The σ–LGOs of a_{1g}, b_{1g} and e_u symmetry will interact with the metal orbitals of the same symmetry to produce the σ–MO diagram (Fig. 3.17.6.3).

For the d^8 metal ions, the total valence electron is 16 (= 8 + 8) in which the four ligands contribute eight electrons. These 16 electrons can be placed in the bonding and nonbonding MOs.

$$\underbrace{\left(a_{1g}\right)^2 \left(b_{1g}\right)^2 \left(e_u\right)^4}_{\sigma\text{-BMOs}} \quad \underbrace{\left(b_{2g}\right)^2 \left(e_g\right)^4}_{\text{NBMOs (in }\sigma\text{-only diagram)}} \quad \underbrace{\left(a_{1g}\right)^2}_{\text{Weakly bonding}}$$

If the total valence electron count exceeds 16, then the additional electrons are to be placed in the antibonding MOs and this will destabilise the system. **This is why, the square planar complex of d^8–system (*e.g.* Ni^{II}) is more stable than that of d^9–system (*e.g.* Cu^{II}).** If the valence electron count is sufficiently small to keep the bonding MOs vacant, then stability of the system is also reduced. In fact, if the valence electron count is less than 16, then the stability of the system is reduced. **This explains why the d^8–system is the most suitable one to generate the square planar complex.**

Interpretation of 16e rule in square planar complexes: The above discussion shows that if the total valence electrons (**TVE**) exceeds 16, then the additional electrons are to be placed in the strongly antibonding MOs. It will destabilise the system.

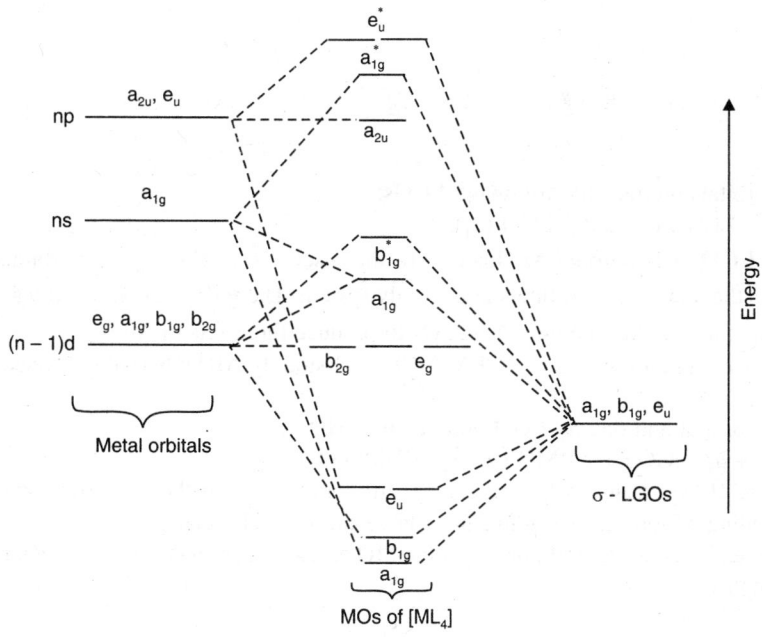

Fig. 3.17.6.3 Sigma only MO energy diagram for a square planar complex [ML$_4$].

Pi bonding in a square planar complex: The p_z, d_{xz}, d_{yz} and d_{xy} orbitals of the metal centre cannot make any net σ–overlap in the xy–plane and these metal orbitals remain as the nonbonding ones in a σ–only system. But these orbitals can participate in π–type interaction with the π–LGOs of the ligands (Fig. 3.17.6.4). The p_x and p_y orbitals may also participate in π–bonding but these orbitals are mainly involved in forming the stronger σ–bonds.

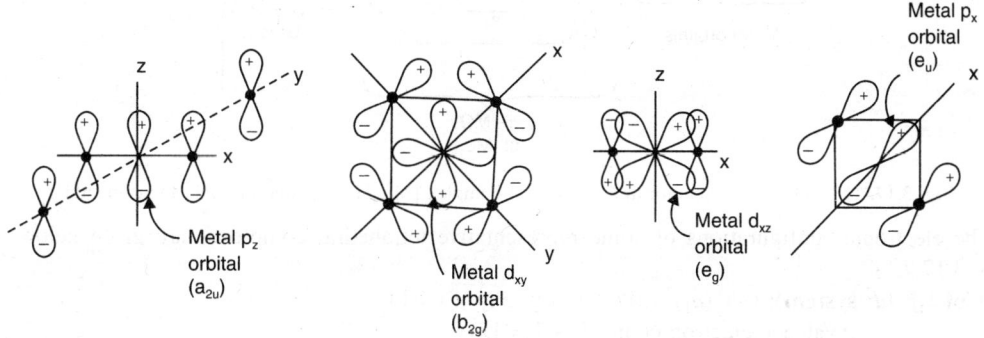

Fig. 3.17.6.4 Formation of the π–LGOs in a square planar complex by considering the π–overlap. $p_y(e_u)$ is similar to p_x; and d_{yz} is similar to d_{xz}.

3.17.7 Molecular Orbital Energy Diagram of a Tetrahedral Complex

In the tetrahedral complex $[ML_4]$ (T_d symmetry), the metal orbitals available are as follows:

Atomic orbitals	Symmetry
s	a_1
p	t_2
$d_{z^2}, d_{x^2-y^2}$	e
d_{xy}, d_{yz}, d_{zx}	t_2

The ligand orbitals produce the following LGOs:

$$3 \text{ LGO } (t_2) \text{ and } 1 \text{ LGO } (a_1).$$

There is no LGO of e–symmetry. This is why, the e–set of (*i.e.* d_{z^2}, $d_{x^2-y^2}$) orbitals of the metal centre remain as the nonbonding orbitals in a σ–only system. The six t_2 orbitals (both p and d_{xy}, d_{yz}, d_{zx}) of the metal can interact with the three t_2–LGOs to produce three sets of MOs.

● t_2 (metal) + t_2 –LGO ⇒ 3 strongly BMOs (t_2) + 3 stongly ABMOs (t_2) + 3 weakly ABMOs or NBMOs (t_2)

One a_1 metal orbital and one a_1–LGO produce two MOs.

● a_1 (metal) + a_1–LGO ⇒ 1 BMO (a_1) + 1 ABMO (a_1).

● The Δ_t, (*i.e.* $10Dq_t$) is given by the energy difference between the e–orbital (nonbonding) and slightly antibonding t_2^*–orbital (residing just above the e–level) relatively enriched with the metal t_2-orbital character. Thus the central portion of the MO energy diagram describes the CFT model of the tetrahedral complex.

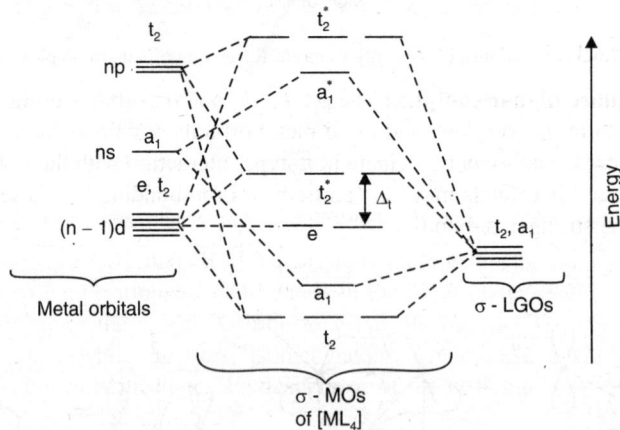

Fig. 3.17.7.1 MO energy diagram in a tetrahedral complex $[ML_4]$ by considering only the σ–bonding.

The electronic configurations of some representative tetrahedral complexes are given below (*cf.* Fig. 3.17.7.1).

$[CoCl_4]^{2-}$ (d^7 **system**): $(t_2)^6 (a_1)^2 (e)^4 (t_2^*)^3$, (*cf.* $e^4 t_2^3$ of CFT)
        (Valence electron count: 8 + 7 = 15)

$[NiCl_4]^{2-}$ (d^8 **system**): $(t_2)^6 (a_1)^2 (e)^4 (t_2^*)^4$, (*cf.* $e^4 t_2^4$ of CFT)
        (Valence electron count: 8 + 8 = 16)

[FeCl$_4$]$^-$ (d^5 system): $(t_2)^6 (a_1)^2 (e)^2 (t_2^*)^3$, (cf. $e^2 t_2^3$ of CFT)

(Valence electron count: $5 + 8 = 13$, P $\rangle$ Δ_t)

Pi–bonding in a tetrahedral complex: The $d_{x^2-y^2}$ and d_{z^2} orbitals (*i.e.* nonbonding e–set of the σ–only system) can efficiently participate in π–bonding with the ligands. Participation of both π-t_2-LGO and σ-t_2-LGO with the t_2-metal orbitals complicates the MO energy diagram. MO energy diagram indicating both σ- and π-bonding in a tetrahedral complex is shown in Fig. 3.17.7.2.

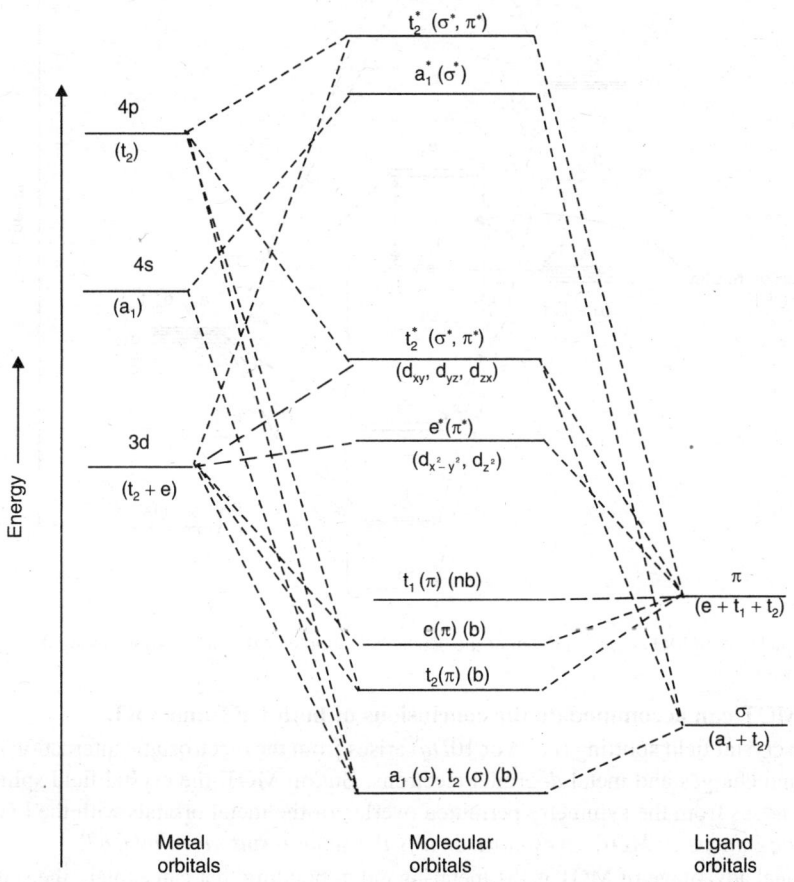

Fig. 3.17.7.2 MO energy diagram (indicating both σ– and π–bonding) in a tetrahedral complex; (b) for BMO and (nb) for NBMO.

3.18 COMPARISON OF VBT, CFT AND MOT OF BONDING IN THE OCTAHEDRAL COORDINATION COMPOUNDS

If we consider the octahedral complex, then the d^2sp^3 hydridisation (*i.e.* inner orbital complex) of *VBT* requires the participation of ns, np and $(n-1)d_{x^2-y^2}$ and $(n-1)d_{z^2}$ orbitals in bonding and the remaining d–orbitals, *i.e.* $(n-1)d_{xy}$, $(n-1)d_{yz}$, $(n-1)d_{xz}$ exist as the nonbonding orbitals. If we consider the bottom portion of the MOT diagram (cf. Fig. 3.18.1) then the MOs, *i.e.* $\sigma - a_{1g}$, $\sigma - t_{1u}$, $\sigma - e_g$ and nonbonding t_{2g}, give the qualitative description of the VBT. The central portion of the diagram consisting of t_{2g} (nonbonding) and $\sigma-e_g^*$ gives the CFT description.

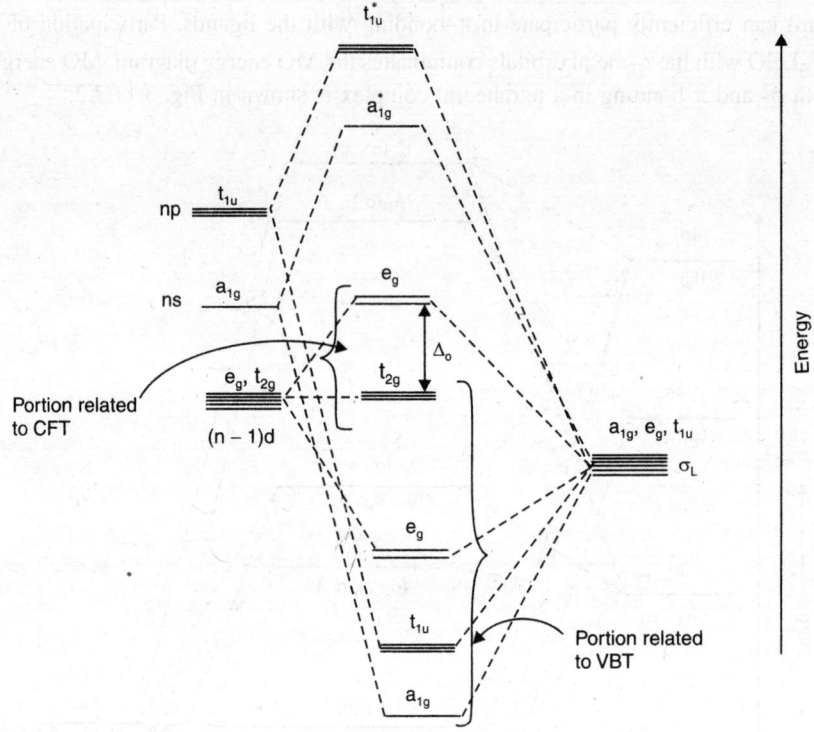

Fig. 3.18.1 Sigma bonding MO energy diagram of the octahedral complex ML_6 and relation among the MOT, VBT and CFT conclusions.

Thus the MOT can accommodate the conclusions of both CFT and VBT.

In CFT, the crystal field splitting (*i.e.* Δ or $10Dq$) arises from the electrostatic interaction between the point-like ligand charges and metal d–orbital electrons. But, in MOT, the crystal field splitting energy automatically arises from the symmetry permitted overlap of the metal orbitals with the LGOs. *In fact, only σ–bonding scheme of MOT, gives more or less the same result given by CFT.*

The additional advantage of MOT is the metal-ligand π–bonding that can explain the stability of the complexes and the spectrochemical series. The MOT is more helpful to explain the magnetic and spectral properties of the complexes.

EXERCISE 3

A. General Type Questions

1. Discuss the role to ionic potential in determining the complexing power of the metal ions.
2. (a) What do you mean by the effective atomic number (EAN) rule? What are the limitations of the rule?
   (b) Verify the applicability of EAN rule in the following complexes.

$[Fe(CO)(\eta^5-C_5H_5)(\eta^3-C_7H_7)]$, $[Cr(\eta^6-C_6H_6)_2]$, $[Mn(CO)_4(\eta^3-C_3H_5)]$,

$[Fe(CO)_2(NO)_2]$, $[Ir(CO)Cl(PPh_3)_2]$, $[Ni(CN)_4]^{2-}$, $[Fe(CN)_6]^{3-}$,

$[Fe(CO)_2(\eta^1-Cp)(\eta^5-Cp)]$, $[Cr(CO)_3(\eta^6-C_8H_8)]$.

(c) Predict the value of x (see Sec. 12.8.5, Vol. 3 of Fundamental Concepts of Inorganic Chemistry).

$[Fe(CO)_3(\eta^x-C_7H_8)]$, $[Cr(CO)_3(\eta^x-C_8H_8)]$,

$[Mn(CO)_4(\eta^x-C_3H_5)]$, $[Fe(CO)_x(\eta^1-Cp)(\eta^5-Cp)]$,

$[Cr(CO)_x(\eta^6-C_7H_8)]$.

3. Discuss the VBT to explain the stereochemistry of complexes.
4. What do you by mean the inner and outer orbital complexes? How can you distinguish them? Do you find any justification in naming the outer orbital complexes as the ionic complexes and the inner orbital complexes as the covalent complexes?
5. Discuss the conditions favouring the outer and inner orbital complexes in terms of VBT.
6. Discuss and illustrate the electroneutrality principle to explain the stability of metal complexes. Comment on the stability of the following complexes in terms of the electroneutrality principle. $[Be(NH_3)_6]^{2+}$, $[Mg(OH_2)_6]^{2+}$, $[Be(OH_2)_6]^{2+}$, $[Be(NH_3)_4]^{2+}$, $[Fe(NH_3)_6]^{3+}$, $[Fe(OH_2)_6]^{3+}$, $[Al(NH_3)_6]^{3+}$, $[Fe(bpy)_3]^{3+}$, $[Fe(CN)_6]^{3-}$.
7. Discuss the limitations of VBT in explaining the bonding in the coordination compounds.
8. Discuss the basic assumptions of crystal field theory (CFT) in understanding the bonding in the coordination compounds.
9. Discuss and compare the crystal field splitting of the d–orbitals in the octahedral, tetrahedral and square planar crystal fields.
10. Compare the splitting of d–orbitals in the (a) octahedral, tetrahedral and cubic geometries, (b) octahedral and square pyramidal complexes, (c) octahedral and tetragonally distorted complexes, (d) trigonal bipyramidal and trigonal planar complexes, (e) trigonal bipyramidal and square pyramidal complexes, (f) octahedral and linear complexes.
11. Explain the origin of low-spin and high-spin complexes:
12. How can you explain the spin-isomerism by CFT? Give the examples to illustrate the spin-isomerism.
13. Discuss the factors controlling the crystal field splitting energy (Δ).
14. Give the splitting of the (i) p–orbitals in the linear and octahedral fields, (ii) p-orbitals in the square planar field, (iii) f–orbitals in the octahedral field.
15. What are the important factors to determine the magnitude of pairing energy (P)? Illustrate with examples.
16. Give the energy of different d–orbitals in different crystal fields in terms of Dq_o.
17. Comment on the possibilities of the formation of low-spin complexes in the octahedral, tetrahedral, trigonal bipyramidal, square pyramidal and square planar complexes.
18. Pairing energy does not appear in the cfse (crystal field stabilisation energy) of the high-spin complexes but of the low-spin complexes—explain.
19. Explain the origin of J.T. distortion by CFT. What are the conditions of J.T. distortion in the octahedral and tetrahedral complexes. Illustrate with examples? How can you predict the z–out or z–in distortion?
20. In some cases, the predicted J.T. distortion is not observed—explain and illustrate.
21. Discuss the effect of J.T. distortion on the structure, thermodynamic and kinetic properties of the complexes.

22. What do you mean by the static and dynamic J.T. distortion? Illustrate with examples.

23. Comment on the possibility of square planar complex formation in the d^8–system. Illustrate with suitable examples.

24. Give some experimental evidences to support CFT.

25. Explain the following curves:
    (i) Hydration energy *vs.* M^{2+} or M^{3+} of 1st transition series. (ii) octahedral or tetrahedral ionic radius (in oxide lattice) *vs.* M^{2+}(of 1st transition series). (iii) lattice energy *vs.* M^{3+} of 1st transition series (assuming both the cases—low-spin and high-spin).

26. Comment on the contribution of cfse in determining the stability of metal complexes and hydration energy.

27. Comment on the effect of electron occupancy of the orbitals to influence the metal-ligand bond length in (i) linear geometry, (ii) square planar geometry, (iii) TBP geometry, (iv) square pyramidal geometry and (v) octahedral geometry.

28. Discuss the term, octahedral site preference energy (OSPE) with respect to the tetrahedral site. Give the most favourable conditions for forming the tetrahedral complexes.

29. Comment on the relative stabilities of $[Ni(CN)_4]^{2-}$ (square planar), $[Ni(CN)_5]^{3-}$ (TBP) and $[Ni(CN)_5]^{3-}$ (square pyramidal) in terms of CFT.

30. Critically comment on the stabilities and structural characteristics of the following complexes:
    (i) $[Ni(CN)_6]^{4-}$, (ii) $[NiCl_4]^{2-}$, (iii) $[Cu(NH_3)_6]^{2+}$, (iv) $[Co(CN)_6]^{4-}$, (v) $[CuCl_6]^{4-}$, (vi) $[CuCl_5]^{3-}$, (vii) $[Co(NO_2)_6]^{4-}$, (viii) $[Pt(NH_3)_6]^{2+}$, (ix) $[Ni(NH_3)_6]^{2+}$, (x) $[CuCl_4]^{2-}$, (xi) $[CoCl_4]^{2-}$, (xii) $[FeCl_4]^{-}$

31. Discuss the spinel and inverse spinel structures of the mixed oxides in terms of CFT.

32. How can you explain the variation of $E°$ values of the Co^{III}/Co^{II} couple and Fe^{III}/Fe^{II} couple for the various types of ligands?

33. Discuss the merits and demerits of CFT.

34. Give the experimental evidences in favour of metal-ligand overlap.

35. What do you mean by nephelauxetic effect and nephelauxetic series? What is the effect of this phenomenon on pairing energy and Racah prameter (B)?

36. Discuss the ESR spectra of $[IrCl_6]^{2-}$. How can you explain it in terms of metal-ligand overlap? Comment on the spectra by considering the possibility of static and dynamic J.T. distortion.

37. Sometimes, the weak field ligands with a high nephelauxetic effect can lead to the low-spin complexes — explain and illustrate.

38. Give the examples of tetrahedral complexes experiencing the J.T. distortion.

38. Only in some limited cases, the low-spin tetrahedral complexes have been found—comment.

39. The d^6–configuration is the most suitable one for the low-spin octahedral complex formation while the d^4–configuration is the most suitable one for the low-spin tetrahedral complex formation—explain with examples.

40. Compare the splitting of f–orbitals in the octahedral and tetrahedral complexes.

41. Compare the splitting of d–orbitals in different 8–coordinate complexes.

42. Compare the splitting of d–orbitals in the octahedral and trigonal prismatic complexes.

43. Give the σ–bonding MO–energy diagram for the $[ML_6]$ (octahedral), $[ML_4]$ (tetrahedral) and $[ML_4]$ (square planar) complexes.

44. Give the π–bonding MO–energy diagram for the $[ML_6]$ (octahedral) complex.

45. How can you find the conclusions of VBT and CFT from the MOT?

46. What is Δ_o in terms of MOT? How does the π–bonding influence the Δ_o value? How do the CFT and MOT differ to predict the Δ_o value?

47. How can you justify the 18e and 16e rule from MOT?
48. The d^8–system is the most ideal one for the square planar complex formation—explain in terms of MOT.

B. Justify the following statements

1. Ionic potential is a good parameter to measure the complexing power of metal ions.
2. Effective atomic number (EAN) rule is very often described as the 18–electron rule.
3. There are some stable complexes disobeying the 18e–rule (*i.e.* EAN rule).
4. Some metal centres prefer the 16e rule rather than the 18e rule.
5. In the electron count process for the complexes bearing the ligands, $C_5H_5^-$, $C_3H_5^-$, C_6H_6, C_4H_6, C_4H_4, C_7H_8, C_8H_8, NO, it is important to know their mode of coordination.
6. Dimerisation of $[Co(CO)_4]$ and $[Mn(CO)_5]$ are expected from the EAN rule.
7. The stability order: $[V(CO)_6]^- \rangle [V(CO)_6]$; $[Mn(CO)_5]^- \rangle [Mn(CO)_5]$ can be explained by considering the 18e rule.
8. $[Fe(CO)_5]$ itself satisfies the 18e rule but it may undergo polymerisation by losing some CO groups to satisfy the 18e rule.
9. For the sp^3d^2 or d^2sp^3 hybridisation in the octahedral geometry, the required d–orbitals are $d_{x^2-y^2}$ and d_{z^2}. On the other hand, in the d^3s hybridisation of tetrahedral geometry, the required d–orbitals are d_{xy}, d_{yz} and d_{xz}.
10. In terms of VBT, the outer orbital complexes were described as the ionic compounds while the inner orbital complexes were described as the covalent compounds. These were done by considering the μ_{spin} only values.
11. Actually, the outer orbital complexes are the high-spin complexes while the inner orbital complexes are the low-spin complexes.
12. The $(n-1)d^{0-3}$ systems always form the inner-orbital octahedral complexes.
13. For the $(n-1)d^{4-7}$ systems, both the possibilities of outer and inner orbital octahedral complexes exist.
14. For the d^7–system, the inner orbital octahedral complexes are prone to oxidation.
15. $[Co(NH_3)_6]^{2+}$ can be readily oxidised to $[Co(NH_3)_6]^{3+}$. This can be explained by VBT.
16. The $(n-1)d^{8-10}$ systems generally form the outer orbital octahedral complexes.
17. The dsp^2 hybridisation of the square planar geometry involves the $d_{x^2-y^2}$ orbital.
18. For the d^8–system, the inner orbital square planar complex can be formed, but for d^9–system, both the inner and outer orbital square planar complexes are possible.
19. The square planar complexes of Cu(II) are not susceptible to oxidation. It can be explained by considering the outer orbital complex formation.
20. Measurement of magnetic moment can distinguish between the outer and inner orbital complexes.
21. Strongly electronegative donor sites (*e.g.* O, F) generally favour the outer orbital complex formation while the low electronegative donor sites (*e.g.* C, P, etc) favour the inner orbital complex formation. These can be rationalised, at least partly, by VBT.
22. The heavier congeners prefer the inner orbital complex formation compared to the outer orbital complex formation. This can be explained by VBT.
23. The higher oxidation state of the metal centre favours the participation of $(n-1)d$ orbitals over the nd orbitals in bonding. This can be explained by VBT.

24. The magnetic moment measurement cannot distinguish between the square planar and tetrahedral complexes of Cu(II), but the magnetic moment value can distinguish between the square planar and tetrahedral complexes of Ni(II).

25. The magnetic moments of many Co(II) and Ni(II) complexes cannot be explained by VBT. In such cases, the observed magnetic moments differ significantly from the μ_{spin} only values.

26. *The electroneutrality principle* does not allow the accumulation of negative charge on the metal centre of the complexes.

27. The electroneutrality principle can explain the following stability order of the complexes:

   (i) $[Be(OH_2)_4]^{2+} \gg [Be(NH_3)_4]^{2+} \gg [Be(OH_2)_6]^{2+}$

   (ii) $[Al(OH_2)_6]^{3+} \gg [Al(NH_3)_6]^{3+}$

   (iii) $[Fe(OH_2)_6]^{3+} \gg [Fe(NH_3)_6]^{3+}$

   Given: $\chi(Be) = 1.5$, $\chi(O) = 3.5$, $\chi(N) = 3.0$, $\chi(H) = 2.1$, $\chi(Al) = 1.5$,

29. Be^{2+} gives $[Be(OH_2)_4]^{2+}$ while Mg^{2+} gives $[Mg(OH_2)_6]^{2+}$. This can be rationalised by considering the Pauling's electroneutrality principle.   **(Hints: $\chi(Be) \rangle \chi(Mg)$)**

30. In some complexes, the metal to ligand π–bonding is important to maintain the electroneutrality principle.

31. The number and nature of d–orbitals available for π–bonding in the octahedral, square planar and tetrahedral complexes are different.

32. The properties (specially with respect to the ease of oxidation) of the square planar Cu(II)–complexes, and the octahedral Co(II)–complexes cannot be clearly explained by VBT.

33. By considering the VBT–concepts of the ionic (*i.e.* outer orbital) and covalent (*i.e.* inner orbital) bonding, the Brönsted acidity sequence of the aqua complexes, *i.e.* $[Cr(OH_2)_6]^{3+}$, $[Fe(OH_2)_6]^{3+}$ and $[Co(OH_2)_6]^{3+}$ cannot be explained.

34. In the basic assumptions of CFT, the ligand electrons and metal electrons are treated in some sort of self-contradictory ways.

35. In terms of CFT, the $(n-1)d$ orbitals of the transition metal ions are affected most.

36. In terms of CFT, the $(n-2)f$ orbitals of the actinides are affected more by the ligands compared to the case of lanthanides.

   ● The $4f$ orbitals of the lanthanides are less affected than the $5f$ orbitals of the actinides by the ligand field.

37. In the linear crystal field, the p–orbitals are split but these are not split in the octahedral crystal field.

38. In the linear crystal field, the d–orbitals are split into three sets while in the octahedral field, these are split into two sets.

39. In a square planar crystal field, degeneracy of the p–orbitals is lifted but is retained in an octahedral field.

40. In the crystal field splitting of orbitals, the barycentre rule translates the principle of conservation of energy.

41. In the crystal field splitting of d–orbitals in an octahedral field, the energy of a t_{2g} electron is not $-4Dq_0$; similarly, energy of an e_g electron is not $+6Dq_0$.

42. In the octahedral crystal field, apparently, it is difficult to consider the $d_{x^2-y^2}$ and d_{z^2} orbitals as the equivalent orbitals.

43. The C_3–symmetry operation on the orbitals of the e_g set of an octahedron produces the orbitals which are the linear combinations of the starting orbitals.

44. Crystal field stabilisation energy (cfse) is only one of the many contributing factors to determine the overall metal-ligand bond energy.

45. An octahedral crystal field splits the f–orbitals into three different sets.
46. When a tetrahedron is generated from a cube, the C_4 axes of the cube are converted into the C_2–axes of the tetrahedron.
47. Crystal field splitting of the d–orbitals is more pronounced in the octahedral field than in the tetrahedral field.
48. The d_{xy}, d_{yz} and d_{xz} orbitals are stabilised and the $d_{x^2-y^2}$ and d_{z^2} orbitals are destabilised in an octahedral crystal field while the reverse is true in a tetrahedral crystal field.
49. The crystal field splitting (Δ) parameters are related as follows:

$$\Delta_{cubic} \approx \frac{8}{9}\Delta_{octahedral}; \ \Delta_{tetrahedral} \approx \frac{4}{9}\Delta_{octahedral}$$

$$\Delta_{square\ planar} > \Delta_{octahedral}$$

50. The limiting situation of z–out tetragonal distortion leads to a square planar geometry.
51. At any stage of tetragonal distortion of an octahedral complex, the energy difference between the $d_{x^2-y^2}$ and d_{xy} orbitals is given by Δ_o (*i.e.* octahedral crystal field splitting parameter).
52. 'Compression of a tetrahedron produces a square planar geometry'. This concept can lead to the splitting pattern of the d-orbitals in a square planar geometry, from the knowledge of splitting pattern of the d–orbitals in the starting tetrahedron.
53. Limiting situations of the tetragonal elongations of an octahedron and the compression of a tetrahedron lead to the common geometry. This idea can correlate the splitting patterns of the d–orbitals in the three geometries.
54. Crystal field splitting pattern of the d–orbitals in a square pyramidal geometry is similar to that in a tetragonally elongated octahedral geometry.
55. In a trigonal bipyramidal crystal field, the d–orbitals are split into three different sets of orbitals.
56. The z–out distortion of a trigonal bipyramidal geometry leads to a trigonal planar geometry. This concept can correlate the splitting patterns of d–orbitals in the two geometries.
57. The crystal field splitting (Δ) increases with the increase of oxidation state of the metal centre.
58. The crystal field splitting (Δ) increases for the heavier congeners.
   ● High-spin complexes are rarely found for the heavier congeners.
59. Spectrochemical series is constructed by some overlaping segments of the series. The segments are constructed from the data of spectral shift.
60. The crystal field splitting power of different ligands depends on the nature of the donor sites:

$$\underrightarrow{\underset{\text{Crystal field splitting power}}{\text{Crystal field strength: } I< Br<Cl<S<F<O<N<C}}$$

61. The very high crystal field splitting by the π–acid ligands can be realised by considering the effect of effective positive charge on the metal centre.
62. The pairing energy (P) varies in the folllowing sequences:

$$M^{n+1} \rangle M^{n+}; \ M^{n+} (3d) \rangle M^{n+} (4d) \rangle M^{n+} (5d).$$

   M^{n+} (gaseous) $\rangle$ M^{n+} (complex); M—L (σ–donor) > M—L (π–acceptor)
63. The pairing energy sequence: $d^5 \rangle d^4 \rangle d^7 \rangle d^6$ can be rationalised in terms of the loss of exchange energy during the pairing process.
64. For Co(III), there are only very few complexes that are the high spin complexes.
65. For the d^5 and d^4 systems, only very strong field ligands can lead to the low-spin complexes.

On the other hand, it is relatively easier to produce the low-spin octahedral complexes for the d^6 system.

● The d^4–system is the most suitable one for the low-spin tetrahedral complex formation.

66. In the high-spin complexes, there is no contribution of pairing energy in the cfse value but the pairing energy is a contributing factor in the cfse value of their low spin complexes.

67. The heavier congeners generally form the low spin complexes because of their higher crystal field splitting power and lower pairing energy.

68. The critical $10Dq_o$ value can lead to the situation of spin isomerism.

69. Energy of both the high and low spin systems decreases with the increase of Δ_o but it decreases more steeply for the low-spin complexes.

70. In the range of critical $10Dq_o$, the observed magnetic moment increases with the increase of temperature.

71. The S–donor sites are not the strong field donor sites but they can lead to spin pairing in many cases.

72. $Fe\left(\begin{array}{c} S \\ \vdots \\ S \end{array} \right)_3 C\!\!-\!\!NR_2$ shows the spin-state isomerism.

● The observed magnetic moment depends on the temperature and nature of the R–group.

● The pairing energy (P) for Fe(III) (d^5–system) is very high but the S–donor sites can favour pairing to establish the spin-state equilibrium.

73. Most of the tetrahedral complexes are the high-spin complexes.

74. In some rare cases, the low-spin tetrahedral complexes are produced for the d^4 and d^5–systems.

75. The most favourable situation of the low-spin octahedral complex formation arises for the d^6–system while the d^4–system gives the most favourable situation for the low-spin tetrahedral complex formation.

76. The bulky and anionic ligands favour the tetrahedral complex formation.

77. The d^n ($n = 1, 2, 3, 4, 9, 10$) systems give one possibility in each case in TBP geometries for all types of ligands but for the d^n ($n = 5, 6, 7, 8$) systems, both the high-spin and low-spin TBP complexes may be formed depending on the crystal field splitting power of the ligands.

78. The d^5, d^6, d^7 and d^8 systems can give both the low-spin and high-spin TBP complexes depending on the ligand field strength.

79. $[Ni(CN)_5]^{3-}$ can give both the low-spin trigonal bipyramidal and square pyramidal complexes of comparable energies.

80. In the trigonal bipyramidal $[Ni(CN)_5]^{3-}$ complex, the axial bonds are slightly shorter than the average equatorial bonds while in the square pyramidal $[Ni(CN)_5]^{3-}$ complex, the axial bond is significantly longer than than the average equatorial bond.

81. In trigonal bipyramidal $[CuCl_5]^{3-}$, the axial bond is relatively longer than than the average equatorial bond.

82. The crystal field splitting patterns of the d–orbitals in $[Co(NH_3)_6]^{3+}$ and trans–$[CoF_2(NH_3)_4]^+$ are different.

83. $[VO(OH_2)_5]^{2+}$ is a tetragonally compressed cation.

● The axial 'V—O' bond is shorter than the equatorial V—O bond.

84. The z–out and z–in distortions in the tetrahedral geometry give the elongated and flattened tetrahedrons respectively.

85. The J.T. distortion is not applicable for the $[VO(OH_2)_5]^{2+}$, $[CoCl(NH_3)_5]^{2+}$ and $[CoF_2(NH_3)_4]^+$ complexes.
86. In an octahedral system, if the origin of J.T. distortion lies in the e_g–set, then the extent of distortion is more severe compared to the cases where the J.T. distortion originates from the t_{2g} level.
87. The driving force of J.T. distortion in the octahedral and tetrahedral systems is the gain of additional cfse.
88. In a tetrahedral system, if the origin of J.T. distortion lies at the t_2–set, then it becomes quite significant.
89. CFT can predict the z–out or z–in distortion.
90. $[CuCl_4]^{2-}$ gives a flattened structure. It can be explained by considering the J.T.distortion.
91. For the d^9–octahedral complexes, in terms of cfse, both the z–out and z–in distortions are equally probable. But in reality, in most of the cases, the z–out distortion prevails.
92. In $[Ti(OH_2)_6]^{3+}$ or $[TiCl_6]^{3-}$, the origin of J.T. distortion lies at the t_{2g}–set, but its effect on the nature of the d–d absorption band is pronounced significantly.
93. The broad peak in the electronic spectrum of $[Ti(OH_2)_6]^{3+}$ is due to the J.T. distortion.
94. In $K_2[MnF_5(OH_2)]$, there are two types of Mn–F bond lengths.
95. For $[CoF_6]^{3-}$, the ligand field peak is split into two components.
96. $[Co(CN)_6]^{4-}$ does not survive.
97. $[Co(NO_2)_6]^{4-}$ shows two types of Co—N bond lengths.
98. $[Mn(dmso)_6]^{3+}$ shows the absorption band having three components instead of the single component.
99. In $[Mn(dmso)_6]^{3+}$, there are two different types of ν_{S-O} stretching frequencies in the ir-spectra.
100. $[Cu(OH_2)_6]^{2+}$ survies but $[Cu(NH_3)_6]^{2+}$ cannot survive in an aqueous system.
101. $[CuCl_6]^{4-}$ does not exist.
102. Sometimes, the chelate ring can avoid the J.T. distortion.
103. In $[Cu(en)_3]^{2+}$, there are two types of Cu—N bond lengths.
104. In $[Cu(bpy)(hfa)_2]$, the axial Cu—O bond is longer than the equatorial Cu—O bond. (hfa = hexafluoroacetylacetonate)
105. $[Cu(OMPA)_3]^{2+}$ is an example of undistorted chelate in spite of the expected inherent J.T. distortion. (OMPA = octamethylpyrophosphoramide)
106. In the spinel structure of $NiCr_2O_4$, the 'NiO_4' unit possesses an elongated tetrahedral structure.
107. Cu(II) is unexpected to produce the symmetrical octahedral complex, but in some cases, it forms the perfectly symmetrical octahedral complexes.
108. There are two types of J.T. distortions—static and dynamic.
109. For Cu(II), in many cases, the room temperature crystallographic data indicate the symmetrical octahedral structure but the spectroscopic data are in favour of their distorted structure.
110. For Cu(II), sometimes, the room temperature crystallographic data indicate the undistorted structure but the crystallographic data at low temperature indicate their distorted structure.
111. The structural features of $[Mn(acac)_3]$ can be explained by considering the dynamic J.T. distortion.
    ● The room temperature crystallographic data and spectroscopic data of $[Mn(acac)_3]$ give the different conclusions regarding the structure of the complex.
112. Structure of the anion $[Cu(NO_2)_6]^{4-}$ differs from crystal to crystal.
113. Ni(II) can form perfectly the octahedral complexes in many cases but its heavier congeners Pd(II) and Pt(II) fail to form the octahedral complexes.
114. For the d^8–system, no J.T. distortion is expected in the octahedral complexes. However, the J.T. distortion can be imposed under the conditions of strong crystal field splitting and low pairing energy.

115. Ni(II) generally forms the octahedral complexes but in the presence of ligand CN⁻, it forms the square planar complex.
    ● Very strong field ligands direct Ni(II) to adopt the square planar structure rather than the octahedral structure.
    ● Sometimes, fairly weak field ligands but possessing a high nephelauxetic effect can favour Ni(II) to adopt the square planar structure.
116. Ni(II) can form the perfectly octahedral complexes but it cannot form the perfectly tetrahedral complexes.
117. Cu(II) is expected to give neither the perfectly octahedral nor the perfectly tetrahedral complexes.
118. Only some special types of ligands (showing the very high crystal field splitting power or high nephelauxetic effect) can generate the square planar complexes with Ni(II) but for Pd(II) and Pt(II), almost all types of ligands are suitable to generate their square planar complexes.
119. With the ligands like Cl⁻, NH₃, etc. Ni(II) forms the octahedral complexes but Pd(II) and Pt(II) form the square planar complexes with these ligands.
    (**Hints:** Crystal field splitting power: Pt(II) ⟩ Pd(II) ⟩ Ni(II); P, *i.e.* pairing energy: Ni(II) ⟩ Pd(II) ⟩ Pt(II)).
120. With CN⁻, Ni(II) and Ni(III) give the different complexes—[Ni(CN)₄]²⁻ and [Ni(CN)₅]³⁻ respectively.
121. Cu(II) is stable but its heavier congener Au(II) is not stable with respect to its disproportionation.
122. The d^7, d^8 and d^9 systems can form the square planar complexes.
    ● Compared to the d^8, d^9 systems, it is more difficult to produce the square planar complexes at the d^7-system
123. Co(II) can produce only a few square planar complexes.
124. Low spin Mn(III) (d^4) experiences the J.T. distortion in the t_{2g} set but it fails to give the square planar complexes even with the very strong field ligands like CN⁻.
    ● With CN⁻, Mn(III) forms low-spin [Mn(CN)₆]³⁻ rather than [Mn(CN)₄]⁻.
125. Hydration and lattice energy of the transition metal ions very often differ from the calculated values obtained by using their ionic radii of gaseous states.
    ● For the transition metal ions, the observed lattice and hydration energy conform to the calculated values, if their crystal radii are used instead of their ionic radii of gaseous state.
126. For the bivalent metal ions (M²⁺) of 1ˢᵗ transition series, if the hydration energy or lattice energy (of oxides or halides) is plotted against the d^n–configuration, a double hamped curve is obtained.
127. Periodic trends of ionic radii (in gaseous state) of the transition metal ions are different from their trends of crystal radii.
128. For the M²⁺ or M³⁺ ions of the 1ˢᵗ transition series, the plot of ionic radius (h.s. octahedral) versus the d^n configuration gives a double humped curve. For the low-spin octahedral ionic radii, it produces a single humped curve instead of the double humped curve.
129. The plot of ionic radius (crystal radius) *vs.* the d^n configuration finds a correlation with the plot of hydration energy *vs.* the d^n configuration.
130. For the transition metal ions, the calculated lattice energy (by Born-Lande equation) needs the correction factor—contribution of cfse.
131. In the capped octahedron, [M(N)₇] system, there are two types of M—N bond distances even in the systems which do not experience the J.T. distortion. These two types of bond length vary in different ways with the d^n–configurations.
132. The plot of tetrahedral ionic radius *vs.* d^n for the M²⁺ ions of 1ˢᵗ transition series, shows two minima.

133. In terms of the effective nuclear charge, the Cu(II)–L bond distance is expected to be shorter than the Ni(II)–L bond distance, but in their comparable square planar complexes, the reverse is true.

134. The ionic radius of the metal ions for the square planar complexes should vary in the sequence: $d^{10} \rangle d^9 \rangle d^8 \rangle d^7$

135. The ionic radius of Ag^+ is larger in a square planar complex compared to that in a tetrahedral complex.

136. The ionic radius of Ni^{2+} varies with its crystal geometries as follows:

$r_{Ni^{2+}}$: Octahedral > tetrahedral > square planar.

(**Hints:** Consider both orbital occupancy factor and ligand-ligand repulsion, *i.e.* effect of coordination number).

137. The weak field and bulky anionic ligands generally favour the tetrahedral complex formation over the octahedral complex formation.

138. The tetrahedral complexes are generally produced for the d^0, d^1, d^2, d^6 and d^7 systems.

139. $[Cu(NH_3)_4]^{2+}$ is square planar while $[Zn(NH_3)_4]^{2+}$ is tetrahedral.

140. $[FeCl_4]^-$ is perfectly tetrahedral but $[CuCl_4]^{2-}$ is flattened tetrahedral.

141. $[FeCl_4]^-$, $[CoCl_4]^{2-}$ are tetrahedral.

142. The d^3, d^4, d^8 and d^9 systems are reluctant to adopt the tetrahedral structure.

143. $\left[Co(OH_2)_6 \right]^{2+} \xrightarrow{\text{conc. HCl}} \underbrace{\left[CoCl_4 \right]^{2-}, \left[CoCl_3(OH_2) \right]^-}_{\text{Tetrahedral}}$
(aqueous solution)

$\left[Ni(OH_2)_6 \right]^{2+} \xrightarrow{\text{conc. HCl}} \left[NiCl_2(OH_2)_4 \right]$
(aqueous solution)

● Co(II) can easily adopt the tetrahedral structure with the bulky and anionic ligands like Cl^-, but Ni(II) is highly reluctant to adopt the tetrahedral structure under the identical conditions, *i.e.* Ni(II) resists strongly to change its octahedral geometry.

144. $[NiCl_4]^{2-}$ is tetrahedral, $[Ni(NH_3)_6]^{2+}$ is octahedral, $[Ni(CN)_4]^{2-}$ is square planar.

145. In the crystal of $[Ni(acac)_2]$, Ni(II) attains the octahedral symmetry through polymerisation.

146. $[Co(acac)_2]$ can exist as both the monomers and polymers to attain the tetrahedral and octahedral structure respectively.

147. The **Lifschitz Salts** of Ni(II) are quite interesting in terms of their structural and magnetic properties.

148. With CN^-, Ni(II) can be trapped as $[Ni(CN)_4]^{2-}$ (square planar), $[Ni(CN)_5]^{3-}$ (TBP) and $[Ni(CN)_5]^{3-}$ (square pyramidal) depending on the conditions.

149. The preference for the normal or inverse spinel structure can be predicted by considering the contribution of cfse.

150. For AB_2O_4 (*i.e.* mixed oxides), the inverse spinel structure is favoured for d^5–B(III) and d^0–B(III), and $d^{7, 8, 9}$–A(II).

151. Mn_3O_4 adopts the normal spinel structure while Fe_3O_4, $CoFe_2O_4$, $NiAl_2O_4$ adopt the inverse spinel structure.

152. Co_3O_4 adopts the normal spinel structure. It can be explained by considering the critical $10Dq_o$ value for Co(III) in the oxide crystal.

153. The reduction potential of the Co(III)/Co(II) couple decreases roughly with the increase of ligand field strength.

● The variation of reduction potential of the Co(III)/Co(II) couple with the ligands does not exactly repeat the spectrochemical series.

154. The Co(III)–aqua complex slowly undergoes redox decomposition to give the Co(II)–aqua complex and oxygen, but in presence of 'en', Co(II) is slowly oxidised to $[Co(en)_3]^{3+}$ by aerial oxygen.

- $$[Co(aq)]^{3+} \xrightarrow[-O_2]{standing} [Co(aq)]^{2+} \xrightarrow[\text{(aerial oxidation)}]{+en} [Co(en)_3]^{3+}.$$

155. Between en and phen ligands, phen is a much stronger field ligand but the E^0 value of the Co^{III}/Co^{II} couple is more negative for the en ligand than that for the phen ligand.

156. $$[Co(aq)]^{2+} + CN^- \xrightarrow[\text{(slowly)}]{-H_2} [Co(CN)_6]^{3-}$$

157. Besides the cfse factor, many other factors are important to determine the E^0 value of a redox couple.

158. The superhyperfine structure of the ESR spectrum of $[IrCl_6]^{2-}$ indicates the overlap between the Ir(IV)–orbital and Cl$^-$–ligand orbital.

159. The nature of ESR spectra of $[IrCl_6]^{2-}$ depends on whether the Ir(IV) centre experiences the static or dynamic J.T. distortion.

160. The PMR signals for the ring protons of $[Al(acac)_3]$ and $[V(acac)_3]$ are of different types.

161. The B value (Racah parameter) always decreases in complexes.

$$\beta = \frac{B_{complex}}{B_{gaseous}} \le 1;$$ in most of the cases, β is less than unity.

B: $Cr^{3+} \rangle Cr^{2+}$; $Ni^{2+} \rangle Pd^{2+} \rangle Pt^{2+}$;

     $Cu^{2+} \rangle Ni^{2+} \rangle Co^{2+} \rangle Fe^{2+} \rangle Mn^{2+} \rangle Cr^{2+} \rangle V^{2+}$

162. The greater nephelauxetic effect favours the low-spin complex formation.

163. The nephelauxetic effect depends on the nature of both the metal centre and ligand.

164. The nephelauxetic series of the ligands is different from the spectrochemical series of the ligands.

165. The β value is close to unity for the hard and nonpolarisable ligands (*e.g.* O, F donors) but it is deviated significantly from unity for the soft and polarisable ligands (*e.g.* I$^-$, Br$^-$, CN$^-$, etc.)

166. The spectral properties like intensity, charge transfer band, etc. are the indirect evidences of the overlap between the metal and ligand orbitals.

167. The antiferromagnetic coupling through the superexchange process supports the overlap between the metal and ligand orbitals.

- In MOT, for σ–bonding with the ligands, in the octahedral complexes, the metal centre can provide 6 orbitals while for π–bonding, it can provide three orbitals.

168. In the MO energy diagrams of the octahedral complexes, the order of MOs can be roughly predicted.

- The bonding MOs generated from the $d_{x^2-y^2}$ and d_{z^2} orbitals of the metal centre are of relatively higher energy compared to the bonding MOs generated from the s– and p–orbitals of the metal centre.

169. The central part of the MO energy diagram of σ–bonding consisting of nonbonding the t_{2g} and antibonding σ_g^* orbitals mimicks the crystal field splitting of the d–orbitals (in terms of CFT) in the octahedral systems.

170. The π–LGOs (t_{1u}, t_{1g} and t_{2u}) generally remain as the nonbonding orbitals in the octahedral complexes.

171. The σ–bonding scheme of MOT can explain the major conclusions of VBT and CFT.

172. The π–donor ligands reduces the $10Dq_o$ value while the π–acceptor ligands enhances the $10Dq_o$ value.

173. The crystal field strength runs as: $F^- \rangle Cl^- \rangle Br^- \rangle I^-$;
$NH_3 \rangle H_2O$; $H_2O \rangle OH^-$; donor sites: $C \rangle N \rangle O \rangle F$ (halogen); $NCS^- \rangle NH_3$
174. The better σ–donor ligands produce the higher $10Dq_o$ value. This can be explained by MOT.
175. Both py and bpy use the $sp^2 - N$ donor sites but the crystal field splitting power of bpy is higher than that of py.
176. The 18e rule is strictly followed for the very strong field ligands only in the octahedral complexes.
   - For the very weak field ligands, the valence electron count may lie in the range $12 - 22$.
   - For the moderately strong field ligands, the valence electron count may lie in the range $12 - 18$.
177. The d^8–system gives the most favourable situation for the square planar complex formation.
178. The metal centre can effectively use the following d–orbitals for π–bonding:

   d_{xy}, d_{yz}, d_{zx} (for octahedral); $d_{x^2-y^2}, d_{z^2}$ (for tetrahedral); d_{xz}, d_{yz}, d_{xy} (for square planar).

179. The bottom and central portion of the MO energy diagram of σ–bonding in $[Co(NH_3)_6]^{3+}$ represent the VBT and CFT description respectively.
180. The σ–bonding scheme of MOT can explain the major conclusions of CFT.
181. To explain the components in the esr signal of $[IrCl_6]^{2-}$, some authors consider the hyperfine splitting while some authors consider the superhyperfine splitting.

CHAPTER 4

Thermodynamic Aspects of Stability of Metal Complexes

4.1 STABILITY OF THE COMPLEXES: THERMODYNAMIC STABILITY AND KINETIC STABILITY (*cf.* Sec. 5.4)

There are two terms *thermodynamic stability and kinetic stability* to express the stability of a coordination compound (say ML).

- **Thermodynamic Stability** measures the strength of the metal-ligand bond in the complex (ML). Thus it gives the measure of the equilibrium constant in the metal-ligand interaction leading to the complex (ML).

  The term, *stability* simply refers to its thermodynamic stability. Thus the terms *stable* and *unstable* complexes bear the thermodynamic sense.

- **Kinetic Stability** of a complex refers to its stability in ligand substitution/dislodgement reactions in terms of rate constant. The *kinetically stable* (*i.e.* **inert**) complexes show *a high activation energy* and they only slowly react while the *kinetically unstable* (*i.e.* **labile**) complexes show *a low activation energy* and they react rapidly.

- The terms **labile and inert** bear the *kinetic sense* while the terms **stable and unstable** bear the thermodynamic sense.

Let us consider the following simple metal-ligand interaction:

$$M + L \rightleftharpoons ML$$

If the equilibrium constant (*i.e. formation constant*) of the above equilibrium is high than we can say that the complex ML is *stable* (thermodynamic term) and this stability is determined by the *reaction energy* (*i.e.* energy difference between the reactants and products). The lability or inertness of the complex depends on *how fast* the above equilibrium is attained. It depends on the *activation energy* (*i.e.* energy difference between the reactants and activated complex produced in the act leading to the product). If the activation energy is high, the rate process is slow (*i.e. inert system*) and if the activation energy is less, the rate is fast (*i.e. labile system*). These are illustrated schematically in Fig. 4.1.1.

How far and How fast?

Stability indicates *how far* a reaction goes on and lability indicates *how fast* the reaction goes on.

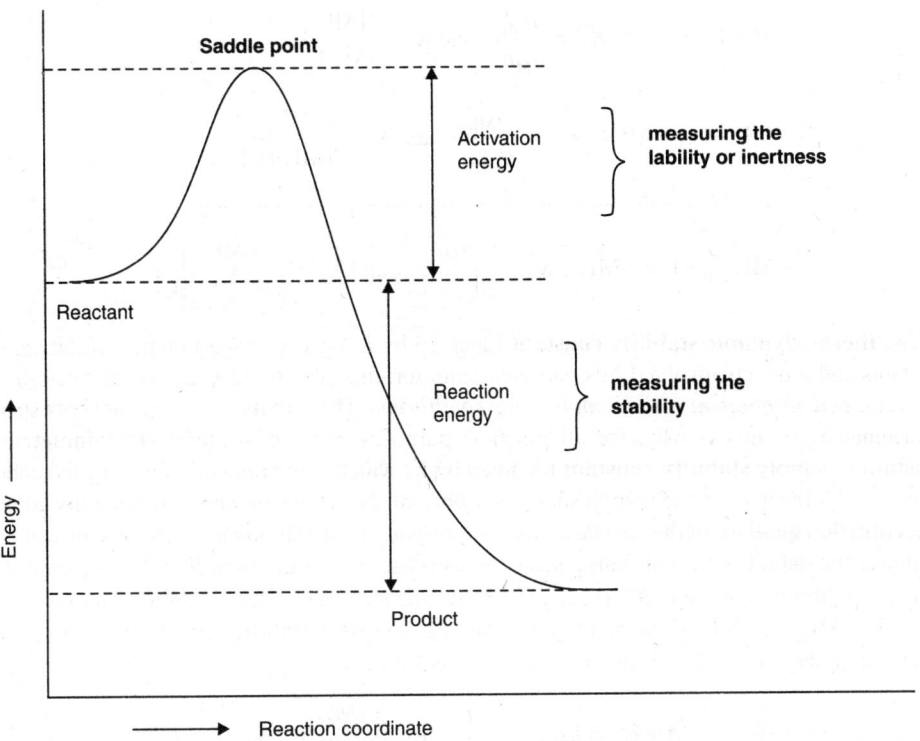

Fig. 4.1.1 Schematic representation of activation energy (measuring the lability or inertness) and reaction energy (measuring the stability).

A stable complex may be labile (*i.e.* kinetically unstable) or inert (*i.e.* kinetically stable) depending on the properties of the complex. The different aspects of lability or inertness will be discussed separately in Chapter 5.

4.2 STEPWISE AND OVERALL STABILITY CONSTANTS OF THE METAL COMPLEXES

It is believed that formation of a complex in solution proceeds by the *stepwise replacement* (*i.e.* substitution) of the coordinated solvent molecules by the ligands. In aqueous media, these may be simply represented (without showing any charge for the sake of simplicity) as follows (L stands for a monodentate ligand):

$$\left[M(OH_2)_n \right] + L \rightleftharpoons \left[M(OH_2)_{n-1} L \right] + H_2O; \; K_1^0 \text{ and } K_1$$

$$\left[M(OH_2)_{n-1} L \right] + L \rightleftharpoons \left[M(OH_2)_{n-2} L_2 \right] + H_2O; \; K_2^0 \text{ and } K_2$$

....

$$\left[M(OH_2)L_{n-1} \right] + L \rightleftharpoons \left[ML_n \right] + H_2O, \; K_n^0 \text{ and } K_n$$

Very often, the solvent molecules (*i.e.* H_2O in aqueous media) are not shown. Then the above equilibria may be simply represented as follows:

$$M + L \rightleftharpoons ML; \quad K_1^0 = \frac{a_{ML}}{a_M a_L} \quad \text{and} \quad K_1 = \frac{[ML]}{[M][L]}$$

$$ML + L \rightleftharpoons ML_2; \quad K_2^0 = \frac{a_{ML_2}}{a_{ML} a_L} \quad \text{and} \quad K_2 = \frac{[ML_2]}{[ML][L]}$$

....

$$ML_{n-1} + L \rightleftharpoons ML_n, \quad K_n^0 = \frac{a_{ML_n}}{a_{ML_{n-1}} a_L} \quad \text{and} \quad K_n = \frac{[ML_n]}{[ML_{n-1}][L]}$$

The **thermodynamic stability constant** (denoted by K^0) is expressed in terms of the *activity* of the reactants and products involved in a particular equilibrium. Thus the thermodynamic stability constant (K^0) is the *activity quotient of the complexation equilibrium*. The activity terms (a's) are not experimentally determinable and this is why, for all practical purposes, we use the term **stoichiometric stability constant** or **simply stability constant** (denoted by K) which is expressed in terms of the concentration of the species involved in the complexation equilibrium. Thus the stoichiometric stability constant is the **concentration quotient** of the *complexation equilibrium*. The stoichiometric stability constant is generally written as the stability constant. This constant is also described as the *conditional concentration quotient.*

The equilibrium constants $K_1, K_2, K_3, ..., K_n$ for the *stepwise formation of the successive complexes* $ML, ML_2, ML_3, ..., ML_n$ respectively are called the **stepwise stability constants.**

The stability constants can also be expressed as follows:

$$M + L \rightleftharpoons ML, \qquad \beta_1 = \frac{[ML]}{[M][L]} = K_1$$

$$M + 2L \rightleftharpoons ML_2, \qquad \beta_2 = \frac{[ML_2]}{[M][L]^2} = K_1 K_2$$

$$M + 3L \rightleftharpoons ML_3, \qquad \beta_3 = \frac{[ML_3]}{[M][L]^3} = K_1 K_2 K_3$$

....

$$M + nL \rightleftharpoons ML_n, \qquad \beta_n = \frac{[ML_n]}{[M][L]^n} = K_1 K_2 K_3 ... K_{n-1} K_n$$

Thus we can write:

$$\beta_n = \frac{[ML]}{[M][L]} \cdot \frac{[ML_2]}{[ML][L]} \cdot \frac{[ML_3]}{[ML_2][L]} \cdots\cdots\cdots \frac{[ML_n]}{[ML_{n-1}][L]}$$

$$= K_1 \cdot K_2 \cdot K_3 .... K_{n-1} \cdot K_n = \prod_{i=1}^{n} K_i$$

and $\qquad\qquad\qquad\qquad \log_{10}\beta_n = \log_{10}K_1 + \log_{10}K_2 + ... + \log_{10}K_n$

i.e. the **overall stability constant (β_n)** is the product of the stepwise stability constants K_n ($n = 1, 2, 3, ...$)

Determination of β_n^0 from β_n

From the definition we can write:

$$M + nL \rightleftharpoons ML_n;\ \beta_n^0 = \frac{a_{ML_n}}{a_M a_L^n} \text{ and } \beta_n = \frac{[ML_n]}{[M][L]^n}$$

The activity (a) and concentration (C) parameters are related as follows:

$$a = C \times f;\ (a = \text{activity};\ C = \text{concentration};\ f = \text{activity coefficient})$$

It leads to:

$$\beta_n^0 = \frac{[ML_n]}{[M][L]^n} \times \frac{f_{ML_n}}{f_M \times f_L^n} = \beta_n \times \frac{f_{ML_n}}{f_M f_L^n}$$

Similarly, it can be shown:

$$K_n^0 = K_n \times \frac{f_{ML_n}}{f_{ML_{n-1}} f_L}$$

Thus from the knowledge of 'f', we can compute β_n^0 or K_n^0 from the corresponding *stoichiometric stability constants*. The activity coefficient (f) tends to be unity at very dilute solution. In fact, f depends on the I (ionic strength) of the medium. In practice, by using some *excess inert electrolyte* (e.g. KNO_3, $NaClO_4$, etc.), the stability constants are determined at different fixed ionic strength values. The extrapolated value of β_n obtained at $I = 0$ gives the *thermodynamic constant*. In fact, β_n^0 and β_n are related as follows in terms of I.

$$\log \beta_n^0 = \log \beta_n - \frac{A\sqrt{I}}{(1 + B\sqrt{I})} \approx \log \beta_n - A\sqrt{I},\ (1 >> B\sqrt{I})$$

i.e.

$$\log \beta_n \approx \log \beta_n^0 + A\sqrt{I},\ (A = \text{constant})$$

(The above relation is obtained from *Debye-Huckel theory*).

The experimentally determined β_n values at different ionic strengths (I) are plotted as $\log\beta_n$ vs. $\sqrt{I}$ and from the intercept (*i.e.* $\sqrt{I} \approx 0$) of the plot, β_n^0 is obtained. In the same way, K_n^0 can also be determined. From the knowledge of β_n^0 or K_n^0, the standard values of the thermodynamic parameters can be obtained.

$$\Delta G^0 = -RT\ln\beta_n^0 = \Delta H^0 - T\Delta S^0$$

i.e.

$$2.303R\log\beta_n^0 = \Delta S^0 - \frac{\Delta H^0}{T} \quad (i.e.\ \log\beta_n^0 \text{ vs. } \frac{1}{T} \text{ gives a linear plot})$$

and

$$\ln\frac{\beta_{n(2)}^0}{\beta_{n(1)}^0} = -\frac{\Delta H^0}{R}\left(\frac{1}{T_2} - \frac{1}{T_1}\right)$$

Note: β_n^0 or K_n^0 does not depend on the ionic strength of the medium while the corresponding stoichiometric stability constant depends on the ionic strength of the medium. It may be noted that both the thermodynamic and stoichiometric stability constants depend on temperature.

For the didentate ligands (L—L), entry of one ligand needs the replacement of two water molecules. For an *octahedral system*, these are shown below.

$$\left[M(OH_2)_6\right]+L-L \rightleftharpoons \left[M(OH_2)_4(L-L)\right]+2H_2O, \ K_1$$

$$\left[M(OH_2)_4(L-L)\right]+L-L \rightleftharpoons \left[M(OH_2)_2(L-L)_2\right]+2H_2O, \ K_2$$

$$\left[M(OH_2)_2(L-L)_2\right]+L-L \rightleftharpoons \left[M(L-L)_3\right]+2H_2O, \ K_3$$

In terms of overall stability constants (β), we can write.

$$\left[M(OH_2)_6\right]+L-L \rightleftharpoons \left[M(OH_2)_4(L-L)\right]+2H_2O, \ \beta_1$$

$$\left[M(OH_2)_6\right]+2L-L \rightleftharpoons \left[M(OH_2)_2(L-L)_2\right]+4H_2O, \ \beta_2$$

$$\left[M(OH_2)_6\right]+3L-L \rightleftharpoons \left[M(L-L)_3\right]+6H_2O, \ \beta_3$$

It can be readily shown:

$$\beta_1 = K_1, \ \beta_2 = K_1K_2, \ \beta_3 = K_1K_2K_3$$

Table 4.2.1 Stepwise stability constants ($I \approx 0.1 - 0.5$ mol dm^{-3}) of some representative metal complexes (~25° C).

System	$logK_1$	$logK_2$	$logK_3$	$logK_4$	$logK_5$	$logK_6$
Cu^{2+}–NH$_3$	4.15	3.50	2.90	2.13	–	–
Cu^{2+}–en	10.70	9.30	–1.0	–	–	–
Cu^{2+}–phen	9.25	6.75	5.35	–	–	–
Ni^{2+}–NH$_3$	2.80	2.25	1.73	1.19	0.75	0.03
Ni^{2+}–en	7.65	6.40	4.55	–	–	–
Zn^{2+}–NH$_3$	2.37	2.45	2.50	2.15	–	–
Zn^{2+}–en	5.92	5.15	1.86	–	–	–
Zn^{2+}–phen	6.43	5.72	4.85	–	–	–
Ag$^+$–NH$_3$	3.14	3.82	–	–	–	–
Hg^{2+}–CN$^-$	18.0	16.70	3.83	2.98	–	–
Hg^{2+}–I$^-$	12.87	10.95	3.67	2.37	–	–

Stability Constant in Terms of Rate Constants

If the following *ML* complex formation and its dissociation occur in *one-step processes* (*i.e.* **elementary reactions**),

$$M + L \underset{k_d}{\overset{k_f}{\rightleftharpoons}} ML$$

then at equilibrium, we have: forward rate ($k_f[M][L]$) = backward rate ($k_d[ML]$)

i.e. $$K_1 = \frac{k_f}{k_d} = \frac{[ML]}{[M][L]}$$

Thus, if k_f and k_d can be determined independently from the kinetic studies, then the corresponding formation constant is obtained. In fact, *this method is quite helpful, when the equilibrium is attained very slowly.*

4.3 CONDITIONAL STABILITY CONSTANTS

In a real situation, in solution, besides the metal-ligand equilibria, many other equilibria may co-exist. These are the *hydrolytic equilibria of the metal ion, protonation and deprotonation equilibria of the ligand, etc.* Thus for the metal ions, the following equilibria (charges are not shown for the sake of simplicity) are quite relevant.

$$M + L \rightleftharpoons ML$$
$$ML + L \rightleftharpoons ML_2$$
$$\cdots\cdots\cdots\cdots$$
$$\underbrace{ML_{n-1} + L \rightleftharpoons ML_n}_{\text{Metal-ligand equilibria}}$$

and

$$M(OH_2)_x \rightleftharpoons M(OH)(OH_2)_{x-1} + H^+$$
$$M(OH)(OH_2)_{x-1} \rightleftharpoons M(OH)_2(OH_2)_{x-2} + H^+$$
$$\cdots\cdots\cdots\cdots\cdots\cdots\cdots\cdots\cdots\cdots\cdots$$
$$\underbrace{M(OH)_{x-1}(OH_2) \rightleftharpoons M(OH)_x + H^+}_{\text{Hydrolytic equilibria of the metal ion.}}$$

Similarly, the protonation-deprotonation process of the ligand may be represented (without showing any charge) as follows:

$$L + H^+ \rightleftharpoons LH, \ LH + H^+ \rightleftharpoons LH_2, \ LH_2 + H^+ \rightleftharpoons LH_3, \ LH_{m-1} + H^+ \rightleftharpoons LH_m$$

(For the sake of simplicity, in the above equilibria, charges are ignored).

In such cases, the stability constants are better defined in another way referred to as the **conditional stability constant (β_n') or the apparent stability constant or the effective stability constant.**

$$\beta_n' = \frac{(\text{Concentration of the complex})}{\begin{pmatrix} \text{Concentration of the metal} \\ \text{not bound to the ligand} \end{pmatrix} \times \begin{pmatrix} \text{Concentration of the ligand} \\ \text{not bound to the metal.} \end{pmatrix}}$$

We have:

T_M (concentration of the total metal)

= concentration of the metal not bound to the ligand + concentration of the metal bound to the ligand.

$$= \left\{ [M] + [M(OH)] + [M(OH)_2] + \cdots\cdots + [M(OH)_x] \right\} + \left\{ [ML] + [ML_2] + [ML_3] + \cdots\cdots + [ML_n] \right\}$$

$$= \sum_{j=0}^{x} [M(OH)_j] + \sum_{i=1}^{n} [ML_i], \ M(OH) \text{ denotes actually } M(OH)(OH_2)_{x-1}$$

or, $$T_M - \sum_{i=1}^{n} [ML_i] = \sum_{j=0}^{x} [M(OH)_j] = \text{concentration of the metal not bound to the ligand.}$$

Similarly, T_L (total ligand concentration) can be expressed as follows:

$$T_L = \left\{ [L] + [LH] + [LH_2] + \cdots\cdots + [LH_m] \right\}$$

$$+ \left\{ 1[ML] + 2[ML_2] + 3[ML_3] + \cdots\cdots + n[ML_n] \right\}$$

$$= \sum_{j=0}^{m} [LH_j] + \sum_{i=1}^{n} i[ML_i]$$

Thus, concentration of the ligand not bound to the metal is given by:

$$T_L - \sum_{i=1}^{n} i[ML_i] = \sum_{j=0}^{m} [LH_j]$$

Thus, the conditional stability constant (β_n') is given by:

$$\beta_n' = \frac{[ML_n]}{\left\{T_M - \sum_{i=1}^{n}[ML_i]\right\}\left\{T_L - \sum_{i=1}^{n} i[ML_i]\right\}}$$

It is illustrated by taking the following representative case.

$$M^{2+} + edta^{4-} \rightleftharpoons [M(edta)]^{2-}$$

The ligand-proton equilibria in the system are:

$$H_4edta \rightleftharpoons H_3edta^- + H^+, (K_1)$$

$$H_3edta^- \rightleftharpoons H_2edta^{2-} + H^+, (K_2)$$

$$H_2edta^{2-} \rightleftharpoons Hedta^{3-} + H^+, (K_3)$$

$$Hedta^{3-} \rightleftharpoons edta^{4-} + H^+, (K_4)$$

(H_4edta: $pK_1 = 2.0$, $pK_2 = 2.7$, $pK_3 = 6.2$, $pK_4 = 10.3$ at 20° C)

If we ignore the hydrolytic process of the metal centre, we can write:

$$T_M = [M^{2+}] + [M(edta)^{2-}]$$

and

$$T_L = [M(edta)^{2-}] + [H_4edta] + [H_3edta^-] + [H_2edta^{2-}] + [Hedta^{3-}] + [edta^{4-}]$$

or

$$T_L - [M(edta)^{2-}] = [edta^{4-}] + \frac{[edta^{4-}][H^+]}{K_4}$$

$$+ \frac{[edta^{4-}][H^+]^2}{K_3K_4} + \frac{[edta^{4-}][H^+]^3}{K_2K_3K_4} + \frac{[edta^{4-}][H^+]^4}{K_1K_2K_3K_4}$$

$$= [edta^{4-}]\left\{1 + \frac{[H^+]}{K_4} + \frac{[H^+]^2}{K_3K_4} + \frac{[H^+]^3}{K_2K_3K_4} + \frac{[H^+]^4}{K_1K_2K_3K_4}\right\}$$

$$= [edta^{4-}]\alpha_L$$

or,

$$\frac{T_L - [M(edta)^{2-}]}{\alpha_L} = [edta^{4-}]$$

i.e.

$$\alpha_L = \frac{T_L - [M(edta)^{2-}]}{[edta^{4-}]}$$

It leads to:

$$K = \frac{\left[M(edta)^{2-}\right]}{\left[M^{2+}\right]\left[edta^{4-}\right]}$$

$$= \frac{\left[M(edta)^{2-}\right] \times \alpha_L}{\left\{T_M - \left[M(edta)^{2-}\right]\right\}\left\{T_L - \left[M(edta)^{2-}\right]\right\}}$$

$$= K'\alpha_L$$

i.e. $\qquad K/\alpha_L = K'$

where K = stability constant, K' = conditional stability constant; α_L depends on pH and α_L may be obtained from the plot of α_L against pH (*see* Fig. 11.7.3.2). The above relation gives:

$$\log K = \log K' + \log \alpha_L; \text{ or } \log K' = \log K - \log \alpha_L$$

α_L can be computed from the knowledge of the acid dissociation constants of H_4edta and pH of the working medium. This aspect is discussed in detail in Secs. 11.7.3, 4.

If the medium contains any other *competing ligand* (which may be the simple OH^- ligand), it will reduce the concentration of $[M^{2+}]$. If the reduction occurs by the factor α_M which is defined as the ratio of the sum of the concentration of all the metal ion species not complexed with the target ligand (say edta in the present case) and the concentration of the simple M^{2+} species, *i.e.* $[M(OH_2)_6^{2+}]$, then α_M is given by:

$$\alpha_M = \frac{T_M - \left[M(edta)^{2-}\right]}{\left[M^{2+}\right]}$$

In such cases, the conditional stability constant K' is given by:

$$\boxed{\log K' = \log K - \log \alpha_L - \log \alpha_M}$$

4.4 ORDER OF THE STEPWISE STABILITY CONSTANTS

It is evident from Table 4.2.1 that the stepwise stability constants generally run as: $K_1 \rangle K_2 \rangle K_3 \rangle \dots$ The **statistical factors** and **nonstatistical factors** like steric effects, electronegativity neutralization, etc. control the order of stepwise stability constants.

4.4.1 Statistical Factors for the Monodentate Ligands

Let us consider the following simple complex formation reaction:

$$ML_n + L \rightleftharpoons ML_{n+1}$$

The above equilibrium actually involves the substitution of the solvent molecules coordinated to M in ML_n. If N be the coordination number of the metal centre then in aqueous solvent, it should be written as:

$$M(OH_2)_{N-n}L_n + L \underset{k_b}{\overset{k_f}{\rightleftharpoons}} M(OH_2)_{N-n-1}L_{n+1} + H_2O$$

If we consider both the forward and backward reactions as the ***one step reactions***, then we can write:

$$\text{forward rate } (R_f) = k_f[M(OH_2)_{N-n}L_n][L]$$

and, $\qquad \text{backward rate } (R_b) = k_b[M(OH_2)_{N-n-1}L_{n+1}]$

In terms of the statistical effect, the ease of replacement of a coordinated water molecule depends on the number of such water molecules (*i.e.* **reaction sites**) available. *Obviously, the number of reaction sites gradually decreases with the successive addition of the ligand.* Thus it is reasonable to conclude that the forward (*i.e.* formation) rate constant (k_f) will be proportional to the number of available reaction sites in the complex $M(H_2O)_{N-n}L$, *i.e.*

$$k_f \propto (N-n), \text{ i.e. } k_f = k_f'(N-n) \; ; \; k_f' = \text{proportionality constant}$$

Similarly, the backward (*i.e.* dissociation) rate constant (k_b) depends on the number of dissociable ligands, *i.e.*

$$k_b \propto (n+1), \text{ i.e. } k_b = k_b'(n+1); \; k_b' = \text{proportionality constant}$$

Now, the equilibrium constant for a such one-step reaction may be expressed as follows:

$$K_{n+1} = \frac{k_f}{k_b} = \frac{k_f'(N-n)}{k_b'(n+1)}, \text{ i.e. } K_{n+1} \propto \frac{(N-n)}{(n+1)}$$

In the same way, we can have:

$$K_n = \frac{k_f'(N-n+1)}{k_b'n}, \text{ i.e. } K_n \propto \frac{(N-n+1)}{n}$$

It leads to:

$$\boxed{\frac{K_n}{K_{n+1}} = \frac{(n+1)(N-n+1)}{n(N-n)}; \text{ or } \frac{K_{n+1}}{K_n} = \frac{n(N-n)}{(n+1)(N-n+1)}}$$

Thus, statistically, K_n is always greater than K_{n+1}.

By using the relation, $K_n \propto \dfrac{N-n+1}{n}$, we get:

$$K_1 : K_2 : K_3 : \ldots\ldots\ldots : K_n : K_{n+1} : \ldots\ldots : K_N$$

$$= \frac{N}{1} : \frac{N-1}{2} : \frac{N-2}{3} : \ldots\ldots : \frac{(N-n+1)}{n} : \frac{N-n}{n+1} : \ldots\ldots : \frac{1}{N}$$

● For an **octahedral system** (*i.e.* $N = 6$), we have:

$$K_1 : K_2 : K_3 : K_4 : K_5 : K_6 = \frac{6}{1} : \frac{5}{2} : \frac{4}{3} : \frac{3}{4} : \frac{2}{5} : \frac{1}{6}$$

For the octahedral system ($N = 6$ and $n = 1$ to 6), we have:

$$\boxed{\frac{K_{n+1}}{K_n} = \frac{n(N-n)}{(n+1)(N-n+1)} = \frac{n(6-n)}{(n+1)(7-n)}}$$

● For the **square planar** and **tetrahedral systems** (*i.e.* $N = 4$, $n = 1$ to 4) we have:

$$K_1 : K_2 : K_3 : K_4 = \frac{4}{1} : \frac{3}{2} : \frac{2}{3} : \frac{1}{4} = 48 : 18 : 8 : 3$$

Table 4.4.1.1 compares the statistical prediction and experimental observations for the Ni^{2+}/NH_3 system, *i.e.*

$$\left[Ni(NH_3)_n (OH_2)_{6-n}\right]^{2+} + NH_3 \rightleftharpoons \left[Ni(NH_3)_{n+1}(OH_2)_{5-n}\right]^{2+} + H_2O$$

Table 4.4.1.1 Comparison of the experimental and statistical values of the ratio of stepwise stability constants of the complexes formed in the $\left[Ni(OH_2)_6 \right]^{2+} - NH_3$ system.

Ratio	Experimental Values (25° C)	Statistical Value
K_2/K_1	0.28	0.42
K_3/K_2	0.31	0.53
K_4/K_3	0.29	0.56
K_5/K_4	0.36	0.53
K_6/K_5	0.20	0.42

It is evident that the statistically predicted value of the ratio (*cf.* Table 4.4.1.1) is always higher than the corresponding experimental value. Thus, the agreement is poor in a quantitative sense. But, as far as the trend is concerned, the statistical prediction is in good agreement. *The experimental ratio is always less than the predicted value. It indicates the importance of the chemical factors (i.e.* nonstatistical factors) *besides the statistical factors.*

4.4.2 Statistical Factors for the Symmetrical Didentate Ligands (Ref. R. Pizer, *Inorg. Chem.*, 23, 3037-38, 1984)

(a) **Octahedral System:** Three didentate ligands (L—L) can occupy the 6 positions. Because of the strain, a didentate ligand can occupy only two *cis*-positions. In other words, the *trans*-positions cannot be occupied by a didentate ligand.

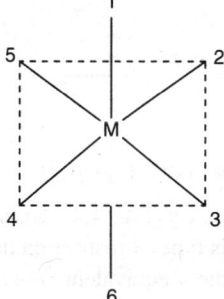

- **6 sites** (*i.e.* vertexes) for the **monodentate ligands**
- **12 edges**; a **didentate ligand** can occupy any one edge;
- 8 trigonal faces; **12 meridians**; a **tridentate ligand** can occupy any one trigonal face or any one meridian.

Fig. 4.4.2.1 Positions of the ligands in an octahedral complex. **Note:** Each C_4-axis can generate 4 meridians, *e.g.* for the C_4-axis passing through 1 and 6, the meridian positions are: 1, 2, 6; 1, 3, 6; 1, 4, 6; 1, 5, 6.

The successive ligand entry processes are:

$$\left[M(OH_2)_6 \right] + L - L \rightleftharpoons \left[M(OH_2)_4 (L-L) \right] + 2H_2O, \ K_1$$

$$\left[M(OH_2)_4 (L-L) \right] \rightleftharpoons \left[M(OH_2)_2 (L-L)_2 \right] + 2H_2O, \ K_2$$

$$\left[M(OH_2)_2 (L-L)_2 \right] \rightleftharpoons \left[M(L-L)_3 \right] + 2H_2O, \ K_3$$

The **first ligand** can be introduced in $^6C_2$ ways, *i.e.* $\dfrac{6!}{2!(6-2)!} = \dfrac{6!}{2!4!} = \dfrac{6 \times 5 \times 4!}{2! \, 4!} = 15$ ways. Out of these 15 ways, the three ways involving the *trans*–positions (*i.e.* 1, 6; 2, 4 and 3, 5) are to be rejected.

Hence, in reality, the first ligand can be introduced in 12(= 15 – 3) ways. Dissociation of the ligand can occur only in one way. It leads to:

$$K_1 \left(= \frac{k_f}{k_b} \right) \propto \frac{12}{1}.$$

For the entry of the **second ligand,** there are four vacant sites. Thus it can be introduced in $^4C_2 = \frac{4 \times 3 \times 2}{2 \times 2}$ = 6 ways, but one *trans*-position is to be rejected. Thus, the number of effective ways to introduce the second ligand is 5, *i.e.* $k_f \propto 5$. If the first ligand is supposed to occupy the 4–5 *cis*–positions, then the second ligand can occupy in 5 ways (2–3, 1–2, 1–3, 6–3, 6–2) giving rise to 4 *cis*-[M(L – L)$_2$(OH$_2$)$_2$] and one *trans*-[M(L – L)$_2$(OH$_2$)$_2$] isomers (Fig. 4.4.2.2). For the dissociation of [M(L – L)$_2$(OH$_2$)$_2$], there are two ligands, *i.e.* $k_b \propto 2$. Therefore, K_2 is given by:

$$K_2 \left(= \frac{k_f}{k_b} \right) \propto \frac{5}{2}$$

The possible structures of [M(L – L)$_2$(OH$_2$)$_2$] are given in Fig. 4.4.2.2:

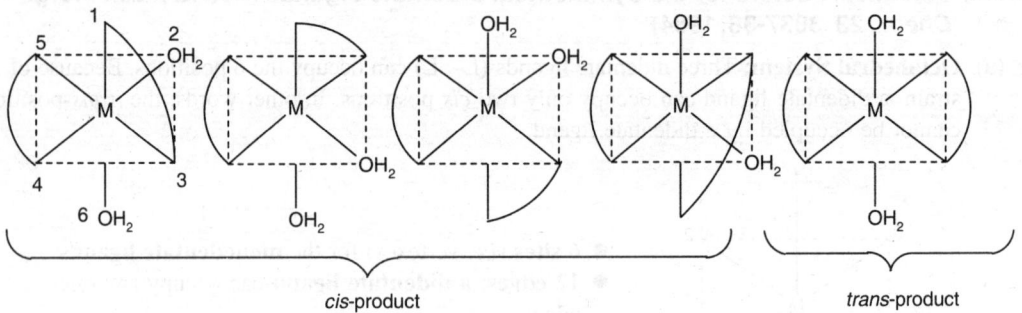

Fig. 4.4.2.2 Five possible structures of the octahedral complex [M(L–L)$_2$(OH$_2$)$_2$].

For the entry of the **third ligand,** out of the above five structures (*cf.* Fig. 4.4.2.2), one structure keeping the H$_2$O molecules at the *trans*–positions is not allowed. This type of restriction does not arise at the K_1 and K_2 steps. The third ligand can add only in *one way* to the 4 equivalent *cis-bis*-complexes (80%, *i.e.* $\frac{4}{5}$th fraction of total *bis*-complexes produced at the K_2 step can accommodate the third ligand). Thus effective k_f is given by: $k_f \propto \frac{4}{5} \times 1$. For the dissociation process, any one of the ligand can come out, *i.e.* $k_b \propto 3$. It leads to:

$$K_3 \left(= \frac{k_f}{k_b} \right) \propto \frac{4}{5} \times \frac{1}{3} = \frac{4}{15}, \text{ i.e. } K_1 : K_2 : K_3 = 12 : \frac{5}{2} : \frac{4}{15}$$

Note: The present problem can be rationalised **in another way.** An octahedron possesses *12 edges* and the incoming first didentate ligand can occupy any one edge, *i.e.* $k_f \propto 12$ and $k_b \propto 1$ (*i.e.* $K_1 \propto 12$). Coordination of this first didentate ligand leaves only **5 edges** for the incoming second didentate ligand, *i.e.* $k_f \propto 5$ and $k_b \propto 2$ (*i.e.* $k_2 \propto 5/2$). For the entry of the third didentate ligand, only **one edge** is effectively available but an **additional statistical problem** arises because the third didentate ligand cannot add if the coordinated two didentate ligands are already in the *trans*-positions, *i.e.* 4 *cis*-isomers can only accommodate the third didentate ligand but the 5th isomer, *i.e.* the *trans*-isomer, *i.e.* *trans*-$[M(L - L)_2(OH_2)_2]$ cannot accommodate the third ligand. Thus for the third incoming didentate ligand, **average number of edges** available in the two types (*i.e. cis*- and *trans*-) of the *bis*-complex is

4/5, *i.e.* $k_f \propto \dfrac{4}{5}$ and $k_b \propto 3 \left(i.e.\ K_3 \propto \dfrac{4}{15} \right)$.

Thus, the ratio of successive stability constants will be:

$$K_1 : K_2 : K_3 = \frac{12}{1} : \frac{5}{2} : \frac{4}{15} = 360 : 75 : 8$$

(b) Square planar complexes: In a square planar complex, there are four equivalent sites.

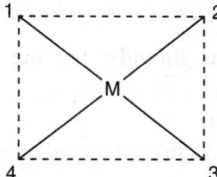

- **4 vertices** for the **monodentate ligands**
- **4 edges**; a **didentate ligand** can occupy any one edge.

Fig. 4.4.2.3 Coordinating sites of a square planar complex.

Thus, the *first didentate ligand* can enter in $^4C_2$ ways $= \dfrac{4!}{(4-2)!\,2!} = \dfrac{4 \times 3 \times 2}{2 \times 2} = 6$ ways. But out of these 6 ways, the two ways involving the *trans*– positions (1–3 and 2–4) are to be rejected. Thus the number of ways for introducing the first ligand is 4 (= 6 – 2). The number of ways for its dissociation is 1. It leads to: $K_1 \left(= \dfrac{k_f}{k_b} \right) \propto \dfrac{4}{1}$

For the entry of the *second ligand*, there is only one way for its entry and there are two possible ways of its dissociation. It leads to: $K_2 \left(= \dfrac{k_f}{k_b} \right) \propto \dfrac{1}{2}$

Thus we can write:

$$K_1 : K_2 = \frac{4}{1} : \frac{1}{2} = 8 : 1$$

Note: The present situation can be analysed in **another way** for a square planar geometry. **4 edges** are available for the first incoming didentate ligand and its dissociation can occur only in one way (*i.e.* $k_f \propto 4$, $k_b \propto 1$; $K_1 \propto 4$). After the addition of one didentate ligand, only **one edge remains for coordination** by the second didentate ligand and the *bis*-complex can dissociate in two ways (*i.e.* $k_f \propto 1$, $k_b \propto 2$, $K_2 \propto 1/2$). It leads to: $K_1 : K_2 = \dfrac{4}{1} : \dfrac{1}{2} = 8 : 1$

(c) Tetrahedral complexes: The first didentate ligand can enter in $^4C_2 (= 6)$ ways in which there is no restriction (*cf.* square planar complex). In other words, the first ligand can occupy any one edge of the available 6 edges. The mono-complex can dissociate in one way, *i.e.* $k_f \propto 6$, $k_b \propto 1$ and $K_1 \propto 6$. For the incoming second didentate ligand, there remains only one edge and the *bis*-complex can dissociate in two ways, *i.e.* $k_f \propto 1$, $k_b \propto 2$, *i.e.* $K_2 \propto \dfrac{1}{2}$. It leads to: $K_1 : K_2 = \dfrac{6}{1} : \dfrac{1}{2} = 12 : 1$

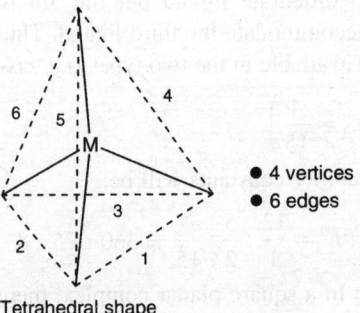

Tetrahedral shape

Fig. 4.4.2.4 Tetrahedral geometry of a complex.

(d) Octahedral complexes involving the symmetrical tridentate ligands: For the **symmetrical tridentate ligand** (*e.g.* dien) forming the octahedral complexes, it can be shown:

$$K_1/K_2 = 16 \text{ (facial)}, 24 \text{ (meridional)}.$$

An octahedron possess 8 faces and 12 meridians. A tridentate ligand can occupy either a trigonal face or a meridian. Thus we have:

$$\textit{fac-}\text{isomer: } K_1/K_2 = \left(\frac{8}{1}\right)\Big/\left(\frac{1}{2}\right) = 16$$

$$\textit{mer-}\text{isomer: } K_1/K_2 = \left(\frac{12}{1}\right)\Big/\left(\frac{1}{2}\right) = 24$$

Statistical Order of the Stepwise Stability Constants

It is evident that the simple statistical factor indicates that the successive stepwise stability constants should decrease gradually. It happens so in general. But, in some cases, involvement of the nonstatistical factors may reverse the situation. These aspects will be discussed later.

4.4.3 Nonstatistical Factors (*i.e.* Chemical Factors)

The order of stepwise stability constants is also determined by different nonstatistical factors. These are discussed below.

(a) **Steric factor:** With the gradual entry of the ligands into the coordination sphere, the steric crowding increases. It operates remarkably when the entering ligands are anionic and sufficiently bulky. This is why, the stepwise stability constants should decrease gradually.

(b) **Electronegativity neutralization:** If the entering ligand is a σ-donor one (*e.g.* NH_3, en, etc.), then the effective positive charge on the metal centre gradually decreases with the successive entry of the ligand. In other words, the electronegativity of the metal centre decreases gradually during complexation. This is why, the metal-ligand interaction measured by the stepwise stability constants also falls in the same way.

(c) **Electronegativity neutralisation opposed by the π-acidity of the ligands:** If the entering ligand is a powerful π-acid ligand (*e.g.* bpy, phen, PR_3, etc.) then the metal→ligand π-back bonding may increase the effective positive charge (*i.e.* electronegativity) on the metal centre. In fact, the π-acid ligands act in two ways: ligand→metal σ-donation, *i.e.* σ-*basicity of the ligands* and metal→ligand π-donation, *i.e.* **π-acidity of the ligand.** This π-bonding interaction opposes the accumulation of negative charge on the metal centre. Consequently, in such cases, during complexation, **the effective positive charge on the metal centre may not drop significantly.**

It has been found that the electronegativity of $[M(OH_2)_6]^{2+}$ and $[M(bpy)(OH_2)_4]^{2+}$ are more or less comparable. While the electronegativity of $[M(OH_2)_6]^{2+}$ is higher than that of $[M(en)(OH_2)_4]^{2+}$ (en acts as a pure σ-donor ligand). *Thus for the π-acid ligands, the stepwise stability constants do not drop significantly on the electronegativity neutralization ground.*

In fact, for many strong π-acid ligands, during the complexation, the effective positive on the metal centre may increase (*i.e.* π-acidity of the ligand is more important than the σ-basicity of the ligand). *In such cases, even the statistical order of stepwise stability constants may be reversed.*

(d) **Ligand-Ligand interaction:** Sometimes, different types of ligand-ligand interactions such as– *stacking interaction, hydrophobic interaction, hydrogen bonding interaction, etc.* may stabilize the higher complexes to such an extent that the statistical stability order may be reversed. These aspects are very much important in the **ternary complexes** (*cf.* Sec. 4.7).

4.4.4 Breaking the Rule: Anomalous Stability Order Including the Reverse Statistical Order

It has been pointed out that if there is *no strong ligand-ligand stabilizing interaction and no enhancement of electron deficiency on the metal centre* due to the π-acidic character of the bound ligand, then both the chemical factors and statistical factors lead to the following sequence:

$$K_1 \rangle K_2 \rangle K_3 \rangle K_4 \rangle \ ...$$

It has been already shown in Table 4.4.1.1 that even in a simple system like $[Ni(OH_2)_6]^{2+}$–NH_3, the experimentally observed ratio of K_{n+1} / K_n is always lower than the statistically predicted value (*cf.* Table 4.4.1.1) However, the predicted trend of stepwise stability constants is in good conformity with the experimentally observed sequence. But, there are cases where even the statistical order of stepwise stability constants is reversed. Sometimes, due to some chemical factors, there may be a remarkably drop in some K-values.

The situations generally arise under the following conditions:

● **change in coordination number** leading to the **change in the state of hybridization** at some stages of successive complexation process;

● **entropic favour or disfavour** at some stages due to the change of coordination number;

● **change in electronic structure** (*e.g.* spin state) of the metal centre at a particular stage of complexation;

● **ligand-ligand steric/cooperative interaction;**

● **enhancement of electron deficiency** due to binding of some powerful **π-acid ligands.**

Some representative examples showing the anomalous stability order are discussed below to illustrate the above fact.

(i) **Hg^{2+}–CN^- or I^- system** (*cf.* Sec. 2.11.4 for details): There is a sudden drop in the value in going from K_2 to K_3 (*cf.* Table 4.2.1). *This abrupt drop is too large to be explained by the statistical consideration only.* This occurs due to the change of linear $[HgX_2]$ to tetrahedral $[HgX_3(OH_2)]^-$. Hg(II) prefers the *linear coordination* (*sp* hybridization with some contribution of the d_{z^2} orbital; *cf.* Sec. 2.11.4) and it is attained in HgX_2. To produce the higher complexes like $[HgX_3(OH_2)]^-$ and $[HgX_4]^{2-}$, it needs the *change of hybridization to sp^3*.

$$\underset{\text{(Linear)}}{[HgX_2]} + X^- + H_2O \rightleftharpoons \underset{\text{(Tetrahedral)}}{[HgX_3(OH_2)]^-} ; \ (K_3)$$

This step (*i.e.* K_3) is strongly disfavoured due to: *strong preference for the linear coordination; reluctance to change the hybridization from sp to sp^3 and loss of entropy due to fixing of a water molecule.*

Note: There is an anomaly in the acid dissociation constants of aqua mercury(II) where the second acid dissociation constant is larger than the first one.

$$\left[Hg(OH_2)_x\right]^{2+} \rightleftharpoons \left[Hg(OH)(OH_2)_{x-1}\right]^+ + H^+, \; pK_{a_1} = 3.4$$

$$\underset{\substack{(\text{Nonlinear, probably} \\ \text{tetrahedral})}}{\left[Hg(OH)(OH_2)_{x-1}\right]^+} \rightleftharpoons \underset{(\text{Linear})}{\left[Hg(OH)_2\right]} + H^+ + (x-2)H_2O, \; pK_{a_2} = 2.6$$

The above mentioned order can also be explained by considering the strong preference of Hg(II) for the *linear geometry and entropic favour due to the release of coordinated water molecules in* K_{a_2}.

(ii) **Ag$^+$–NH$_3$ system** (*cf.* **Sec. 2.11.4 for details**): The reverse statistical order (*i.e.* $K_2 \rangle K_1$) can be explained by considering the *preference for the linear geometry* and *entropic favour.*

$$\underset{(\text{Tetrahedral})}{\left[Ag(NH_3)(OH_2)_3\right]^+} + NH_3 \rightleftharpoons \underset{(\text{Linear})}{\left[Ag(NH_3)_2\right]^+} + 3H_2O, \; K_2$$

Thus aspect has been discussed in Sec. 2.11.4.

(iii) **Zn^{2+}–en system:** There is a **drastic drop in K_3** (*i.e.* abrupt decrease in the K_3/K_2 value). Zn(II) is a d^{10} system and consequently it does not earn any cfse in any geometry. Between the tetrahedral and octahedral geometries, though in the octahedral geometry, there is a larger number of metal-ligand interactions, the steric crowding is more in the octahedral system, *i.e.* [Zn(en)$_3$]$^{2+}$, compared to that in the tetrahedral geometry. In fact, to **avoid the steric crowding**, Zn(II) prefers the tetrahedral geometry which is attained in [Zn(en)$_2$]$^{2+}$. *To accommodate third ligand, it is to change its favoured tetrahedral structure to the disfavoured octahedral structure.* **It explains the drastic drop in K_3.**

$$\underset{(\text{Tetrahedral})}{\left[Zn(en)_2\right]^{2+}} + en \rightleftharpoons \underset{(\text{Octahedral})}{\left[Zn(en)_3\right]^{2+}}, \; K_3$$

(iv) **Cu^{2+}–en system:** There is a huge drop in K_3 compared to that in the other systems like Ni^{2+}–en. The drastic drop in K_3 for the Cu^{2+}–en system arises from the J.T. effect (*i.e.* *z*-out distortion). In fact, for K_3, the Irving-Williams stability order at Cu(II) is not maintained. This aspect has been discussed later (*cf.* Figs. 2.5.11-13).

(v) **Cd^{2+}–Br$^-$:** The stability constant order is: $K_1 \rangle K_2 \rangle K_4 \rangle K_3$. The higher value of K_4 than that of K_3 needs an explanation. It is believed that the aqua-halido complexes are octahedral, *i.e.* [CdBr$_x$(OH$_2$)$_y$]$^{(2-x)+}$ ($x + y = 6$, $y = 1, 2, 3$) but [CdBr$_4$]$^{2-}$ is tetrahedral. Thus the K_3 and K_4 involve the following processes.

$$\left[CdBr_2(OH_2)_4\right] + Br^- \rightleftharpoons \left[CdBr_3(OH_2)_3\right]^- + H_2O, \; K_3$$

$$\underset{(\text{Octahedral})}{\left[CdBr_3(OH_2)_3\right]^-} + Br^- \rightleftharpoons \underset{(\text{Tetrahedral})}{\left[CdBr_4\right]^{2-}} + 3H_2O, \; K_4$$

Thus up to K_3, the octahedral structure is maintained but at K_4 to attain the tetrahedral structure, the coordinated water molecules are released. *This makes the entropy change (ΔS) highly positive.* The attainment of the tetrahedral structure also gives a **steric relaxation** because in the starting octahedral complex, the steric crowding is more. The strong preference for the tetrahedral structure and steric relaxation give also an enthalpic favour for K_4. This **entropic favour** and **enthalpic favour** can overweigh the opposing contributing factors like charge neutralization, statistical factor, etc.

(vi) **Cr(II)–bpy and Fe(II)-bpy systems:** In the Cr(II)-bpy system, the stepwise stability constants are:

$$\log K_1 = 4.5, \log K_2 = 6.0 \text{ and } \log K_3 = 3.5, \text{ i.e. } K_2 \rangle K_1 \rangle K_3.$$

The increase in K_2 value occurs due to the fact: $[Cr(OH_2)_6]^{2+}$ and $[Cr(bpy)(OH_2)_4]^{2+}$ are **the high-spin complexes** $(t_{2g}^3 e_g^1)$ but $[Cr(bpy)_2(OH_2)_2]^{2+}$ and $[Cr(bpy)_3]^{2+}$ are **the low-spin complexes** (*i.e.* $t_{2g}^4 e_g^0$). This spin state change, *i.e. high-spin to low-spin,* occurs during the ligation by the 2nd ligand (*i.e.* K_2-step). This change produces a higher *cfse* (crystal field stabilization energy) for the *bis–* and *tris–* complexes. It explains the sequence: $K_2 \rangle K_1 \rangle K_3$.

In the **Fe(II)-bpy and Fe(II)-phen systems**, the *tris–* complexes are *diamagnetic* and in *low-spin state* (*i.e.* $t_{2g}^6 e_g^0$) while the *bis–* and *mono–* complexes are *paramagnetic* and in *high spin state* (*i.e.* $t_{2g}^4 e_g^2$). At the K_3-step, the spin state change, *i.e.* $t_{2g}^4 e_g^2$ to $t_{2g}^6 e_g^0$ produces an enthalpic favour. This makes K_3 greater than both K_1 and K_2. The step-wise stability constants are:

$$\begin{cases} \textbf{Fe(II)-bpy} : \log K_1 = 4.2, \log K_2 = 3.7, \log K_3 = 9.3 \\ \textbf{Fe(II)-phen} : \log K_1 = 5.85, \log K_2 = 5.30, \log K_3 = 9.85 \end{cases}$$

$$\text{i.e., } K_3 \rangle\rangle K_1 \rangle K_2$$

Here, it is worth mentioning that bipyridine and phenanthroline are the π-acid ligands. Consequently, the **metal $\rightarrow$ bpy/phen π-bonding enchances the positive charge on the metal centre**. This factor in general, leads to increase the stepwise stability constants (*i.e.* $K_3 \rangle K_2 \rangle K_1$). This factor may also partially contribute here. Importance of this factor has been illustrated for the ternary complexes in Sec. 4.7.2.

(**Note:** It may be noted that **Fe(II)-porphyrin, *i.e.* [Fe(P)] complex** is high spin while $[Fe(P)(L)_2]$ having two *trans–* axial ligands (L) is low-spin, *i.e.*

$$\underset{(h.s.)}{\left[Fe(P)\right]} + L \underset{}{\overset{K_1}{\rightleftharpoons}} \underset{(h.s.)}{\left[Fe(P)(L)\right]}, \underset{(h.s.)}{\left[Fe(P)(L)\right]} + L \overset{K_2}{\rightleftharpoons} \underset{(l.s.)}{trans\text{-}\left[Fe(P)(L)_2\right]}$$

It leads to: $K_2 \rangle K_1$)

(vii) **Ni²⁺/Pd²⁺–cysteines:** In such cases, there is an unusual stability order $K_2 \rangle K_1$. Probably, the strong π-acidic character of the –SH group of cysteine and its derivatives, enhances the electron deficiency on the metal centre after ligation by the first ligand. This explains the order $K_2 \rangle K_1$.

(viii) **Hb (hemoglobin)–O₂ (dioxygen) system:** In hemoglobin, per molecule of Hb (which is a tetramer), four O_2 molecules can bind successively as the ligands where the stability constants follow a reverse statistical order, *i.e.* $K_4 \rangle K_3 \rangle K_2 \rangle K_1$. This can be explained in terms of *positive cooperativity* or *homotropic allosteric effect*. This aspect has been discussed in detail in the present author's book, *Bioinorganic Chemistry* (Books & Allied, Kolkata).

4.5 CHEMICAL FACTORS GOVERNING THE STABILITY CONSTANTS

Besides the chemical properties of the metal centre and ligands, many other *external factors* such as— pressure, temperature, dielectric constant of the medium, ionic strength of the medium, etc. can influence the metal-ligand stability constants. The temperature dependence of the stability constants can be expressed as follows:

$$-RT \ln \beta^0 = \Delta G^0 = \Delta H^0 - T \Delta S^0$$

i.e.
$$\ln \beta^0 = -\frac{\Delta H^0}{RT} + \frac{\Delta S^0}{R}$$

From the plot of $\ln \beta^0$ vs. $\frac{1}{T}$ (a linear relationship), ΔH^0 and ΔS^0 may be evaluated. ΔH^0 is directly

obtained from the slope of the plot but ΔS^0 is not directly obtained from the intercept (*cf.* $\frac{1}{T}$ = 0 when

$T = \infty$). ΔS^0 is obtained from the equation by using the value of ΔH^0. In needs the values of β^0 at different temperatures.

4.5.1 Effect of the Properties of the Metal Centres

(A) **Charge, size and ionic potential** (*cf.* Sec. 3.2): If the metal-ligand interaction is considered to be a purely electrostatic one, then with the increase of charge, and decrease of size, the stability should increase. In terms of *ionic potential* (= charge / radius), *with the increase of ionic potential, the stability constant increases.*

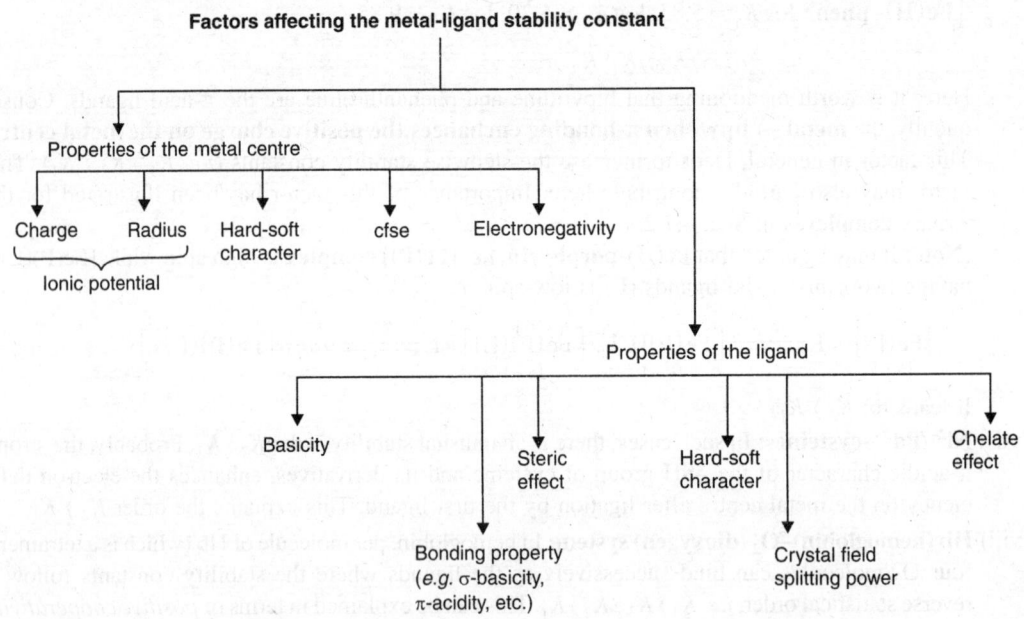

Effect of Ionization Energies and Electronegativity of the Metal Centres

In general, with the increase of ionization energies, the electronegativity and ionic potential of the metal centres increase. Thus, generally, complexing power of the metal centres increases with the increase of ionization energies of the metal centres. In fact, *the parameter ionic potential (cf. Sec. 3.2) takes care of the effects of ionization energies, electronegativities, charge and radius.*

Thus the *complexing power* runs as:

$$Li^+ > Na^+ > K^+ > Rb^+ > Cs^+ \; ; \quad Be^{2+} > Mg^{2+} > Ca^{2+} > Sr^{2+} > Ba^{2+} > Ra^{2+};$$
$$\xrightarrow{\text{(increasing radius)}} \qquad\qquad\qquad \xrightarrow{\text{(increasing radius)}}$$

(**Note:** In many cases, Mg^{2+} forms the stabler complexes than Ca^{2+}, but with edta^{4-}, *Mg^{2+} forms a less stable complex than Ca^{2+}* cf. log$K$: 8.7 vs. 10.7; probably edta^{4-} spans octahedrally better around the larger Ca^{2+} ion)

Stability constant:
$$\underset{\text{(increasing ionic potential)}}{\xleftarrow{\hspace{2cm}} Al^{3+} > Sc^{3+} > Y^{3+} > La^{3+};} \quad \underset{\text{(increasing ionic potential)}}{\xleftarrow{\hspace{1cm}} Fe^{3+} > Fe^{2+}} \; ; \quad \underset{\text{(increasing ionic potential)}}{\xleftarrow{\hspace{1cm}} Co^{3+} > Co^{2+}} \; ; \text{etc.}$$

For the **lanthanides** (Ln^{3+}, *i.e.* La^{3+} to Lu^{3+}), there is a gradual shrinkage in ionic radius due to *thanthanide contraction*. It causes a gradual increase in stability constants from La^{3+} to Lu^{3+}.

For the ions of comparable ionic radius, the **stability constants** follow the sequence as follows:

$$Th^{4+} \rangle Y^{3+} \rangle Ca^{2+} \rangle Na^+; La^{3+} \rangle Sr^{2+} \rangle K^+.$$

Table 4.5.1.1 Shannon-Prewitt crystal ionic radii (for coordination number 6).

Ion	Radius (pm)	Ion	Radius (pm)
Li^+	90	Ti^{2+}	100
Na^+	116	V^{2+}	93
K^+	152	Cr^{2+}	94 (87)*
Rb^+	166	Mn^{2+}	97 (81)*
Cs^+	181	Fe^{2+}	92 (75)*
Be^{2+}	59	Co^{2+}	88 (79)*
Mg^{2+}	86	Ni^{2+}	83
Ca^{2+}	114	Cu^{2+}	87
Sr^{2+}	132	Zn^{2+}	88
Ba^{2+}	149	Sc^{3+}	88
Al^{3+}	67	Ti^{3+}	81
Ga^{3+}	76	V^{3+}	78
In^{3+}	94	Cr^{3+}	75
Tl^{3+}	102	Mn^{3+}	78 (72)*
		Fe^{3+}	78 (69)*
		Co^{3+}	75 (68)*
		Ni^{3+}	74 (70)*

(* Values for the corresponding low spin complexes).

Here it must be mentioned that the above predictions are of importance only *when the metal ions have the similar electronic configurations*, but if they have different types of electronic configuration then the predictions do not work. For example, the ionic radii of the pair Na^+ and Cu^+ or Ca^{2+} and Cd^{2+} are comparable but the stabilities of their complexes differ drastically. In fact, Cu^+ is a better complexing centre than Na^+; similarly, Cd^{2+} is a better complexing centre than Ca^{2+}. In such cases, the *effective positive charge* on the metal centre is to be considered. In Cu^+ or Cd^{2+}, the presence of *poor shielding d-electrons* produces more effective positive charge experienced by the ligands.

- **Inert gas configuration vs. pseudo-inert gas configuration** (*cf.* Sec. 10.6.2, Vol. 2 of **Fundamental Concepts of Inorganic Chemistry**): The cations of **nontransition metals (representative elements)** possess the *inert gas configuration,* $(n-1)s^2p^6$ while the **cations of post-transition and transition metals** possess the *pseudo-inert gas configuration* $(n-1)s^2p^6d^{10}$. For the pseudo-inert gas configuration, the presence of low shielding 10-d electrons produces the *larger effective nuclear charge* experienced by the ligands at the periphery. This is why, the transition and post-transition metal ions form the more stable complexes than the nontransition metal ions even when there is no cfse (as in d^{10} configuration).

- **Ionic potential vs. hard-soft matching:** With the increase of ionic potential, the stability constant increases and it is expected from **the electrostatic model of complex formation.** *This is true when basically, a hard-hard interaction prevails, i.e. covalent interaction is less important.* If the metal-ligand interaction is basically of hard-soft or soft-soft nature, then the covalent interaction is important. In such cases, the effect of ionic potential (*i.e.* effect of charge and radius) is not so straight-forward.

Ionic potential *vs.* Hard-soft interaction

In terms of the property of ionic potential parameter (*cf.* Sec. 3.2), it is expected that the higher ionic potential of the metal centre will attract the more negative charge from the ligand towards the metal centre. This electron donation (*i.e.* flow of negative charge towards the metal centre) will be favoured if the ligand is itself *more polarizable* (*i.e.* more soft). *Thus a complex made by a hard metal centre and a soft ligand should be very stable.* **But in reality, it does not happen so. Because, this will lead to an excessive accumulation of negative charge on the metal centre to lower the stability of the system** (*cf. electroneutrality principle,* Chapter 3). This aspect is also well understood in terms of the HSAB principle.

Size factor for encapsulation by the macrocyclic ligands like crown ethers

It has been mentioned that the smaller size of the metal ion giving rise to the higher ionic potential gives the higher complexing power of the metal ion. *It is true for the monodentate and open chain multidentate ligands.* For the macrocyclic ligands like crown ethers, cryptands, etc. there is a *fixed cavity size and this is why, only the metal ions of appropriate size can sit in the cavity to bring about a strong metal-ligand interaction.* If the metal ion is too small, it cannot interact properly with the ligating/donor sites of the macrocycle. On the other hand, if the metal ion is too large then it cannot enter into the cavity. All these aspects will be discussed separately in Sec. 4.6 in connection with the crown ethers and other macrocyclic ligands.

Thus for the macrocyclic ligands providing a fixed cavity size to accommodate the metal ion, *the matching of size factor overrides the effect of ionic potential and many other factors.*

(B) Electronegativities of the metal ions: With the increase of electronegativity of the metal ion, the stability increases. Thus stability and electronegativity run in the same sequence.

$$Ba^{2+} \langle Ca^{2+} \langle Mg^{2+} \langle Mn^{2+} \langle Co^{2+} \langle Ni^{2+} \langle Cu^{2+}$$

(C) Crystal field effect and Irving-Williams Stability Order (*cf.* **Figs. 4.5.1.1,2**): In this connection, the **Irving-Williams Natural Stability Order** is important. Irving and Williams (1948) noted that stability constants of the octahedral complexes of the bivalent metal ions of 1st transition series with the common ligands having the donor sites like oxygen, sulfur, nitrogen and halogen follow the following general order (*cf.* Figs. 4.5.1.1,2).

$$Mn^{2+} \langle Fe^{2+} \langle Co^{2+} \langle Ni^{2+} \langle Cu^{2+} \rangle Zn^{2+}$$

This series is called the ***Irving-Williams stability order for the octahedral complexes.*** This occurs due to the following contributing factors:

- *decrease of ionic radii, i.e.* increase of electronegativity along the series
- *Increase of cfse (crystal field stabilization energy)* along the series – **the main contributing factor.** *J.T. distortion gives an additional cfse to the Cu(II)-complex.*

(**Note:** Alkanine earth cations may be included in the Irving-Williams series as follows:

$$Ba^{2+} \langle Sr^2 \langle Ca^{2+} \langle Mg^{2+} \langle Mn^{2+} \langle Fe^{2+} \langle Co^{2+} \langle Ni^{2+} \langle Cu^{2+} \rangle Zn^{2+}.)$$

The **decreasing trend of ionic radii** *(cf. Table 4.5.1.1)* *(i.e. increasing electronegativity)* is partially responsible to explain the Irving-Williams series. **The cfse is the main contributing factor** to explain the sequence (*cf.* Sec. 3.14). Let us illustrate for the *high-spin octahedral complexes.*

Sc^{2+} (d^1, $t_{2g}^1 e_g^0$): 4Dq$_o$;   **Ti^{2+}** (d^2, $t_{2g}^2 e_g^0$): 8Dq$_o$;

V^{2+} (d^3, $t_{2g}^3 e_g^0$): 12Dq$_o$;   **Cr^{2+}** (d^4, $t_{2g}^3 e_g^1$): 6Dq$_o$; (additional cfse due to the J.T. distortion

Mn^{2+} (d^5, $t_{2g}^3 e_g^2$): 0;   **Fe^{2+}** (d^6, $t_{2g}^4 e_g^2$): 4Dq$_o$;   is ignored in all cases)

Co^{2+} (d^7, $t_{2g}^5 e_g^2$): 8Dq$_o$;   **Ni^{2+}** (d^8, $t_{2g}^6 e_g^2$): 12Dq$_o$;

Cu^{2+} (d^9, $t_{2g}^6 e_g^3$): 6Dq$_o$;   **Zn^{2+}** (d^{10}, $t_{2g}^6 e_g^4$): 0

(cfse has been calculated in terms of 10Dq$_o$ of the octahedral splitting; cfse change due to the J.T. distortion has been ignored).

(**Note:** For the d^4–d^7 systems, the *low spin complexes* may be obtained. For such systems, cfse may be obtained as follows:

Cr^{2+} (d^4, $t_{2g}^4 e_g^0$): 16Dq$_o$ – P; Mn^{2+} (d^5, $t_{2g}^5 e_g^0$): 20Dq$_o$ – 2P,

Fe^{2+} (d^6, $t_{2g}^6 e_g^0$): 24Dq$_o$ – 2P; Co^{2+}(d^7, $t_{2g}^6 e_g^1$): 18Dq$_o$ – P.)

It is evident that *for the high-spin states*, there is no cfse for the d^0, d^5 and d^{10} systems. For such systems, the stability constant is *mainly determined by the electrostatic interaction* (*i.e.* ionic radius and effective nuclear charge experienced at the periphery). In fact, for Ca^{2+}(d^0), Mn^{2+}(d^5) and Zn^{2+}(d^{10}) having no cfse, their stability constants bear a smooth relationship with their ionic radii as predicted from the pure electrostatic model using the ionic radii of gaseous ions. Thus the higher ionic potential of Zn^{2+} explains its higher complexing power than that of Mn^{2+} and Ca^{2+}. For the bivalent metal ions (Mn^{2+} to Zn^{2+}) of 1st transition series, the cfse increases up to Ni^{2+}, *but if the additional cfse from the J.T. distortion is included, this increasing trend goes up to* Cu^{2+} *and then it drops for* Zn^{2+}. **The Irving-Williams stability order also follows the same trend.**

Deviation from the simple electrostatic model

The simple electrostatic model predicts: higher the ionic potential (*i.e.* high charge and small radius) of a metal ion, higher the stability constant. But in many cases, there is a significant deviation from the above prediction. This deviation can be generally explained by considering the following two factors:

- Consideration of the **contribution of cfse** (*cf.* Sec. 3.14 and Fig. 4.5.1.1)
- Consideration of the **contribution of covalence** in the metal-ligand interaction. It is important mainly for the soft and heavier congeners of the *d*-block elements and for the highly charged *p*-block metal ions, *i.e.* posttransition metal ions of the pseudo-inert gas configuration.
- Fig. 4.5.1.2 indicates that for the O,O donor ligands, the difference in stability constants for different metal ions is very small while it is high for the N,N or N,S donor ligands.

It may be noted that for Sc^{2+} to Mn^{2+}, again the cfse increases up to V^{2+} (ignoring the J.T. effect at Cr^{2+}) and then it decreases up to Mn^{2+} (having no cfse). Thus, in terms of cfse contribution, the stability constants *vs.* the M^{2+} (of 1st transition series) plot should show **two maxima** – one at V^{2+} (d^3) and another at $Ni^{2+}(d^8)$–$Cu^{2+}(d^9)$ region. This trend is also experimentally verified (*cf.* Fig. 4.5.1.1).

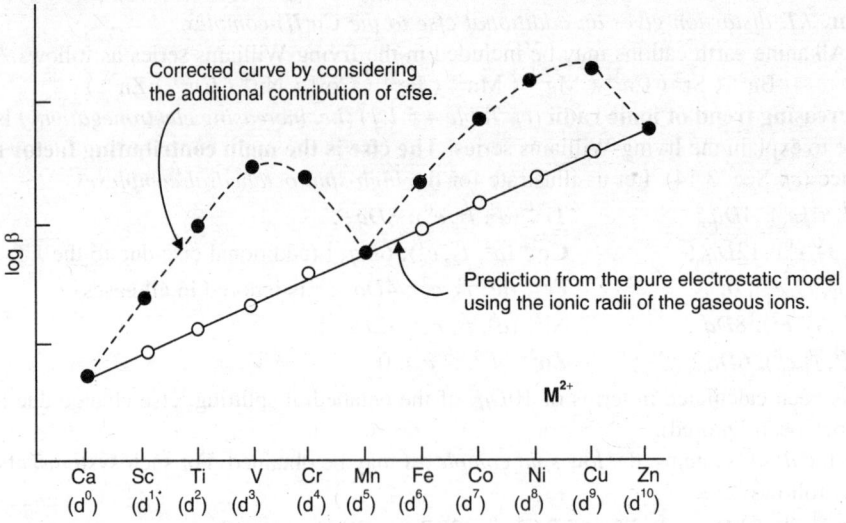

Fig. 4.5.1.1 Effect of cfse (ignoring the effect of J.T. distortion) on the stability constant (β) (qualitative representation) and rationalization of the **Irving-Williams stability order.** O $\Rightarrow$ the value predicted in terms of the ionic radii of gaseous ions without considering the cfse; $\bullet$ $\Rightarrow$ the value predicted after incorporation of cfse.

Here it may be noted that hydration energy (*cf.* Fig. 3.14.6.1) for the M^{2+} ions of 1st transition series also follows the similar trend found in the Irving-Williams series.

(D) **Hard-soft character of the metal ions** (*i.e.* matching of hard-soft character with that of the ligands): In terms of **HSAB theory**, the **hard metal ions** (*i.e. class a* metal ions, *e.g.* Li^+, Na^+, K^+; Be^{2+}, Mg^{2+}, Ca^{2+}, Sr^{2+}; Al^{3+}, Ga^{3+}, Cr^{3+}, Fe^{3+}, Co^{3+}; etc.) form the more stable complexes with the ligands having the *hard donor sites* (*e.g.* O, F, N, etc.). On the other hand, the *soft metal centres* (*i.e. class b* metal ions, *e.g.* Cu^+, Ag^{2+}, Au^+, Tl^+, Hg^{2+}, Pd^{2+}, Pt^{2+}, etc.) prefer to bind with the ligands having the soft donor sites (*e.g.* P, S, As, etc.).

Metal ions

Hard: Li^+, Na^+, K^+, Mg^{2+}, Ca^{2+}, Cr^{3+}, Fe^{3+}, Co^{3+}, etc.
Border line: Mn^{2+}, Fe^{2+}, Co^{2+}, Ni^{2+}, Cu^{2+}, Zn^{2+}, etc.
Soft: Cu^+, Ag^+, Tl^+, Pd^{2+}, Pt^{2+}, Hg^{2+}, etc.

Ligands

Hard: H_2O, OH^-, ROH, NH_3, F^-, PO_4^{3-}, CO_3^{2-}, NO_3^-, etc.
Border line: py, RNH_2, N_2, N_3^-, Br^-, etc.
Soft: RSH, R_2S, RS^-, R_3P, R_3As, CO, CN^-, SCN^-, I^-, etc.

The hardness of a metal centre increases with the increase of its positive charge and decrease of its ionic radius, *i.e.* **hardness increases with the increase of its ionic potential.** For the transition metal ions, the hardness decreases with the increase of the number of d-electrons. **In fact, these d-electrons (which are less penetrating) make the systems more polarizable and it enhances the softness.** Thus hardness varies as follows:

- $Ba^{2+} \langle Sr^{2+} \langle Ca^{2+} \langle Mg^{2+}$
- $Fe^{3+} \rangle Fe^{2+}; Co^{3+} \rangle Co^{2+}.$

 $\}$ (mainly controlled by ionic potential)

- $Mn^{2+} \rangle Fe^{2+} \rangle Co^{2+} \rangle Ni^{2+} \rangle Cu^{2+} \rangle Zn^{2+}$
- $Pt^{2+} \rangle Pd^{2+} \rangle Ni^{2+}; Hg^{2+} \rangle Cd^{2+} \rangle Zn^{2+}.$

 $\}$ (mainly controlled by the number of $d-$ and f-electrons which can be perturbed by the external field easily).

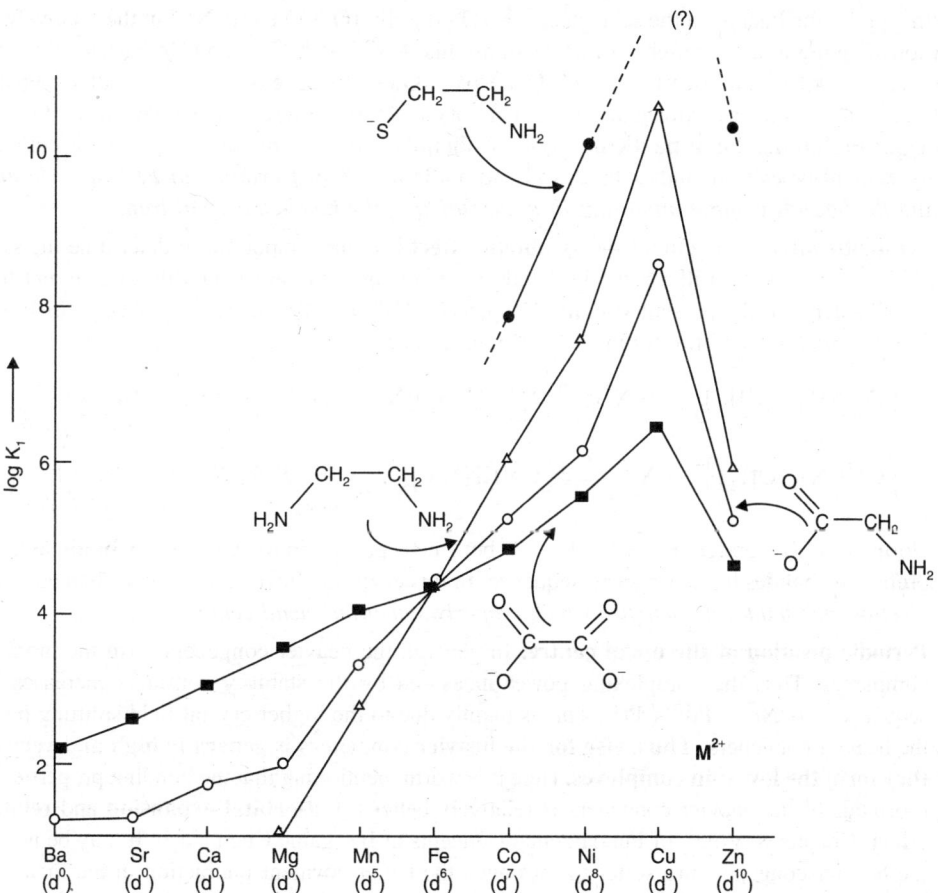

Fig. 4.5.1.2 Variation of $\log K_1 (25° C)$ values for different M^{2+} ions complexing with different didentate ligands varying the donor sites (O,O; O,N; S,N) (An illustration of the effect of hardness-softness factor on Irving-Williams stability order).

- **Ligand preference of the M^{2+} ions:** Let us consider the following didentate ligands in which the hardness gradually decreases from left to right.

 i.e. ox^{2-} (O, O) $\rangle$ gly^- (O, N) $\rangle$ en (N, N) $\rangle$ mercaptoethylamine (N, S)

Hardness:

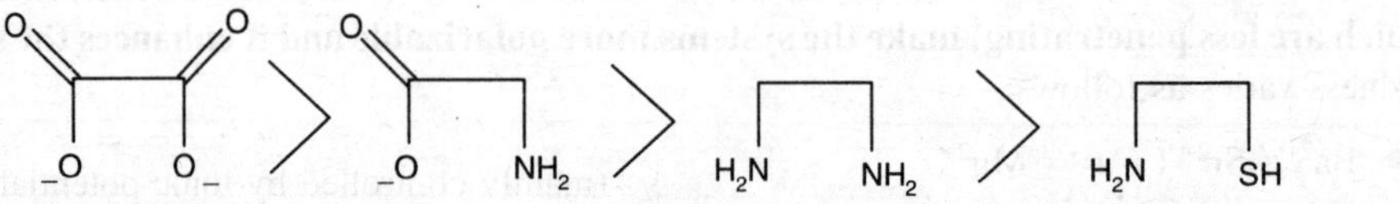

(oxalate; O, O)   (glycinate; O, N)   (ethylenediamine; N, N)   2-aminoethane thiol or (mercaptoethylamine; N, S)

Figure 4.5.1.2 illustrates the fact that *hard-soft interaction* plays a crucial role in determining the stability constants (*cf.* H. Sigel et al, *Acc. Chem. Res.,* **3**, 201, 1970). In each case, the Irving-Williams stability order is maintained. It is evident that the relatively harder metal ions like Ba^{2+}, Sr^{2+}, Ca^{2+}, Mg^{2+} and Mn^{2+} prefer the ligands in the sequence, ox^{2-} (O,O) $\rangle$ gly⁻ (O,N) $\rangle$ en (N,N), but the relatively softer metal ions (having a larger number of *d*-electrons) like Co^{2+}, Ni^{2+}, Cu^{2+} and Zn^{2+} prefer the inverted sequence, *i.e.* (N,S) $\rangle$ en (N,N) $\rangle$ gly⁻ (N,O⁻) $\rangle$ ox^{2-} (O,O). In developing the metalloenzymes and metalloproteins, nature has utilized the above mentioned fact. It is interesting to note that for **Fe^{2+}, there is no rigid preference for a particular type of ligands** and it forms the complexes of comparable stability with all types of ligands – hard, soft and borderline. *This flexibility of Fe^{2+} in preference of selecting the ligands is quite important in understanding the biochemistry of iron.*

● **Symbiotic effect:** Sometimes the **symbiotic effect** becomes important to determine the stability order. This aspect has been illustrated in detail in explaining the relative stability of different linkage isomers (Chapter 2 of the present volume; Chapter 15, Vol. 3 of Fundamental Concepts of Inorganic Chemistry). It is illustrated further in the following examples.

$$\left[Co(NH_3)_5(OH_2)_6\right]^{3+} + X^- \xrightleftharpoons{K} \left[Co(NH_3)_5X\right]^{2+}; \; K \text{ for } X^- : Cl^- > Br^- > I^-$$

$$\left[Co(CN)_5(OH_2)\right]^{2-} + X^- \xrightleftharpoons{K} \left[Co(CN)_5X\right]^{3-}; \qquad \begin{matrix} K \text{ for } X^- : Cl^- \langle Br^- \langle I^- \\ {\scriptstyle(\text{due to the softening effect of the } CN^- \text{ ligands})} \end{matrix}$$

Normally Zn(II) prefers F⁻ $\rangle$ Cl⁻ $\rangle$ Br⁻ $\rangle$ I⁻ but Zn(II) present in the enzyme **carbonic anhydrase** prefers the halides in the opposite sequence. In this enzyme, Zn(II) remains coordinated with two *electron rich imidazole moieties which properly soften the metal centre.*

(E) **Periodic position of the metal centre:** In general, the heavier congeners form the more stable complexes. Thus the complexing power measured by the stability constants increases in the sequence: *e.g.* $Ni^{2+} \langle Pd^{2+} \langle Pt^{2+}$. This is mainly due to the higher crystal field splitting power of the heavier congeners. **Thus, cfse for the heavier congeners is generally high and very often, they form the low spin complexes.** Here it is worth mentioning that the bonding properties of the *d*-orbitals of the heavier congeners is relatively better (*cf.* **d-orbital expansion** and **relativistic effect**; Chapter 8, Vol. 1 of Fundamental Concepts of Inorganic Chemistry). It may be noted that the heavier congeners are softer to introduce the better covalent interaction in the metal-ligand linkages. Sometimes, even for the same ligands, the heavier congeners form different types of complexes compared to the complexes formed by the lighter congeners.

4.5.2 Properties of the Ligands Affecting the Stability Constants

(A) **Charge and size – electrostatic and entropic effect:** According to the **electrostatic consideration,** with the increase of the anionic charge and decrease of the size of the ligating sites, stability of the complexes increases.

If in complexation, **charge neutralization** occurs then the electrorestriction over the movement of solvent molecules decreases and consequently, the randomness among the solvent molecules increases. *This leads to an increase in entropy to favour the process.* Thus in the complexation process with a particular metal ion, M^{n+}, the **entropic favour** runs in the sequence: $(O,O)^{2-}$ (dicarboxylate) ⟩ $(O,N)^-$ (aminocarboxylate) ⟩ (N–N) (diamine). Here it should be mentioned that to determine the stability, both the enthalpy (ΔH^0) and entropy (ΔS^0) parameters are to be considered simultaneously (*cf.* $-RT\ln\beta_n^0 = \Delta H^0 - T\Delta S^0$).

(B) Hard-soft character: The hard metal centres prefer the hard ligands while the soft metal centres prefer the soft ligands. This aspect has been illustrated in Fig. 4.5.1.2.

(C) Ligand basicity: With the increase of basicity of the ligand, the stability constant is expected to increase. *The more basic ligands bind H^+ more firmly and thus the more basic ligands are expected to bind M^{n+} more firmly.*

$$L^- + H^+ \rightleftharpoons LH, \left(K_{LH}\right);$$

$$L^- + M^{n+} \rightleftharpoons ML^{(n-1)+}, \left(K_{ML}\right)$$

In terms of the electrostatic interaction, K_{LH} and K_{ML} should run in parallel. In fact, in many cases, $\log K_{LH}$ and $\log K_{ML}$ bear a linear relationship (Fig. 4.5.2.1). Very often, $pK_a\left(=-\log\dfrac{1}{K_{LH}}\right)$ of the conjugate acids of the ligands are used in such plots. Higher pK_a value indicates a weaker acidic character of the conjugate acid (LH), *i.e.* a stronger basic character of the ligand (L^-). Thus, the plot of pK_a versus $\log K_{ML}$ gives a straight line with a upward slope.

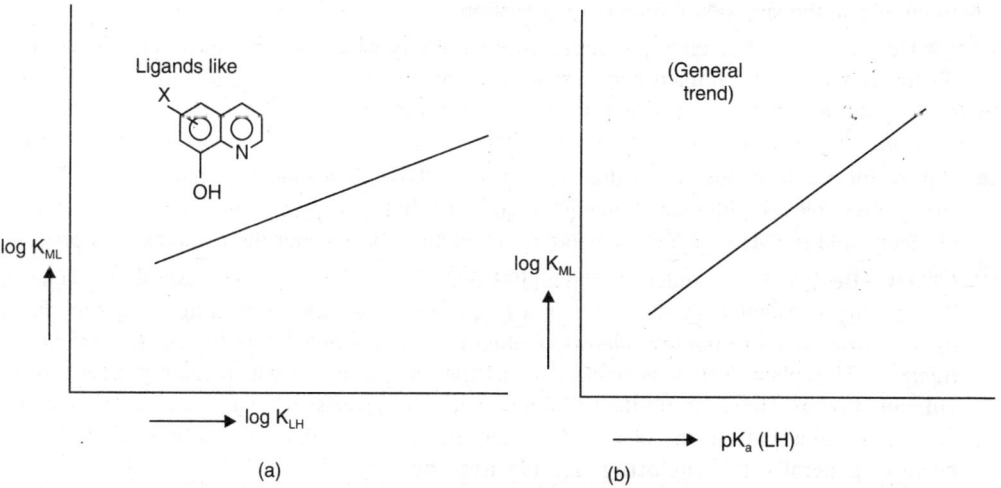

Fig. 4.5.2.1 Relationship between the ligand basicity and metal-ligand stability constant.

● **Restricted conditions for the correlation between pK_a and $\log K$:** The said correlation between pK_a (of the ligand conjugate acid) and $\log K$ (stability constant) is noticed only in some restricted conditions where the ligands are closely related (*i.e.* except basicity, other structural features remain more or less the same). Even for the ligands like ammonia, pyridine and imidazole (all having the N-donor site), no such correlation is noticed for their Cu(II)-complexes.

Ligand (L)

	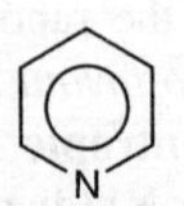 (pyridine)	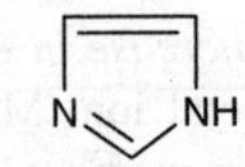 (imidazole)	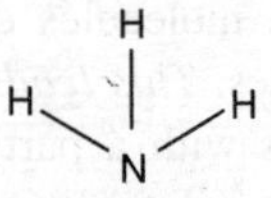 (ammonia)
$pK_a(LH^+)$:	5.2	6.9	9.3
$\log\beta_4$ (for $[CuL_4]^{2+}$) :	6.6	12.5	12.8

Here basicity of NH_3 is greater than that of imidazole by 2.4 log unit while the stability constants (β_4) differ by only 0.3 log unit.

If the donor sites are different as in RNH_2, RCO_2^- and RS^-, then practically no correlationship between pK_a and $\log K$ can be predicted.

The linear relationship between $\log K_{LH}$ and $\log K_{ML}$ (or pK_a and $\log K_{ML}$) has been experimentally verified with various types of ligand like *amines, amino acids, β-diketonates, etc.* (provided the steric factors remain more or less comparable). The secondary and tertiary amines may be more basic than the primary amines but because of steric crowding in the complexes of secondary and tertiary amines, the stability constants are less.

● **Common inorganic anionic ligands:** For many metal ions, the observed metal-ligand stability constant sequence for the **common inorganic ligands** follows their base strength order.

$$F^- \rangle Cl^- \rangle Br^- \rangle I^- \rangle\rangle ClO_4^- \approx BF_4^-$$

ClO_4^- **and** BF_4^- **are the very weak bases** (*i.e.* their conjugate acids are very strong) **and consequently their complexing power is very weak.** This is why, for the studies of complexation reaction in presence of inert electrolytes, perchlorate or fluoroborate salts are very often used. These anions cannot complicate the situation through complexation.

(Note: ● Dependence of K_{ML} on K_{LH} is straight-forward only when the other factors like ligating sites, steric factor, ring size, etc. remain equal, at least comparable. The said prediction, *i.e.* relationship between K_{LH} and K_{ML} has been verified for the amine complexes of Ag^+ and Hg^{2+}. *Steric factor may oppose the basicity factor to reverse the trend of stability constants.* These aspects will be illustrated later.

● **Pipyridine vs. pyridine:** Pipyridine is a stronger base than pyiridine, but pyridine forms the stronger complexes with many transition metal ions. In the M-py complexes, the π-delocalization of charge and $d\pi(M) \to \pi^*(py)$ π-bonding interaction may favour the complexation process.

(D) **Chelate effect:** When the didentate and multidentate ligands form the complexes through chelation, the stability constant increases. *This is why, even for the similar binding sites, the chelating ligands form the more stable complexes compared to the nonchelating ligands (i.e. monodentate ligands).* This phenomenon is referred to as the *chelate effect* which mainly arises from the **entropic favour**. However, for the transition metal ions possessing cfse, the enthalpic factor may also partly contribute to the observed chelate effect. The enthalpic benefit in chelation arises because generally the **chelating ligands are the stronger field ligands compared to their corresponding monodentate ligands** (*cf.* NH_3 vs. en; py vs. bpy in the spectrochemical series). Less **preorganisation energy** for the chelating ligands gives also an **enthalpic benefit** for all types of metal ions. The thermodynamic and kinetic aspects of chelate effect have been discussed separately in detail in Sec. 4.5.3.

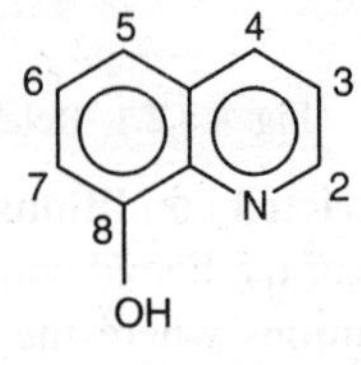

(8-hydroxyquinoline, LH)

Fig. 4.5.2.2 Structure of 8-hydroxyquinoline.

(E) Steric effect: When there is a bulky group attached to the ligating sites or close to the ligating sites, there will be a steric hindrance in the complex and it destabilizes the metal-ligand bond.

- **8-Hydroxyquinoline and its substituted derivatives:** The complexing power of 8-hydroxyquino-line and its monomethyl derivatives (substitution at the 2-, 4-, and 5- positions) may be compared to understand the steric effect on the stability constants.

Substitution at the 2-position causes a more steric crowding than the substitution at the position 5 or 4. Here it may be mentioned that introduction of a Me-group makes the system more basic, but the 2-methyl derivative forms the weaker complexes (*cf.* Table 4.5.2.1).

Table 4.5.2.1 Comparison of complexing power of 8-hydroxyquinoline and its monomethyl derivatives (LH) in 50% dioxane-water medium (20° C) with Cu^{2+}.

LH	pK_{LH}	$logK_1$	$logK_2$
8-Hydroxyquinoline	10.8	13.0	12.4
2-Methyl-8-hydroxyquinoline	11.0	10.2	9.3
5-Methyl-8-hydroxyquinoline	11.1	13.6	12.4

It is evident from Table 4.5.2.1:

Basicity: 5-Methyl-8-hydroxyquinoline $\approx$ 2-Methyl-8-hydroxyquinoline $\rangle$ 8-hydroxyquinoline

Complexing power: 5-Methyl-8-hydroxyquinoline $\rangle$ 8-hydroxyquinoline $\rangle$ 2-Methyl-8-hydroxy-quinoline.

The 5-Me group is far away from the donor site to sterically affect the complex formation. Its higher basicity favours the complexation process.

- **Ethylenediamine and its N-substituted derivatives:** Ethylenediamine very often forms the stabler complexes than the substituted ethylenediamines. If the substitution occurs on the N-sites, then the steric effect becomes very much prominent. Here it is worth mentioning that the substituted ethylene-diamines are relativity more basic but the steric factor very often overrides the basicity factor during the complexation process.

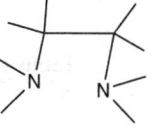

(skeleton of en)

The complexing power of en, N,N'-dimethyl-en and N,N'-dimethyl-en with Ni(II) has been compared in Table 4.5.2.2.

Table 4.5.2.2 Comparison of complexing power (~25 °C) of en and N-substituted ethylendiamine with Ni(II) and Cu(II).

	Ni(II)-en	Ni(II)-N,N'-dimethyl-en	Ni(II)-N,N'-diethyl-en
$logK_1$	7.6	7.1	5.6
$logK_2$	6.5	4.7	3.3
$logK_3$	5.0	1.5	~0
$log\beta_3$	19.1	13.3	~8.9
	Cu(II)-en	Cu(II)-N,N'-dimethyl-en	Cu(II)-N,N'-diethyl-en
$logK_1$	11.3	11.2	10.8
$logK_2$	9.9	8.3	7.8
$log\beta_2$	21.2	19.5	18.6

The steric factor is more pronounced in K_2 and K_3 values (*cf.* Table 4.5.2.2). The K_1-values are more or less comparable because of two opposing factors: basicity and steric factors. **But in the higher complexes, the steric hindrance is too high to be compensated by the slightly favoured basicity of the ligands** by the introduction of alkyl groups. The general trend of β_3 or β_2 is as shown in Fig. 4.5.2.3:

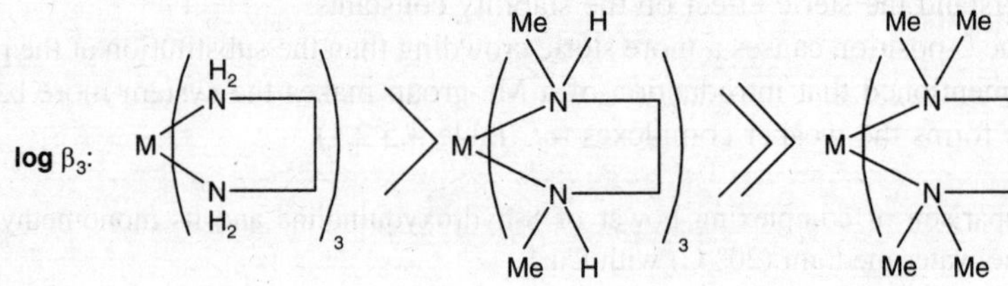

Fig. 4.5.2.3 Effect of steric crowding in the ligand, N-substituted ethylenediamines on the metal-ligand stability constants.

● **Glycine vs. substituted glycine:** Because of the same ground, N-methylglycine (sarcosine) forms the less stable complexes than glycine.

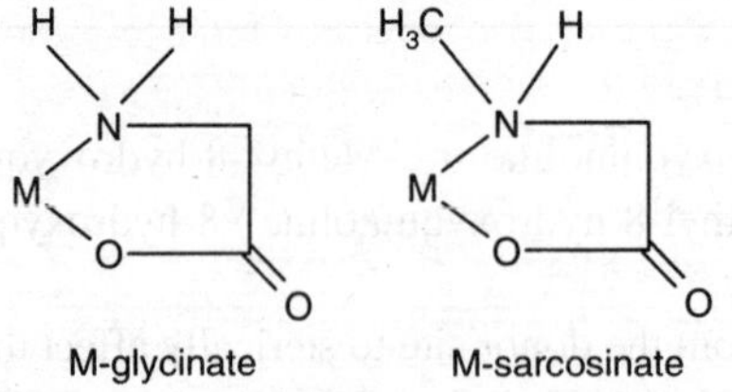

M-glycinate   M-sarcosinate

Fig. 4.5.2.4 Steric crowding in the glycinate and sarcosinate complexes.

Table 4.5.2.3 Comparison of complexing power of glycinate and sarcosinate (25° C)

Ligand (LH$^\pm$)	$pK_a^{LH_2^+}$	$pK_a^{LH^\pm}$	$\log\beta_1$	$\log\beta_2$	$\log\beta_3$
				Ni(II)–L	
Glycine	2.5	9.7	5.6	10.5	14.0
Sarcosine	2.3	10.1	5.2	9.5	12.4

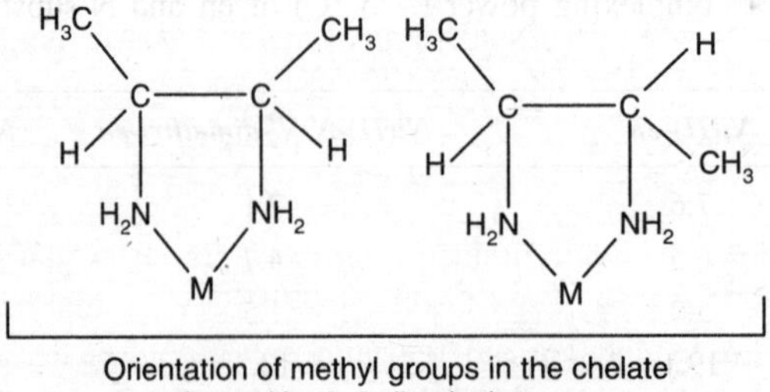

Orientation of methyl groups in the chelate
formed by 2, 3-diaminobutane.

Fig. 4.5.2.5 Steric crowding in the complexes where 2,3-diaminobutane exists in two diastereomeric forms.

● **2,3-diaminobutane:** It can exist in two diastereomeric forms in the metal complexes depending on the *cis-trans* (*meso* and *dl*) orientations of the two Me-groups on the chelate ring. In the *meso-*

isomer, the steric crowding is obviously more than that in the *dl*-isomer. It is reflected in the stability constants (β_1) of Ni(II)-complexes. The values are (25° C): $\log\beta_1 = 7.02$ (*meso*-isomer) and = 7.70 (for *dl*-isomer).

● **Bipyridine and its substituted derivatives:** Because of the steric crowding in the complexes of 6,6′-dimethyl-2,2′-bipyridine (bpy), this substituted dimethyl derivative of bpy can form only the *mono* or *bis*– complexes while the unsubstituted bpy can form the stable *tris*–chelates.

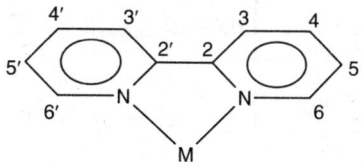

Fig. 4.5.2.6 M-bpy complex

● **Acetylacetone and its substituted derivatives:** Unsubstituted acetylacetone (acacH) can form the stable complexes with different metal ions. Its complexing power dramatically depends on the size of the substituents (*i.e.* alkyl group) at the 3-position. If it is an ethyl group, then it can form a *bis*-complex with Cu(II), but if it is an isopropyl group, *then the steric crowding becomes too high to allow the chelate ring formation.* Here it may be pointed out that the substitution is far away from the O-donor site but, the steric repulsion between the CH_3 group and alkyl group at the 3-position disfavours the chelate ring formation.

[Chemical structures]

Complexes formed   Complex not formed

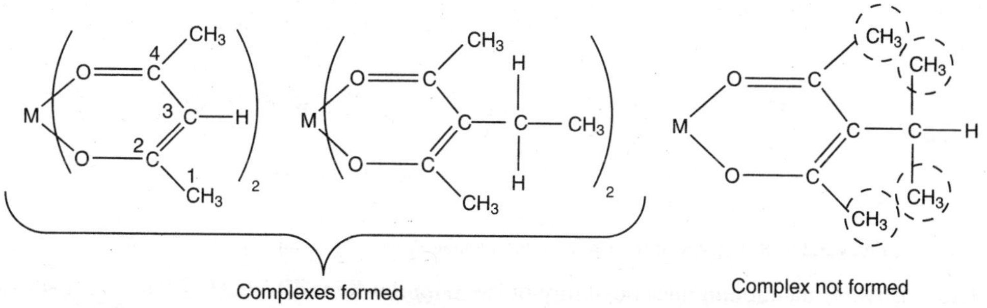

Fig. 4.5.2.7 Steric crowding in the chelate ring formed by acetylacetonate and its derivatives.

● **Ethylenediamine and its C-substituted derivatives:** It has been already pointed out that substitution at the N-sites in the en-skeleton causes the steric crowding severely because the N-sites are the coordinating sites. Substitution on the C-sites also enhances the steric crowding in another way and *it may enforce the system to adopt a different stereochemical arrangement in the higher complexes.*

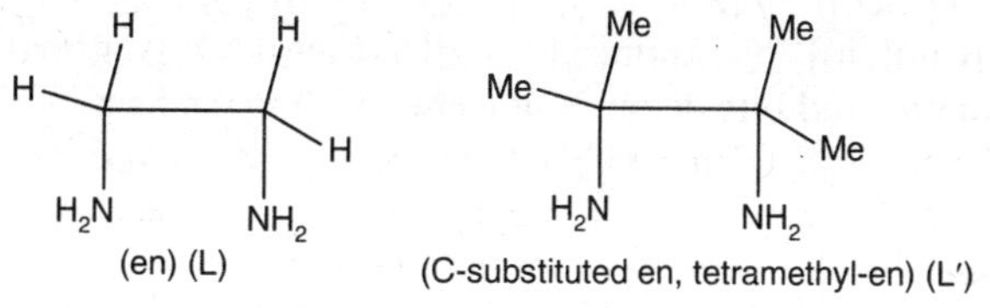

[Chemical structures]

(en) (L)   (C-substituted en, tetramethyl-en) (L′)

Fig. 4.5.2.8 (a) Steric crowding in ethylene diamine and its C-substituted derivatives.

With en (*i.e.* L), Ni(II) forms a ***tris*-complex [Ni(L)₃]²⁺ of octahedral structure (purple coloured and paramagnetic)** but with tetramethyl-en, in the *tris*-complex, the steric crowding will be too high and in fact, it fails to form a *tris*-complex. Rather, with L′, Ni(II) forms a *bis*-complex **[Ni(L′)₂]²⁺ of square planar geometry (yellow coloured and diamagnetic)** in which the ligand-ligand repulsion is optimized.

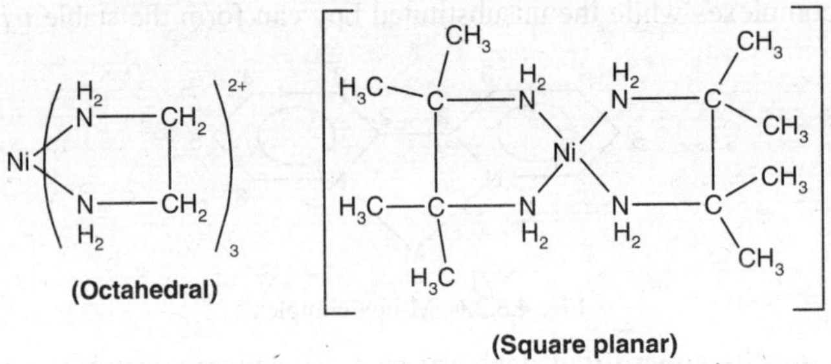

Fig. 4.5.2.8 (b) Nature of the Ni(II)-complexes formed by en and its C-substituted derivatives (*i.e.* illustration of the consequence of steric crowding).

(F) **Stereochemical requirements (tren vs. trien; *cf.* Sec. 2.2.4):** It can be illustrated by considering the isomeric tetradentate ligands *tren* (stereochemically more rigid) and *trien* (more flexible) (*cf.* Sec. 1.7.2, Figs. 2.11.7.8-9).

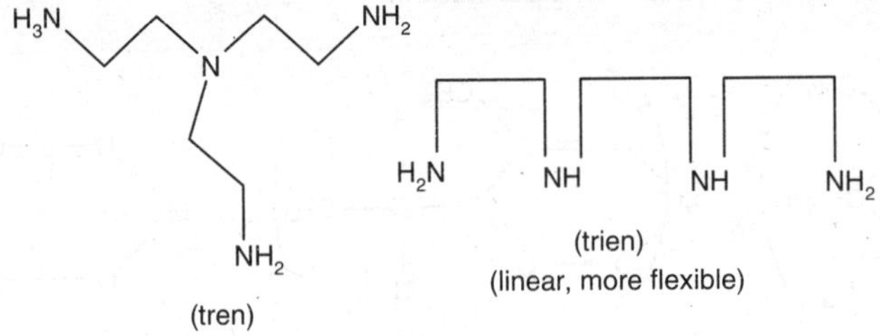

Fig. 4.5.2.9 Structures of the isomeric tetradentate ligands trien and tren (*cf.* Sec. 2.2.4).

Here it is worth mentioning that flexibility of the **tripodal ligand** $N\{(CH_2)_xNH_2\}_3$ depends on the value of x (= 2, 3 etc.; $x = 2$ for *tren*) (*cf.* Fig. 2.11.7.8-9).

The **more or less rigid geometry of *tren*** does not allow it to provide the square planar geometry but it can provide the tetrahedral geometry. The reverse is true for the linear trien. It is more flexible to provide the square planar geometry. Between Zn(II) (d^{10}) and Cu(II) (d^9), Zn(II) having no cfse prefers the tetrahedral geometry to relax the steric crowding, while Cu(II) prefers the square planar geometry to earn the high cfse. ***Thus, tren forms a more stable complex with Zn(II) while trien forms a more stable complex with Cu(II).***

Stability sequence: $[Cu(trien)]^{2+} \rangle [Zn(trien)]^{2+}$ $[Zn(tren)]^{2+} \rangle\rangle [Zn(trien)]^{2+}$

                        $[Cu(tren)]^{2+} \langle [Zn(tren)]^{2+}$ $[Cu(trien)]^{2+} \rangle\rangle [Cu(tren)]^{2+}$

(G) **Jahn-Teller distortion and bite distance of a chelate ring** (*cf.* Secs. 3.11.4-5): This is illustrated for the M²⁺-en system. The didentate ligand like en forms a *chelate ring* that has a preferred *bite distance* (the distance between the coordinating atoms in the chelate). This bite distance is fixed for a particular ligand and it increases with the ring size.

    ● **Stepwise stability constants for the M²⁺-en system:** Among the M²⁺ metal ions (Mn²⁺ to Zn²⁺ in the 1ˢᵗ transition series), Cu²⁺ (d^9) system experiences the *z*-out distortion (*i.e.* bonds along the axial directions are elongated significantly). This J.T. distortion gives an **additional stabilization energy** but the bite distance of en occupying an axial and equatorial position is not sufficiently large to allow the z-out distortion. *Thus, the J.T. effect tends to elongate the axial bonds in the tris-chelate of Cu(II) with en, but the fixed bite distance of en opposes this bond length elongation.*

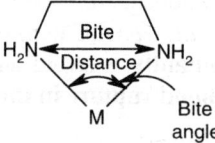

Fig. 4.5.2.10 Representaion of the bite distance in the M(en) chelate.

$$\left[M(OH_2)_6\right]^{2+} + en \rightleftharpoons \left[M(en)(OH_2)_4\right]^{2+} + 2H_2O, \ K_1$$

$$\left[M(en)(OH_2)_4\right]^{2+} \rightleftharpoons \left[M(en)_2(OH_2)_2\right]^{2+} + 2H_2O, \ K_2$$

$$\left[M(en)_2(OH_2)_2\right]^{2+} + en \rightleftharpoons \left[M(en)_3\right]^{2+} + 2H_2O, \ K_3$$

Table 4.5.2.4 Stepwise stability constants for the $[M(OH_2)_6]^{2+}$ -en system (25°)

	Mn^{2+}	Fe^{2+}	Co^{2+}	Ni^{2+}	Cu^{2+}	Zn^{2+}
$\log K_1$	2.73	4.28	5.89	7.52	10.75	5.71
$\log K_2$	2.06	3.25	4.83	6.28	9.25	4.66
$\log K_3$	0.88	1.99	3.10	4.26	-1.0	1.72

For K_1 and K_2, the Irving-Williams stability order is maintained as expected. Thus for K_1 and K_2, the value increases from Mn^{2+} to Cu^{2+} and then drops at Zn^{2+}. But, **for K_3 the value strikingly drops at Cu^{2+}** This large drop in K_3, makes $[Cu(en)_3]^{2+}$ less stable.

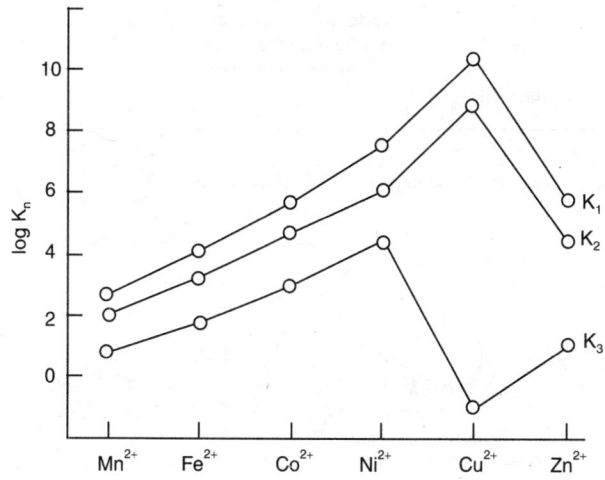

[Cu(en)₃]²⁺: (i) z-out distortion; 2 axial Cu–N bonds longer; 4 equatorial Cu–N bonds shorter; 2 strained chelate rings;

(ii) rupture of the Cu–N bond to form the tris-complex from the predominant species, trans-bis complex.

Fig. 4.5.2.11 Stepwise fromation constants for the mono-, bis-, and tris ethylenediamine complexes of the bivalent metal ions of the first transition series.

Among the given metal ions, Mn^{2+} to Zn^{2+}, the J.T. distortion prevails significantly in the octahedral complex of Cu^{2+} ($t_{2g}^6 e_g^3$). Upto the *bis–* complex of Cu(II), the axial sites are occupied by the monodentate ligand H_2O to avoid any strain (due to the J.T. distortion) in the chelated en-ring. Thus, upto the *bis-* complex in which the equatorial positions are occupied by two en ligands keeping the H_2O molecules at the *trans–*axial positions, no strain is developed. Formation of the *tris*-complex, *i.e.* $[Cu(en)_3]^{2+}$ from the *trans*-configuration of the *bis*-complex, *i.e.* *trans*-$[Cu(en)_2(OH_2)_2]^{2+}$ (keeping the H_2O ligands at the *trans*-positions) **which prevails predominantly** to avoid any strain (due to the J.T. distortion) in the chelated en-ring, **requires the Cu–N bond rupture in the *bis*-complex. This also disfavours the K_3-step.**

For the other metal ions having no preference for a particular configuration, their *bis*-complexes are distributed statistically in the following ratio:

$$cis : trans = 4 : 1 \ (cf. \ \text{Fig. 4.4.2.2})$$

$$cis\text{-}\left[M(en)_2(OH_2)_2\right]^{2+} + en \underset{}{\overset{K_3}{\rightleftharpoons}} \left[M(en)_3\right]^{2+} + 2H_2O;$$

(Predominant form, **80% statistically**)

(No requirement of bond rupture to accomodate the 3rd en ligand)

From the *cis*-configuration, *i.e.* *cis*-$[M(en)_2(OH_2)_2]^{2+}$ (80% statistically) (keeping the two H_2O ligands at the *cis*-positions), the *tris*-complex can be generated **without any M–N bond breaking.** This line of argument indicates that K_3 is relatively disfavoured for Cu(II) compared to other metal ions. *Besides this hurdle of bond rupture in the formation of $[Cu(en)_3]^{2+}$ from trans-$[Cu(en)_2(OH_2)_2]^{2+}$ which predominantly exists denying the statistical demand, the strain developed in $[Cu(en)_3]^{2+}$ also disfavours K_3.* This aspect is discussed below.

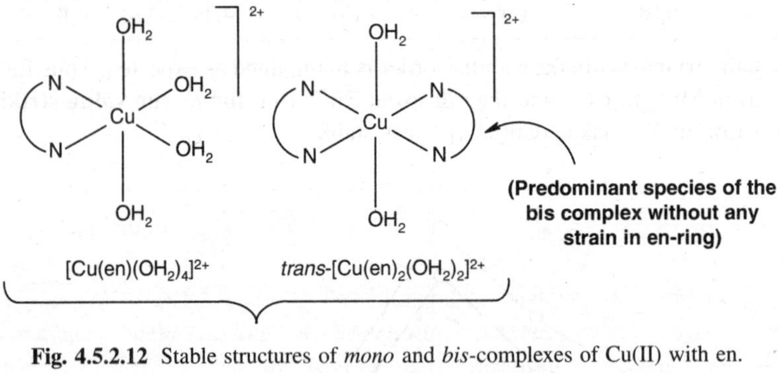

[Cu(en)(OH$_2$)$_4$]$^{2+}$　　　　*trans*-[Cu(en)$_2$(OH$_2$)$_2$]$^{2+}$

(Predominant species of the bis complex without any strain in en-ring)

Fig. 4.5.2.12 Stable structures of *mono* and *bis*-complexes of Cu(II) with en.

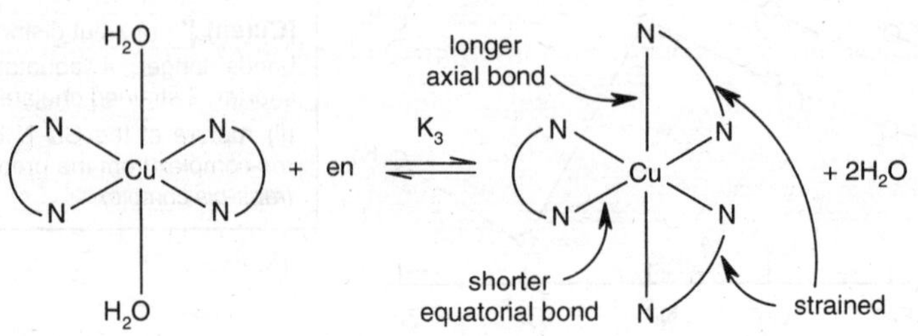

Fig. 4.5.2.13 Formation of **strained** $[Cu(en)_3]^{2+}$ from *trans*-$[Cu(en)_2(OH_2)_2]^{2+}$ through the **rupture of Cu – N bond in the bis-complex.**

In forming the *tris–* chelate, *i.e.* [Cu(en)$_3$]$^{2+}$ from the *bis–* chelate, the **rearrangement** places two en ligands in the strained conditions (where each en ligand spans one equatorial and axial position) but the other en ligand occupying the equatorial *cis–* positions does not experience any such strain.

Thus, in [Cu(en)$_3$]$^{2+}$, two chelate rings are in strain to destabilize the system. In other words, the restricted bite distance of en prevents the elongation of the axial bond, *i.e.* it prevents the earning of additional cfse through the J.T. distortion. **Thus the concept of strain in the chelate rings or prevention of J.T. distortion leading towards the perfectly octahedral geometry will reduce the stability of the system** (*i.e.* drop in K_3). For the other metal ions of Irving-Williams series, no such J.T. effect prevails and the normal stability order is maintained.

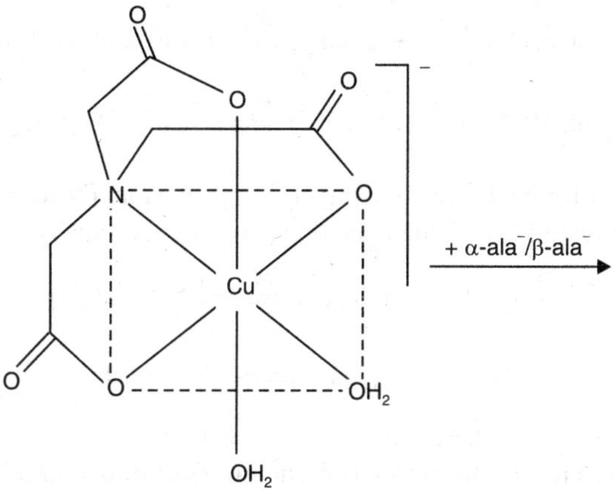

cis-[Cu(nta)(OH$_2$)$_2$]$^-$

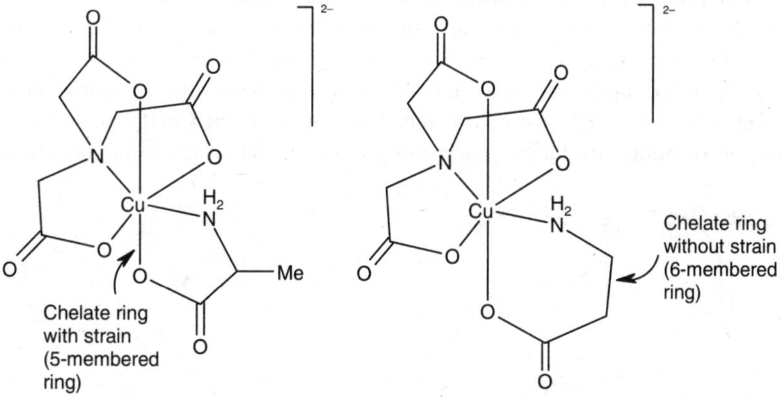

[Cu(α-ala)(nta)]$^{2-}$ [Cu(β-ala)(nta)]$^{2-}$

Fig. 4.5.2.14 Structures of the **ternary complexes**, [Cu(α-ala)(nta)]$^{2-}$ and [Cu(β-ala)(nta)]$^{2-}$ (K_β ⟩ K_α for Cu^{II} but K_β ⟨ K_α for Ni^{II}).

Here it may be noted that for the Cu(II) – en system, though K_3 is dropped significantly, but in terms of the overall stability constant (*i.e.* β_3), the [Cu(en)$_3$]$^{2+}$ complex is still more stable than [Ni(en)$_3$]$^{2+}$ and [Zn(en)$_3$]$^{2+}$ as expected from the Irving-Williams series.

$$\left[\text{M(en)}_3\right]^{2+}: \quad \overline{\begin{matrix} \text{M}^{2+} \\ \text{Ni}^{2+} \quad \text{Cu}^{2+} \quad \text{Zn}^{2+} \end{matrix}}$$

$$\sim \log \beta_3 \,(25°\text{C}): \quad 18.1 \quad < 19.0 >> \quad 12.1$$

- cis-$[\text{M(nta)(OH}_2)_2]^-$–$\alpha$-ala$^-$/$\beta$-ala$^-$ (M = Cu, Ni): It may be noted that in general, the 5-membered chelate ring formed by α-ala$^-$ is more stable than the 6-membered chelate ring formed by β-ala$^-$ (cf. Table 4.5.2.6) but the mixed chelates having these 6-membered rings which can allow the J.T. distortion without any geometric strain are more stable. This is illustrated in the following example (where $K_\beta \rangle K_\alpha$).

$$cis-\left[\text{Cu(nta)(OH}_2)_2\right]^- +\alpha\text{-ala}^- \xrightarrow{K_\alpha} \left[\text{Cu}(\alpha\text{-ala})(\text{nta})\right]^{2-} +2\text{H}_2\text{O},$$

$$cis-\left[\text{Cu(nta)(OH}_2)_2\right]^- +\beta\text{-ala}^- \xrightarrow{K_\beta} \left[\text{Cu}(\beta\text{-ala})(\text{nta})\right]^{2-} +2\text{H}_2\text{O}.$$

Here it may be noted that for **Ni(II) (d^8)** having no J.T. distortion, **$[\text{Ni}(\alpha\text{-ala})(\text{nta})]^{2-}$ is more stable than $[\text{Ni}(\beta\text{-ala})(\text{nta})]^{2-}$ as usual** (i.e. 5-membered chelate ring is more stable), i.e.

$$cis-\left[\text{Ni(nta)(OH}_2)_2\right]^- +\alpha\text{-ala}^- \xrightarrow{K_\alpha} \left[\text{Ni}(\alpha\text{-ala})(\text{nta})\right]^{2-} +2\text{H}_2\text{O},$$

$$cis-\left[\text{Ni(nta)(OH}_2)_2\right]^- +\beta\text{-ala}^- \xrightarrow{K_\beta} \left[\text{Ni}(\beta\text{-ala})(\text{nta})\right]^{2-} +2\text{H}_2\text{O}.$$

where $K_\alpha \rangle K_\beta$ (cf. reverse is true for the Cu^{II} – complex).

- **5-Membered chelate ring vs. 6-membered chelate ring in accomodating the J.T. distortion** (cf. Fig. 4.5.2.15 for **[Cu(bpy)(hfa)$_2$]**): The didentate ligands providing the larger bite distances can stabilize the system experiencing the J.T. distortion. **A 6-membered ring can provide a larger bite distance than a 5-membered ring.** In the mixed chelates (i.e. ternary complexes) of Cu(II), the 5-membered chelate ring forming ligand bpy occupies the cis– equatorial positions while the other two 6-membered chelate forming ligands hfa occupy the remaining positions – where each ligand occupies one axial and one equatorial position to accommodate the J.T. distortion (cf. Fig. 4.5.2.15). Such distorted octahedral complexes of Cu(II) are stable because of the J.T. distortion without introducing much strain in the chelate rings of larger bite distance.

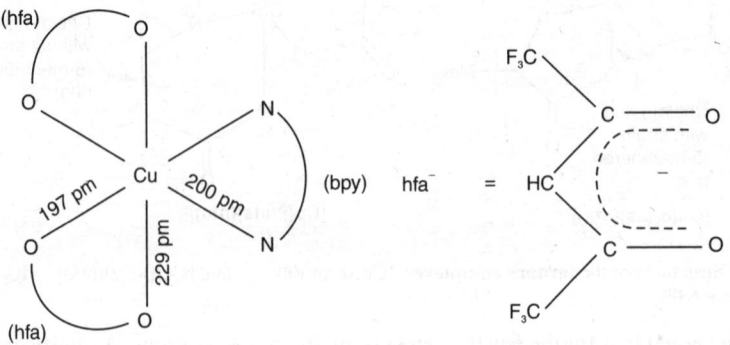

Fig. 4.5.2.15 Structural parameters of [Cu(bpy)(hfa)$_2$].

- **Positions of the didentate ligands in $[Cu(bpy)(hfa)_2]$ and resonance stabilisation *vs.* stabilisation due to the J.T. distortion in the chelate ring formed by hfa:** This complex is an example of distorted octahedral mixed chelate (*i.e.* ternary complex) of Cu(II). In $[Cu(bpy)(hfa)_2]$ where bpy = 2,2′-bipyridine (5-membered chelate forming ligand) occupies the *cis–* equatorial positions and hfa = hexa-fluoroacetylacetonate, *i.e.* $CF_3C(O)CHC(O)CF_3^-$ (6-membered chelate ring forming ligand) occupies the remaining positions to accommodate the J.T. distortion without any strain. The bond length difference by about 32 pm between the equatorial and axial Cu–O bonds indicates the *presence of a strong J.T. distortion in the complex*. In general, the acetylacetonate chelate ring experiences resonance (*i.e.* pseudo-aromaticity) but in the present case because of the unequal Cu–O bond length, *the resonance stabilization energy in the chelate ring formed by hfa is partly reduced*. However, *this is compensated (at least in part) by the J.T. distortion*.

- **(H) Effect of the chelate ring size:** In the coordination compounds, chelate rings of different sizes ranging from 4 to 9 membered rings are known. But except the 5- and 6- membered chelate rings, all other chelate rings are less stable. The 7-9 membered chelate rings are so unstable that they can exist only in the noncoordinating solvents. *In aqueous media, H_2O as ligands can replace such unstable large chelate rings.*

Four membered chelate rings produced by the η^2-didentate ligands like acetate, carbonate, sulfate and nitrate are fairly stable and **they are preferred in the complexes of high coordination number like 8, 10, 12** (*see* Chapter 2).

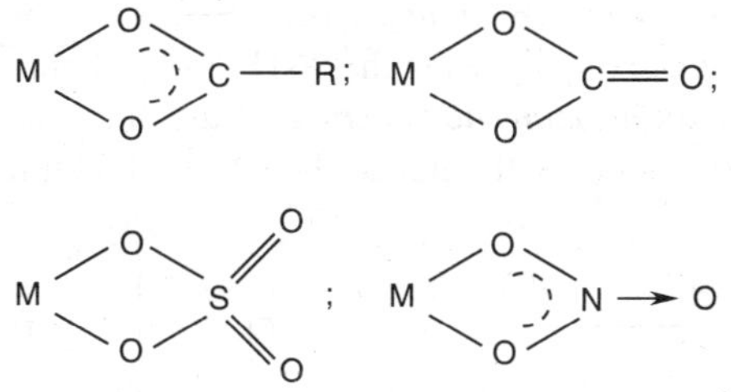

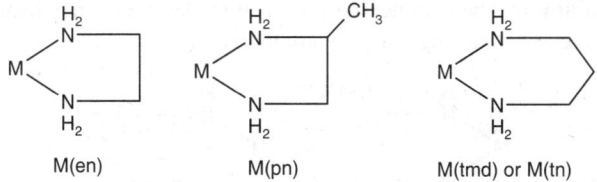

Fig. 4.5.2.16 4-membered chelate rings formed by η^2-oxyacid anions.

The 5-membered chelate rings formed by the saturated aliphatic or amino acids or aminocarboxylates are more stable than the **6-membered chelate** rings formed by the similar ligands. In fact, *in the 6-membered chelate ring, there is a more strain*. This is illustrated in the following cases.

- **5-membered vs. 6-membered chelate rings formed by the aliphatic diamines:** The complexing power of *en* (ethylenediamine), *pn* (propylenediamine or propane-1,2-diamine) and *tmd* or *tn* (trimethylenediamine or propane-1,3-diamine) may be compared.

M(en)   M(pn)   M(tmd) or M(tn)

Fig. 4.5.2.17 5- and 6-membered chelate rings formed by the diamines.

Among these diamines, *en* and *pn* produce the 5-membered chelate rings while *tmd* produces a 6-membered chelate ring. *Among these, though tmd is the strongest base, it forms the weakest complex.*

Table 4.5.2.5 Comparison of the complexing power of *en*, *pn* and *tmd* (values at 25°C).

| Ligand | $pK_a^{LH^+}$ | $pK_a^{LH_2^{2+}}$ | Cu(II)–L | | Ni(II)–L |
			$logK_1$	$logK_2$	$logK_1$
en	10.2	7.4	10.7	9.3	7.7
pn	10.0	7.1	10.8	9.3	7.3
tmd (*i.e. tn*)	10.6	9.1	10.0	7.2	6.3

$$LH_2^{2+} \rightleftharpoons LH^+ + H^+ \left(K_a^{LH_2^{2+}} \right); \ LH^+ \rightleftharpoons L + H^+ \left(K_a^{LH^+} \right)$$

From Table 4.4.2.5, it is evident:

basicity: *tmd* ⟩ *en* ≈ *pn* ; **complexing power:** *en* ≈ *pn* ⟩ *tmd*.

● **Complexing behaviour of *ptn* determined by the chelate ring size:** The complexing behaviour of 1,2,3-triaminopropane (*ptn* be denoted by L) is quite interesting.

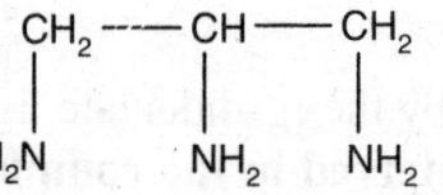

$$CH_2 ---- CH ---- CH_2$$
$$\quad | \qquad | \qquad | \qquad (L) \ i.e., \ ptn$$
$$H_2N \quad NH_2 \quad NH_2$$

The said ligand reacts with $[PtCl_6]^{2-}$ to produce a complex $[PtCl_4L]$ where the triamine ligand *acts as a didendate ligand and the complex is optically active*. This can be explained by considering the formation of a 5-membered chelate ring by using the $-NH_2$ groups bound to the adjacent C-atoms. ***This mode of binding will produce an asymmetric C-centre.*** If the didentate function is expressed through a 6-membered chelate ring formation by the ligand then no such chiral centre can be generated.

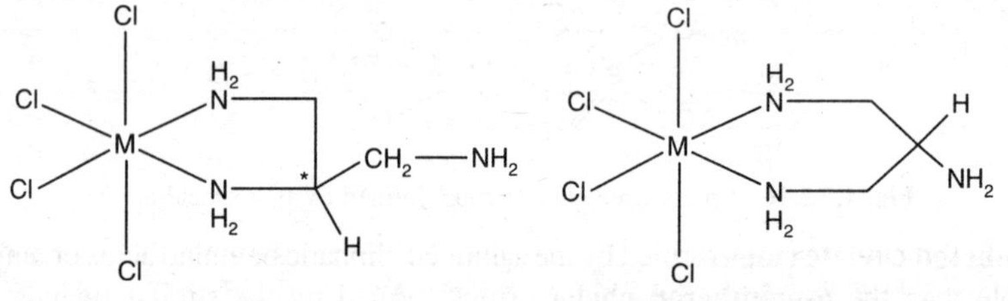

(More stable and optically active, *C denotes the chiral centre)

(Less stable and optically inactive)

Fig. 4.5.2.18 Complexing behaviour of *ptn* determined by the relative stabilities of chelate rings.

In the above Pt(IV)-complex, **the favoured tendency to produce the 5-membered chelate ring is the driving force to generate the origin of optical activity.**

● **α-alanine vs. β-alanine:** The complexing power of α-alanine (5-membered chelate ring) and β-alanine (6-membered chelate ring) may be compared.

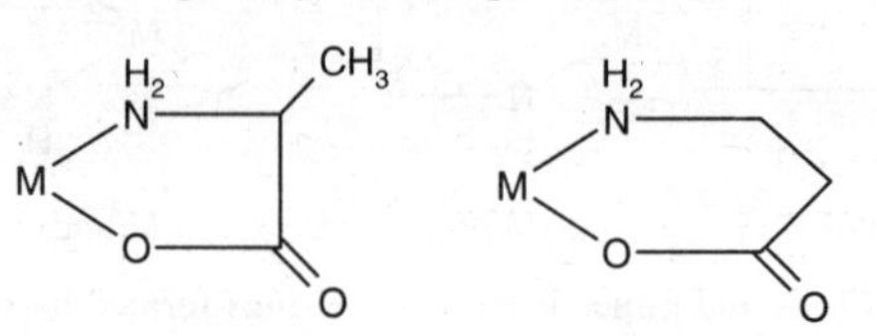

M(α-ala)

M(β-ala)

Fig. 4.5.2.19 Chelate rings formed by α-ala⁻ and β-ala⁻.

Table 4.5.2.6 Comparison of the complexing behaviour of α-ala⁻ and β-ala⁻ (25° C).

Ligand (LH±)	$pK_a^{LH_2^+}$	$pK_a^{LH^\pm}$	Cu(II)–L		Ni(II)–L	
			$logK_1$	$logK_2$	$logK_1$	$logK_2$
α-Hala	2.4	9.8	8.3	6.7	5.5	4.4
β-Hala	3.7	10.2	7.1	5.4	4.7	3.4

From Table 4.5.2.6, it is evident that though β-ala⁻ is more basic but it produces a less stable complex because of the strain in the 6-membered ring produced by β-ala⁻. However, in some 6-coordinate ternary complexes of Cu(II), β-ala⁻ may form the more stable chelates than α-ala⁻ to accommodate the J.T. distortion. This aspect has been already pointed out (*cf.* Fig. 4.5.2.14).

- **Complexing power of edta⁴⁻ and its homologues:** Comparison of the complexing power of edta and its homologues is quite interesting. With the increase of the size of the central chelate ring, the stability decreases (Table 4.5.2.7).

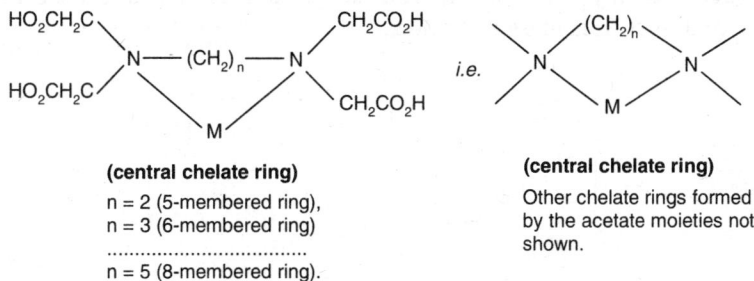

(central chelate ring)
n = 2 (5-membered ring),
n = 3 (6-membered ring)
....
n = 5 (8-membered ring).

(central chelate ring)
Other chelate rings formed by the acetate moieties not shown.

Fig. 4.5.2.20 Chelate ring size for complexation by edta⁴⁻ and its homologues.

Table 4.5.2.7 Comparison of stability constants of 6-coordinate **Ca-complexes** with edta⁴⁻ (*i.e. n* = 2) and its homologues (at 25° C).

n	Central chelate ring size	logK	
2	5	10.5	
3	6	7.1	Decreasing trend
4	7	5.0	
5	8	4.6	

- **Complexing power of oxalate, malonate and succinate:** Comparison of the complexing power of oxalate, malonate and succinate forming the 5-, 6- and 7-membered chelate rings respectively is illustrated here.

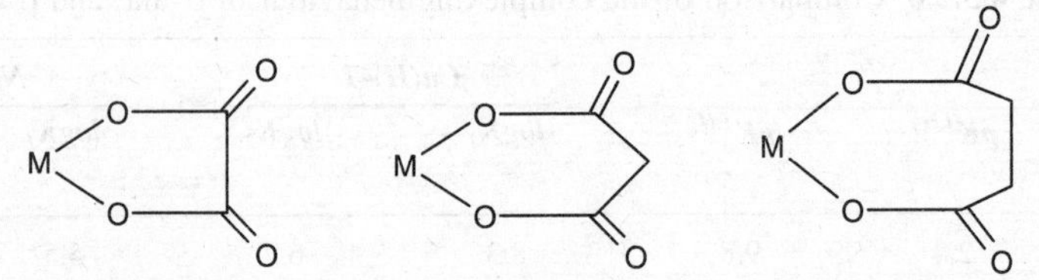

	M(oxalate)	M(malonate)	M(succinate)
log β_1: (with Ni^{II}, 25°C)	5.2	4.1	2.3

Decreasing trend →

Fig. 4.5.2.21 Chelate ring size for complexation by oxalate, malonate and succinate and their relative complexing power.

- **Complexing power of the polydentate ligands having both the 5- and 6-membered ring forming capacity (2, 2, 2-tet *vs.* 2, 3, 2-tet):** For these polydentate ligands, the complexes containing both the 5- and 6-membered rings (both are saturated rings) are more stable due to the *greater flexibility in the ring systems* compared to the similar complexes bearing only the 5-membered rings. It is illustrated below.

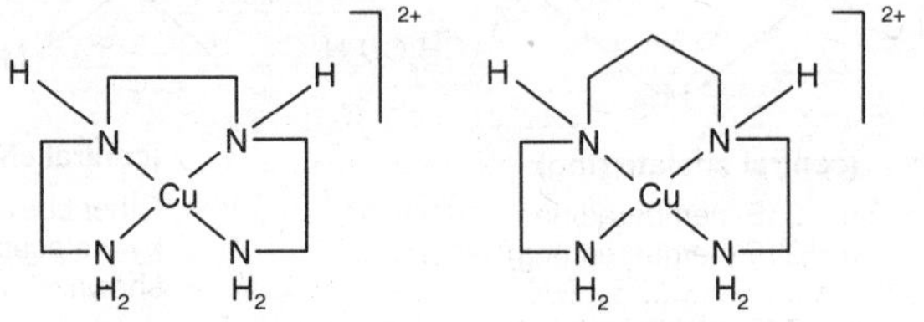

i.e. [Cu(2, 2, 2)tet]$^{2+}$ *i.e.* [Cu(2, 3, 2)tet]$^{2+}$

logK (25° C):   20.1   23.9

(2,2,2)tet = 1,4,7,10-tetraazadecane (*i.e.* trien); **(2,3,2)tet** = 1,4,8,11-tetraazaundecane

Fig. 4.5.2.22 Comparison of the complexing power of 2,2,2-tet and 2,3,2-tet.

6-Membered Aromatic Chelate Ring *vs.* 5-Membered Nonaromatic Chelate Ring

Sometimes, presence of the conjugated double bonds in the chelate ring may produce the aromaticity. **This aromaticity may be extended by using the d-orbitals of the metal centre.** *To maintain the aromaticity (i.e. resonance through the conjugated linkages), the chelate ring needs an even number of atoms in the unsaturated chelate ring.* Obviously, such a conjugation is possible in the 6-membered chelate rings not in the 5-membered chelate rings. *Consequently, for such aromatic chelate rings, the six membered chelate ring is more stable than the five membered chelate ring.* On the other hand, for the nonaromatic chelate rings, the 5-membered chelate ring is more *stable* than the 6-membered chelate ring.

- *In general, any chelate ring less than 5-membered or more than 6-membered will be less stable. Between the 5- and 6-membered chelate rings, if the conjugation leading to an aromaticity is possible then the 6-membered chelate ring will be more stable, otherwise the 5-membered chelate ring will be more stable.*

(J) *Aromaticity, i.e. resonance stabilisation in the chelate rings:* Aromaticity (*i.e.* pseudoaromaticity) or resonance is possible in many 6-membered chelate rings through the participation of metal $d\pi$-orbitals and such chelate rings are highly stable. This is possible for even number of atoms in the **unsaturated** ring. *This resonance or delocalization of charge stabilizes the 6-membered chelate rings.* This is illustrated in the examples involving acetylacetonate, biguanide, and dimethylglyoximate as the chelate forming didentate ligands.

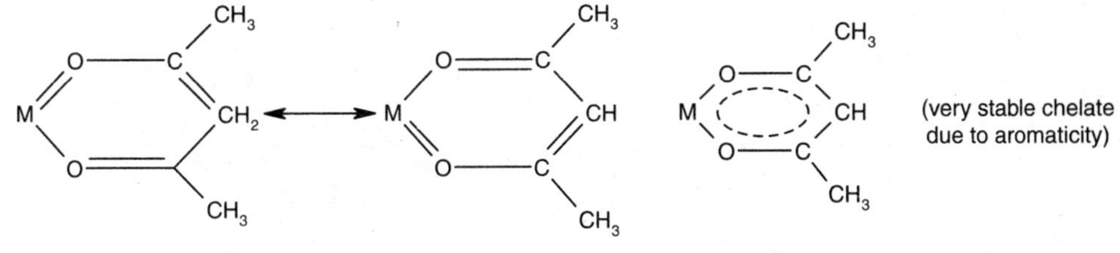

M(acac), *i.e.* chelate ring formed by acetylacetonate

Fig. 4.5.2.23 Aromaticity in the chelate ring formed by acetylacetonate (acac⁻).

By considering the participation of metal d-orbitals, the resonance may be extended throughout the ring (*cf.* Fig. 4.5.2.23).

● **Acetylacetonate complexes:** Acetylacetonate exists in a keto-enol tautomeric equilibrium and the enol form participates in complexation (Fig. 4.5.2.24). The enolic-H is a very weak acid but the deprotonation is favoured due to the formation of a very stable chelate.

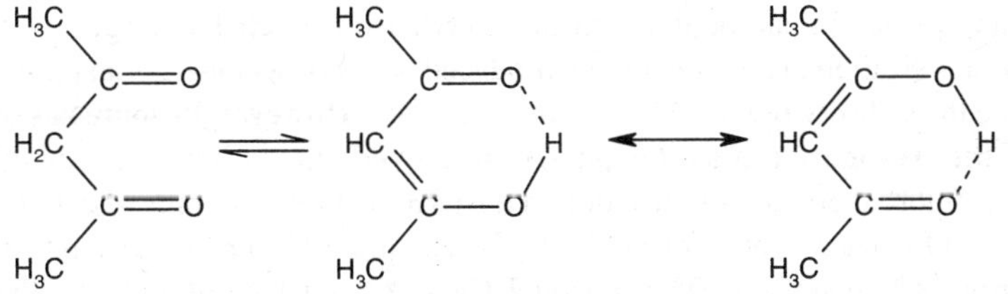

$CH_3 \!-\! C(OH)CHC(O)CH_3 \rightleftharpoons [CH_3 \!-\! C(O)CHC(O)CH_3]^- + H^+$, $pK_a \approx 10$ (*i.e.* very weak acidic character)

i.e.

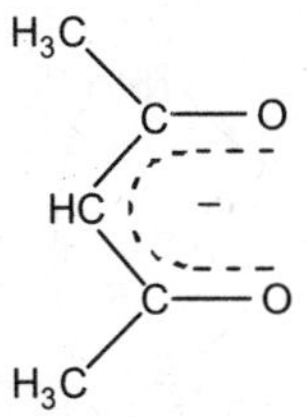

(acac⁻)

$Fe^{3+} + 3acac^- \rightleftharpoons [Fe(acac)_3]$, $\log K = 26$

Fig. 4.5.2.24 Complexation by acetylacetonate.

The **pseudoaromaticity** (*i.e.* resonance stabilization) in the acetylacetonate chelate ring is evidenced by the fact that *at the central C-centre of the chelate ring, an* **electrophilic substitution** *can occur.*

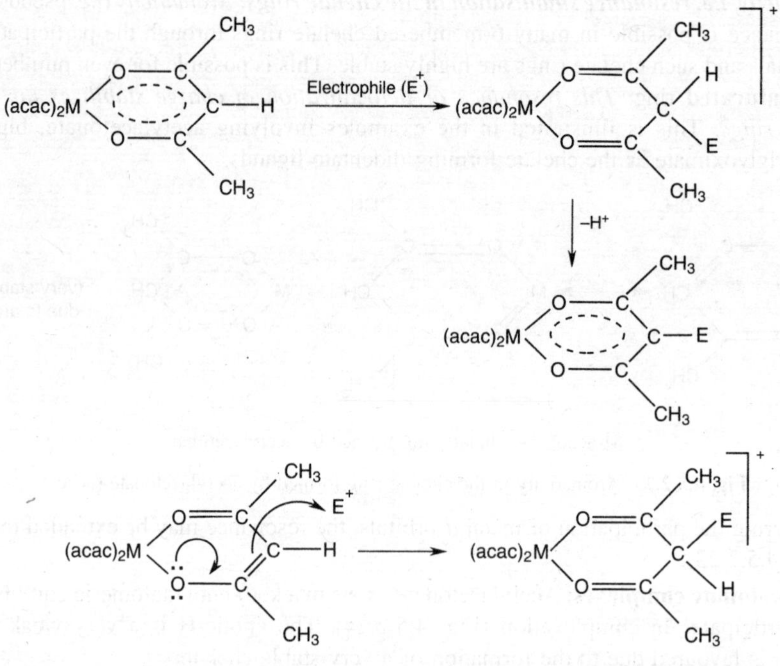

Fig. 4.5.2.25 Electrophilic substitution on the central C-atom of the acetylacetonate chelate ring.

Thus halogenation reaction can be done on the acetylacetonate chelate ring.

(**Note:** Resonance stabilization in the chelate ring produced by acetylacetonate is an important contributing factor to determine the stability of the acetylacetonate chelate. **However, in some cases, because of the J.T. distortion, this resonance may be partly destroyed.** In the complex, $[Cu(bpy)(hfa)_2]$, *i.e.* bipyridinebis(hexafluoroacetylacetonato)copper(II) (where bpy occupies the two equatorial *cis*–positions), the Cu—O axial and equatorial bonds are unequal (229 pm *vs.* 197 pm, *cf.* Fig. 4.5.2.15) because of the *z*-out J.T. distortion. This unequal Cu—O bond lengths indicate that the resonance stabilization in the chelate is partly lost due to the J.T. distortion. (This aspect has been already discussed.)

● **Biguanide complexes:** As in the acetylacetonate complexes, in the **biguanide complexes**, the pseudoaromatic character is maintained to stabilize the system (Fig. 4.5.2.26).

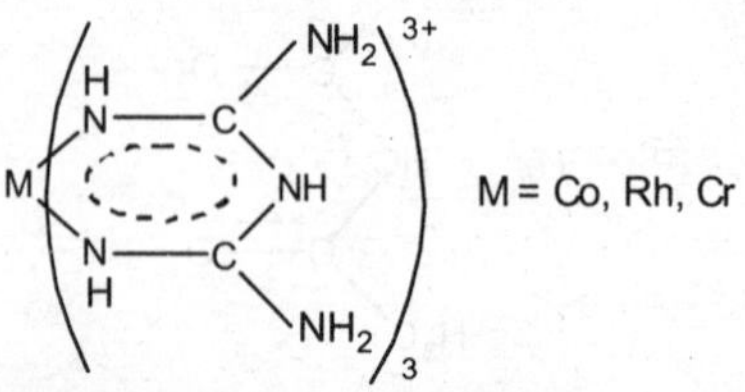

Fig. 4.5.2.26 Pseudoaromaticity in a biguanide chelate.

● **Dimethylglyoximate complexes:** Here it may be mentioned that dimethylglyoximate forms a 5-membered unsaturated chelate ring where the electron delocalization may occur through the H-bonded segments to stabilize the system. This is illustrated for the square planar complexes, $[M(dmgH)_2]$ (M = Ni, Pd, etc.)

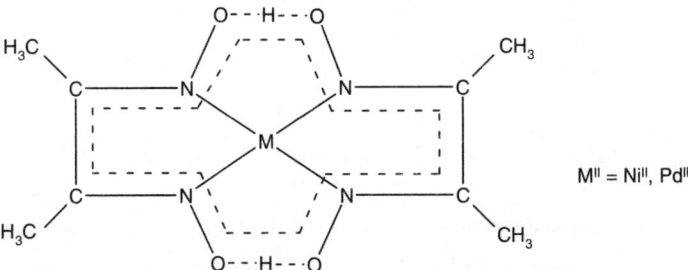

Fig. 4.5.2.27 Electron delocalisation in the dimethylglyoximato complexes like [M(dmgH)$_2$].

(K) **Effect of the bonding property of the ligand:** Depending on this property, position of a particular ligand in the spectrochemical series is mainly determined. If it is a **π-acid ligand** (*i.e.* strong field ligand), then it can stabilize the low oxidation states of the transition metal ions through complexation. On the other hand, to stabilize the higher oxidation states, hard σ-donor (without any π-acidic character) ligands are suitable. *The spectrochemical position of a particular ligand is of an important consideration to determine the metal-ligand stability constant.*

(L) **π-acidity of the ligands:** The metal $\rightarrow$ ligand π-bonding can enhance the electron deficiency on the metal centre and consequently it may favour the binding of the next incoming ligand. *Sometimes, it may reverse the statistical order of the stepwise stability constants, i.e.* $K_1 \langle K_2 \langle K_3 \langle$. This aspect is well documented for the ternary complexes (*cf.* Sec. 4.7.2).

4.5.3 Chelate Effect: Thermodynamic and Kinetic Aspects

(A) Illustration of Chelate Effect: It is well known that complexes containing the chelate rings are more stable than the complexes containing the similar but nonchelating ligands. This phenomenon is described as the **chelate effect**. This is illustrated in the following representative examples.

- **M(II) – en system *vs.* M(II) – NH$_3$ system:**

$$\left[Ni(OH_2)_6\right]^{2+} + 6NH_3 \rightleftharpoons \left[Ni(NH_3)_6\right]^{2+} + 6H_2O, \quad \log\beta_6 = 8.6 \ (25°C)$$

$$\left[Ni(OH_2)_6\right]^{2+} + 3en \rightleftharpoons \left[Ni(en)_3\right]^{2+} + 6H_2O, \quad \log\beta_3 = 18.4 \ (25°C)$$

It indicates that compared to [Ni(NH$_3$)$_6$]$^{3+}$, the complex [Ni(en)$_3$]$^{3+}$ having three chelate rings is about **10^{10} times more stable**. It can be illustrated by taking the **Cd(II)-ammine** complexes.

	[Cd(NH$_3$)$_4$]$^{2+}$	[Cd(en)$_2$]$^{2+}$
$\log\beta$ $\approx$	7.5	10.6

Thus [Cd(en)$_2$]$^{2+}$ is about **10^3 times more stable** than [Cd(NH$_3$)$_4$]$^{4+}$. **It is evident that the chelate effect is more pronounced for Ni(II) having the cfse contribution compared to that for Cd(II) having no cfse contribution** (*cf.* ligand field strength: en $\rangle$ NH$_3$).

- **Stability constant and degree of chelation:** In general, with the increase of the **degree of chelation**, the stability increases. It is illustrated in the Cu(II)-amine complexes.

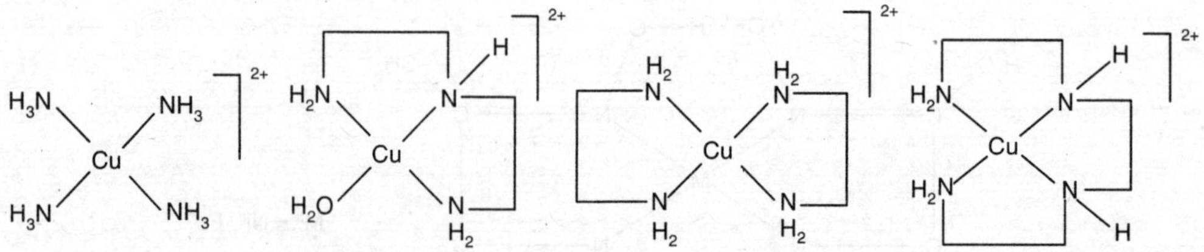

i.e. [Cu(NH₃)₄]²⁺   i.e. [Cu(dien)(OH₂)]²⁺   i.e. [Cu(en)₂]²⁺   i.e. [Cu(trien)]²⁺

$\log\beta_4 \approx 12.7$ $\log\beta_1 = 16.0$ $\log\beta_2 = 20$ $\log\beta_1 \approx 20.5$

● **Stability constant and the number of chelate rings:** That the stability increases with the increase of the number of chelate ring is illustrated for the Ni(II)-amine complexes.

Complex:	$[Ni(NH_3)_4(OH_2)_2]^{2+}$	$[Ni(en)_2(OH_2)_2]^{2+}$	$[Ni(OH_2)_2(trien)]^{2+}$	$[Ni(dien)_2]^{2+}$
No. of chelate rings:	0	2	3	4
$\log\beta$:	7.8	13.8	14.0	18.9

Because of the same fact, the stability order for the aminopolycarboxylates increases in the order: $gly^- \langle\ ida^{2-} \langle\ nta^{3-}$

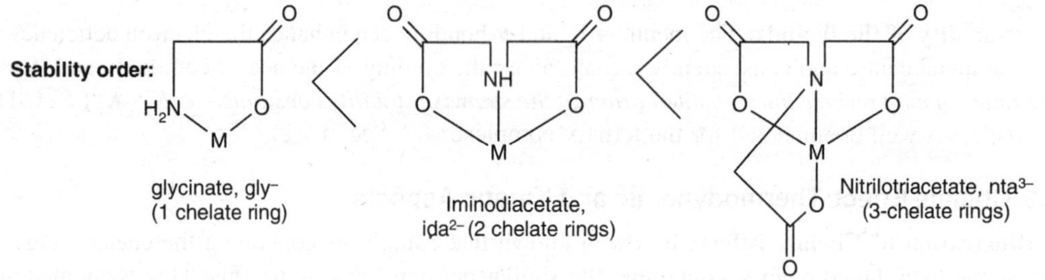

Stability order:

glycinate, gly⁻ (1 chelate ring) $\langle$ Iminodiacetate, ida²⁻ (2 chelate rings) $\langle$ Nitrilotriacetate, nta³⁻ (3-chelate rings)

● **Chelate effect (NH₃ vs. en) – quantitative relationships:** Chelate effect in the en-complexes compared to those of the analogous monodentate ligand complexes, *i.e.* NH₃– complexes, may be expressed by the following relationship experimentally verified.

$$\log\beta_1(en) = 1.152\ \log\beta_2(NH_3) + \log 55.5\ (\text{entropic factor}) = 1.152 \times 4.8 + 1.74 = 7.26$$

● **Chelate effect (NH₃ vs. en; py vs. bpy) - both the enthalpic and entropic effect:** Taking pyridine (py) as the monodentate ligand and 2,2′-bipyridine (bpy) as the corresponding didentate ligand (*cf.* NH₃ vs. en), the stability constants calculated for the bpy-complexes from the stability constants of the py-complexes by using the above equation are less than the experimental values by about three log units.

$$\log\beta_1(bpy) = 1.152\log\beta_2(py) + \log 55.5(\text{entropic factor}) + 3(\text{enthalpic factor})$$

It indicates that the chelate effect is more pronounced when there is a pyridyl group. **This may be explained by considering the relative positions of the ligands, i.e. NH₃ vs. en and py vs. bpy in the spectrochemical series.** The crystal field splitting power of bpy is much higher than that of py while in the pair, *en* and *NH₃*, the splitting power of en is not so higher than that of NH₃. It indicates: the ligand field strength difference between *py* and *bpy* $\rangle\rangle$ ligand field strength difference between *NH₃* and *en*. Thus in the observed chelate effect, the entropic factor is comparable in all cases but the enthalpic factor from cfse is more important in the bpy complexes than that in the en-complexes compared to their respective complexes formed by the monodentate ligands, *i.e. py* and NH₃ respectively. It suggests the sequence of **cfse value difference:** bpy-complex – py-complex $\rangle\rangle$ en-complex – NH₃-complex.

● **Chelate effect for the transition *vs.* nontransition metals and enthalpy factor *vs.* entropy factor:** Here it may be pointed out that chelate effect is more pronounced for the transition metal ions compared to that for the non-transition metal ions. It is illustrated in the following system.

$$\left[M(NH_3)_2(OH_2)_4\right]^{2+} + en \xrightleftharpoons{K} \left[M(en)(OH_2)_4\right]^{2+} + 2NH_3$$

For Cu^{2+}: $\Delta H^0 = -8$ kJ mol^{-1}, $\Delta S^0 = +31.0$ J K^{-1} mol^{-1}

$M^{2+} =$	Co^{2+}	Ni^{2+}	Cu^{2+}	Zn^{2+}	Cd^{2+}
$\log K$ (25° C)	= 2.1	3.0	3.2	1.4	1.1

Here it may be pointed out that for the transition metal complexes, in chelate effect both the enthalpic (i.e. cfse and preorganisation energy) and entropic factor contribute where the entropic factor is the predominant one. For the non-transition metal ions having no cfse, chelate effect mainly arises from the entropic factor and partly from the preorganisation energy.

Examples Opposing the Chelate Effect

It happens so for the following system:

$$\left[Ag(NH_3)_2\right]^+ + en \xrightleftharpoons{K} \left[Ag(en)\right]^+ + 2NH_3, \ \log K \approx -2.6 \ (25°C)$$

Here the chelate effect is denied. Ag(I) prefers the **linear coordination** and it is attained in $[Ag(NH_3)_2]^+$ but not in the $[Ag(en)]^+$ complex. Because of the geometric constraints in the en-chelate ring, it cannot provide the required linear coordination. This is why, $[Ag(en)]^+$ is destabilized compared to $[Ag(NH_3)_2]^+$. Sometimes, the J.T. effect may oppose the chelate effect. It happens so in the process:

$$trans\text{-}\left[Cu(en)_2(OH_2)_2\right]^{2+} + en \xrightleftharpoons{K_3} \left[Cu(en)_3\right]^{2+} + 2H_2O, \ \log K_3 = -1.0$$

● **Chelate effect in the metal-edta complex:** The hexadentate ligand edta^{4-} forms very stable complexes with a large variety of metal ions.

$$\left[M(OH_2)_6\right]^{n+} + edta^{4-} \rightleftharpoons \left[M(edta)\right]^{(4-n)-} + 6H_2O$$

$M^{n+} =$	Mg^{2+}	Ca^{2+}	Sr^{2+}	Ba^{2+}	Fe^{2+}	Co^{2+}	Ni^{2+}	Cu^{2+}	Fe^{3+}
$\log K$ (20° C) =	8.7	10.7	8.6	7.8	14.3	16.3	18.6	18.8	25.1

The high stability in these complexes arises mainly from the chelate effect.

(B) Thermodynamic Aspects of the Chelate Effect: In order to understand the thermodynamic aspects of chelate effect, we are to recall the following relationship.

$$-RT\ln\beta^0 = \Delta G^0 = \Delta H^0 - T\Delta S^0$$

During the chelation, there is an increase in the number of unbound (*i.e.* uncoordinated) molecules. It is illustrated in the following reactions.

$$\left[M(OH_2)_6\right]^{2+} + 6NH_3 \rightleftharpoons \left[M(NH_3)_6\right]^{2+} + 6H_2O$$

$$\left[M(OH_2)_6\right]^{2+} + 3en \rightleftharpoons \left[M(en)_3\right]^{2+} + 6H_2O$$

$$\left[M(OH_2)_6\right]^{2+} + 2dien \rightleftharpoons \left[M(dien)_2\right]^{2+} + 6H_2O$$

$$\left[M(OH_2)_6\right]^{2+} + \text{edta}^{4-} \rightleftharpoons \left[M(\text{edta})\right]^{2-} + 6H_2O$$

Here, 6 monodentate NH_3 ligands replace 6 H_2O molecules; 3 molecules of the didentate en ligands replace 6 H_2O molecules; 2 molecules of the tridentate dien ligands replace 6 H_2O molecules; and one molecule of the hexadentate edta^{4-} ligand replaces 6 H_2O molecules.

Thus, in the above reactions leading to the formation of octahedral complexes, in moving from monodentate (*i.e.* NH_3) to didentate (*i.e.* en), to tridentate (*i.e.* dien), to hexadentate (*i.e.* edta^{4-}) there is a net increase of 0, 3, 4 and 5 unbound molecules respectively.

Such an increase in the number of unbound molecules during the complexation will lead to an increase in entropy (ΔS^0) from the increase in *translational randomness* to favour the process. ***Thus chelation is always accompanied with an entropic favour. This entropic favour is the basic reason for the chelate effect. Here it may be pointed out that cyclisation of a chelating ligand during the complexation causes a loss of entropy.*** But the increase of entropy due to the net increase in the number of unbound molecules during the chelation can overweigh the decrease of entropy due to cyclization. Thus, the net effect is the increase of entropy in the chelation process. In the above examples, for edta^{4-}, there is a charge neutralisation during the complexation. It causes an **additional entropic favour** due to the reduction in the **electrorestriction** over the dipolar solvent molecules.

However, the **enthalpic factor (as a minor factor)** may also contribute in part for the observed chelate effect.

● **Entropic favour:** During the chelation, the **increase of entropy** occurs due to the increase in the number of unbound molecules in solution. It is given by (in aqueous solution):

$$\Delta S = xR\ln 55.5 = 33.4x \text{ J K}^{-1} \text{ mol}^{-1}$$

(x = number of net increase of the unbound molecules during chelation).

$T\Delta S$ gives the contribution to the free energy change and it becomes about 10.0 kJ mol^{-1} per chelate ring at 300 K. For the pair of complexes: $[Cu(NH_3)_4]^{2+}$ and $[Cu(trien)]^{2+}$, $x = 3$; $[Cu(NH_3)_4]^{2+}$ and $[Cu(en)_2]^{2+}$; $x = 2$; $[Cu(NH_3)_4]^{2+}$ and $[Cu(cyclam)]^{2+}$, $x = 3$ (*cf.* Sec. 4.5.4, additional entropy effect for a macrocyclic ligand).

Illustration: $\left[Cu(NH_3)_4\right]^{2+} + \text{trien} \rightleftharpoons \left[Cu(\text{trien})\right]^{2+} + 4NH_3$; $x = 5 - 2 = 3$

This theoretical prediction on the magnitude of **entropic favour** (= 33.4x J K^{-1} mol^{-1}) in the chelation process has been experimentally verified in many cases.

● **Enthalpic favour:** Enthalpy effect may have some contribution to the observed chelate effect. This may be understood by considering the following structures:

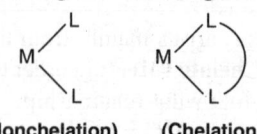

(Nonchelation)   (Chelation)

In the chelation, the binding sites occupy the adjacent sites of the metal centre and for this act of the event, the didentate chelating ligand is **preorganized or preformed** to some extent. If two unidentate ligands are supposed to occupy the similar adjacent positions, then some energy must be spent to manage the steric hindrance, repulsion between the dipoles at the adjacent positions, etc. This is not required to that extent for the didentate ligand which is already preorganized to some extent for the purpose (*cf.* $2NH_3$ vs. en; $3NH_3$ vs. dien; $4NH_3$ vs. trien; etc.). *This is the common origin of the enthalpic contribution to the chelate effect for both the transition and nontransition metal complexes.* This concept of *preorganization energy* is more pronounced in the *macrocyclic effect* (*i.e.* macrocyclic *vs.* open chain polydentate ligand) and in the exceedingly high stability of the *spherand complexes* (compared to the crown ether complexes). These aspects will be discussed later (*cf.* Sec. 4.6).

Spherands⟩ Cryptands (macrobicylcic ligand)⟩ Crown ether

$\overset{\longleftarrow}{(\text{monocyclic macrocycle})⟩ \text{Open chain polydentate ligand}}$

Increasing trend of preorganisation

Besides the concept of *preorganization energy* leading to an enthalpic favour in the chelate effect, **the cfse may become important for the transition metal ions.** Because, very often, the chelating ligand produces a higher **crystal field splitting power** (*cf.* NH_3 *vs.* en; py *vs.* bpy; etc.). Sometimes, the **difference in the basicities** of the chelating and nonchelating ligands may contribute to the enthalpic factor. For example, en is more basic than NH_3 because of the electron pushing inductive effect of the methylene groups. This also makes the en-complexes more stable than the NH_3-complexes because of an enthalpic favour.

Thus, the **enthalpic favour** in the chelation originates in general from the following factors:
preorganised structure of the chelating ligands; higher ligand field strength of the chelating ligand; higher basicity of the chelating ligand.

- **Thermodynamic parameters of the chelatic effect for the nontransition metal ions:** For the **non-transition metal ions** like Zn^{2+}, Cd^{2+}, etc. having no cfse, *the chelate effect mainly arises from the entropic effect.* It is illustrated below:

$$\left[Cd(OH_2)_4\right]^{2+} + 4CH_3NH_2 \rightleftharpoons \left[Cd(NH_2CH_3)_4\right]^{2+} + 4H_2O, \text{ (nonchelation)}$$

$\Delta H^0 = -57.33 \text{ kJ mol}^{-1}$, $\Delta S^0 = -67.0 \text{ J K}^{-1} \text{ mol}^{-1}$, $\Delta G^0 = -37.41 \text{ kJ mol}^{-1}$, $\log \beta_4^0 = 6.55 \ (25°C)$

$$\left[Cd(OH_2)_4\right]^{2+} + 2en \rightleftharpoons \left[Cd(en)_2\right]^{2+} + 4H_2O, \text{ (chelation)}$$

$$\Delta H^0 - -56.5 \text{ kJ mol}^{-1}, \Delta S^0 = 13.7 \text{ J K}^{-1} \text{ mol}^{-1}$$

$$\Delta G^0 = -60.7 \text{ kJ mol}^{-1}, \log \beta_2^0 = 10.62 \ (25° C)$$

The above data indicate that the enthalpy factor remains more or less the same for both the chelation and nonchelation process, but for the chelation process, there is *a huge entropic favour.* The chelate effect in terms of increase of entropy = $+13.7 - (-67.0) = 80.7 \text{ J K}^{-1}\text{mol}^{-1}$. The theoretically predicted value is: $2 \times 33.4 = 66.8 \text{ J K}^{-1} \text{ mol}^{-1}$

- **Thermodynamic parameters of the chelate effect for the transition metal ions (for NH_3 *vs.* en; py *vs.* bpy):** For the **transition metal ions** excepting those of the d^0, d^5 (high spin) and d^{10} systems, *the enthalpic factor from the cfse contribution also partly contributes along with the main contributing factor, entropic favour.* It is illustrated below:

$$\left[Cu(OH_2)_6\right]^{2+} + 2NH_3 \rightleftharpoons \left[Cu(NH_3)_2(OH_2)_4\right]^{2+} + 2H_2O$$

$$\Delta H^0 = -46.45 \text{ k J mol}^{-1}, \Delta S^0 = -8.37 \text{ J K}^{-1} \text{ mol}^{-1},$$

$$\Delta G^0 = -43.94 \text{ kJ mol}^{-1}, \log \beta_2^0 = 7.65$$

$$\left[Cu(OH_2)_6\right]^{2+} + en \rightleftharpoons \left[Cu(en)(OH_2)_4\right]^{2+} + 2H_2O,$$

$$\Delta H^0 = -54.4 \text{ kJ mol}^{-1}, \Delta S^0 = +22.6 \text{ J K}^{-1} \text{mol}^{-1},$$

$$\Delta G^0 = -61.0 \text{ kJ mol}^{-1}, \log \beta_1^0 = 10.64$$

Chelate effect in terms of gain of entropy $= +22.6 - (-8.37) = 30.97 \text{ J K}^{-1} \text{mol}^{-1}$ (*cf.* $33.4 \times 1 \text{ J K}^{-1} \text{mol}^{-1}$)

In the pair: $[Cu(NH_3)(OH_2)_4]^{2+}$ and $[Cu(en)(OH_2)_4]^{2+}$, for the observed chelate effect, besides the entropic favour ($\approx 31 \text{ J K}^{-1} \text{mol}^{-1}$), there is an enthalpic favour given by: $-54.4 - (-46.45) = -7.95 \text{ kJ mol}^{-1}$. **This enthalpic favour is more pronounced compared to that in the Cd(II)-system where there is no cfse effect.**

In the Ni(II)-complexes, a similar observation is noted.

$$\left[Ni(OH_2)_6 \right]^{2+} + 4NH_3 \rightleftharpoons \left[Ni(NH_3)_4 (OH_2)_2 \right]^{2+} + 4H_2O$$

$$\Delta H^0 = -65 \text{ kJ mol}^{-1}, \Delta S^0 = -63 \text{ J K}^{-1} \text{mol}^{-1}$$

$$\left[Ni(OH_2)_6 \right]^{2+} + 2en \rightleftharpoons \left[Ni(en)_2 (OH_2)_2 \right]^{2+} + 4H_2O$$

$$\Delta H^0 = -77 \text{ kJ mol}^{-1}, \Delta S^0 = 12 \text{ J K}^{-1} \text{mol}^{-1}$$

Thus the **entropic favour** in the above chelation is given by $= 12 - (-63) = 75 \text{ J K}^{-1} \text{mol}^{-1}$ (*cf.* $33.4 \times 2 \text{ J K}^{-1} \text{mol}^{-1}$) and the **enthalpic favour** in chelation is given by: $-77 - (-65) = -12 \text{ kJ mol}^{-1}$ (more pronounced than that in the Cd^{II}-complexes).

Compared to the Cd(II)-system, the enthalpic favour in the chelation process with the Ni(II) or Cu(II) centres is more pronounced. Because, for Ni(II) and Cu(II), besides the preorganization favour, cfse contributes in the chelation process. In the *spectrochemical series,* en lies right to NH$_3$ (*i.e.* crystal field splitting power of en is higher than that of NH$_3$). **This is why, in the en-complex, the cfse is higher than that in the NH$_3$-complex.** It gives the enthalpic favour. The higher basicity of en compared to that of NH$_3$ due to the presence of the electron pushing methylene group in en makes the en-complexes more stable. This also favours the enthalpy. Thus, for the chelate effect, both the *enthalpic factor and entropic factor contribute simultaneously but the entropic effect is more predominant. The enthalpic factor becomes appreciable for the transition metal ions having the cfse.* This aspect is further illustrated in the following reactions:

$$\left[Cu(NH_3)_2 (OH_2)_4 \right]^{2+} + en \xrightleftharpoons{K} \left[Cu(en)(OH_2)_4 \right]^{2+} + 2NH_3, \ x = 1$$

$$\Delta H^0 = -7.95 \text{ kJ mol}^{-1}, \Delta S^0 = +31 \text{ J K}^{-1} \text{mol}^{-1}, \ \log K(25°C) = 3.0$$

$$\left[Ni(NH_3)_6 \right]^{2+} + 3en \xrightleftharpoons{K} \left[Ni(en)_3 \right]^{2+} + 6NH_3, \ x = 3$$

$$\Delta H^0 = -12.2 \text{ kJ mol}^{-1}, T\Delta S^0 = +55.5 \text{ kJ mol}^{-1}, \ \log K(25°C) = 9.7$$

In the above reactions, the entropic favour is quite expected (where increase in the number of unbound molecules $= x = 1, 3$). The enthalpic favour also drives the equilibrium to the right side. The enthalpic favour arises from the higher basicity of en and more gain of cfse in the en-complex (*cf.* **crystal field strength:** en $\rangle$ NH$_3$). This is explained by considering the relative positions of NH$_3$ and en in the spectrochemical series.

This enthalpic contribution is more pronounced if we consider the stabilities of py and bpy complexes.

$$\left[M(py)_6 \right]^{2+} + 3bpy \rightleftharpoons \left[M(bpy)_3 \right]^{2+} + 6py, \ x = 3$$

The entropic favour to drive the above equilibrium will arise as usual but because of much higher crystal field splitting power of *bpy* than that of *py*, there will be a significant amount of gain of cfse in producing the *bpy*-complex from the *py*-complex. This gives the contribution of enthalpic factor in the observed chelate effect for the *bpy*-complexes. *This is why, the chelate effect is more pronounced in the bpy-complexes than in the en-complexes where the ligand field strengths of en and NH_3 do not differ so widely.*

(C) **Sidgwick's Qualitative Explanation of the Chelate Effect:** Sidgwick attempted to explain the phenomenon *in terms of the ease and difficulty in dissociation of the ligands.* This is illustrated by taking a didentate ligand and a monodentate ligand. For a didentate ligand (say, *en*), if one M—N bond is ruptured, the ligand is **not completely dissociated** from the coordination sphere as it is still remaining coordinated with the metal centre through the remaining M—N bond. Thus, the partially dissociated chelating ligand remains still within the coordination sphere and it can easily reestablish the broken bond. On the other hand, for the monodentate ligand (say NH_3), if the metal-ligand bond is broken then it comes out from the coordination sphere (*i.e.* **it gets completely dissociated**) and consequently, the possibility of reestablishing the broken metal-ligand bond does not arise. **In general, with the increase of the degree of chelation, the difficulty in dislodging the ligand gradually increases.** This explains the higher stability of the chelated compounds (*cf.* $\beta = k_f / k_d$ where k_f and k_d are the formation and dissociation rate constants of the complex; decrease in k_d makes β higher). Interpretation of the chelate effect in terms of k_d has been given later in details (see 'D' of this Section).

Distortion in a geometry due to the low bite angle and low bite distance

In the chelate, the two terms *bite distance* and *bite angle* are quite important. When the J.T. distortion elongates the axial bond length as in the Cu(II)-complexes, the low bite distance in a chelate occupying the two *cis*– equatorial and axial positions causes a strain (*cf.* Figs. 4.5.2.11, 13; [Cu(en)$_3$] complex where K_3 is drastically dropped). The low bite angle can also distort the geometry. In fact, [Ru(bpy)$_3$]$^{2+}$ having the ~ 90° bite angle possesses a distorted octahedral geometry.

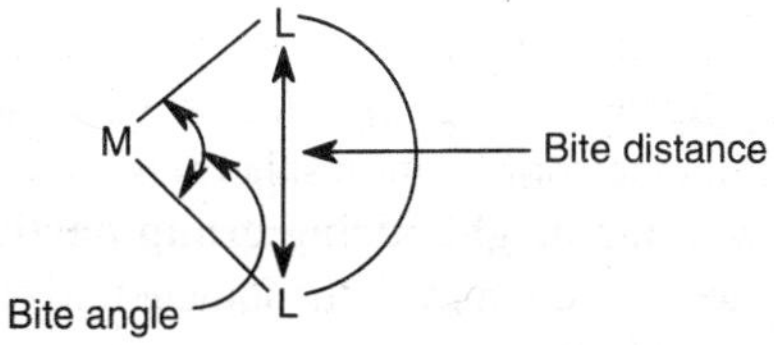

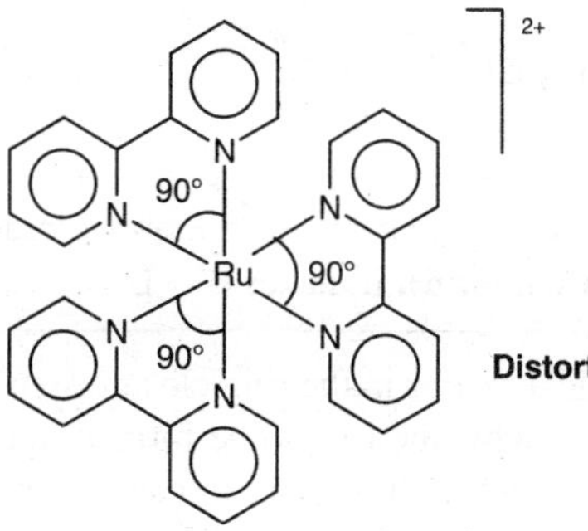

Bite distance

Bite angle

Distorted octahedral geometry

D. Kinetic Aspects of the Chelate Effect: This phenomenon can be interpreted in terms of the kinetic and mechanistic aspects by considering the following simple reaction scheme (by taking the didentate and monodentate ligands).

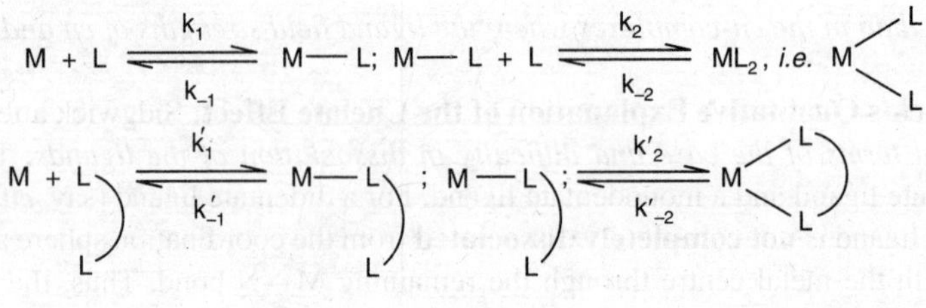

By considering the above **single step reactions**, the stability constants may be expressed as follows:

$$\beta = \frac{k_1 k_2}{k_{-1} k_{-2}}, \left(\text{for the monodentate ligand L}\right)$$

$$\beta' = \frac{k_1' k_2'}{k_{-1}' k_{-2}'}, \left(\text{for the didentate ligand L} - \text{L}\right)$$

If the ligands are having the similar binding sites (*e.g.* NH_3 *vs.* en), then it is reasonable to assume $k_1 \approx k_1'$ and $k_{-1} \approx k_{-1}'$. To explain the chelate effect, *i.e.* $\beta' \rangle \beta$, we should have the following condition:

$$\frac{k_2'}{k_{-2}'} \rangle \frac{k_2}{k_{-2}}$$

For the monodentate ligands, the k_2-step represents a *simple bimolecular step* while for the didentate ligand, the k_2'-step represents a **unimolecular step. Thus, statistically, k_2' is favoured over k_2.** This *statistical favour* arises from the ring closure step by the partially ligated ligand that lies within the coordination sphere. The statistical favour in k_2' can only be counterbalanced in the complexation process by the monodentate ligand by increasing the concentration of the monodentate ligand. In other words, for the didentate chelating ligand, *in the ring closure step*, the **effective concentration of the nucleophile** (which is already within the coordination sphere) is increased. This phenonmenon (*i.e.* ring closure step) may be compared with the **neighbouring group participation (NGP) phenomenon** in the nucleophilic substitution reaction (a well known phenomenon in organic chemistry). This makes the following situation causing the chelate effect.

$$\frac{k_2'}{k_{-2}'} \rangle \frac{k_2}{k_{-2}} \; (i.e. \; \beta' > \beta)$$

In the light of **Sidgwick's argument** discussed earlier, in the above reaction, $k_{-2}' < k_{-2}$ (*i.e.* **dechelation is easier for the monodentate ligand**). It makes $\beta' > \beta$.

Here it must be mentioned that the statistical favour in the ring closure step is significant only when there is no steric strain in the process. This is why, the ring to be formed must be of proper size.

In terms of *entropy of activation* $(\Delta S^{\neq})$, the favour in k_2' over k_2 can also be rationalized. The parameter $\Delta S^{\neq}$ is more positive for the **unimolecular step** (*i.e.* ring closure step in the chelation process) than in the **bimolecular step** (for the monodentate ligand).

In the unimolecular step, there is a net increase by one in the number of unbound molecule while in the bimolecular step, there is no change in the number of unbound molecule. Here it may be mentioned that **cyclisation during the chelation makes a loss of entropy** but this effect is overweighed by the net increase in the number of unbound molecules (*cf.* $\Delta S = nR\ln 55.5 = 33.4n$ J K^{-1} mol^{-1}).

4.5.4 Macrocyclic Effect and Cryptate Effect (*i.e.* Macrobicyclic effect) – An Extension of the Chelate Effect: Preorganization Energy (Enthalpic Factor) and Entropic Factor

The closed ring macrocyclic ligands from the more stable (both thermodynamically and kinetically) complexes than the corresponding open-chain polydentate ligands having the similar binding sites. This phenomenon is called the *macrocyclic effect which is an extension of the chelate effect.* Thus the metal-ligand stability constant for the *n*-unidentate, *n*-dentate open chain ligand, and *n*-dentate macrocyclic ligand varies in the following sequence (all having the comparable binding sites):

n-dentate macrocyclic ligand ⟩ n-dentate open chain ligand ⟩ n-unidentate ligands.

● The representative macrocyclic ligands like cyclam, crown ether, spherand, cryptand, etc. have been discussed in Secs. 1.8, 4.6.

● The metal complexes of the macrocyclic ligands experience the *macrocyclic effect* characterized by the *greater thermodynamic stability* and *kinetic stability* (with respect to the ligand dissociation).

Because of this macrocyclic effect, even the normally labile metal centres produce the kinetically stable complexes with the macrocyclic ligands. **This kinetic stability arises from the disfavoured metal-ligand dissociation process.**

● The enhanced thermodynamic stability due to the macrocyclic effect is illustrated in the following examples.

[Zn(3, 2, 3-tet)]$^{2+}$　　[Zn(cyclam)]$^{2+}$　　[Cu(2, 2, 2-tet)]$^{2+}$　[Cu(2, 3, 2-tet)]$^{2+}$　　[Cu(tet-a)]$^{2+}$
log $K \approx 11.3$　　　log $K \approx 15.3$　　　log $K = 20.1$　　log $K = 23.9$　　　log $K = 28.0$

3,2,3-*tet* = 1,5,8,12-tetraazadodecane (open chain ligand)
2,2,2-*tet (trien)* = 1,4,7,10-tetraazadecane (open chain ligand)
2,3,2-*tet* = 1,4,8,11-tetraazaundecane (open chain ligand)
cyclam =1,4,8,11-tetraazacyclotetradecane (macrocyclic ligand)
　tet-a = *meso*-1,4,8,11-tetraaza-5,5,7,12,12,14-hexamethylcyclotetradecane (macrocyclic ligand) (corresponding *d, l, –*form is *tet-b*)

(A) Thermodynamic Aspects of the Macrocyclic Effect: We have already pointed out that the macrocyclic effect is *an extension of the chelate effect*. In the chelate effect involving the **didentate** or **linear polydentate ligands**, the main thermodynamic favour arises from the **entropic favour** given by $xR\ln 55.5$ (where x = number of increase of the unbound molecules) and the **enthalpic favour** makes a minor contribution to the chelate effect. The enthalpic favour has been explained by considering *the less preorganization energy* required for the chelating ligands and *higher crystal field splitting power* of the chelating ligand compared to that of the corresponding monodentate ligand. Sometimes, the **difference in ligand basicity** may also contribute to the enthalpic factor. This *enthalpic favour* has been found more pronounced in the chelate effect for the transition metal complexes (compared to the nontransition metal ions and transition metal ions having no cfse).

● **Entropic favour:** If we consider the complexation, **n-dentate macrocyclic vs. n-unidentate ligands,** then the entropic favour to the macrocyclic effect is expressed in terms of the relative increase in the number of unbound molecules during the complexation and it is given by $xR\ln 55.5$ J K^{-1} mol^{-1} ($x = n - 1$ = number of increase of the unbound molecules, n = number of chelate rings produced by the **macrocyclic ligand**; *cf.* for the **linear polydentate ligand,** $x = n - 1$ = number of chelate rings formed.)

$$\left[M(OH_2)_y\right] + nL \rightleftharpoons \left[M(L)_n(OH_2)_{y-n}\right] + nH_2O; \ (x = 0)$$

$$\left[M(OH_2)_y\right] + L_n \text{ (open chain)} \rightleftharpoons \left[M(L_n)(OH_2)_{y-n}\right] + nH_2O; \ (x = n - 1)$$

$$\left[M(OH_2)_y\right] + L_n \text{ (macrocycle)} \rightleftharpoons \left[M(L_n)(OH_2)_{y-n}\right] + nH_2O; \ (x = n - 1)$$

If we consider the complexation, **n-dentate macrocyclic ligand** *vs.* **n-dentate polydentate open chain ligand,** then in terms of the relative increase in the number of unbound molecules, there is no entropic favour for the macrocycle. Because, in both the cases, $x = n - 1$ but **cyclization of a linear polydentate ligand during the complexation leads to a loss of entropy (*i.e.* entropic disfavour).** On the other hand, the macrocyclic ligand is already cyclized and there is no such loss of entropy during the complexation. This gives an entropic benefit to the macrocycle compared to the linear polydentate ligand. Here it is worth mentioning that **desolvation of the macrocycle** during the complexation experiences a less entropic benefit but a more enthalpic benefit. But this entropic factor is relatively less important. These aspects are discussed later. *In summary, the resultant entropic disfavour (relatively) for the open chain ligand (compared to a macrocycle) arises from the cyclisation.*

Here the **enthalpic favour** is also quite important. This enthalpic favour can be explained by considering the concept of *preorganization energy* and *desolvation energy of the ligand.*

● **Preorganization of the ligand:** In a complex of a *polydentate ligand* (both the open-chain and macrocycle), the conformation of the ligand is different from that of the ligand in its free state (*i.e.* uncomplexed state). Thus in the complexation process, the ligand is to change its conformation for the chelation. *This process is called preorganization of the ligands.* For this preorganisation of the ligands, it requires some energy to be spent. This is called **preorganisation energy.** *Thus a more preorganised ligand experiences a less conformational change during the complexation.* Compared to an open-chain multidentate ligand, a macrocyclic ligand is more preorganized to encapsulate the metal centre through the complexation. Here it must be pointed out that the conformation of a macrocyclic ligand is not exactly the same as in its complex. For the crown ether ligand, in a hydrophilic solvent, the O-sites with their lone pairs are projected outwards to produce a hydrophobic core. But during the complexation, these O-sites get projected towards the core where the metal ion resides. **For the open chain ligand, a more structural rearrangement is needed to bring it in a state present in the complex. This less preorganization energy for a macrocyclic ligand gives the enthalpic favour in the observed macrocyclic effect.**

Degree of preorganisation: 18-crown-6 $\rangle\rangle$ pentaglyme (EG5)

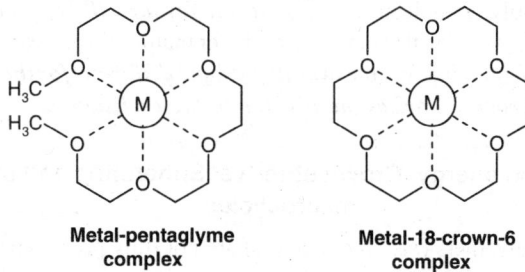

Metal-pentaglyme
complex

Metal-18-crown-6
complex

Fig. 4.5.4.1 Schematic representation of the metal complexes of pentaglyme and 18-crown-6 ligands.

That the preorganized nature of a macrocycle gives an enthalpic favour is understood by comparing the thermodynamic parameters of the complexation process of 18–crown–6 and its linear analogue pentaglyme, $CH_3(OCH_2CH_2)_5OCH_3$, *i.e.* pentaethyleneglycol dimethyl ether (EG5) (an example of **podand**) with some nontransition metal ions *where no cfse contributes.*

Table 4.5.4.1 Thermodynamic aspects of complexation of 18–crown–6 and pentaglyme with K$^+$ and Ba^{2+} in methanol.

	ΔH^0 (kJ mol^{-1})	ΔS^0 (J K^{-1} mol^{-1})	$logK$
K$^+$–18–crown–6:	−56.0	−71	6.05
K$^+$–pentaglyme:	−36.5	−84	2.1
Ba^{2+}–18-crown–6:	−43.5	−13.0	7.0
Ba^{2+}–pentaglyme:	−23.8	−33.0	2.3

It is evident that in the complexation process, for the macrocyclic ligand, 18-crown-6, the entropy change is more positive as expected. Besides this, there is an **enthalpic favour** for the macrocycle. It is given by: −56 − (−36.5) = −19.5 kJ mol^{-1} (for K$^+$) and −43.5 − (−23.8) = −19.7 kJ mol^{-1}(for Ba^{2+}). *This enthalpic favour originates from the more preorganised nature of the macrocycle.* Here it may be noted that the basicity of the said ligands are comparable and there is no cfse. ***Thus, the basicity effect and cfse effect do not complicate the interpretation.***

Here it may be pointed out that in the observed enthalpic favour, besides the preorganization energy, the *solvation-desolvation process* also contributes in part. This aspect is discussed below.

● **Desolvation of the ligand:** The complexation process may be described as follows:

$$M_{solvated} + L_{solvated} \rightleftharpoons [ML_{solvated}], (charges\ not\ shown)$$

Desolvation of the metal ion is the common factor for both the macrocycle and open-chain ligand. Solvation of the complex is more or less comparable for the both types of ligands. Here, during the complexation, the ligand must be desolvated and the **extent of solvation is different for the macrocycle and its corresponding open-chain ligand.** This difference will lead to the different contributions to the values of ΔH^0 for these two types of ligands.

Due to the **structural rigidity in the free macrocyclic ligand,** in aqueous media, it is less hydrated than the corresponding open chain ligand. In fact, a macrocyclic ligand can accommodate relatively less number of H_2O molecules through H-bonding. **Thus the energy required for desolvation of a macrocyclic ligand is less.** This gives a contribution to the observed enthalpic favour in the macrocyclic effect. Here it is worth mentioning that because of the less solvation of the free macrocycle, during the desolvation (*i.e.* release of solvent molecules), the *entropic favour will be less compared to the case of an open-chain ligand.* However, for the macrocyle, the enthalpic favour can overweigh this entropic disfavour. Here it is worth mentioning that the *overall entropic disfavour for the open chain polydentate ligand (compared to a macrocycle) arises mainly due to the cyclisation.*

Preorganization energy: Crown ether *vs.* Spherands and other related macrocycles

The macromonocyclic polyethers (*i.e.* crown ethers) are not fully preorganized to encapsulate the metal ion. They get fully organized through the conformational change when the metal ions are placed in their cavities. *The spherands, almost fully preorganized macromonocyclic polyethers,* can provide a spherical cavity and the O-donor sites are properly positioned to produce a cavity of fixed size. In fact, the O-donor sites in spherands are present in the *intraannular substituents and thus the donor sites are pointed towards the cavity.* The macrocyclic oligomers of methoxytolune and its derivatives produce the well known spherands.

Thus from the structural features of spherands, it is evident that the coordinating sites of the spherands are positioned strictly towards the cavity without any other degree of freedom. *This preorganized nature* of the spherands, *gives an* **enthalpic favour** *in the complexation process.* Consequently, the spherands very often form more stable complexes than the corresponding *crown ethers provided the metal ions fit properly in the cavity.* This property, consequently, provides the more **cation selectivity** for the spherands. The degree of preorganisation (*i.e.* the ease of encapsulation of a metal ion measured by stability constant) runs in the following sequence:

spherands > cryptaspherands > cryptands (*i.e.* macrobiocyclic ligands) > hemispherands > corands (*i.e.* monomacrocyclic ligands like the crown ethers) > *podands* (*i.e.* open chain polydentate ligands).

(B) Kinetic Aspects of the Macrocyclic Effect: The stability constant for the metal-ligand interaction can be approximately expressed as follows:

$$M + L \underset{k_d}{\overset{k_f}{\rightleftharpoons}} ML, \quad K \text{ or } \beta = \frac{k_f}{k_d}$$

Compared to the linear open-chain ligand, for the macrocyclic ligand, the rates of formation and dissociation of the macrocyclic complex are slower due to the excessive geometric strain. *But, compared to k_f (i.e. formation process), k_d (i.e. dissociation process) is drastically reduced to enhance the stability constant K.* For the dissociation, all the metal-ligand bonds are to be ruptured and the partially dissociated ligand must rotate to take away the dissociated coordinating sites away from the coordination sphere. *Because of the geometrical rigidity in the macrocyclic complex, the dissociated coordinating sites experience a hindered rotation. On the other hand, for the complexes of the open-chain ligand, there is no such geometrical rigidity to hinder the rotation of the dissociated sites.* The prediction has been experimentally verified in many cases, *e.g.*

$$k_d: [M(trien)]^{2+} \rangle\rangle [M(cyclam)]^{2+}, (M = Cu, Co, etc.)$$

(C) Macrobicyclic Effect, *i.e.* **Cryptate Effect:** The crown ethers are macromonocyclic while the cryptands are macrobicyclic. The cryptands form generally the more stable complexes than the macromonocycles. The general trend of increase in the stability constant on going from a macro-monocyclic ligand to a macrobicyclic ligand (*i.e.* cryptand) is described as the *macrobicyclic effect or cryptate effect which is a special case of macrocyclic effect.*

4.6 STABILITIES OF THE METAL COMPLEXES OF CROWN ETHER, SPHERANDS AND CRYPTANDS: ALKALI METAL COMPLEXES

4.6.1 Crown Ether Complexes: Structure (*see* Sec. 13.5, Vol. 3 of Fundamental Concepts of Inorganic Chemistry)

A. Characteristic features:

- **Structure:** The monomacrocyclic polyethers, *i.e.* *cyclic polymers of ethylene glycol,* $(OCH_2CH_2)_n$ are called the **crown ethers.** These monocyclic macrocycles are described as the *corands* in the terminology of **supramolecular chemistry.**

- **Crown ether – why named so?** In such cyclic polyethers, the ether O-atoms are separated by two methylene groups. These O-sites are the coordinating sites. In the complexes of such **cyclic polyethers**, the organic linkages or segments, *i.e.* $-(CH_2CH_2)-$ joining the O-atoms, are puckered to produce a *'crown'* **arrangement** while the coordinating O-sites with their lone pairs (projected towards the inner hole of the macrocycle) are arranged **in a nearly planar arrangement** about the metal ion positioned at the centre of the cavity. The **crown like arrangement of the organic segments of the ligands** in their complexes makes the name *crown ether.*

- **Lariat crown ethers:** They contain the additional coordinating sites in side chain (called **pendant arm**). Thus the lariat crown ethers can bind the metal ions more strongly (*see* Sec. 13.5, Vol. 3 of Fundamental Concepts of Inorganic Chemistry).

- **Naming the crown ethers:** Pedersen made a significant contribution in understanding the properties of the crown ether complexes. The nomenclature of such organic ring system is complicated. Pedersen introduced a simpler system for naming the crown ethers as *m-crown-n* where *m* stands for the total number of the atoms present in the ring and *n* denotes the number of O-atoms. Thus the **cyclic hexamer** of **ethyleneglycol** is called **18-crown-6** which may be further abbreviated as **18C6**. The benzo analogue of 18C6 is dibenzo-18-crown-6, where the C_6H_4 moiety is fused to the crown-18 ring at the 2,3 and 11,12 positions. Structures of some representative crown ethers are shown in Fig. 4.6.1.1.

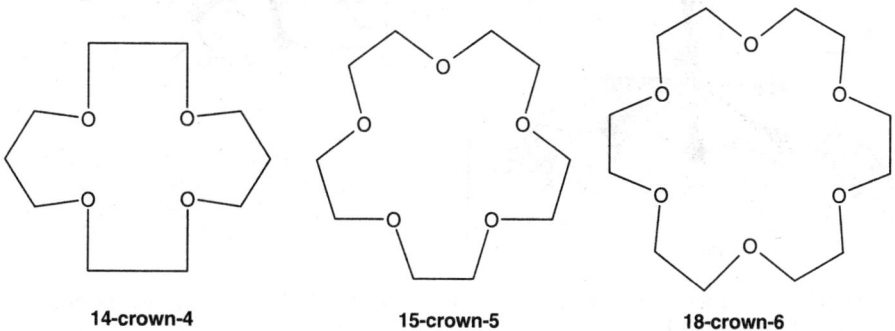

14-crown-4   **15-crown-5**   **18-crown-6**

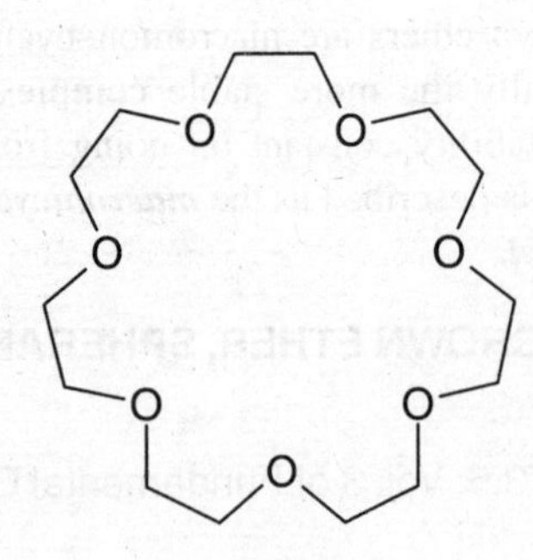

21-crown-7 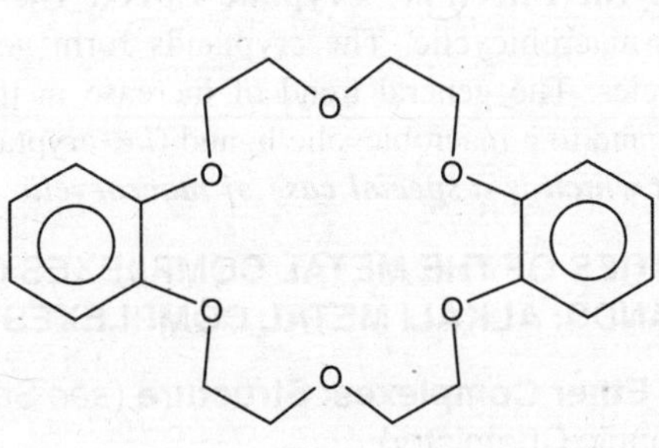   **dibenzo-18-crown-6**

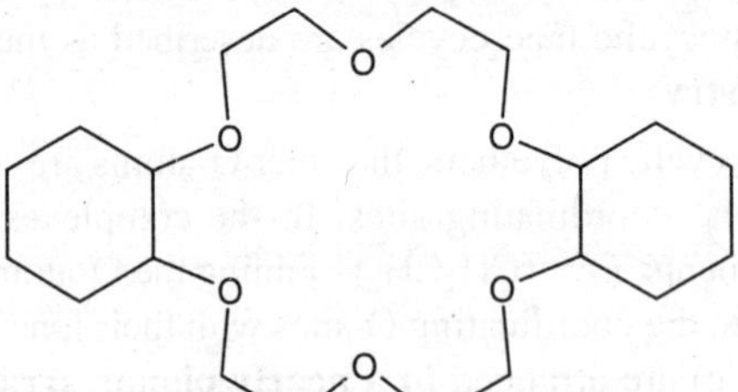

dicyclohexyl-18-crown-6 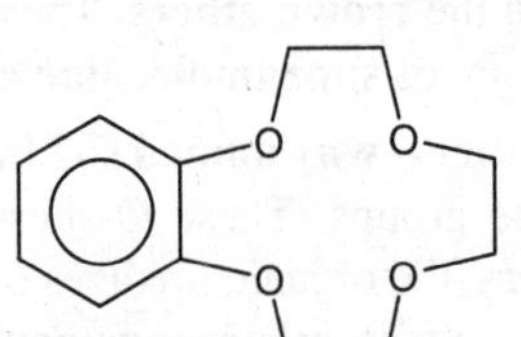   **benzo-12-crown-4**

Fig. 4.6.1.1 Structures of some representative crown ethers.

● **Nature of the donor sites:** In the crown ethers, the coordinating sites are the **hard O-donor sites**. This is why, they generally prefer selectively the alkali and alkaline earth metal ions which are hard. They also form complexes with the other metal ions like Ln^{3+} (lanthanides), Ag^+, Au^+, Tl^+, Cd^{2+}, Hg^{2+}, etc. and some limited number of transition metal ions. The structure of [Rb(dibenzone-18-crown-6)(NCS)] is shown in Fig. 4.6.1.2.

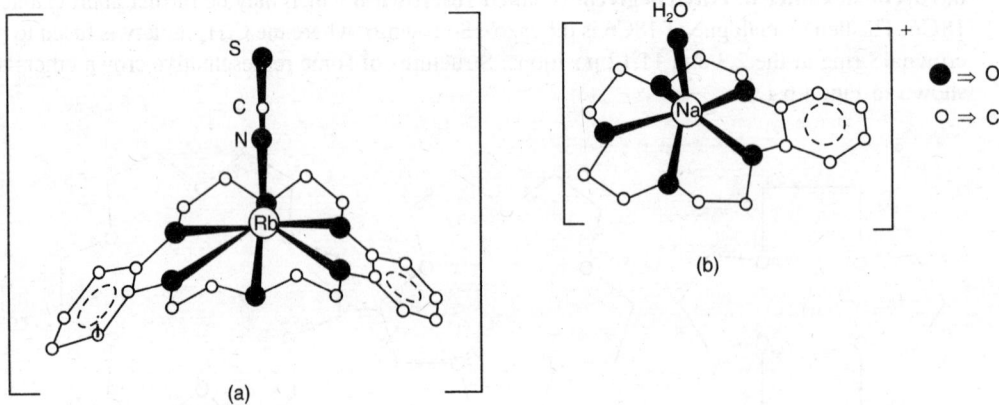

Fig. 4.6.1.2 (a) Structure of [Rb(dibenzone-18-crown-6)(NCS)] (b) Pentagonal pyramidal structure of [Na(benzo-15-crown-5)(OH₂)]⁺.

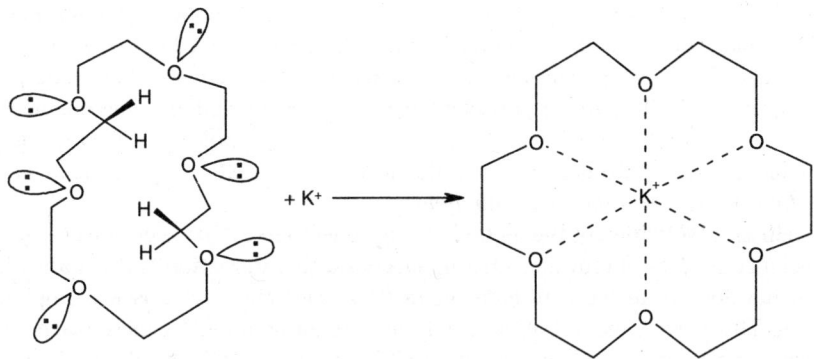

Fig. 4.6.1.3 Change of conformation of 18-crown-6 during complexation.

● **Nature of the metal-ligand interaction:** In the crown ether complexes, the bonding interaction to trap the cation is **basically electrostatic in nature.** This is why, the d-block metal centres are not preferred.

● **Crown ethers mimicking the ionophores:** The periphery of the crown ether complexes are hydrophobic while the interior portion of the cavity where the metal centre is placed is hydrophilic. Because this **hydrophobic or nonpolar nature of the periphery of such complexes,** they are soluble in organic solvents. In this respect, these macrocycles can mimic the properties of ionophores that carry the ions through the biological lipid membrane.

● **Conformations of the crown ethers:** The cyclic polyethers may also accommodate the nonpolar guests within the cavity in a hydrophilic medium. The conformation to be adopted by these cyclic polyethers depend on the nature of the solvent or medium (Fig. 4.6.1.4). In a hydrophilic medium, (Fig. 4.6.1.4), the O-sites are projected outwards (*i.e.* towards the hydrophilic surrounding) to

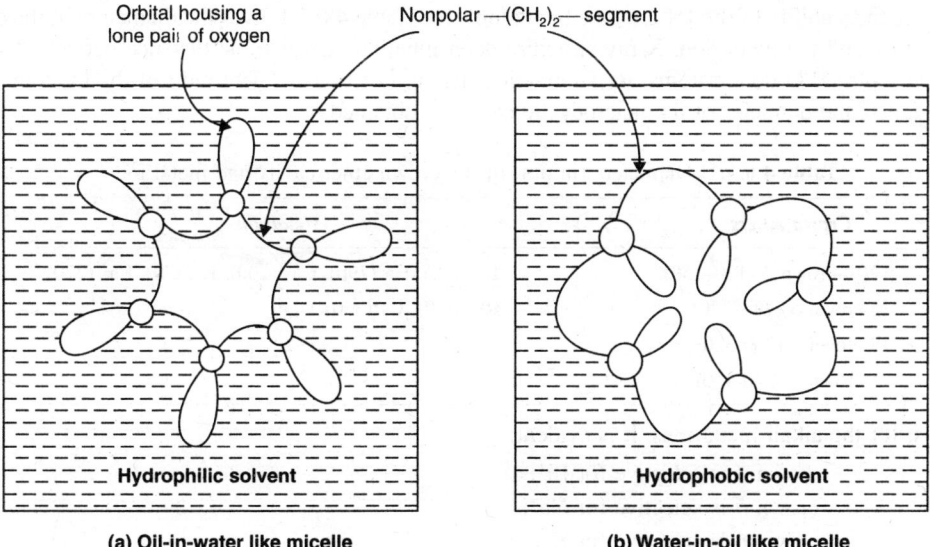

(a) Oil-in-water like micelle   (b) Water-in-oil like micelle

Fig. 4.6.1.4 (a) Qualitative representation of different conformation of 18-crown-6 in (a) hydrophilic and (b) hydrophobic solvent.

produce a **hydrophobic core.** Thus it produces a situation of **oil-in water like micelles** and within the hydrophobic core, the nonpolar molecules may be accommodated as the guest molecules. On the other hand, in a hydrophobic medium, a different conformation of the cyclic polyethers is adopted. Then the O-sites will be projected towards the cavity and the hydrophobic CH_2-groups will remain projected outwards. Thus, the core becomes hydrophilic while the periphery becomes hydrophobic mimicking the situation of **water-in-oil like micelles.** *In fact, the metal complexes of these ligands represent the same situation.*

● **Size matching, desolvation of the metal ions and selectivity:** It has been already mentioned that in the metal complexes of crown ethers, **the electrostatic interaction is the main contributing factor.** It develops from the hard-hard interaction. The role of cfse is less important in this interaction. This is why, the transition metal ions are not preferred by these macrocycles.

In terms of the electrostatic interaction, the complexing power of the alkali metal ions should run as follows:

$$\overset{\text{increasing ionic potential}}{\underset{\longleftarrow}{Li^+ \rangle Na^+ \rangle K^+ \rangle Rb^+ \rangle Cs^+}}$$

This trend is experimentally verified with many **hard donor noncyclic ligands.** But astonishingly, for the crown ethers, the above trend is not followed always. **Here matching between the cavity size and metal ion size is of prime importance.** If the size of the dehydrated metal ion is too small with respect to that of the cavity, then it cannot bring about the appropriate metal-oxygen interaction. On the other hand, if the size of the dehydrated metal ion is larger than that of the cavity, then it cannot fit into the cavity of the macrocycle. *However, because of the **flexibility of the crown ethers,** the ligand can fold to trap the metal ion when there is a mismatch in size factor. But it will reduce the stability of the complex to some extent (cf.* Table 4.6.1.2).

In fact, the **size matching factor** can override the other factor like **ionic potential which is important for the noncyclic ligands.** Thus the *size factor (i.e.* matching in size) is the most important factor to determine the stability of the 1:1 crown ether complexes. Table 4.6.1.1 gives the diameters of the crown ether cavity and the metal ion. X-ray structure determination can give the distance between the two diagonally placed O-donor atoms. By subtracting the sum of two covalent radii of the O-atoms from this distance, the diameter of the macrocyclic cavity is obtained.

Table 4.6.1.1 Diameters (in pm) of the crown ether cavity and metal ions.

Crown ethers	*Metal ion*
12/14-crown-4 (120–150)	Li^+(140), Na^+(190), Ca^{2+}(200), K^+(270), Ba^{2+}(270),
15-crown-5 (180–220)	Rb^+(300), Cs^+(330).
18-crown-6 (260-320)	
21-crown-7 (340-430)	

Thus, for the alkali metal ions, it is evident:

> crown-4 is selective for Li^+;
> crown-5 is selective for Na^+;
> crown-6 is selective for K^+;
> crown-7 is selective for Rb^+;
> crown-8 is selective for Cs^+

The **selectivity sequence** of different crown ethers for the alkali metal ions (in methanol) is given below:

$$\textbf{crown-5: } Na^+ \rangle K^+ \rangle Cs^+ \rangle Li^+$$
$$\textbf{crown-6: } K^+ \rangle Rb^+ \rangle Cs^+ \geq Na^+ \rangle Li^+$$

This selectivity principle has been utilized by the ***naturally occurring ionophores in transporting the ions through the cell membranes***. In fact, in the living bodies, there is a preferential transportation of K^+ from the extracellular fluid into the intracellular fluid. Na^+ remains concentrated in the extracellular fluid. Thus the concentration gradient of the Na^+ and K^+ ions across the cell membrane is maintained and it is important for the transmission of nerve impulse.

(**Note:** The selectivity sequence of the crown ethers in gas phase is different from the sequence in solution phase (*cf.* selectivity sequence for **crown-5 in gas phase:** $Li^+ \rangle\rangle Na^+ \rangle K^+ \rangle Cs^+$). The above sequences are maintained in solution phase. *Thus, the solvent effect is quite important.*)

Table 4.6.1.2 gives the stability constants (log K) of some metal-crown ether complexes.

The important observations are:

(i) *all the crown ethers prefer K^+ to some extent;*

(ii) *the 18-crown-6 and its derivatives bind almost all the alkali metal cations more strongly than the other crown ethers, but among the alkali metal ions, K^+ binds most strongly with the 18-crown-6.*

Probably the *flexible nature* of the crown ethers allows them to accomodate a wide variety of the metal ions, irrespective of their ionic radii but the best fit arises for K^+ with the crown-6. Moreover, the **ease of desolvation** ($K^+ \rangle Na^+ \rangle Li^+$) favours K^+ (which requires less energy for desolvation) to bind with the crown ether.

Table 4.6.1.2 Stability constants (log K, 25° C, in methanol solvent) of some representative metal-crown ether complexes

Crown ethers	log K		
	Na⁺	K⁺	Cs⁺
Dicyclohexyl-14-crown-4	2.2	1.3	–
18-crown-6	4.3	6.1	4.6
Cyclohexyl-18-crown-6	4.1	5.9	4.3
Dicyclohexyl-18-crown-6	4.1	6.0	4.6
Dibenzo-18-crown-6	4.4	5.0	3.5
Dibenzo-21-crown-7	2.4	4.3	4.2
Dibenzo-24-crown-8	–	3.5	3.8

● **Macrocyclic effect:** The metal-crown ether complexes experience the **macrocyclic effect** as usual. These complexes are about 10^2–10^3 times more stable than the complexes of the analogous open-chain ligands, **glymes**, $CH_3(OCH_2CH_2)_nOCH_3$ (*cf.* Table 4.5.4.1). The thermodynamic aspects (*i.e.* enthalpic and entropic favour) and kinetic aspects (*i.e.* hindered metal-ligand dissociation process) of the macrocyclic effect have been discussed in detail in Sec. 4.5.4.

● **Preorganisation energy** (crown ether *vs.* corresponding open chain ligand): This factor is important to understand the macrocyclic effect and this aspect has been illustrated in Sec. 4.5.4.

● **Basicity of the O-donor sites:** Besides the matching between the cavity size and metal ion size, other factors to influence the crown ether-metal ion (1:1) stability constants are: *number of the O-donor sites and their basicities.* Larger number of the metal—O interactions will favour the stability constant provided there is no mismatch in the size factor. If in the complex, the coordinated

O-sites are arranged symmetrically and the macrocycle becomes free from strain, then the stability constant increases. Increase of the basicity of the O-sites will favour the stability constant. *The O-atom attached to the aliphatic carbon is more basic than the one attached to an aromatic system.*

● **Desolvation of the metal ion and ligand during the complexation** (*cf.* Sec. 4.5.4):

$$M_{solvated} + L_{solvated} \rightleftharpoons [ML]_{solvated}$$

Thus during the complexation, desolvation of the metal ion is important. In fact, the less solvated metal centre gives an enthalpic favour to the stability constant (*cf.* **ease of desolvation:** $K^+ \rangle Na^+ \rangle Ca^{2+}$, *i.e.* less energy is required to desolvate K^+ in order to bind it). In fact, to determine the enthalpic favour, *basicity of the crown ether-O vs. H_2O* is of an important consideration.

B. Synthesis of crown ethers and thiacrowns:

(i) 18-crown-6 is prepared by reacting triethylene glycol with its corresponding dichloride in presence of aqueous KOH. Actually, initially the K^+-complex of 18-crown-6 is obtained and then from this complex, the free ligand may be isolated. It represents the **kinetic template effect.** Through the complexation with the starting reactants, K^+ brings the reactive moieties in a proper orientation required for the cyclisation. In fact, *if an organic base like NEt_3 is used instead of KOH, the crown ether is not produced.*

(triethylene glycol)   18-crown-6

(ii) The reaction between free catechol and dichloroether in presence of NaOH in *n*-butanol solvent gives the Na^+ complex of dibenzo-18-crown-6 from which the free ligand may be isolated. The reaction enjoys the **kinetic template effect** through the complexation of the partially formed ligand with Na^+.

(Catechol)   (Dichloroether)   Dibenzo-18-crown-6

Note: The above route of synthesis of 18-crown-6 was an accidental discovery, *i.e.* a case of scientific **serendipity** (in 1967 by Charles Pedersen) which ultimately brought the **Nobel Prize (1987) in Chemistry**. The 1987 Nobel Prize in Chemistry was jointly won by D.J. Cram, C.J. Pedersen and J.M. Lehn for their development of chemistry using the macrocyclic ligands like crown ethers, cryptands, etc. Pedersen attempted to prepare a **linear diol** (an open chain polydentate ligand) by using a catechol derivative in which one phenolic OH group was protected by a tetrahydropyran ring to prevent its participation in the reaction.

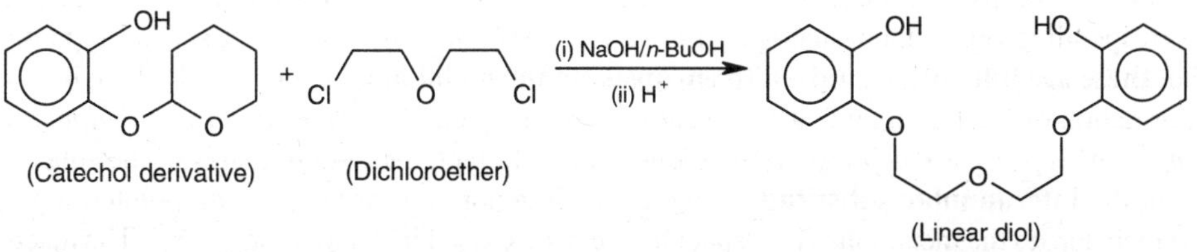

(Catechol derivative)   (Dichloroether)   (i) NaOH/n-BuOH   (ii) H$^+$ (Linear diol)

Unknowingly, the starting material (*i.e.* catechol derivative) was slightly contaminated by free catechol which produced the **undesired product dibenzo-18-crown-6** at trace quantities (**ca. 0.4% yield**). It is a great tribute to Pedersen's skill that the *said trace quantity of the undesired product could not escape the attention of Pedersen's skill*. This accidental discovery of crown ether gave the birth of a new discipline of chemistry called **supramolecular chemistry** (honoured by the 1987 Nobel Prize). Here, it is worth mentioning that synthesis of the said 18-crown-6 occurred due to the **kinetic template effect** of Na$^+$ which led to the cyclisation through the coordination with the hard donor O-atoms of the reaction intermediate, a linear polydentate ligand. In fact, if other bases like NEt$_3$ instead of NaOH or KOH are used, then the cyclisation does not occur. *It supports the role of Na$^+$ or K$^+$ ion.*

(iii) In *thiacrowns*, the sulfur atoms takes the positions of oxygens. In thiacrowns, the coordinating sites are the *S*-sites and these are the soft donor sites. 14-thiacrown-4 can be obtained in the following way.

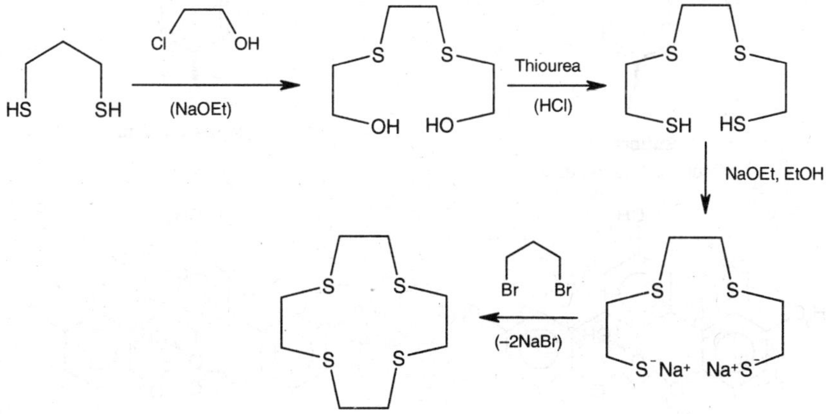

14-thiacrown-4

Metallacrown and Inverse Crown Ethers (*i.e.* Anticrown Ethers)

The metallacrowns incorporate the metals into the crown structure. The polymetallic cationic octagonal rings (called **inverse crown ethers**) are the novel **anion receptors**. These are discussed in Sec. 13.5.19, Vol. 3 of Fundamental Concepts of Inorganic Chemistry.

4.6.2 Spherands: The Preorganized Macrocycles

The crown ethers need the conformational changes during the complexation because they are not fully preorganized to encapsulate the metal ion. But definitely, the crown ethers are more preorganized than their corresponding open-chain analogue, *i.e.* glymes. *Spherand* is basically a special type of crown ether but **these are fully preorganized to encapsulate the metal ion**, *i.e.* practically there is no change of conformation of the free ligand at the event of complexation. *This gives a more enthalpic favour in the complexation process compared to the crown ethers.* In fact, the O-donor sites of the spherands are present in the **interannular substituents** and these O-donor sites are remaining pointed towards the cavity which houses the metal ion. This aspect has been discussed in detail in Sec. 4.5.4. The macrocyclic oligomers of **methoxytoluene** give the common examples of spherands. Because of this enthalpic favour, the spherands form very stable complexes when the metal ions properly fit in the cavity size of the spherands. The spherand (shown in Fig. 4.6.2.1) form the complexes with Li$^+$ and Na$^+$ with the stability constants 10^{16} and 10^{14} respectively. Because of the size factor, the corresponding K$^+$ complex is not stable ($\log K \approx 2.0$). Thus the selectivity of this spherand for Na$^+$ is about 10^{12} times as that for K$^+$.

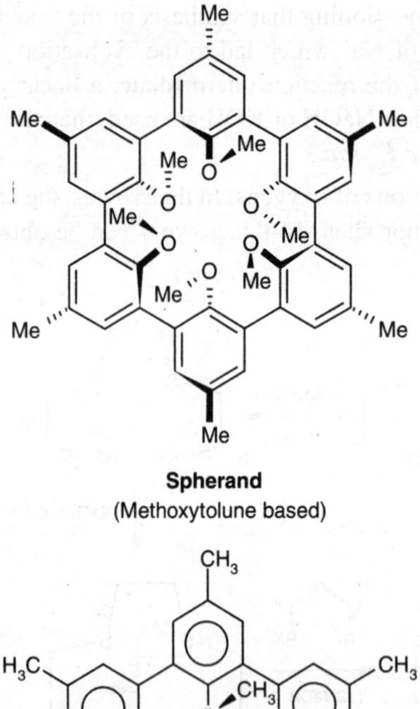

Spherand
(Methoxytolune based)

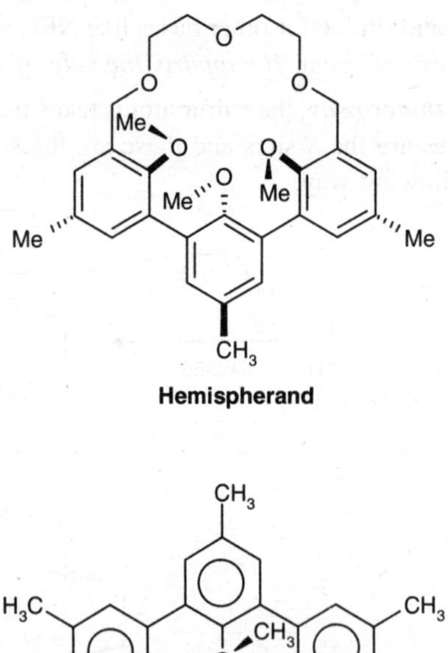

Hemispherand

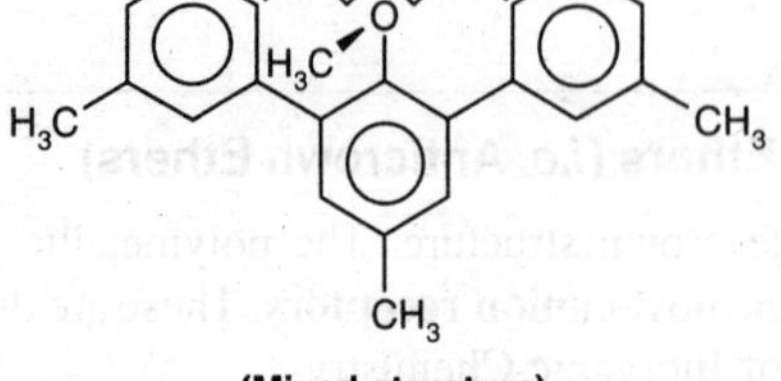

(Mixed structure)

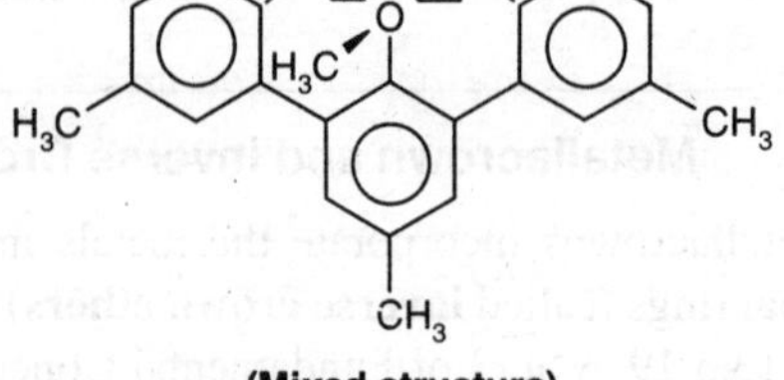

(Mixed structure)

Fig. 4.6.2.1 Structures of some representative methoxytolune based spherands and hemispherands.

Obviously, the *hemispherands* (having a mixed structure of crown ethers and spherands) are less preorganised than the spherands and the hemispherands form the less stable complexes than the spherands.

Naming of the ligands and their metal complexes

The ending -*and* denotes the free ligand while the ending -*ate* denotes the metal complex. It is illustrate below.

Podand (open chain ligand) → **Podate** (metal complex of a podand)

Corand (monocyclic ligand) → **Corate** (metal complex of a corand)

Cryptand → **Cryptate** (metal complex of a cryptand)

The suffix -*ate* is generally used for the anions. Thus the above procedure for describing the **guest-host complexes** is misleading and it has been suggested that -*ate* should be replaced by -*plex*, *e.g.* corand → *coraplex*, spherand → *spheraplex*.

Torands

Torands represent a series of two dimensional **extremely rigid** (*cf.* spherands) macrocyclic ligands constituted by the fused **aromatic pyridyl rings.** The structural rigidity arises from the extensive electron delocalisation over the conjugated double bonds. The **softer pyridyl-N donor atoms** bind strongly different softer transition metals. Generally, the pyridyl-N donor sites do not strongly bind the hard alkali metal ions but the torands can bind strongly the alkali metal ions, if the size factor matches. The presence of *n*-butyl substituents makes torands soluble in organic solvents. A representative torand is shown below.

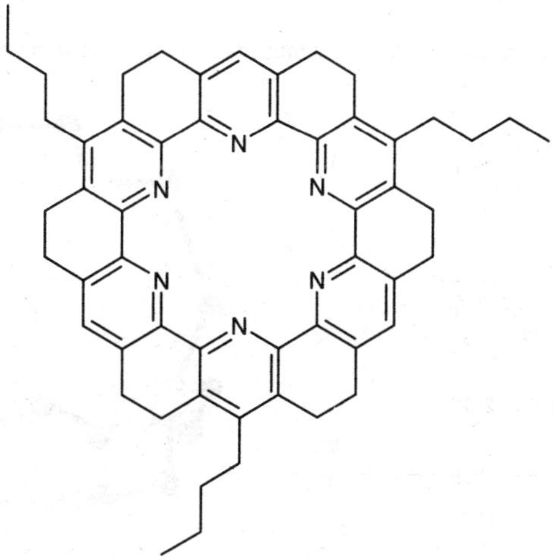

(A representative torand)

4.6.3 Cryptands – Multicyclic Macrocycles: Football Ligands (*see* Sec. 13.5, Vol. 3 of Fundamental Concepts of Inorganic Chemistry)

- **Structural features:** The crown ethers are monocyclic while the cryptands are **bicyclic or tricyclic.** The multicyclic cryptands are basically the *amino ethers* where the N-sites act as the **bridgehead atoms.** The cryptands can provide the cage structure to accommodate the metal ion and other suitable guest species. In the cage structure, they can provide the *bridgehead N-atoms* and ether O-atoms as the donor sites. Thus these multicyclic ligands provide the **heterocyclic rings (N,O-generally)** to encapsulate the metal ion in a **cage-like structure**. This is why, the cryptate complexes are described as the **clathrochelates**. The heterocyclic cryptand rings may contain the N and S donor sites in place of the O-donor sites.

The **degree of preorganisation** runs as:

spherands 〉 cryptaspherands (mixed structure of cryptands and spherands 〉 cryptands 〉 hemispherands (mixed structure of spherands and crown ethers) 〉 crown ethers.

It explains the order of **macrocyclic effect.**

● **Naming of the cryptands:** The cryptands are named starting with the chain offering the highest number of O-donors. These are illustrated in Figs. 4.6.3.1, 2.

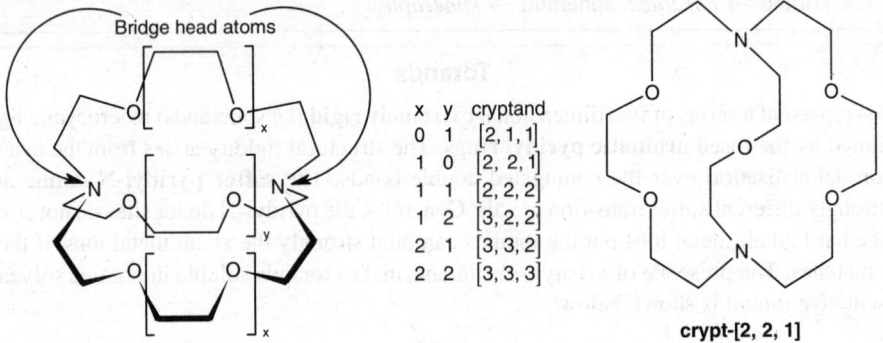

x	y	cryptand
0	1	[2, 1, 1]
1	0	[2, 2, 1]
1	1	[2, 2, 2]
1	2	[3, 2, 2]
2	1	[3, 3, 2]
2	2	[3, 3, 3]

crypt-[2, 2, 1]

Fig. 4.6.3.1 Illustration of naming of the cryptands and examples of some representative cryptands.

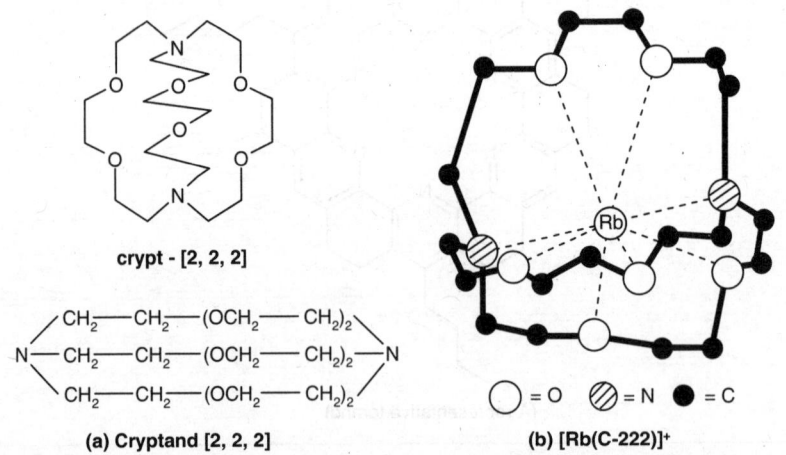

crypt - [2, 2, 2]

(a) Cryptand [2, 2, 2]   **(b) [Rb(C-222)]⁺**

$\bigcirc$ = O $\oslash$ = N   ● = C

Fig. 4.6.3.2 Structural representation of crypt–[2, 2, 2] and its Rb⁺-complex.

● **Representation of cryptands:** For the sake of simplicity, representation of the cryptands has been simplified, *e.g.* crypt-[2,2,1] is denoted by C-221; thus C-222 stands for crypt-[2,2,2], etc.

● **Cryptates – why named so?** The metal complexes of cryptands are called the **cryptates**. The name *cryptate* arises from the Greek word meaning *hidden*. Just like the crown ether complexes, the exterior part of the complex is hydrophobic while the interior part is hydrophilic (*i.e.* mimicking the situation *water-in-oil*). Thus like the naturally occurring ionophores, the cryptands can also hide the metal ions in their cavities and pass through the nonpolar membranes.

- **C-222 described as a football ligand:** The bicyclic cryptands like $N(CH_2CH_2OCH_2CH_2OCH_2CH_2)_3N$, *i.e.* crypt–[2,2,2] have been described as the **football ligands** because in such ligands, the polyether linkages between the bridgehead N-atoms look like the seams of a football. These football ligands (*i.e.* **bicyclic cryptands**) show a high degree of selectivity for the alkali metal ions.

- **Synthesis of cryptands:** The cryptands can be synthesized and it is illustrated (*cf.* Fig. 4.6.3.3) for the synthesis of crypt–[2,2,2], etc. **Template synthesis** of the cryptands in presence of the suitable alkali metal ions like K^+ (available from the base K_2CO_3 used) is now being widely practised.

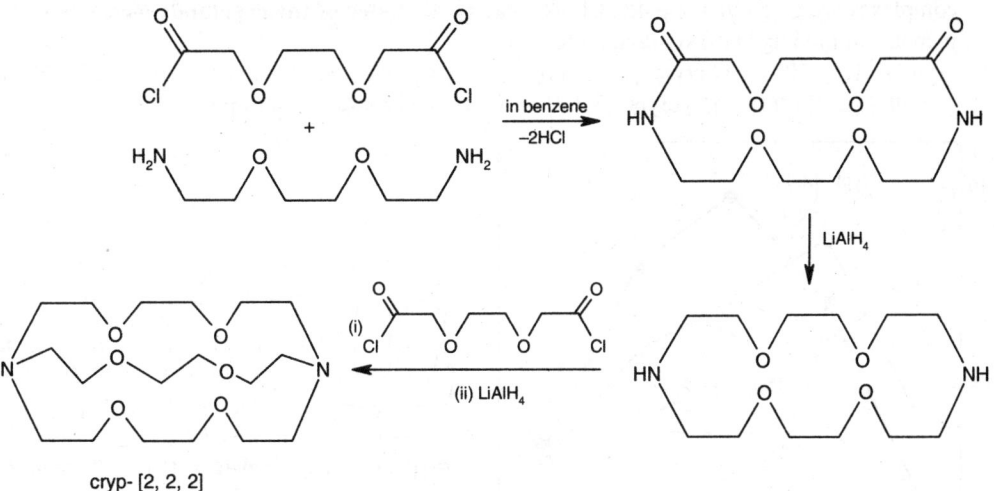

Fig. 4.6.3.3 Scheme for the synthesis of crypt-[2, 2, 2].

Encapsulation of Cl⁻: Anion Coordination: Spherical Recognition by the Soccerball Cryptand (*cf.* Fig. 4.6.3.5 and Chapter 13, Vol. 3 of Fundamental Concepts of Inorganic Chemistry)

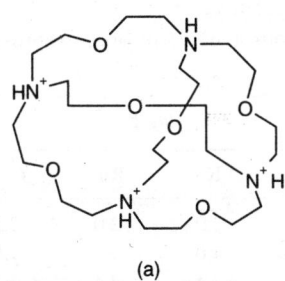

(a)
Protonated macropolycyclic
ligand: a tricyclic cryptand
called the **soccerball cryptand**
(LH_4^{4+}) - **Anion Receptor**

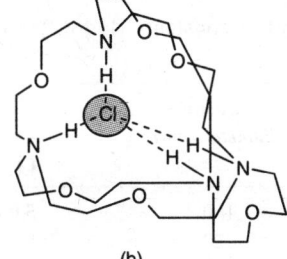

(b)
Encapsulation of spherical Cl⁻ by LH_4^{4+}
through H-bonding

- **Cryptate effect – a special type of macrocyclic effect:** In the cryptate complexes, the metal ions are placed at the centres of the ligands. The cryptate complexes are more stable than the macromonocyclic crown ether complexes and the cryptate complexes are more stable (by a factor about $10^5 - 10^6$) compared to the complexes formed by the corresponding open-chain polydentate ligands. The enhanced stability of the cryptate complexes is called the ***cryptate effect*** or ***macro-multicyclic effect*** (commonly called the **macrobicyclic effect** considering the bicyclic cryptands). In fact, the general trend of increase in stability constant on going from a **macromonocyclic ligand** to a **macropolycyclic ligand** (*i.e.* cryptand) is described as the ***cryptate effect***. *This enhanced stability, i.e. cryptate effect arises from the complete encapuslation of the metal ion in a cage-like structure. The structure* $[Rb(C-222)]^+$ *is shown in Fig. 4.6.3.2.*

- **Cavity size of the cryptands:** The cavity diameter and stability constants of some cryptate complexes are given in Table 4.6.3.1. The **cavity diameter of the cryptands** increases with the increase in the number of O-donor sites.

  crypt–[1,1,1] (100 pm) ⟨ crypt–[2,1,1] (160 pm) ⟨ crypt–[2,2,1] (220 pm) ⟨ crypt–[2,2,2] (280 pm) ⟨ *crypt*–[3,2,2] (360 pm) ⟨ crypt–[3,3,2] (420 pm) ⟨ crypt–3,3,3 (480 pm).

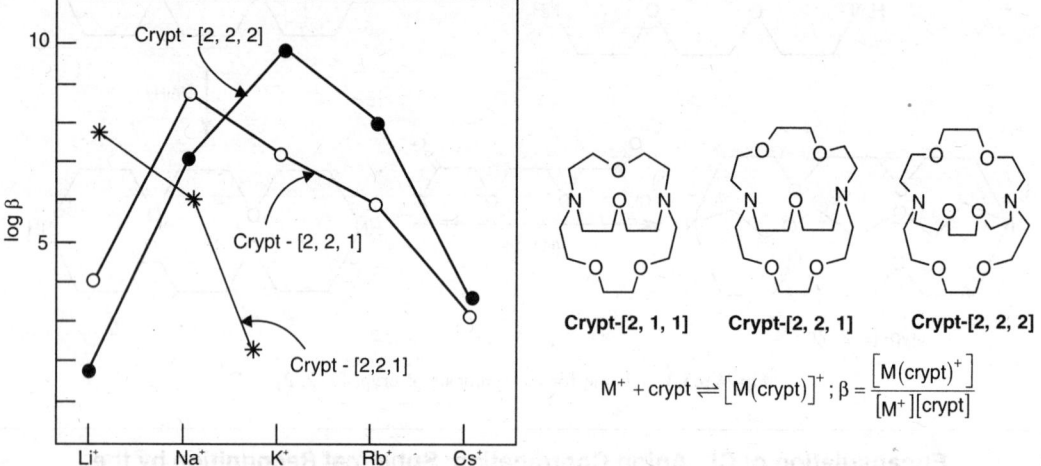

Fig. 4.6.3.4 Variation of the stabilities of alkali metal ion complexes with crypt–[2,1,1], crypt–[2,2,1] and crypt–[2,2,2] in methanol-aqueþus (95:5 V/V) media at 25° C.

Table 4.6.3.1 Stability constants (logβ, 25° C, in aqueous media) of some representative cryptate complexes (1:1).

Cryptand (bicyclic)	Cavity diameter (pm)*	log β					
		Li⁺	Na⁺	K⁺	Rb⁺	Ca²⁺	Ba²⁺
[2,1,1]	160	**5.0**	2.8	2.0	2.0	2.5	2.0
[2,2,1]	220	2.5	**5.4**	4.0	2.6	**7.0**	6.3
[2,2,2]	280	2.0	3.9	**5.4**	**4.4**	4.4	**9.5**
[3,2,2]	360	2.0	1.8	2.2	2.0	2.0	6.0
[3,3,2]	420	2.0	2.0	2.0	0.8	2.0	3.7

* Ionic diameters of the metal ions are given in Table 4.5.1.1.

It is evident (*cf.* Table 4.6.3.1) that for Li$^+$, Na$^+$ and K$^+$, the most suitable cryptands are crypt–[2,1,1], crypt–[2,2,1] and crypt–[2,2,2] respectively.

● **Steric conformations of the cryptands:** By considering the dispositions of the lone pairs on the bridgehead N-sites, there may be three *steric conformations* of the bicyclic cryptands.

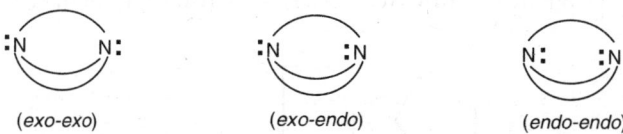

| (*exo-exo*) | (*exo-endo*) | (*endo-endo*) |

The *bicyclic cryptands* from the 1:1 metal complexes, where the bridgehead N-sites acts as the donor sites and the metal ion is positioned at the centre of the cage. All the N– and O– donor sites are more or less equidistant from the metal centre. *Thus in the metal complex, the ligand possesses the endo-endo conformation.*

● **Tricyclic cryptands and cryptaspherands:** The tricyclic cryptands are also known (Fig. 4.6.3.5). The cryptaspherands (Fig. 4.6.3.5) are having the mixed structure of the crown ether and cryptand.

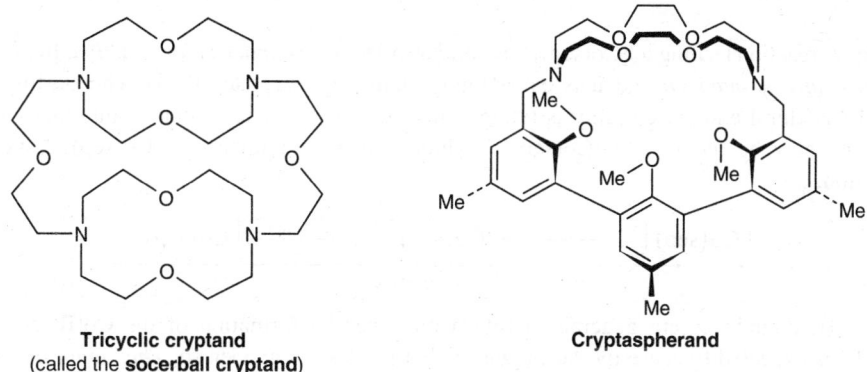

Tricyclic cryptand
(called the **socerball cryptand**)

Cryptaspherand

Fig. 4.6.3.5 Structures of a representative tricyclic cryptand and a cryptaspherand.

Notes: ● **Inorganic Cryptates:** In some hetropolytungstates, the alkal metal centre may get completely encaged giving rise to the **inorganic cryptates.**

● **Nonmacrocyclic octopus ligands:** The benzene derivatives bearing 2 to 6 pendant mercaptopolyether groups may form the stable complexes with the alkali metal ions. Such ligands, $C_6H_{6-x}R_x$ (R = –SCH$_2$CH$_2$OCH$_2$CH$_2$OMe, R = –SCH$_2$CH$_2$OCH$_2$CH$_2$OCH$_2$CH$_2$O–Bu, etc.) are *highly flexible* to form the complexes with the alkali metal ions. Such **nonmacrocyclic ligands** are described as the **octopus ligands** because their flexible character can completely encapsulate the metal centre.

4.6.4 Sepulchrate: Football Ligands

The sepulchrate macrobicyclic ligands structurally resemble the bicyclic cryptand ligands, *i.e.* football ligands. The sepulchrate (sep) ligand is:

$$N \begin{cases} CH_2-NH-CH_2-CH_2-NH-CH_2 \\ CH_2-NH-CH_2-CH_2-NH-CH_2 \\ CH_2-NH-CH_2-CH_2-NH-CH_2 \end{cases} N$$

Sepulchrate (sep)

Thus sepulchrate is closely related to crypt–[2,2,2] and it can encapsulate the complexed metal ion in a cage like structure. The ligand in a complex was synthesized in the following reaction (Sargeson et al, 1977).

$$\left[Co(en)_3\right]^{3+} + 6HCHO + 2NH_3 \longrightarrow \left[Co(sep)\right]^{3+} + 6H_2O$$

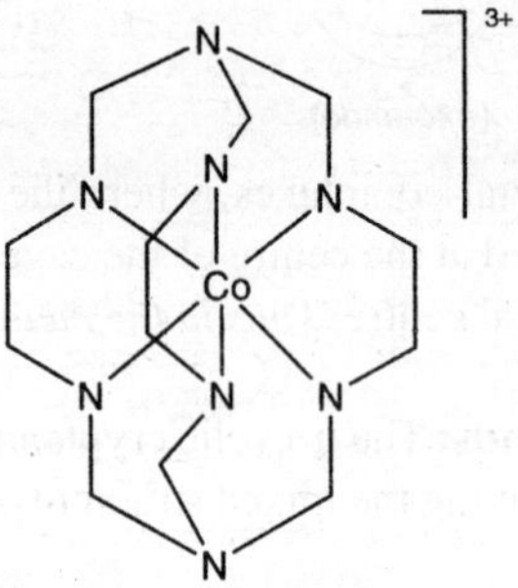

i.e. **[Co(sep)]³⁺.**

The above reaction leading to [Co(sep)]³⁺ is similar to *Mannich condensation reaction.* In the complex, *the metal centre is encaged* and it is exceedingly stable (*cf.* cryptate effect). The starting complex [Co(en)₃]³⁺ is chiral and the chirality is retained in [Co(sep)₃]³⁺. A particular enantiomer (Δ or Λ) of [Co(sep)₃]³⁺ may be reduced to [Co(sep)]²⁺ which can be reoxidized to [Co(sep)]³⁺ without any racemization, *i.e.*

$$\underbrace{\left[Co\left(sep\right)\right]^{3+} \xrightarrow{\text{reduction}} \left[Co\left(sep\right)\right]^{2+} \xrightarrow{\text{oxidation}} \left[Co\left(sep\right)\right]^{3+}}_{\text{Retention of chirality}}$$

The Co(II) complexes are generally labile. Consequently, formation of the Co(II)-complex, *i.e.* [Co(sep)]²⁺ is expected to cause the racemization, but it does not happen so. *This again indicates both the kinetic and thermodynamic stability provided by the bicyclic cryptand like ligands.*

4.6.5 Application of the Macrocyclic Ligands Related to the Crown Ethers (*see* Sec. 13.5, Vol. 3 of Fundamental Concepts of Inorganic Chemistry)

(a) **Solubilization of the alkali metals:** The alkali metals dissolve in liquid NH_3 but not in ether. But in presence of a suitable cryptand or crown ether, the alkali metals dissolve in ether. It is due to stabilization of the alkali metal ions through complexation.

$$K + \text{Crown-6} \xrightarrow{Et_2O} \left[K\left(\text{crown-6}\right)\right]^+ + \left[e\left(\text{crown-6}\right)\right]^-$$

The K⁺ ion is stabilized in the complex and the released electron is sheltered and stabilized in the cavity of the macrocycles. Such complexes like [K(crown-6)]⁺e⁻ are called the **electrides.**

(b) **Formation of the alkalides and electrides:** Stabilization of the alkali metal ion (M⁺) through the complexation can give the **alkalides** (where counter anions are the M⁻ ions of alkali metals, *e.g.* Na⁻ **sodide,** more correctly **natride;** K⁻; Rb⁻; Cs⁻) and the **electrides** (where the solvated electron balances the charge of M⁺).

Examples:
$$\underbrace{\left[K\left(\text{crown-6}\right)\right]^+ e, \left[Cs\left(\text{crown-6}\right)_2\right]^+ e, \text{etc.}}_{\textbf{Electrides}}$$

$$\left[Na(crown\text{-}6)\right]^+ Na^-;\ \left[Na(C\text{-}222)\right]^+ Na^-;$$
$$\underbrace{\left[Na(macrocycle)\right]^+ Au^-;\ \left[K(crown\text{-}6)\right]^+ Na^-}_{\textbf{Alkalides and aurides}}$$

In these unusual compounds, *i.e. electrides* and *alkalides*, the large macrocyclic ligands around M^+ shield the cation effectively to prevent the transfer of an electron from M^- (*i.e.* alkalide) or the solvated electron (*i.e.* electride) to the complexed M^+ ion. *The stabilization of M^+ and shielding effect on M^+ by the macrocycle can explain the stability of these unusual compounds.*

In general, **the excess ligand will lead to the electrides while the excess metal will lead to the alkalides in the following reactions:**

$$Cs^+ + 18\text{-crown-6 (excess)} \longrightarrow \left[Cs(18\text{-crown-6})_2\right]^+ e^-$$
$$\text{(Sandwich type)}$$

$$\left.\begin{array}{l} 2Na + 18\text{-crown-6} \longrightarrow \left[Na(18\text{-crown-6})\right]^+ Na^- \\[2mm] 2Na + C\text{-222} \longrightarrow \left[Na(C\text{-222})^+\right]^+ Na^- \end{array}\right\};\ \text{Metal: Ligand} = 2:1$$

$$Na + K + 18\text{-crown-6} \longrightarrow \left[K(18\text{-crown-6})^+\right]Na^- \Big\};\ Na:K:\text{Ligand} = 1:1:1$$

In the last reaction, a **mixed metal compound is produced**. The metal ion which will be present in the cationic part is determined by two factors: *ionization energy* and *selectivity of the macrocycle.* In the above reaction, K is preferably ionized because of its lower ionization energy and the crown-6 shows the more selectivity for K^+ than for Na^+. This is why, **$[K(crown\text{-}6)]^+Na^-$** is preferably produced rather than $[Na(crown\text{-}6)]^+K^-$.

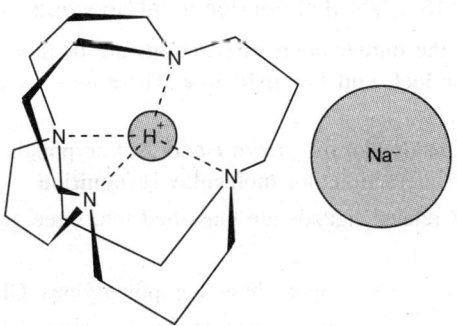

$H^+(3^6\text{-adamanzane})Na^-$, *i.e.* **Inverse sodium hydride**

Inverse sodium hydride consisting of Na^- and H^+ ion (encapsulated in 3^6-adamanzane) has been synthesised.

(c) **Solubilization of salts in the nonpolar solvents:** The macrocycles can solubilize the salts in the organic solvents through encapsulation of the metal ions in their cavities. The exterior part of the complex remains hydrophobic (*cf.* **water-in-oil micelles, ionophores**).

$$KF \xrightarrow[\text{(benzene)}]{\text{crown ether}} \text{solubilised, } i.e. \left[K(Crown)\right]^+ F^-\ ;$$

$$KMnO_4 \xrightarrow[\text{(benzene)}]{\text{crown ether}} \text{solubilised, } i.e. \left[K\left(Crown\right)\right]^+ MnO_4^-.$$

In such cases, F^- or MnO_4^- remain almost **unsolvated** and these **naked anions** are quite reactive. It may be noted that cryptand can solubilize $BaSO_4$ in water through the complexation of Ba^{2+}.

(d) **Phase transfer catalyst:** When KF is solubilized in the organic solvents in presence of 18-crown-6 or cryptand like C222, K^+ ion is stabilised as $[K(macrocycle]^+$ and **the unsolvated F^- ion acts as a powerful nucleophile.** It is illustrated in the following reaction.

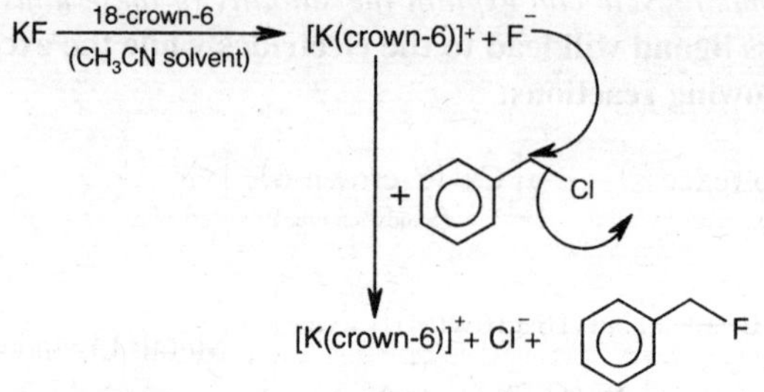

Similarly, $KMnO_4$ solubilized in benzene in presence of crown-6 or suitable cryptand, produces a purple solution called the **purple benzene** which can act as a **powerful oxidizing agent** because of the presence of unsolvated MnO_4^-.

(e) **Other applications:** They can be used for developing the ion selective electrodes, separation of the chemically similar metal ions (*e.g.* lanthanides, alkali metal ions, etc.) based on the size-factor (*i.e.* matching of size in the cavity), etc.

4.6.6 Crown Ether Related Macrocycles and Supramolecular System and Molecular Recognition (*see* Sec. 13.5, Vol. 3 of Fundamental Concepts of Inorganic Chemistry)

In the supramolecular systems, the **noncovalent interactions** are predominant where the **molecular recognition** occurs based on the **lock and key principle**. These noncovalent interactions lead to the **guest-host type complexes**.

The synthetic macrocyclic ligands like the crown ethers and cryptands are used as the *molecular receptors* in the supramolecular interactions for **molecular recognition.**

These macrocycles and their related ligands are classified into three groups from the *topological point of view*. These are:

podands	(open-chain, *e.g.* polyglymes, $CH_3(OCH_2CH_2)_xOCH_3$
corands	(monomacrocyclic, *e.g.* crown ethers)
cryptands	(multicyclic giving rise to the spherical shape)

Thus the monocyclic crown ethers are the corands.

The metal complexes are described as: *podates, corates, cryptates,* etc. (*i.e.* the ending *-and* of the free ligands is replaced by *-ate*). The ending *-ate* is generally used for the naming of the anions. This is why, it has been suggested that the suffix *-ate* should be replaced by *-plex, e.g.* coraplex, spheraplex.

All these aspects of **Supramolecular Systems** have been discussed in the present author's book, "Fundamental Concepts of Inorganic Chemistry, Vol. 3"-2nd Edition, CBS Publishers and Distributors, New Delhi.

4.6.7 Ionophores Resembling the Crown Ethers and Naturally Occurring Antibiotics: Alkali Metal Complexes of the Naturally Occurring Ionophores.

The ionophores are biologically very much important for the transport of different metal ions through the lipid bilayer membrane. The **cyclic ionophores** are: *valinomycin, enniatin A, enniatin B*, etc. The **open chain ionophores** are: *monensin, nigericin, dianemycin*, etc. These are discussed in detail in the author's book ***Bioinorganic Chemistry*** (Books and Allied, Kolkota).

The exterior portion of the metal complexes of the ionophores is nonpolar. This is why, such metal complexes can pass through the nonpolar lipid membrane. Thus the hydrophilic metal ions are transported through the lipid membrane. Based on the structural properties, the naturally occurring ionophores are classified as follows:

Naturally occurring ionopohores

Cyclic Ionophores (*i.e.* Class-I antibiotics)

Open-chain ionophores
(*i.e.* Class-II antibiotics)
(having a free –CO$_2$H group; *i.e.* RCO$_2$H;
capable of cyclising through H-bonding).
(*e.g.* monensin, nigericin, dianemycin).

Macrotetrolides
(*e.g.* actins)

Cyclodepsipeptides
(*i.e.* condenstation products of aminoacids and hyrdoxy acids)
(*e.g.* valinomycin, enniatin A, enniatin B, etc.)

The structural features of the naturally occurring ionophores are given in Fig. 4.6.7.1.

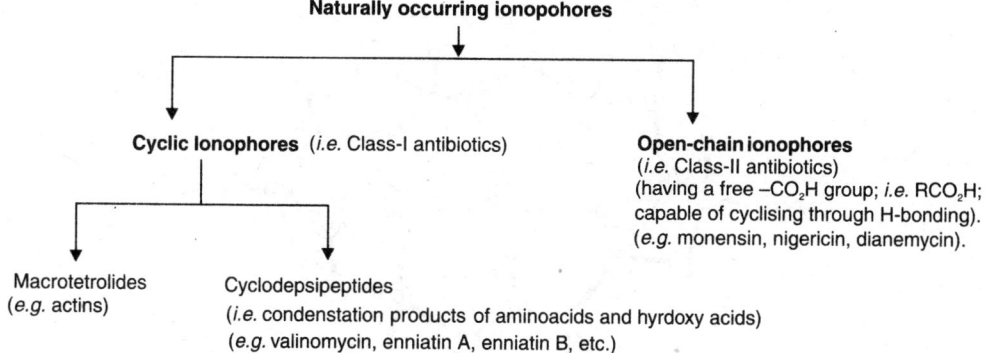

Valinomycin

Actins

$R_1 = R_2 = R_3 = R_4 = CH_3$, **Nonactin**,

$R_1 = R_2 = R_3 = CH_3$ and $R_4 = C_2H_5$, **Monactin**

$R_1 = R_3 = CH_3$ and $R_2 = R_4 = C_2H_5$, **Dinactin**

$R_1 = CH_3$ and $R_2 = R_3 = R_4 = C_2H_5$, **Trinactin**

$R_1 = R_2 = R_3 = R_4 = C_2H_5$, **Tetranactin**

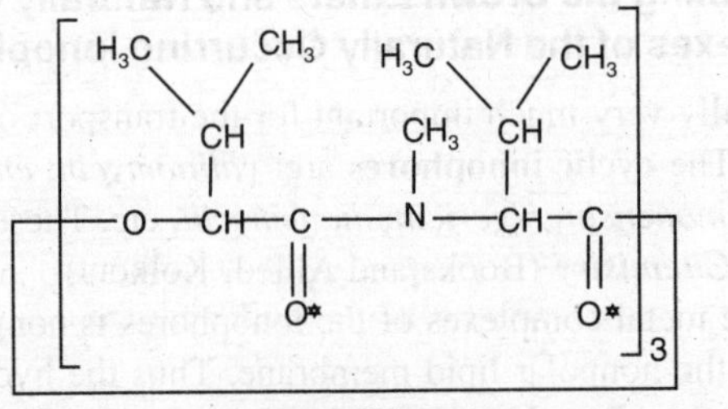

Enniatin B

i.e. [D-α-hydroxyisovaleric acid-N-methyl-L-valine]₃

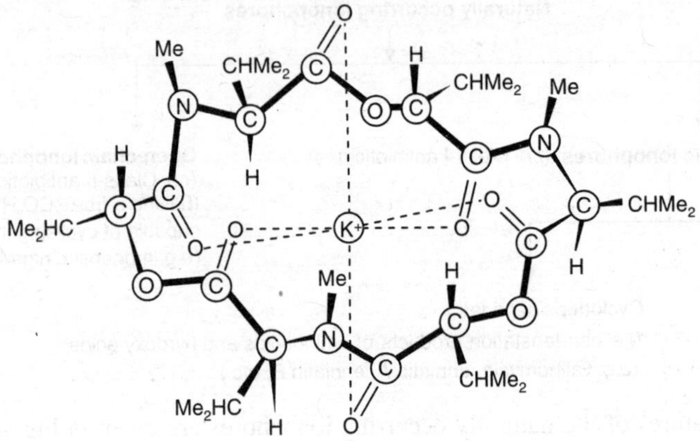

K⁺-complex of enniatin-B

Fig. 4.6.7.1 Structures of some representative naturally occurring cyclic ionophores, *i.e.* antibiotics of class-I. (*denotes the donor sites)

The alkali metal complexes of the above mentioned naturally occurring ionophores have been extensively studied. The structures of some representative alkali metal complexes of these ligands are shown in Figs. 4.6.7.1,3,4. Biochemical effects of these complexes on mitochondria have also been studied. *The antibiotic activity* of *valinomycin develops from its ability to induce the K⁺ -transport-process across the lipophilic membrane.* In fact, *valinomycin selectively binds the K⁺ -ion.* For the stability of such alkali metal complexes, the following factors are important as usual.

• **size matching** with the macrocyclic cavity;

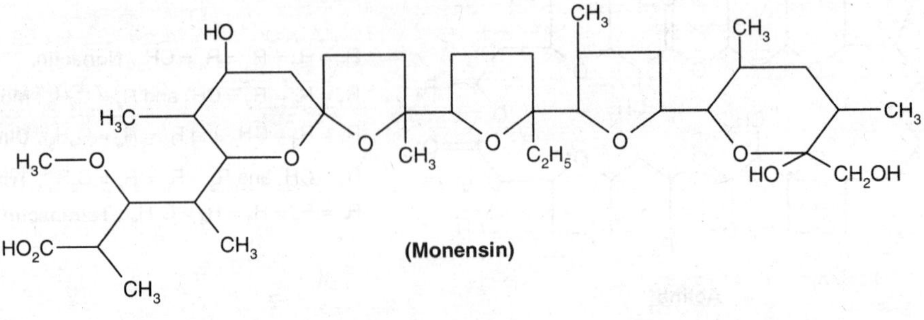

(Monensin)

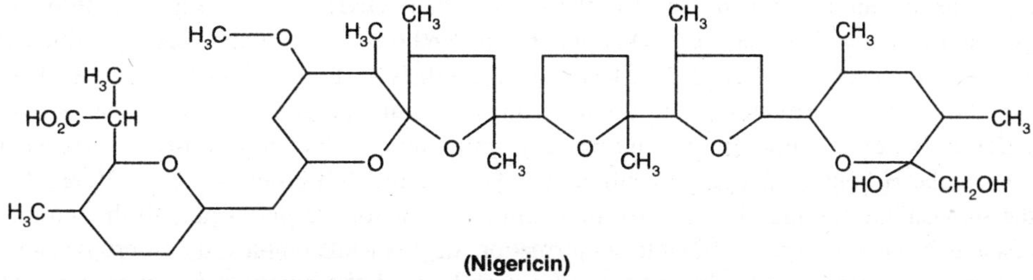

Fig. 4.6.7.2 Structures of some representative open-chain naturally occurring carboxylic acid ionophores, *i.e.* antibiotics of class-II.

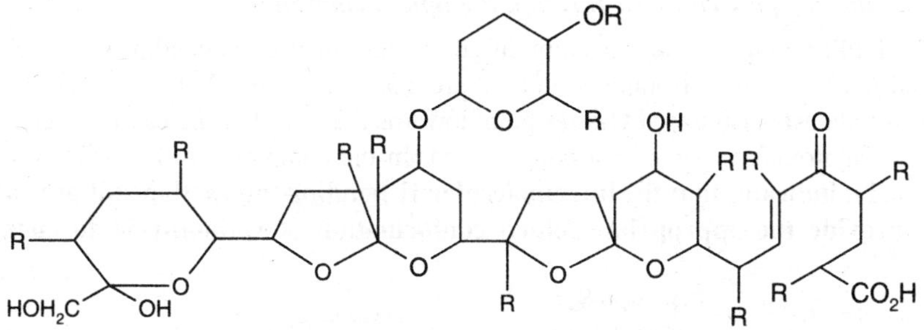

- **desolvation energy of the metal ions** (*i.e.* highly solvated metal ions form the weak complexes); *cf.* ease of dehydration: $K^+ \rangle Na^+ \rangle Li^+$.
- **desolvation energy of the ligand** and **preorganization energy** needed to change the conformation of the ligand for complexation.

Selectivity sequence (*i.e.* stability order) of the alkali metal complexes with the naturally occurring ionophores is as follows:

Valinomycin:	$Rb^+ \rangle K^+ \rangle Cs^+ \rangle Na^+$
Actins:	$K^+ \rangle Rb^+ \rangle Cs^+ \rangle Na^+ \rangle Li^+$
Nigericin:	$K^+ \rangle Rb^+ \rangle Na^+ \rangle Li^+$
Monensin:	$Na^+ \rangle K^+ \rangle Rb^+$
Dianemycin:	$Na^+, K^+, Rb^+ \rangle Li^+$

The members of the **actins** are: nonaction, monactin, dinactin, trinactin and tetranactin. These are differentiated by the gradual stepwise replacement of the Me-groups by the Et-groups. Thus, the hydrophobic character increases from nonactin to tetranactin. For a particular alkali metal ion, the stability constant in a polar solvent like MeOH runs as:

tetranactin $\rangle$ trinactin $\rangle$ dinactin $\rangle$ monactin $\rangle$ nonactin.

With the increase of hydrophobic character in the ligand, the ligand gets less solvated in a polar solvent, *i.e.* it needs **less desolvation energy** during the complexation process. Thus, tetranactin (having the most hydrophobic character) needs the minimum desolvation energy and it forms the most stable complex.

Valinomycin consists of three repeated sequence, [L-lactic acid (P), L-valine (Q), D-α-hydroxyisovaleric acid (R) and D-valine (S)]. Thus, the macrocycle consists of alternate residues of the hydroxy acids and amino acids to generate the six peptide (CONH) groups. It gives a **36-membered flexible macrocycle** with 12 carbonyl O-atoms (6 ester and 6 peptide groups). Each peptide carbonyl O is H-bonded to the closest $\rangle$NH group. The amide to amide N —H$\cdots$OC H-bonding network depends on the polarity of the surroundings. In the nonpolar media, the predominant form bears the 6 intra-molecular H-bonding while in the moderately polar media, the predominant form bears the 3 intramolecular H-bonding keeping free three $\rangle$C = O and three $\rangle$NH groups for H-bonding interaction with the surrounding (*cf.* Fig. 4.6.7.3c, d). In a strongly polar solvent like water, all the intramolcular H-bonds may be broken to give a flexible conformation. In the metal complex, the isopropyl and methyl groups remain projected outwards and *thus the periphery of the complex becomes nonpolar (i.e. lipophilic) and the complex can pass through the lipid membrane.*

In fact, this **folding macrocycle** can adopt different conformations depending on the extent of intra-molecular and intermolecular H-bonding and the presence and absence of metal coordination. In the K$^+$-complex, the **six ester carbonyl O-sites** point towards the centre of the cavity where K$^+$ is accom-modated for an approximate octahedral coordination. In the complex, no free $\rangle$NH ($v = 3395$ cm^{-1}) group exists and **it indicates that the intramolecular H-bonding interaction still prevails in the K$^+$-complex to provide the appropriate folded conformation of valinomycin to encapsulate K$^+$.**

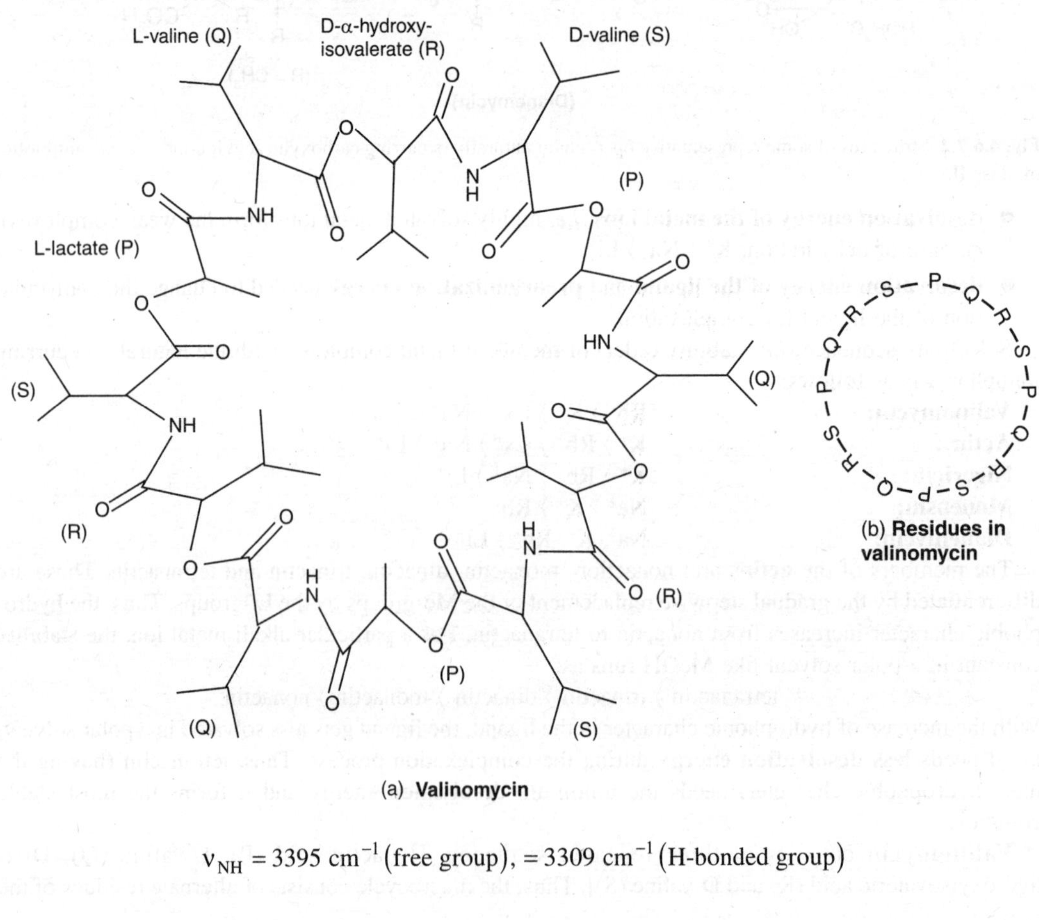

(a) **Valinomycin**

(b) **Residues in valinomycin**

$$v_{NH} = 3395 \text{ cm}^{-1} (\text{free group}), = 3309 \text{ cm}^{-1} (\text{H-bonded group})$$

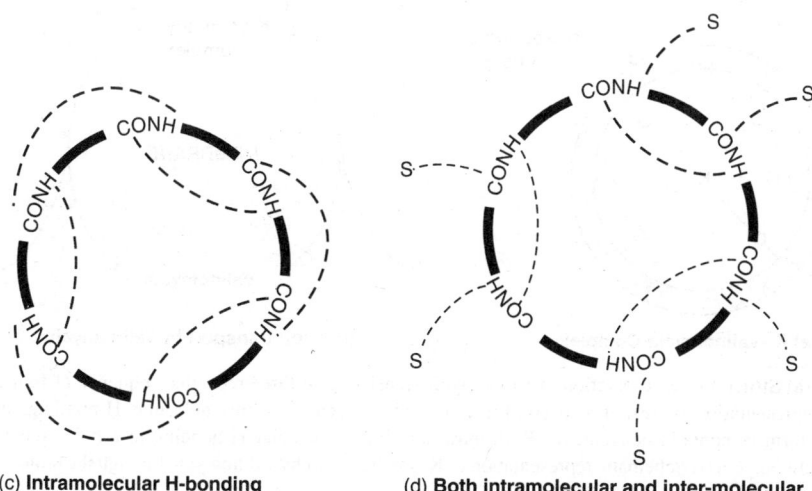

(c) **Intramolecular H-bonding**

(d) **Both intramolecular and inter-molecular H-bonding (S = polar solvent)**

Isopropyl group

Ester carbonyl-O

Methyl group

Peptide-O

K⁺

Intramolecular amide to amide H-bonding to preorganise the macrocycle to encapsulate K⁺

(e) **K⁺-valinomyein complex**

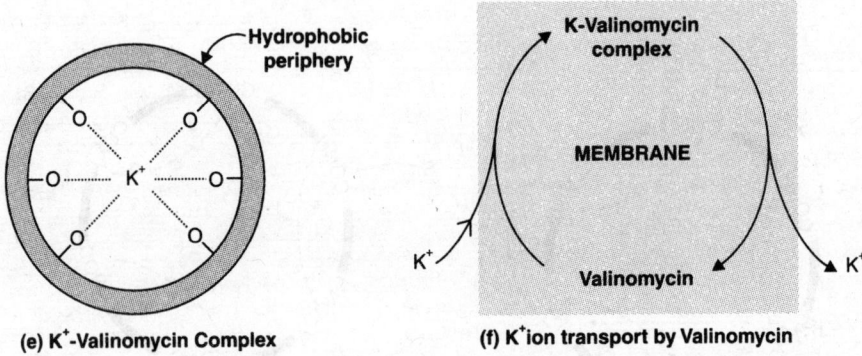

(e) K⁺-Valinomycin Complex   **(f) K⁺ion transport by Valinomycin**

Fig. 4.6.7.3 (a) Structural representation of valinomycin consisting of three repeating sequence of four residues (P, Q, R, S). (b) Representation of residues to constitute the valinomycin; (c) Intramolecular H-bonding in valinomycin; predominant form in nonpolar solvent; (d) Both intra- and inter-molecular H-bonding in a polar solvent (S); (e) K⁺-complex of valinomycin (f) Schematic representation of K⁺ ion transport by valinomycin through the biological membrane.

Thus in the complex, the peptide groups still retain the intramolecular H-bonding (indicated by the presence of v_{NH} = 3309 cm⁻¹). This H-bonding interaction generating the right information of the macrocycle is important for all cyclodesipeptides acting as the ionophores, *i.e.* ion carriers. It is evidenced by the fact that on *N-methylalation of valinomycin, its antibiotic activity (i.e. K⁺ -binding power) is lost.*

Valinomycin binds K⁺ more tightly than Na⁺ by a factor about 10³. This selectivity can be explained by considering the factors like size-matching and desolvation energy required during the complexations. These are discussed in detail in the author's book, **Bioinorganic Chemistry**. By using this selectivity towards the K⁺ ion, the K⁺-selective membrane electrode has been developed.

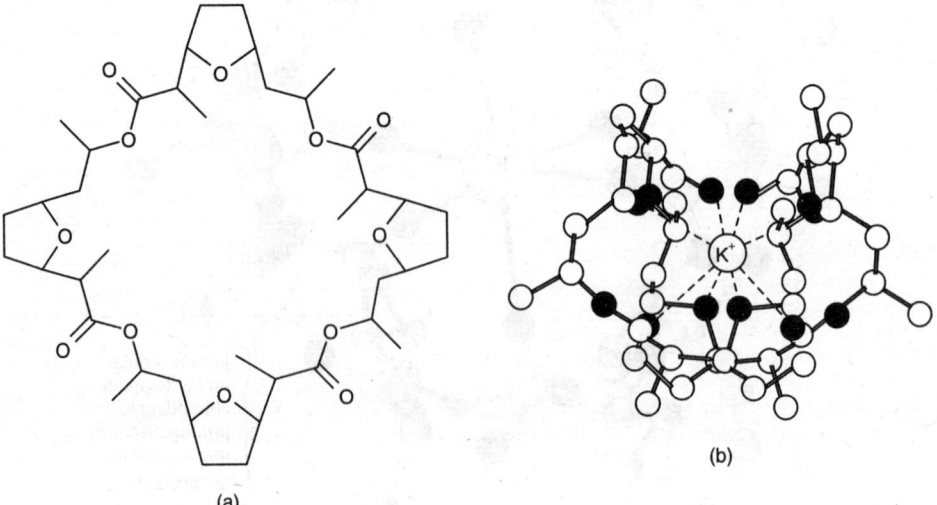

Fig. 4.6.7.4 (a) Structure of nonactin; (b) Structure of K⁺-complex of nonactn (● denotes oxygen).

4.7 TERNARY COMPLEXES: FACTORS AFFECTING THEIR STABILITIES
(*cf.* A.K. Das, *Transition Met Chem.*, **14**, 66-68, 1989; *Int. J. Chem. Kinet.*, **28**, 275-82, 1996; *Chemistry Education*, **4 (2)**, 44-51, 1987).

In the *mixed ligand complexes*, more than one type of ligands except the solvents are present. Thus in a mixed ligand complex, besides the coordinated solvent, at least there must be two different types of ligands. Such complexes containing two different types of ligands (besides the solvents) are called the **ternary complexes.** The ternary complex may be simply represented by *MAB* where the two different ligands *A* and *B* are coordinating the metal centre *M*. The complexes containing the same ligands are called the **binary complexes,** *e.g.* MA_2, MB_2, etc.

In the ternary complexes, even the same ligand with different chiralities may exist. The situation happens for the optically active amino acids like *L*-his and *D*-his. (*cf.* [M(L-his)(D-his)]). Sometimes, the same ligand binding through the different coordinating sites may lead to the ternary complexes. The ligating property of histidine depends on pH and in the mixed ligand complexes like [Cu(Hhis)(his)]$^+$, [Cu(his)$_2$], etc. the two histidine ligands bind in different ways. These aspects have been illustrated in Figs. 1.7.2.3, 4.7.2.1.

The mixed ligand complexes are quite relevant in the understanding of the **activity of metalloenzymes** (EM) where the proteins are providing the coordinating sites to produce the metalloenzyme (*cf.* **Importance of Ternary Complex Formation in Biological Systems,** A.K. Das, **Chemistry Education, 4(2),** 44-51, 1987). In the enzymatic activity, EM binds with the substrate (S) to produce a mixed ligand complex EMS that ultimately reacts to give the products.

$$EM + S \xrightleftharpoons{K} E - M - S \xrightarrow{k} \text{Product}$$

Thus the enzymatic activity largely depends on the success of the formation of the mixed ligand complex EMS. These are illustrated in the author's book, **Bioinorganic Chemistry.**

4.7.1 Thermodynamic Parameters Measuring the Stabilities of the Ternary Complexes

In the mixed ligand complex (*MAB*), the ligands may be described as the **primary ligands** and the **secondary** or **auxiliary ligands** depending on the sequence of their entry in the complex.

$$MA + B \rightleftharpoons MAB$$

Here, the ligand *A* which is already present is called the *primary ligand* while the ligand *B* that enters at the next step is called the *secondary ligand*. *MBA* is formed in the following reaction.

$$MB + A \rightleftharpoons MBA, \ (B = \text{primary ligand}, \ A = \text{secondary ligand}).$$

In the labile systems, the ternary complexes *MAB* and *MBA* are indistinguishable.

In the binary complexes like MA_2 or MB_2, generally $K_1 \rangle K_2$, *i.e.* *MA* shows a reduced affinity for binding another *A*. But during the formation of a ternary complex, *i.e.* ligation by *B* to *MA*, this step may show an increased affinity for the secondary ligand *B*. **Thus the ternary complexes are very often more stable than the corresponding binary complexes.** The stabilities of the ternary complexes may be expressed in terms of different thermodynamic parameters. A ternary complex *MAB* may be formed in different ways and the corresponding equilibria are:

$$M + A \rightleftharpoons MA, \qquad K_{MA}^{M} = \frac{[MA]}{[M][A]}$$

$$MA + A \rightleftharpoons MA_2, \qquad K_{MA_2}^{MA} = \frac{[MA_2]}{[MA][A]}$$

$$MA + B \rightleftharpoons MAB, \qquad K_{MAB}^{MA} = \frac{[MAB]}{[MA][B]}$$

$$M + A + B \rightleftharpoons MAB, \qquad \beta_{MAB}^{M} = \frac{[MAB]}{[M][A][B]}$$

$$M + B \rightleftharpoons MB, \qquad K_{MB}^{M} = \frac{[MB]}{[M][B]}$$

$$MB + B \rightleftharpoons MB_2, \qquad K_{MB_2}^{MB} = \frac{[MB_2]}{[MB][B]}$$

$$MB + A \rightleftharpoons MBA, \qquad K_{MBA}^{MB} = \frac{[MBA]}{[MB][A]}$$

$$2MAB \rightleftharpoons MA_2 + MB_2 \qquad K_D = \frac{[MA_2][MB_2]}{[MAB]^2}$$

Stabilities of the ternary complexes (*e.g. MAB*) are generally expressed in terms of **$\Delta \log K_M$ and $\log K_D$.** (where K_D denotes the disproportionation constant) which are related with the stability constants of the binary and ternary complexes.

● **$\Delta \log K_M$** is defined as follows (assuming MAB and MBA identical):

$$\Delta \log K_M = \log K_{MAB}^{MA} - \log K_{MB}^{M} = \log\left(\frac{[MAB]}{[MA][B]} \times \frac{[M][B]}{[MB]}\right) = \log\left(\frac{[MAB][M]}{[MA][MB]}\right)$$

$$= \log K_{MBA}^{MB} - \log K_{MA}^{M} = \log\left(\frac{[MBA]}{[MB][A]} \times \frac{[M][A]}{[MA]}\right) = \log\left(\frac{[MBA][M]}{[MB][MA]}\right)$$

$$= \log \beta_{MAB}^{M} - \log K_{MA}^{M} - \log K_{MB}^{M} = \log\left(\frac{[MAB]}{[M][A][B]} \times \frac{[M][A]}{[MA]} \times \frac{[M][B]}{[MB]}\right)$$

$$= \log\left(\frac{[MAB][M]}{[MA][MB]}\right)$$

Actually $\Delta \log K_M$ corresponds to $\log K_{eq}$ of the following process:

$$MA + MB \xrightleftharpoons[\quad]{K_{eq}} MAB + M, \quad K_{eq} = \frac{[MAB][M]}{[MA][MB]} = 10^{\Delta \log K_M}$$

$$K_{eq} = \frac{[MAB][M]}{[MA][MB]} = \frac{[MAB]}{[MA][B]} \times \frac{[M][B]}{[MB]} = K_{MAB}^{MA} \times \frac{1}{K_{MB}^{M}}$$

i.e. $$\qquad \log K_{eq} = \log K_{MAB}^{MA} - \log K_{MB}^{M} = \Delta \log K_M$$

In the same way, K_{eq} may be expressed in terms of K_{MBA}^{MB} or β_{MBA}^{MB}. *Thus $\Delta \log K_M$ gives a measure of the coordinating capacity of the secondary ligands (say B) towards the binary complex, MA relative to their coordinating tendencies towards M as described by the equilibrium, $M + B \rightleftharpoons MB$.*

In a **binary system** *(i.e. A = B), statistically* $\Delta \log K_M \left(= \log K_{MA_2}^{MA} - \log K_{MA}^{M} = \log K_2 - \log K_1 \right)$ values range from -0.5 to -0.8 for the monodentate ligands. For the symmetrical didentate ligands, statistically $\Delta \log K_M$ values may vary in the range -0.6 to -1.2 for different geometries. *Less negative value (compared to the statistical value) or more positive value of $\Delta \log K_M$ indicates the enhanced stability of the ternary complex, i.e.* the equilibrium, $MA + MB \rightleftharpoons MAB + M$ will be favoured to produce the ternary complex.

Table 4.7.1.1 $\Delta \log K_M$ values (25° C) for some representative ternary complexes.

M^{2+}	A (primary ligand)	B (secondary ligand)	$\Delta \log K_M$
Cu^{2+}	en	gly⁻	-0.80
		ox²⁻	-0.79
		en	-1.50
	bpy	gly⁻	-0.35
		ox²⁻	0.70
		en	-1.29
	nta³⁻	py	0.54
		im	0.25
		NH₃	-0.39
Ni^{2+}	bpy	en	-0.18
		gly⁻	0.21
	nta³⁻	py	0.31
		im	0.02
		NH₃	-0.20

- **log K_D** for the disproportionation process, $2MAB \rightleftharpoons MA_2 + MB_2$ is always -0.6 **(statistical value).** *More negative value of* log K_D *indicates the enhanced stability of the ternary complex.*
- log K_D can be obtained in the following way:

$$K_D = \frac{[MA_2][MB_2]}{[MAB]^2} = \frac{[MA_2]}{[M][A]^2} \times \frac{[MB_2]}{[M][B]^2} \times \left\{ \frac{[M][A][B]}{[MAB]} \right\}^2 = K_{MA_2}^{M} \times K_{MB_2}^{M} \times \left(K_{MAB}^{M} \right)^{-2}$$

i.e. $\quad \log K_D = \log K_{MA_2}^{M} + \log K_{MB_2}^{M} - 2 \log K_{MAB}^{M} = -0.6$ (statistical value)

- The **overall stability constant** (β_{MAB}^{M} or K_{MAB}^{M}) can be calculated theoretically based on the statistical consideration *(Ref. J. Chem. Edu,* **46**, 506, 1969) as follows:

$$\log \beta_{MAB\,(calcd)}^{M} = \frac{1}{2} \log K_{MA_2}^{M} + \frac{1}{2} \log K_{MB_2}^{M} + \log S$$

where S is the **statistical factor**, $\log S = 0.3$. Thus the calculated value can be compared with the experimentally observed value to measure the *contribution of the nonstatistical factors* to stabilize the ternary complex.

$$\Delta \log K_i = \log \beta^M_{MAB(obsd)} - \log \beta^M_{MAB(calcd)}$$

*The positive value of $\Delta \log K_i$ gives a quantitative measure of the **nonstatistical ligand enhancement factor** while its negative value gives the measure of the **nonstatistical ligand destabilization factor**.*

Parameters measuring the stabilities of the ternary complexes

- Less negative or more positive values of $\Delta \log K_M$, *i.e.* more stability of the ternary complex.
- More negative values of $\log K_D$, *i.e.* more stability of the ternary complex.
- More positive values of $\Delta \log K_i$, *i.e.* more stability of the ternary complex.

4.7.2 Different Factors Determining the Stabilities of the Ternary Complexes

It has been pointed out that the less negative or more positive values of $\Delta \log K_M$, more negative values of $\log K_D$ and positive value of $\Delta \log K_i$ *indicate the enhanced stabilities in the ternary complexes (MAB) over the binary MA_2 and MB_2 complexes.*

Besides the coulombic and steric factors as in the binary complexes, the **direct or indirect interaction between the ligands A and B in the ternary complex can explain the enhanced stability in the ternary complexes.** These are discussed below.

(i) **Effect of the presence of a π-acid ligand in the ternary complex:** If in a ternary complex (MAB), one ligand (say A) is a π-acidic ligand and the other ligand (*i.e.* B) is σ-basic, then stability of the ternary complex is enhanced due to the synergistic interaction.

$$A \underset{\pi}{\overset{\sigma}{\rightleftharpoons}} M \overset{\sigma}{\longleftarrow} B \quad ; \quad A \overset{\sigma}{\longrightarrow} M \overset{\sigma}{\longleftarrow} B$$

$$\underbrace{}_{\text{Synergistic stabilising combination}} \qquad \underbrace{}_{\text{No synergistic stabilising combination}}$$

For the π-acid ligands (A) like heteroaromatic N-bases (*e.g.* bpy, phen, etc.), the $M \rightarrow A$ π-back bonding leads to the withdrawal of electron cloud from the metal d-orbitals. Thus, the **electron deficiency produced on the metal centre** favours the binding of the σ-donor ligand (*e.g.* NH_3). If A is itself a σ-donor ligand (*e.g.* en, nta^{3-}, etc.), then no such electron deficiency on the metal centre can be created to favour the binding of another σ-donor ligand.

Based on the above consideration, we can reasonably expect:

$$K^{MA}_{MAB} \Big\rangle K^{MA'}_{MA'B}, \qquad \begin{aligned} & A = \pi\text{-acid ligands (}e.g.\text{ bpy, phen, etc.)} \\ & A' = \sigma\text{-basic ligand (}e.g.\text{ en, } nta^{3-}\text{)} \\ & B = \sigma\text{-basic ligand (}e.g.\text{ } NH_3\text{)} \end{aligned}$$

It has been illustrated (*cf.* Table 4.7.1.1) for the positive or more positive $\Delta \log K_M$ values of the complexes [Cu(bpy)(B)] and negative $\Delta \log K_M$ values of the [Cu(en)(B)] complexes ($B = gly^-$, ox^{2-}, en).

- The relative stabilities, $[M(nta)(py)] \rangle [M(nta)(im)] \rangle [M(nta)(NH_3)]$ in terms of $\Delta \log K_M$ (*cf.* Table 4.7.1.1) indicates that pyridine (*py*) and imidazole (*im*) can act as the π-acid ligands, but the π-acidic character of imidazole moiety is less pronounced compared to that of pyridine. In $[M(nta)(NH_3)]$ both the ligands are the σ-donor ligands (without any π-acidic character) and consequently, the complex is relatively less stable.

- Presence of a π-acid ligand (*A*) in the ternary complex introduces the **discriminating properties** in selecting the other ligand (*B*). This unique property to discriminate the ligands is very much important and nature has utilized this **property of selectivity** in different metalloenzymatic process. From the Δlog K_M values (*cf.* Table 4.7.1.1), selectivity of $[Cu(bpy)]^{2+}$ for the *B* ligands runs in the order:

$$ox^{2-}\left(O^- - O^-\right) \;\rangle\; gly^-\left(^-O - N\right) \;\rangle\; en\,(N - N)$$

$$\Delta \log K_M \quad = +0.70 \qquad -0.35 \qquad -1.29$$

It indicates that presence of a π-acidic ligand like bpy favours the O-donor ligands than the N-donor ligands. It is believed that the electron withdrawal by the π-acid ligand from the metal centre *increases the* ***hardness of the metal centre*** and *this is why, the metal centre prefers the hard donor secondary ligands.*

It is also suggested that presence of the *O*-donor ligands (*e.g.* ox^{2-}) as in **[Cu(bpy)(ox)]**, ***there will be an extensive electron delocalization involving the vacant π*-MO of bpy, metal d-orbitals and the filled p-orbitals of the O-donor sites.*** If the secondary ligand (*i.e. B*) uses the sp^3-nitrogen (as in *en*) donor sites, then no such delocalization is possible as the sp^3-*N*-sites of en do not have any such orbital that can be used for the delocalization purpose.

Synergistic action to stabilize the ternary complexes

Δlog K_M becomes more positive when one of the ligands is π-acidic (*i.e.* electron withdrawal through the *M→L*, π-bond) and the other ligand is a strong donor ligand (*L→M*, σ or σ and π bonding).

- **Relative stabilities of the binary and ternary complexes, $[Cu(his)_2]$:** The **ligand selectivity** in the ternary complex to facilitate the *synergistic bonding interaction* is well illustrated in nature for the $[Cu(his)_2]$, *i.e.* $[Cu(histidinate)_2]$ complex in solution (pH ≥ 6.5, pK_a of protonated imidazole ≈ 6.0).

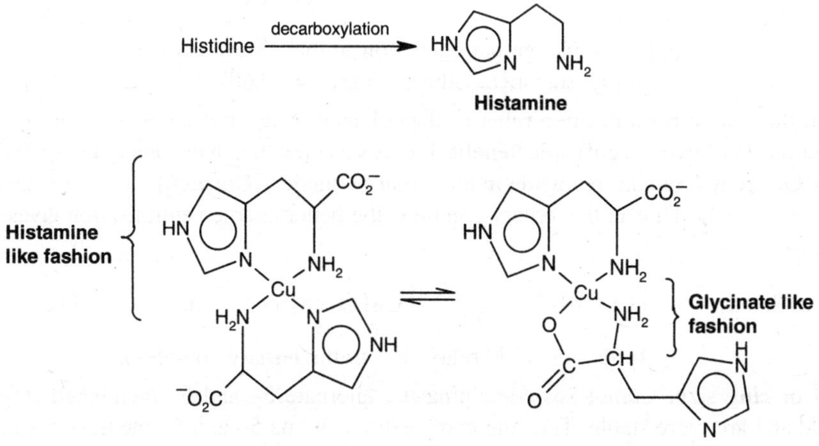

(Binary complex) (I) (Ternary complex) (II)

Fig. 4.7.2.1 Ligating behaviour of histidine in the binary and ternary complexes of Cu(II), (*cf.* Fig. 1.7.2.3).

At pH ≥ 6.5, histidine (Hhis) can provide three different binding sites: N(amino), N(imidazole) and O(carboxyl) (*cf.* Fig. 1.7.2.3). [Cu(his)$_2$] remains in solution in an equilibrium between the two forms **I** and **II**. In I, both the ligands bind through the amino-N and imidazole-N sites while in II, one ligand binds through the amino-N and imidazole-N and the other ligand binds through the amino-N and carboxylate-O. Thus, in I, *both the histidine ligands bind in a **histamine-like fashion** while in II, one histidine binds in a histamine-like fashion and the other histidine binds in a **glycinate-like fashion**.* The equilibrium is shifted towards II. Imidazole can act as a π-acid ligand. Hence to stabilize the mixed ligand complex, one histidine binds as a π-acid ligand (*i.e.* histamine-like behaviour) while the other ligand should bind as a σ-donor ligand without any π-acidic character (*i.e.* glycinate-like behaviour). Better charge neutralisation in the ternary complex II gives also an **entropic benefit** discussed later.

To favour the above equilibrium towards the right side, another factor, *i.e.* **ring-size**, also operates. Complexes having one 5– and one 6– membered rings are more stable than those having two 5– or two 6– membered rings. *In fact, existence of the alternate 5– and 6– membered rings in a complex gives a more flexibility to stabilize the system.* Moreover, in the ternary complex, [Cu(his)$_2$] there is a **statistical entropic benefit** while in the binary complex no such benefit exists. This fact will be further illustrated later.

Biological importance of the imidazole moiety in metalloproteins and metalloenzymes

There are many evidences that histidine binding through the imidazole-N can act as a π-acid ligand. This π-acidic character of histidine can introduce the special properties like the enhanced stability, enhanced Brönsted acid strength of the metal bound H_2O molecule, selectivity, etc. in the metalloprotein complex.

(ii) **Effect of charge neutralization and relaxation of steric crowding:** Both these factors will favour the stability of the system.

$$\left[Cu(en)_2\right]^{2+} + \left[Cu(ox)_2\right]^{2-} \underset{K_D}{\rightleftharpoons} 2\left[Cu(en)(ox)\right], \log K_D = -0.95$$

(effect of charge neutralization in the ternary complex;
cf. statistical value of $\log K_D = -0.60$).

In fact, the charge neutralization releases the solvated water molecules held through ion-dipole interaction. This gives an **entropic benefit**. It may be noted that in the binary complex, **[Cu(his)$_2$]**, two $-CO_2^-$ groups remain free while in the ternary complex **[Cu(his)$_2$]**, one $-CO_2^-$ group remains free (Fig. 4.7.2.1). Thus in the ternary complex, the better charge neutralisation gives an entropic benefit.

$$\left[Cu(Et_2en)_2\right]^{2+} + \left[Cu(en)_2\right]^{2+} \underset{K_D}{\rightleftharpoons} 2\left[Cu(Et_2en)(en)\right]^{2+}, \log K_D = -1.83$$

(effect of steric relaxation in the ternary complex).

(iii) **Effect of ring size:** Complexes containing the alternate 5– and 6– membered rings are more flexible and are more stable. Thus the complexes with one 5– and 6– membered rings are more stable than the complexes containing two 5– or 6– membered rings. This is illustrated here.

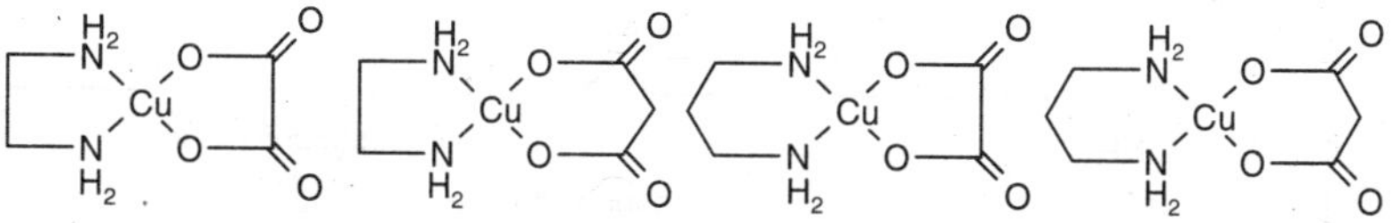

[Cu(A)(B)]:	[Cu(en)(ox)]	[Cu(en)(mal)]	[Cu(ox)(pn)]	[Cu(mal)(pn)]
Ring size:	(5,5)	(5,6)	(6,5)	(6,6)
log K_D:	−0.94	−2.31	−3.14	−2.55

The similar effect in the tetraamine complexes like [Cu(2,2,2-tet)]$^{2+}$ (log$K$ = 20.1) and [Cu(2,3,2-tet)]$^{2+}$ (log$K$ = 23.9) has been already illustrated (cf. Fig. 4.5.2.22).

(iv) **Intermolecular ligand-ligand interactions:** It may be of different types. The interaction between A and B in MAB is represented, in general, as follows:

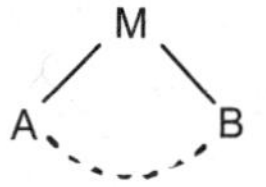

$$
\begin{array}{c}
M \\
\diagup \quad \diagdown \\
A \cdots\cdots B
\end{array}
$$

(a) **Intermolecular stacking interaction between the aromatic rings:** If the bound ligands possess the aromatic rings (i.e. delocalized π-electron clouds) and the ligands are sufficiently flexible then the aromatic rings may be stacked in parallel planes to allow the overlapping between their π-electron clouds in a lateral way (cf. graphite structure). Actually, the electrons from the **HOMO of an electron rich aromatic ring** flow towards the **LUMO of an electron deficient aromatic ring.** It represents basically an electrostatic attraction. Besides this, the **London dispersion force** may also contribute to the **π-π-stacking interaction.** In fact, there is a debate over the nature of π-π interaction but this interaction will give an additional stabilization.

To allow the stacking interaction, the following condition is to be maintained:

"*Aromatic rings of one ligand will be kept fixed in a plane and the other ligand will bear the aromatic rings as the nonbonding moieties in a flexible manner so that these rings can be stacked above the fixed rings of the other ligand.*"

The heteroaromatic ligands like bpy, phen, etc. can bind the metal centre keeping their aromatic rings fixed in a particular plane. Then the aromatic rings as the nonbonding moieties in **a flexible tail** of the other ligands, may be stacked over the aromatic rings of the heteroaromatic ligands.

In the mixed ligand complexes of Mn(II), Cu(II) or Zn(II) containing bpy and ATP (adenosine triphosphate), the additional stabilization (indicated by the positive value of Δlog K_M) is induced due to *stacking of the purine moiety* (**nonbonding tail**) *of ATP over the aromatic π-system of bpy.* It may be argued that positive value of Δlog K_M (i.e. additional stabilization of the ternary complex) is due to *synergistic stabilization of the metal-ligand bonds, i.e.* π-acidic property of bpy and σ-basic property of the phosphate moiety of ATP.

$$(ATP)P-O \xrightarrow{\sigma} M \underset{\sigma}{\overset{\pi}{\rightleftharpoons}} bpy$$

But in the above mentioned transition metal complexes, the positive value of Δlog K_M (e.g. 0.50 for the [Cu(ATP)(bpy)]$^{2-}$ complex) is not only *due to the synergistic stabilization of the metal-ligand bonds.* It is evidenced by the fact that the analogous complexes like [Mg(ATP)(phen)]$^{2-}$, [Ca(ATP)(phen)]$^{2-}$, etc. where M→bpy/phen π-interaction is absent (because of the d^0-system), also show the comparable positive Δlog K_M values. The **bpy-purine stacking interaction** is schematically shown in Fig. 4.7.2.2. *Here it may be pointed out that the N-7 site of adenine moiety of ATP fails to coordinate in [Cu(ATP)(bpy)]$^{2-}$ in presence of the strongly coordinating ligand bpy.* It gives here both the stacking

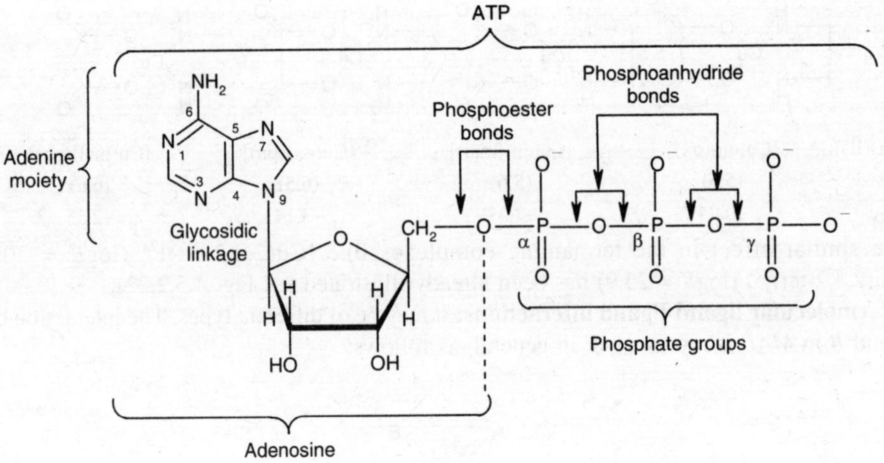

ATP

Adenine
moiety

NH₂

Phosphoester
bonds

Phosphoanhydride
bonds

Glycosidic
linkage

Phosphate groups

Adenosine

(a) **ATP**

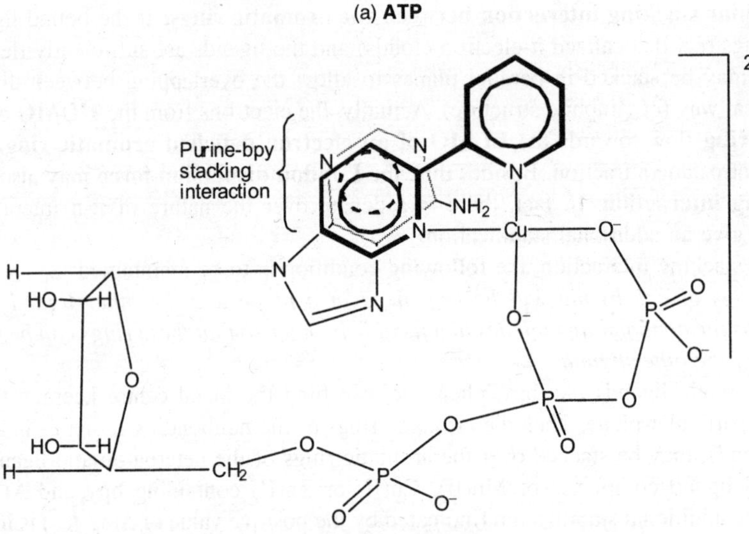

Purine-bpy
stacking
interaction

(b) **[Cu(ATP)(bpy)]²⁻**

Fig. 4.7.2.2 Structures of ATP and [Cu(ATP)(bpy)]²⁻

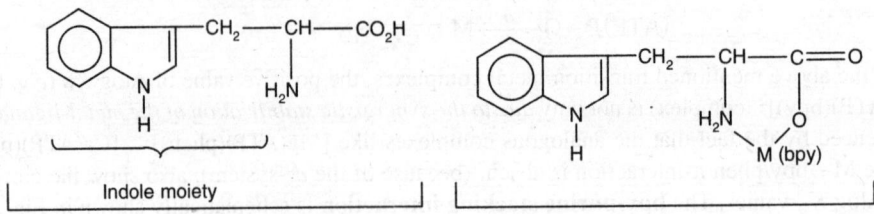

Indole moiety

Tryptophen
(*i.e.* α-amino-β-indole propionic acid)

In this ternary complex, tryptophenate
binds in a glycinate-like fashion

interaction effect and synergistic effect of stabilization. The synergistic stabilization has been discussed in [Cu(his)$_2$] (*cf.* Fig. 4.7.2.1).

In the mixed ligand complexes of M(II) containing tryptophenate and ATP, the stacking interaction between the *indole ring* of tryptophenate and *purine ring* of ATP stabilizes the systems. In fact, the **indole-bpy, indole-phen stacking interactions** are well documented. The mixed ligand complexes bearing *bpy* or *phen* and phenoxyacetic acid derivatives as the ligands can also earn the stacking interaction by placing the benzene ring on the *bpy* or *phen* ring.

(b) **Hydrophobic interaction:** This interaction originates from (*cf.* Chapter 13, Vol. 3 of Fundamental Concepts of Inorganic Chemistry) the **London dispersion** force among the nonpolar, *i.e.* hydrophobic moieties and **entropic favour** arising from the association of the nonpolar moieties in aqueous media. This association produces a **less structurisation of the water molecules** to cause the entropic favour. This hydrophobic interaction may occur between the aromatic moieties of one ligand (*e.g.* bpy, phen) and noncoordinating nonaromatic side chains of the other ligand. Such an interaction between the phenanthroline ring and the isopropyl residue of leucine in [Zn(leu)(phen)]$^+$ is established. Thus in the mixed ligand complexes [M(A)(B)] where A = bpy or phen, B = an amino acid with a noncoordinating nonaromatic side chain, the hydrophobic interaction stabilizes the system.

(c) **Covalent interaction:** In some cases, the covalent bond formation among the ligands may occur. Unusually high positive values of $\Delta \log K_M$ for the ternary complexes, [M(gly$^-$)(sal$^-$)] (sal$^-$ = salicyl-aldehydate and gly$^-$ = glycinate) of Mn(II) (2.45), Ni(II) (1.50), Cu(II) (2.70) and Zn(II) (1.90) result from the Schiff base, salicylidineglycinate (salgly^{2-}) formation through the condensation reaction between the $\rangle$C = O group of the coordinated sal$^-$ and –NH$_2$ group of the coordinated gly$^-$. It leads to the covalent — CH $=$ N — bond formation.

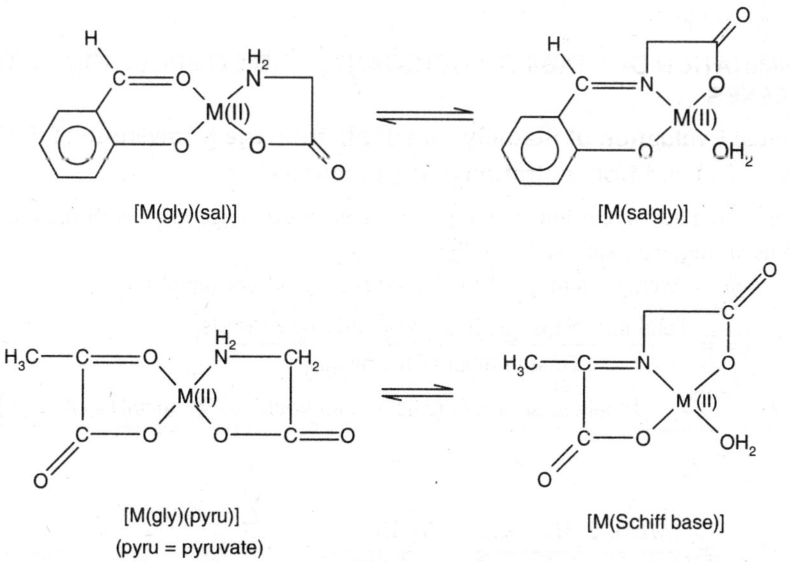

Fig. 4.7.2.3 Covalent interaction between the coordinated ligands of the ternary complexes giving rise to the polydentate Schiff base complex (*cf.* Template effect).

Similarly, pyruvate and glycinate coordinated in the mixed ligand complexes of M(II) (M = Co, Mn, Zn) from the Schiff bases within the coordination sphere characterized by the highly positive value of $\Delta \log K_M$. In fact, these covalent interactions lead to the *template synthesis of polydentate ligands*.

(d) **Ionic interaction:** This type of interaction may arise when the noncoordinating oppositely charged side chains of the bound ligands are brought in contact. In the complexes like [M(α-amino acid$_1$) (α-aminoacid$_2$)], having free $-NH_3^+$ and $-CO_2^-$ groups in the side chains of the amino acids may establish *an electrostatic interaction* between these two free groups. This interaction can arise under a *particular stereospecific condition, i.e.* two *L*-amino acids are coordinated *cis–* to one another.

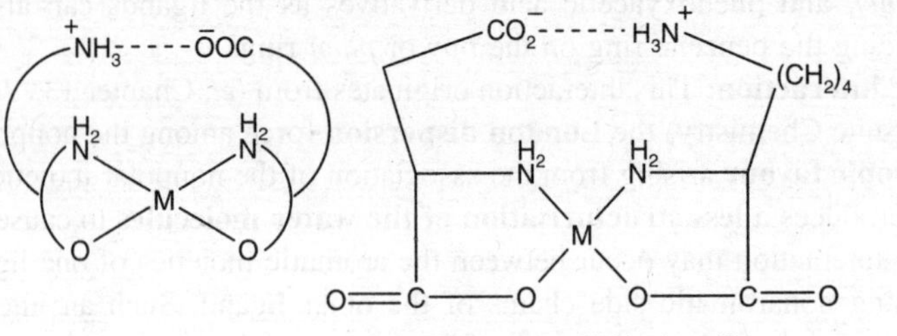

[M(aspartate)(H-lysinate)]

Fig. 4.7.2.4 Ionic interaction between the oppositely charged noncoordinating moieties of the coordinated ligands in a ternary complex.

Such a favourable electrostatic interaction occurs in [Cu(aspartate)(H-lysinate)] from the (aspartate)COO$^-$····$^+$H$_3$N(H-lysinate) bridging. A similar interaction may occur in [Cu(L-A)(L-his)] complexes (A = lysine, arginine) from the bridging between the protonated free basic site of A and free $-CO_2^-$ group of histidine.

(e) **H-bonding interaction:** Sometimes, the H-bonding interaction among the bound ligands can stabilize the ternary complexes.

4.8 DETERMINATION OF STABILITY CONSTANTS AND COMPOSITION OF THE COMPLEXES

4.8.1 Graphical Evaluation of Stability Constants from the Knowledge of Formation Function ($\bar{n}$) and Concentration of Free Ligand ([L])

The parameter, $\bar{n}$ (called **formation function**) represents the average number of the ligand bound per metal ion. Thus $\bar{n}$ may be expressed as follows:

$\bar{n}$ = Average number of the ligand (*L*) bound per metal ion.

$$= \frac{\text{Total number of the ligands bound to the metals}}{\text{Total number of the metals}}$$

$$= \frac{T_L - (\text{concentration of the ligand not bound to the metal})}{T_M} = \frac{T_L - [L]}{T_M}$$

$$= \frac{[ML] + 2[ML_2] + \dots\dots + N[ML_N]}{[M] + [ML] + [ML_2] + \dots\dots + [ML_N]} = \frac{\displaystyle\sum_{n=1}^{N} n[ML_n]}{[M] + \displaystyle\sum_{n=1}^{N} [ML_n]}$$

Here T_L and T_M denote the total analytical concentrations of the ligand and the metal respectively in solution. Here all the probable complexes are considered.

$\bar{n}$ can be expressed in terms of stability constants and [L] (= concentration of the free ligand) as follows:

$$\bar{n} = \frac{\beta_1 [M][L] + 2\beta_2 [M][L]^2 + \ldots\ldots + N\beta_N [M][L]^N}{[M] + \beta_1 [M][L] + \beta_2 [M][L]^2 + \ldots\ldots + \beta_N [M][L]^N}$$

$$= \frac{\displaystyle\sum_{n=1}^{N} n\beta_n [L]^n}{1 + \displaystyle\sum_{n=1}^{N} \beta_n [L]^n}$$

$$= \frac{\beta_1 [L] + 2\beta_2 [L]^2 + \ldots\ldots + N\beta_N [L]^N}{1 + \beta_1 [L] + \beta_2 [L]^2 + \ldots\ldots + \beta_N [L]^N}$$

Rearrangement of the expression gives:

$$\frac{\bar{n}}{(1-\bar{n})[L]} = \beta_1 + \frac{(2-\bar{n})}{(1-\bar{n})}\beta_2 [L] + \ldots\ldots + \frac{(N-\bar{n})}{(1-\bar{n})}\beta_N [L]^{N-1} \qquad \ldots (4.8.1.1)$$

The above general equation involving $\bar{n}$, stability constants and $[L]$ is very much important. Free ligand concentration, *i.e.* $[L]$, can be determined experimentally by different techniques: emf measurement (if a suitable reversible electrode exists), by pH measurement (if the protonation constant of the ligand is known, *cf.* **Bjerrum's method**), etc. Then from the knowledge of $[L]$, $\bar{n}$ can be evaluated by using the following relation.

$$\bar{n} = \frac{T_L - [L]}{T_M} \text{ ; assuming no protonated form of the ligand.}$$

By using the different T_L values, a set of $\bar{n}$ and $[L]$ values can be obtained and then by making the use of Eqn. 4.8.1.1, stability constants can be evaluated graphically. The use of a computer is a great promise in such solutions.

● Application of Eqn 4.8.1.1 in a simple case where only ML and ML_2 complexes exist can be illustrated. If no higher complex than ML_2 exists, then the above equation is simplified as follows:

$$\bar{n} = \frac{\beta_1 [L] + 2\beta_2 [L]^2}{1 + \beta_1 [L] + \beta_2 [L]^2}, \text{ or } \bar{n} + \bar{n}\beta_1 [L] + \bar{n}\beta_2 [L]^2 = \beta_1 [L] + 2\beta_2 [L]^2$$

or, $$\bar{n} = \beta_1 [L](1 - \bar{n}) + \beta_2 [L]^2 (2 - \bar{n})$$

or, $$\frac{\bar{n}}{(1-\bar{n})[L]} = \beta_1 + \frac{(2-\bar{n})}{(1-\bar{n})}\beta_2 [L]$$

Then the plot of $\dfrac{\bar{n}}{(1-\bar{n})(L)}$ vs. $\dfrac{(2-\bar{n})}{(1-\bar{n})}[L]$ gives a straight line having **slope** $= \beta_2$ and **intercept** $= \beta_1$.

● An equation similar to Eqn. 4.8.1.1, may be used for evaluation of protonation constants of a ligand. It can be illustrated as follows:

$$L + H^+ \xrightleftharpoons{K_1^H} LH^+, \ LH^+ + H^+ \xrightleftharpoons{K_2^H} LH_2^{2+}.$$

For the above system, $\bar{n}_H$ can be defined and expressed as follows:

$$\bar{n}_H = \text{Average number of } H^+ \text{ ions bound per ligand species} = \frac{T_{H^+} - [H^+]}{T_L}$$

$$= \frac{[LH^+] + 2[LH_2^{2+}]}{[L] + [LH^+] + [LH_2^+]}$$

$$= \frac{K_1^H [L][H^+] + 2K_1^H K_2^H [L][H^+]^2}{[L] + K_1^H [L][H^+] + K_1^H K_2^H [L][H^+]^2}$$

$$= \frac{K_1^H [H^+] + 2K_1^H K_2^H [H^+]^2}{1 + K_1^H [H^+] + K_1^H K_2^H [H^+]^2}$$

or, $$\frac{\bar{n}_H}{(1 - \bar{n}_H)[H^+]} = K_1^H + \frac{2 - \bar{n}_H}{1 - \bar{n}_H} K_1^H K_2^H [H^+]$$

Free hydrogen ion concentration, *i.e.* [H^+] can be obtained from the pH measurement and then $\bar{n}_H$ can be evaluated from its definition.

$$\bar{n}_H = \frac{T_{H^+} - [H^+]}{T_L}$$

T_{H^+} = total analytical concentration of H^+, T_L = total analytical concentration of the ligand.

Thus $\bar{n}_H$ and [H^+] are the experimental quantities and from the linear plot of $\dfrac{\bar{n}_H}{(1 - \bar{n}_H)[H^+]}$ *vs.* $\dfrac{2 - \bar{n}_H}{1 - \bar{n}_H}[H^+]$,

the protonation constants may be obtained (intercept = K_1^H and slope = $K_1^H K_2^H$).

In general, $\bar{n}_H$ and [H^+] are correlated in terms of the ligand protonation constants as follows:

$$\frac{\bar{n}_H}{(1 - \bar{n}_H)[H^+]} = K_1^H + \frac{2 - \bar{n}_H}{1 - \bar{n}_H} K_1^H K_2^H [H^+] + \ldots\ldots + \frac{(N - \bar{n}_H)}{(1 - \bar{n}_H)} K_1^H K_2^H \ldots\ldots K_N^H [H^+]^{(N-1)}$$

4.8.2 Formation Curves and Bjerrum's Half $\bar{n}$-Method

The metal-ligand stability constants and protonation constants of the ligands can be determined from the *formation curves*, *i.e.* plot of $\bar{n}$ *vs.* pL (= $-\log [L]$) for the metal-ligand stability constants, and $\bar{n}_H$ *vs.* $pH(= -\log[H^+])$ for the ligand protonation constants. If the successive constants differ widely, *i.e.* $\dfrac{K_n}{K_{n+1}} \geq 10^{2.5}$ (for metal-ligand stability constants) and $\dfrac{K_n^H}{K_{n+1}^H} \geq 10^{2.5}$ (for protonation constants of the ligand), then the formation curves will have the **wave-like nature** (Fig. 4.8.2.1). In such cases, according to Bjerrum's **half $\bar{n}$-method,** we have:

$\log K_1 = (pL)_{\bar{n}=0.5}$, $\log K_2 = (pL)_{\bar{n}=1.5}$, $\log K_3 = (pL)_{\bar{n}=2.5}$, etc.

Similarly for the protonation constants we have: $\log K_1^H = (pH)_{\bar{n}_H=0.5}$, $\log K_2^H = (pH)_{\bar{n}_H=1.5}$; and so on.

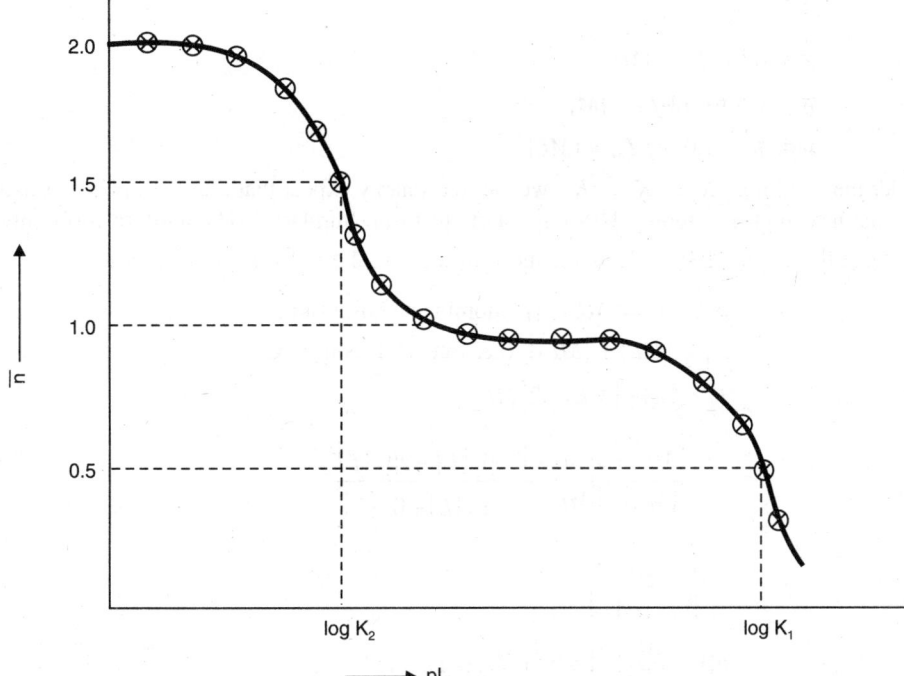

Fig. 4.8.2.1 Schematic representation of determination of metal-ligand stability constants (K_1, K_2) from the formation curve by **half $\bar{n}$-method**.

The above mentioned conclusions can be arrived by considering the following reasonable approximations. When the step-wise stability constants differ widely (*i.e.* after almost complete formation of a particular complex, the next higher complex starts to form), at the smaller values of $\bar{n}$ ($= 0$ to 1), there will be practically no ML_2 complex in solution and we can write:

$$T_M = [M] + [ML], \quad (\bar{n} = 0 \text{ to } 1 \text{ range})$$

and

$$\bar{n} = \frac{[ML]}{[M]+[ML]} = \frac{\beta_1 [L]}{1+\beta_1 [L]}; \quad \text{or, } \bar{n} + \bar{n}\beta_1 [L] = \beta_1 [L]$$

or, $\dfrac{\bar{n}}{(1-\bar{n})} = \beta_1 [L]$; or $\log \dfrac{\bar{n}}{(1-\bar{n})} = \log \beta_1 + \log [L]$;

or, $pL = -\log[L] = \log K_1 + \log \dfrac{(1-\bar{n})}{\bar{n}}$, $(K_1 = \beta_1)$.

It leads to: $pL = \log K_1$ for $\bar{n} = 0.5$

i.e. $\qquad (pL)_{\bar{n}=0.5} = \log K_1$

For $\bar{n} = 0 - 1$, we can write:

$$T_M = [M] + [ML], \quad T_L = [L] + [ML], \quad [\text{ML}_2] \approx 0 \text{ as } K_1 \gg K_2$$

and

$$\bar{n} = \frac{T_L - [L]}{T_M}$$

It leads to: $\quad \bar{n} = 0$ for $T_M = [M]$

$$\bar{n} = 0.5 \text{ for } [ML] = [M]$$

and, $\bar{n} = 1.0$ for $T_M = [ML]$

● Under the condition, $K_1 \gg K_2 \gg K_3$, we can reasonably expect that after complete formation of ML, the next higher complex ML_2 only starts to form. Similarly, ML_3 starts to form only after complete formation of ML_2. Thus for the values of $\bar{n} : 2 \geq \bar{n} \geq 1.0$, we can write:

$$ML + L \rightleftharpoons ML_2, \text{ (predominant equilibrium)}$$

$$T_M = [ML] + [ML_2], \text{ (\textit{i.e.} free } M \text{ does not exist)}$$

$$T_L = [L] + [ML] + 2[ML_2]$$

and

$$\bar{n} = \frac{[ML] + 2[ML_2]}{[ML] + [ML_2]} = \frac{\beta_1[L] + 2\beta_2[L]^2}{\beta_1[L] + \beta_2[L]^2}$$

$$= \frac{\beta_1 + 2\beta_2[L]}{\beta_1 + \beta_2[L]}$$

or, $\quad \bar{n}\beta_1 + \bar{n}\beta_2[L] = \beta_1 + 2\beta_2[L]$

or, $\quad (\bar{n} - 1)\beta_1 = \beta_2[L](2 - \bar{n})$

or, $\quad \dfrac{\bar{n} - 1}{2 - \bar{n}} = \dfrac{\beta_2}{\beta_1}[L] = K_2[L], \quad \beta_2 = K_1K_2$

or $\quad \log\dfrac{(\bar{n} - 1)}{(2 - \bar{n})} = \log K_2 + \log[L]$

or $\quad pL = -\log[L] = \log K_2 + \log\dfrac{(2 - \bar{n})}{(\bar{n} - 1)}$

or, $\quad (pL)_{\bar{n}=1.5} = \log K_2 \, (cf. \text{ Fig. 4.8.2.1})$

For the range: $2.0 \geq n \geq 1.0$ we have:

$$\bar{n} = 1.5, \text{ for } [ML] = [ML_2]$$

and, $\quad \bar{n} = 2.0$ for $T_M = [ML_2]$

● For the range, $\bar{n} = 0-2$ (\textit{i.e.} complex upto ML_2 exists) we can write (\textit{cf.} Eqn. 4.8.1.1).

$$\bar{n} = \frac{\beta_1[L] + 2\beta_2[L]^2}{1 + \beta_1[L] + \beta_2[L]^2}$$

or,　　　　$$\bar{n} + \bar{n}\beta_1[L] + \bar{n}\beta_2[L]^2 = \beta_1[L] + 2\beta_2[L]^2$$

or,　　　　$$\bar{n} + (\bar{n} - 1)\beta_1[L] + (\bar{n} - 2)\beta_2[L]^2 = 0$$

At $\bar{n} = 1$ (*i.e.* at the condition of $[ML]_{max}$), the above relation reduces to:

$$\beta_2[L]^2 = 1$$

or,　　　　$$\log\beta_2 + 2\log[L] = 0$$

or,　　　　$$2pL = \log\beta_2 = \log K_1 K_2 = \log K_1 + \log K_2$$

● The successive protonation constants of the ligand can be evaluated in the same way by the $\bar{n}$-method.

Note: If the successive stability constants differ widely, *i.e.* $K_1/K_2 \geq 10^{2.5}$, *i.e.*

$\left[(pL)_{\bar{n}=0.5} - (pL)_{n=1.5} = \log K_1 - \log K_2 \right] \geq 2.5$, then the half $\bar{n}$-method is suitable. Because in this method, it has been assumed that after complete formation of one complex, the next higher complex starts to form, *i.e.* in the solution only ML_x and ML_{x+1} can coexist at the same time. If this condition is not maintained then K_1, K_2... are determined graphically (*cf.* Sec. 4.8.1))

4.8.3 Determination of Stability Constants from the Knowledge of [M], *i.e.* Free Metal Concentration

In the earlier sections, the procedures for the stability constant determination from the knowledge of $[L]$ have been described. *If it is easier to determine $[M]$ than $[L]$,* then a different equation may be used for the stability constant determination. We can write:

$$\frac{T_M}{[M]} = \frac{[M] + [ML] + [ML_2] + \dots\dots + [ML_N]}{[M]}$$

$$= \frac{1}{[M]}\left\{ [M] + \beta_1[M][L] + \beta_2[M][L]^2 + \dots\dots + \beta_N[M][L]^N \right\}.$$

$$= 1 + \beta_1[L] + \beta_2[L]^2 + \dots\dots + \beta_N[L]^N$$

or,　　　　$$\frac{T_M}{[M]} = 1 + \sum_{n=1}^{N} \beta_n[L]^n \qquad \qquad \dots(4.8.3.1)$$

If $T_L \rangle\rangle T_M$, then we can reasonably write:

$$T_L = [L] + [ML] + 2[ML_2] + \dots + N[ML_N]$$
$$\approx [L] \approx \text{constant}$$

Under the above condition, determination of $[M]$ for various T_M values can evaluate β's by making the use of Eqn. 4.8.3.1.

4.8.4 pH-Metric Determination of the Stability Constant under the Competition between Hydrogen Ion and Metal Ion for the Ligand

There is a competition between H^+ and M (charge not shown for the sake of simplicity) for the ligand.

$$L + H^+ \rightleftharpoons LH^+, \; K^H = \frac{1}{K_a} = \frac{[LH^+]}{[L][H^+]}$$

K^H = protonation constant of L; K_a = acid dissociation constant of the conjugate acid LH^+.

$$L + M \rightleftharpoons ML, \; K \text{ or } \beta \; (= \text{stability constant}) = \frac{[ML]}{[L][M]}$$

Thus to determine K, we need the values of $[ML]$, $[L]$ and $[M]$. These can be calculated by using $[H^+]$ obtained from the *pH* measurement provided the acid dissociation constant of LH^+ is known.

If the total analytical concentrations of H^+, ligand and metal are T_H, T_L and T_M respectively, then we have the following relations:

$$T_H = [H^+] + [LH^+]; \; T_L = [L] + [LH^+] + [ML]$$

and $$T_M = [M] + [ML]$$

$$T_H = [H^+] + K^H[L][H^+] \quad \text{or} \quad [L] = \frac{T_H - [H^+]}{K^H[H^+]} \qquad \qquad \text{...(a)}$$

$$[ML] = T_L - [L] - [LH^+] = T_L - [L] - K^H[L][H^+]$$

$$= T_L - [L]\{1 + K^H[H^+]\}$$

$$= T_L - \frac{(T_H - [H^+])}{K^H[H^+]}\{1 + K^H[H^+]\} \qquad \qquad \text{...(b)}$$

and $[M] = T_M - [ML]$ $\qquad \qquad \qquad \qquad \qquad \qquad \qquad \qquad \qquad \text{...(c)}$

It is evident that from the values of $[H^+]$ (which is an experimental quantity), the values of $[ML]$, $[L]$ and $[M]$ can be calculated by making the use of Eqns. (a), (b) and (c).

In practice, different solutions containing various amounts of initial concentrations of T_H, T_L and T_M are prepared. Generally, for all the solutions, a fixed ionic strength is maintained. Then for each solution, *pH* is measured at a particular temperature. From the measured *pH* value, the corresponding $[H^+]$ is obtained. Then by using Eqns. (a), (b) and (c), the values of $[L]$, $[ML]$ and $[M]$ (*i.e. equilibrium concentrations of the species*) are obtained to evaluate the stability constant.

● If in the system, more than one complex is formed then graphical method by using the concept of $\bar{n}$ (*cf.* Eqn. 4.8.1.1) is applied. The measurement of *pH* leads to the value of $[L]$ as discussed above. Then $\bar{n}$ is calculated for the different solutions from the knowledge of $[L]$.

$$\bar{n} = \frac{T_L - [L] - [LH^+]}{T_M} = \frac{T_L - [L] - K^H[L][H^+]}{T_M} = \frac{T_L - [L](1 + K^H[H^+])}{T_M}$$

If only two complexes, *i.e. ML* and ML_2 (*e.g.* **Ag$^+$ –NH$_3$, Cu^{2+}–glycine**, etc.) are present in solution, then the corresponding equation for graphical solution is:

$$\frac{\bar{n}}{(1 - \bar{n})[L]} = K_1 + \frac{(2 - \bar{n})}{(1 - \bar{n})} K_1 K_2 [L]$$

(*i.e.*, linear plot of $\dfrac{\bar{n}}{(1-\bar{n})[L]}$ *vs.* $\dfrac{2-\bar{n}}{(1-\bar{n})}[L])$

If K_1 and K_2 differ widely $\left(i.e.,\ \dfrac{K_1}{K_2}\geq 10^{2.5}\right)$ then, the half $\bar{n}$ method may be used to evaluate K_1 and K_2 from the formation curve (*i.e.* $\bar{n}$ vs. pL)

Ag⁺–NH₃ system:

The relevant equilibria are:

$$NH_3 + H^+ \rightleftharpoons NH_4^+\ \left(K_1^H\right),\ i.e.\ L = NH_3\ \text{and}\ LH^+ = NH_4^+$$

$$Ag^+ + NH_3 \rightleftharpoons \left[Ag(NH_3)\right]^+ (K_1);\left[Ag(NH_3)\right]^+ + NH_3 \rightleftharpoons \left[Ag(NH_3)_2\right]^+ (K_2)$$

The relevant mass-balance relations are:

$$T_H = [H^+] + [NH_4^+] = [H^+] + K_1^H [NH_3][H^+]$$

i.e. $[L] = [NH_3] = \dfrac{T_H - [H^+]}{K_1^H [H^+]}$ and $\bar{n} = \dfrac{T_L - [L] - [LH^+]}{T_M} = \dfrac{T_L - [L]\left(1 + K_1^H [H^+]\right)}{T_M}$

$$T_L = [NH_3] + [NH_4^+] + \left[Ag(NH_3)^+\right] + 2\left[Ag(NH_3)_2^+\right]$$

$$= [L] + [LH^+] + [ML] + 2[ML_2]$$

$$T_M = [Ag^+] + \left[Ag(NH_3)^+\right] + \left[Ag(NH_3)_2^+\right]$$

$$= [M] + [ML] + [ML_2]$$

Cu²⁺–gly⁻ system:

The relevant equilibria are:

$$gly^- + H^+ \rightleftharpoons Hgly, \left(K_1^H\right);\ Hgly + H^+ \rightleftharpoons H_2gly^+, \left(K_2^H\right)$$

$$Cu^{2+} + gly^- \rightleftharpoons \left[Cu(gly)\right]^+, (K_1); \left[Cu(gly)\right]^+ + gly^- \rightleftharpoons \left[Cu(gly)_2\right], (K_2)$$

To find $[L] = [gly^-]$, the following mass-balance relationship is to be used.

$$T_H = [H^+] + [Hgly] + 2\left[H_2gly^+\right]$$

$$= [H^+] + K_1^H \left[gly^-\right][H^+] + 2K_1^H K_2^H \left[gly^-\right][H^+]^2$$

i.e.
$$\left[gly^-\right] = \frac{T_H - [H^+]}{K_1^H[H^+] + 2K_1^H K_2^H [H^+]^2}$$

$\bar{n}$ can be calculated as follows:

$$\bar{n} = \frac{T_L - [L^-] - [LH] - \left[LH_2^+\right]}{T_M} \text{ where } L^- = gly^-,\ LH = Hgly,\ LH_2^+ = H_2gly^+ \text{ and } T_M = T_{Cu}$$

● **Evaluation of protonation constant(s) of the ligand:** The protonation constant (K^H) can be evaluated by treating a certain amount of ligand with variable amounts of $[H^+]$. Then evaluation of

$$\bar{n}_H \left(= \frac{T_H - [H^+]}{T_L}\right) \text{ and use of the following equation,}$$

$$\frac{\bar{n}_H}{(1 - \bar{n}_H)[H^+]} = K^H[H^+],\ (\textit{cf. Sec. 4.8.1})$$

allows us to estimate K^H. If there is a one protonation constant for the ligand (*e.g.* NH_3) then the above simple equation is applicable. For the ligands like glycinate, there are two successive protonation constants, *i.e.* K_1^H and K_2^H

$$gly^- + H^+ \underset{}{\overset{K_1^H}{\rightleftharpoons}} Hgly; \quad Hgly + H^+ \underset{}{\overset{K_2^H}{\rightleftharpoons}} H_2gly^+,\ T_L = \left[gly^-\right] + \left[Hgly\right] + \left[H_2gly^+\right]$$

Under these conditions, we are to use the following equation for the graphical evaluation of K_1^H and K_2^H.

$$\frac{\bar{n}_H}{(1 - \bar{n}_H)[H^+]} = K_1^H + \frac{2 - \bar{n}_H}{1 - \bar{n}_H} K_1^H K_2^H [H^+]$$

i.e.

plot of $\dfrac{\bar{n}_H}{(1 - \bar{n}_H)[H^+]}$ *vs.* $\dfrac{2 - \bar{n}_H}{1 - \bar{n}_H}[H^+]$ gives: intercept $= K_1^H$ and slope $= K_1^H K_2^H$, *i.e.* $\dfrac{\text{slope}}{\text{intercept}} = K_2^H$

4.8.4 Spectrophotometric Method of Determination of Stability Constant

If the complex formed absorbs at a particular wavelength at which the free ligand and metal ion do not absorb then from the measured optical density (A), concentration of the complex at equilibrium can be determined by using the Beer's law:

$A = \varepsilon l C$ (where A = optical density; ε = molar extinction coefficient of the complex; C = concentration of the complex (in moles / litre), l = optical path length)

Under the condition $T_M \gg T_L$, only the monocomplex is formed, *i.e.*

$$M^{n+} + L \rightleftharpoons ML^{n+},\ K(=\beta) = \frac{[ML^{n+}]}{[M^{n+}][L]}$$

$$C = [ML^{n+}] = A/\varepsilon l$$

$$T_M = [M^{n+}] + [ML^{n+}],\ \textit{i.e. } [M^{n+}] = T_M - [ML^{n+}] = T_M - A/\varepsilon l$$

$$T_L = [L] + [ML^{n+}], \ i.e. \ [L] = T_L - [ML^{n+}] = T_L - A/\varepsilon l \ ; \ \text{(considering no protonation}$$
$$\text{of the ligand)}$$

$$K = \frac{A/\varepsilon l}{(T_M - A/\varepsilon l)(T_L - A/\varepsilon l)}$$

Now, if ε is known, then by using the other experimental quantities, K can be evaluated. In practice, to simplify the process, some approximations are taken into consideration. If T_M is taken in large excess over T_L, then *formation of the noncomplex is assured* and we can have:

$$[M^{n+}] \approx T_M \ \text{ when } T_M \gg T_L$$

Thus we can write:

$$K = \frac{A/\varepsilon l}{T_M (T_L - A/\varepsilon l)} = \frac{A}{T_M (\varepsilon l T_L - A)}$$

$$\text{or,} \quad \frac{1}{K} = \frac{T_M (\varepsilon l T_L - A)}{A} = \frac{T_M \varepsilon l T_L}{A} - T_M$$

$$\text{or,} \quad \frac{1}{K} + T_M = \frac{T_M \varepsilon l T_L}{A}$$

$$\text{or,} \quad \frac{1}{A} = \frac{1}{T_L \varepsilon l} + \frac{1}{K T_L \varepsilon l} \times \frac{1}{T_M}$$

If a number of solutions are prepared having different T_M values at a fixed T_L value, maintaining the condition, $T_M \gg T_L$, then the plot of $\dfrac{1}{A}$ vs. $\dfrac{1}{T_M}$ will give a straight line having:

$$\text{Intercept} = \frac{1}{T_L \varepsilon l} \ \text{ and slope} = \frac{1}{K T_L \varepsilon l}$$

and
$$K (= \beta) = \frac{\text{intercept}}{\text{slope}}.$$

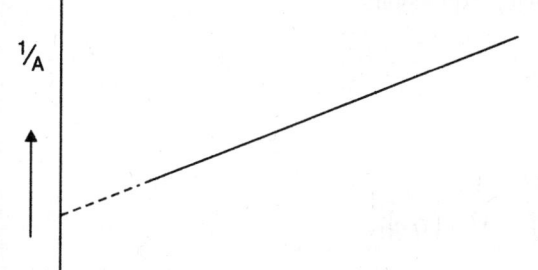

A representative experiment

$$Fe(aq)^{3+} + SCN^- \rightleftharpoons Fe(aq)(SCN)^{2+} \ ; \ \lambda = 460 \text{ nm}$$

T_L (fixed) $= 2 \times 10^{-4} \text{ mol dm}^{-3}$; $[H^+] = 0.4 \text{ mol dm}^{-3}$ (fixed);

$T_{Fe} = 2 \times 10^{-3}$ to 2×10^{-2} mol dm^{-3} (at least 6 sets).

Fig. 4.8.4.1 Spectrophotometric determination of formation constant of a monocomplex.

The above method can be applied for the determination of formation constant of the $[Fe(SCN)]^{2+}$ complex under the condition $T_{Fe} \gg T_{NCS}$ to ensure the formation of the monocomplex only. The bright red $[Fe(SCN)]^{2+}$ complex is having $\lambda_{max} = 450$ nm at which Fe^{3+} is almost nonabsorbing.

- **Determination of the formation constant from a reaction releasing protons** (*cf.* Asim K. Das, *Bull. Chem. Soc. Jpn*, **65**, 2205, 1992; *Indian J. Chem.*, **33A**, 740, 1994): The following reaction may be considered as a representative example.

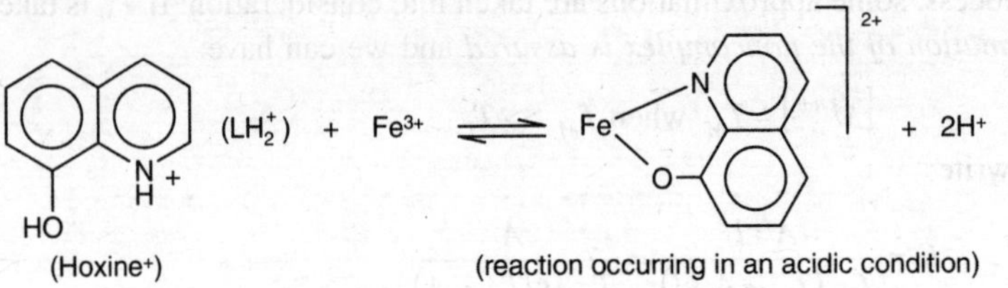

(Hoxine+)   (reaction occurring in an acidic condition)

i.e.
$$Fe^{3+} + LH_2^+ \rightleftharpoons FeL^{2+} + 2H^+, \ (Q)$$

In the acidic condition, the ligand will remain predominantly as a protonated species and hydrolysis of Fe^{3+} may be ignored.

Under the condition, $T_{Fe} \gg T_L$ (*i.e.* to ensure the formation of the monocomplex only).

We can write: $T_{Fe} = [Fe^{3+}] + [FeL^{2+}] \approx [Fe^{3+}]$

$$T_L = [L^-] + [LH] + \left[LH_2^+\right] + \left[FeL^{2+}\right]$$

$$= \frac{K_{a(1)}K_{a(2)}\left[LH_2^+\right]}{\left[H^+\right]^2} + \frac{K_{a(1)}\left[LH_2^+\right]}{\left[H^+\right]} + \left[LH_2^+\right] + \left[FeL^{2+}\right]$$

$$\approx \left[LH_2^+\right] + \left[FeL^{2+}\right]; \text{ when } \left[H^+\right] \gg K_{a(1)} \left(\approx 10^{-5}\right),\ K_{a(2)} \left(\approx 10^{-10}\right)$$

where $K_{a(1)}$ and $K_{a(2)}$ are the successive acid dissociation constants of LH_2^+; $(K_{a(1)} \gg K_{a(2)})$]

$$Q = \frac{\left[FeL^{2+}\right]\left[H^+\right]^2}{\left[Fe^{3+}\right]\left[LH_2^+\right]},\ \left[FeL^{2+}\right] = A/\varepsilon l$$

From the above relations, we can write the following expression:

$$Q = \frac{(A/\varepsilon l)\left[H^+\right]^2}{T_{Fe}\left(T_L - \dfrac{A}{\varepsilon l}\right)} = \frac{A\left[H^+\right]^2}{T_{Fe}\left(T_L \varepsilon l - A\right)}$$

or,
$$\frac{1}{Q} = \frac{T_{Fe}T_L \varepsilon l}{A\left[H^+\right]^2} - \frac{T_{Fe}}{\left[H^+\right]^2},\ \text{ or } \frac{T_{Fe}T_L \varepsilon l}{A\left[H^+\right]^2} = \frac{1}{Q} + \frac{T_{Fe}}{\left[H^+\right]^2}$$

or,
$$\frac{1}{A} = \frac{\left[H^+\right]^2}{T_{Fe}\left(QT_L \varepsilon l\right)} + \frac{1}{T_L \varepsilon l}$$

From the plot of $\dfrac{1}{A}$ vs. $\left[H^+\right]^2/T_{Fe}$, we can determine Q as follows:

$$\text{intercept} = \frac{1}{T_L \varepsilon l}; \ \text{slope} = \frac{1}{Q T_L \varepsilon l} \ \text{and} \ Q = \frac{\text{intercept}}{\text{slope}}$$

To determine the **formation constant** (*i.e.* stability constant) K from Q, we need the values of the acid dissociation constants of LH_2^+.

$$Fe^{3+} + L^- \rightleftharpoons FeL^{2+}, \ K(=\beta) = \frac{[FeL^{2+}]}{[Fe^{3+}][L^-]}$$

$$LH_2^+ \rightleftharpoons LH + H^+, \ K_{a(1)} = \frac{[LH][H^+]}{[LH_2^+]}, \ pK_{a(1)} \approx 5.0$$

$$LH \rightleftharpoons L^- + H^+, \ K_{a(2)} = \frac{[L^-][H^+]}{[LH]}, \ pK_{a(2)} \approx 9.8$$

$$Q = \frac{[FeL^{2+}][H^+]^2}{[Fe^{3+}][LH_2^+]} \times \frac{[L^-]}{[L^-]} = \frac{[FeL^{2+}]}{[Fe^{3+}][L^-]} \times \frac{[L^-][H^+]^2}{[LH_2^+]} = K \times K_{a(1)} K_{a(2)}$$

i.e.
$$K = \frac{Q}{K_{a(1)} K_{a(2)}}$$

4.8.5 Determination of Stability Constant by Ion Exchange Method

If a cation exchange resin is allowed to remain in contact with the metal ion M^{n+} then an equilibrium between the M^{n+} in resin and in solution will be attained.

$$K_r = \frac{[M^{n+}]_{\text{resin}}}{[M^{n+}]_{\text{soln.}}}$$

In presence of a suitable ligand (L), the following complex may be formed with stability constant (K).

$$M^{n+} + L \rightleftharpoons \underset{(\text{soluble})}{ML^{n+}}, \ K(=\beta) = \frac{[ML^{n+}]}{[M^{n+}][L]}$$

Because of the formation of the complex, concentration of M^{n+} will decrease in the solution phase and to maintain the constancy in K_r, concentration of the metal ion in the resin phase will also decrease. Under the condition, the *distribution coefficient* (D) is given by:

$$D = \frac{[M^{n+}]_{\text{resin}}}{[M^{n+}]_{\text{soln.}} + [ML^{n+}]_{\text{soln.}}}$$

$$= \frac{[M^{n+}]_{\text{resin}}}{[M^{n+}]_{\text{soln.}} + K[M^{n+}]_{\text{soln.}} [L]_{\text{soln.}}}$$

$$\frac{1}{D} = \frac{[M^{n+}]_{\text{soln.}}}{[M^{n+}]_{\text{resin}}} + \frac{K[M^{n+}]_{\text{soln.}} [L]_{\text{soln.}}}{[M^{n+}]_{\text{resin}}} = \frac{1}{K_r} + \frac{K[L]_{\text{soln.}}}{K_r} \qquad \qquad \text{...(a)}$$

If in presence of an excess ligand, *i.e.* $T_L \rangle\rangle T_M$, only one complex is produced then it gives: $[L]_{soln.} \approx T_L$. The plot of $\dfrac{1}{D}$ *vs.* $[L]_{soln.}$ gives a straight line (D, a measurable quantity; $[L]_{soln.} \approx T_L$, known values) having the slope = K/K_r and intercept = $1/K_r$, *i.e.* K = slope/intercept.

Illustration for the Ca(II) - citrate complex: Ca^{2+} ion is allowed to chelate with citrate which is added in the form of Na-citrate. In presence of excess Na-citrate, Na^+ ion will also be distributed between the resin and solution and a new equilibrium will be established.

$$Ca^{2+}_{resin} + 2Na^+_{soln.} \rightleftharpoons Ca^{2+}_{soln.} + 2Na^+_{resin}$$

Under the condition, we will have:

$$K'_r = \frac{[Na^+]^2_{soln.} [Ca^{2+}]_{resin}}{[Na^+]^2_{resin} [Ca^{2+}]_{soln.}}$$

or

$$K''_r = \frac{[Ca^{2+}]_{resin}}{[Ca^{2+}]_{soln.}}, \left(\begin{array}{l} \text{when } T_{Na^+} >> T_{Ca^{2+}}, \ i.e. \ \text{concentration} \\ \text{of } Na^+ \ \text{remains constant in two phases.} \end{array} \right)$$

Under the condition, distribution of Ca^{2+} in the two phases in presence of excess Na-citrate allowing the formation of Ca-citrate complex in solution will give a similar equation given by (a) and K_r will be replaced by K''_r, *i.e.*

$$\frac{1}{D} = \frac{1}{K''_r} + \frac{K[\text{citrate}]_{soln.}}{K''_r} \ ; \ [\text{citrate}]_{soln.} \approx T_{Na\text{-citrate}}$$

D can be measured at different concentrations of citrate and then the linear plot of $\dfrac{1}{D}$ *vs.* $[\text{citrate}]_{soln.}$ will give the stability constant K as follows:

$$\text{Intercept} = \frac{1}{K''_r}, \ \text{slope} = \frac{K}{K''_r}, \ i.e. \ K(=\beta) = \frac{\text{slope}}{\text{intercept}}$$

(**Note:** If the metal ion is radioactive, then concentration of the metal ion in two phases can be easily determined by measuring its radioactivity and the method becomes convenient.)

4.8.6 Polarographic Method of Determination of Stability Constants and Composition of the Complex (*see* Chapter 16, Vol. 3 of Fundamental Concepts of Inorganic Chemistry for details)

Due to the complexation, *the half-wave potential* $(E_{1/2})$ of a redox couple is altered. Generally, due to the complexation, $E_{1/2}$ (for the reduction process) becomes more negative. This shifting of $E_{1/2}$ in presence of ligands indicates the complex formation and from this shifting of $E_{1/2}$, the formation constant may be obtained.

Let us consider the following redox couple to generate the polarographic wave.

$$M^{n+} + ne \rightleftharpoons M$$

The reduction occurs on the *dropping mercury electrode* (DME) and thus the process may be represented as:

$$M^{n+} + ne + Hg \rightleftharpoons M(Hg)$$

According to Nernst Eqn., the electrode potential (*i.e.* reduction potential) of the couple is given by:

$$E = E^0 + \frac{RT}{nF} \ln \frac{[Ox]}{[Red]} = E^0 + \frac{RT}{nF} \ln \frac{[M^{n+}]}{[M(Hg)]}$$

$$= E^0 + \frac{0.059}{n} \log \frac{[M^{n+}]}{[M(Hg)]}, \quad \text{(at 25°C)} \qquad \text{...(i)}$$

In terms of $E_{1/2}$ (**half-wave potential**), we can write:

$$E_{1/2} = E^0 + \frac{0.059}{n} \log\left(\frac{k}{K}\right) - E_{ref}, \qquad \begin{array}{l}\text{(see Sec. 16.22.1, Vol. 3 of Fundamental} \\ \text{Concepts of Inorganic Chemistry)}\end{array}$$

k and K denote the diffusion factors of the reductant and oxidant respectively.
In presence of a suitable ligand (L), the complexation reaction will occur.

$$M^{n+} + xL \rightleftharpoons ML_x^{n+}, \ \beta = \frac{[ML_x^{n+}]}{[M^{n+}][L]^x}$$

$$\textit{i.e.} \quad [M^{n+}] = \frac{[ML_x^{n+}]}{\beta[L]^x} \qquad \text{...(ii)}$$

Using Eqn. (ii) in Eqn (i), we get:

$$E = E^0 + \frac{0.059}{n} \log \frac{[ML_x^{n+}]}{\beta[L]^x [M(Hg)]}$$

or,

$$E = E^0 + \frac{0.059}{n} \log \frac{[ML_x^{n+}]}{[M(Hg)]} - \frac{0.059}{n} \log \beta - \frac{0.059}{n} \log [L]^x$$

$$= E_f^0 + \frac{0.059}{n} \log \frac{[ML_x^{n+}]}{[M(Hg)]}$$

Concept of $E_{1/2}$ (*see* Vol. 3 of Fundamental Concepts of Inorganic Chemistry for details of expression of $E_{1/2}$)

E_{appl} represents the half-wave potential ($E_{1/2}$) when i (current) $= \dfrac{1}{2} i_d$, (i_d = limiting current).

$E_{appl} = E_{1/2} = E^0 + \dfrac{0.059}{n} \log\left(\dfrac{k}{K}\right) - E_{ref}$; k and K are the diffusion factors of the reductant and oxidant respectively.

It leads to: $E_{appl} = E_{1/2} + \dfrac{0.059}{n} \log\left(\dfrac{i_d - i}{i}\right) = E_{1/2} - \dfrac{0.059}{n} \log\left(\dfrac{i}{i_d - i}\right).$

where $E_f^0 = E^0 - \dfrac{0.059}{n}\log\beta - \dfrac{0.059}{n}\log[L]^x$, denotes the *formal reduction potential* of the couple

$M^{n+}/M(Hg)$ on the DME (dropping mercury electrode) surface in presence of the ligand (L). Now, $E_{1/2}$ in presence of the ligand (L) is given by (assuming no change in the diffusion factor K due to complexation):

$$\left(E_{1/2}\right)_c (say) = E_f^0 + \frac{0.059}{n}\log\left(\frac{k}{K}\right) - E_{ref}$$

Thus the shifting of $E_{1/2}$ due to complexation is given by:

$$\Delta E_{1/2} = \left(E_{1/2}\right)_s - \left(E_{1/2}\right)_c = E^0 - E_f^0 = \frac{0.059}{n}\log\beta + \frac{0.059}{n}x\log[L]$$

$(E_{1/2})_s$ and $(E_{1/2})_c$ denote the half-wave potentials for the reduction of simple metal ion (M^{n+}) and complexed metal ion (ML_x^{n+}) on DME respectively.

Thus the plot $\Delta E_{1/2}$ vs. log $[L]$ gives a straight line with:

$$\text{intercept} = \frac{0.059}{n}\log\beta, \text{ slope} = \frac{0.059x}{n}$$

When 'n' is known, from the intercept, β is obtained, and x is obtained from the slope. Otherwise, β may be obtained as follows:

$$\frac{\text{intercept}}{\text{slope}} = \frac{\log\beta}{x}.$$

4.8.7 Determination of Composition of Complexes by the Method of Continuous Variation (Job's Method) and Determination of Stability Constant from the Job's Plot

(A) Determination of the composition of a complex: This method measures a physical property called the *indicative property* (generally, absorbance or optical density) which changes due to the formation of *a stable complex* in the reaction.

$$M + nL \rightleftharpoons ML_n, \ \beta = \frac{[ML_n]}{[M][L]^n}$$

- For this purpose, a series of solutions containing the varying amounts of M and L keeping the total amount constant in each solution, *i.e.* $T_M + T_L$ = constant, is prepared but each solution has the different values of T_L and T_M. It is generally done by mixing the equimolar solutions of the metal ion and ligand to vary the composition of the mixture keeping $V_M + V_L$ = constant where V_M and V_L denote the volumes of the equimolar metal ion solution and ligand solution respectively. It is illustrated in the following series of solution having $V_M + V_L = 10$ mL

Solution No:	1	2	3	4	5	6	7	8	9	10	11
V_M (mL):	0	1	2	3	4	5	6	7	8	9	10
V_L (mL):	10	9	8	7	6	5	4	3	2	1	0

If 'C' be the molar concentration of both the metal and ligand solution used to prepare the series of solutions then at a particular solution, we have:

$$T_M = \frac{V_M C}{V_M + V_L} = \frac{V_M C}{10} \text{ and } T_L = \frac{(10 - V_M)C}{10}$$

and

$$T_M + T_L = \frac{V_M C}{10} + \frac{(10 - V_M)C}{10} = C \text{ (constant)}$$

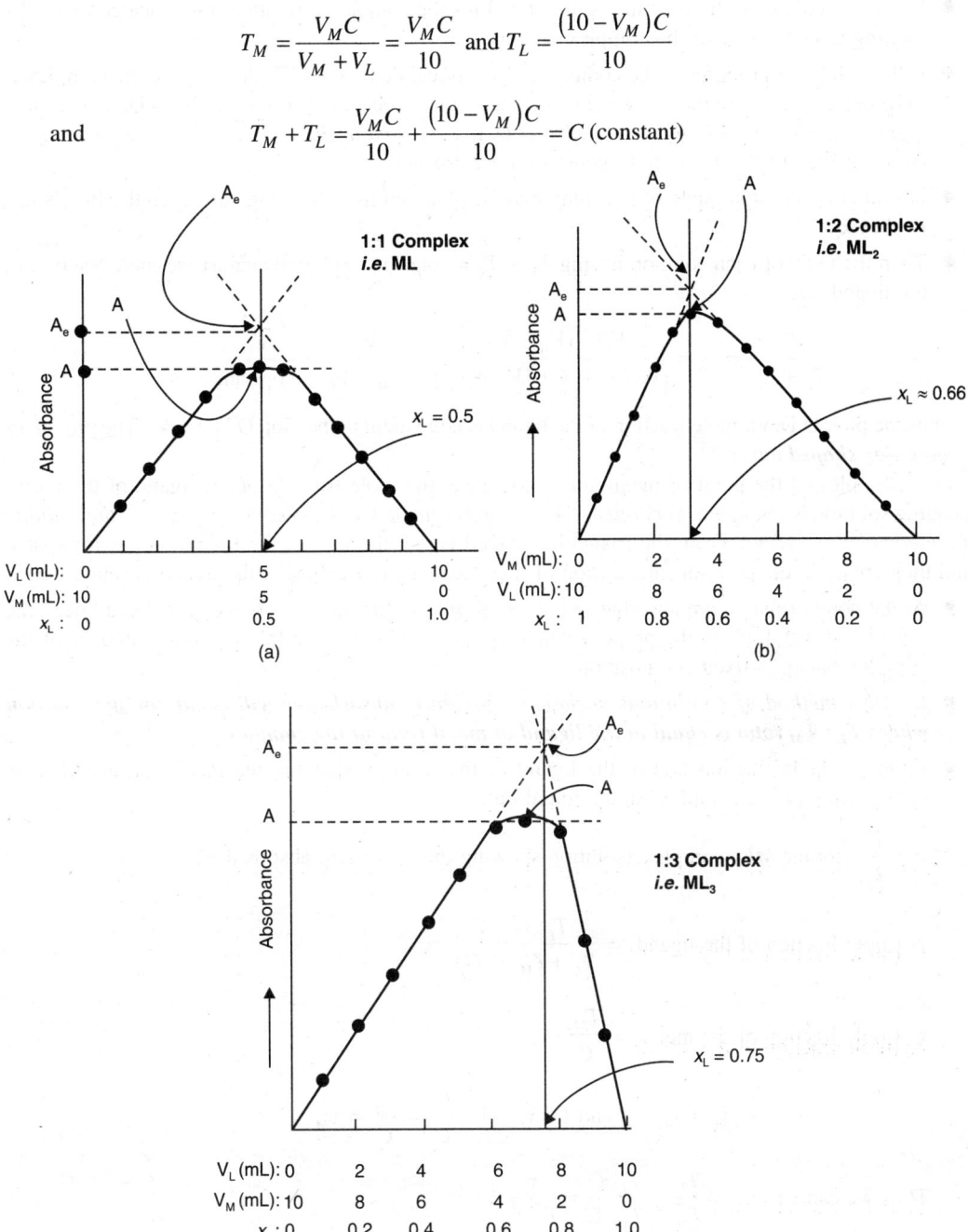

Fig. 4.8.7.1 Job's plot when only one stable complex is formed. (a) for 1:1 complex, *i.e.* *ML*; (b) for 1:2 complex, *i.e.* ML_2; (c) for 1:3 complex, *i.e.* ML_3. **Note:** For ML_n, the maximum appears at x_L (mole fraction of the ligand) given by $n = x_L/(1 - x_L)$, *i.e.* $x_L = n/(1 + n)$.

- Now it is required to find out the optimum pH for the complex formation. It is done generally by plotting O.D. (at λ_{max} of the complex) against pH.
- If the indicative property of the complex is the optical density (O.D.) then λ_{max} of the complex is to be determined from the full wavelength spectrum of the complex. Generally, O.D. is measured at the λ_{max}. *But to avoid the complexity, O.D. is measured at the wavelength where the complex is absorbing but the metal and ligand are nonabsorbing.*
- Useful concentration range of the study may be obtained from the range in which the Beer's law is obeyed.
- Then the O.D. of each solution having $T_M + T_L$ = constant is plotted against the mole fraction of the ligand, *i.e.*

$$\frac{T_L}{T_L + T_M} = x_L = \frac{V_L C / (V_M + V_L)}{(V_M C + V_L C)/(V_M + V_L)} = \frac{V_L}{V_M + V_L} = \frac{V_L}{\text{constant}}$$

Thus the plot, O.D. *vs.* mole fraction of the ligand is equivalent to the plot, O.D. *vs.* V_L. The plot gives a *triangular shaped curve.*

From the plot, at the point of maximum absorbance, the mole fraction of the ligand of the corresponding solution is found out. *Very often, the peak gets rounded and it becomes sharper as the stability of the complex increases.* From the triangular shaped curve, the legs of the triangle are extrapolated until they cross. At the point of intersection of these legs, mole fraction of the ligand is found out.

- At the point of maximum absorbance (*i.e.* point of intersection of the two extrapolated lines), the ligand and metal are in the proper relative concentrations to give the maximum amount of the complex having a fixed composition.
- *In Job's method of continuous variation, maximum absorbance will occur for the solution where $T_L : T_M$ ratio is equal to the ligand to metal ratio in the complex.*
- From the mole fraction (x_L) of the ligand of the solution showing the maximum absorbance, composition of the complex can be found out.

$$n = \frac{T_L}{T_M} \text{ for the } ML_n \text{ complex, (solution showing the maximum absorbance)}$$

$$x_L \text{ (mole fraction of the ligand)} = \frac{T_L}{T_L + T_M} = \frac{T_L}{C}$$

$$x_M \text{ (mole fraction of the metal)} = \frac{T_M}{C}$$

$$x_L + x_M = 1 \text{ and } 1 - x_L = 1 - \frac{T_L}{C} = \frac{T_M}{C} = x_M$$

Thus we can write: $n = \dfrac{T_L}{T_M} = \dfrac{T_L/C}{T_M/C} = \dfrac{x_L}{1 - x_L}$;

$$1 + n = 1 + \frac{x_L}{1 - x_L} = \frac{1}{1 - x_L} = \frac{1}{x_M}; \text{ i.e. } x_M = \frac{1}{1 + n}$$

● The condition $n = \dfrac{T_L}{T_M}$ (for the complex ML_n) where T_L and T_M denote the respective concentrations of the ligand and metal of the solution showing the highest absorbance, can be realized as follows:

$$M + nL \rightleftharpoons ML_n, \; \beta = \frac{[ML_n]}{[M][L]^n};$$
(assuming only the ML_n complex to remain under the experimental condition)

or, $$[M][L]^n = \frac{1}{\beta}[ML_n]$$ ...(i)

$$T_M = [M] + [ML_n]$$ ...(ii)

$$T_L = [L] + n[ML_n]$$ ...(iii)

$$T_M + T_L = C = \text{constant}$$ ...(iv)

Differentiating the Eqns. (i)-(iv) with respect to $[L]$ we get:

$$\frac{d[M]}{d[L]}[L]^n + n[M][L]^{n-1} = \frac{1}{\beta}\frac{d[ML_n]}{d[L]}$$ ...(v)

$$\frac{d[T_M]}{d[L]} = \frac{d[M]}{d[L]} + \frac{d[ML_n]}{d[L]}$$ ...(vi)

$$\frac{d[T_L]}{d[L]} = 1 + n\frac{d[ML_n]}{d[L]}$$ ...(vii)

$$\frac{d[T_M]}{d[L]} + \frac{d[T_L]}{d[L]} = 0$$

or $$\frac{d[T_M]}{d[L]} = -\frac{d[T_L]}{d[L]}$$ ...(viii)

Under the condition of maximum complex formation, we have the condition:

$$\frac{d[ML_n]}{d[L]} = 0$$

By using the above condition in Eqns. (v)-(vii), we get:

$$n[M][L]^{n-1} = -\frac{d[M]}{d[L]}[L]^n$$ ...(ix)

$$\frac{d[T_M]}{d[L]} = \frac{d[M]}{d[L]}$$ ...(x)

$$\frac{d[T_L]}{d[L]} = 1$$ ...(xi)

Thus we can write from Eqns. (viii) and (xi):

$$\frac{d[T_M]}{d[L]} = -\frac{d[T_L]}{d[L]} = -1 \qquad \qquad \ldots(\text{xii})$$

From the Eqn. (ix), we have:

$$n[M][L]^{n-1} = -\frac{d[M]}{d[L]}[L]^n$$

$$= -\frac{d[T_M]}{d[L]}[L]^n, \ (cf. \ \text{Eqn. x})$$

$$= [L]^n, \ (cf. \ \text{Eqn. xii})$$

or, $\quad n = \dfrac{[L]^n}{[M][L]^{n-1}} = \dfrac{[L]}{[M]} = \dfrac{T_L - n[ML_n]}{T_M - [ML_n]}, \ \left(cf. \ T_L = [L] + n[ML_n] \ \text{and} \ T_M = [M] + [ML_n]\right)$

or, $\quad nT_M - n[ML_n] = T_L - n[ML_n]$

or, $\quad n = \dfrac{T_L}{T_M}$ (at the condition of maximum ML_n complex formation)

Thus, maximum absorbance will occur for the solution where $T_L : T_M$ ratio is equal to the ligand to metal ratio in the complex.

● The method can be generalized for the system.

$$mM + nL \rightleftharpoons M_m L_n$$

● **Assumptions and requirements of the method:** These are given below:
  (i) Only one type complex having a fixed composition is formed in the system.
  (ii) The complex must be very stable, *i.e.* extent of dissociation of the complex is negligibly small.
  (iii) There must have a suitable indicative property (measurable) which is a linear function of the concentration of the complex.

(B) Evaluation of stability constant (β) from the Job's plot: Let us illustrate for the 1:1 complex (*i.e.* ML). Generally, the peak of the curve is deviated from the point of intersection of the two extrapolated lines of the triangular shaped curve. *The degree of this deviation depends on the stability constant. The deviation is less for a more stable complex.* Thus, the degree of deviation gives the measure of stability constant.

If the extrapolated absorbance is A_e (which is obtained from the point of intersection of the two extrapolated lines, *cf.* Fig. 4.8.7.1) and A be corresponding absorbance obtained from the curve, then the difference $(A_e - A)$ arises due to the dissociation of the complex. Thus we can write:

$A \propto [ML]$ (actually present in solution).

$A_e \propto$ expected $[ML]$ considering no dissociation (*i.e.* $\beta \to \infty$) $\propto$ total concentration of the metal or ligand, whichever is the limiting concentration to determine $[ML]$ at the point considered (let it be $C_{\lim}$).

i.e. $\qquad\qquad \dfrac{A}{A_e} = \dfrac{[ML]}{C_{\lim}}$ where $C_{\lim} = T_M$ or T_L which is the limiting factor.

or $\qquad\qquad [ML] = \left(\dfrac{A}{A_e}\right) C_{\lim}$

Actual concentration of *ML* in terms of *A* and *A_e*

Actually, $A_e = \varepsilon C_{\text{lim}} l$, l = path length, ε = molar extinction coefficient of the complex, *i.e.* $\varepsilon = \dfrac{A_e}{C_{\text{lim}} l}$.

Then from the measured absorbance (A) of a solution, actual concentration of *ML* present in the

solution can be determined, $[ML] = \dfrac{A}{\varepsilon l} = \dfrac{A}{A_e} \times C_{\text{lim}}$.

For 1:1 complex,

$$\beta_1 = \frac{[ML]}{[M][L]} = \frac{[ML]}{\left(T_M - [ML]\right)\left(T_L - [ML]\right)} = \frac{\left(A/A_e\right)C_{\text{lim}}}{\left[T_M - \left(A/A_e\right)C_{\text{lim}}\right]\left[T_L - \left(A/A_e\right)C_{\text{lim}}\right]}$$

For 1:2 complex,

$$M + 2L \rightleftharpoons ML_2, \beta_2 = \frac{[ML_2]}{[M][L]^2}$$

If the limiting concentration (C_{lim}) is determined by T_L then $C_{\text{lim}} = T_L/2$; if C_{lim} is determined by T_M, then $C_{\text{lim}} = T_M$. From the knowledge of C_{lim} and A_e (extrapolated value of absorbance), ε value of the complex is determined. Then from the measured absorbance, $[ML_2]$ is obtained for a particular solution. Thus β_2 is obtained as follows:

$$\beta_2 = \frac{[ML_2]}{\left(T_M - [ML_2]\right)\left(T_L - 2[ML_2]\right)^2}; \ \left(\text{assuming, } T_M = [M] + [ML_2], T_L = [L] + 2[ML_2]\right)$$

where $[ML_2] = \left(\dfrac{A}{A_e}\right)C_{\text{lim}}$.

In the same way, β_n for $M + nL \rightleftharpoons ML_n$ can be determined.

4.8.8 Determination of Composition of a Complex by Mole Ratio Method

Due to the formation of the complex, $M_m L_n$ having a very high stability constant, in the following reaction,

$$mM + nL \rightleftharpoons M_m L_n, \text{ where } [M_m L_n] = \frac{T_L}{n} \text{ or } \frac{T_M}{m} \text{ depending on the condition.}$$

the physical property, *i.e.* indicative property (*e.g.* conductance, optical density, acidity, electrode potential) of the solution will change with the change of the ratio of molar concentration of the metal and ligand provided the physical property is a linear function of the concentration of the complex. If the complex is of a high stability constant, the equilibrium will be shifted to the right hand side and *the ligand to metal ratio in the complex is equal to the molar ratio of the ligand and the metal in the solution* having the minimum (T_L/T_M) ratio (at a fixed T_M value) but *showing the maximum change of the indicative property* (under the experimental condition discussed below) (*cf.* Fig. 4.8.8.1).

In this method, a number of solutions are prepared in which concentration of the metal (T_M) is kept fixed while concentration of the ligand (T_L) is varied. Generally, for preparing such sets of solutions, *equimolar solutions* of the metal and ligand are used. It is illustrated below for a representative set. If the ML_3 complex is formed, then its concentration, *i.e.* $[ML_3]_{max}$ will gradually change with the ratio $(T_L/T_M)_{T_M}$ and then it will remain constant after a certain value of $(T_L/T_M)_{T_M}$.

Solution No:	1	2	3	4	5	6	7	8	9	10	11	12	13
V_M (mL):	1	1	1	1	1	1	1	1	1	1	1	1	1
V_L (mL):	0.2	0.4	0.6	0.8	1.0	1.25	1.50	1.75	2.0	2.5	3.0	3.5	4.0
$(T_L/T_M)_{T_M}$:	0.2	0.4	0.6	0.8	1.0	1.25	1.50	1.75	2.0	2.5	3.0	3.5	4.0

$$[ML_3]_{max} \propto \quad \frac{0.2}{3} \quad \frac{0.4}{3} \quad \frac{0.6}{3} \quad \frac{0.8}{3} \quad \frac{1.0}{3} \quad \frac{1.25}{3} \quad \frac{1.50}{3} \quad \frac{1.75}{3} \quad \frac{2.0}{3} \quad \frac{2.5}{3} \quad \underbrace{\frac{\mathbf{3.0}}{\mathbf{3}} \quad \frac{1.0}{1} \quad \frac{1.0}{1}}_{\substack{\text{Constant}}}$$

determined by T_L/n determined by T_M/m

Here it may be noted that for the above mentioned solutions, a *fixed total volume*, *constant acidity* (optimum value for complex formation) and *constant ionic strength* will be maintained.

Now the change of a suitable physical property (say optical density measured at a particular wavelength where the complex shows a significant absorbance while the metal and ligand are particularly nonabsorbing) which is a linear function of the concentration of the complex is measured for all the solutions. If the metal and ligand are nonabsorbing at the wavelength of measurement of the optical density, then the measured optical density gives the required change of optical density, *i.e.* absorbance (A). These changes are plotted against the molar ratio of T_L and T_M at fixed T_M, *i.e.* $\left(T_L/T_M\right)_{T_M}$. For the M_mL_n complex, A will linearly increase with the ratio $(T_L/T_M)_{T_M}$ upto its value n/m and then A will remain constant for the further increase of the ratio $(T_L/T_M)_{T_M}$.

- The plot A vs. $\left(T_L/T_M\right)_{T_M}$ will show a number of linear segments intersecting at such values of $\left(T_L/T_M\right)_{T_M}$ representing the molar ratio of the ligand to metal in the complexes existing under the experimental conditions.

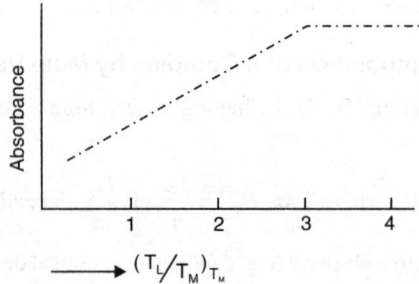

Fig. 4.8.8.1 Mole ratio plot for **1:3 complex, *i.e.* ML_3.**

- If only one complex say M_mL_n of sufficient stability is formed in the system, then two linear segments will intersect at $\left(T_L/T_M\right)_{T_M} = \dfrac{n}{m}$. Thus if the only one complex, say ML is formed in the system, then the two linear segments will intersect at $\left(T_L/T_M\right)_{T_M} = 1$. If ML_3 only exists in the condition, then two linear segments will intersect at $\left(T_L/T_M\right)_{T_M} = 3.0$.

- If under the experimental condition, the complex ML_3 only exists, then its concentration will gradually increase with the increase of the ratio $(T_L/T_M)_{T_M}$ upto its value $= 3.0$ and then $[ML_3]$ will remain constant assuming the very high value of the stabilty constant. In reality, $[ML_3]$ will increase very slightly beyond the ratio $(T_L/T_M)_{T_M} > 3.0$.
- If in the system, two different complexes, say ML and ML_2 are formed, then three linear segments will intersect at the mole ratio 1 and 2 respectively. Here it is assumed that the stepwise stability constants K_1 and K_2 are widely different.
- *If the successive stability constants do not differ widely, then the different linear segments will not develop properly. This is why, each complex should be sufficiently stable, otherwise the method is not applicable.*

4.8.9 Determination of Composition of Complexes by Slope Ratio Method

Let us consider the following complexation reaction leading to a sufficiently stable M_mL_n complex.

$$mM + nL \rightleftharpoons M_mL_n, \quad \beta = \frac{[M_mL_n]}{[M]^m[L]^n}$$

If β is sufficiently high then it is reasonable to assume that in presence of large excess of either M or L, dissociation of the complex will be negligibly small. **Under this condition, at equilibrium, concentration of the complex will be proportional to the analytical concentration of the other reactant L or M which is not in excess.**

- Thus under the condition, $T_M \gg T_L$, $[M_mL_n] = \dfrac{T_L}{n}$ *i.e.* $[M_mL_n] \propto T_L$

    and under the condition, $T_L \gg T_M$, $[M_mL_n] = \dfrac{T_M}{m}$ *i.e.* $[M_mL_n] \propto T_M$

    where T_M and T_L denote the total analytical concentrations of the metal and ligand respectively and $[M_mL_n]$ gives the concentration of the complex at equilibrium in the mixture.
- If the complex is coloured and its absorbance is sufficiently high at a particular wavelength where the ligand and metal are practically nonabsorbing, then measurement of absorbance (A) at that particular wavelength gives the concentration of the complex according to Beer's law, *i.e.*

$$A = \varepsilon[M_mL_n]l \text{ or; } [M_mL_n] = \frac{A}{\varepsilon l}$$

where ε = molar extinction coefficient of the complex; l = path length.
- For a series of solutions having a **constant large excess of T_M** and varying the concentrations of T_L, the absorbance (A_M) is given by:

$$A_M = \varepsilon[M_mL_n]l = \varepsilon(T_L/n)l, \ T_M \gg T_L \qquad \qquad \dots\text{(a)}$$

Plot of A_M vs. T_L gives a straight line with a slope given by: $(\text{slope})_M = \dfrac{\varepsilon l}{n} \qquad \dots\text{(b)}$

Similarly, for a series of solutions having a **constant large excess of L** and varying the concentrations of T_M, the absorbance (A_L) is given by:

$$A_L = \varepsilon[M_mL_n]l = \varepsilon(T_M/m)l, \ T_L \gg T_M \qquad \qquad \dots\text{(c)}$$

Plot of A_L vs. T_M gives a straight line with a slope given by: $(\text{slope})_L = \dfrac{\varepsilon l}{m} \qquad \dots\text{(d)}$

● The ratio of the slopes gives the following result.

$$\frac{(\text{slope})_M}{(\text{slope})_L} = \frac{\varepsilon l}{n} \times \frac{m}{\varepsilon l} = \frac{m}{n}$$

It indicates that for the 1:1 complex (*i.e.* $m = n = 1$), the two straight lines (*i.e.* A_L vs. T_M and A_M vs. T_L) will run parallel with the same slopes. For the 1:2 complex (*i.e.* $m = 1$ and $n = 2$), $(\text{slope})_L = 2(\text{slope})_M$; similarly for 1:3 complex (*i.e.* $m = 1, n = 3$), $(\text{slope})_L = 3(\text{slope})_M$. These are illustrated in Fig. 4.8.9.1.

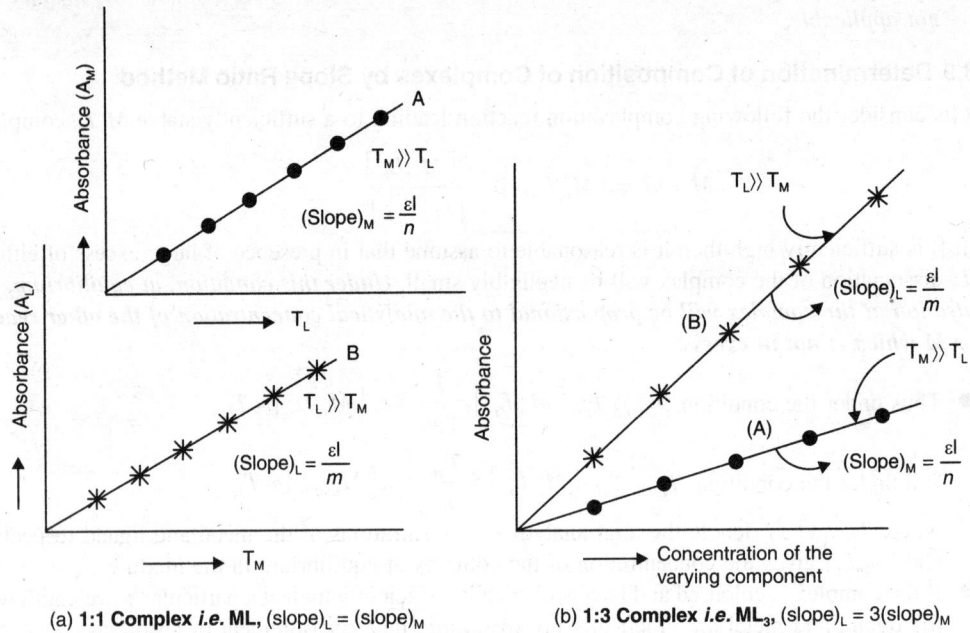

(a) **1:1 Complex *i.e.* ML**, $(\text{slope})_L = (\text{slope})_M$ (b) **1:3 Complex *i.e.* ML$_3$**, $(\text{slope})_L = 3(\text{slope})_M$

Fig. 4.8.9.1 Representation of the slope ratio method for 1:1 and 1:3 complex. One plot: Absorbance *vs.* T_L under the condition $T_M \gg T_L$ keeping T_M fixed; Other plot: Absorbance *vs.* T_M under the condition $T_L \gg T_M$ keeping T_L fixed.

4.9 STABILIZATION OF UNUSUAL (OR LESS COMMON) OXIDATION STATES OF METALS

For the transition metal ions, the incompletely filled *d*-electrons can participate in bonding to different extents depending on the chemical environment around the metal centre. This is why, the variable oxidation states are well documented in the transition metal complexes. But, for a particular metal, all the oxidation states are not equally stable. Possible oxidation states of the first transition series elements are given in Table 4.9.1.

In Table 4.9.1, the oxidation states other than those characterized by * and ** are generally referred to as the ***unusual oxidation states.*** The chemical environment required to stabilize the high oxidation states is different from that required in stabilizing the low oxidation states. These are discussed in Secs. 4.9.1 and 4.9.2.

Thermodynamic stability *vs.* kinetic stability: Thermodynamic stability can be measured from different parameters like *stability constant* (β), *reduction potential* (E^0), *solubility product* (K_{sp}) of the insoluble compounds, etc.

Table 4.9.1 Possible oxidation states of first transition series elements

Element	Oxidation States								
	+7*	+6*	+5*	+4*	+3**	+2	+1	0	-1
Ti				+4*	+3**	+2		0	-1
V			+5*	+4**	+3	+2	+1	0	-1
Cr		+6*	+5	+4	+3*	+2	+1	0	
Mn	+7*	+6	+5	+4	+3	+2*	+1	0	-1
Fe		+6	+5	+4	+3*	+2*	+1	0	
Co				+4	+3*	+2*	+1		-1
Ni				+4	+3	+2*	+1	0	
Cu					+3	+2*	+1**		

$* \Rightarrow$ most stable oxidation state; $** \Rightarrow$ stable but not so stable designated by $*$

- Higher β-value signifies higher stability.
- Higher reduction potential signifies lower stability of the higher oxidation state.
- Lower solubility product (K_{sp}) of an insoluble compound signifies its higher stability.

However, thermodynamic stability cannot always give the real picture. For example, from the E^0-value, MnO_4^- should oxidize water in acidic condition, *i.e.* MnO_4^- is thermodynamically unstable in aqueous acidic media. But in reality, under the condition of volumetric analysis, loss of MnO_4^- through the oxidation of water can be ignored because of the very slow rate of the process. *Thus, the kinetic aspect (i.e.* rate of the reaction) *is of an important point of consideration to define the stability.*

Complexation can lead to both the *thermodynamic and kinetic stabilization.* These aspects have been discussed separately.

- Formation of the insoluble compounds can stabilize a particular oxidation state, *e.g.* $NaBiO_3$ (Bi in +5 oxidation state), CuI, etc. are stable because of their insoluble character. Cu^+ disproportionates in aqueous solution to Cu^{2+} and Cu^0 but it is stabilized in insoluble CuI.

Enhanced Stability of the Higher Oxidation States for the 4d and 5d Series

This occurs due to many factors: lower ionization potentials, high *Dq* value, relativistic effect causing an expansion of *d*-orbitals (*see* Chapter 8, Vol. 1).

4.9.1 Stabilization of Higher Oxidation States (*cf.* Sec. 3.14.16)

Metal ions in higher oxidation states become oxidizing (*i.e.* electron seeker). This is why, the surrounding environment must be resistant to defy the oxidation. Thus the ligands must be *nonoxidizable* and *nonpolarizable* (*i.e.* hard in terms of HSAB concept).

- **Nonoxidisable and nonpolarisable anionic ligands:** F^- and O^{2-} are **nonoxidizable and nonpolarizable** and they can stabilize many higher oxidation states, *e.g.* K_2MF_6 (M = Co, Ni, etc. in +4 oxidation state), PtF_6 (+6 oxidation state), $O_2^+ PtF_6^-$ (Pt in +5 state), Ni_2O_3 (Ni in +3 state), FeO_4^{2-} (Fe in +6 state), etc. O^{2-} can stabilize the highest possible oxidation states of V, Cr and Mn as VO_4^{3-}, CrO_4^{2-} and MnO_4^- respectively.

- **Nonoxidisable bulky anionic ligands with high negative charges producing an entropic favour:** Periodate (IO_6^{5-}) and tellurate (TeO_6^{6-}) are *very much resistant to oxidation.* In these species, the respective central element exists at its highest possible oxidation state, *i.e.* I in +7 state; Te in +6 state. These ligands are promising to stabilize the higher oxidation states of metals. The well documented examples are:

$$M_7^I[Ag^{III}(IO_6)_2].nH_2O \; ; \qquad\qquad M_9^I[Ag^{III}(TeO_6)_2].nH_2O \; ;$$

$$M_9^I[Cu^{III}(TeO_6)_2].nH_2O \; ; \qquad\qquad M^I[Ni^{IV}(IO_6)].nH_2O.$$

Stabilization of the species may be explained at least in part by considering the **entropic favour** in the formation of the compounds. The higher valent cations like M^{III}, M^{IV} are strongly solvated but combination with the anions leads to the neutralization of charge. It releases the solvent molecules held around the cation. Solvent molecules held with the anions through the ion-dipole interaction are also released due to the charge neutralisation in the complex formation. *Thus the entropy increases* and it is favoured for the large anions with high charge.

- **Alkaline condition:** The metal ions in high oxidation states are electron deficient and **consequently they are highly acidic.** *This is why, the high oxidation states are frequently stabilized in strongly alkaline media* and the synthesis of such compounds is very often carried out in an alkaline media. For example, NaOBr in alkaline media can oxidize Ni(II) to Ni(III) as Ni_2O_3 and Ni(IV) as $[Ni^{IV}(dmg)_2]$ (H_2dmg = dimethylglyoxime); anodic oxidation of Fe(III) in alkaline condition leads to ferrate FeO_4^{2-}. Formation of the higher oxidation state in an alkaline condition is favoured and this aspect can be rationalized by considering the following equilibria.

$$\left[M(OH_2)_6\right]^{n+} \underset{H^+}{\overset{OH^-}{\rightleftharpoons}} \left[M(OH)(OH_2)_5\right]^{(n-1)+} \underset{H^+}{\overset{OH^-}{\rightleftharpoons}} \left[M(OH)_2(OH_2)_4\right]^{(n-2)+} \rightleftharpoons$$

Formation of the hydroxo-species (*i.e. less positively charged or anionic species*) is favoured in an alkaline condition and **from such species electron removal leading to the higher oxidation gets favoured.** On the other hand, *in an acidic condition, more positively charged aqua species predominantly exist* and electron removal from such species is a more difficult task. In fact, transition metal hydroxides are readily oxidized in an alkaline medium. This is well known from the formation of the oxo species of Cr(VI), Mn(VI), Fe(VI) in the **oxidative fusion** of their lower valent compounds in an alkaline medium.

Stabilization of the higher oxidation states in an alkaline condition is evidenced from the decrease of reduction potential. It is illustrated in the following redox couples.

$$\begin{cases} Mn(aq)^{3+} + e \rightleftharpoons Mn(aq)^{2+}, & E^0 = +1.51V \\ Mn(OH)_3 + e \rightleftharpoons Mn(OH)_2 + OH^-, & E^0 = +0.15V \end{cases}$$

$$\begin{cases} Fe(aq)^{3+} + e \rightleftharpoons Fe(aq)^{2+}, & E^0 = +0.76V \\ Fe(OH)_3 + e \rightleftharpoons Fe(OH)_2 + OH^-, & E^0 = -0.56V \end{cases}$$

It happens so due to the lower solubility product of the hydroxides of higher oxidation state:

K_{sp}:	$Mn(OH)_2$	$Mn(OH)_3$	$Fe(OH)_2$	$Fe(OH)_3$
	$\sim 10^{-13}$	$\sim 10^{-36}$	$\sim 10^{-15}$	$\sim 10^{-38}$

These aspects have been discussed and illustrated in *Chapter 16, Vol. 3 of Fundamental Concepts of Inorganic Chemistry.*

- **Nonoxidisable organic chelating ligands:** The nonoxidizable organic chelating ligands having *the good σ-donor sites* can stabilize the higher oxidation states. *The metal ions in higher oxidation* *states are electron deficient and favour the good* σ-*donor sites.* Such metal centres are reluctant in participating the $M \rightarrow L$ π-back bonding that occurs for the π-acid ligands.

In this regard, ***ethylenedibiguanide*** is highly promising. It can stabilize Ni(III), Ag(III), Mn(III) and Mn(IV). The complex of Ag(III) is shown below (**P. Ray**).

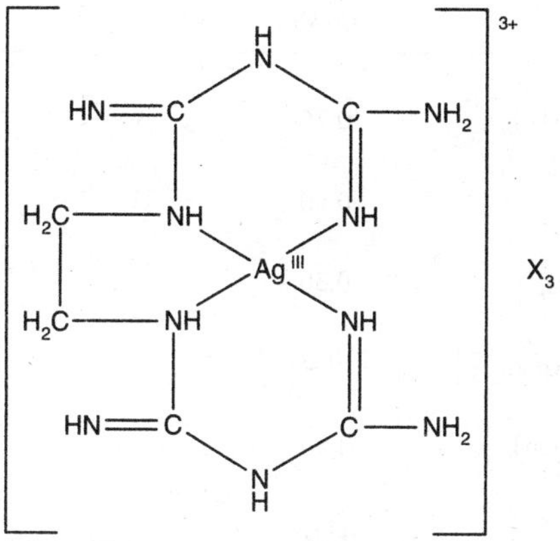

$$X^- = NO_3^-, \; ClO_4^-, \; \frac{1}{2} SO_4^{2-}$$

Pyridine and its derivatives, *e.g.* Hpic (pyridine–2–carboxylic acid, *i.e.* picolinic acid), bpy, phen, etc. can stabilize Ag(II) through the complexation, *e.g.* $[Ag(pic)_2]$, $[Ag(py)_4]^{2+}$, $[Ag(bpy)_2]^{2+}$, etc. These are prepared by predisulfate oxidation. Here it is worth mentioning that bpy, phen can also participate in the $M \rightarrow L$ π-back bonding. But in the present case, probably this is not important because of the high oxidation state of the metal centre.

Stabilization of the C^{4+}-state

R.S. Nyholm (1964) became successful in preparing $[C(\textbf{diars})_2]^{4+}$ in the reaction between diars and CCl_4.

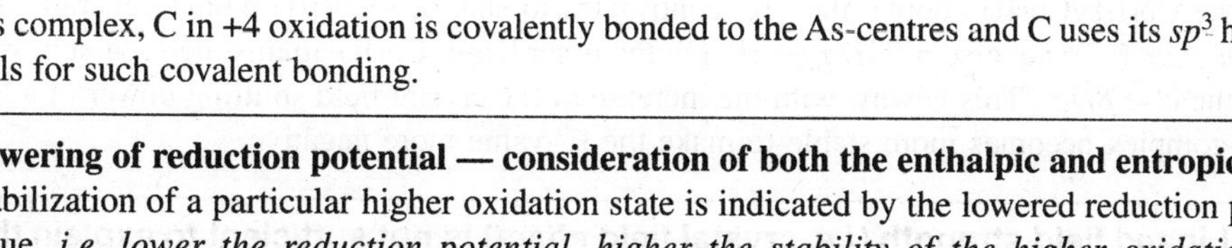

In this complex, C in +4 oxidation is covalently bonded to the As-centres and C uses its sp^3 hybrid orbitals for such covalent bonding.

● **Lowering of reduction potential — consideration of both the enthalpic and entropic factor:**
Stabilization of a particular higher oxidation state is indicated by the lowered reduction potential value, *i.e. lower the reduction potential, higher the stability of the higher oxidation state.* Complexation may stabilize the higher and lower oxidation states of a particular redox couple to different extents. If the higher oxidation state is stabilized more than the lower oxidation state, the reduction potential decreases. On the other hand, if the lower oxidation state is stabilized more,

the reduction potential increases. This aspect is illustrated by taking the Fe^{3+} / Fe^{2+} couple and Co^{3+}/Co^{2+} couple (*cf.* Chapter 16, Vol. 3 of Fundamental Concepts of Inorganic Chemistry).

Table 4.9.1.1 Effect of complexation on the E^0 value of Fe(III)/Fe(II) couple

Couple	$E^0(V)$	Conclusion, stability sequence
$\left[Fe(OH_2)_6\right]^{3+} + e \rightleftharpoons \left[Fe(OH_2)_6\right]^{2+}$	0.76	–
$\left[FeF_6\right]^{3-} + e \rightleftharpoons \left[FeF_6\right]^{4-}$	0.40	Fe(III) ⟩ Fe(II)
$\left[Fe(CN)_6\right]^{3-} + e \rightleftharpoons \left[Fe(CN)_6\right]^{4-}$	0.36	Fe(III) ⟩ Fe(II)
$\left[Fe(oxinate)_3\right] + e \rightleftharpoons \left[Fe(oxinate)_3\right]^{-}$	−0.20	Fe(III) ⟩ Fe(II)
$\left[Fe(phen)_3\right]^{3+} + e \rightleftharpoons \left[Fe(phen)_3\right]^{2+}$	+1.14*	Fe(II) ⟩ Fe(III)
$\left[Fe(bpy)_3\right]^{3+} + e \rightleftharpoons \left[Fe(bpy)_3\right]^{2+}$	0.96*	Fe(II) ⟩ Fe(III)

* The π-acid ligands like bpy and phen stabilise the lower oxidation state better.

Table 4.9.1.2 Effect of complexation on the E^0 value of Co(III)/Co(II) couple (cf. Sec. 3.14.16)

Couple	$E^0(V)$	Stability sequence
$\left[Co(OH_2)_6\right]^{3+} + e \rightleftharpoons \left[Co(OH_2)_6\right]^{2+}$	+1.84	
$[Co(edta)]^{-} + e \rightleftharpoons [Co(edta)]^{2-}$	0.60	Co(III) ⟩ Co(II)
$\left[Co(phen)_3\right]^{3+} + e \rightleftharpoons \left[Co(phen)_3\right]^{2+}$	0.42	Co(III) ⟩ Co(II)
$\left[Co(NH_3)_6\right]^{3+} + e \rightleftharpoons \left[Co(NH_3)_6\right]^{2+}$	0.10	Co(III) ⟩ Co(II)
$\left[Co(en)_3\right]^{3+} + e \rightleftharpoons \left[Co(en)_3\right]^{2+}$	−0.26	Co(III) ⟩ Co(II)
$\left[Co(CN)_6\right]^{3-} + e \rightleftharpoons \left[Co(CN)_5\right]^{3-} + CN^{-}$	−0.83	Co(III) ⟩ Co(II)

For the Co(III)/Co(II) couple, the *cfse* contributes to stabilize Co(III) which generally attains the *low-spin state* (t_{2g}^6), i.e. cfse = $24Dq_0 - 2P$. On the other hand, Co(II) attains the high-spin state ($t_{2g}^5 e_g^2$) having the cfse $8Dq_0$. This is why, with the increase of the crystal field splitting power of a ligand, the Co(III)-complex becomes more stable to make the E^0-value more negative.

Ligand field strength (*i.e.* crystal field effect) is not sufficient to explain the variation of E^0 value of the Co(III) /Co(II) couple (*cf.* Sec. 3.14.16)

The above aspect is illustrated by taking the ligands en, phen, and CN^{-}.

- Considering the high-spin Co(II) (*i.e.* $t_{2g}^5 e_g^2$) and low spin Co(III) (*i.e.* $t_{2g}^6 e_g^0$) complexes, the **gain of cfse** (an enthalpic factor) in moving from the Co(II) complex to the Co(III) complex is: $(24Dq_{Co(III)}$ $- 2P_{Co(III)}) - (8Dq_{Co(II)}) \approx 28Dq_{Co(II)} - 2P_{Co(III)}$ taking $24Dq_{Co(III)} \approx 36Dq_{Co(II)}$. Thus the calculated value of the gain of cfse becomes *ca.* 336 kcal mol^{-1} (for en), 358 kcal mol^{-1} (for phen). This is expected as phen is a stronger field ligand than en. Thus, consideration of cfse only indicates that stabilization of Co(III) relative to that of Co(II) is more in the case of phenanthroline than that in the case of *en*. *It should make the E^0-value (i.e. standard reduction potential) more negative for phen than that for en.* But the reverse is true.

  In fact, for the phen ligand, Co(II) is stabilized to some extent. Because of the π-acidic character (*cf.* en does not have any π-acidic character) of phen, it can stabilize Co(II) better than Co(III), because Co(III) is a poorer π-donor for its higher positive charge. **Thus the metal → ligand π-bonding effect opposes the cfse effect in determining the relative stabilities of Co(III) and Co(II).** Because of this fact, stabilization of Co(III) relative to that of Co(II) is less than 358 kcal mol^{-1} for the phen ligand. In fact, en stabilizes Co(III) relative to Co(II) more than phen. This is why, E^0 value of Co(III)/Co(II) is more negative for the en-complex than that for the phen complex.

- CN$^-$ is a very strong field ligand and both Co(III) and Co(II) are in the low spin state. Thus the **gain of cfse** in the corresponding Co(III)-complex is: $(24Dq_{Co(III)} - 2P_{Co(III)}) - (18Dq_{Co(II)} - P_{Co(II)})$. It leads to about 398 kcal mol^{-1}. Again, CN$^-$ a strong π-acid ligand prefers the lower oxidation state Co(II) more. But compared to phen (neutral π-acid ligand), the anionic π-acid ligand CN$^-$ can stabilise better the higher oxidation state Co(III). Moreover, the **entropic favour** for the oxidation of Co(II) to Co(III) makes the E^0 value more negative for the CN$^-$ ligand.

$$\left[Co(CN)_5 \right]^{3-} + CN^- \rightleftharpoons \left[Co(CN)_6 \right]^{3-} + e, \; \Delta S \; (\text{due to the electrorestriction effect}) = \text{Positive}$$

In the above equilibrium, an electrorestriction over the solvent is more by the reactants than by the product (*cf.* electrorestriction effects of 4 negative charge *vs.* 3 negative charge on the dipolar solvents). Thus the **cfse factor** (enthalpic factor) and **entropy factor** are jointly responsible for this highly negative E^0 value (*cf.* Sec. 3.14.16).

> **Cyanido complex of Fe(III) is more stable than the cyanido complex of Fe(II). While for bpy or phen, the reverse is true – why? : Enthalpy factor (cfse and π-bonding) and Entropy factor**

- The above conclusions are supported by the E^0 values of the relevant redox couples (Table 4.9.1.1). The involved ligands, *i.e.* CN$^-$, bpy and phen are the π-acid ligands. The Fe(II) (*i.e.* lower oxidation state) centre is expected to be the better centre than the Fe(III) centre (*i.e.* higher oxidation state) to favour the metal→ligand π-back bonding. Thus, the lower oxidation state (*i.e.* Fe(II)-centre) should form the more stable complexes than the higher oxidation state (*i.e.* Fe(III)-centre). Here it may be pointed out that effect of the charge on the metal centre causing the electrostatic or ion-dipole attraction in the metal-ligand interaction will oppose the stability sequence as expected from the effect of the metal→ ligand π-bonding. Considering the low-spin complexes of both Fe(II) (t_{2g}^6) and Fe(III) (t_{2g}^5), the relative gain of cfse in the Fe(III) complex is: $20Dq_{Fe(III)} - 2P_{Fe(III)} - (24Dq_{Fe(II)} - 2P_{Fe(II)})$. **This gain of cfse in the Fe(III) complexes is opposed by the predominant π-bonding effect present in the Fe(II) complexes.**

- For **bpy and phen**, Fe(II) forms the more stable complexes than Fe(III). It indicates that for these π-acid ligands, the metal-ligand π-bonding interaction is the most predominant factor. *For the cyanide which is also a π-acid ligand, the reverse stability order needs some explanation.* Actually in the cyanido complex, the π-bonding effect is also important. It is supported by the fact that the Fe(II)–C bond (192 pm) is shorter than the Fe(III)–C bond (195 pm). In terms of the electrostatic interaction and ionic radii $\left(r_{Fe^{2+}} > r_{Fe^{3+}} \right)$, the Fe(III)–C bond should be shorter. In fact, in the aqua complex, the Fe(II)–OH$_2$ bond length is longer than the Fe(III)–OH$_2$ bond length. *Thus the observed bond length order strongly indicates the stronger metal → ligand π-bonding interaction in [Fe(CN)$_6$]$^{4-}$ compared to that in [Fe(CN)$_6$]$^{3-}$.* To find the answer for the reverse order of the E^0-values for the bpy/phen and CN$^-$ complexes, we are to consider the **entropic factor,**

Charge annihilation: $\left[Fe(L)_3 \right]^{3+} + e \rightleftharpoons \left[Fe(L)_3 \right]^{2+}$, ΔS (due to electrorestriction effect) = +ve,

$$L = bpy, \ phen, \ E^0 \approx 1.0 \, V$$

Charge creation: $\left[Fe(CN)_6 \right]^{3-} + e \rightleftharpoons \left[Fe(CN)_6 \right]^{4-}$, ΔS (due to electrorestriction effect) = –ve,

$$E^0 = 0.36 \ V$$

For the bpy and phen complexes, the Fe(II) complex is less charged while for the CN$^-$ complex, the Fe(II) complex is more charged. In above redox equilibria, for L = bpy and phen, the oxidant is more solvated while for L = CN$^-$, the reductant is more solvated. *Thus reduction of [FeL$_3$]$^{3+}$ causes an increase of entropy due to the release of the bound solvent molecules while reduction of [Fe(CN)$_6$]$^{3-}$ causes a decrease of entropy due to the more solvation effect of the product (cf.* the electrorestriction effect of: 2 units of positive charge *vs.* 3 units of positive charge and 3 units of negative charge *vs.* 4 units of negative charge, on the dipolar solvent). Thus both the *enthalpic factor* (*i.e.* effect of metal → ligand π-bonding) and *entropic factor* favour the above redox equilibrium for L = bpy, phen. On the other hand, for L = CN$^-$, though the enthalpic factor (*i.e.* metal → ligand π-bonding effect) favours the equilibrium as usual but the entropic factor disfavours the equilibrium. ***This entropic disfavour is probably more important than the said effect of enthalpic favour for the CN$^-$ ligand.***

4.9.2 Stabilization of Low Oxidation States

- **Nonoxidising environment and reducing ligands:** The very low oxidation states (even zero or negative oxidation states) are very much sensitive to oxidation by the atmospheric oxygen or ligands. This is why, synthesis of such complexes is very often carried out in an oxygen free atmosphere and in deaerated solvents like tetrahydrofuran, dichloromethane, etc. The chosen ligands should provide a *reducing environment* (*i.e.* nonoxidizing environment) around the metal centre. The large ligands like iodide, sulfide which are reducing in character are suitable to stabilize the low oxidation states. *It may be noted that oxides, fluorides, etc. which are poor the reducing agents are not suitable to stabilize the low oxidation states.*

- **Softer ligands:** The metal centres in lower oxidation states are expected to be *softer* and *this is why they generally favour the softer ligands.*

- **Preference for the 3d-metal centres:** Lower oxidation states are not favoured for the heavier congeners (*i.e.* 4d and 5d series element) and these states are more documented for the 3d – series. The heavier congeners can easily attain the higher oxidation states due to relativistic effect (*d-orbital expansion*; *cf.* Chapter 8. Vol. 1 of Fundamental Concepts of Inorganic Chemistry).

- **Pi-acid ligands:** To stabilze the lower oxidation states, the π-acid ligands are extremely important. Due to the σ-donation to the metal centre by the ligands, it will enhance the electron density on the metal centre. *This accumulation of electron density leads to an unfavourable situation when* the metal centre is already in a low oxidation state. To avoid this unfavourable condition, there should be some π-acid ligands which can reduce the electron density on the metal centre through the metal→ligand π bonding interaction. **This is why, the π-acid ligands are extremely important to stabilize the low oxidation states of the metal centres.** The representative examples of such π-acid ligands are: CO, CN^-, NO^+, PR_3, RNC, bpy, phen, etc. It may be noted that CO, CN^- and NO^+ are the *isoelectronic species.*

- **Examples of complexes bearing the low oxidation states:**

Ti:	$[Ti(bpy)_3]$, (0)	$[Ti(bpy)_3]^-$, (−1)	$[Ti(CO)_6]^{2-}$ (−2)		
V:	$[V(bpy)_3]^+$, (+1)	$[V(bpy)_3]$, (0)	$[V(bpy)_3]^-$, (−1)	$[V(CO)_6]^{3-}$ (−3)	
Cr:	$[Cr(bpy)_3]^+$, (+1)	$[Cr(bpy)_3]$, (0)	$[Cr(CO)_6]$, (0)	$[Cr(C_6H_6)_2]$, (0)	$[Cr(C_6H_6)_2]^+$ (+1)
	$[Cr(bpy)_3]^-$, (−1)	$[Cr(CO)_5]^{2-}$, (−2)	$[Cr(bpy)_3]^{3-}$, (−3)	$[Cr(CO)_4]^{4-}$ (−4)	
Mn:	$[Mn(CN)_5]^{5-}$, (+1)	$[Mn(bpy)_3]$, (0)	$[Mn(bpy)_3]^-$, (−1)	$[Mn(CO)_5]^-$, (−1)	$[Mn(CO)_4]^{3-}$ (−3)
Fe:	$[Fe(NO)(OH_2)_5]^{2+}$, (+1)*	$[Fe(bpy)_3]$, (0)	$[Fe(CO)_5]$, (0)	$[Fe(bpy)_3]^-$, (−1)	$[Fe(CO)_4]^{2-}$ (−2)
Co:	$[Co(bpy)_3]^+$, (+1)	$[Co(bpy)_3]$, (0)	$[Co_2(CO)_8]$, (0)	$[Co(CO)_4]^-$ (−1)	
Ni:	$[NiBr(PPh_3)_3]$, (+1)	$[Ni(CN)_4]^{4-}$, (0)	$[Ni(CO)_4]$, (0)	$[Ni(PF_3)_4]$, (0)	$[Ni(bpy)_2]$, (0)
	$[Ni(CNR)_4]$, (0)	$[Ni_2(CO)_6]^{2-}$ (−1)			
Cu:	$[Cu(CN)_2]^-$ (+1)	$[Cu(CN)_4]^{3-}$ (+1)			

---

* *See* Sec, 16, Vol. 3 of Fundamental Concepts of Inorganic Chemistry regarding the controversy on the oxidation state of iron in the **brown ring complex.**

Note: Ambiguity in the Oxidation States – Representative Examples

- **$[Fe(NO)(OH_2)_5]^{2+}$ (brown ring complex):** There is a controversy whether 'Fe' is in +1 (*i.e.* NO as NO^+) or +3 (*i.e.* NO as NO^-) oxidation state (*see* Chapter 16, Vol. 3 of Fundamental Concepts of Inorganic Chemistry for discussion).

- **$[Ti(bpy)_3]$:** It may have Ti(0) or Ti(II) having two bpy^- and one neutral bpy.

- **Carbonyl nitrosyls:** The oxidation state depends on the state of 'NO' species in the complex.

- **$[M(R_2C_2S_2)_3]$ (trigonal prismatic complex)** (*cf.* Sec. 2.11.8): The organic ligand may remain in two possible forms. For the neutral form of the ligand, oxidation state of M becomes 0 (zero) while for the dianionic form of the ligand, oxidation state of the metal centre becomes +6. These aspects have been discussed in Sec. 2.11.8F.

$R_2C_2S_2$:

or

- **Nitrosyl complexes:** Nitric oxide after loosing one electron can produce NO^+ which is isoelectronic with CO. In fact, the unpaired electron residing in the π^*–MO of NO is easily lost. NO^+ can act as a π-acid ligand. Thus apparently NO acts as a 3e donor (1e is donated from the π^*–MO, other 2e are donated through the usual σ-donation) ligand towards the metal centre and the metal centre will move towards the lower oxidation state. *This is why, the complexes (in low oxidation) having only the NO-moieties are not known.* Rather, they form the **mixed nitrosyl-carbonyl** complexes, *e.g.* $[Mn(CO)(NO)_3]$, $[Fe(CO)_2(NO)_2]$ and $[Co(CO)_3(NO)]$ having -3, -2 and -1 oxidation states of the metals respectively in the complexes. It may be noted that these are the *isoelectronic nitrosyl carbonyl complexes.*
- **Dinitrogen complexes (Sec. 9.11):** N_2 (dinitrogen) is isoelectronic with CO but in N_2 the HOMO (σ–MO) is stabilized (*i.e.* poor σ-donor) and LUMO (π^*–MO) is destabilized (*i.e.* poor π-acceptor). Thus N_2 is a poor π-acid ligand. However in some complexes of low oxidation states, N_2 is present as a π-acid ligand. The representative examples are:

$[Ru(N_2)(NH_3)_5]^{2+}$, $[(H_3N)_5Ru-N_2-Ru(NH_3)_5]^{4+}$,

$[Co(H)(N_2)(PPh_3)_3]$, $[FeH_2(N_2)(Ph_2EtP)_3]$, etc.

EXERCISE 4

A. General Type Questions

1. What do you mean by the thermodynamic and kinetic stability of a complex? What are the factors to determine these parameters?
2. Define and illustrate: stepwise and overall stability constants; thermodynamic and stoichiometric stability constants; conditional stability constants; binary and ternary complex; chelate effect; torand; corand; podand; reverse crown ether; molecular recognition; anion coordination; macrocyclic effect; cryptate effect; preorganized macrocycles; football ligands; crown ethers and spherands; cryptands and sepulchrate; hydrophobic and stacking interaction in the ternary complexes; formation curves in terms of $\bar{n}$; effect of chelate ring size on the metal-ligand stability; steric factors on the metal-ligand stability.
3. Discuss the statistical factors in controlling the stepwise stability constants in the octahedral, tetrahedral and square planar complexes for the monodentate and didentate (symmetrical) chelating ligands.
4. Discuss the conditions under which statistical stability order can be denied.
5. What are the chemical factors (*i.e.* nonstatistical factors) to determine the metal-ligand stability constants?
6. Discuss the entropic effect, enthalpic effect and preorganization energy in understanding the chelate effect and macrocyclic effect.
7. Discuss the factors to control the stability of the complexes formed by the crown ethers and cryptands. What are the important applications of such macrocyclic ligands?

8. What do you mean by alkalides and electrides?
9. Discuss the different factors to determine the stability of the ternary complexes.
10. How can you determine the stepwise stability constants and protonation constants (of the ligands) graphically by using the experimental quantities $\bar{n}$ and $\bar{n}_H$ respectively?
11. How can you determine the stability constants by (a) Bjerrum's half $\bar{n}$-method; (b) by spectro-photometric method; (c) Job's method; (d) ion exchange method; (e) pH-metric method in terms of $\bar{n}$; (f) polarographic method.
12. Discuss the principles of Job's method, slope ratio method and mole ratio method to determine the composition of a complex.
13. Discuss the conditions to favour to stabilize the higher and low oxidation states of the metal centres.

B. Explain /Justify the following statements

1. Stability and lability characterize the different properties of a complex.
2. Overall stability constant can be expressed in terms of stepwise stability constants.
3. Thermodynamic stability constants and stoichiometric stability constants are different, but ther-modynamic stability constant may be evaluated from the stoichiometric stability constants deter-mined at different ionic strength values.
4. Stability constant can be expressed in terms of rate constants.
5. Stability constant and conditional stability constants are different but they can be correlated.
6. Statistical factors suggest: $K_1 \rangle K_2 \rangle K_3 \rangle \ldots$
7. Very often, the statistical prediction of the ratio of K_{n+1}/K_n differs from the experimental value.
8. For the octahedral complexes involving M^{n+} and monodentate ligand (L), the statistical ratio of the stepwise stability constants is:

$$K_1: K_2: K_3: K_4: K_5: K_6 = 6 : \frac{5}{2} : \frac{4}{3} : \frac{3}{4} : \frac{2}{5} : \frac{1}{6}.$$

9. For the octahedral complexes involving M^{n+} and symmetrical didentate ligands $(L-L)$, the statistical ratio of the stepwise stability constants is: $K_1: K_2: K_3 = 12 : \frac{5}{2} : \frac{4}{15}$.

10. For the square planar complexes involving M^{n+} and symmetrical didentate ligands $(L-L)$, the statistical ratio of the stepwise stability constants is: $K_1: K_2 = 8 : 1$
   - For the tetrahedral complexes the ratio is: $K_1: K_2 = 12 : 1$
   - For the octahedral complexes involving the symmetrical tridentate ligands, the statistical ratio of the stepwise stability constants is: $K_1: K_2 = 16 : 1$ (*fac*-isomer) and $K_1: K_2 = 24 : 1$ (*mer*-isomer).

11. For the π-acid ligands, the stepwise stability constants do not drop significantly.
12. Statistical stability order is reversed in the following systems:
   (a) Ag^+-NH_3: $K_2 \rangle K_1$
   (b) Cd^+-Br^-: $K_1 \rangle K_2 \rangle K_4 \rangle K_3$
   (c) $Cr^{2+}-bpy$: $K_2 \rangle K_1 \rangle K_3$
   (d) $Fe^{2+}-bpy$: $K_1 \rangle K_3 \rangle K_2$
   (e) $Pd^{2+}-cysteine$: $K_2 \rangle K_1$
   (f) $Hb-O_2$: $K_4 \rangle K_3 \rangle K_2 \rangle K_1$; (Hb = hemoglobin)

13. In the Hg^+–CN^- or I^- system, there is a drastic drop in going from K_2 to K_3.

14. In the M^{2+}–*en* (M = Zn, Cu) system, there is a drastic in K_3 (*i.e.* abrupt decrease in the ratio K_3/K_2).

15. For $Hg(aq)^{2+}$, the Brönsted acid dissociation constants follow a reverse statistical order.

16. Effect of electronegativity and cfse are important to rationalize the Irving-Williams stability order.

17. Hard-hard interaction leads to increase the stability constants with the increase of ionic potential of the metal centre.

18. Interaction between the hard metal ion and soft ligand (*i.e.* hard-soft interaction) should give a favourable situation to stabilize the metal-ligand interaction but it does not happen so in reality.

19. The parameter, ionic potential, cannot always explain the stability order for the interactions of alkali metal ions with the crown ethers, though the interaction is basically electrostatic in nature.

20. The hard-soft interaction is of an important consideration to determine the stability constant.

21. The harder metal ions prefer the didentate ligands in the sequence: (O,O) ⟩ (O,N) ⟩ (N,N) ⟩ (N,S) while the softer metal ions prefer the reverse sequence.

22. In the metal-ligand interaction, charge neutralization favours the complexation process due to the both favoured electrostatic interaction and entropic benefit.

23. In many cases, the ligand basicity (measured by pK_a) and stability constant (measured by $\log K$) bear a linear relationship.

24. Though the basicity of NH_3 and imidazole differs widely but the stability constants of their Cu(II)-complexes differ slightly.

25. BF_4^- and ClO_4^- show only the very poor complexing power.

26. The complexing power of 8-hydroxyquinoline and its monomethyl derivatives (at 2–, 4–, 5– positions) differs significantly.

27. The basicity order and complexing power order are not the same in the followings:
{ *Basicity*: 5-methyl-8-hydroxyquinoline ≈2-methyl-8-hydroxyquinoline ⟩ 8-hydroxyquinoline
  Complexing power: 5-methyl-8-hydroxyquinoline ⟩ 8-hydroxyquinoline ⟩ 2-methyl-8-hydroxy-quinoline.
{ *Basicity*: tmd or tn ⟩ pn
  Complexing power: pn ⟩ tmd or tn.

28. 2,3-diaminobutane can produce two isomeric complexes differing in their stability constants.

29. In the Ni(II)–en and Ni(II)– N,N′-dimethyl-en systems, the K_3 values differ more drastically than the K_1-values.

30. bpy can produce easily the *tris*– chelates while 6,6′-dimethyl-bpy fails to produce the *tris*– chelates.

31. The complexing power of *acac* depends drastically on the size of substituents (*i.e.* alkyl group) at the 3-position of the *acac*-skeleton.

32. Ni(II) + [H₂N— / H₂N—] (excess) ⟶ purple coloured and paramagnetic complex

Ni(II) + [H₂N— / H₂N—] (excess) ⟶ yellow coloured and diamagnetic complex

33. *trien* forms a more stable complex with Cu(II) than with Zn(II) while the reverse is true for the isomeric tetradentate ligand *tren*.

34. In the M(II)–en system, Irving-Williams stability order is maintained for K_1 and K_2 but not for K_3.

35. α-*ala* forms a more stable complex with $[Ni(nta)(OH_2)_2]^-$ than β-*ala* while the reverse is true for $[Cu(nta)(OH_2)_2]^-$.

36. In the octahedral complex $[Cu(bpy)(hfa)_2]$, bpy preferably occupies the equatorial positions and the resonance stabilization in the chelate ring formed by hfa is drastically reduced.

$$hfa = \text{hexafluoroacetylacetonate.}$$

37. Complexing power of *en*, *pn* and *tn* differs.

38. The Pt(IV) complex, $[PtCl_4(ptn)]$ can produce two isomeric compounds and one of them is optically active which is more stable while the other is optically inactive and less stable.

$$ptn = \quad \begin{array}{ccc} & \!\!\!\!\!\!\text{—CH—} & \\ \text{H}_2\text{N} & \text{NH}_2 & \text{NH}_2 \end{array}$$

40. Complexing power of the ligand, $(HO_2CH_2C)_2N–(CH_2)_n–N(CH_2CO_2H)_2$ depends on the values of n that may have the values 2, 3, 4, 5, ...

41. Among the didentate chelating ligands oxalate, malonate and succinate, oxalate forms the most stable complex while succinate forms the least stable complex.

42. Sometimes, the 5– membered chelate ring is more stable and sometimes the 6-membered chelate ring is more stable.

43. In the free ligand, acac, electrophilic substitution at the 3-position is a difficult task but in its metal chelate, it can occur more easily.

44. The stability order runs as: $[Cu(2,2,2)tet]^{2+} \langle [Cu(2,3,2)tet]^{2+}$

45. In the chelation process, cyclization of an open-chain multidentate ligand causes a loss of entropy but the process goes on with a resultant entropic favour.

46. The square planar complex $[Pd(edta)]^{2-}$ can exist in two geometrical isomers in which one is optically active while the other is optically inactive.

47. Chelate effect in py *vs.* bpy system is more pronounced than in the system, NH_3 *vs.* en.

48. Chelate effect is denied in $[Ag(en)]^+$.

49. For the observed chelate effect, the entropic favour is the main contributing factor for the complexes of nontransition metal ions but for the complexes of transition metal ions, the enthalpic favour may also contribute to some extent to the observed chelate effect.

50. Both the entropy and enthalpy factor contribute to the observed chelate effect.

51. Chelate effect in the system, py *vs.* bpy is more pronounced than in the system, NH_3 *vs.* en. This can be explained by considering the enthalpy factor to the chelate effect.

52. By considering the dissociation process of the ligands, the chelate effect may be explained.

53. Chelate effect may be explained by considering the kinetic aspects of formation and dissociation of the complexes.

54. Macrocyclic effect is an extension of the chelate effect.

55. In the macrocyclic effect, both the entropic and enthalpic factor contribute.

56. To understand the macrocyclic effect, the concept of preorganization energy, is very much important.
   - Degree of preorganisation in the free ligands runs in the sequence:
     Spherands ⟩ cryptaspherands ⟩ cryptands ⟩ hemispherands ⟩ crown ethers.
   - Stability of the complexes formed by the above ligands follows the same sequence.

57. Stability of the complexes runs as follows for the ligands: n-dentate macrocyclic ligand $\rangle$ n-dentate open chain ligand $\rangle$ n-unidentate ligands.

58. Entropic favours in the following complexation processes are different:

$$\left[M\left(OH_2\right)_x\right] + nL\,(\text{monodentate ligands}) \rightleftharpoons \left[M(L)_n\left(OH_2\right)_{x-n}\right] + nH_2O$$

$$\left[M\left(OH_2\right)_x\right] + L_n\,(n\text{-dentate open chain ligand}) \rightleftharpoons \left[M(L_n)(OH_2)_{x-n}\right] + nH_2O$$

$$\left[M\left(OH_2\right)_x\right] + L_n\,(n\text{-dentate macrocyclic ligand}) \rightleftharpoons \left[M(L_n)(OH_2)_{x-n}\right] + nH_2O$$

59. $\left[Zn(3, 2, 3\text{-tet})\right]^{2+} + \text{cyclam} \rightleftharpoons \left[Zn(\text{cyclam})\right]^{2+} + 3, 2, 3\text{-tet}, \; K = 10^4$

60. The stability of the complexes runs as:
$$[Cu(tet\text{-}a)]^{2+} \rangle [Cu(2,3,2\text{-tet})]^{2+} \rangle [Cu(2,2,2\text{-tet})]^{2+}$$

61. Macrocyclic effect in the system, $CH_3(OCH_2CH_2)_5OCH_3$ $vs.$ crown-6 originates from both the enthalpic and entropic factor.

62. Spherands generally form the more stable complexes than crown ethers.

63. The macrobicyclic ligands like cryptands form the more stable complexes than the macromonocyclic ligands like crown ethers.

64. Stability of the complexes formed by the macrocyclic ligands like crown ethers, cryptands, etc. mainly depends on the size-matching factor.

65. Crown ethers adopt different conformations in the uncomplexed and complexed state.
   - In synthesis of crown ethers, NaOH or KOH is used instead of the organic base like NEt$_3$.
   - In general, all crown ethers prefer K$^+$.
   - Preferential sequences of the alkali metal ions for a particular crown ether are different in gas phase and aqueous phase.
   - For the alkali metal-crown ether complexes, the stability constant depends not only on the size matching factor but also on the desolvation energy of the alkali metal ions.
   - 18-crown-6 and its derivatives bind all the alkali metal ions more strongly than the other crown ethers.

66. Free crown ether adopts different conformations in hydrophobic and hydrophilic solvents.

67. Crown ethers and cryptands can mimic the properties of ionophores.

68. Aliphatic crown ethers form the more stable complexes than the aromatic crown ethers.

69. The bicyclic cryptands are described as the football ligands.

70. The metal complexes of bicyclic cryptands adopt the endo-endo conformation of the cryptand.

71. Sepulchrate is related to $crypt$-[2,2,2].

72. Sepulchrate complexes are extremely stable. Even the [Co(sep)]$^{2+}$ complex can deny the racemization.

73. Retention of chirality occurs in the following sequence of reactions.

$$\left[Co(sep)\right]^{3+} \xrightarrow{\;\text{reduction}\;} \left[Co(sep)\right]^{2+} \xrightarrow{\;\text{oxidation}\;} \left[Co(sep)\right]^{3+}$$

74. The unusual compounds like electrides, alkalides are formed in the presence of suitable macrocycles of crown ethers and cryptands.

75. F$^-$ is a poor nucleophile but KF dissolved in benzene in presence of crown ethers provides a powerful F$^-$ nucleophile.

76. *Purple benzene* is a powerful oxidizing agent for the organic substrates.

77. [Cu(his)$_2$] and [Cu(his)(Hhis)]$^+$ are the examples of ternary complexes.

78. $\Delta \log K_M$ is a good parameter to measure the stability of a ternary complex.

79. The stability constants of the ternary complexes run as follows:

   (a) $K^{M(nta)}_{M(nta)(py)} > K^{M(nta)}_{M(nta)(im)} > K^{(nta)}_{M(nta)(NH_3)}$ $\left(M = Cu^{II}, Ni^{II} \right).$

   (b) $K^{Cu(bpy)}_{Cu(bpy)(ox)} > K^{Cu(bpy)}_{Cu(bpy)(gly)} > K^{Cu(bpy)}_{Cu(bpy)(en)}$

   (c) $K^{Cu(bpy)}_{Cu(bpy)(ox)} > K^{Cu(en)}_{Cu(en)(ox)}$ *but* $K^{Cu(bpy)}_{Cu(bpy)(en)} \approx K^{Cu(en)}_{Cu(en)_2}$

   (d) $K^{Cu(nta)}_{Cu(nta)(\alpha\text{-}ala)} < K^{Cu(nta)}_{Cu(nta)(\beta\text{-}ala)}$ *but* $K^{Ni(nta)}_{Ni(nta)(\alpha\text{-}ala)} > K^{Ni(nta)}_{Ni(nta)(\beta\text{-}ala)}$

80. The parameters $\Delta \log K_M$, $\log K_D$ and $\Delta \log K_i$ are the good parameters to measure the stability of ternary complexes.

81. In [Cu(his)$_2$], one histidinate prefers to bind in a histamine-like fashion while the other ligand binds in a glycinate-like fashion.
   - The ternary complex [Cu(his)$_2$] is more stable than the binary complex.
   - The ternary complex [Cu(his)$_2$] enjoys both the enthalpic and entropic favour.
   - The [M(bpy)] complex possesses a **ligand discriminating property.** It prefers the ligand in the order: (O, O) ⟩ (O, N) ⟩ (N, N).

82. Nature uses the imidazole moiety as a potential ligand in different metalloproteins because of its π-acidic character.

83. $\Delta \log K_M$ value for [M(ATP)(L)]$^{2-}$ (L = bpy or phen, M = Mg, Mn, Cu, etc.) is positive indicating the stability of the ternary system.

84. Ternary complexes bearing bpy or phen and tryptophenate are quite stable.

85. $\Delta \log K_M$ value for [M(sal)(gly)] or [M(pyruvate)(gly)] is highly positive (sal = salicylaldehydate).

86. Different types of intermolecular ligand-ligand interactions in the ternary complexes are important to stabilize the ternary complexes.

87. Presence of a π-acid ligand in a ternary complex can introduce a discriminating property in selecting the other ligands.

88. Hydrophobic and stacking interaction are important in many cases to enhance the stability of ternary complexes.

89. In the ternary complexes of amino acids, electrostatic interaction between the noncoordinating side chains can stabilize the system.

90. $\log K_1 = \left(pL \right)_{\bar{n}=0.5}$, $\log K_2 = \left(pL \right)_{\bar{n}=1.5}$ and $\log K_3 = \left(pL \right)_{\bar{n}=2.5}$

   (*cf.* Bjerrum's half $\bar{n}$-method).

91. For $\bar{n} = 0\text{--}2$ range, at the condition of [ML]$_{max}$, $\log K_1 K_2 = 2pL$, $(pL = -\log [L])$.

92. Protonation constants, K_1^H and K_2^H of a ligand (L) are related with $\bar{n}_H$ as follows:

$$\frac{\bar{n}_H}{\left(1 - \bar{n}_H \right)\left[H^+ \right]} = K_1^H + \frac{2 - \bar{n}_H}{1 - \bar{n}_H} K_1^H K_2^H \left[H^+ \right].$$

93. $M + LH_2 \rightleftharpoons ML + 2H^+, Q$

$$\frac{1}{A} = \frac{\left[H^+ \right]^2}{T_{Fe} \left(Q T_L \varepsilon l \right)} + \frac{1}{T_L \varepsilon l}, \quad \text{(under the condition } T_M \gg T_L \text{)}.$$

(A = absorbance of the complex ML at a particular wavelength where the ligand and metal are nonabsorbing).

94. Shifting of $E_{1/2}$ (half wave potential) due to complexation may be utilized to determine the stability constant.

95. In Job's method (continuous variation method), the peak point of the curve is not sharp always and it is rounded very often.

96. In Job's method, both the composition and stability of a complex can be determined.

97. In Job's method, a triangular shaped curve is obtained and the peak of the curve is deviated from the point of intersection of two extrapolated legs of the curve. This degree of deviation gives the measure of the stability of the complex.

98. In the mole ratio method, different linear segments of different slopes are obtained.

99. Pi-acid ligands are suitable for stabilization low oxidation states of metals.

100. Nonpolarizable and hard ligands are needed to stabilize the higher oxidation states of metals.

101. For the d-block elements, higher oxidation states are favoured for the heavier congeners.

102. Periodate and tellurate can stabilize Cu(III), Ni(IV), Ag(III).

103. Generation of higher oxidation states is favoured in an alkaline condition.

104. Stabilization of higher oxidation states can be done by the nonpolarizable anionic ligands.

105. Though py, pic, bpy are known as the π-acid ligands, they can stabilize the higher oxidation states like Ag^{2+}.

106. The π-acid ligands like bpy, phen, CN^- are expected to stabilize the lower oxidation state of a redox couple, $e.g.$ Fe(III)/Fe(II) to enhance the E^0 value (reduction potential). It happens so in the Fe(III)/Fe(II) couple for bpy and phen but for CN^-, the E^0 value decreases for the Fe(III)/Fe(II) couple.

107. Pure nitrosyl complexes are less known than the mixed ligand complexes ($e.g.$ nitrosyl-carbonyl complexes).

108. Oxidation states of the metal centres in the complexes like [Ti(bpy)$_3$], [M(R$_2$C$_2$S$_2$)] and brown ring complex are in ambiguity.

109. In [Cu(ATP)(bpy)]$^{2-}$ there are two possible favourable factors, $i.e.$ stacking interaction and synergistic bonding while in [Mg(ATP)(bpy)]$^{2-}$ only stacking interaction is possible, but for the both complexes $\Delta \log K_M$ values are comparable.

110. The E^0 value of the Co(III)/Co(II) couple is more negative for the en-ligand than for the phen ligand, through phen is a stronger field ligand than en.
   - Highly negative value of E^0 for the Co(III)/Co(II) couple in presence of CN^- can be explained by both cfse and entropic factor.
   - The lower E^0 value of the Co(III)/Co(II) couple for L = en than that for L = edta can be explained in terms of cfse ($i.e.$ enthalpic factor).
   - The order of E^0 value, [Fe(bpy)$_3$]$^{3+}$/[Fe(bpy)$_3$]$^{2+}$ $\rangle$ [Fe(CN)$_6$]$^{3-}$/[Fe(CN)$_6$]$^{4-}$ can be explained by considering both the enthalpic and entropic factor.

APPENDIX I
Units and Conversion Factors

Table 1: Basic physical quantities in SI units

Physical quantity	Symbol for quantity	Name of unit	Symbol for unit
Length	l	metre	m
Mass	m	kilogram	kg
Time	t	second	s
Electric current	I	ampere	A
Temperature	T	kelvin	K
Amount of substance	n	mole	mol
Luminous intensity	I_v	candela	cd

Table 2: Derived physical quantities in SI units

Physical quantity	Name of SI unit	Symbol and definition of SI unit	Named after
Force, weight	newton (N)	$N = kg\ m\ s^{-2} = J\ m^{-1}$	Isaac Newton (1642–1727)
Work, energy, quantity of heat	joule (J)	$J = N\ m$	James Prescott Joule (1818–1889)
Power	watt (W)	$W = J\ s^{-1}$	James Watt (1737–1819)
Pressure	pascal (Pa)	$Pa = N\ m^{-2}$	B. Pascal (1623–1662)
Electrical charge	coulomb (C)	$C = A\ s$	Charles Auguste de Coulomb (1736–1806)
Electrical potential	volt (V)	$V = kg\ m^2\ s^{-3}\ A^{-1}$ $= J\ A^{-1}\ s^{-1}$ $= J\ C^{-1} = W\ A^{-1}$	Allesandro Volta (1745–1827)
Electric capacitance	farad (F)	$F = C\ V^{-1}$	Michael Faraday (1791–1867)
Electric resistance	ohm (Ω)	$\Omega = V\ A^{-1}$	Georg Simon Ohm (1787–1854)
Electric conductance	siemens (S)	$S = \Omega^{-1}$	W. von Siemens (1816–1892)
Magnetic flux	weber (Wb)	$Wb = V\ s$	W. Weber (1804–1891)
Magnetic flux density	tesla (T)	$T = Wb\ m^{-2}$	Nikola Tesla (1856–1943)
Inductance	henry (H)	$H = V\ s\ A^{-1}$	Joseph Henry (1799–1878)
Frequency	hertz (Hz)	$Hz = s^{-1}$	Heinrich Hertz (1857–1894)
Radioactivity	becquerel (Bq)	$Bq = s^{-1}$	A. Henri Becquerel (1852–1908)
Absorbed dose of radiation	gray (Gy)	$Gy = m^2\ s^{-2}$ $= J\ kg^{-1}$	Louis H. Gray (1905–1965)
Area (A)	square metre	m^2	
Volume (V)	cubic metre	m^3	
Density (ρ)	kilogram per cubic metre	$kg\ m^{-3}$	
Velocity (u, v, w, c)	metre per second	$m\ s^{-1}$	
Acceleration	metre per square second	$m\ s^{-2}$	
*Concentration (c)	mole per cubic metre	$mol\ m^{-3}$	

*for concentration (c) in mol per litre, it is given by mol dm^{-3}

Table 3: SI prefixes

Fraction	Prefix	Symbol	Multiple	Prefix	Symbol
10^{-1}	deci	d	10	deka	da
10^{-2}	centi	c	10^2	hecto	h
10^{-3}	milli	m	10^3	kilo	k
10^{-6}	micro	μ	10^6	mega	M
10^{-9}	nano	n	10^9	giga	G
10^{-12}	pico	p	10^{12}	tera	T
10^{-15}	femto	f	10^{15}	peta	P

Table 4: CGS and SI units for some common physical quantities

| Physical quantity | CGS units | | SI Units | |
	Name	Symbol	Name	Symbol
Length	centimetre	cm	metre	m
	Angstrom (10^{-8} cm)	Å		
Mass	gram	g	kilogram	kg
Time	second	sec	second	s
Temperature	celsius	°C	kelvin	K
	kelvin	°K		
Energy	calorie	cal	joule	J
	kilocalorie	kcal	kilojoule	kJ
	litre-atmosphere ergs	lit-atm erg		

* tonne (t) = 10^3 kg, lb = 453.59 g, Btu = 1.055×10^{10} erg = 252 cal

Table 5: Conversion of CGS units to SI units

Physical quantity	Relation between the units
Length	Angstrom (Å) = 10^{-10} m = 10^{-1} nm = 10^2 pm
	micron (μ) = 10^{-6} m
Volume	llitre (L) = dm^3
Force	dyne = 10^{-5} N
Energy*	erg = 10^{-7} J
	cal = 4.185 J
	eV = 1.602×10^{-19} J; eV/molecule = 96.484 kJ mol^{-1}
Pressure	atm = 1.013×10^6 dyne cm^{-2} = 101.33 kN m^{-2}
	atm = 760 torr
	mm Hg or torr = 133.32 N m^{-2}
	bar (10^6 dynes/cm^2) = 10^5 N m^{-2}
Viscosity	poise = 10^{-1} kg m^{-1} s^{-1}
Magnetic flux density	gauses = 10^{-4} T

* 1 cm^{-1} (wave number) = 2.86 cal mol^{-1}, reciprocal centimetre is known as **Kayser (K)** 1 kK (kilo Kayser) = 1000 cm^{-1}, 1 eV per molecule = 23.06 kcal mol^{-1}, 1 erg = 2.39×10^{-11} kcal. 1 eV = (charge of an electron) × 1 V = 1.6×10^{-19} C × V = 1.6×10^{-19} J,

$$1 \text{ eV/molecule} = \frac{1.6 \times 10^{-19} \times 6.02 \times 10^{23}}{1000} \text{ kJ/mole} = 96.488 \text{ kJ mol}^{-1}.$$

$$1 \text{ kK molecule}^{-1} = 10^3 \text{ cm}^{-1} \text{ molecule}^{-1}$$

$$= hc \times 10^3 \text{ cm}^{-1} \text{ molecule}^{-1} \left(cf. \ E = h\nu = \frac{hc}{\lambda} \right)$$

$$= 6.62 \times 10^{-34} \text{ J s} \times 2.99 \times 10^8 \text{ m s}^{-1} \times 10^5 \text{ m}^{-1} \text{ molecule}^{-1}$$

$$= 1.985 \times 10^{-20} \text{ J molecule} = 11.96 \text{ kJ mol}^{-1}$$

Table 6: Atomic Units (see text, Vol. 1, Chapter 1, Sec. 1.8.2)

APPENDIX II
Some Physical and Chemical Constants

Table 1: Values of some physical and chemical constants

Constant	CGS and other units	SI units
Acceleration due to gravity (g)	980.665 cm sec^{-2}	9.80665 m s^{-2}
Avogadro's number (N_A)	6.02217×10^{23} molecules mole^{-1}	6.02217×10^{23} mol^{-1}
Bohr magneton (μ_B)	9.274×10^{-21} erg gauss^{-1}	9.274×10^{-24} J T^{-1}
Nuclear Magneton (μ_N)	5.04×10^{-24} erg G^{-1}	5.04×10^{-27} J T^{-1}
Bohr radius (a$_o$)	0.52918 Å	5.2918×10^{-11} m
Boltzmann constant (k_B or k)	1.3806×10^{-16} erg (degree K)$^{-1}$	1.3806×10^{-23} J K^{-1}
Debye	10^{-18} esu cm	3.3356×10^{-30} C m
Electronic charge (e)	4.80298×10^{-10} esu	1.60216×10^{-19} C
Electronic rest mass (m_e)	9.10953×10^{-28} g	9.10953×10^{-31} kg
Faraday (F)	96487 coulomb equiv^{-1}	9.6487×10^4 C mol^{-1}
Gas constant (R)	8.3144×10^7 erg (degree K)$^{-1}$ mole^{-1}	8.3144 J K^{-1} mol^{-1}
	8.31441 Joules (degree K)$^{-1}$ mole^{-1}	8.3144 N m K^{-1} mol^{-1}
	82.053×10^{-3} litre-atm (degree K)$^{-1}$ mole^{-1}	8.3144 Pa m^3 K^{-1} mol^{-1}
	1.987 cal (degree K)$^{-1}$ mole^{-1}	
Permitivity of vacuum (ε_o)		8.854188×10^{-12} C V^{-1} m^{-1}
		(C V^{-1} m^{-1} = C^2 N^{-1} m^{-2})
Permeability of vacuum (μ_o)		$4\pi \times 10^{-7}$ H m^{-1}
Planck's constant (h)	6.62618×10^{-27} erg sec	6.62618×10^{-34} J s
Rest mass of proton (m_p)	1.672648×10^{-24} g	1.672648×10^{-27} kg
Rest mass of neutron (m_n)	1.674954×10^{-24} g	1.674954×10^{-27} kg
Atomic mass unit (u)	1.660565×10^{-24} g	1.660565×10^{-27} kg
(1 u = 10^{-3} kg mol^{-1}/N_A)		($\equiv$ 931.48 MeV)
Vacuum speed of light (c)	2.99793×10^{10} cm sec^{-1}	2.99793×10^8 m s^{-1}
Standard atmospheric pressure	76 cm Hg	101.325 kPa
	760 mm Hg (or torr)	1.01325 bar
	1.1032×10^6 dyne cm^{-2}	
Hartree energy (E_h)		4.3597×10^{-18} J
Zero of the Celsius scale	0°C	273.15 K

APPENDIX III
Wavelength and Colours

Table 1: Spectral wavelength and colours

Range of the wavelength absorbed (nm)	Colour of the radiation absorbed	Complementary colour observed
400–450	Violet	Green-yellow
450–470	Indigo	Yellow
470–490	Blue	Orange
490–500	Blue-green	Red
500–560	Green	Purple
560–575	Green-yellow	Violet
575–590	Yellow	Indigo
590–640	Orange	Blue
640–730	Red	Blue-green

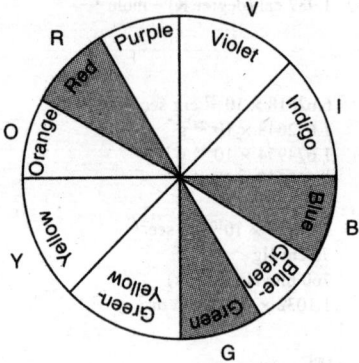

Artist's Colour Wheel: Colours and complementary colours.

Colours that stand opposite each other on the colour wheel are called as the **complimentary colours.** The amount of a colour in a particular image can be decreased by increasing its complimentary colour and vice-versa.

RGB colour model: In this model, **red (R), green (G), and blue (B)** are considered as the **three primary colours** that can generate all the colours of the visible spectrum when mixed together in different proportions. Mixing of these three primary colours in equal parts gives **white** and complete absence of these three colours results in **black**. Computer monitor devices use these **three additive primaries** to produce the different colour shades.

APPENDIX IV

Names, symbols, atomic numbers and atomic weights* of the elements

Element	Symbol	Atomic number	Atomic weight	Element	Symbol	Atomic number	Atomic weight
Actinium	Ac	89	227.03	Lead	Pb	82	207.19
Aluminium	Al	13	26.98	Lithium	Li	3	6.94
Americium	Am	95	(243)	Lutetium	Lu	71	174.97
Antimony	Sb	51	121.75	Magnesium	Mg	12	24.31
Argon	Ar	18	39.95	Manganese	Mn	25	54.94
Arsenic	As	33	74.92	Mendelevium	Md	101	(258)
Astatine	At	85	(210)	Mercury	Hg	80	200.59
Barium	Ba	56	137.34	Molybdenum	Mo	42	95.94
Barkelium	Bk	97	(247)	Neodymium	Nd	60	144.24
Beryllium	Be	4	9.01	Neon	Ne	10	20.18
Bismuth	Bi	83	208.98	Neptunium	Np	93	237.05
Boron	B	5	10.81	Nickel	Ni	28	58.71
Bromine	Br	35	79.91	Niobium	Nb	41	92.91
Cadmium	Cd	48	112.40	Nitrogen	N	7	14.01
Calcium	Ca	20	40.08	Nobelium	No	102	(259)
Californium	Cf	98	(251)	Osmium	Os	76	190.2
Carbon	C	6	12.01	Oxygen	O	8	16.00
Cerium	Ce	58	140.12	Palladium	Pd	46	106.4
Cesium	Cs	55	132.91	Phosphorus	P	15	30.97
Chlorine	Cl	17	35.45	Platinum	Pt	78	195.09
Chromium	Cr	24	52.01	Plutonium	Pu	94	(244)
Cobalt	Co	27	58.93	Polonium	Po	84	(209)
Copper	Cu	29	63.54	Potassium	K	19	39.10
Curium	Cm	96	(247)	Praeseodymium	Pr	59	140.91
Dysprosium	Dy	66	162.50	Promethium	Pm	61	(145)
Einsteinium	Es	99	(252)	Protactinium	Pa	91	231.04
Erbium	Er	68	167.26	Radium	Ra	88	226.03
Europium	Eu	63	151.96	Radon	Rn	86	(222)
Fermium	Fm	100	(257)	Rhenium	Re	75	186.2
Fluorine	F	9	19.00	Rhodium	Rh	45	102.91
Francium	Fr	87	(223)	Rubidium	Rb	37	85.47
Gadolinium	Gd	64	157.25	Ruthenium	Ru	44	101.07
Gallium	Ga	31	69.72	Samarium	Sm	62	150.35
Germanium	Ge	32	72.59	Scandium	Sc	21	44.96
Gold	Au	79	196.97	Selenium	Se	34	78.96
Hafnium	Hf	72	178.49	Silicon	Si	14	28.09
Helium	He	2	4.00	Silver	Ag	47	107.87
Holmium	Ho	67	164.93	Sodium	Na	11	22.99
Hydrogen	H	1	1.008	Strontium	Sr	38	87.62
Indium	In	49	114.82	Sulfur	S	16	32.06
Iodine	I	53	126.90	Tantalum	Ta	73	180.95
Iridium	Ir	77	192.2	Technetium	Tc	43	(99)
Iron	Fe	26	55.85	Tellurium	Te	52	127.60
Krypton	Kr	36	83.80	Terbium	Tb	65	158.92
Lanthanum	La	57	138.91	Thallium	Tl	81	204.37
Lawrencium*	Lr	103	(260)	Thorium	Th	90	232.04

(Contd.)

* Previously it was represented by Lw. Now as per IUPAC recommendation, it is represented by Lr.

Element	Symbol	Atomic number	Atomic weight	Element	Symbol	Atomic number	Atomic weight
Thulium	Tm	69	168.93	Xenon	Xe	54	131.30
Tin	Sn	50	118.69	Ytterbium	Yb	70	173.04
Titanium	Ti	22	47.90	Yttrium	Y	39	88.91
Tungsten	W	74	183.85	Zinc	Zn	30	65.37
Uranium	U	92	238.03	Zirconium	Zr	40	91.22
Vanadium	V	23	50.94				

* Values are based on the presently accepted *physical scale* of ^{12}C = 12.00000 as the standard. The ratio of these atomic weights to those on older *chemical scale* assuming the value 16.00000 for the atomic weight of natural oxygen (composition : ^{16}O, 99.757%; ^{17}O, 0.039%; ^{18}O, 0.204%) is 1.000279. The values given within the parentheses are the mass numbers of the most commonly known isotopes.

The true atomic weight of natural oxygen is given by : $16 \times 0.99757 + 17 \times 0.00039 + 18 \times 0.00204 = 16.00447$.

$$\text{It gives}: \frac{\text{Physical atomic weight}}{\text{Chemical atomic weight}} = \frac{16.00447}{16.00000} = 1.000279$$

Presently the term *atomic weight* is replaced by *atomic mass* or *relative atomic mass* (RAM). These are expressed in terms of *amu* or *u* or *Da* (Dalton). These aspects have been discussed in Sec. 5.2.3.

** For the *translawrencium* elements with $Z > 103$, the currently used names (cf. Chemistry of the Elements by N.N. Greenwood and A. Earnshaw and IUPAC, 1997) are : $_{104}Rf$ (rutherfordium), $_{105}Db$ (dubnium), $_{106}Sg$ (seaborgium), $_{107}Bh$ (bohrium), $_{108}Hs$ (hassium), $_{109}Mt$ (meitnerium). However, other names to these elements have also been suggested from time to time. These are ; $Z = 104$ (kurchatovium, *Ku*; dubnium, *Db*), $Z = 105$ (hahnium, *Ha*, neilsbohrium, *Ns*, joliotium, *Jl*), $Z = 106$ (rutherfordium, *Rf*), $Z = 107$ (neilsbohrium, *Ns*), $Z = 108$ (hahnium, *Hn*). The IUPAC (1977) recommendation, i.e. a *hybrid Latin-Greek numerical method* (cf. Sec. 8.6), is free from such confusions. The elements with $Z = 110, 111, 112, 114, 116, 118$ have been characterised.

APPENDIX V
Some Useful Mathematical Relationship

(A) Differentials

(u and v are the functions of x; a and m are constants).

(i) $\dfrac{dx}{dx} = 1$

(ii) $\dfrac{d}{dx}(au) = a\dfrac{du}{dx}$

(iii) $\dfrac{d}{dx}(u+v) = \dfrac{du}{dx} + \dfrac{dv}{dx}$

(iv) $\dfrac{d}{dx}x^m = m\,x^{m-1}$

(v) $\dfrac{d}{dx}\ln x = \dfrac{1}{x}$

(vi) $\dfrac{d\ln u}{dx} = \dfrac{1}{u}\dfrac{du}{dx}$

(vii) $\dfrac{d}{dx}(e^x) = e^x$

(viii) $\dfrac{d}{dx}(e^u) = e^u\dfrac{du}{dx}$

(ix) $\dfrac{d}{dx}(uv) = u\dfrac{dv}{dx} + v\dfrac{du}{dx}$

(x) $\dfrac{d}{dx}\left(\dfrac{u}{v}\right) = \dfrac{v\left(\dfrac{du}{dx}\right) - u\left(\dfrac{dv}{dx}\right)}{v^2}$

(xi) $\dfrac{d}{dx}(\sin x) = \cos x$

(xii) $\dfrac{d}{dx}(\cos x) = -\sin x$

(xiii) $\dfrac{d}{dx}(\tan x) = \sec^2 x$

(xiv) $\dfrac{d}{dx}(\sin u) = \cos u\dfrac{du}{dx}$

(xv) $\dfrac{d}{dx}(\cos u) = -\sin u\dfrac{du}{dx}$

(B) Integrals

(u and v are the functions of x; a and m are constants; for the indefinite integrals, there should be an addition of *arbitrary constant of integration*).

(i) $\int dx = x$

(ii) $\int au\,dx = a\int u\,dx$

(iii) $\int (u + v)dx = \int u\,dx + \int v\,dx$

(iv) $\int x^m\,dx = \dfrac{x^{m+1}}{m+1}, (m \neq -1)$

(v) $\int \dfrac{dx}{x} = \ln x$

(vi) $\int e^x dx = e^x$

(vii) $\int e^{ax}\,dx = \dfrac{1}{a}e^{ax}$

(viii) $\int \ln ax\,dx = x\ln ax - x$

(ix) $\int \sin x\,dx = -\cos x$

(x) $\int \cos x\,dx = \sin x$

(xi) $\int uv\,dx = u\int v\,dx - \int \left[\dfrac{du}{dx}\int v\,dx\right] dx$, (by parts)

(xii) $\int u\,dv = uv - \int v\,du$

(xiii) $\int\limits_0^\infty x^n e^{-ax} = \dfrac{n!}{a^{n+1}}$, ($n$ positive integer)

(C) Trigonometric relationship

(i) $\sin(90° - \theta) = \cos \theta$

(ii) $\cos(90° - \theta) = \sin \theta$

(iii) $\exp(\pm im\theta) = \cos(m\theta) \pm i\sin(m\theta)$, (*Demovier's theorem*)

$$\cos(m\theta) = \dfrac{1}{2}[\exp(im\theta) + \exp(-im\theta)] \qquad \sin(m\theta) = \dfrac{1}{2i}[\exp(im\theta) - \exp(-im\theta)]$$

(iv) $\sin^2\theta + \cos^2\theta = 1$

(v) $\sin 2\theta = 2\sin\theta\cos\theta$

(vi) $\cos 2\theta = \cos^2\theta - \sin^2\theta = 2\cos^2\theta - 1 = 1 - 2\sin^2\theta$

(vii) $\sin(\alpha \pm \beta) = \sin\alpha\cos\beta \pm \cos\alpha\sin\beta$

(viii) $\cos(\alpha \pm \beta) = \cos\alpha\cos\beta \mp \sin\alpha\sin\beta$

(ix) $\sin\alpha \pm \sin\beta = 2\sin\dfrac{1}{2}(\alpha \pm \beta)\cos\dfrac{1}{2}(\alpha \mp \beta)$

(x) $\cos\alpha + \cos\beta = 2\cos\dfrac{1}{2}(\alpha + \beta)\cos\dfrac{1}{2}(\alpha - \beta)$

(xi) $\cos\alpha - \cos\beta = -2\sin\dfrac{1}{2}(\alpha + \beta)\sin\dfrac{1}{2}(\alpha - \beta)$

(xii) $\sin\alpha\sin\beta = \dfrac{1}{2}\cos(\alpha - \beta) - \dfrac{1}{2}\cos(\alpha + \beta)$

(xiii) $\cos\alpha\cos\beta = \dfrac{1}{2}\sin(\alpha - \beta) + \dfrac{1}{2}\cos(\alpha + \beta)$

(xiv) $\sin\alpha\cos\beta = \dfrac{1}{2}\sin(\alpha + \beta) + \dfrac{1}{2}\sin(\alpha - \beta)$

(D) Some Useful Expansions ($x^2 \leq 1$)

$(1 + x)^{-1} = 1 - x + x^2 - x^3 + \ldots$ $(1 - x)^{-1} = 1 + x + x^2 + x^3 + \ldots$ $(1 - x)^{-2} = 1 + 2x + 3x^2 + 4x^3 + \ldots$

$$\ln(1 + x) = x - \dfrac{x^2}{2} + \dfrac{x^3}{3} - \dfrac{x^4}{4} + \ldots$$

$$\ln(1 - x) = -x - \dfrac{x^2}{2} - \dfrac{x^3}{3} - \dfrac{x^4}{4} \ldots$$

$$\exp(x) \text{ or } e^x = 1 + x + \dfrac{x^2}{2!} + \dfrac{x^3}{3!} + \dfrac{x^4}{4!} + \ldots, \text{ (for all } x)$$

$$\exp(-x) \text{ or } e^{-x} = 1 - x + \dfrac{x^2}{2!} - \dfrac{x^3}{3!} + \ldots$$

$$\sin x = x - \frac{x^3}{3!} + \frac{x^5}{5!} - \frac{x^7}{7!} + ..., \text{(for all } x)$$

$$\cos x = 1 - \frac{x^2}{2!} + \frac{x^4}{4!} - \frac{x^6}{6!} + ..., \text{(for all } x)$$

$\ln x! = x \ln x - x$ or $x! = \dfrac{x^x}{e^x}$ (**Stirling approximation**, 1730)

(for $x > ca.10$)

(E) Progressions

(i) **Arithmatic progression :** $a, a + d, a + 2d, ...$

n-th term $(T_n) = a + (n - 1)d;$

Sum of first n terms $(S_n) = \dfrac{n}{2}(T_1 + T_n) = \dfrac{n}{2}\{2a + (n-1)d\}$

Sum of first n-natural numbers $= 1 + 2 + 3 + ... + n = \dfrac{n(1+n)}{2}$

(ii) **Geometric progression :** $a, ar, ar^2, ...$

n-th term $(T_n) = ar^{n-1};$ Sum of first n terms $(S_n) = \dfrac{a(1-r^n)}{(1-r)}$ when $r < 1$

$S_n = \dfrac{a(r^n - 1)}{(r - 1)}$ when $r > 1$; sum to infinity $(S_\infty) = \dfrac{a}{1-r}$ when $|r| < 1$

(iii) **Useful results :**

$$\left. \begin{array}{l} \displaystyle\sum_{x=1}^{n} x^2 = \frac{n}{6}(n+1)(2n+1) \\[2mm] \displaystyle\sum_{x=1}^{n} x^3 = \left[\frac{n}{2}(n+1)\right]^2 \end{array} \right\} \text{For first } n \text{ natural numbers, } 1, 2, 3, ..., n$$

(F) Permutations and combinations

$$^nP_r = \frac{n!}{(n-r)!}; \quad ^nP_n = \frac{n!}{0!} = \frac{n!}{1} = n!; \quad ^nC_r = \frac{n!}{r!(n-r)!}; \quad ^nP_r = r! \, ^nC_r; \quad ^nC_r = \, ^nC_{n-r}; \quad ^nC_n = \, ^nC_0 = 1$$

$$n! = 1 \times 2 \times 3 \times ...... \times (n-2) \times (n-1) \times n; \, 0! = 1$$

(G) Logarithms

$\log(mn) = \log m + \log n; \log\left(\dfrac{m}{n}\right) = \log m - \log n,$

$\log(m^n) = n \log m; \log 1 = 0; \log_a a = 1; \ln(x) = 2.3 \log x$

Natural logarithm (base e); common logarithm (base 10).

(H) Mathematical Constants

$\pi = 3.14159265....., e = 2.71828...., \ln x = 2.30258 ... \log_{10}{}^x$

APPENDIX VI
Books Consulted

Akhmetov, N.: General and Inorganic Chemistry
Akitt, J.W.: NMR and Chemistry
Anantharaman, R.: Fundamentals of Quantum Chemistry
Aruldhas, G.: Molecular Structure and Spectroscopy
Ballhausen, C.J.: Introduction to Ligand Field Theory
Banerjea, D.: Inorganic Chemistry—Elements and Compounds
Banerjea, D.: Coordination Chemistry
Banerjee, D.: Inorganic Chemistry (Principles)
Banerjea, D.: Fundamental Principles of Inorganic Chemistry
Banwell, C.N.: Fundamentals of Molecular Spectroscopy
Berrow, G.M.: Introduction to Molecular Spectroscopy
Basolo, F. and Pearson, R.G.: Mechanisms of Inorganic Reactions
Bertini, I., Gray, H.B., Lippard, S.J. and Valentine, J.S.: Bioinorganic Chemistry
Carlin, R.L.: Magnetochemistry
Chanda, M.: Atomic Structure and Chemical Bonding
Chatwal, G.R. and Anand, S.K.: Spectroscopy
Connelly, G., Damhus, T. (Editors): Nomenclature of Inorganic Chemistry: IUPAC Recommendations 2005.
Cotton, F.A. and Wilkinson, G.: Advanced Inorganic Chemistry
Cotton, F.A. and Wilkinson, G.: Basic Inorganic Chemistry
Cotton, F.A. and Wilkinson, G., Murillo, C.A. and Bochmann, M.: Advanced Inorganic Chemistry
Cotton, F.A.: Chemical Applications of Group Theory
Das, Asim K.: Bioinorganic Chemistry
Das, Asim K.: Inorganic Chemistry: Biological and Environmental Aspects
Das, Asim K.: Environmental Chemistry with Green Chemistry
Das, A.K.: A Textbook on Medicinal Aspects of Bioinorganic Chemistry
Day (Jr), D.A. and Underwood, A.L.: Quantitative Analysis
Day, M.C. and Selbin, J.: Theoretical Inorganic Chemistry
Dorain, P.B.: Symmetry in Inorganic Chemistry
Douglas, B.E. McDaniel, D.H. and Alexander, J.J.: Concepts and Models of Inorganic Chemistry
Drago, R.S,: Physical Methods in Chemistry
Drago, R.S.: Physical Methods for Chemists
Dyer, J.R.: Applications of Absorption Spectroscopy of Organic Compounds
Dutta, P.K.: General and Inorganic Chemistry
Dutta, R.L. and Syamal, A.: Elements of Magnetochemistry
Dutta, R.L.: Inorganic Chemistry (Part. I and II)
Earnshaw, A.: Introduction to Magnetochemistry
Ebsworth E.A.V., Rankin, D.W.H. and Cradock, S.: Structural Methods in Inorganic Chemistry:
Emeleus, H.J. and Anderson, J.S.: Modern Aspects of Inorganic Chemistry
Figgis, B.N.: Introduction to Ligand Fields
Glasstone, S.: Textbook of Physical Chemistry
Griffith J.S.: The Theory of Transition Metal Ions
Greenwood, N.N. and Earnshaw, A.: Chemistry of the Elements

Huheey, J.E.: Inorganic Chemistry—Principles of Structure and Reactivity
Huheey, J.E., Keiter, E.A., Keiter, R.L. and O.K. Mehdi: Inorganic Chemistry—Principles of Structure and Reactivity
Jolly, W.L.: Principles of Inorganic Chemistry
Jorgensen, C.K.: Absorption spectra and Chemical Bonding in Complexes
Kalshi, P.S.: Spectroscopy of Organic Compounds
Kapoor, K.L.: A Textbook of Physical Chemistry (Vol. 1-4)
Kettle, S.F.A.: Coordination Compounds
Langford, C.H. and Gray, H.B.: Ligand Substitution Processes
Laidler, K.J. and Meiser, J.H.: Physical Chemistry
Lee, J.D.: Concise Inorganic Chemistry
Lever, A.B.P.: Inorganic Electronic Spectroscopy
Lippard, S.J. and Berg, J.M.: Principles of Bioinorganic Chemistry
Mackay, K.M. and Mackay, R.A.: Introduction of Modern Inorganic Chemistry
Mahan, B.H.: University Chemistry
Miessler, G.L. and Tarr, D.A.: Inorganic Chemistry
Mohan, J.: Organic Spectroscopy: Principles and Applications.
Mukherjee, G.N. & Das, A.: Elements of Bioinorganic Chemistry
Nakamoto, K.: Infrared Spectra of Inorganic and Coordination Compounds
Orchin, M. and Jaffe, H.H.: The Importance of Antibonding Orbitals
Pavia, D.L., Lampman, G.M. and Kriz, G.S. : Introduction to Spectroscopy
Potterfied, W.W.: Inorganic Chemistry: A Unified Approach
Purcell, K.F. and Kotz, J.C.: Inorganic Chemistry
Rakshit, P.C.: Physical Chemistry
Rao, C.N.R.: Chemical Applications of Infrared Spectroscopy
Rao, C.N.R.: Ultraviolet and Visible Spectroscopy
Ray, R.K.: Electronic Spectra of Transition Metal Complexes
Reddy, K.V.: Symmetry and Spectroscopy of Molecules
Shriver, D.F., Atkins, P.W. and Langford, C.H.: Inorganic Chemistry
Sarkar, R.: General and Inorganic Chemistry (Part I and II)
Satyanaryana, D.N.: Electronic Absorption Spectroscopy and Related Techniques
Satyanarayana, D.N.: Vibrational Spectroscopy – Theory and Applications
Sharpe, A.G.: Inorganic Chemistry
Skoog, D.A., West, D.M., Holler, F.J. and Crouch, S.R.: Fundamental of Analytical Chemistry
Solomon, E.I. and Lever, A.B.P.: Inorganic Electronic Structure and Spectroscopy
Steed, J.W. and Atwood, J.L.: Supramolecular Chemistry
Topping, J.: Errors of Observation and Their Treatment
Vogel, A.I.: A Textbook of Quantitative Inorganic Analysis
Wilkins, R.G.: The study of Kinetics and Mechanism of Reactions of Transition Metal Complexes
Wulfsberg, G.: Inorganic Chemistry

Acknowledgement

The above listed references and sources have been freely consulted to borrow their views and ideas. The present authors are indebted to all these authors to an endless extent. The present author expresses his heartiest thanks and grateful acknowledgement to all of them.

Subject Index

Reader's Note

Reader's Note

Reader's Note